Finite and Infinite Combinatorics in Sets and Logic

NATO ASI Series

Advanced Science Institutes Series

A Series presenting the results of activities sponsored by the NATO Science Committee, which aims at the dissemination of advanced scientific and technological knowledge, with a view to strengthening links between scientific communities.

The Series is published by an international board of publishers in conjunction with the NATO Scientific Affairs Division

A Life Sciences	Plenum Publishing Corporation
B Physics	London and New York
C Mathematical	Kluwer Academic Publishers
and Physical Sciences	Dordrecht, Boston and London
D Behavioural and Social Sciences	
E Applied Sciences	
F Computer and Systems Sciences	Springer-Verlag
G Ecological Sciences	Berlin, Heidelberg, New York, London,
H Cell Biology	Paris and Tokyo
I Global Environmental Change	

NATO-PCO-DATA BASE

The electronic index to the NATO ASI Series provides full bibliographical references (with keywords and/or abstracts) to more than 30000 contributions from international scientists published in all sections of the NATO ASI Series.
Access to the NATO-PCO-DATA BASE is possible in two ways:

– via online FILE 128 (NATO-PCO-DATA BASE) hosted by ESRIN,
Via Galileo Galilei, I-00044 Frascati, Italy.

– via CD-ROM "NATO-PCO-DATA BASE" with user-friendly retrieval software in English, French and German (© WTV GmbH and DATAWARE Technologies Inc. 1989).

The CD-ROM can be ordered through any member of the Board of Publishers or through NATO-PCO, Overijse, Belgium.

Series C: Mathematical and Physical Sciences - Vol. 411

Finite and Infinite Combinatorics in Sets and Logic

edited by

N.W. Sauer,

R.E. Woodrow

and

B. Sands
Department of Mathematics and Statistics,
The University of Calgary,
Calgary, Alberta, Canada

Kluwer Academic Publishers

Dordrecht / Boston / London

Published in cooperation with NATO Scientific Affairs Division

o5590o5x

MATH-STAT.

Proceedings of the NATO Advanced Study Institute on
Finite and Infinite Combinatorics in Sets and Logic
Banff, Alberta, Canada
April 21 - May 4, 1991

A C.I.P. Catalogue record for this book is available from the Library of Congress.

ISBN 0-7923-2422-6

Published by Kluwer Academic Publishers,
P.O. Box 17, 3300 AA Dordrecht, The Netherlands.

Kluwer Academic Publishers incorporates the publishing programmes of
D. Reidel, Martinus Nijhoff, Dr W. Junk and MTP Press.

Sold and distributed in the U.S.A. and Canada
by Kluwer Academic Publishers,
101 Philip Drive, Norwell, MA 02061, U.S.A.

In all other countries, sold and distributed
by Kluwer Academic Publishers Group,
P.O. Box 322, 3300 AH Dordrecht, The Netherlands.

Printed on acid-free paper

CONTENTS

v

BACK ROW (LEFT TO RIGHT) P. Ille, G. Cherlin, B. Rothschild, W.S. Shi, R. Diestel, S. Starchenko, C. Laflamme, M. Rabus, R. Halin, N. Sauer, B. Hart.

THIRD ROW (LEFT TO RIGHT) W. Brierley, A. Mekler, P. Winkler, C. Lam, B. Li, H. Wang, J. Truss, D. Lascar, R. Nowakowski, D. Evans, J-M. Brochet, M. Giraudet, D. Delhomme.

SECOND ROW (LEFT TO RIGHT) I. Rival, R. Schipperus, J. Schönheim, B. Sands, A. Carson, H. Harborth, E.C. Milner, P. Erdős, A. Hajnal, D. Haskell, S. Gunay, M. Pouzet, T. Goddard.

FIRST ROW (LEFT TO RIGHT) D. Duffus, H. Kierstead, N. Zaguia, R. Bonnet, S. Shelah, D. Gunderson, J. Baumgartner, J. Siráň, C. Toffalori.

PREFACE

This volume contains the accounts of papers delivered at the Nato Advanced Study Institute on Finite and Infinite Combinatorics in Sets and Logic held at the Banff Centre, Alberta, Canada from April 21 to May 4, 1991. As the title suggests the meeting brought together workers interested in the interplay between finite and infinite combinatorics, set theory, graph theory and logic. It used to be that infinite set theory, finite combinatorics and logic could be viewed as quite separate and independent subjects. But more and more those disciplines grow together and become interdependent of each other with ever more problems and results appearing which concern all of those disciplines. I appreciate the financial support which was provided by the N.A.T.O. Advanced Study Institute programme, the Natural Sciences and Engineering Research Council of Canada and the Department of Mathematics and Statistics of the University of Calgary.

The meeting on Finite and Infinite Combinatorics in Sets and Logic followed two other meetings on discrete mathematics held in Banff, the Symposium on Ordered Sets in 1981 and the Symposium on Graphs and Order in 1984. The growing inter-relation between the different areas in discrete mathematics is maybe best illustrated by the fact that many of the participants who were present at the previous meetings also attended this meeting on Finite and Infinite Combinatorics in Sets and Logic.

The main topics of this meeting illustrating the cohesiveness of finite and infinite discrete mathematics dealt with finite and infinite graphs, partitions of infinite graphs and rays, random graphs, random structures, homogeneous structures and graphs, the partition calculus and Ramsey theory, partition regular systems, coloring and partition problems, finite and infinite partial orders, the use of general logical constructions in discrete mathematics and stability theory, relational structures and Boolean algebras, pseudo finite structures, cardinal arithmetic and representation.

The papers in this volume give a good account of the topics mentioned in the previous paragraph. They mirror the talks presented during the conference and highlight some new research, but are also intended to allow non-specialists an entrance into the area under discussion and are generally of a more survey type nature. For students interested in topics somewhere in between the theory of relations, logic, infinite set theory and finite combinatorics the present proceedings provide good first reading on those topics and demonstrate, due to the interconnectedness of the articles, that here is a cohesive subject very worthwhile of further study.

There were lively problem and discussion sessions in the evening which led to several new results published in this volume. I am very glad and proud that we could include the results of successful cooperation initiated by the opportunity for discourse presented at the meeting. Of course this necessitated on the other hand that we had to be forgiving with deadlines and may partially explain the late publication of this volume.

E.C. Milner has always been interested in both finite and infinite combinatorial mathematics. In this sense the topic of this conference and the participants have been old friends of his in whose honor we organized this meeting. I know of many mathematicians who are grateful for the unselfish help he has provided in improving their results both scientifically and stylistically. I certainly count myself as one of them.

I am particularly grateful to R. Woodrow and B. Sands, the co-editors of this volume, for their help without which I would have found it almost impossible to organize the meeting and prepare the proceedings for publication. I also received help from the Scientific Affairs Division of N.A.T.O. and support in the selection and notification of the main participants from M. Pouzet and A. Lachlan. Several of the participants lent a helping hand, just to mention a few of them, R.K. Guy, D. Duffus and E. Milner. The help of G. Vezina, R. Maltby and J. Longworth in typing as well as the very enthusiastic help of D. Gunderson is greatly appreciated. Special thanks go to Bianca Sauer for enduring my preoccupation for long periods of time.

Calgary, Canada, March 5, 1993   Norbert Sauer

LIST OF PARTICIPANTS

Brian Alspach (Canada)

Joseph Barback (U.S.A.)

James E. Baumgartner (U.S.A.)

Arie Bialostocki (U.S.A.)

Robert Bonnet (France)

William Brierley (Canada)

Jean-Michel Brochet (France)

Andrew B. Carson (Canada)

Gregory L. Cherlin (U.S.A.)

Jacinta Covington (Australia)

Walter Deuber (Germany)

Christian Delhomme (France)

Reinhard Diestel (Germany)

Dwight Duffus (U.S.A.)

Paul Erdős (Hungary)

David M. Evans (U.K.)

Michèle Giraudet (France)

Ted Goddard (Canada)

Suleyman Gunay (Turkey)

David Gunderson (U.S.A.)

Richard K. Guy (England)

Gena Hahn (Canada)

Andras Hajnal (Hungary)

Rudolf Halin (Germany)

Heiko Harborth (Germany)

Bradd Hart (Canada)

Egbert Harzheim (Germany)

Deirdre Haskell (U.K.)

Katherine Heinrich (Canada)

Peter Horak (Canada)

Ehud Hrushowski (U.S.A.)

Pierre Ille (France)

Heinz Jung (Germany)

Hal A. Kierstead (U.S.A.)

Claude Laflamme (Canada)

Clement Lam (Canada)

Jean A. Larson (U.S.A.)

Daniel Lascar (France)

Hanno Lefmann (Germany)

Boyu Li (Canada)

Dugald Macpherson (U.K.)

Adrian R.D. Mathias (U.S.A.)

Alan Mekler (Canada)

Cheryl Miller (U.S.A.)

Eric C. Milner (Canada)

William Mitchell (U.S.A.)

Rognvaldur G. Möller (U.K.)

Charles Morgan (U.S.A.)

Richard J. Nowakowski (Canada)

Norbert Polat (France)

Maurice Pouzet (France)

Mariusz Rabus

Ivan Rival (Canada)

Bruce Rothschild (U.S.A.)

Bill Sands (Canada)

Norbert Sauer (Canada)

Rene Schipperus (Canada)

Todd J. Schneider (U.S.A.)

Jochanan Schönheim (Israel)

Saharon Shelah (Israel)

Jozef Širáň (Czechoslovakia)

Sergei Starchenko (U.S.S.R.)

Juris Steprans (Canada)

Michael G. Stone (U.S.A.)

Stevo Todorčević (Canada)

Carlo Toffalori (Italy)

J.K. Truss (U.K.)

Boban Veličković (U.S.A.)

Edward T.H. Wang (Canada)

Hong Wang (Canada)

Shang-zhi Wang (Canada)

Peter M. Winkler (U.S.A.)

K.C. Wong (Canada)

Carol Wood (U.S.A.)

Robert E. Woodrow (Canada)

Nejib Zaguia (Canada)

PROGRAMME FOR MONDAY APRIL 22.

9:25 *a.m.* N. Sauer, Opening comments

9:30–10:30 *a.m.* R. Halin—*Tree Partitions of Infinite Graphs*

11:00–12:00 *p.m.* P. Winkler—*Random Structures*

2:00–4:00 *p.m.* **Special Session:** *Relational Structures* (Chairman: M. Pouzet)

    2:00–2:30 *p.m.* P. Ille—*Indecomposable Relations*

    2:40–3:10 *p.m.* D. Evans—*Finite Covers of $\aleph_0$-Categorical Structures*

    3:20–3:50 *p.m.* D. Haskell—*A Transfer Theorem in Constructive p-adic Algebra*

4:30–5:30 *p.m.* G. Cherlin—*Classification of Homogeneous Directed Graphs*

PROGRAMME FOR TUESDAY APRIL 23.

9:30–10:30 *a.m.* R. Halin—*Lattices of Cuts in Graphs*

11:00–12:00 *p.m.* P. Winkler—*Random Structures*

2:00–4:00 *p.m.* **Special Session:** *Graphs* (Chairman: G.W. Sands)

    2:00–2:20 *p.m.* B.L. Rothschild—*Numbers of Vertices and Edges Which Must Occur in Induced Subgraphs*

    2:25–2:45 *p.m.* J. Širáň—*Discrepancy Versus Size*

    2:50–3:20 *p.m.* R. Möller—*Ends of Graphs*

    3:25–3:45 *p.m.* G. Hahn—*A Nice Proof of Halin's Theorem*

    3:50–4:10 *p.m.* R. Nowakowski—*Vertex to Vertex Pursuit on Disjoint Sets of Edges*

4:30–5:30 *p.m.* G. Cherlin—*Classification of Homogeneous Directed Graphs*

8:00–9:30 *p.m.*—Problem session in the Solarium bar: *Infinite Graphs* (Chairman: M. Pouzet)

PROGRAMME FOR WEDNESDAY APRIL 24.

9:30–10:30 *a.m.* J. Baumgartner—*Aspects of the Partition Calculus*

11:00–12:00 *p.m.* I. Rival—*Order, Ice Flow and Invariance*

12:00—Conference Photo

2:00–4:00 *p.m.* **Special Session:** *Partial Orders* (Chairman: D. Duffus)

    2:00–2:40 *p.m.* H.A. Kierstead—*Recursive and/or On-line Coloring of Co-comparability Graphs*

    2:45–3:25 *p.m.* N. Zaguia—*On Order Preserving Maps and Perpendicular Orders*

    3:30–4:10 *p.m.* S. Todorčević—*Conjectures of Rado and Chang*

PROGRAMME FOR THURSDAY APRIL 25.

9:30–10:30 *a.m.* J. Baumgartner—*Aspects of the Partition Calculus*

11:00–12:00 *p.m.* M. Pouzet—*From Relational Structures to Ordered Sets*

2:00–4:00 *p.m.* **Special Session:** *Homogeneous Structures* (Chairman: N.W. Sauer)

    2:00–2:30 *p.m.* J. Covington—*Maximal Subgroups of Infinite Symmetric Groups*

    2:40–3:10 *p.m.* J.K. Truss—*Generic Automorphisms of Homogeneous Structures*

    3:20–3:50 *p.m.* C. Wood—*Partially Homogeneous Posets*

4:30–5:30 *p.m.* Lascar D.—*Automorphism Groups of Relational Structures*

8:00–9:30 *p.m.*—Problem session in the Solarium bar: *Finite Combinatorics* (Chairman: H. Lefmann)

PROGRAMME FOR FRIDAY APRIL 26.

9:30–10:30 *a.m.* M. Pouzet—*From Relational Structures to Ordered Sets*

11:00–12:00 *p.m.* R. Bonnet—*On Superatomic Boolean Algebras*

2:00–4:00 *p.m.* **Special Session:** *Algebraic Structures* (Chairman: M.G. Stone)

    2:00–2:25 *p.m.* A. Carson—*The Characterization of Function Rings*

    2:30–2:55 *p.m.* E.T.H. Wang—*Expressing Positive Integers as Sums of Consecutive Positive Integers*

    3:00–3:25 *p.m.* C. Laflamme—*A Few Sigma Ideals of Measure Zero Sets*

    3:30–3:55 *p.m.* B. Veličković—*On the Deterministic Approximation of DNF Formulas*

4:30–5:30 *p.m.* D. Lascar—*Automorphism Groups of Relational Structures*

PROGRAMME FOR SATURDAY APRIL 27.

9:30–10:30 *a.m.* E. Hrushovski—*Pseudo Finite Structures*

11:00–12:00 *p.m.* H. Lefmann—*On Canonical Ramsey Numbers for Coloring 3-Element Sets*

2:00–4:00 *p.m.* **Special Session:** *Infinite Combinatorics* (Chairman: E.C. Milner)

    2:00–2:30 *p.m.* C. Morgan—*Morasses and Augmentations*

    2:35–3:05 *p.m.* J. Steprans—*Cardinal Invariants Associated With Splitting Finite Sets*

    3:05–3:35 *p.m.* J-M. Brochet—*Infinite Ordered Sets Spanned by Finitely many Chains*

4:30–5:30 *p.m.* H.D. Macpherson—*Combinatorial Questions on Infinite Permutation Groups*

6:00–9:30 *p.m.* **Banquet**

PROGRAMME FOR SUNDAY APRIL 28.

There were three hikes organized by M.G. Stone, D. Macpherson and R.K. Guy and an excursion to the icefields.

PROGRAMME FOR MONDAY APRIL 29.

9:30–10:30 *a.m.* E.C. Milner—*Cardinal Representation*

11:00–12:00 *p.m.* P. Erdős—*Some of my Favourite Problems in Finite and Infinite Combinatorics*

2:00–4:00 *p.m.* **Special Session:** *Finite Structures* (Chairman: R.K. Guy)

   2:00–2:30 *p.m.* N. Polat—*A Compactness Theorem for Perfect Matching in Matroids*

   2:40–3:10 *p.m.* J. Schönheim—*Covering Groups by Cosets, by Conjugacy Classes*

   3:20–3:50 *p.m.* C. Delhomme—*A Projection Property for Graphs of Girth Larger Than or Equal to Five*

4:30–5:30 *p.m.* H.D. Macpherson—*Combinatorial Questions on Infinite Permutation Groups*

8:00–9:30 *p.m.*—Problem session in the Solarium bar: *Homogeneous Structures and Permutation Groups* (Chairman: D. Evans)

PROGRAMME FOR TUESDAY APRIL 30.

9:30–10:30 *a.m.* E.C. Milner—*Cardinal Representation*

11:00–12:00 *p.m.* R. Diestel—*Graph Decomposition: A Study in Infinite Graph Theory*

2:00–4:00 *p.m.* **Special Session:** *Ramsey Theory* (Chairman: H. Lefmann)

   2:00–2:30 *p.m.* A. Bialostocki—*Zero Sum Trees: A Survey of Results and Open Problems*

   2:35–3:05 *p.m.* H. Harborth—*Ramsey Numbers For Five Vertex Graph Sets with Fixed Number of Edges* .

   3:10–3:40 *p.m.* D. Duffus—*2-Coloring Maximal Chains in Ordered Sets*

4:30–5:30 *p.m.* S. Shelah—*Cardinal Arithmetic*

PROGRAMME FOR WEDNESDAY MAY 1.

9:30–10:30 *a.m.* A. Hajnal—*True Embedding Partition Relations*

11:00–12:00 *p.m.* S. Shelah—*Cardinal Arithmetic*

2:00–4:00 *p.m.* **Special Session:** *Logic and Stability* (Chairman: R.E. Woodrow)

   2:00–2:30 *p.m.* S. Starchenko—*On Strongly Abelian Stable Varieties*

   2:40–3:10 *p.m.* B. Hart—*Superstable Quasi-varieties*

   3:20–3:50 *p.m.* T.J. Schneider—*Stable Theories of Order*

8:00–9:30 *p.m.*—Problem session in the Solarium bar. *Miscellaneous*
                (Chairman: H. Harborth)

PROGRAMME FOR THURSDAY MAY 2.

9:30–10:30 *a.m.* A. Hajnal—*True Embedding Partition Relations*

11:00–12:00 *p.m.* W.A. Deuber—*On Partition Regular Systems of Equations*

2:00–4:00 *p.m.* **Special Session:** *Infinite Permutation Groups*
            (Chairman: H.D. Macpherson)

   2:00–2:30 *p.m.* O.V. Belegradek—*Model Theory of Unitriangular Groups Over Rings*

   2:40–3:10 *p.m.* M. Giraudet—*Automorphism Groups of Cyclic Orders*

   3:20–3:50 *p.m.* A.R.D. Mathias—*The Status of the Axiom of Determinacy*

4:30–5:30 *p.m.* J. Larson—*Ordinal Partition Relations and Infinite Exponents*

PROGRAMME FOR FRIDAY MAY 3.

9:30–10:30 *a.m.* A. Mekler—*Homogeneous Posets*

11:00–12:00 *p.m.* W.A. Deuber—*On Partition Regular Systems of Equations*

5:00–9:30 *p.m.* **Barbecue.**

Extensions of the Erdős–Rado Theorem

JAMES E. BAUMGARTNER*
Mathematical Institute
Dartmouth College
Hanover, NH 03755 USA

ANDRÁS HAJNAL†
Institute of Mathematics
Hungarian Academy of Sciences
Budapest, Hungary

STEVO TODORČEVIĆ‡
Matematicki Institut
Kneza Mihaila 35
11001 Beograd, p.p. 367
Jugoslavija

Dedicated to Eric Milner on the occasion of his third coming of age

0   Statement of results

We find here some extensions of the Erdős–Rado Theorem that answer some longstanding problems. Ordinary partition relations for cardinal numbers are fairly well understood (see [5]), but for ordinal numbers much has been open, and much remains open. For example, any proof of the simplest version of the Erdős–Rado Theorem seems to yield

$$\text{For any regular cardinal } \kappa, \text{ if } \mu < \kappa \text{ then } (2^{<\kappa})^+ \to (\kappa + 1)^2_\mu,$$

but to replace $\kappa + 1$ by $\kappa + 2$ seems quite a nontrivial problem. In this paper we will prove that if κ is regular and uncountable, then

Theorem 3.1. $\forall k < \omega \ \forall \xi < \log \kappa \ (2^{<\kappa})^+ \to (\kappa + \xi)^2_k.$

Theorem 4.1. $\forall n, k < \omega \ (2^{<\kappa})^+ \to (\rho, (\kappa + n)_k)^2$, where $\rho = \kappa^{\omega+2} + 1.$

*The preparation of this paper was partially supported by National Science Foundation grant number DMS–8906946.
†Research supported by Hungarian National Science Foundation OTKA grant 1908.
‡Research supported by the Science Fund of Serbia grant number 0401A

1

N.W. Sauer et al. (eds.), Finite and Infinite Combinatorics in Sets and Logic, 1–17.
© 1993 *Kluwer Academic Publishers. Printed in the Netherlands.*

Theorem 5.1. $\forall n < \omega \ (2^{<\kappa})^+ \cdot \omega \to (\kappa \cdot n)_2^2$.

The actual version of 5.1 is slightly stronger. Here $\log \kappa$ is the least cardinal μ such that $2^\mu \geq \kappa$, the exponentiation in 4.1 is ordinal exponentiation, and the products in 5.1 represent ordinal multiplication. For $\kappa = \omega$, 3.1, 4.1, and 5.1 all follow from the known result $\omega_1 \to (\alpha)_k^2$ for all $\alpha < \omega_1$ and all $k < \omega$ (see [2]), so the uncountable case is the interesting one.

The proofs make use of elementary substructures of structures of the form $(H(\mu), \in)$ where $H(\mu)$ is the set of all sets hereditarily of cardinality $< \mu$, and of ideals on ordinals generated from such elementary substructures. It is possible to recast our arguments in such a way as to fit them under the heading of ramification arguments but we choose not to do so, both because the proofs remain clearer this way, and because we believe that this may generally be a better approach to ramification arguments.

The proof of 4.1 also uses a metamathematical trick seen in [2] for countable ordinals. Our approach is to prove 4.1 in a generic extension of the universe obtained by κ-closed forcing, and then to argue that it must therefore be true.

Several open problems remain. The simplest versions are whether any of the following are provable from GCH:

$$\omega_3 \to (\omega_2 + 2)_\omega^2$$

$$\omega_3 \to (\omega_2 + \omega_1, \omega_2 + \omega)^2$$

$$\omega_2 \to (\omega_1^{\omega+2} + 2, \omega_1 + 2)^2$$

or whether the following is provable from CH:

$$\omega_3 \to (\omega_1 + \omega)_3^2.$$

1 Terminology

Our partition calculus notation is standard. If x is a set and λ is a cardinal then $[x]^\lambda = \{y \subseteq x : |y| = \lambda\}$, $[x]^{<\lambda} = \bigcup\{[x]^\kappa : \kappa < \lambda\}$, and $[x]^{\leq\lambda} = [x]^{<\lambda^+}$. Usually, but not always, such a λ is finite. The ordinary (ordinal) partition relation $\alpha \to (\beta)_\mu^2$ means that $\forall f : [\alpha]^2 \to \mu$ $\exists X \subseteq \alpha$ X has order type β and is homogeneous for f, i.e., f is constant on $[X]^2$. Here α and β may be ordinals; μ is always taken to be a cardinal. If $f(x) = i$ for all $x \in [X]^2$ then we say X is *i-homogeneous*. The unbalanced relation

$$\lambda \to (\gamma_0, \gamma_1, \ldots, \gamma_{n-1})^2$$

means that $\forall f : [\lambda]^2 \to n \; \exists X \subseteq \lambda \; \exists i < n \; X$ has order type γ_i and is i-homogeneous. We abbreviate

$$\lambda \to (\alpha, \beta, \beta, \beta, \ldots, \beta)^2$$

by $\lambda \to (\alpha, (\beta)_n)^2$ if there are n occurrences of β.

If κ and λ are cardinals, then $\lambda^{<\kappa} = \sum \{ \lambda^\mu : \mu < \kappa \}$. Clearly $2^{<\kappa} \geq \kappa$. If κ is regular then $\kappa^{<\kappa} = 2^{<\kappa}$.

If $(N, \in)$ is an elementary substructure of $(H(\lambda), \in)$ (i.e., $(N, \in) \prec (H(\lambda), \in)$) then we often abbreviate this by $N \prec H(\lambda)$. The $\in$-relation is always understood. Suppose κ is regular, $N \prec H(\kappa^{++})$, $|N| = \kappa$ and $\alpha = N \cap \kappa^+ < \kappa^+$. Then we may define an ideal I on α as $\{ X \subseteq \alpha : \exists A \in N \; \alpha \notin A$ and $X \subseteq N \}$. Clearly I contains all bounded subsets of α. If in addition $2^{<\kappa} = \kappa$ then it is quite possible that $[N]^{<\kappa} \subseteq N$, and we generally work with sets N with this property. Note in that case that I is κ-closed, i.e., if $\mu < \kappa$ and $X_i \in I$ for $i < \mu$, then $\bigcup \{ X_i : i < \mu \} \in I$. This follows from the fact that if $A_i \in N$ for each $i < \mu$ and $\alpha \notin A_i$, then $\langle A_i : i < \mu \rangle \in N$ since $[N]^{<\kappa} \subseteq N$ so $A = \bigcup \{ A_i : i < \mu \} \in N$ and $\alpha \notin A$.

If I is an ideal on α then $I^+ = \{ X \subseteq \alpha : X \notin I \}$ and $I^* = \{ X \subseteq \alpha : \alpha - X \in I \}$. Note that if I is not proper, then $I^+ = 0$ and $I^* = I$. If $I(\alpha, \beta)$ is an ideal then we write $I^+(\alpha, \beta)$ instead of $I(\alpha, \beta)^+$.

If $N \prec H(\kappa^{++})$ as above, $S \in N$ and $\alpha \in S$, then $S \cap \kappa^+$ must be cofinal in κ^+. If not, an upper bound is definable from S and κ^+, both of which lie in N ($\kappa^+ \in N$ since it is the largest cardinal in N), so the upper bound must be $< \alpha$, which is impossible. It follows that $S \cap \alpha$ is cofinal in α. In fact, $S \cap \kappa^+$ is actually stationary in κ^+, since otherwise there would be closed unbounded $C \in N$ with $S \cap C = 0$. But then $C \cap \alpha$ is unbounded in α so $\alpha \in C$, which is impossible.

We often use variations of the ideal I to produce large homogeneous sets for a partition function $f \in N$. Such arguments can also be made using ramification ideas. A very good introduction to the ideas we plan to use is in Section 2, where a simple version of the Erdős–Rado Theorem is proved with these techniques.

In Section 4 we will use the notions of stationary and closed unbounded subsets of $[X]^\kappa$, where X is a set such that $|X| > \kappa^+$. We assume the reader is familiar with these notions.

2 The Erdős–Rado Theorem

In this section we present a slightly nonstandard proof of the Erdős–Rado Theorem, and an associated result, using elementary substructures. Our goal is to prepare the way for an extension of the Erdős–Rado Theorem to be proved in the next section.

Theorem 2.1. (Erdős–Rado [4]) *Let κ be a regular cardinal and let $\lambda = (2^{<\kappa})^+$. Then $\lambda \to (\kappa+1)^2_\mu$ for all $\mu < \kappa$.*

Proof. Fix $f : [\lambda]^2 \to \mu < \kappa$. Choose $N \prec H(\lambda^+)$ such that $f \in N$, $|N| = 2^{<\kappa}$, $[N]^{<\kappa} \subseteq N$ and $N \cap \lambda = \alpha < \lambda$. (Note that this implies cf $\alpha = \kappa$. If cf $\alpha < \kappa$ then some cofinal subset x of α would be an element of N, and $\alpha = \bigcup x$ so $\alpha \in N$, contradiction.) Let $I_\alpha = \{ X \subseteq \alpha : \exists A \in N \ \alpha \notin A \text{ and } X \subseteq A \}$. Then I_α is a κ-complete ideal on α since $[N]^{<\kappa} \subseteq N$.

For $i < \mu$ let $H_i = \{ \beta < \alpha : f\{\beta,\alpha\} = i \}$. Then $\bigcup\{ H_i : i < \mu \} = \alpha$ so for some i $H_i \in I_\alpha^+$.

Claim. *If $H_i \in I_\alpha^+$, $X \subseteq H_i$ and $X \in I_\alpha^+$ then $\exists Y \subseteq X$ Y is i-homogeneous of order type κ.*

This will suffice since then $Y \cup \{\alpha\}$ is i-homogeneous of type $\kappa + 1$.

It will suffice to show

Lemma 2.2. *If $X \subseteq H_i$, $X \in I_\alpha^+$, $x \subseteq X$, $|x| < \kappa$ and x is i-homogeneous, then $\exists \beta \in X$ $\beta > \sup x$ and $x \cup \{\beta\}$ is i-homogeneous.*

Proof. Let $Z = \{\gamma < \lambda : \forall \delta \in x f\{\delta,\gamma\} = i\}$. Then $Z \in N$ and $\alpha \in Z$ so $\alpha - Z \in I_\alpha$. Since $X \in I_\alpha^+$ we have $X \cap Z \in I_\alpha^+$ as well. Now choose $\beta \in X \cap Z$, $\beta > \sup x$. Using the same methods we can do a little better.

Theorem 2.3. *As in Theorem 2.1, let κ be regular, $\lambda = (2^{<\kappa})^+$, $\mu < \kappa$ and $f : [\lambda]^2 \to \mu$. Then $\exists A \subseteq \lambda \ \exists i < \mu$ A is i-homogeneous and either $|A| = \lambda$ and $i = 0$ or else A has type $\kappa + 1$ and $i > 0$. (In short notation $\lambda \to (\lambda, (\kappa+1)_{\mu-1})^2$.)*

Proof. We use the notation of the previous proof. If $H_i \in I_\alpha^+$ for some $i > 0$ we are done, so assume $H_i \in I_\alpha$ for all $i > 0$. Then $\bigcup\{ H_i : i > 0 \} \in I_\alpha$ so for some $X \in N$ we have $\alpha \in X$ and $X \cap \alpha \subseteq H_0$. If in N we define $Y = \{ \beta \in X : \forall \gamma \in X \cap \beta f\{\gamma,\beta\} = 0 \}$, then $\alpha \in Y$. Hence $|Y| = \lambda$ and is clearly 0-homogeneous.

Remark. Theorem 2.3 is a well known result of Erdős, Dushnik and Miller. See [3], for example. To get the rest of the Erdős–Rado Theorem with this approach (i.e., the version for partitions of n-element sets) simply proceed via end-homogeneous sets, an approach which is itself highly adaptable—and well known—in the context of elementary substructures. Details are left to the reader.

3 A balanced extension of the Erdős–Rado Theorem.

In this section we seek an improvement of the Erdős–Rado Theorem, Theorem 2.1. This will require a clearer analysis of the ideals I_α defined in Section 2.

Recall that $\log \kappa$ is the least cardinal μ such that $2^\mu \geq \kappa$.

Theorem 3.1. *Let κ be a regular uncountable cardinal, and let $\lambda = (2^{<\kappa})^+$. Then $\forall k < \omega$*
$$\forall \xi < \log \kappa \quad \lambda \to (\kappa + \xi)^2_k.$$

Some parts of this theorem are already known. In unpublished work, Hajnal proved Theorem 3.1 for $\kappa = \omega_1$ and $k = 2$ about thirty years ago. Shelah proved Theorem 3.1 for the case $k = 2$ in [Sh1, §6].

The rest of this section will be devoted to the proof of Theorem 3.1.

Fix $f : [\lambda]^2 \to k$, and let $\langle N_\alpha : \alpha < \lambda \rangle$ be a continuous sequence of elementary substructures of $H(\lambda^+)$ such that $f \in N_0$, $|N_\alpha| = 2^{<\kappa}$, $N_\alpha \in N_{\alpha+1}$ and $[N_\alpha]^{<\kappa} \subseteq N_{\alpha+1}$. We call a sequence continuous if each limit point is the union of the preceding elements. We may assume that $\{\alpha : N_\alpha \cap \lambda = \alpha\}$ is closed and unbounded in λ. Note that if $\alpha \in S_0 = \{\alpha : N_\alpha \cap \lambda = \alpha$ and $\mathrm{cf}\,\alpha = \kappa\}$ then $[N_\alpha]^{<\kappa} \subseteq N_\alpha$.

If $\alpha \in S_0$ then let $I_\alpha = \{X \subseteq \alpha : \exists A \in N_\alpha \; \alpha \notin A$ and $X \subseteq A\}$.

Lemma 3.2. *Let $S \subseteq S_0$ be stationary. Then there is closed unbounded $C \subseteq \lambda$ such that*
$$\forall \alpha \in C \cap S \quad S \cap \alpha \in I_\alpha^+.$$

We refer to α as a *reflection point* of S.

Proof of Lemma 3.2. Let $\langle M_\alpha : \alpha < \lambda \rangle$ be a continuous sequence of elementary substructures of $H(\lambda^+)$ such that $N_\alpha \subseteq M_\alpha$, $|M_\alpha| = 2^{<\kappa}$ and $S \in M_0$. Let $N = \bigcup \{N_\alpha : \alpha < \lambda\}$ and let $C = \{\alpha : M_\alpha \cap N = N_\alpha\}$. If $\alpha \in C \cap S$ then $S \cap \alpha \in I_\alpha^+$ since otherwise there is $A \in N_\alpha \subseteq M_\alpha$ such that $S \subseteq A$ and $\alpha \notin A$, which is impossible. [Note: we really only know that $S \cap \alpha \subseteq A \cap \alpha$, but since $S, A \in M_\alpha$ we must have $M_\alpha \models S \subseteq A$ and since $M_\alpha \prec H(\lambda^+)$, it is true that $S \subseteq A$. We will leave such remarks tacit in future.]

For $i < k$ let $f_i(\alpha) = \{\beta < \alpha : f\{\beta, \alpha\} = i\}$.

Next, if σ is a sequence of length n of elements of k we define an ideal $I(\alpha, \sigma)$ for $\alpha \in S_n$, where $S_{n+1} = \{\alpha \in S_n : \alpha$ is a limit point of $S_n\}$. The definition is by induction on $n = \mathrm{length}(\sigma)$. Let $I(\alpha, \langle\rangle) = I_\alpha$. If $\sigma = \tau^\frown \langle i \rangle$ then put

$$X \in I(\alpha, \sigma) \quad \text{iff} \quad \{\beta < \alpha : X \cap f_i(\alpha) \cap \beta \in I^+(\beta, \tau)\} \in I_\alpha.$$

Notice that some of the $I(\alpha, \sigma)$ may not be proper. As part of the proof of the Erdős–Rado Theorem we essentially made the easy observation that there exists i such that $I(\alpha, \langle i \rangle)$ is proper.

Lemma 3.3. *We always have* $I_\alpha \subseteq I(\alpha, \sigma)$.

Proof. Once again we proceed by induction on length(σ). Suppose $\sigma = \tau^\frown \langle i \rangle$. We may assume $I(\alpha, \sigma)$ is proper. Let $X \in I_\alpha$. Then $\exists A \in N_\alpha$ $X \subseteq A \cap \alpha$ and $\alpha \notin A$. We may as well assume $X = A \cap \alpha$. Since α is limit and $A \in N_\alpha$ we know $\exists \gamma < \alpha$ $A \in N_\gamma$. Suppose $\beta < \alpha$, $\gamma < \beta$, and $A \cap \beta \in I^+(\beta, \tau)$. Then by the inductive hypothesis $A \cap \beta \in I_\beta^+$ so $\beta \in A$. Thus

$$\{ \beta < \alpha : (A \cap \alpha) \cap f_i(\alpha) \cap \beta \in I^+(\beta, \tau) \} \subseteq \{ \beta < \alpha : A \cap \beta \in I^+(\beta, \tau) \} \subseteq A \cap \alpha \in I_\alpha,$$

so $A \cap \alpha \in I(\alpha, \sigma)$ as desired.

Now we extend an observation made in the proof of the Erdős–Rado Theorem.

Lemma 3.4. *If* $X \subseteq \alpha$ *and* $X \in I^+(\alpha, \sigma)$ *then* $\forall j \in \text{range}\,\sigma$ $\exists W \subseteq X$ W *is* j-*homogeneous and* $|W| = \kappa$.

Proof. The proof is by induction on length(σ). We may assume $\sigma = \tau^\frown \langle i \rangle$. If $j \in \text{range}(\tau)$ then we may apply the inductive hypothesis to $X \cap \beta \in I^+(\beta, \tau)$ for some $\beta < \alpha$. So suppose $j = i$. As in Lemma 2.2 let us argue that if $x \subseteq X$, $|x| < \kappa$ and $x \cup \{\alpha\}$ is i-homogeneous then $\exists \xi \in X$ $\xi > \sup x$ and $x \cup \{\xi\} \cup \{\alpha\}$ is i-homogeneous. Let $A = \{\beta : x \cup \{\beta\}$ is i-homogeneous $\}$. Then $A \in N_\alpha$ and $\alpha \in A$ so $\alpha - A \in I_\alpha$. By Lemma 3.3 $\alpha - A \in I(\alpha, \sigma)$, so $A \cap X \in I^+(\alpha, \sigma)$. Hence there is $\beta < \alpha$, $\beta > \sup x$, such that $A \cap X \cap f_i(\alpha) \cap \beta \in I^+(\beta, \tau)$. Now just choose $\xi \in A \cap X \cap f_i(\alpha)$ with $\xi > \sup x$.

Given an ordinal ρ and a finite sequence σ of elements of k we define what it means for $x \subseteq \lambda$ to be (ρ, σ)-*good*. x is $(\rho, \langle \rangle)$-good iff x is a singleton. If $\sigma = \tau^\frown \langle i \rangle$ then x is (ρ, σ)-good iff $x = \bigcup \{ x_\xi : \xi < \rho \}$ where each x_ξ is (ρ, τ)-good, $\xi < \eta < \rho$ implies $\sup x_\xi < \inf x_\eta$, and whenever $\xi < \eta < \rho$, $\gamma \in x_\xi$ and $\delta \in x_\eta$ then $f\{\gamma, \delta\} = i$. Note that any (ρ, σ)-good set must have order type ρ^n where $n = \text{length}(\sigma)$.

Lemma 3.5. *If* x *is* (ρ, σ)-*good then* $\forall i \in \text{range}\,\sigma$ $\exists y \subseteq x$ y *is* i-*homogeneous and has order type* ρ.

Proof. Easy by an induction on length(σ).

Lemma 3.6. *If* $X \in I^+(\alpha, \sigma)$ *then* $\forall \rho < \kappa$ $\exists x \subseteq X$ x *is* (ρ, σ)-*good.*

Proof. Say $\sigma = \tau^\frown \langle i \rangle$. We construct $x = \bigcup \{ x_\xi : \xi < \rho \} \subseteq X \cap f_i(\alpha)$. By induction on ξ we obtain (ρ, τ)-good $x_\xi \subseteq X \cap f_i(\alpha)$. Suppose x_η has been obtained for $\eta < \xi$. Since $\rho < \kappa$ we know $x_\eta \in N_\alpha$ for each $\eta < \xi$ and hence $\langle x_\eta : \eta < \xi \rangle \in N_\alpha$. Thus $A = \{\beta : \forall \eta < \xi \, \forall \gamma \in x_\eta \, f\{\gamma, \beta\} = i\} \in N_\alpha$. Moreover, $\alpha \in A \in I_\alpha^* \subseteq I^*(\alpha, \sigma)$, so

$X \cap A \in I^+(\alpha, \sigma)$. Find $\beta < \alpha$ such that $X \cap A \cap f_i(\alpha) \cap \beta \in I^+(\beta, \tau)$ and choose (ρ, τ)-good $x_\xi \subseteq X \cap A \cap f_i(\alpha) \cap \beta$. We may assume β is large enough so that we may choose x_ξ with $\sup x_\eta < \min x_\xi$ for all $\eta < \xi$. The rest is easy.

Recall that an *indecomposable* ordinal is an ordinal power of ω. An indecomposable ordinal is characterized by the property that whenever $\rho = A \cup B$, either A or B has order type ρ.

Lemma 3.7. *If ρ is indecomposable, $x \subseteq \lambda$ is (ρ, σ)-good, $m < \omega$ and $x = \bigcup \{ t_j : j < m \}$ then $\exists j\ \exists y \subseteq t_j\ y$ is (ρ, σ)-good.*

Proof. Let $\sigma = \tau^\frown \langle i \rangle$ and let $\langle x_\xi : \xi < \rho \rangle$ witness (ρ, σ)-goodness of x. By inductive hypothesis for each $\xi < \rho$ there is $j(\xi) < m$ and $y_\xi \subseteq x_\xi \cap t_{j(\xi)}$ such that y_ξ is (ρ, τ)-good (this is trivial if $\tau = \langle \rangle$). Since ρ is indecomposable there is j such that $\{ \xi : j(\xi) = j \}$ has order type ρ. Let $y = \bigcup \{ y_\xi : j(\xi) = j \}$. Then $y \subseteq t_j$ and y is (ρ, σ)-good.

If σ is a finite sequence of elements of k let $f_\sigma(\alpha) = \bigcup \{ f_i(\alpha) : i \in \text{range}\,\sigma \}$.

Recall that we defined $S_{n+1} = \{ \alpha \in S_n : \alpha \text{ is a limit point of } S_n \}$. Thus $S_\omega = \bigcap \{ S_n : n < \omega \}$ differs from S_0 by a nonstationary set, and $I(\alpha, \sigma)$ is defined for all σ whenever $\alpha \in S_\omega$.

If $T \subseteq S_\omega$ is stationary and $\alpha \in T$, let $\Sigma(\alpha, T) = \{ \sigma : \sigma \text{ is one-to-one, } \text{length}(\sigma) \geq 1, \text{ and } T \cap \alpha \in I^+(\alpha, \sigma) \}$. Note that if α is a reflection point of T then $\Sigma(\alpha, T) \neq 0$.

Lemma 3.8. *There is stationary $S \subseteq S_\omega$ and Σ such that for all stationary $T \subseteq S$ there is a closed unbounded set C such that $\forall \alpha \in T \cap C\ \Sigma(\alpha, T) = \Sigma$.*

Proof. Note that if $\alpha \in T \subseteq S$ then $\Sigma(\alpha, T) \subseteq \Sigma(\alpha, S)$. Choose descending sequences

$$S_\omega = T_0 \supseteq T_1 \supseteq T_2 \supseteq \dots \quad \text{and} \quad \Sigma_0 \supseteq \Sigma_1 \supseteq \Sigma_2 \supseteq \dots$$

such that $\forall n\ \forall \alpha \in T_{n+1}\ \Sigma(\alpha, T_n) = \Sigma_n$, the T_n are stationary and Σ_{n+1} is a proper subset of Σ_n. Since each Σ_n is finite this process must end after finitely many steps. Let S and Σ be the final elements of each sequence.

Clearly $\Sigma \neq 0$. Choose $\sigma \in \Sigma$ maximal with respect to inclusion.

Lemma 3.9. *There are $\alpha \in S$ and stationary $T \subseteq S$ such that $\forall \beta \in T\ \alpha - f_\sigma(\beta) \in I(\alpha, \sigma)$.*

Proof. Suppose the lemma is false. Then for each $\alpha \in S$ there is closed unbounded C_α such that $\forall \beta \in C_\alpha \cap S\ \alpha - f_\sigma(\beta) \in I^+(\alpha, \sigma)$.

Let $C = \{ \beta : \forall \alpha \in \beta \cap S\ \beta \in C_\alpha \}$, the diagonal intersection of the C_α. Then C is closed unbounded. Let $\alpha \in S$ be a reflection point of $C \cap S$ with $\sigma \in \Sigma(\alpha, S)$. For all

$\beta \in C \cap S \cap \alpha$ there must be $i \notin \text{range}(\sigma)$ such that $\beta \cap f_i(\alpha) \in I^+(\beta, \sigma)$. Fix $i < k$ such that $\{\beta \in C \cap S \cap \alpha : \beta \cap f_i(\alpha) \in I^+(\beta, \sigma)\} \in I_\alpha^+$. But now clearly if $\tau = \sigma^\frown\langle i\rangle$ then $S \cap \alpha \in I^+(\alpha, \tau)$, contradicting maximality of σ.

Now we are ready to put all this material together to complete the proof of Theorem 3.1.

Fix an indecomposable ordinal $\rho < \log \kappa$. It will suffice to find a homogeneous set of order type $\kappa + \rho$. Let α and T be as in Lemma 3.9. Choose $\beta > \alpha$ such that $\Sigma(\beta, T) = \Sigma$. Since $T \cap \beta \in I^+(\beta, \sigma)$ there is $x \subseteq T$ that is (ρ, σ)-good. We may assume $\min x > \alpha$. Also, since $I(\alpha, \sigma)$ is κ-complete we know $F = \bigcup\{\alpha - f_\sigma(\beta) : \beta \in x\} \in I(\alpha, \sigma)$. For each $\gamma \in \alpha - F$ define $g_\gamma : x \to \text{range}(\sigma)$ by $g_\gamma(\beta) = f\{\gamma, \beta\}$. Since $|x| < \log \kappa$ there is g such that $\{\gamma : g_\gamma = g\} \in I^+(\alpha, \sigma)$. Also, by Lemma 3.7 $\exists i \; g^{-1}(i)$ is (ρ, σ)-good.

But now by Lemma 3.4 $\exists W \subseteq \{\gamma : g_\gamma = g\}$ such that W is i-homogeneous, and by Lemma 3.5 we may find $y \subseteq g^{-1}(i)$ such that y is i-homogeneous. Also, if $\gamma \in W$ and $\beta \in y$ than $f\{\gamma, \beta\} = g_\gamma(\beta) = g(\beta) = i$ so $W \cup y$ is i-homogeneous of order type $\kappa + \rho$, and the proof is complete.

4   An unbalanced extension of the Erdős–Rado Theorem.

It is natural to ask if Theorem 3.1 can be improved in the same way that Theorem 2.1 was improved to Theorem 2.3. For example, is it true under CH that $\omega_2 \to (\omega_2, \omega_1 + 2)^2$? Hajnal showed in [6] that if GCH holds and κ is regular, then $\kappa^+ \nrightarrow (\kappa^+, \kappa + 2)^2$, and Todorčević has shown in unpublished work that this remains the case when κ is singular. Thus the best we could hope to prove under CH would be $\omega_2 \to (\alpha, \omega_1 + 2)^2$ for all $\alpha < \omega_2$. In this section we will prove (a generalization of) an improvement of part of this.

Theorem 4.1. *Suppose κ is regular and $\lambda = (2^{<\kappa})^+$. Then $\lambda \to (\rho, (\kappa + n)_k)^2$ for all $n, k < \omega$, where $\rho = \kappa^{\omega+2} + 1$.*

Here $\kappa^{\omega+2}$ represents ordinal exponentiation. Recall that the partition relation means that if $f : [\lambda]^2 \to k + 1$, then either there is a 0-homogeneous set of order type ρ or else there is an i-homogeneous set of type $\kappa + n$ for some $i > 0$.

The rest of this section will be devoted to the proof of Theorem 4.1. The strategy of the proof is to derive Theorem 4.1 from the auxiliary assumption $Q(\kappa)$, which asserts that $2^{<\kappa} = \kappa$ and in addition that

$(*)$ $\forall\langle f_\alpha : \alpha < \kappa^+\rangle \subseteq {}^\kappa\kappa \; \exists g \in {}^\kappa\kappa \; \forall\alpha \; \exists\eta < \kappa \; \forall\xi > \eta \; f_\alpha(\xi) < g(\xi)$.

Then, rather as in [2], we observe that the assumption $Q(\kappa)$ is unnecessary, and therefore that Theorem 4.1 holds in ZFC.

Let us deal with this latter observation first.

Let P_0 be the natural κ-closed ordering for making $2^{<\kappa} = \kappa$. Then in V^{P_0} we have $\lambda = \kappa^+$. Working in V^{P_0} and using a standard iterated forcing argument (as in [1]) we can force (∗) to be true via a partial ordering P_1 that is κ-closed and has the λ-chain condition. Let $P = P_0 * P_1$. Then P is κ-closed and in V^P $\lambda = \kappa^+$ and $Q(\kappa)$ holds. Note that in V^P we will have $2^\kappa > \kappa^+$ (since for one thing that is implied by (∗)).

Assuming we have proved Theorem 4.1 under the assumption $Q(\kappa)$ we may assume it holds in V^P. Assume $f : [\lambda]^2 \to k + 1$, where $f \in V$. Then in V^P there is $A \subseteq \lambda$ such that either (a) A is 0-homogeneous of type ρ or (b) A is i-homogeneous of type $\kappa + n$, where $i > 0$. Suppose (a) holds. Note that $\kappa^{\omega+2} + 1$ is the same whether computed in V or in V^P. Let $h : \kappa \to \rho$ be a bijection with $h \in V$. In V^P fix an order-isomorphism $j : \rho \to A$. Now, working in V, find a decreasing sequence $\langle p_\xi : \xi < \kappa \rangle$ of elements of P and a sequence $\langle \alpha_\xi : \xi < \kappa \rangle$ of elements of λ such that $\forall \xi\ p_\xi \Vdash j(h(\xi)) = \alpha_\xi$. This is easy to do by induction on ξ, using the fact that P is κ-closed. But now it is clear that $\{ \alpha_\xi : \xi < \kappa \} \in V$ has order type ρ and is 0-homogeneous. Case (b) may be handled the same way.

From now on, assume $Q(\kappa)$ holds (so $\lambda = \kappa^+$).

We begin with a couple of observations about order types. The only use of $Q(\kappa)$ will be in Corollary 4.3 below.

Lemma 4.2. *Let $n < \omega$, $n \geq 1$, $\alpha < \kappa^+$, and assume α has a cofinal subset of type κ^n. Then for every $f : \kappa \to \kappa$ there is a set $A(f, n, \alpha)$ of order type κ^n cofinal in α such that*

(i) *if A is cofinal in α of type κ^n then $\exists f\ A \subseteq A(f, n, \alpha)$*

(ii) *$\forall f_1, f_2$ if $\forall \xi < \kappa\ f_1(\xi) \leq f_2(\xi)$ then $A(f_1, n, \alpha) \subseteq A(f_2, n, \alpha)$.*

Proof: The proof is by induction on n. Suppose $n = 1$. Let $\langle \alpha_\xi : \xi < \kappa \rangle$ be an increasing, continuous sequence cofinal in α with $\alpha_0 = 0$. Let $A_\xi = \alpha_{\xi+1} - \alpha_\xi$ and let $h_\xi : \kappa \to A_\xi$ be onto. Define $A(f, n, \alpha) = \bigcup\{ h_\xi \text{``} f(\xi) : \xi < \kappa \}$. It is easy to see that this works.

Now suppose $n > 1$. Let $\langle \alpha_\xi : \xi < \kappa \rangle$ be as above. If $\alpha_\xi < \beta \leq \alpha_{\xi+1}$ then for $1 \leq i < n$ let $A'(f, i, \beta) = A(f, i, \beta) \cap (\alpha_{\alpha+1} - \alpha_\xi)$ if $A(f, i, \beta)$ is defined; let $A'(f, i, \beta) = \{\beta\}$ otherwise. Let $A(f, 0, \beta) = \{\beta\}$. Now let $f : \kappa \to \kappa$. We must define $A(f, n, \alpha)$. We may regard f as being defined on $\kappa \times \kappa$, i.e., $f : \kappa \times \kappa \to \kappa$, and we define f_ξ by $f_\xi(\eta) = f(\xi, \eta)$ for $\xi, \eta < \kappa$. Let $h : \kappa \to \alpha$ be a bijection. Now let

$$A(f, n, \alpha) = A(f_0, 1, \alpha) \cup \bigcup\{ A'(f_\xi, i, h(\xi)) : h(\xi) \in A(f_0, 1, \alpha), i < n \}.$$

Since $A(f_0, 1, \alpha)$ has type κ, it is clear that $\bigcup \{ A'(f_\xi, i, h(\xi)) : h(\xi) \in A(f_0, 1, \alpha) \cap (\alpha_{\eta+1} - \alpha_\eta) \}$ has type $< \kappa^n$ for fixed η. Thus $A(f_0, 1, \alpha)$ has type at most κ^n.

It is clear that condition (ii) is satisfied. Let us check (i). Fix $A \subseteq \alpha$ cofinal of type κ^n. For $\eta < \kappa$ let $A_\eta = A \cap (\alpha_{\eta+1} - \alpha_\eta)$. Then A_η has type $< \kappa^n$ so there is a family F_η such that $|F_\eta| < \kappa$, $A_\eta = \bigcup F_\eta$, every $B \in F_\eta$ has type κ^i for some $i < n$ (possibly $i = 0$), and if $B, C \in F_\eta$ and $B \neq C$ then $\sup B < \inf C$ or $\sup C < \inf B$. Let $H_\eta = \{ \sup B : B \in F_\eta \}$. Now choose $f : \kappa \times \kappa \to \kappa$ so that $\bigcup \{ H_\eta : \eta < \kappa \} \subseteq A(f_0, 1, \alpha)$ and whenever $B \in F_\eta$, B has type κ^i for $i > 0$, and $\sup B = \beta = h(\xi)$, then $B \subseteq A(f_\xi, i, \beta)$. But now it is clear that $A \subseteq A(f, n, \alpha)$.

There may be special circumstances in which the $A(f, n, \alpha)$ as defined above have order type $< \kappa^n$. This defect may be remedied by choosing a fixed set of order type κ^n cofinal in α and adjoining its elements to every $A(f, n, \alpha)$.

Corollary 4.3. $(Q(\kappa))$ *Let $\alpha < \kappa^+$ and let A_γ, $\gamma < \kappa^+$, be a sequence of sets of order type κ^n cofinal in α, where $1 \leq n \leq \omega$. Then we may write $\kappa^+ = \bigcup \{ X_\xi : \xi < \kappa \}$ where $\bigcup \{ A_\gamma : \gamma \in X_\xi \}$ has order type κ^n for each ξ.*

Proof. First suppose $n < \omega$. For each γ choose $f_\gamma : \kappa \to \kappa$ such that $A_\gamma \subseteq A(f_\gamma, n, \alpha)$. Use $Q(\kappa)$ to find $f : \kappa \to \kappa$ such that $\forall \gamma \, \exists \xi_\gamma \, \forall \eta > \xi_\gamma \, f_\gamma(\eta) < f(\eta)$. Thus there is $s_\gamma : \xi_\gamma \to \kappa$ such that if $g_\gamma = s_\gamma \cup f|(\kappa - \xi_\gamma)$ then $\forall \eta \, f_\gamma(\eta) < g_\gamma(\eta)$. By $\kappa^{<\kappa} = \kappa$ there are only κ functions s_γ so we may write $\kappa^+ = \bigcup \{ X_\xi : \xi < \kappa \}$ where if $\gamma, \delta \in X_\xi$ then $s_\gamma = s_\delta$, so $g_\gamma = g_\delta$. But then

$$\bigcup \{ A_\gamma : \gamma \in X_\xi \} \subseteq \bigcup \{ A(f_\gamma, n, \alpha) : \gamma \in X_\xi \} \subseteq A(g_{\gamma_0}, n, \alpha)$$

by Lemma 4.2(ii) where γ_0 is any member of X_ξ.

Now suppose $n = \omega$. For each $\gamma < \kappa^+$ let $g_\gamma(0) = 0$ and if $1 \leq m < \omega$ let $g_\gamma(m)$ be the supremum of the first κ^m elements of A_γ and let $A_\gamma^m = A_\gamma \cap [g_\gamma(m-1), g_\gamma(m))$. Since $\kappa^\omega = \kappa$ we may write $\kappa^+ = \bigcup \{ Y_\xi : \xi < \kappa \}$ where for each $\gamma, \delta \in Y_\xi$ we have $g_\gamma = g_\delta$. And now by the earlier case, for each m we may write $Y_\xi = \bigcup \{ Y_{\xi\eta}^m : \eta < \kappa \}$, where $\bigcup \{ A_\gamma^m : \gamma \in Y_{\xi\eta}^m \}$ has order type κ^m for each ξ and η. For each γ let $h_\gamma(m) = (\xi, \eta)$ if $\gamma \in Y_{\xi\eta}^m$. Finally put $\kappa^+ = \bigcup \{ X_\xi : \xi < \kappa \}$ where for $\gamma, \delta \in X_\xi$ we have $h_\gamma = h_\delta$. It is easy to check that this works.

This is the only use of $Q(\kappa)$ that we need.

For the main proof let us fix a partition function $f : [\lambda]^2 \to k + 1$, and assume that there is no i-homogeneous set of type $\kappa + n$, where $i > 0$. We must find a 0-homogeneous set of type ρ.

If $H(\lambda^+)$ is the collection of sets hereditarily of cardinality $\leq \lambda$, then $f \in H(\lambda^+)$. Let $S = \{N : N \prec H(\lambda^+), |N| = \kappa, N \cap \lambda \in \lambda, f \in N, \text{ and } [N]^{<\kappa} \subseteq N\}$. Then S is a stationary subset of $[H(\lambda^+)]^\kappa$.

Define $\pi : S \to \lambda$ by $\pi(N) = N \cap \lambda$. If $T \subseteq S$ is stationary then let $J_T = \{X \subseteq \lambda : \pi^{-1}(X) \cap T \text{ is nonstationary}\}$. It is easy to see that J_T is a normal ideal on λ.

Claim 4.4. *Let $\xi < \omega + 2$. Then*

   (a) *if $T \subseteq S$ is stationary, then there is a set C closed and unbounded in $[H(\lambda^+)]^\kappa$*
      such that for all $N \in C \cap T$, if $A \in N$ and $\pi(N) \in A$ then there is $x \in N$ such
      that $x \subseteq A \cap \pi(T)$, x is of order type κ^ξ, and $x \cup \{\pi(N)\}$ is 0-homogeneous;

   (b) *if $T \subseteq S$ is stationary then there is stationary $U \subseteq T$ and $x \subseteq \pi(T)$ such that x*
      is 0-homogeneous of type κ^ξ and $\forall \alpha \in x \; \forall \beta \in \pi(U) \; f\{\alpha, \beta\} = 0$.

Let us observe that Claim 4.4(a) for $\xi = \omega + 1$ will complete the proof. Fix $N \in C \cap S$. By induction on $\gamma < \kappa$ we define $A_\gamma \in N$ with $\pi(N) \in A_\gamma$ and $x_\gamma \subseteq A_\gamma$ such that x_γ has order type $\kappa^{\omega+1}$ and $x_\gamma \cup \{\pi(n)\}$ is 0-homogeneous. Let A_0 be arbitrary and choose x_0 as in 4.4(a). Suppose A_γ, x_γ have been determined for $\gamma < \delta$. Since $N \in S$ we know $[N]^{<\kappa} \subseteq N$ so $\langle x_\gamma : \gamma < \delta \rangle \in N$. Thus

$$A_\delta = \left\{\alpha : \bigcup\{x_\gamma : \gamma < \delta\} \cup \{\alpha\} \text{ is 0-homogeneous and } \sup\bigcup\{x_\gamma : \gamma < \delta\} < \alpha\right\} \in N$$

and $\pi(N) \in A_\delta$. Choose $x_\delta \subseteq A_\delta$ as in 4.4(a). This completes the construction. And now $\bigcup\{x_\gamma : \gamma < \kappa\} \cup \{\pi(N)\}$ is 0-homogeneous and has order type $\kappa^{\omega+1} \cdot \kappa + 1 = \kappa^{\omega+2} + 1 = \rho$, as desired.

Thus we may devote the rest of the section to Claim 4.4. The proof is by induction on ξ.

Let us begin the induction by showing that (b) holds for $\xi = 0$. If there is a $\alpha \in \pi(T)$ such that $X_\alpha = \{\beta : f\{\alpha, \beta\} > 0\} \in J_T^+$, then let $x = \{\alpha\}$ and $U = T \cap \pi^{-1}X_\alpha$. If not, then the diagonal intersection $X = \{\beta : \forall \alpha < \beta \; \beta \notin X_\alpha\}$ belongs to J_T^+, hence has cardinality λ. But now by Theorem 3.1 there is an i-homogeneous set of type $\kappa + n$ for some $i > 0$, contrary to hypothesis.

Next we show (b) implies (a). If (a) is false, then there is stationary $S' \subseteq S$ such that the assertion in (a) fails for all $N \in S'$. By normality of the nonstationary ideal on $[H(\lambda^+)]^\kappa$ there is a single set A and stationary $T \subseteq S'$ such that A is a counterexample to (a) for all $N \in T$. Since $\pi(N) \in A$ for $N \in T$ we have $\pi(T) \subseteq A$. Let x and U be as in (b). Then

$x \subseteq \pi(T) \subseteq A \cap \pi(S')$ and x is of type κ^ξ. Since $x \in H(\lambda^+)$ there must be $N \in U$ with $x \in N$. Moreover $x \cup \{\pi(N)\}$ is 0-homogeneous, so A is not in fact a counterexample for N, contradiction.

We now concentrate on showing that (a) for ξ implies (b) for $\xi + 1$, provided $\xi \leq \omega$.

Let T be as in (b), let C be as in (a), and let $\langle N_\alpha : \alpha < \lambda \rangle$ be an increasing continuous sequence of elements of C with $[N_\alpha]^{<\kappa} \subseteq N_{\alpha+1}$. Let $N_\lambda = \bigcup \{ N_\alpha : \alpha < \lambda \}$. We may assume that for all $N \in T$, if $\pi(N) = \alpha$ then $N \cap N_\lambda = N_\alpha$. (If $(N \cap N_\lambda) - N_{\pi(N)} \neq 0$ for stationarily many N, then for stationarily many N there is a common element of $(N \cap N_\lambda) - N_{\pi(N)}$, and this is impossible.) We also assume that if $N \in T$ and $N \cap N_\lambda = N_\alpha$, then $N_\alpha \cap \lambda = \alpha$.

Let Σ_0 be the set of all one-to-one sequences σ of elements of $k+1$ such that $\sigma(0) = 0$. For each $\alpha \in \pi(C \cap T)$ and each $\sigma \in \Sigma_0$ we will define an ideal $I(\alpha, \sigma)$ on α. Each $I(\alpha, \sigma)$ will contain all bounded subsets of α but will not necessarily be closed under countable unions.

First suppose $\sigma = \langle 0 \rangle$. Choose $N \in T \cap C$ with $\pi(N) = \alpha$. Let $\langle A_\gamma : \gamma < \kappa \rangle$ enumerate $\{ A \in N : \alpha \in A \}$. We define $\langle x_\gamma : \gamma < \kappa \rangle$ by induction on γ so that $x_\gamma \in N$, x_γ has type κ^ξ and $x_\gamma \cup \{\alpha\}$ is 0-homogeneous. Suppose x_γ, $\gamma < \delta$, has been obtained. Let

$$A = \bigcap \{ A_\gamma : \gamma < \delta \} \cap \Big\{ \beta : \beta > \sup\{ x_\gamma : \gamma < \delta \} \text{ and } \forall \gamma < \delta \; \forall \beta' \in x_\gamma \; f\{\beta, \beta'\} = 0 \Big\}.$$

Note that $A \in N$ since it is definable from $\langle A_\gamma : \gamma < \delta \rangle$ and $\langle x_\gamma : \gamma < \delta \rangle$, both of which are in N, and clearly $\alpha \in A$. By 4.4(a) choose $x_\delta \in N$, 0-homogeneous of type κ^ξ, such that $x_\delta \subseteq A \cap \pi(T)$. Thus $\bar{x}(\alpha) = \bigcup \{ x_\gamma : \gamma < \kappa \}$ is 0-homogenous of type $\kappa^{\xi+1}$, and is cofinal in α since for any $\beta < \alpha \; \lambda - \beta = A_\gamma$ for some γ. Let $I(\alpha, \langle 0 \rangle) = \{ X \subseteq \alpha : X \cap \bar{x}(\alpha) \text{ has type } < \kappa^{\xi+1} \}$.

Now suppose $\tau \in \Sigma_0$ and $\sigma = \tau^\frown \langle i \rangle$. Let I_α be the ideal on α generated by $\{ A \cap \alpha : A \in N_\alpha, \alpha \notin A \}$, as in the previous section. Recall that $f_i(\alpha) = \{ \beta < \alpha : f\{\beta, \alpha\} = i \}$. For $X \subseteq \alpha$, put $X \in I(\alpha, \sigma)$ iff $\{ \beta < \alpha : X \cap f_i(\alpha) \cap \beta \in I^+(\beta, \tau) \} \in I_\alpha$.

Next we prove analogues of Lemmas 3.3, 3.4, and 3.9.

Lemma 4.5. $I_\alpha \subseteq I(\alpha, \sigma)$ for $\alpha \in \pi(C \cap T)$, $\sigma \in \Sigma_0$.

Proof. Let $X \in I^+(\alpha, \sigma)$. If $\sigma = \langle 0 \rangle$, then by construction $X \cap A \neq 0$ for all $A \in N_\alpha$ with $\alpha \in A$. Thus $X \in I_\alpha^+$. Suppose $\sigma = \tau^\frown \langle i \rangle$, where $\tau \in \Sigma_0$. Let $A \in N_\alpha$, $\alpha \in A$. Since α is limit $\exists \gamma < \alpha \; A \in N_\gamma$. Choose β such that $\gamma < \beta < \alpha$, $\beta \in A$, and $X \cap \beta \in I^+(\beta, \tau)$. By inductive hypothesis $X \cap \beta \in I_\beta^+$, so $X \cap \beta \cap A \neq 0$. Thus $X \cap A \neq 0$ so $X \in I_\alpha^+$.

Lemma 4.6. If $X \in I^+(\alpha, \sigma)$, then for all $j \in \mathrm{range}\,\sigma$, if $j > 0$ then X contains a j-homogeneous set of type κ.

Proof. We proceed by induction. Say $\sigma = \tau^\frown\langle i\rangle$ where $\tau \in \Sigma_0$. The induction will handle the case $j \in \text{range}\,\tau$. It will suffice to show, as in Section 3, that if $x \cup \{\alpha\}$ is i-homogeneous and $|x| < \kappa$ then $\exists \beta \in X - x$ $x \cup \{\beta, \alpha\}$ is i-homogeneous. Let $A = \{\gamma : x \cup \{\gamma\}$ is i-homogeneous $\}$. Then $A \in N_\alpha$ and $\alpha \in A$. Since $X \in I^+(\alpha,\sigma)$ there is $\gamma \in A$ such that $X \cap f_i(\alpha) \cap \gamma \in I^+(\gamma,\tau)$. We may choose γ large enough that $A \in N_\gamma$ and $\gamma > \sup x$. Thus $X \cap f_i(\alpha) \cap \gamma \cap A \neq 0$ and we may choose $\beta \in X \cap f_i(\alpha) \cap \gamma \cap A$, $\beta > \sup x$.

For $\beta < \lambda$ and $\sigma \in \Sigma_0$, let $f_\sigma(\beta) = \bigcup\{ f_i(\beta) : i \in \text{range}\,\sigma\}$.

Lemma 4.7. *There are $\sigma \in \Sigma_0$, $\alpha \in \pi(C \cap T)$, and stationary $T' \subseteq T$ such that $\alpha \in I^+(\alpha,\sigma)$ (i.e., $I(\alpha,\sigma)$ is proper) and $\forall\beta \in \pi(T')\ \alpha - f_\sigma(\beta) \in I(\alpha,\sigma)$.*

Proof. For $\alpha \in \pi(C \cap T)$ we know $\pi(T) \in I^+(\alpha,\langle 0\rangle)$. Choose $\sigma_\alpha \in \Sigma_0$ maximal with respect to inclusion such that $\pi(T) \in I^+(\alpha,\sigma)$, and let $U \subseteq C \cap T$ be stationary such that $\forall\alpha \in \pi(U)\ \sigma_\alpha = \sigma$.

Suppose the lemma is false. Then for each $\alpha \in \pi(U)$ there is $C_\alpha \in J_U^*$ such that $\forall\beta \in C_\alpha$ $\alpha - f_\sigma(\beta) \in I^+(\alpha,\sigma)$. (Recall that $J_U = \{ X \subseteq \lambda : \pi^{-1}(X) \cap U$ is stationary $\}$.) Since J_U is normal, the set $C_\lambda = \{\beta : \forall\alpha < \beta\ \beta \in C_\alpha\} \in J_U^*$ also. By Lemma 3.2 there is closed unbounded $D \subseteq \lambda$ such that $\forall\alpha \in D \cap \pi(U)\ \pi(U) \cap \alpha \in I_\alpha^+$. Fix $\alpha \in D \cap C_\lambda \cap \pi(U)$.

If $\beta < \alpha$ and $\beta \in \pi(U)$ then $\alpha \in C_\beta$ so there is i, $i \notin \text{range}\,\sigma$, such that $f_i(\alpha) \cap \beta \in I^+(\beta,\sigma)$. Since k is finite we may fix i such that $\{\beta < \alpha : \beta \in \pi(U)$ and $f_i(\alpha) \cap \beta \in I^+(\beta,\sigma)\} \in I_\alpha^+$. But this means that $I(\alpha,\sigma^\frown\langle i\rangle)$ is proper contrary to the assumption about the maximality of σ.

One more lemma will allow us to complete this part of the proof of Claim 4.4.

Lemma 4.8. *Let $\alpha < \lambda$ and $\sigma \in \Sigma_0$ be arbitrary. If $X \in I^+(\alpha,\sigma), Y \in I(\alpha,\sigma)$ and $Y \subseteq X$, then $\exists\beta \leq \alpha$ $\bar{x}(\beta) \cap X$ has type $\kappa^{\xi+1}$ and $Y \cap \bar{x}(\beta)$ has type $< \kappa^{\xi+1}$.*

Proof. If $\sigma = \langle 0\rangle$ this is clear. Just take $\beta = \alpha$. Suppose $\sigma = \tau^\frown\langle i\rangle$ for $\tau \in \Sigma_0$. Then we can find $\beta < \alpha$ such that $X \cap f_i(\alpha) \cap \beta \in I^+(\beta,\tau)$ and $Y \cap \beta \in I(\beta,\tau)$, so we are done by inductive hypothesis.

Finally, let α, σ and T' be as in Lemma 4.7. For $\beta \in \pi(T')$ let $g(\beta) = \bigcup\{ f_i(\beta) \cap \alpha : i \in \text{range}\,\sigma, i > 0\}$. Let $F \subseteq \pi(T')$ be maximal such that $\{\bigcap\{ g(\beta) : \beta \in x\} : x \in [F]^{<\omega}\} \subseteq I^+(\alpha,\sigma)$. (It is possible that $F = 0$.)

First, suppose $|F| = \lambda$. Let G be an ultrafilter on α such that $\{ g(\beta) : \beta \in F\} \subseteq G \subseteq I^+(\alpha,\sigma)$. Thus for each $\beta \in F$ there is i_β such that $f_{i_\beta}(\beta) \cap \alpha \in G$. Find i so that $F' = \{\beta \in F : i_\beta = i\}$ has cardinality λ. We may assume we are proving Theorem 4.1 by induction on k, so by $\lambda \to (\lambda,\kappa)^2$, which follows from Theorem 2.3, we may assume there is $x \subseteq F'$ i-homogeneous of cardinality n. But now the set $\bigcap\{ f_i(\beta) \cap \alpha : \beta \in x \}$ belongs to $I^+(\alpha, \sigma)$ so by Lemma 4.6 must contain an i-homogeneous set H of type κ. And now $H \cup x$ is i-homogeneous of type $\kappa + n$, contrary to our hypothesis about the partition function f. Thus $|F| < \lambda$.

From the maximality of F it follows that if $N \in T'$ and $\pi(N) \notin F$ then there is a finite set $y_N \subseteq F$ such that $\bigcap\{ g(\beta) : \beta \in y_N \} - f_0(\pi(N)) \in I(\alpha, \sigma)$. Since $|F| < \lambda$ there is y such that $T'' = \{ N \in T' : y_N = y \}$ is stationary. For $N \in T''$ let $a_N = \bigcap\{ g(\beta) : \beta \in y \} - f_0(\pi(N))$. By Lemma 4.8 there is $\beta_N \leq \alpha$ such that $\bar{x}(\beta_N) \cap \bigcap\{ g(\beta) : \beta \in x \}$ has type $\kappa^{\xi+1}$ and $\bar{x}(\beta_N) \cap a_N$ has type $< \kappa^{\xi+1}$. There is stationary $T''' \subseteq T''$ and β such that $T''' = \{ N \in T'' : \beta_N = \beta \}$.

By thinning out T''' we may assume without loss of generality that each a_N for $N \in T'''$ is such that $a_N \subseteq (\bar{x}(\beta) \cap \zeta) \cup b_N$ where $\zeta < \sup \bar{x}(\beta)$ and b_N has type κ^η for some $\eta \leq \xi$, and that η and ζ are the same for all N. Now it follows from Corollary 4.3 that $T''' = \bigcup\{ T_\delta : \delta < \kappa \}$ where for each δ, $\bigcup\{ a_N : N \in T_\delta \}$ has type $< \kappa^{\xi+1}$. Pick such δ for which T_δ is stationary. Let $U = T_\delta$ and let $x = \bar{x}(\beta) \cap \bigcap\{ g(\beta) : \beta \in y \} - \bigcup\{ a_N : N \in T_\delta \}$. Then x is 0-homogeneous of type $\kappa^{\xi+1}$ and $\forall \alpha \in x \; \forall \beta \in \pi(U) \; f\{\alpha, \beta\} = 0$.

This completes the proof of Claim 4.4(b) for $\xi + 1$ from Claim 4.4(a) for ξ, provided $\xi \leq \omega$.

To complete the proof of Claim 4.4, and hence Theorem 4.1, let us observe that 4.4(a) for $\xi = \omega$ follows from 4.4(a) for $\xi < \omega$. Let C_n be a closed unbounded set as in 4.4(a) for $\xi = n$, and let $C = \bigcap\{ C_n : n < \omega \}$. Let $N \in C \cap T$, $A \in N$, and $\pi(N) \in A$. By induction on n choose $x_n \in N$ as follows. Let x_0 be as in 4.4(a) for $\xi = 0$. Given x_m for $m < n$, let

$$A_n = \{ \beta : \bigcup\{ x_m : m < n \} \cup \{\beta\} \text{ is 0-homogeneous } \}.$$

Then $A_n \in N$ and $\pi(N) \in A$, so we may find $x_n \in N$ as in 4.4(a) with A replaced by A_n. But now $x = \bigcup\{ x_n : n < \omega \} \in N$ and satisfies 4.4(a) with $\xi = \omega$.

5   An extension of the Erdős–Rado Theorem to ordinals

In this section we prove a theorem implying that if CH holds then for any $n < \omega$ there is $m < \omega$ such that $\omega_2 \cdot m \to (\omega_1 \cdot n)_2^2$. This is an improvement of a result of Shelah [9] that implies under similar circumstances $\omega_2^m \to (\omega_1 \cdot n)_2^2$.

Let A be a set and let $k < \omega$. A *set-mapping* on A of order k is a function $p : A \to [A]^{<k}$ such that $a \notin p(a)$ for all $a \in A$. Given such a mapping p, a set $F \subseteq A$ is said to be *free* for p if $\forall a, b \in F \; a \notin p(b)$. It is well-known that if $n, k < \omega$ then there is $g(n, k) < \omega$ such that any set-mapping on $g(n,k)$ of order k has a free set of cardinality n. (This is easy to see using elementary Ramsey theory, for example.)

Let $n,k < \omega$.

We define $f(n,k)$ by induction on k. Let $f(n,0) = 1$, and let $f(n,k+1) = g(n,f(n,k))$.

Theorem 5.1. *Let κ be regular and let $\lambda = (2^{<\kappa})^+$. Then for all $n,k < \omega$ we have*

$$\lambda \cdot f(n,k) \to (\kappa \cdot n, \kappa \cdot (k+1))^2.$$

Proof. The proof is by induction on k. If $k = 0$ then $f(n,k) = 1$ so we must show $\lambda \to (\kappa \cdot n, \kappa)^2$, and this follows easily from Theorem 2.3, which implies $\lambda \to (\lambda, \kappa)^2$.

So we may assume $k > 0$. Let $m = f(n,k)$, and suppose $h : [m \times \lambda]^2 \to 2$. Note that with the lexicographic order, $m \times \lambda$ has order type $\lambda \cdot m$.

Choose $N_0 \prec N_1 \prec \cdots \prec N_{m-1} \prec H(\lambda^+)$ such that $N_i \in N_{i+1}$, $|N_i| = 2^{<\kappa}$, $N_i \cap \lambda \in \lambda$, and $[N_i]^{<\kappa} \subseteq N_i$ for all i. For each i let $\beta_i = N_i \cap \lambda$.

Let $A_i = \{i\} \times \lambda$. Without loss of generality we may assume that for each i there is n_i such that A_i contains no 1-homogeneous set of type $\kappa \cdot (n_i + 1)$ but for all $B \subseteq A_i$, if $|B| = \lambda$ then B contains a 1-homogeneous set of type $\kappa \cdot n_i$. Of course if $n_i \geq k+1$ we are done, so we may assume $n_i \leq k$ for each i.

By induction on $\xi < \kappa$ we will define $a_{i\xi} \in A_i \cap N_0$ for each $i < m$, and we will obtain a set-mapping g_ξ on m.

Fix i, and let $X(i,\xi) = \{a \in A_i : \forall \eta < \xi \; \forall j < m \; f\{a,a_{j\eta}\} = f\{(i,\beta_i),a_{j\eta}\}\}$. Since each $a_{j\eta} \in N_0$ and $[N_0]^{<\kappa} \subseteq N_0$, we have $X(i,\xi) \in N_0$ (note that the function assigning $a_{j\eta}$ to $f\{(i,\beta_i),a_{j\eta}\}$ must be in N_0). Since $(i,\beta_i) \in X(i,\xi)$ it is clear that $|X(i,\xi)| = \lambda$.

Now by $\lambda \to (\lambda,\kappa)^2$ we may assume $X(i,\xi)$ contains a 1-homogeneous set Y of order type κ (so $n_i \geq 1$ also), and we may assume $Y \in N_0$ since $N_0 \prec H(\lambda^+)$ and the statement that such a Y exists is true in $H(\lambda^+)$. If $Y \in N_0$, then in particular $Y \subseteq N_0$ as well.

We define $Y = Y_0 \supseteq Y_1 \supseteq \cdots \supseteq Y_m$ with $|Y_j| = \kappa$, as follows. Given Y_j, let $Y_{j+1} = \{y \in Y_j : f\{(j,\beta_j),y\} = 0\}$ if this set has cardinality κ; otherwise set $Y_{j+1} = Y_j$ and put $j \in g_\xi(i)$. Finally, choose $a_{i\xi} \in Y_m$. Note that $a_{i\xi} \in N_0$.

For the following lemmas let us assume for convenience that Theorem 5.1 is false for k.

Lemma 5.2. $i \notin g_\xi(i)$.

Proof. Suppose $i \in g_\xi(i)$. Let $C = \{y \in Y_i : f\{(i,\beta_i),y\} = 0\}$. Since $|C| < \kappa$ we have $C \in [N_0]^{<\kappa} \subseteq N_0$. Now (i,β_i) belongs to $Z = \{b \in A_i : \forall y \in Y_i - C \; f\{b,y\} = 1\}$. But $Y_i \in N_i$ since it is defined using Y and the β_j for $j < i$ and all these belong to N_i. Thus $Z \in N_i$ and since $(i,\beta_i) \in Z$ we must have $|Z| = \lambda$. By our hypothesis on n_i there is $B \subseteq Z$,

1-homogeneous of type $\kappa \cdot n_i$, and we may assume $\min B > \max Y_i$. But now $(Y_i - C) \cup B$ is 1-homogeneous of type $\kappa \cdot (n_i + 1)$, contradiction. Hence $i \notin g_\xi(i)$.

Lemma 5.3. $|g_\xi(i)| < f(n, k-1)$.

Proof. Again, suppose otherwise. For each $j \in g_\xi(i)$, choose a set $C_j \in [Y_j]^{<\kappa}$ such that $(j, \beta_j) \in Z_j = \{ b \in A_j : \forall y \in Y_j - C_j \ f\{b, y\} = 1 \}$. Then $Z_j \in N_j$, and $|Z_j| = \lambda$.

By the inductive hypothesis for Theorem 5.1 we know $\lambda \cdot f(n, k-1) \to (\kappa \cdot n, \kappa \cdot k)^2$. If we apply this to $\bigcup \{ Z_j : j \in g_\xi(i) \}$, which has order type $\geq \lambda \cdot f(n, k-1)$, either this set contains a 0-homogeneous set of type $\kappa \cdot n$ (in which case we have established Theorem 5.1) or else it contains a 1-homogeneous set B of type $\kappa \cdot k$. But then $B \cup Y_m - \bigcup \{ C_j : j \in g_\xi(i) \}$ is 1-homogeneous of type $\kappa \cdot (k+1)$ and again we are done. Thus $|g_\xi(i)| < f(n, k-1)$.

Since there are only finitely many possibilities for g_ξ we may assume that they are all the same. Say $g_\xi = g$ for all $\xi < \kappa$. By the definition of $m = f(n, k)$ and Lemmas 5.2 and 5.3 we know there is a free set F for g of cardinality n.

Lemma 5.4. If $\xi \neq \eta$ and $j_1, j_2 \in F$ then $f\{a_{j_1\xi}, a_{j_2\eta}\} = 0$.

Proof. Assume $\xi < \eta$, $i = j_1$, $j = j_2$. Since $j \notin g(i)$ we must have had $a_{i\xi} \in Y_m \subseteq Y_{j+1} = \{ y \in Y_j : f\{(j, \beta_j), y\} = 0 \}$ during the choice of $a_{i\xi}$. Hence $f\{(j, \beta_j), a_{i\xi}\} = 0$, so necessarily $f\{a_{j\eta}, a_{i\xi}\} = 0$ also.

And now if $F = \{i_0, \ldots, i_{n-1}\}$ and $X_0, \ldots, X_{n-1}$ are disjoint subsets of κ, each of cardinality κ, then $\{ a_{i_l\xi} : l < n \text{ and } \xi \in X_l \}$ is 0-homogeneous of type $\kappa \cdot n$. This completes the proof of Theorem 5.1.

Corollary 5.5. $\lambda \cdot \omega \to (\kappa \cdot n)_2^2$ for all $n < \omega$.

It is natural to ask whether the subscript in Theorem 5.1 can be improved. By Theorem 3.1 we know that if $\rho < \log \kappa$ then $\lambda \to (\kappa + \rho)_3^2$. If $\log \kappa < \kappa$ then a κ-closed forcing construction yields the consistency of $\lambda \not\to (\kappa + \log \kappa)_2^2$ as in [7] (and we may arrange $2^\kappa > \lambda$ as well), and Corollary 5.5 shows $\lambda^+ \to (\kappa \cdot n)_2^2$ for all $n < \omega$. What about $\lambda^+ \to (\kappa + \log \kappa)_3^2$? No ordinal below λ^+ will work in view of the following. Throughout, we assume κ is regular and $\lambda = (2^{<\kappa})^+$.

Proposition 5.6. If $\lambda \not\to (\kappa + \log \kappa)_2^2$ then $\forall \alpha < \lambda^+ \ \alpha \not\to (\kappa + \log \kappa, \kappa + \log \kappa, \omega)^2$.

Proof. Suppose $f : [\lambda]^2 \to 2$ witnesses $\lambda \not\to (\kappa + \log \kappa)_2^2$. Define $g : [\alpha]^2 \to 3$ as follows. Let $\pi : \alpha \to \lambda$ be a bijection (we assume $\lambda \leq \alpha < \lambda^+$). If $\xi < \eta < \alpha$ then let $g\{\xi, \eta\} = 2$ if $\pi(\xi) > \pi(\eta)$ and let $g\{\xi, \eta\} = f\{\pi(\xi), \pi(\eta)\}$ otherwise. It is easy to see that this works.

Proposition 5.7. $\lambda^+ \to (\kappa \cdot n, \kappa \cdot n, \kappa + 1)^2$ *for all* $n < \omega$.

Proof. A proof exactly similar to Theorem 2.3 shows that $\lambda^+ \to (\lambda^+, \kappa + 1)^2$. Thus if $f : [\lambda^+]^2 \to 3$, either there is a 2-homogeneous set of type $\kappa + 1$ or else there is a set X of cardinality λ^+ such that $f''[X]^2 \subseteq \{0, 1\}$. And now in the latter case Theorem 5.1 may be applied to find a subset of X homogeneous of type $\kappa \cdot n$.

Question. Does $\lambda^+ \to (\kappa + \log \kappa, \kappa + \log \kappa, \kappa + 2)^2$? Assuming CH, the simplest nontrivial case is $\omega_3 \to (\omega_1 + \omega, \ \omega_1 + \omega, \ \omega_1 + 2)^2$ (and $2^{\omega_1} \geq \omega_3$).

References

[1] J. Baumgartner, Iterated forcing, *Surveys in Set Theory*, A. R. D. Mathias, ed., *London Math. Soc. Lecture Note Series* **87** (1983), 1–59.

[2] J. Baumgartner and A. Hajnal, A proof (involving Martin's Axiom) of a partition relation, *Fund. Math.* **78** (1973),193–203.

[3] B. Dushnik and E. W. Miller, Partially ordered sets, *Amer. J. of Math.* **63** (1941), 605.

[4] P. Erdős and R. Rado, A partition calculus in set theory, *Bull. Amer. Math. Soc.* **62** (1956), 427–489.

[5] P. Erdős, A. Hajnal, A. Máté and R. Rado, *Combinatorial Set Theory: Partition Relations for Cardinals*, North-Holland, 1984.

[6] A. Hajnal, Some results and problems in set theory, *Acta Math. Acad. Sci. Hung.* **11** (1960), 277–298.

[7] K. Prikry, On a problem of Erdős, Hajnal and Rado, *Discrete Math.* **2** (1972), 51–59.

[8] S. Shelah, Notes on combinatorial set theory, *Israel J. Math.* **14** (1973), 262–277.

[9] S. Shelah, On CH $+ 2^{\aleph_1} \to (\alpha)^2_2$ for $\alpha < \omega_2$, to appear.

Zero Sum Trees: A Survey of Results and Open Problems

A. BIALOSTOCKI

Mathematics and Statistics

University of Idaho

Moscow, ID 83843

U.S.A.

Abstract

This is an expository paper which surveys results related to the recent confirmation of the zero-sum-tree conjecture by Z. Füredi and D. Kleitman. The author believes that trees play a major role in zero-sum Ramsey theory. He makes some observations that may lead in the far future to very general theorems and presents a collection of fifteen conjectures attributed to various people. Some of the conjectures have already appeared in print but most of them are new.

1 Introduction

Let Z_m denote the additive group of residues modulo m and let $0, 1 \in Z_m$. By a Z_m-coloring ($\{0,1\}$-coloring) of a set S we mean a function $c : S \to Z_m$ (a function $c : S \to \{0,1\}$). The following theorem of P. Erdős, A. Ginzburg and A. Ziv, [14], is the cornerstone of "Zero Sum Ramsey Theory" a new developing direction in combinatorial number theory.

Theorem 1.1 (E.G.Z.) *In every Z_m-coloring c of a set S of cardinality $2m-1$, there exists a subset T of S of cardinality m such that $\sum_{a_i \in T} c(a_i) = 0$.*

If we replace the Z_m-coloring in the E.G.Z. Theorem by a $\{0,1\}$-coloring, then its conclusion follows from the pigeon hole principle. Thus the E.G.Z. Theorem can be viewed as a generalization of the pigeon hole principle. Moreover, observing a set S of cardinality $2m - 2$ and coloring half of its elements by 0 and the other half by 1, we see that $2m - 1$ is the smallest integer for which the conclusion of the E.G.Z. Theorem holds.

The "Zero Sum Ramsey Theory" develops the E.G.Z. Theorem in the same way that ordinary "Ramsey Theory" was developed from the pigeon hole principle. Over 15 papers have been written in this direction. Among the authors are N. Alon, A. Bialostocki, Y. Caro, P. Dierker, Z. Füredi, J. Kahn, D. Kleitman, M. Lotspeich, A. Schrijver, P. Seymour and Y. Roditty. However, most of the papers have not yet appeared in print.

In the present paper we shall focus on some past and hopefully future developments concerning the recent confirmation by Z. Füredi and D. Kleitman of the following conjecture of [4].

N.W. Sauer et al. (eds.), Finite and Infinite Combinatorics in Sets and Logic, 19–29.
© 1993 *Kluwer Academic Publishers. Printed in the Netherlands.*

Conjecture. *In every Z_m-coloring c of the edges of K_{m+1}, the complete graph on $m + 1$ vertices, there exists a spanning tree of K_{m+1}, say T, with edges $e_1, e_2, \ldots, e_m$ satisfying*

$$\sum_{i=1}^{m} c(e_i) = 0.$$

This paper is of expository nature. In Section 2, we shall review some generalizations by Z. Füredi & D. Kleitman, A. Schrijver & P. Seymour, and A. Bialostocki & P. Dierker. In Section 3, we shall introduce the notion of a *proper Ramsey type theorem* and make a few observations. Section 4 contains a collection of fifteen conjectures concerning zero sum problems which involve trees and forests.

2 The zero-sum tree theorem and its generalizations

We start with some definitions. For a graph G, we denote by $e(G)$ the edge set of G.

Definitions.

(i) Let $\mathcal{G}_1, \mathcal{G}_2, \ldots, \mathcal{G}_k$ be k sets of graphs. We denote by $R(\mathcal{G}_1, \mathcal{G}_2, \ldots, \mathcal{G}_k)$ the minimal n such that for every k-coloring of $e(K_n)$, there exists in K_n an i-monochromatic copy of some $G \in \mathcal{G}_i$ for some $i \in \{1, 2, \ldots, k\}$. If all the $\mathcal{G}_i$'s are equal to $\mathcal{G}$, then we shall use the notation of $R(\mathcal{G}, k)$.

(ii) Let $\mathcal{G}$ be a set of graphs each having m edges and let $\Gamma = (\Gamma, +)$ be a group of order m. We denote by $R(\mathcal{G}, \Gamma)$ the minimal integer n such that for every Γ-coloring of $e(K_n)$, say c, there is in K_n a copy of some $G \in \mathcal{G}$ whose edges can be ordered $e_1, e_2, \ldots, e_m$ such that

$$\sum_{e_i \in e(G)} c(e_i) = 0.$$

Remarks.

(i) The above definitions can be easily modified to uniform r-graphs (hypergraphs); in this case we shall use the notation of $R^{(r)}(\mathcal{G}_1, \mathcal{G}_2, \ldots, \mathcal{G}_k)$ and $R^{(r)}(\mathcal{G}, \Gamma)$ respectively.

(ii) A singleton set will be denoted by G instead of $\{G\}$.

(iii) It is clear that if $\Gamma = Z_m$, then

$$R(\mathcal{G}, 2) \leq R(\mathcal{G}, Z_m) \leq R(\mathcal{G}, m).$$

The most common set of graphs considered throughout this paper is the set of all trees on s vertices and it will be denoted by τ_s. The definition of a tree can be easily modified to an r-tree, namely; it is a connected r-graph on $m(r - 1) + 1$ vertices with m edges. We shall denote by $\tau_{m(r-1)+1}^{(r)}$ the set of all r-trees on $m(r - 1) + 1$ vertices with m edges.

In [15], Z. Füredi and D. Kleitman confirmed the zero sum tree conjecture by proving the following stronger theorem.

Theorem 2.1 *If Γ is a group of order m, then $R(\tau_{m+1}, \Gamma) = m + 1$.*

Later, A. Schrijver & P. Seymour proved in [18] the following theorem.

Theorem 2.2 *If Γ is an abelian group of order m, then $R^{(r)}(\tau^{(r)}_{m(r-1)+1}, \Gamma) = m(r-1) + 1$.*

To the best of our knowledge there is no proof of a theorem that combines Theorem 2.1 and Theorem 2.2.

Definition. Let $\tau^r_{m(r-1)+1}$ be as defined before and let t be a positive integer. We denote by $t\tau^{(r)}_{m(r-1)+1}$ an r-graph that consists of t vertex-disjoint r-trees each belonging to $\tau^{(r)}_{m(r-1)+1}$.

Using the result of Theorem 2.2, A. Bialostocki and P. Dierker proved in [5] the following theorem.

Theorem 2.3 $R^{(r)}(t\tau^{(r)}_{m(r-1)+1}, Z_{tm}) = t(m(r-1)+2) - 1$.

If in Theorem 2.3 we have $t = 1$ and $r = 2$, then the confirmation of the zero sum tree conjecture follows. Moreover, if $m = 1$, then a known theorem of [6] concerning matchings is implied.

Restricting our consideration to the group Z_p, where p is a prime, A. Schrijver and P. Seymour proved in [19] the following theorem.

Theorem 2.4 *Let G be a connected graph and let c be a Z_p-coloring of $e(G)$, where p is prime. If every cut-set of G has two edges e_1 and e_2 such that $c(e_1) \neq c(e_2)$, then:*

$$\left| \left\{ \sum_{e \in e(T)} c(e) \middle| T \text{ is a spanning tree of } G \right\} \right| \geq \min\{p, |v(G)|\}.$$

It is worthwhile to mention that the proof of Theorem 2.4 is in terms of matroids and it implies the Cauchy-Davenport Theorem, the Erdős-Ginzburg-Ziv Theorem in the prime case, and the Zero Sum Tree Theorem again only in the prime case.

3 Proper Ramsey Theorems, examples and observations

The notion of a proper Ramsey type theorem followed by six observations make up the heart of this section. But first, following the spirit of [16], we introduce the following definitions.

Definitions.

(i) Let A be a set and let $\mathcal{B}$ be a family of subsets of A each of cardinality m. If for every $\{0,1\}$-coloring of the elements of A, say c, there exist a $B = \{b_1, b_2, \ldots, b_m\} \in \mathcal{B}$ such that $\sum_{i=1}^{m} c(b_i) = 0$ (i.e. all the elements of B have the same color), then we say that the pair $(A, \mathcal{B})$ is 2-Ramsey.

(ii) Let A and B be as above and let $\Gamma = (\Gamma, +)$ be a group of order m. If for every Γ-coloring of the elements of A, say c, there exists a $B \in \mathcal{B}$ whose elements can be ordered $b_1, b_2, \ldots, b_m$ such that $\sum_{i=1}^{m} c(b_i) = 0$, then we say that the pair (A, B) is ZS-Ramsey with respect to Γ (ZS stands for zero sum). The set B will be called a zero sum set.

(iii) Let the conclusion of a Ramsey type theorem T state that under certain assumptions the pair (A, B) is 2-Ramsey. If in theorem T the $\{0, 1\}$-coloring is extended to a Z_m-coloring but all the other assumptions stay and it follows that the pair (A, B) is ZS-Ramsey with respect to Γ, then we say that the theorem T is Γ-proper.

(iv) We shall use the notion of a proper Ramsey type theorem if it is Γ-proper for $\Gamma = Z_m$.

In informal words, if a Ramsey type theorem can be generalized from a $\{0, 1\}$-coloring to a Z_m-coloring, then it will be called proper. We are aware that the notion of a Ramsey type theorem is not well defined, but every attempt to define it may limit the imagination of the reader. We shall proceed with some examples of proper and non-proper Ramsey type theorems.

The following are five examples of proper Ramsey type theorems.

Examples.

(i) In every 2-coloring of a set S of cardinality $2m - 1$, there exists a monochromatic subset T of S of cardinality m. See [14].

(ii) $R(K_{1,m}, 2) = \begin{cases} 2m & \text{if } m \text{ is odd;} \\ 2m - 1 & \text{if } m \text{ is even.} \end{cases}$ See [6].

(iii) $R(mK_2, 2) = 3m - 1$. See [6].

(iv) $R(\tau_{m+1}, 2) = m + 1$. See [15] and [18].

(v) All theorems which determine $R(T, 2)$ where T is a tree with 3 or 4 edges. See [7].

The following are two examples of non-proper Ramsey type theorems.

(i) $R(K_3, 2) = 6$ is a non-proper Ramsey type theorem since $R(K_3, Z_3) = 11$. See [7].

(ii) $R(K_{1,3} \cup K_2, 2) = 7$ is a non-proper Ramsey type theorem since $R(K_{1,3} \cup K_2, Z_4) = 8$. See [7].

Next, we proceed with six observations concerning proper Ramsey numbers which inspire some of the conjectures in Section 4.

Observation 3.1 Observing the examples above of *proper Ramsey type theorems*, it turns out that whenever m is involved in the graph (or graphs) under consideration, the corresponding Ramsey numbers are linear in m. Moreover, the Ramsey numbers $R(K_{1,m}, k)$,

$R(mK_2, k)$ and $R(\tau_{m+1}, k)$ are linear in k. Thus, it seems that linearity plays some role in proper Ramsey type theorems.

Observation 3.2 Observing Theorem 2.1 and the theorem of J. Olson [17] which generalizes the E.G.Z. Theorem for every group, it might be the case that if a Ramsey type theorem is Z_m-proper, then it is Γ-proper for every group Γ of order m. However, the proofs for general groups are much more complicated.

Observation 3.3 Observing the proof of the E.G.Z. Theorem for a prime p, [14], we see that actually the following stronger theorem is proved, which is not true for a non-prime.

Theorem *Let $S = \{a_1, a_2, \ldots, a_{2p-1}\}$ be a set of cardinality $2p - 1$, where p is a prime, and let c be a Z_p-coloring of S. If x is an arbitrary element of S then either*

(i) *there is a monochromatic subset T of S of cardinality p, or*

(ii) *the set of all the p-sums of the form*

$$\{c(a_{i_1}) + c(a_{i_2}) + \cdots + c(a_{i_{p-1}}) + c(x) | a_{i_j} \in S\}$$

(i.e. which include $c(x)$) equals to Z_p.

In addition, observing Theorem 2.4, it seems that while generalizing a proper Ramsey theorem to a prime p, a much stronger result holds than for a general m. The transition from a prime to a non-prime is often of major difficulty. In the case of the E.G.Z. Theorem, the transition was possible because the above theorem was weakened while in the proof of Theorem 2.2 the transition was possible because the theorem was generalized to r-graphs.

Observation 3.4 Proper Ramsey theorems tend to have a further natural generalization to a Z_k-coloring where $k|m$. See for instance [11] and [12]. We shall state this generalized version of the E.G.Z. Theorem.

Theorem *In every Z_k-coloring of a set S of cardinality $m + k - 1$, there exists a subset T of S of cardinality m such that $\sum_{a_i \in T} c(a_i) = 0$ (in Z_k).*

Observation 3.5 Consider the following definition and theorem which appears in [9].

Definition. Let k and m be positive integers such that $k \leq m$. Denote by $g(m, k)$ the minimum integer g for which the following holds: if S is a set of cardinality g and c is a Z_m-coloring of S such that it has at least k distinct elements of Z_m in its image, then there is a subset T of S of cardinality m such that $\sum_{a_i \in T} c(a_i) = 0$.

Theorem

(i) $g(m, 2) = 2m - 1$.

(ii) $g(m, 3) = 2m - 2$.

(iii) $g(m, 4) = 2m - 3$.

Our observation is: In the Z_m-generalized form of a proper Ramsey type theorem, if one insists on having at least k elements in the image of the Z_m-coloring and $k \geq 2$, then the larger k is, the more likely it is to find a zero sum set. In the above case the cardinality of the set S can be decreased in the assumption, obtaining the same conclusion.

Observation 3.6 In view of the results so far despite the fact that $\dot{R}(K_{1,3} \cup K_{2,2}) = 7$ is a non-proper Ramsey type theorem, we believe that Ramsey type theorems about trees tend to be proper. Special proper behavior is expected from the family τ_s.

4   Conjectures

Inspired by Observation 3.6, we start with a conjecture of [7] followed by a conjecture of [5].

Conjecture 4.1

$$\lim_{n \to \infty} \max_G \frac{R(G, Z_m)}{R(G, 2)} = 1,$$

where the maximum is taken over all forests with m edges.

Conjecture 4.2 *If G is a forest with m edges, then there is a t_0 such that*

$$R(tG, Z_{tm}) = R(tG, 2) \text{ for all } t \geq t_0.$$

It is worthwhile noting that $R(tK_3, 2) = 5t$, see [10]; but $6t + 3 \leq R(tK_3, Z_{3t}) \leq 6t + 5$. The upper bound appears in [5] and the lower bound is due to Y. Caro. It is remarkable that the graph tK_3 does not have a proper behavior despite being very sparse.

Next, we continue with two new conjectures concerning subfamilies of τ_s of bounded diameter. If these conjectures will be proved, then we shall have a further generalization of the Zero Sum Tree Theorem and a nice bridge between the E.G.Z. Theorem and the Zero Sum Tree Theorem as well.

Definition. $\tau_n(k) = \{T | T$ is a tree on n vertices and $\text{diam}(T) \leq k\}$.

It is clear that since $\tau_n(2) = K_{1,n-1}$, we have $R(\tau_{m+1}(2), 2) = R(\tau_{m+1}(2), Z_m)$.

Conjecture 4.3   (i) $R(\tau_{m+1}(3), 2) = R(\tau_{m+1}(3), Z_m)$.

(ii) $R(\tau_{4k-r+1}(3), 2) = 5k - r$ *where* $r = 1, 2, 3, 4$.

Conjecture 4.4 $R(\tau_{m+1}(4), 2) = R(\tau_{m+1}(4), Z_m)$.

We note that it is not difficult to prove that $R(\tau_{m+1}(4), 2) = R(\tau_{m+1}, 2) = m + 1$. This fact actually was the motivation for Conjecture 4.4.

Generalized Turán numbers and zero sum Turán numbers were introduced in [3]. First, we recall the definition and then we shall state two conjectures of [3].

Definition. Let $\mathcal{G}$ be a set of graphs, each having m edges; then

(i) $T(n, \mathcal{G}, k)$ is the maximum possible number of edges in a k-edge-colored graph on n vertices, avoiding a monochromatic copy of $G \in \mathcal{G}$;

(ii) $T(n, \mathcal{G}, Z_m)$ is the maximum possible number of edges in a Z_m-edge-colored graph on n vertices, avoiding a zero sum copy of $G \in \mathcal{G}$.

We note that if $k = 1$ and $\mathcal{G} = \{G\}$, then $T(n, G, 1)$ is the well known Turán number $T(n, G)$. Moreover, it is clear that

$$T(n, \mathcal{G}, 2) \leq T(n, \mathcal{G}, Z_m) \leq T(n, \mathcal{G}, m).$$

Conjecture 4.5 $T(n, \tau_{m+1}, 2) = T(n, \tau_{m+1}, Z_m)$.

Conjecture 4.6 $T(n, sK_2, 2) = T(n, sK_2, Z_m)$.

In [3], Conjecture 4.5 was proved in the case where $n \geq m^2$ and $n \equiv 0 \pmod{m}$ and an outline of a proof was given for the fact that Conjecture 4.6 holds asymptotically for large n. Recently, Y. Caro noted that, using a theorem of Sauer and Spencer instead of the Hajnal-Szemerédi Theorem, the result of [3] regarding Conjecture 4.5 can be proved under a weaker assumption than stated above. However, Conjecture 4.5 is still open in the case where $n - m$ is small.

We proceed with a conjecture of [2] concerning zero sum trees in complete bipartite graphs. First we need the following definition.

Definition. Let H be a graph and let $\mathcal{G}$ be a set of graphs, each having m edges.

(i) We write $H \xrightarrow{2} \mathcal{G}$ to denote that in every 2-coloring of the edges of H, there is in H a monochromatic copy of some $G \in \mathcal{G}$.

(ii) We write $H \xrightarrow{Z_m} \mathcal{G}$ to denote that in every Z_m-coloring, say c, of the edges of H, there is in H a copy of some $G \in \mathcal{G}$ with edges $e_1, e_2, \ldots, e_m$ such that $\sum_{i=1}^{m} c(e_i) = 0$.

Conjecture 4.7 *If $K_{m,n}$ denotes the complete bipartite graph with two sets of vertices of cardinalities m and n, then*

$$K_{m,n} \xrightarrow{Z_s} \tau_{s+1}, \quad \text{where} \quad s+1 = \left\lceil \frac{m}{2} \right\rceil + \left\lceil \frac{n}{2} \right\rceil.$$

We note that it is easy to prove that $K_{m,n} \xrightarrow{2} \tau_{s+1}$, where $s+1 = \lceil m/2 \rceil + \lceil n/2 \rceil$. Using the results of [9], among them part (iii) of the theorem quoted in Observation 3.5, we can also prove that $K_{3,2m-3} \xrightarrow{Z_m} \tau_{m+1}$ but not more.

The next two conjectures are due to P. Seymour and A. Schrijver. They were able to prove them for m being a prime, see [19]. Their paper uses the more general notion of matroids and they actually proved a somewhat stronger result.

Conjecture 4.8 *If G is an m-connected graph with $|v(G)| \equiv 1$ (mod m), then for every Z_m-coloring of $e(G)$ there is a zero sum spanning tree of G.*

Conjecture 4.9 *If G is a graph such that every $\{0,1\}$-coloring of its edges implies a monochromatic spanning tree of G, then every Z_m-coloring of its edges implies a zero sum spanning tree of G.*

Next, we proceed with the notion of multiplicity.

The following multiplicity conjecture regarding the E.G.Z. Theorem was stated in [2]: In every Z_m-coloring c of a set S of cardinality n, there exist at least $\binom{\lceil n/2 \rceil}{m} + \binom{\lfloor n/2 \rfloor}{m}$ subsets T of S such that $\sum_{a_i \in T} c(a_i) = 0$ for every T.

Recently, M. Kisin proved the above conjecture for m a power of a prime or if $m = p^\alpha q$ where p and q are distinct primes and $\alpha \geq 1$. Moreover, Z. Füredi and D. Kleitman proved the conjecture for a fixed m and asymptotically for n. Thus, a significant progress toward confirming the above conjecture has been made. Motivated by this progress we introduce the following definition and conjecture.

Definition. Let $\mathcal{G}$ be a family of graphs, each having m edges.

(i) $M(n, G, k)$ is the minimum number of monochromatic copies of $G \in \mathcal{G}$ assured in every k-coloring of $e(K_n)$, and

(ii) $M(n, G, Z_m)$ is the minimum number of zero sum copies of $G \in \mathcal{G}$ assured in every Z_m-coloring of $e(K_n)$.

It is clear that

$$M(n, \mathcal{G}, m) \leq M(n, \mathcal{G}, Z_m) \leq M(n, \mathcal{G}, 2).$$

Conjecture 4.10 $M(n, \tau_{m+1}, Z_m) = M(n, \tau_{m+1}, 2)$.

To the best of our knowledge, not very much is known about $M(n, \tau_{m+1}, 2)$.

Motivated by Observation 3.5, we state the following conjecture which appears in [2].

Conjecture 4.11 *If c is a Z_m-coloring of $e(K_{m+1} \backslash \lfloor \frac{m+1}{2} \rfloor K_2)$ such that the image of c contains at least 3 elements of Z_m, then $K_{m+1} \backslash \lfloor \frac{m+1}{2} \rfloor K_2$ contains a zero sum spanning tree.*

Next, we introduce some definitions that will hopefully enable us in the future to generalize some Ramsey type theorems that involve more than 2 colors to zero-sum Ramsey theorems.

Definition. Let $\mathcal{G}_1$ and $\mathcal{G}_2$ be sets of graphs where $\mathcal{G}_2$ consists of graphs having m edges each. Denote by $R(\mathcal{G}_1, \mathcal{G}_2; \{\infty\} \cup Z_m)$ the minimum n such that in every $(\{\infty\} \cup Z_m)$-coloring of $e(K_n)$, say c, there is in K_n either

(i) a ∞-monochromatic copy of some $G \in \mathcal{G}_1$, or

(ii) a copy of some $G \in \mathcal{G}_2$ such that $\sum_{e_i \in e(G)} c(e_i) = 0$.

It is clear that

$$R(\mathcal{G}, \mathcal{H}, \mathcal{H}) \leq R(\mathcal{G}, \mathcal{H}; \{\infty\} \cup Z_m) \leq (\mathcal{G}, \underbrace{\mathcal{H}, \mathcal{H}, \ldots, \mathcal{H}}_{m \text{ times}}).$$

Very recently, we have learned from J. Kahn that, confirming a conjecture of the author, Kim Jeong-Han proved the following theorem.

Theorem $R(sK_2, \tau_{m+1}, \tau_{m+1}) = \max\{2s, s+m\}$.

Motivated by this theorem, we make the following two conjectures.

Conjecture 4.12 $R(sK_1, \tau_{m+1}, \tau_{m+1}) = R(sK_2, \tau_{m+1}; \{\infty\} \cup Z_m)$.

Conjecture 4.13 $R(\tau_{m+1}, 3) = R(\tau_{m+1}, \tau_{m+1}; \{\infty\} \cup Z_m)$.

Definition. We denote by $Z_m^{(1)} \cup Z_m^{(2)}$ a union of two disjoint copies of Z_m. Let $\mathcal{G}$ be a set of graphs having m edges each. Denote by $R(\mathcal{G}, Z_m^{(1)} \cup Z_m^{(2)})$ the minimum n such that in every $(Z_m^{(1)} \cup Z_m^{(2)})$-coloring of $e(K_n)$ there is, in K_n, a zero sum copy of a $G \in \mathcal{G}$ (by a zero sum we mean a zero sum in either $Z_m^{(1)}$ or in $Z_m^{(2)}$).

It is clear that

$$R(\mathcal{G}, 4) \leq R(\mathcal{G}, Z_m^{(1)} \cup Z_m^{(2)}) \leq R(\mathcal{G}, 2m).$$

Conjecture 4.14 $R(\tau_{m+1}, 4) = R(\tau_{m+1}, Z_m^{(1)} \cup Z_m^{(2)})$.

We note that $R(\tau_{m+1}, 4)$ was determined in [11] and [8].

Observing the last two definitions that enable us to generalise Ramsey theorems for 3 and 4 colors to zero sum Ramsey theorems, it is not difficult to introduce the right definitions that will enable us to generalize Ramsey theorems for any number of colors. We note that one has to distinguish between two cases where the number of colors is even or odd. Going one step further we can easily define a proper Ramsey theorem that involves any number of colors. However, not very much is known about Ramsey numbers involving many colors. Among the little that is known is the following deep theorem of N. Alon, P. Frankl and L. Lovász, see [1], which determines the Ramsey number of k copies of r-matching for a t-coloring of the complete uniform r-graph. Let $K_n^{(r)}$ denote the complete r-graph on n vertices.

Theorem (N. Alon, P. Frankl and L. Lovász)

$$R^{(r)}(kK_r^{(r)}, t) = kr + (t-1)(k-1).$$

The natural question that arises is: is the above theorem proper? We believe that it is, but prefer not to state it explicitly as a conjecture.

We conclude with an unpublished conjecture of Y. Caro that is supported by the confirmation of some small cases.

Conjecture 4.15 *Let T be an arbitrary tree with m edges. If G is a graph whose minimal degree is $2m-1$, then for every Z_m-coloring of $e(G)$ there is in G a zero sum copy of T.*

ACKNOWLEDGEMENT

The author would like to thank Z. Füredi and D. Kleitman for a helpful conversation.

References

[1] N. Alon, P. Frankl and L. Lovász, The chromatic number of Kneser hypergraphs, *Trans. Amer. Math. Soc* **298** (1986) 359–370.

[2] A. Bialostocki, Some combinatorial number theory aspects of Ramsey theory, research proposal (1989).

[3] A. Bialostocki, Y. Caro, and Y. Roditty, On zero sum Turán numbers, *Ars Combinatoria*, **29A** (1990) 117–127.

[4] A. Bialostocki and P. Dierker, Zero sum Ramsey theorems, *Congressus Numerantium* **70** (1990) 119–130.

[5] A. Bialostocki and P. Dierker, On zero sum Ramsey numbers: multiple copies of a graph, to appear in *J. Graph Theory*.

[6] A. Bialostocki and P. Dierker, On the Erdős–Ginzburg–Ziv Theorem and the Ramsey numbers for stars and matchings, to appear in *Discrete Math.*

[7] A. Bialostocki and P. Dierker, On zero sum Ramsey numbers: small graphs, *Ars Combinatoria* **29A** (1990) 193–198.

[8] A. Bialostocki and P. Dierker, Monochromatic connected subgraphs in a multicoloring of the complete graph, manuscript.

[9] A. Bialostocki and M. Lotspeich, Some developments of the Erdős–Ginzburg–Ziv Theorem I, manuscript submitted.

[10] S.A. Burr, P. Erdős, and J.H. Spencer, Ramsey theorems for multiple copies of graphs, *Trans. Amer. Math. Soc.* **209** (1975) 87–99.

[11] J. Bierbauer and A. Gyárfás, On (n, k) colorings of complete graphs, *Congressus Numerantium* **58** (1987) 123–139.

[12] Y. Caro, On zero-sum Ramsey numbers — stars, to appear in *Discrete Math.*

[13] Y. Caro, On zero-sum Turán numbers — stars and cycles, to appear in *Ars Combinatoria*.

[14] P. Erdős, A. Ginzburg and A. Ziv, Theorem in additive number theory, *Bull. Research Council Israel* **10F** (1961) 41–43.

[15] Z. Füredi and D.J. Kleitman, On zero-trees, to appear in *J. Graph Theory*.

[16] R.L. Graham, B.L. Rothschild, and J.H. Spencer, *Ramsey Theory*, John Wiley & Sons, New York–Chichester–Brisbane–Toronto, 1980.

[17] J.E. Olson, On a combinatorial problem of Erdős, Ginzburg and Ziv, *J. Number Theory* **8** (1976) 52–57.

[18] A. Schrijver and P.D. Seymour, A simpler proof and a generalization of the zero-trees theorem, to appear in *J. Combinatorial Theory, Series A.*

[19] A. Schrijver and P.D. Seymour, Spanning trees of different weights, in "Polyhedral Combinatorics", in *DIMACS Series in Discrete Math. and Theoret. Comp. Sc.* **1** (1990) (eds. W. Cook and P.D. Seymour).

On Superatomic Boolean algebras

Université d'Aix-Marseille

Abstract

Let us recall that a Boolean algebra B is *superatomic* if every homomorphic image
is atomic. A Boolean algebra B is an *interval algebra* if there is a chain C having a first
element, say 0^C, such that B is isomorphic to the subalgebra of $\wp(C)$ generated by the
intervals of the form $[a, b)$ $(a < b$ in $C)$. The typical example of superatomic interval
algebra is an *ordinal algebra* (i.e. an interval algebra of an ordinal). We will survey
different generalizations of the properties of ordinal algebras.

The first one concerns the fact that every ordinal algebra $B(\alpha)$ is generated by a well-
founded sublattice (namely $\{[0, \gamma) : \gamma < \alpha\}$). We develop results obtained on Boolean
algebras generated by a well-founded sublattice. This notion of well-generatedness arises
when one seeks analogues of the notions of well-founded partially ordered sets for the
context of Boolean algebras. A well-generated algebra must be superatomic. There
are more canonical notions of well-generatedness arising in this investigation: we define
the notion of canonically well-generatedness and of well-founded tree algebra (B is a
well-founded tree algebra if there is $T \subseteq B$, which satisfies: (1) T generates B, (2) T
is well-founded for $\leq^B$ and (3) for $s, t \in T$, $s \cdot t = 0^B$ or s and t are comparable).
A Boolean algebra is a well-founded tree algebra if and only if it is a superatomic
subalgebra of an interval algebra. (This result is a corollary of two results: one due to
U. Avraham and the author, the second one obtained in collaboration with M. Rubin
and H. Si-Kaddour. A self-contained proof of this result is given in the last part of this
work.)

Every ordinal algebra has the property that every homomorphic image is isomorphic
to a factor. We investigate also what are the other algebras having this property (so-
called CO algebras): the results are obtained in collaboration with M. Bekkali and M.
Rubin; and with S. Shelah. Now, if we regard the subalgebra A generated by the atoms
of $B(\lambda)$, where λ is an infinite cardinal, then A has the property that each subalgebra of
A of cardinality λ is isomorphic to A (algebra having the ISAP). We study the different
candidates of such algebras (results in collaboration with M. Rubin).

1 Survey of the results

For classical notions and results on Boolean algebras, we refer to Koppelberg [32]. In
a Boolean algebra B we denote by 0^B and 1^B, respectively, the smallest and the largest
elements of B. For x and y in B, we denote by $x +^B y$ and $x \cdot^B y$, respectively, the supremum
and the infimum of x and y in B, by $-^B x$ the complement of x in B, by $x -^B y \overset{\text{def}}{=} x \cdot^B (-^B y)$

*Supported by the CNRS (UPR 9016), and NSERC grants 69-0259, 69-3378 and 69-1325.

N.W. Sauer et al. (eds.), Finite and Infinite Combinatorics in Sets and Logic, 31–62.
© 1993 *Kluwer Academic Publishers. Printed in the Netherlands.*

(so $-^B x = 1^B -^B x$). Also $x \leq^B y$ if $x \cdot^B y = x$ and $x\Delta^B y \stackrel{\text{def}}{=} (x -^B y) +^B (y -^B x) = (x +^B y) -^B (x \cdot^B y)$. If B is understood from the context, we omit the superscript ^{B}.

For $0^B \neq a \in B$, we denote by $B\restriction a$ the Boolean algebra induced by B on the set $\{t \in B : t \leq a\}$. So $1_{B\restriction a} = a$ and the complement of t in $B\restriction a$ is $a - t$ (such an algebra is called a *factor of B*).

For a Boolean algebra B, we denote by $\text{At}(B)$ the set of atoms B. For a subset D of B, we denote by $\text{cl}_B(D)$ the subalgebra of B generated by D, and by $\text{cl}_B^{\text{Id}}(D)$ the ideal of B generated by D.

Let I be an ideal of B. Then $\text{cl}_B(I) = I \cup -I$, where $-I = \{-x \in B : x \in I\}$. For $x \in B$, we set $x/I \stackrel{\text{def}}{=} \{y \in B : x\Delta y \in I\}$, and for a subset S of B, let $S/I \stackrel{\text{def}}{=} \{x/I : x \in S\}$ (so B/I is the quotient algebra and $S/I \subseteq B/I$).

Let B be a Boolean algebra. We denote by $\text{Ult}(B)$ its Boolean space, i.e. the space of all ultrafilters of B.

Let $(B_i)_{i \in I}$ be a family of Boolean algebras.

(1) $\prod_{i \in I} B_i$ denotes the *product* of the algebras B_i (as usual, $\prod_{i<2} B_i$ is denoted by $B_0 \times B_1$). If I is finite, then $\text{Ult}(\prod_{i \in I} B_i)$ is homeomorphic to the direct sum of the $\text{Ult}(B_i)$'s.

(2) $\prod_{i \in I}^w B_i \stackrel{\text{def}}{=} \{(b_i)_{i \in I} : \{i \in I : b_i \neq 0\}$ is finite or $\{i \in I : b_i \neq 1\}$ is finite$\}$ denotes the *weak product* of the algebras B_i. $\text{Ult}(\prod_{i \in I}^w B_i)$ is homeomorphic to the one-point compactification (so called Alexandroff compactification) of the $\text{Ult}(B_i)$'s.

(3) $\bigoplus_{i \in I} B_i$ denotes the *free product* (also called the *direct sum* or the *tensor product*) of the algebras B_i. $\text{Ult}(\bigoplus_{i \in I} B_i)$ is homeomorphic to $\prod_{i \in I} \text{Ult}(B_i)$.

A Boolean algebra B is said to be *superatomic* if every quotient of B is atomic.

Day [19] (see also Koppelberg [32]) has shown the following result:

Proposition 1.1 *Let B be a Boolean algebra. The following properties are equivalent:*
    (i) *B is superatomic;*
    (ii) *every subalgebra of B is atomic;*
    (iii) *there is no embedding from the atomless countable algebra into B;*
    (iv) *the rational chain $\mathbb{Q}$ is not embeddable in the partial ordering $\langle B, \leq^B \rangle$.*   □

Let B be a Boolean algebra. For every ordinal α we define the ideal $I_\alpha(B)$ of B. $I_0(B) = \{0^B\}$ and $I_1(B)$ is the ideal of B generated by $\text{At}(B)$. If δ is a limit ordinal, then $I_\delta(B) = \bigcup_{\alpha<\delta} I_\alpha(B)$. Suppose that $I_\alpha(B)$ has been defined, and we define $I_{\alpha+1}(B)$. Let φ_α^B be the canonical homomorphism from B onto $B/I_\alpha(B)$, then $I_{\alpha+1}(B) = (\varphi_\alpha^B)^{-1}[I_1(B/I_\alpha(B))]$. The algebra $D_\alpha(B) \stackrel{\text{def}}{=} B/I_\alpha(B)$ is called the *α-th Cantor–Bendixson derivative of B*. Trivially, $(I_\alpha(B))_{\alpha \in \text{Ord}}$ is an increasing sequence of ideals of B (Ord denotes the class of ordinals). We set $I_\infty(B) \stackrel{\text{def}}{=} \bigcup_{\alpha \in \text{Ord}} I_\alpha(B)$. We denote by $\infty(B)$ the first ordinal such that $I_{\infty(B)}(B) = I_\infty(B)$.

If B is superatomic, then $1^B \in I_\infty(B)$, and then $\infty(B)$ is a successor ordinal, say $\text{rk}(B)+1$, and $\text{rk}(B)$ is called the *rank of B*. Note that $D_{\text{rk}(B)}(B)$ is a non-trivial finite algebra, isomorphic to $\wp(n)$ ($n > 0$ integer). Let $I(B)$ and $D(B)$ denote $I_{\text{rk}(B)}(B)$ and $B/I(B)$ respectively. If $n = 1$, then $I(B)$ is a maximal ideal of B. Let $\widehat{\text{At}}_\alpha(B) \stackrel{\text{def}}{=} (\varphi_\alpha^B)^{-1}[\text{At}(B/I_\alpha(B))]$, and $\widehat{\text{At}}(B) \stackrel{\text{def}}{=} \bigcup_{\alpha \leq \text{rk}(B)} \widehat{\text{At}}_\alpha(B)$. Next, if B is not superatomic, then $1^B \notin I_\infty(B)$: this is so, since $B/I_\infty(B)$ is an infinite atomless algebra. Therefore we have:

B is superatomic if and only if $1^B \in I_\alpha(B)$ for some α, and in such a case,
$1^B \in I_{rk(B)+1}(B) - I_{rk(B)}(B)$.

Definition 1.2 (a) Let B be a superatomic Boolean algebra, and $b \in B$, $b \neq 0^B$. Let γ be the first ordinal α such that $b \in I_\alpha(B)$. Clearly, γ is a successor ordinal, say $\alpha + 1$, and we set $rk^B(b) = \alpha$. So $b \in I_{rk^B(b)+1}(B) - I_{rk^B(b)}(B)$. (In Lemma 1.3(a), we will see that $rk^B(b) = rk(B\lceil b)$.) For instance $rk^B(b) = 0$ for $b \in At(B)$; and $rk^B(1^B) = rk(B)$.

Let B be a superatomic Boolean algebra. We say that H is a *complete set of representatives of generalized atoms of B*, whenever $H = \bigcup\{H_\alpha : \alpha \leq rk(B)\}$, where $H_\alpha \subseteq \widehat{At}_\alpha(B)$ satisfies: for every $a \in \widehat{At}_\alpha(B)$, there is a unique $h \in H_\alpha$ such that $rk^B(a\Delta h) < rk^B(a)$ $(= rk^B(h))$.

(b) The *cardinal sequence* $(\gamma_\alpha)_{\alpha \leq \rho}$ of a superatomic Boolean algebra B is defined as follows: $\rho = rk(B)$ and for every $\alpha \leq \rho$, $\gamma_\alpha = |At(D_\alpha(B))|$. So a cardinal sequence verifies $\gamma_\alpha \geq \aleph_0$ for $\alpha < \rho$ and $1 \leq \gamma_\rho < \omega$.

A Boolean algebra B is a *thin-tall* algebra if $rk(B) = \omega_1$, $|At(D_\alpha(B))| = \aleph_0$ for $\alpha < \omega_1$ and $D_{\omega_1}(B) = \{0, 1\}$. So its cardinal sequence is

$$\overbrace{\langle\aleph_0, \aleph_0, \ldots, \aleph_0, \ldots}^{\omega_1 - times}, 1\rangle.$$

(c) If E is a set, then $\wp(E)$, the power set of E, is regarded as a Boolean algebra. Hence, in $\wp(E)$, the operations $+$ and $\cdot$ are denoted by $\cup$ and $\cap$ respectively, and $a \leq b$ means $a \subseteq b$. Moreover $FC(E)$ denotes the subalgebra of $\wp(E)$ of finite or cofinite subsets of E.

(d) A partial ordering P is *scattered* (resp. *well-founded*) if the rational chain (resp. the chain of non-positive integers) is not embeddable in P. A partial ordered set C is a *chain* if every pair of members of C are comparable. We consider ordinals as chains.

(e) Let C be a chain with a first element denoted by 0^C (if C has no first element, then we must add one). Let $C^+ \overset{def}{=} C \cup \{\infty^C\}$ be the chain obtained by adding to C a greatest element ∞^C. We denote by $B(C)$ the subalgebra of $\wp(C)$ generated by the set of $[a, b)$ for $a \in C$ and $b \in C^+$, i.e. $cl_{\wp(C)}(\{[a, b) : a \in C \text{ and } b \in C^+\})$. $B(C)$ is called the *interval algebra on C* (see [32]).

Because $B(C) \subseteq \wp(C)$, for $a, b \in B(C)$, $a + b = a \cup b$ and $a \cdot b = a \cap b$.

An *ordinal algebra* is an interval algebra of a well-ordered chain.

From (d) and Proposition 1.1, it follows that the following holds. *A Boolean algebra is superatomic if and only if $\langle B, \leq^B \rangle$ is scattered.* Concerning the ideals $I_\alpha(B)$ and properties of ranks, the following result seems to be well-known. For completeness, we will prove it at the end of Section 1.5.

Lemma 1.3 (a) *Let B be a Boolean algebra and $b \in B$, $b \neq 0^B$. Then for every ordinal α:*
(1) $I_\alpha(B\lceil b) = I_\alpha(B) \cap (B\lceil b)$;
(2) $D_\alpha(B) \to (D_\alpha(B\lceil b)) \times (D_\alpha(B\lceil - b))$ *defined by:*

$$a/I_\alpha(B) \mapsto \langle a \cdot b/I_\alpha(B\lceil b), a \cdot -b/I_\alpha(B\lceil - b)\rangle (a \in B)$$

is an onto isomorphism;
(3) $rk^B(b) = rk(B\lceil b) = rk(cl_B(B\lceil b))$;

(4) $I_\infty(B) = \{a \in B : B{\upharpoonright} a$ is a superatomic Boolean algebra$\}$;

(5) $\mathrm{cl}_B(I_\infty(B))$ *is a superatomic subalgebra of B.*

(b) *Let A and B be superatomic Boolean algebras, and f be a homomorphism from B onto A. Then for every ordinal α the following hold.*

(1) $f[I_\alpha(B)] \subseteq I_\alpha(A)$.

(2) *f induces a homomorphism f_α from $D_\alpha(B)$ onto $D_\alpha(A)$ defined by $f_\alpha(b/I_\alpha(B)) = f(b)/I_\alpha(A)$ (for $b \in B$).*

(3) $\mathrm{rk}(A) \le \mathrm{rk}(B)$.

(4) *Let A be a subalgebra of B, and $a, b \in B$ such that $a \le b$. Then $\mathrm{rk}(A{\upharpoonright} a) \le \mathrm{rk}(A{\upharpoonright} b) \le \mathrm{rk}(A) = \mathrm{rk}^A(1^A)$.*

(c) *Let A and B be superatomic Boolean algebras, and f be a one-to-one homomorphism from A into B. Then for every ordinal α the following hold.*

(1) $A \cap f^{-1}[I_\alpha(B)] \subseteq I_\alpha(A)$.

(2) $\mathrm{rk}(A) \le \mathrm{rk}(B)$.

In fact, it is easy to show also, by induction on α, that if $H = \bigcup_{\alpha \le \mathrm{rk}(B)} H_\alpha$ is a complete set of representatives of generalized atoms of a superatomic Boolean algebra B, then:

$$I_\alpha(B) = \mathrm{cl}_B^{\mathrm{Id}}\Big(\bigcup_{\beta < \alpha} H_\beta\Big) \subseteq \mathrm{cl}_B\Big(\bigcup_{\beta < \alpha} H_\beta\Big) = I_\alpha(B) \cup -I_\alpha(B).$$

1.1 Characterization of superatomic subalgebras of an interval algebra

Let T be a subset of a Boolean algebra B.

(1) T is *well-founded* if T has no strictly decreasing sequence for $\le^B$.

(2) T is a *tree* if for every $s, t \in T$: either $s \cdot t = 0^B$ or s and t are comparable.

(3) T is a *disjointed tree* if T is a tree contained in $B - \{0^B\}$.

(4) B is called a *well-founded tree algebra*, if B contains a generating subset T which is a well-founded tree (T is called a *well-founded generating tree of B*).

(5) B is called a *disjointed tree algebra*, if B contains a generating subset T which is a disjointed tree (T is called a *generating tree of B*).

Comment 1.4 (1) Let B be a Boolean algebra. The following properties are equivalent:

(i) B is a disjointed tree algebra.

(ii) B is generated by a tree, i.e. there is a subset T of B such that for every $s, t \in T$: either $s \cdot t = 0^B$ or s and t are comparable.

This is so, since if T is a tree contained in B, then $\mathrm{cl}_B(T) = \mathrm{cl}_B(T - \{0^B\})$.

About disjointed tree algebras, some closure properties are quite obvious to prove: for instance the fact that "disjointed tree algebra" is closed under homomorphic image (Claim 2.6(a)). But our definition of a well-founded disjointed tree algebra is unusual (see also 1.9–1.12 and 2.2(c)): let T be a subset of a Boolean algebra B.

(2) Suppose that T is a disjointed tree. Then T ordered by its converse ordering (i.e. $t' \le t''$ if $t' \ge^B t''$) is a pseudo-tree in the sense of Bekkali [6] (cf. also [33]).

(3) Suppose that T is a disjointed tree, and that $\langle T, \ge^B {\upharpoonright} T \rangle$ is well-founded. Then T is a tree in the usual meaning ([28], [32], [34].)

The following result was proved by Bonnet, Rubin and Si-Kaddour [14] (for (ii) $\Leftrightarrow$ (iv)), and Avraham and Bonnet [1] (for (i) $\Leftrightarrow$ (ii) $\Leftrightarrow$ (iii)).

Theorem 1.5 (Avraham, Bonnet, Rubin and Si-Kaddour) *Let B be a Boolean algebra. The following properties are equivalent.*

(i) *B is a superatomic Boolean algebra, embeddable in an interval algebra.*

(ii) *B is embeddable in an ordinal algebra.*

(iii) *B is isomorphic to a subalgebra $\underline{B}$ of an interval algebra $B(C)$ of a well-founded chain C, such that $\mathrm{At}(\underline{B}) = \mathrm{At}(B(C))$.*

(iv) *B is generated by a well-founded disjointed tree.*

(v) *B is a well-founded tree algebra.*

(vi) *There is a well-founded generating tree T of B which is a complete set of representatives of generalized atoms of B.*

Note that (iii) $\Rightarrow$ (ii) $\Rightarrow$ (i) and (vi) $\Rightarrow$ (v) $\Rightarrow$ (iv) are trivial. In Sections 2.4, 2.5, 2.6, we will prove that (iv) $\Rightarrow$ (ii), (i) $\Rightarrow$ (iii) and (ii) $\Rightarrow$ (vi) respectively. Moreover, the proof of (iv) $\Rightarrow$ (ii) shows that the embedding f from B into the ordinal algebra has the following property:

(ii$^+$): f *is an embedding from B in an ordinal algebra $B(\alpha)$ such that the image under f of every member t of T is a half-open interval $[\beta^t, \gamma^t)$ of α ($\gamma^t \in \alpha \cup \{\infty^\alpha\}$).*

In this paper, we give a self-contained proof of Theorem 1.5.

For example, let $\alpha = \omega^\gamma \cdot q$ be an ordinal and $B \overset{\mathrm{def}}{=} B(\alpha)$. We set $\underline{0} = 1$. Let $\beta < \alpha$, $\beta \neq 0$. Then, we have the (Cantor) normal form (see Fraïssé [24], Kuratowski and Mostowski [35] or Rosenstein [54]): $\beta = \omega^{\beta_0} \cdot p_0 + \omega^{\beta_1} \cdot p_1 + \cdots + \omega^{\beta_{\ell-1}} \cdot p_{\ell-1}$, with $\beta_0 > \beta_1 > \cdots > \beta_{\ell-1}$ ($\ell > 0$) and for $i < \ell$: $0 < p_i < \omega$ ($\omega^0 = 1$). We denote by $\Gamma(\beta)$ the set of $\beta + \omega^\gamma$ for $0 \leq \gamma \leq \beta_{\ell-1}$. So $\Gamma(0) = \{0\}$, $\Gamma(\omega + 1) = \{\omega + 1, \omega + \omega\}$, and $\Gamma(\omega^2) = \{\omega^2 + 1, \omega^2 + \omega, \omega^2 + \omega^2\}$. Then the set

$$G \overset{\mathrm{def}}{=} \{[0, \delta) : \delta < \alpha \text{ and } \delta \text{ is indecomposable }\} \cup \bigcup_{\beta < \alpha} \{[\beta, \underline{\beta}) : \underline{\beta} \in \Gamma(\beta)\}$$

satisfies Theorem 1.5(iv).

The class of scattered chains is obtained from the class of well-ordered chains by converse order operation and lexicographic sum over well-ordered chains (see for example [24] or [54]). Theorem 1.5 arises as an analogue of the scattered chain result in the context of superatomic algebras: we can see (ii) as an external approach, and (iv) as an internal approach.

Obviously (i) $\Leftrightarrow$ (ii) are equivalent for a countable algebra B, since B is isomorphic to an interval algebra (S. Mazurkiewicz and W. Sierpinski [40]).

Remark 1.6 Let B be a superatomic subalgebra of an interval algebra $B(C)$. Is there a superatomic interval algebra A such that $B \subseteq A \subseteq B(C)$? The answer is negative: let $2 \cdot \mathbf{R}$ be the chain obtained from $\mathbf{R}$ (chain of real numbers) by replacing each real number by the 2-element chain. Let $B \overset{\mathrm{def}}{=} \mathrm{cl}_{B(2 \cdot \mathbf{R})}(\mathrm{At}(B(2 \cdot \mathbf{R})))$. Let $A \overset{\mathrm{def}}{=} B(D)$ be a superatomic interval algebra containing B. First, because $\mathrm{At}(B)$ is uncountable, the scattered chain D is uncountable. By a theorem of Hausdorff (see [54], Theorem 5.28), D contains a copy of the chain ω_1 or $\omega_1^\star$. For a contradiction, let us suppose that $B(D) \subseteq B(2 \cdot \mathbf{R})$. Because $B(D)$ satisfies: there is an uncountable family of pairwise disjoint elements such that each element contains infinitely many atoms, the algebra $B(2 \cdot \mathbf{R})$ has the same property. Now, we obtain a contradiction with the fact that, in $\mathbf{R}$, every family of pairwise disjoint non-trivial intervals is at most countable.

Let us state some direct consequence of Theorem 1.5.

Corollary 1.7 (M. Rubin and S. Shelah [57]) (a) *Let B be an infinite superatomic Boolean algebra, embeddable in an interval algebra. Then* At(B) *and B have the same cardinality.*

(b) *In particular, a thin-tall Boolean algebra is not embeddable in an interval algebra.*

Proof. (a) For an infinite well-ordered chain C, $|$At$(B(C))| = |B(C)|$. With the notations of 1.5(iii), (a) follows from the following string: $|$At$(\underline{B})| \leq |\underline{B}| \leq |B(C)| = |At(B(C))| = |At(\underline{B})|$.

(b) is a consequence of (a) and of the definition of a thin tall algebra.   □

Finally, for an interval algebra: every element has a unique decomposition as a finite union of half-open intervals (see Definition 1.2(e)). We have a similar decomposition theorem in the case of superatomic subalgebras of an interval algebra. Rewriting the proof of Lemma 16.6 of [32], we obtain:

Proposition 1.8 *Let B be a superatomic subalgebra of an interval algebra, and let T be a well-founded tree of B. Let $b \in B$ be such that $b \neq 0^B$. Then b has a unique decomposition of the form:*

$$b = \sum_{i<n}\Big(g_i - \sum_{k \in F_i} g_{i,k}\Big)$$

where $n < \omega$, each F_i is a finite set, $g_i \in T \cup \{1^B\}$, $g_{i,k} \in T$ for every $n < \omega$ and $k \in F_i$, satisfying the following conditions:

(1) $(g_i - \sum_{k \in F_i} g_{i,k}) \cdot (g_j - \sum_{k \in F_j} g_{j,k}) = 0^B$ *for distinct $i, j < n$;*

(2) $g_{i,k} < g_i$ *for all $g_{i,k} \in F_i$;*

(3) *the members of each F_i are pairwise disjoint;*

(4) $g_i \notin F_j$ *for all $i, j < n$.*   □

Comment 1.9 Let us make a digression, concerning our definition of "disjointed tree algebra" and "well-founded tree algebra". We begin by a definition:

Definition 1.10 Let P be a partial ordering.

(a) We say that P is a *pseudo-tree* if for every $p \in P$, $\{q \in P : q \leq^P p\}$ is a chain.

(b) For a pseudo-tree P, we define the *pseudo-tree algebra* $B(P)$ of P as follows: $B(P)$ is the subalgebra of $\wp(P)$, generated by the set of $\{q \in P : q \geq^P p\}$ for $p \in P$. $B(P)$ is called the *pseudo-tree algebra* of P.

In the literature ([28], [32], [34]) a *tree* is a well-founded pseudo-tree P. The notion of tree algebra $B(P)$ (that is a pseudo-tree algebra of a tree P) was introduced by G. Brenner ([17], [18]) to construct a rigid Boolean algebra (see also [32]). Pseudo-tree algebras were developed by M. Bekkali [5]. Let us remark that every pseudo-tree algebra is a disjointed tree algebra: let $B(P)$ be the pseudo-tree of P. Then $T \overset{\text{def}}{=} \{-p : p \in P\}$ is a disjointed tree generating $B(P)$.

A pseudo-tree generates a unique pseudo-tree (up to isomorphy), i.e. if two partial orderings are isomorphic, then the corresponding pseudo-tree algebras are isomorphic. For disjointed tree algebras, the situation is different: the algebras $\wp(\{a, b, c\})$ and $\wp(\{a, b\})$ are not isomorphic and are algebras generated by the disjointed tree $\{a, b, \{a, b\}\}$.

M. Bekkali ([5], [6]) has proved (see also [33]: Theorem 3.1):

Proposition 1.11 (Bekkali) *Every pseudo-tree algebra is embeddable in an interval algebra.*
□

Using Bekkali's ingredients of the original proof of 1.11, we can also show that: every pseudo-tree algebra is embeddable in an interval algebra ([16]). But the converse of 1.11 is unknown:

Question 1.12 (a) *Is every subalgebra of an interval algebra isomorphic to a disjointed tree algebra (a pseudo-tree algebra)?*

(b) *In particular, is every subalgebra of the interval algebra of* R *(*R *denotes the real chain) isomorphic to a disjointed tree algebra (a pseudo-tree algebra)?*

Question 1.12(a) was put by Bekkali, Koppelberg and Monk. Koppelberg and Monk ([33]: Theorem 3.2(b)) give a partial answer to Question 1.12(a) with the following result. Let C be a chain, and A be a subalgebra of $B(C)$, generated by a family of half-open intervals of C. Then A is isomorphic to a pseudo-tree algebra. It is easy to check that if B and T are as in 1.5(vi), then B is the pseudo-tree algebra $B(T)$ of T. Therefore Proposition 1.11 gives another proof of 1.5(vi) $\Rightarrow$ (i). Moreover, we can add to the different characterizations of superatomic subalgebras of an interval algebra the following result due to M. Bekkali ([7]). Let T be a pseudo-tree. The following properties are equivalent:

(i) the pseudo-tree-algebra $B(T)$ is superatomic;

(ii) $B(T)$ is embeddable in a superatomic interval algebra;

(iii) T is a scattered partial ordering and T does not contain a copy of the dychotomic tree.

1.2 Well-generated, canonically well-generated Boolean algebras. Poset Boolean algebras

Every ordinal algebra $B(\alpha)$ is generated by $\{[0, \gamma) : \gamma < \alpha\}$, that is a well-founded lattice. So we can introduce the notion of well-generatedness as follows.

Let B be a Boolean algebra.

(a) We say that B is a *well-generated algebra* if B has a well-founded sublattice which generates B.

(b) A Boolean algebra B is a *canonically well-generated algebra* if there is a complete set H of representatives of generalized atoms of B such that $cl_B(H) = B$ and the sublattice generated by H is a well-founded lattice.

First, let us remark that: well-founded treeness $\Rightarrow$ canonically well-generatedness $\Rightarrow$ well-generatedness $\Rightarrow$ superatomicity (cf. [14]).

Before giving closure properties concerning (canonically) well-generatedness, let us state a characterization of canonically well-generatedness in an algebraic way, and in a topological way. We denote by $Y^{(\alpha)}$ the α-th Cantor–Bendixson derivative of a space Y. In [14], we prove the following results.

(a) Let B be a superatomic Boolean algebra and $S = \mathrm{Ult}(B)$ be its Boolean space. Then B is canonically well-generated if and only if there is a family $(U_x)_{x \in S}$ of clopen subsets of S such that the following hold:

(1) for every $x \in S$, there is α_x such that $U_x^{(\alpha_x)} = \{x\}$;

(2) if $y \in U_x$, then $U_y \subseteq U_x$.

Moreover B is a well-founded tree algebra if and only if there is a family $(U_x)_{x \in S}$ of clopen sets of X such that (1), (2) and

(3) if $U_x \cap U_y \neq \emptyset$, then $U_y \subseteq U_x$ or $U_x \subseteq U_y$

hold.

(b) Let B be a Boolean algebra. The following statements are equivalent:

(i) B is a canonically well-generated algebra;

(ii) B has a set H of representatives of generalized atoms such that the intersection of two elements of H is a finite union of elements of H;

(iii) B has a set H of representatives of generalized atoms such that for every a and b in H, if a is almost contained in b (that means $\mathrm{rk}^B(a - b) < \mathrm{rk}^B(a)$ or equivalently $\mathrm{rk}^B(a \cdot b) = \mathrm{rk}^B(a)$), then $a \leq^B b$.

(c) Let $H \subseteq B$ satisfy (b) (iii).

(1) Let $a, b \in H$. Then either $a \leq b$, $b \leq a$, or $\mathrm{rk}^B(a \triangle b) < \min(\mathrm{rk}^B(a), \mathrm{rk}^B(b))$ (that means by definition: a and b are almost disjoint).

(2) H verifies the fact that the intersection of two elements of H is a finite union of elements of H.

The following results concern closure properties of (canonically) well-generated algebras, and state some examples of well-generated algebras ([14]).

Theorem 1.13 (a) (1) *Let B be a (canonically) well-generated Boolean algebra. Every factor of B is (canonically) well-generated.*

(2) *Every homomorphic image of a well-generated algebra is well-generated.*

(3) *Every weak product of well-generated algebras is well-generated. In particular a finite product of well-generated algebras is well-generated.*

(4) *Every weak product of canonically well-generated algebras is canonically well-generated. In particular a finite product of canonically well-generated algebras is canonically well-generated.*

(5) *A finite free product of well-generated algebras is well-generated.*

(6) *A finite free product of canonically well-generated algebras is canonically well-generated.*

Note that, by Part (1), if a finite product (resp. a weak product, a finite free product) of Boolean algebras B_i's is (canonically) well-generated, then each B_i is (canonically) well-generated.

(7) (i) *If B is a well-generated algebra, then there is a sublattice G which well-generates B with $1^B \in G$.*

(ii) *If B is a canonically well-generated algebra, then there is a canonical subset B which well-generates B with $1^B = \sum \{h \in H : \mathrm{rk}^B(h) = \mathrm{rk}(B)\}$.*

(b) (1) *Every subalgebra of a well-founded tree-generated algebra is canonically well-generated (Theorem 1.5(vi)).*

(2) *Every algebra is of rank ≤ 2. (This is so, since any complete set of representatives of generalized atoms generates a canonically well-generating set of the algebra.)*

(3) *There is a thin tall algebra which is canonically well-generated.* □

By Theorem 1.13, we have large classes of well-generated algebras. However, there are superatomic non well-generated algebras:

Theorem 1.14 (a) (1) *Assume that B is a superatomic algebra with only countably many atoms, and B has a homomorphic image having a chain of order-type ω_1. Then B cannot be well-generated.*

(2) *There is a superatomic Boolean algebra A which is not embeddable in a well-generated algebra.*

(3) *There is a non-well-generated thin-tall Boolean algebra.*

(b) *There is a non-well-generated Boolean algebra of cardinal sequence $\langle \aleph_0, \aleph_0, 2^{\aleph_0}, 1 \rangle$, and there is a non-well-generated Boolean algebra of cardinal sequence $\langle \aleph_1, \aleph_0, \aleph_1, 1 \rangle$.*   □

Note that the algebras mentioned in (a2) and (a3) are algebras of rank ω_1, and their corresponding proofs use (a1).

(a3) is related to possible homomorphic images of thin-tall superatomic Boolean algebras. A. Dow and P. Simon [21], and independently J. Roitman, have shown the following unpublished result. *Let A be a superatomic Boolean algebra of rank $\mathrm{rk}(A) \leq \omega_1$. We suppose that for every $b \in A$, if $\mathrm{rk}^A(b) < \mathrm{rk}(A)$, then $A{\restriction}\, b$ is countable, and $D_{\mathrm{rk}(A)}(A) = \{0,1\}$.* (For instance $B(\omega_1)$ has these properties.) *Then A is a homomorphic image of a thin-tall Boolean algebra.*

Because every algebra of rank ≤ 2 is canonically well-generated, 1.14(b) gives minimal "cardinal sequences" for which there are superatomic Boolean algebras which are not well-generated. But in [14], it is proved that if we assume (MA plus $\aleph_1 < 2^{\aleph_0}$), then "every Boolean algebra of cardinal sequence $\langle \aleph_0, \aleph_0, \aleph_1, 1 \rangle$ (resp. $\langle \aleph_0, \aleph_1, \aleph_1, 1 \rangle$) is canonically well-generated".

As a direct consequence of the above results, we obtain:

Theorem 1.15 *The following statements are independent in ZFC.*
(a) *Every Boolean algebra of cardinal sequence $\langle \aleph_0, \aleph_0, \aleph_1, 1 \rangle$ is canonically well-generated.*
(b) *Every Boolean algebra of cardinal sequence $\langle \aleph_0, \aleph_1, \aleph_1, 1 \rangle$ is canonically well-generated.*
The same results hold whenever we replace canonically well-generated by well-generated.   □

More generally, in [14], we prove the following result. The following statements are independent in ZFC: Let $\alpha, \beta, \gamma, \delta$ and ε be countable ordinals such that $\alpha \geq 2$ and $\gamma \geq 1$. Every Boolean algebra of cardinal sequence

$$\langle \overbrace{\aleph_0, \aleph_0, \ldots, \aleph_0, \ldots}^{\alpha-\text{times}}, \aleph_1, \overbrace{\aleph_0, \aleph_0, \ldots, \aleph_0, \ldots}^{\beta-\text{times}}, 1 \rangle$$

$$(resp. \quad \langle \overbrace{\aleph_0, \aleph_0, \ldots, \aleph_0, \ldots}^{\gamma-\text{times}}, \aleph_1, \overbrace{\aleph_0, \aleph_0, \ldots, \aleph_0, \ldots}^{\delta-\text{times}}, \aleph_1, \overbrace{\aleph_0, \aleph_0, \ldots, \aleph_0, \ldots}^{\varepsilon-\text{times}}, 1 \rangle \,)$$

is well-generated. The same result holds whenever we replace well-generated by canonically well-generated.

We don't have an answer to the following problem:

Question 1.16 *Is it consistent with (ZFC plus $\aleph_2 < 2^{\aleph_0}$), that every Boolean algebra of cardinal sequence $\langle \aleph_0, \aleph_1, \aleph_2, 1 \rangle$ is well-generated (canonically well-generated)? Same question for cardinal sequence $\langle \aleph_0, \aleph_2, \aleph_1, 1 \rangle$; $\langle \aleph_0, \aleph_2, \aleph_2, 1 \rangle$.*

Now, we will define the notion of poset Boolean algebras. For a partial ordering P, we denote by $L(P)$ the free Boolean algebra generated by the set $\{x_p : p \in P\}$. Let $E(P)$ be the ideal of $L(P)$ generated by the equations of the form $x_a \cdot x_b = x_a$ in $L(P)$ if and only if $a \leq b$ in P.

Definition 1.17 A Boolean algebra B is called a *poset algebra* if there is a partial ordering P such that B is isomorphic to $F(P) \overset{\text{def}}{=} L(P)/E(P)$.

We denote by i^P the one-to-one order-preserving mapping from P into $F(P)$ ($i^P(p) \overset{\text{def}}{=} x_p/E(P)$). $F(P)$ is characterized by the following universal property. Let A be a Boolean algebra. Every order-preserving mapping f from P into A can be extended to a unique homomorphism $\bar{f}$ from $F(P)$ into A, i.e. $\bar{f} = f \circ i^P$.

Let us remark that if C is a chain with a first element 0^C, then $F(C - \{0^C\})$ is isomorphic to $B(C)$. Namely the mapping $a \mapsto x_a \overset{\text{def}}{=} [0^C, a)$ extends to a homomorphism from $L(C - \{0^C\})$ onto $B(C)$ whose kernel is $E(P)$ giving the indicated isomorphism, as is easily verified.

A partial ordering P has the *finite antichain condition* if P has no infinite subset of pairwise incomparable elements. A partial ordering W is a *well-quasi-ordering* if W has no strictly decreasing sequence and W satisfies the finite antichain condition. M. Pouzet [46] (see Bonnet and Rubin [16]), has proved:

Theorem 1.18 (Pouzet) *Let P be a partial ordering. P is scattered and satisfies the finite antichain condition if and only if $F(P)$ is superatomic.* □

Bonnet and Rubin [16] have shown:

Theorem 1.19 *Let B be a superatomic poset algebra. Then B is a homomorphic image of a subalgebra of a poset algebra $F(W)$, where W is a quasi-well-ordering.* □

Concerning well-generated poset algebras, we have only the following partial result. Assume that P satisfies one of the following conditions:

(1) P is a well-quasi ordering, or

(2) P is a finite scattered partial ordering of finite width (i.e. there is $p < \omega$ such that every set of pairwise incomparable elements of P has at most p elements).

Then $F(P)$ is well-generated.

Let $\omega_1 \sqcup \omega_1$ be the direct union of two copies of ω_1: so $\omega_1 \sqcup \omega_1 = \omega_1 \times \{0\} \cup \omega_1 \times \{1\}$ with $(\alpha, 0)$ and $(\beta, 1)$ incomparable for $\alpha, \beta < \omega_1$. It is easy to see that $B \overset{\text{def}}{=} B(\omega_1 \sqcup \omega_1)$ is isomorphic to $B(\omega_1) \oplus B(\omega_1)$. In [16], we show that there is a subalgebra A of $B(\omega_1 \sqcup \omega_1)$ which is not well-generated (namely the subalgebra of $B(\omega_1 \times \{0\} \cup \omega_1 \times \{1\})$) generated by $\{x_{(\alpha,0)} - x_{(\beta,1)} : \alpha \leq \beta < \omega_1\}$).

Question 1.20 *Let $F(P)$ be a superatomic Boolean algebra.*

(a) Is $F(P)$ embeddable in $F(W)$, where W is a quasi-well-ordering?

(b) Is $F(P)$ a well-generated algebra?

(c) Is $F(P)$ a canonically well-generated algebra?

1.3   CO and HCO algebras: uncountable algebras for which each homomorphic image is isomorphic to a factor

Let α be an ordinal, and $B \overset{\text{def}}{=} B(\alpha)$. Then B has the following property: if A is a homomorphic image of B, then A is isomorphic to a factor of B. This is so, since if we denote by π the homomorphism from B onto A, then $\pi[\alpha]$ is well-ordered, and thus isomorphic to some $\beta \le \alpha$. A and $B(\beta)$ are isomorphic algebras (because $\pi[\alpha]$ generates A). It is easy to check that $B(\beta)$ is isomorphic to a factor of B: if $\beta = \alpha$, then there is nothing to prove, and if $\beta < \alpha$, then $B(\alpha)$ is isomorphic to $B(\beta) \times B(\alpha - \beta)$ $(\alpha - \beta = [\beta, \alpha))$. So, we introduce the following definition.

Definition 1.21 Let B be a Boolean algebra. We say that B is a *CO algebra*, if for every ideal I of B, there is an algebra A such that B is isomorphic to $(B/I) \times A$, and B is a *hereditarily CO algebra (HCO algebra)*, if every homomorphic image of B is a CO algebra.

Let X be a topological space. We say that X is a *CO space* if every closed subspace is homeomorphic to an open subspace. From the definitions, it follows that B *is a CO algebra if and only if its Boolean space* $\mathrm{Ult}(B)$ *is a CO space.*

The definition of CO space is related to the following notion. Let X be a T_2 space (i.e. a Hausdorff space). We say that X is a *Toronto space* if every subspace of X of the same cardinality, is homeomorphic to X. For example, there is only one countable Toronto space (up to homeomorphy), namely the discrete topology on a countable set. We don't know if there an uncountable non-discrete Toronto space.

Every ordinal algebra is a CO algebra, and because every homomorphic image of such an algebra is an ordinal algebra, such an algebra is a HCO algebra. We recall that if L and K are linear orderings, then $L^{\star}$, $L + K$, $L \cdot K$ denote respectively the reverse ordering of L, the ordered sum of L and K and the lexicographic order on $L \times K$ (so $\omega \cdot 2 = \omega + \omega$ and $2 \cdot \omega = \omega$). Cardinals are initial ordinals, and thus are considered as chains. In Bekkali, Bonnet and Rubin [8] and [9], we show:

Theorem 1.22 *Let B be an interval Boolean algebra. Then B is a CO algebra if and only if B is isomorphic to an interval algebra $B(C)$, where C is of order-type $\alpha + \sum_{i<n}(\kappa_i + \lambda_i^{\star})$, where α is any ordinal, $n \in \omega$, for every $i < n$, κ_i, λ_i are regular cardinals and $\kappa_i \ge \lambda_i \ge \omega$, and if $n > 0$, then $\alpha \ge \max(\{\kappa_i : i < n\}) \cdot \omega$.* $\square$

The class of CO algebras seems to be interesting. We expect that for various special classes of CO algebras we have a meaningful structure theorem (like in the case of interval algebras). It is indicated by Bonnet and Shelah [13] that there does not exist a structure theorem for the whole class of CO algebras (see Theorem 1.26). The construction uses $(\Diamond_{\aleph_1})$. As is usual in such constructions, it is possible to construct many such CO algebras which are very non-isomorphic.

Moreover there is a superatomic CO algebra B which is not of the form described in Theorem 1.22. Let $A = \mathrm{FC}(X)$ where $|X| = \aleph_1$, and $B = A \times B(\omega^2)$. Then B is a superatomic CO algebra, but does not have the form described in 1.22. We do not know, however, whether X is a sporadic example, or whether it leads to a new class of superatomic CO algebras. In spite of the above example, the CO property seems to be very restrictive.

But:

(1) Every infinite CO algebra has infinitely many atoms.

(2) If a CO algebra is atomic, then by the definition, it is superatomic.

Accordingly we have the following conjecture.

Question 1.23 (a) *Is every CO algebra superatomic?*

(b) *Describe the superatomic CO algebras.*

Shelah [59] proved the following fact related to Question 1.23(a). Let B be a Boolean algebra. We recall that a subset S of B is *dense* in B if for every $0^B \neq b \in B$, there is $a \in S$ such that $0 < a \leq b$. Let $d(B) = \min(\{|S| : B \text{ has a dense set } S \subseteq B\})$.

Let $(A^\star)$ denote the following axiom of set theory:

$$\text{for every infinite ordinal } \alpha, \ 2^{|\alpha|} > \aleph_\alpha.$$

Theorem 1.24 (Shelah) *Assume $A^\star$. Let B be a CO algebra. For $n \in \omega$, let $U_n(B)$ be the set of $a \in B$ such that $d(B \restriction a) < \aleph_n$. Then for some $n \in \omega$, $U_n(B)$ is dense in X.* □

Note that *for every $n < \omega$, $U_n(B)$ is an ideal of B, and B is atomic if and only if $U_0(B)$ is dense in B*. The meaning of this theorem is that under axiom $(A^\star)$, a CO algebra has a certain property which is a weak notion of scatteredness.

In Bonnet and Shelah [13], we show:

Theorem 1.25 (Bonnet) (a) *Every HCO subalgebra of an interval algebra is isomorphic to an ordinal algebra.*

(b) *If B is an HCO algebra of countable Cantor–Bendixson rank, then B is isomorphic to a countable ordinal algebra.* □

Let us remark that the proof of Part (a) of this result uses Theorem 1.5. However, we describe an HCO algebra, which is not, by Corollary 1.7, embeddable in an interval algebra:

Theorem 1.26 (Shelah) *Assume $\diamondsuit_{\aleph_1}$. Then there is an HCO interval algebra B of Cantor–Bendixson rank ω_1 and of cardinality $\aleph_1$ such that:*

(1) *B is atomic with only countable many atoms;*

(2) *every countable homomorphic image of B is isomorphic to a factor of B, and every uncountable homomorphic image of B is isomorphic to B;*

(3) *B is retractive, that means that for every ideal J of B there is a subalgebra A of B such that $J \cap A = \{0\}$ and $J \cup A$ generates B.*

Hence, B is a thin-tall algebra. □

J. Roitman ([52]) has constructed, under CH, a Boolean algebra satisfying the properties (1) and (2) of 1.26. G. Gruenhage has pointed out that it is an easy consequence of a work of Balogh, Dow, Fremlin and Nyikos [2] that, under PFA, every HCO algebra of cardinality $\aleph_1$ is isomorphic to $B(\omega_1)$. The following problem is open.

Question 1.27 *Is it consistent with ZFC that every HCO algebra of cardinality $\geq \aleph_2$ is isomorphic to an interval algebra (and thus, by 1.25(a), to an ordinal algebra)?*

Related to the properties of the Boolean algebra described in Theorem 1.26 above, M. Weese [68] has shown:

Theorem 1.28 *Let B be a Boolean algebra of cardinality $\aleph_1$. We suppose that for every ideal J of B for which B/J is uncountable, B/J is isomorphic to B. Then B is superatomic. Moreover the cardinal sequence of B has the form*

$$\overbrace{\langle \aleph_0, \aleph_0, \ldots, \aleph_0, \ldots, 1 \rangle}^{\omega_1-\text{times}} \;,\; \overbrace{\langle \aleph_1, \aleph_1, \ldots, \aleph_1, \ldots, 1 \rangle}^{\omega_1-\text{times}} \; \text{ or } \; \overbrace{\langle \aleph_1, \aleph_1, \ldots, \aleph_1, \ldots, 1 \rangle}^{\alpha-\text{times}}$$

where α is a countable indecomposable ordinal (so $\alpha = \omega^\gamma$ for some $\gamma < \omega_1$). □

Theorem 1.26 shows the existence of such algebras. But the following question is still open:

Question 1.29 *Is there a Boolean algebra B such that:*
(1) for every ideal J of B for which B/J is uncountable, B/J is isomorphic to B;
(2) B is of cardinal sequence $\overbrace{\langle \aleph_1, \aleph_1, \ldots, \aleph_1, \ldots, 1 \rangle}^{\omega_1-\text{times}}$ (resp. $\overbrace{\langle \aleph_1, \aleph_1, \ldots, \aleph_1, \ldots, 1 \rangle}^{\alpha-\text{times}}$) where α is an infinite countable indecomposable ordinal)?

1.4 Boolean algebras for which each uncountable subalgebra is isomorphic to the algebra: algebras having the ISAP

Let $B = \mathrm{FC}(\lambda)$ where λ is any infinite cardinal (that is, B is a subalgebra of $B(\lambda)$). Every subalgebra of B of cardinality λ is isomorphic to B. The converse holds for $\lambda = \aleph_0$. We will analyse the situation for $\lambda = \aleph_1$. So we introduce the following definition.

Let B be an uncountable Boolean algebra. We say that B *has the isomorphic subalgebra property (ISAP)* if each uncountable subalgebra A of B is isomorphic to B.

So $\mathrm{FC}(\omega_1)$ has the ISAP. Trivially, an algebra satisfying the ISAP is of cardinality $\aleph_1$ and has infinitely many atoms. In Bonnet and Rubin [15], we show that: if B is an algebra satisfying the ISAP and is embeddable in an interval algebra, then B is isomorphic to $\mathrm{FC}(\omega_1)$. We show also the following:

Theorem 1.30 *Assume $\diamondsuit_{\aleph_1}$. There is a Boolean algebra B that has the following properties.*
(1) B is a thin-tall Boolean algebra.
(2) Every countable subalgebra of B is isomorphic to a factor of B.
(3) Every uncountable subalgebra of B is isomorphic to B (i.e. B has the ISAP).
(4) Every uncountable subalgebra of B contains an uncountable ideal.
(5) If I is an uncountable ideal in B, then B/I is countable.
(6) If A is an uncountable subalgebra of B, then there are an uncountable ideal J of B and a countable set $D \subseteq A$ such that $A = \mathrm{cl}^B(J \cup D)$.
(7) If A is an uncountable subalgebra of B, then there is a countable set $E \subseteq B$ such that $B = \mathrm{cl}_B(A \cup E)$.
(8) Every chain and every antichain (set of pairwise incomparable elements) of B is countable.
(9) If I is an ideal in B such that B/I is uncountable, then B/I has properties (1)–(7). □

In fact (4) and (7) are consequences of (6) and (5). Note that $FC(\omega_1)$ satisfies (3) but not the other properties. By (1) and Corollary 1.7, B is not embeddable in an interval algebra.

The proof of Theorem 1.30, uses the techniques of "nowhere dense sets" introduced by M. Rubin [55] and developed by S. Shelah [58].

Question 1.31 (a) *Is it consistent with ZFC that there is an uncountable Boolean algebra B having the ISAP such that every uncountable quotient of B is isomorphic to B, and every countable quotient of B is isomorphic to a factor of B?*

(b) *Is it consistent with ZFC that there is a retractive Boolean algebra that has the ISAP? (B is retractive means that for every ideal I of B, there is a subalgebra C of B such that $C \cap I = \{0\}$ and $C \cup I$ generates B.)*

Concerning Theorem 1.30, Weese ([68]) has shown the following result.

(a) *Let B be a Boolean algebra of cardinality $\aleph_1$ satisfying ISAP. Then B is superatomic.*

(b) *Moreover, if we suppose that $2^{\aleph_0} < 2^{\aleph_1}$, then a Boolean algebra of cardinality $\aleph_1$ satisfying the ISAP is either isomorphic to $FC(\omega_1)$ or is thin-tall.*

The following result is still open:

Question 1.32 *Let B be a Boolean algebra of cardinality $\kappa \geq \aleph_2$. Suppose that every subalgebra of B of cardinal κ is isomorphic to B. Is B superatomic?*
Characterize such algebras.

Related to this topic, Shelah and Steprans [60] proved the following result.

Theorem 1.33 *The following statement is consistent. There is a superatomic Boolean algebra B of cardinal sequence $\langle \aleph_0, \aleph_1, 1 \rangle$ such that every uncountable subalgebra A of B satisfying $\mathrm{At}(A) = \mathrm{At}(B)$ is isomorphic to B.*   □

Note that a Boolean algebra of cardinal sequence $\langle \aleph_0, \aleph_1, 1 \rangle$ can be regarded as the subalgebra of $\wp(\omega)$ generated by an almost disjoint family of subsets of ω of cardinality $\aleph_1$. (We recall that $\mathcal{A} \subseteq \wp(\omega)$ is an *almost disjoint family* if every element of $\mathcal{A}$ is infinite, and the intersection of two distinct members of $\mathcal{A}$ is finite.)

Question 1.34 (a) *Is it consistent with ZFC that there is an algebra B of cardinal sequence $\langle \aleph_0, \aleph_1, 1 \rangle$ satisfying the ISAP?*

(b) *Is it consistent that all Boolean algebras of cardinal sequence $\langle \aleph_0, \aleph_1, 1 \rangle$ are isomorphic?*

Note that in a universe of ZFC in which (b) is true, (a) is also true.

1.5   Complements

Let us complete some of the results concerning superatomic Boolean algebras. For developments on this topic, we refer to [32] and more precisely to [51]: we don't try to develop deep results concerning thin very tall or very thin thick superatomic algebras. More of the recent developments on this subject are obtained by Baumgartner, Just, Roitman, Simon, Shelah, Weese... (see the bibliography). First La Grange [36] has characterised cardinal sequences for which there is a superatomic algebra of this cardinal sequence. Let B be a superatomic Boolean algebra of cardinal sequence $(\gamma_\alpha)_{\alpha\leq\rho}$. Then (1) $\gamma_\alpha \geq \aleph_0$ for $\alpha < \rho$, and (2) $1 \leq \gamma_\rho < \omega$. La Grange has shown that: if ρ is a countable ordinal, and $(\gamma_\alpha)_{\alpha\leq\rho}$ is a cardinal sequence satisfying (1) and (2), plus (3) if $\beta \leq \alpha \leq \rho$ then $\gamma_\alpha \leq \gamma_\beta^{\aleph_0}$, then $(\gamma_\alpha)_{\alpha\leq\rho}$ is the cardinal sequence of a supertomic Boolean algebra B. By Theorem 1.5(vi), it follows that the cardinal sequence of any superatomic subalgebra of an interval algebra is decreasing. Let $H = \bigcup_{\alpha\leq\mathrm{rk}(B)} H_\alpha$ be a strongly canonically well-generated set of representatives of generalized atoms of B. Then clearly, for $\alpha < \beta \leq \mathrm{rk}(B)$ the function $\varphi_{\beta,\alpha} : H_\alpha \to H_\beta$ defined by $\varphi_{\beta,\alpha}(g)$ is the unique member of H_β such that $\varphi_{\beta,\alpha}(g) \geq g$ $(g \in H_\alpha)$ is onto.

Now, let us give some information concerning thin-tall algebras. The question of the existence of a thin-tall algebra was asked by Tegárski [64]. It was answered by Hujhász and Weiss [29] and under $(\diamondsuit_{\aleph_1})$ by Ostaszewski [44]. Let us state only one recent classical result on thin-tall algebras. Let B be a superatomic Boolean algebra. We denote by $\mathrm{Aut}(B)$ the group of automorphisms of B. Every automorphism f of B must carry each $I_\alpha(B)$ onto itself, and induces an automorphism f_α of $D_\alpha(B)$. Let us say that B is *almost-rigid* if for every automorphism f of B, there is $\alpha < \omega_1$ such that f_α is the identity on $D_\alpha(B)$ (that implies that for $\alpha \leq \beta \leq \omega_1$, f_β is the identity on $D_\beta(B)$). Let $N(B)$ be the normal subgroup of $\mathrm{Aut}(B)$ of all almost-rigid automorphisms of B, and $G(B) \overset{\mathrm{def}}{=} \mathrm{Aut}(B)/N(G)$. Assuming CH, M. Weese constructed a thin-tall algebra which has only quasi-rigid automorphisms (i.e. $|G(B)| = 1$). Assuming CH, S. Koppelberg has shown that every countable group can be a $G(B)$ for some thin-tall algebra B. J. Roitman has shown that CH is unnecessary, and A. Dow and P. Simon [63] have shown that: every countable group can be a $G(B)$ for $2^{\aleph_1}$ thin-tall algebras B.

For the completeness of basic results, we will prove Lemma 1.3.

Proof of Lemma 1.3. (a) Simultaneously, we will prove by induction: (a1) and (a2) and,

$(\star)_\alpha$: $\{a \in B\restriction b : a/I_\alpha(B\restriction b) \in \mathrm{At}(D_\alpha(B\restriction b))\} = \{a \in B\restriction b : a/I_\alpha(B) \in \mathrm{At}(D_\alpha(B))\}$.

We prove the three conditions simultaneously by induction on α; except that $(\star)_\alpha$ and $(a2)_\alpha$ are consequences of $(a1)_\alpha$. We prove the last fact first. For $(a2)_\alpha$, define for any $a \in B$, $h(a) = \langle a \cdot b/I_\alpha(B\restriction b), a \cdot -b/I_\alpha(B\restriction -b)\rangle$. Thus h is a homomorphism from B onto $D_\alpha(B\restriction b) \times D_\alpha(B\restriction -b)$, with kernel $\{a \in B : a \cdot b \in I_\alpha(B\restriction b)$ and $a \cdot -b \in I_\alpha(B\restriction -b)\}$, which is by $(a1)_\alpha$ $\{a \in B : a \cdot b \in I_\alpha(B)$ and $a \cdot -b \in I_\alpha(B)\} = I_\alpha(B)$. The existence of the indicated isomorphism is hence clear. For $(\star)_\alpha$, suppose that $c \in B\restriction b$ is such that $c/I_\alpha(B\restriction b) \in \mathrm{At}(B\restriction b/I_\alpha(B\restriction b))$, and hence by $(a2)_\alpha$, $c/I_\alpha(B) \in \mathrm{At}(D_\alpha(B))$, as desired; the other inclusion in $(\star)_\alpha$ is proved similarly.

Now, we turn to the inductive step. The case $\alpha = 0$ is obvious. Suppose the conditions for α. Suppose that $c \in I_{\alpha+1}(B\restriction b)$. There is a finite subset $\{d_i : i < n\}$ of $B\restriction b$ such that $d_i/I_\alpha(B\restriction b) \in \mathrm{At}(D_\alpha(B\restriction b))$ $(i < n)$ and $c/I_\alpha(B\restriction b) = \sum\{d_i/I_\alpha(B\restriction b) : i < n\}$. It follows that $c\triangle(\sum\{d_i : i < n\}) \in I_\alpha(B\restriction b) \subseteq I_\alpha(B)$ by $(a1)_\alpha$. By $(a2)_\alpha$, $d_i/I_\alpha(B) \in \mathrm{At}(D_\alpha(B))$ for $i < n$, and thus $c \in I_{\alpha+1}(B) \cap (B\restriction b)$. Conversely, suppose that $c \in I_{\alpha+1}(B) \cap (B\restriction b)$. Then we write $c/I_\alpha(B) = \sum\{d_i/I_\alpha(B) : i < n\}$. Now $c \leq b$, so $d_i/I_\alpha(B) \leq c/I_\alpha(B) \leq b/I_\alpha(B)$. Hence we may assume that $d_i \leq b$ for $i < n$, and the argument goes as above (but backwards). The limit step of the induction is obvious.

(a3) follows directly from (a1), the definitions of $\mathrm{rk}^B(b)$ and of $\mathrm{rk}(B\restriction b)$, and the fact that $\mathrm{cl}_B(B\restriction b) = B\restriction b \cup -B\restriction b$ is isomorphic to $(B\restriction b) \times \{0,1\}$.

(a4) Let $b \in B$. We have $(B{\restriction} b) \cap I_\infty(B) = I_\infty(B{\restriction} b)$. So the following properties are equivalent (i): $I_\infty(B{\restriction} b) = B{\restriction} b$; (ii): $(B{\restriction} b) \cap I_\infty(B) = I_\infty(B{\restriction} b) = B{\restriction} b$; (iii): $B{\restriction} b \subseteq I_\infty(B)$; and (iv): $b \in I_\infty(B)$. Hence $b \in I_\infty(B)$ if and only if $B{\restriction} b$ is superatomic.

(a5) follows from (a4), and the fact that either $I_\infty(B) = B$, or $I_{\infty(B)}(B) = I_\infty(B)$ is a maximal ideal of $\mathrm{cl}_B(I_\infty(B))$, i.e. $D_{\infty(B)}(B) = \{0,1\}$.

(b1) is proved by induction. The case $\alpha = 0$ or α limit are trivial. Now, suppose that $f[I_\alpha(B)] \subseteq I_\alpha(A)$ and f induces an homomorphism f_α from $D_\alpha(B)$ onto $D_\alpha(A)$ defined by $f_\alpha(b/I_\alpha(B)) = f(b)/I_\alpha(A)$ (for $b \in B$). Let $a \in f[I_{\alpha+1}(B) - I_\alpha(B)]$. So $a = \sum_{i<n} f(b_i)$, where the $b_i/I_\alpha(B)$'s are atoms of $D_\alpha(B)$. It suffices to show that $f(b_i) \in I_{\alpha+1}(A)$ for all $i < n$. Let $i < n$. Let $\langle a_i^0, a_i^1 \rangle$ be a partition of $f(b_i)$. Hence there are $b_i^\ell \in B$ such that $f(b_i^\ell) = a_i^\ell$ for $\ell = 0, 1$. There is no loss in assuming that $\langle b_i^0, b_i^1 \rangle$ is a partition of b_i. Because $b_i/I_\alpha(B)$ is an atom, $b_i^\ell/I_\alpha(B) = 0$ for some ℓ, say ℓ'. Hence, $a_i^{\ell'}/I_\alpha(A) = 0$ by the induction hypothesis. So $a \in I_{\alpha+1}(A)$.

(b2) follows from (b1). (b3) follows from (b1) and (b2). For (b4), we recall that $A{\restriction} a$ is a homomorphic image of A.

(c1) is proved by induction on α and (c2) is a consequence of (c1). $\square$

Comment 1.35 (1) For the reader who knows the Cantor–Bendixson derivation, the proofs of (a)–(c) are more clear. We recall that for a space Y, we denote by $Y^{(\alpha)}$ the α-th Cantor–Bendixson derivative subspace of Y: $Y^{(0)} = Y$ and $Y^{(1)}$ is the closed subset of limit points of Y, $Y^{(\alpha+1)} \stackrel{\mathrm{def}}{=} (Y^{(\alpha)})^{(1)}$ and $Y^{(\lambda)} = \bigcap_{\alpha<\lambda} Y^{(\alpha)}$ for limit λ (see, for example [32] or Pierce [45]). Let B be a Boolean algebra. Then the Boolean space of $D_\alpha(B)$ is canonically homeomorphic to $\mathrm{Ult}(B)^{(\alpha)}$; B is superatomic if and only if there is an ordinal β such that $(\mathrm{Ult}(B))^{(\beta)} = \emptyset$; and $\mathrm{rk}(B)+1 = \inf\{\beta : (\mathrm{Ult}(B))^{(\beta)} = \emptyset\}$. Let us recall that if C and D are Boolean algebras, any homomorphism from C into D induces a continuous mapping from $\mathrm{Ult}(D)$ into $\mathrm{Ult}(C)$. A compact space Z is *scattered* if for some ζ, $Z^{(\zeta)} = \emptyset$. For such a space, the first ordinal ρ such that $Z^{(\rho)} = \emptyset$ is a successor ordinal; say $\alpha+1$, and $\mathrm{rk}(Z) \stackrel{\mathrm{def}}{=} \alpha$. Moreover, for every $t \in Z$, there is a unique $\tau \le \mathrm{rk}(Z)$ such that $t \in Z^{(\tau)} - Z^{(\tau+1)}$; τ is denoted by $\mathrm{rk}^Z(t)$. Moreover, a Boolean algebra B is superatomic if and only if $\mathrm{Ult}(B)$ is scattered, and in this case $\mathrm{rk}(B) = \mathrm{rk}(\mathrm{Ult}(B))$.

We denote by X and Y scattered compact spaces, U a clopen subset of X, and f a continuous mapping from X into Y. First we claim that:

$(\star)$ *Let Z be a subspace of X. Then $Z^{(1)} \subseteq X^{(1)}$, i.e. if $y \in Z$ is a limit point of elements of Z, then y is a limit point of X. Moreover if Z is open, then $Z^{(1)} = Z \cap X^{(1)}$.*

$(\star\star)$ *Let U be a closed subspace of X and V be a closed subspace of Y. Suppose that $f[U] \supseteq V$. Then $f[U^{(1)}] \supseteq V^{(1)}$.*

First $(\star)$ is trivial. For $(\star\star)$, let v be a limit point of V. There is a net, i.e. an up-directed family, $(v_i)_{i \in I}$ converging to v ($v \ne v_i \in V$) —see Kelley [31] or Dugundji [22]. For $i \in I$, choose $u_i \in U$ such that $f(u_i) = v_i$. By compactness, there is u such that u is a cluster point of $(u_i)_{i \in I}$ (that means that for every neighborhood W of u and every $i_0 \in I$, there is $i \ge i_0$ such that $u_i \in W$). So, $u \in U^{(1)}$ and, by continuity, $f(u) = v$. That proves $(\star\star)$.

(a) Because U is clopen and $(\star)$ holds, we have $U^{(\alpha)} = U \cap X^{(\alpha)}$ (see also [45], Proposition 1.7.1). Therefore, (a) is clear.

(b) Let us suppose that f is one-to-one. Then, by induction on α, $f[X^{(\alpha)}] \subseteq Y^{(\alpha)}$ follows from $(\star)$. By duality, we easily finish.

(c) Let us suppose that f is onto. We will prove by induction on α that $f[X^{(\alpha)}] \supseteq Y^{(\alpha)}$. For a non-limit ordinal, this follows from the induction hypothesis and the fact that $(\star\star)$ holds. Now, suppose that α is limit. Let $v \in Y^{(\alpha)}$ and V be a clopen subset of Y such that $V^{(\alpha)} = \{v\}$. For $\beta < \alpha$, let $v_\beta \in V \cap Y^{(\beta)} = V^{(\beta)}$. Because for every clopen neighborhood W of v contained in V has $\mathrm{rk}(V - W) < \alpha$, $(v_\beta)_{\beta<\alpha}$ is an up-directed family converging to v. By induction hypothesis, for $\beta < \alpha$, let $u_\alpha \in X^{(\alpha)}$ such that $f(u_\alpha) = v_\alpha$. By compactness, there is u such that u is a cluster point of $(u_\beta)_{\beta<\alpha}$. So, $u \in U^{(\alpha)}$ and $f(u) = v$. That finishes the proof of our claim. Now, (c) is obvious.

(2) Note that (c) is also a corollary of a result of M. Rubin [56] (§ 7). *Let X and Y be compact scattered spaces and let f be a continuous function from Y onto Z. Let $\gamma = \sup\{\mathrm{rk}(f^{-1}(y)) : y \in Y\} + 1$.*

(i) *For every $y \in Y$, $\mathrm{rk}^Y(y) \leq \sup(\{\mathrm{rk}_X(x) : f(x) = y\})$.*

(ii) *For $x \in X$, $\mathrm{rk}^X(x) \leq \gamma \cdot \mathrm{rk}(f^{-1}(f(x))) + \mathrm{rk}^{f^{-1}(f(x))}(x)$ (· denotes the lexicographic product; so $\omega \cdot 2 = \omega + \omega$ and $2 \cdot \omega = \omega$).*

(iii) $\mathrm{rk}(Y) \leq \mathrm{rk}(X) \leq \gamma \cdot (\mathrm{rk}(Y) + 1)$.

(3) Let A and B be Boolean algebras. Then $\mathrm{rk}(A \times B) = \max(\mathrm{rk}(A), \mathrm{rk}(B))$ follows from Lemma 1.3. Let $A \oplus B$ be the free product of A and of B: the Boolean space $\mathrm{Ult}(A \oplus B)$ of $A \oplus B$ is the product $\mathrm{Ult}(A) \times \mathrm{Ult}(B)$ of the Boolean spaces of A and B (see Koppelberg [32], ch 4, § 11). Tegárski [64] (see also [45],§ 2.21) has shown that $\mathrm{rk}(A \oplus B) = \mathrm{rk}(A) \oplus \mathrm{rk}(B)$. (Let us recall that for ordinals α and β, $\alpha \oplus \beta$ denotes the Hessenberg (or polynomial) sum of α and β: if $\alpha = \omega^{\gamma_1} \cdot m_1 + \cdots + \omega^{\gamma_k} \cdot m_k$ and $\beta = \omega^{\gamma_1} \cdot n_1 + \cdots + \omega^{\gamma_k} \cdot n_k$ with $\gamma_1 > \cdots > \gamma_k$ and $m_i, n_i < \omega$, then $\alpha \oplus \beta = \omega^{\gamma_1} \cdot (m_1 + n_1) + \cdots + \omega^{\gamma_k} \cdot (m_k + n_k)$.) More precisely Tegárski has proved: *Let X and Y be compact scattered spaces. Then for every ordinal γ, $(X \times Y)^{(\gamma)} = \bigcup\{X^{(\alpha)} \times Y^{(\beta)} : \alpha \oplus \beta = \gamma\}$.*

The rest of this work is devoted to a self-contained proof of Theorem 1.5.

2   Proof of Theorem 1.5

2.1   Chains

Definition 2.1 Let C be a chain, $u \in C$ and D be a subset of C.

(a) C is a *complete chain* if every subset of C has a supremum and an infimum.

(b) C is a *relatively complete chain* if every bounded non-empty subset A of C (i.e. there are b and c in C such that $b \leq^C a \leq^C c$ for every $a \in A$) has a supremum and an infimum in C.

(c) Assume that C is complete, and $D \subseteq C$. We denote by $c(D, C)$, or more simply $c(D)$, the closure of D in C by supremum and infimum.

(d) Every chain C is embeddable in a complete chain C^d, namely its *Dedekind completion* (completion by cuts) which satisfies: for every $c \in C^d$, if c is not the first element of C^d, then $c = \sup\{p \in C : p \leq^C c\}$, and if c is not the last element of C^d, then $c = \inf\{p \in C : p \geq^C c\}$ (see [54]).

(e) u is a *predecessor in C* if there is an (unique) element $u^+ \in C$ such that $u^+ >^C u$ and $[u, u^+) = \{u\}$. We denote by $\mathrm{Pred}(C)$ the set of predecessors of C.

(f) C is *totally disconnected* if for every $v <^C w$ in C, we have $[v, w) \cap \mathrm{Pred}(C) \neq \emptyset$.

(g) If $(C_d)_{d \in D}$ is a family of chains indexed by a chain D, then $\sum_{d \in D} C_d$ denotes the *lexicographic sum of the C_d over D*, i.e. $\sum_{d \in D} C_d = \bigcup_{d \in D} C_d$, and $u \leq v$ if u, v are in the same C_d and $u \leq^{C_d} v$ in C_d; or $u \in C_{d'}$, $v \in C_{d''}$ with $d' <^D d''$.

As usual, the order relation $\leq^C$ on a chain C is denoted by $\leq$.

Consequently $B(C)$ is atomic if and only if C is totally disconnected. The term "totally disconnected" comes from the fact that a chain C is totally disconnected if and only if C, endowed with the interval topology, is a totally disconnected space.

Let $\cong$ be an equivalence relation on C. For $a \in C$, we denote by $a/\cong$ its equivalence class, i.e. $a/\cong \overset{\text{def}}{=} \{a' \in C : a' \cong a\}$. We suppose that each equivalence class is an interval of C. For two equivalence classes $a'/\cong$, $a''/\cong$ of $C/\cong$, we set $a'/\cong < a''/\cong$ if for every $a' \in a'/\cong$ and $a'' \in a''/\cong$, we have $a' < a''$. If each equivalence class is an interval of C, then $\langle C/\cong, \leq \rangle$ is a chain. Moreover if $\langle C, \leq \rangle$ is a complete chain, then $\langle C/\cong, \leq \rangle$ is too.

2.2 Partial orderings and trees

Definition 2.2 (a) Let P be a poset, and X a subset of P. We say that:
(1) X is *cofinal in P* if for every $p \in P$, there is $q \in X$ such that $q \geq^P p$;
(2) X is an *initial interval of P* if $p \leq^P q$ and $q \in X$ implies $p \in X$;
(3) X is an *ideal of P* if X is an initial interval of P such that for every $p, q \in X$, there is $r \in X$ such that $r \geq^P p, q$.

We denote by Maxideal(P) the set of maximal ideals of P.
(4) Let $p, q \in X$. q is a *successor of p in X* if $p <^P q$ and there is no $r \in X$ such that $p <^P r <^P q$. We also say that p and q are *consecutive*. Moreover, if p has a unique successor, it is denoted by p_X^+ or more simply p^+.

(b) We recall that P is a *partial well-founded set* if P has no strictly decreasing sequence. For such a P we define, by induction, subsets P_α. Let P_0 be the set of minimal elements of P. If $(P_\beta)_{\beta < \alpha}$ are defined, let P_α be the set of minimal elements of $P - \bigcup\{P_\beta : \beta < \alpha\}$. Trivially, for some ρ, $P_\rho = \emptyset$. Now, rk^P is the function with domain P defined by $\text{rk}^P(p) = \alpha$ if and only if $p \in P_\alpha$, and $\text{rk}(P) \overset{\text{def}}{=} \sup\{\text{rk}^P(p) + 1 : p \in P\} = \min\{\rho : P_\rho = \emptyset\}$.

(c) Let P be a poset. P is a *tree* if P is a partial well-founded set such that for every $p \in P$, $\{q \in P : q \geq p\}$ is a chain.

Note that this definition is unusual if we compare to [28], [32], [34]: in fact the exact term for a tree must be *well-founded dual pseudo-tree*: P is well-founded and P with its converse order relation is a pseudo-tree.

The following result gives some information on Theorem 1.5(iv).

Claim 2.3 (a) *Let P be a partial ordering. Every element of P is a member of a maximal ideal of P.*

(b) *Let P be a partial ordering. We suppose that P satisfies the following property:*

$$(\bullet) \quad \textit{for every } p \in P, \ \{q \in P : q \geq p\} \textit{ is a chain.}$$

Then:

(1) *if I is an ideal of P and D is a maximal chain of I, then D is cofinal in I;*

(2) *if I is an ideal of P and D is a maximal chain of I, then I is the ideal generated by D (i.e. if $x \in I$, then there is $d \in D$ such that $x \leq d$);*

(3) *two ideals of P are either comparable or disjoint;*

(4) *two distinct maximal ideals of P are disjoint;*

(5) *every element of P is contained in a unique maximal ideal of P.*

(c) *Let B and T be as in Theorem 1.5(iv) (so B is superatomic). We consider T as a partial ordering.*

(1) *T is a tree (i.e. T is well-founded, satisfies the property ($\bullet$), and $\mathrm{cl}_B(T) = B$).*

(2) *If D is a subchain of T, and if $d \in D$ is not a maximal element of D, then d has a unique successor d_D^+ in D.*

(3) *The elements of Maxideal(T) are pairwise disjoint.*

(4) *If $g \in T$, there is a unique member of Maxideal(T), say K_g, such that $g \in K_g$. If g_0, g_1 are distinct members of T such that $g_0 \cdot g_1 \neq 0^B$, then $K_{g_0} = K_{g_1}$.*

(5) *If a is minimal in T, then $C_a \stackrel{\text{def}}{=} \{g \in T : g \geq a\}$ is a maximal chain in T, and C_a generates K_a.*

Proof. (a) is a direct consequence of Zorn's lemma.

(b1) Let $p \in I$, and $d \in D \subseteq I$. There is $q \in I$ such that $q \geq p, d$. Because $L \stackrel{\text{def}}{=} \{t \in I : t \geq d\}$ is a chain, we have $D \supseteq L \ni q$. So $q \in D$ and $q \geq p$.

(b2) is trivial.

(b3) Let I_0 and I_1 be ideals of P. Suppose that $d \in I_0 \cap I_1$. For $\ell = 0, 1$, let D_ℓ be a maximal chain of I_ℓ such that $d \in D_\ell$. We can suppose that

$$D_0 \cap \{p \in I_0 : p \leq d\} = D_1 \cap \{p \in I_1 : p \leq d\} .$$

Then, by ($\bullet$), $D_0 \subseteq D_1$ or $D_1 \subseteq D_0$; say $D_0 \subseteq D_1$. Now, if $x \in I_0$, then $x \leq e \in D_0$ for some e, by (b1). Hence $e \in D_1$, so $x \in I_1$. This proves that $I_0 \subseteq I_1$.

(b4) is a consequence of (b3), and (b5) is a consequence of (a) and (b4).

(c) is a consequence of (a) and (b).   □

2.3 Boolean algebras

Every finite product of interval algebras is isomorphic to an interval algebra, and thus *if B is a subalgebra of a finite product of interval algebras, then B is embeddable in an interval algebra.*

More precisely, let us recall the following fact concerning chains and Boolean algebras (see Proposition 15.11 in [32]). Let C_1 and C_2 be chains with first elements 0^{C_1} and 0^{C_2} respectively. Let $C = C_1 + C_2$ be the chain which is lexicographic sum of C_1 and C_2 (so $c_1 < c_2$ for $c_1 \in C_1$ and $c_2 \in C_2$). Note that C has a first element, namely 0^{C_1}. A canonical isomorphism f from $B(C)$ onto $B(C_1) \times B(C_2)$ is obtained by letting: $f(c) = \langle c \cap C_1, c \cap C_2 \rangle$. Let us remark that we identified ∞^{C_1} with 0^{C_2}. $B(C_1), B(C_2)$ are factors of $B(C)$; and by identification, $B(C) \restriction C_1 = B(C_1)$ and $B(C) \restriction C_2 = B(C_2)$.

An element a of an interval algebra $B(C)$, different from 0^B $(= \emptyset)$, has a unique decomposition (called the *canonical decomposition*), under the form: $a = \bigcup \{[a_{2i}, a_{2i+1}) : i < n\}$ where $0 < n < \omega$, $0^C \leq a_0 < a_1 < a_2 < \ldots < a_{2n-1} \leq \infty^C$ and $a_k \in C^+$, $(k = 0, 1, \ldots, 2n-1)$. The integer n is called the *length* of a, and is denoted by $\ell(a)$.

Let D be a subset of a chain C, containing 0^C. Hence D is a chain with a first element. We denote by $B_C(D)$ the subalgebra of $B(C)$ consisting of those elements a such that $\sigma(a) \subseteq D \cup \{\infty^C\}$. Let us remark that the Boolean algebras $B_C(D)$ and $B(D)$ are isomorphic.

Now, we are ready to prove Theorem 1.5.

Trivially, (vi) $\Rightarrow$ (v) $\Rightarrow \ldots \Rightarrow$ (i). So it suffices to prove that (i) $\Rightarrow$ (iii) and (ii) $\Rightarrow$ (iv).

2.4   Part A: *Proof of (iv) $\Rightarrow$ (ii): If B is generated by a well-founded disjointed tree, then B is embeddable in an ordinal algebra*

First let us introduce a definition:

Definition 2.4 Let T be a tree (and thus T is well-founded).
   (a) $g', g'' \in T$ are said to be *disjoint* if there is no $g \in T$ such that $g \le g'$ and $g \le g''$. We denote by $g' \cdot g'' = 0$ (or $g' \cap g'' = \emptyset$) this fact.
   (b) $\langle W, (\varphi_l, \varphi_r) \rangle$ is said to be a *representation of T* if W is a well-founded chain, and φ_l and φ_r are functions from T into W such that:
   (1) for $g \in T$, $\varphi_l(g) < \varphi_r(g)$;
   (2) if $g \in T$ is minimal, then $[\varphi_l(g), \varphi_r(g)) = \{\varphi_l(g)\}$;
   (3) for $g', g'' \in T$, $g' \le g''$ if and only if $[\varphi_l(g'), \varphi_r(g'')) \subseteq [\varphi_l(g'), \varphi_r(g''))$;
   (4) for $g', g'' \in T$, $g' \cdot g'' = 0$ if and only if $[\varphi_l(g'), \varphi_r(g'')) \cap [\varphi_l(g'), \varphi_r(g'')) = \emptyset$.
   (c) Let $\langle W, (\varphi_l, \varphi_r) \rangle$ be a representation of T. We define a function $\vec{\varphi}$ from T into $B(W)$ by $\vec{\varphi}(g) = [\varphi_l(g), \varphi_r(g))$. Then $\vec{\varphi}$ defines φ_l and φ_r, and thus we say that $\langle W, \vec{\varphi} \rangle$ is a *representation of T* too.

Lemma 2.5 Let T be a tree. Then T has a representation $\langle W, \vec{\varphi} \rangle$.

Lemma 2.5 is a Boolean result, in the sense that it is not extendable to partial ordering. The following result is obvious. Let C be a chain of order-type $\omega \cdot \omega^*$. Then C has no representation, i.e. C cannot be representable as set of intervals of a well-founded chain. Other partial orderings (which are well-founded) with no representation (as set of intervals of a well-founded chain) are: $P = \{a_i, b_i : 0 \le i < 3\}$, ordered by $a_i \ne a_j$, $b_i \ne b_j$ and $a_i < b_j$, for $i \ne j$; the set $\omega \times \omega$, with the cartesian product order relation; and the dychotomic tree.

Proof of (ii) as a consequence of Lemma 2.5. We first show:

Claim 2.6 Let B be a Boolean algebra generated by a disjointed tree T.
   (a) *Every homomorphic image is generated by a disjointed tree.*
   (b) *In particular, if $b \in B - \{0^B\}$, then $B \!\restriction\! b$ is generated by a disjointed tree.*
   (c) *B is atomic.*
   (d) *$T \cup At(B)$ is a disjointed tree generating B.*

Proof. (a) Let T be a disjointed tree generating B, and $f : B \to A$ be a homomorphism onto. Then $f[T]$ is a tree generating A, and thus $f[T] - \{0^A\}$ is a disjointed tree generating A.

(b) follows from (a). Note that by (a), $T_b \stackrel{\text{def}}{=} \{b \cdot t : t \in T\} - \{0^B\}$ is a disjointed tree generating $B \restriction b$.

(c) Let $b \in B - \{0\}$. Let a be a minimal element of T_b. If $v \in T_b$, then $v \cdot a = 0$ or $a \leq v$. Therefore a is an atom of $B \restriction b$.

(d) Trivially $T \cup \text{At}(B)$ generates B. Let $a \in T \cup \text{At}(B)$. It is easy to check that $\{x \in T \cup \text{At}(B) : x \geq a\}$ is a well-founded chain.   □

Next, let B be a Boolean algebra generated by a disjointed tree T. By Claim 2.6(d), we can suppose that $\text{At}(B) \subseteq T$. Let $\langle W, \vec{\varphi} \rangle$ be a representation of T. We set $W_0 = \bigcup_{t \in T} \vec{\varphi}(t) \subseteq W$. W_0 is a well-founded chain, and $\langle W_0, \vec{\varphi} \rangle$ be a representation of T. Therefore we can suppose that $W = W_0$. We claim that $\vec{\varphi} : T \to B(W)$ is extendable to an embedding φ from B into $B(W)$. But that can be seen as a consequence of the Sikorski extension criterion (see Theorem 5.5 in [32]), because $\vec{\varphi}$ takes minimal elements of T to atoms of $B(W)$, and thus φ takes atoms to atoms. Hence φ exists and is one-to-one.

So it suffices to prove Lemma 2.5.

Proof of Lemma 2.5. By induction on the rank $\text{rk}(T)$ of the tree T (which is a well-founded partial ordering). If $\text{rk}(T) = 1$, then T is a non-empty antichain. Let $\{g_\nu : \nu < \gamma\}$ be an enumeration of the elements of T, $W = \gamma$, and for $\nu < \gamma$, $\vec{\varphi}(g_\nu) = [\nu, \nu + 1)$. Trivially $\langle W, \vec{\varphi} \rangle$ is a representation of T. Now, let $\alpha > 1$ be given. Assume that every tree of rank $< \alpha$ has a representation. Let T be a tree of rank α. We consider T as a partial ordering.

CASE 1: T IS AN IDEAL WITH A GREATEST ELEMENT g^∞.

Let $\underline{T} = T - \{g^\infty\}$. So $\text{rk}(\underline{T}) < \text{rk}(T) = \alpha$. By the induction hypothesis, $\underline{T}$ has a representation $\langle \underline{W}, \vec{\underline{\varphi}} \rangle$. Let $W = \underline{W} + \{\infty\}$ and $0 \stackrel{\text{def}}{=} \min(W) = \min(\underline{W})$. We define $\vec{\varphi}$ with domain T by $\vec{\varphi}(g) = \vec{\underline{\varphi}}(g)$ for $g \in \underline{T}$ and $\vec{\varphi}(g^\infty) = [0, \infty)$. Obviously $\langle W, \vec{\varphi} \rangle$ is a representation of T.

CASE 2: T IS AN IDEAL WITH NO GREATEST ELEMENT.

Let D be a maximal chain of T, and $\{d_\nu : \nu < \rho\}$ be the canonical enumeration of elements of D. By Claim 2.3(b1), D is cofinal in T, i.e. if $g \in T$, there is $\nu < \rho$ such that $g \leq d_\nu$. For $\mu < \rho$, $\mu > 0$, let

$$H_\mu = \{g \in T : g < d_\mu \text{ and } (\forall \nu)(\nu < \mu \Rightarrow g \cdot d_\nu = 0)\}$$

with the induced order relation. Let us remark that:

(a) for $0 < \mu < \rho$, H_μ is a tree (note that H_μ is an initial interval of T);
(b) for $0 < \mu < \rho$, $\text{rk}(H_\mu) < \text{rk}^T(d_\mu) < \text{rk}(T)$;
(c) $T = D \cup \bigcup \{H_\mu : 0 < \mu < \rho\}$;
(d) for $\mu < \nu < \rho$, $H_\mu \cap H_\nu = \emptyset$ and $H_\mu \cap D = \emptyset$;
(e) for $\mu < \nu < \rho$ and $g' \in H_\mu$, $g'' \in H_\nu$, $g' \cdot g'' = 0$;
(f) for $\mu < \nu < \rho$ and $g \in H_\nu$, $g \cdot d_\mu = 0$;
(g) for $\mu \leq \nu < \rho$ and $g \in H_\mu$, $g \leq d_\nu$.

Hence, by the induction hypothesis, for $0 < \mu < \rho$, H_μ has a representation $\langle W_\mu, \vec{\varphi}_\mu \rangle$. Let $a_\mu, b_\mu \notin W_\mu$. We set $W_\mu^+ = \{a_\mu\} + W_\mu + \{b_\mu\}$ with $a_\mu < x < b_\mu$ for $x \in W_\mu$, and let $W = \{a_0\} + \sum_{0 < \mu < \rho} W_\mu^+$ be the lexicographic sum of the W_μ^+'s (over the ordinal ρ) plus a new first element a_0. We define $\vec{\varphi} : T \to W$. For $g \in T$, let

$$\vec{\varphi}(g) = \begin{cases} \vec{\varphi}_\mu(g) & \text{if } g \in H_\mu \text{ with } 0 < \mu < \rho \\ [a_0, a_1] & \text{if } g = d_0 \\ [a_0, b_\mu] & \text{if } g = d_\mu \text{ with } 0 < \mu < \rho. \end{cases}$$

We show that $\langle W, \vec{\varphi} \rangle$ is a representation of T.

(1) Let $g \in T$ be minimal. If $g = d_0$, then $\vec{\varphi}(d_0) = \{a_0\}$. If $g \neq d_0$, then $\vec{\varphi}(g) = \vec{\varphi}_\mu(g)$ where μ is the unique ordinal such that $g \in H_\mu$.

(2) Let $g' \le g''$ in T. If g', g'' are in the same H_μ, then $\vec{\varphi}(g') = \vec{\varphi}_\mu(g') \subseteq \vec{\varphi}_\mu(g'') = \vec{\varphi}(g'')$. If $g' \in H_\mu$, $g'' = d_\nu$, then $\mu \le \nu$ and $\vec{\varphi}(g') = \vec{\varphi}_\mu(g') \subseteq W_\mu \subseteq [a_0, b_\nu] = \vec{\varphi}(d_\nu)$. If $g' = d_\mu$ and $g'' = d_\nu$, then $\mu \le \nu$ and $\vec{\varphi}(d_\mu) \subseteq \vec{\varphi}(d_\nu)$.

(3) Let $g', g'' \in T$ be such that $g' \cdot g'' = 0$. If g', g'' are in the same H_μ, then $\vec{\varphi}(g') \cap \vec{\varphi}(g'') = \vec{\varphi}_\mu(g') \cap \vec{\varphi}_\mu(g'') = \emptyset$. If $g' = d_\mu$ and $g'' \in H_\nu$, then $\mu < \nu$ and thus $\vec{\varphi}(d_\mu) \cap \vec{\varphi}(g'') \subseteq [a_0, b_\mu] \cap [a_\nu, b_\nu] = \emptyset$.

CASE 3: GENERAL CASE.

If T is an ideal, then T has a representation by Case 1 or Case 2. Now, suppose that T is not an ideal. Let $(J_\xi)_{\xi < \lambda}$ be an enumeration of maximal ideals of the partial ordering T. So $T = \bigcup_{\xi < \lambda} J_\xi$ and because T is a tree, by Claim 2.3 for $\zeta < \xi < \lambda$, $J_\zeta \cap J_\xi = \emptyset$, and hence $d' \cdot d'' = 0$ for $d' \in J_\zeta$ and $d'' \in J_\xi$. Moreover $\mathrm{rk}(J_\xi) \le \mathrm{rk}(T) = \alpha$. By Case 1 and Case 2, for $\zeta < \lambda$, let $\langle W_\zeta, \vec{\varphi}_\zeta \rangle$ be a representation of J_ζ. Let $W = \sum_{\zeta < \lambda} W_\zeta$ and $\vec{\varphi} = \bigcup_{\zeta < \lambda} \vec{\varphi}_\zeta$. It is easy to check that $\langle W, \vec{\varphi} \rangle$ is a representation of T. □

2.5　Part B: *Proof of (i) $\Rightarrow$ (iii): If B is a superatomic Boolean algebra, embeddable in an interval algebra, then B is isomorphic to a subalgebra $\underline{B}$ of an interval algebra $B(C)$ generated by a well-founded chain C such that $\mathrm{At}(\underline{B}) = \mathrm{At}(B(C))$*

Let B be a subalgebra of an algebra A, and $c \in A$. We denote by $B \restriction c$ the set of $b \cdot c$ for $b \in B$. We regard $B \restriction c$ as a quotient of B, as a subalgebra of the factor $A \restriction c$ of A, and thus as a Boolean algebra. Note that if $c \in B$, then $B \restriction c$ is a factor of B.

The following result is due to M. Rubin and S. Shelah [57], and was one of the ingredients of the original proof of Corollary 1.7.

Proposition 2.7 (Rubin and Shelah [57]). *Let B be an atomic subalgebra of an interval algebra. There are a totally disconnected complete chain C and an embedding ϕ from B into $B(C)$ such that $B(C)$ is an atomic algebra and $\mathrm{At}(\phi(B)) = \mathrm{At}(B(C))$.*

Note that the property of C implies that $c(\bigcup \mathrm{At}(B(C)), C) = C$. The proof of Proposition 2.7 needs some preliminary results. Let C be a chain such that $B \subseteq B(C)$.

Claim 2.8 *We can suppose that C satisfies (1): C is a complete chain, and (2): every atom of B is a finite subset of C.*

Proof. We can suppose that C is a complete chain (consider its Dedekind completion). Let $\underline{C} = \bigcup \{\sigma(a) : a \in B\}$ be the set of end-points of elements of B. We set $\underline{C}^c = c(\underline{C}, C)$. The function ϕ from B into the subalgebra $B_{\underline{C}^c}(\underline{C})$ of $B(\underline{C}^c)$ defined by $\phi(b) = b \cap \underline{C}$ is trivially a one-to-one homomorphism, and $\underline{C}^c$ is as required (note that by the construction, (2) is satisfied). □

Claim 2.9 *We can suppose that C satisfies (1) and (3): every atom of B is a singleton of C.*

Proof. For every $a \in \text{At}(B)$, we have $a = \bigcup_{i<\ell(a)}[a_{2i}, a_{2i+1}) = \bigcup_{i<\ell(a)}\{a_{2i}\}$ with $a_k \in C^+ = C \cup \{\infty^C\}$, and a_{2i}, a_{2i+1} consecutive in C. Let $\sim$ be the equivalence on C defined by $a_{2i} \sim a_{2i+1}$ for $0 < i < l(a)$ and $a \in \text{At}(B)$. Let $\underline{C} = C/\sim$. Hence $\underline{C}$, with the linear order induced by C, is a complete chain. Let φ be the function from $B(C)$ into $B(\underline{C})$ defined as follows: if $b = \bigcup_{i<\ell(b)}[b_{2i}, b_{2i+1})$, then $\varphi(b) = \bigcup_{i<\ell(b)}[b_{2i}/\sim, b_{2i+1}/\sim)$. Obviously φ is a homomorphism from $B(C)$ onto $B(\underline{C})$. It suffices to show that $\varphi(a) = [a_0/\sim, a_1/\sim) = \{a_0/\sim\} \neq 0$ for $a \in \text{At}(B)$, and φ restricted to B is one-to-one. But this is trivial. $\square$

Proof of Proposition 2.7. To prove the proposition, there is no loss in assuming that B and C satisfy the assumptions (1) and (3) of Claim 2.9. Let $\equiv$ be the equivalence on C defined by $x \equiv y$ if $x = y$, or $x \leq y$ and $[x, y)$ does not contain an atom of B, or $y \leq x$ and $[y, x)$ does not contain an atom of B. Then the quotient chain $\underline{C} \overset{\text{def}}{=} C/\equiv$ is complete and totally disconnected. Let ρ be the canonical increasing function from C onto $\underline{C}$. Note that if $x < y$ in C are such that $\rho(x) < \rho(y)$ in $\underline{C}$, then there is $a \in \text{At}(B)$ such that $a \subseteq [x, y)$. This shows that the function ϕ from B into $B(\underline{C})$, defined by $\phi(a) = \bigcup\{[\rho(a_{2i}), \rho(a_{2i+1})) : i < \ell(a)\}$ for $a = \bigcup\{[a_{2i}, a_{2i+1}) : i < \ell(a)\}$ in $B \subseteq B(C)$, is as required, and satisfies $\text{At}(\phi(B)) = \text{At}(B(\underline{C}))$. That finishes the proof of Proposition 2.7. $\square$

To prove that (i) $\Rightarrow$ (iii), there is no loss in assuming that (1): B satisfies both the premises and the conclusions of Proposition 2.7, and (2): $I(B) = I_{\text{rk}(B)}(B)$ is a maximal ideal of B. We denote by $\cong$ the equivalence relation on C defined by $x \cong y$ if $x \leq y$ and there is $b \in B$, with $\text{rk}^B(b) < \text{rk}^B(1^B)$, containing $[x, y)$, or $y \leq x$ and there is $b \in B$, with $\text{rk}^B(b) < \text{rk}^B(1^B)$, containing $[y, x)$. We will show:

Lemma 2.10 (a) *Each equivalence class is an interval of C.*
  (b) *Let $b \in B$ be such that $\text{rk}^B(b) < \text{rk}^B(1^B)$. Then there is a finite set $a_0/\cong, a_1/\cong, \ldots, a_{n-1}/\cong$ of equivalence classes such that $b \subseteq \bigcup_{k<n} (a_k/\cong -\{\sup_{C^+}(a_k/\cong)\})$.*

Proof. The first part of the claim is trivial. Let us show the second part. Let $b = \bigcup\{[b_{2i}, b_{2i+1}) : i < n\}$. It suffices to show that $b_{2i} \cong b_{2i+1}$ for $i < n$. But this is a trivial consequence of the definition of $\cong$. $\square$

The following two claims are obvious.

Claim 2.11 *If $a \in \text{At}(B)$, then a is contained in an equivalence class.* $\square$

Claim 2.12 *Let $a/\cong$ be an equivalence class. If $a/\cong$ has a last element v, then $v \notin \text{Pred}(C)$.* $\square$

Definition 2.13 Let C be a complete totally disconnected chain, B a superatomic subalgebra of $B(C)$, λ an ordinal, and ψ a function from B into $B(\lambda)$. We say that (B, C, λ, ψ) is a *good system* if $\text{At}(B) = \text{At}(B(C))$, ψ is a one-to-one homomorphism from B into the interval algebra $B(\lambda)$ and the restriction $\psi \upharpoonright \text{At}(B)$ of ψ on $\text{At}(B)$ is a one-to-one function from $\text{At}(B)$ onto $\text{At}(B(\lambda))$. We say that there is a *good system for* (B, C) if there are λ and ψ such that (B, C, λ, ψ) is a good system.

Note that $At(\psi(B)) = \psi(At(B))$. Equivalently a good system is (B, C, λ, ψ_0) where $At(B) = At(B(C))$, and ψ_0 is a one-to-one function from $\mathrm{Pred}(C)$ into the chain λ such that the function $\underline{\psi}_0$ from $At(B(C))$ into $At(B(\lambda))$, defined by $\underline{\psi}_0([u, u^+)) = [\psi_0(u), \psi_0(u) + 1)$ for $u \in \mathrm{Pred}(C)$, can be extended to an embedding ψ from B into $B(\lambda)$.

We will prove by induction on α that the following statement holds:

$\mathrm{Th}(\alpha)$: *For every chain C and for every superatomic subalgebra B of $B(C)$, such that $rk(B) \le \alpha$ and $At(B) = At(B(C))$, there is a good system for (B, C).*

$\mathrm{Th}(0)$ and $\mathrm{Th}(1)$ hold. Indeed B is isomorphic to the Boolean algebra $FC(X)$ of finite or cofinite subsets of a set X, where $X = At(B(C))$ (since $I(B) = I_{rk(B)}(B)$ is a maximal ideal of B). Consider λ to be the (initial) ordinal corresponding to the cardinality of the set $At(B(C))$.

In what follows, we suppose that $rk(B) \ge 2$.

Let $a \in C$. Let $\underline{a} = a/\cong$, $a^- = \inf_C(\underline{a})$, $a^+ = \sup_{C^+}(\underline{a})$ and $\tilde{a} = [a^-, a^+)$.
Note that $a^- \notin a/\cong$ if and only if $a/\cong$ has no first element.

Claim 2.14 *Let $a \in C$. There is a good system for $(B \restriction \tilde{a}, \tilde{a})$.*

Proof. By induction. If $a/\cong = \{a\}$, then $\tilde{a} = \emptyset$ and there is nothing to prove. Assume $|a/\cong| \ge 2$. Let $e \in \mathrm{Pred}(a/\cong)$, $\tilde{a}^+ = \{x \in \tilde{a} : x > e\}$ and $\tilde{a}^- = \{x \in \tilde{a} : x \le e\}$. Note that $\tilde{a}$ is the lexicographic sum $\tilde{a}^- + \tilde{a}^+$, and $\tilde{a}^+$ has a first element (the successor of e). If $\tilde{a}^-$ has no first element, then we must add one, namely a^-. Suppose that $(B \restriction \tilde{a}^+, \tilde{a}^+, \lambda^+, \psi^+)$ and $(B \restriction \tilde{a}^-, \tilde{a}^-, \lambda^-, \psi^-)$ are good. Let $(B \restriction \tilde{a}, \tilde{a}, \lambda^- + \lambda^+, \psi)$, where ψ is defined in the following way: for $b \in B \restriction \tilde{a}$, we have $b = \langle b^-, b^+ \rangle \in B(\tilde{a}^-) \times B(\tilde{a}^+)$ and we set $\psi(b) = \langle \psi^-(b^-), \psi^+(b^+) \rangle \in B(\lambda^-) \times B(\lambda^+)$ (that is identified with $B(\lambda^- + \lambda^+)$). Trivially, $(B \restriction \tilde{a}, \tilde{a}, \lambda^- + \lambda^+, \psi)$ is as required. So it suffices to prove Claim 2.14, whenever $\tilde{a}^+ = \tilde{a}$ or $\tilde{a}^- = \tilde{a}$.

First suppose that $\tilde{a}^+ = \tilde{a}$. Note that $\tilde{a}$ satisfies: $\tilde{a}$ has a first element, denoted by e, and for every element x of $\tilde{a}$, we have $x \cong e$.

CASE 1: $a/\cong$ HAS A LAST ELEMENT a^+.

So $\tilde{a} = a/\cong -\{a^+\}$. We have $a^+ \cong a$ and $\tilde{a} = [e, a^+)$ (that is the case of Remark 1.6). Let $b \in B$ be such that $\tilde{a} \subseteq b$, and $rk_{B \restriction b}(1_{B \restriction b}) = rk^B(b) < rk^B(1^B)$. Let $B \restriction \tilde{a} \overset{\mathrm{def}}{=} \{c \cap \tilde{a} : c \in B\}$. Note that $B \restriction \tilde{a} = (B \restriction b) \restriction \tilde{a}$. We regard $B \restriction \tilde{a}$ as a Boolean algebra. By the definition, $B \restriction \tilde{a}$ is a homomorphic image of $B \restriction b$. From the fact that for every superatomic Boolean algebra A and every ideal I of A, we have $rk(A/I) \le rk(A)$, it follows that $rk(B \restriction \tilde{a}) \le rk(B \restriction b) = rk^B(b) < rk^B(1^B) = rk(B)$. By the induction hypothesis there is a good system for $(B \restriction \tilde{a}, \tilde{a})$.

CASE 2: $a/\cong$ HAS NO LAST ELEMENT, I.E. $\tilde{a} = a/\cong$.

Let $(e_\alpha)_{\alpha < \sigma}$ be a strictly increasing sequence, cofinal in $\tilde{a}$. We can suppose that $e_0 = e$, and $e_\beta = \sup\{e_\alpha : \alpha < \beta\}$ for every limit ordinal $\beta < \sigma$ (because $\tilde{a}$ is relatively complete). Let $\alpha < \sigma$ be given. Let $b_\alpha \in B$ be such that $rk^B(b_\alpha) < rk^B(1^B)$ and $[e_\alpha, e_{\alpha+1}) \subseteq b_\alpha$. We set $B_\alpha \overset{\mathrm{def}}{=} B \restriction [e_\alpha, e_{\alpha+1})$. From Lemma 1.3, we have $rk(B_\alpha) \le rk(B \restriction b_\alpha) = rk^B(b_\alpha) < rk^B(1^B) = rk(B)$. Applying the induction hypothesis to $(B_\alpha, [e_\alpha, e_{\alpha+1}))$, there is a good system $(B_\alpha, [e_\alpha, e_{\alpha+1}), \mu_\alpha, \psi_\alpha)$. Hence $\psi_\alpha([e_\alpha, e_{\alpha+1})) = \mu_\alpha$. Let $\mu = \sum_{\alpha < \sigma} \mu_\alpha$, and $\psi =$

$\bigcup\{\psi_\alpha : \alpha < \sigma\}$. We have $\mathrm{Pred}(\mu) = \mu$. We extend ψ to a one-to-one homomorphism $\underline{\psi}$ from $B\!\restriction\tilde{a}$ into $B(\mu)$: let $b \in B$. We set:

$$\underline{\psi}(b) = \bigcup\{\mu_\alpha : [e_\alpha, e_{\alpha+1}) \subseteq b\,\} \cup \bigcup\{\psi_\alpha(b \cap [e_\alpha, e_{\alpha+1})) : b \cap [e_\alpha, e_{\alpha+1}) \neq 0_{B_\alpha}, 1_{B_\alpha}\}.$$

We must remark that $\underline{\psi}$ is well defined, because b is a finite union of half-open intervals and thus $\{\alpha < \sigma : [e_\alpha, e_{\alpha+1}) \subseteq b\} \in B(\sigma)$, and the set $\{\alpha < \sigma : b \cap [e_\alpha, e_{\alpha+1}) \neq 0_{B_\alpha}, 1_{B_\alpha}\}$ is finite. Consequently $\underline{\psi}(b)$ is a finite union of half-open intervals of μ, and thus $(B\!\restriction\tilde{a}, \tilde{a}, \mu, \underline{\psi})$ is a good system. That finishes the proof of Claim 2.14 in the case where $\tilde{a} = \tilde{a}^+$.

Let us show Claim 2.14 in the case $\tilde{a} = \tilde{a}^-$. First suppose that $a/ \cong$ has a first element, namely a^-. So $\tilde{a} = [a^-, e)$, and we conclude as in Case 1 above. Now suppose that $a/ \cong$ has no first element. So $a^- \notin a/ \cong$ and $\tilde{a} = \{a^-\} \cup (a/ \cong)$. Let $A_0 \overset{\mathrm{def}}{=} \{b \cap (a/ \cong) : b \in B\}$ and $A_1 \overset{\mathrm{def}}{=} \{b \cap \tilde{a} : b \in B\}$ (that is the homomorphic images of B) and $\varphi : A_1 \to A_0$ be defined by $\varphi(b') = b' - \{a^-\}$. Clearly φ is a homomorphism onto. Now φ is one-to-one follows from the fact that if $b \in B$, then $b \subseteq C$, and the fact that $a^- \not\subseteq b$ if and only if $\mathrm{rk}^B(b) < \mathrm{rk}^B(1^B)$. Now this case is similar to Case 2 by reversing the order relation. Consider a strictly decreasing coinitial sequence $(e'_\alpha)_{\alpha < \sigma}$ in $a/ \cong$ such that $e'_0 < e$, $e'_\beta = \inf\{e'_\alpha : \alpha < \beta\}$ for limit $\beta < \sigma$ and $a^- = \inf\{e'_\alpha : \alpha < \sigma\}$. For $\alpha < \sigma$, let $b_\alpha \in B$ be such that $\mathrm{rk}^B(b_\alpha) < \mathrm{rk}^B(1^B)$ and $[e'_{\alpha+1}, e'_\alpha) \subseteq b_\alpha$. Let $B_\alpha = B\!\restriction [e'_{\alpha+1}, e'_\alpha)$. By the induction hypothesis there is a good system $(B_\alpha, [e'_{\alpha+1}, e'_\alpha), \mu_\alpha, \psi_\alpha)$. We conclude as in Case 2, using the fact that $B\!\restriction\tilde{a} \overset{\mathrm{def}}{=} A_1$ is isomorphic to A_0. $\qquad\Box$

End of the proof of (i) $\Rightarrow$ (iii). Let $(a_\zeta / \cong)_{\zeta < \theta}$ be an enumeration of the set of equivalence classes. By Claim 2.14, for $\zeta < \theta$, let $(B\!\restriction\tilde{a}_\zeta, \tilde{a}_\zeta, \lambda_{\tilde{a}_\zeta}, \psi_{\tilde{a}_\zeta})$ be a good system. Let $\lambda = \sum_{\zeta < \theta} \lambda_{\tilde{a}_\zeta}$. Each $\lambda_{\tilde{a}_\zeta}$ is an interval of λ. Now, let ψ be the function from $\mathrm{At}(B)$ into λ defined by $\psi(a) = \psi_{\tilde{a}_\zeta}(a)$ where $\tilde{a}_\zeta$ is the unique class such that $a \in \mathrm{At}(B) \cap B\!\restriction\tilde{a}_\zeta$. Let $b \in B$. First, suppose that $\mathrm{rk}^B(b) < \mathrm{rk}^B(1^B)$. There is a finite subset $\{\tilde{a}_0, \tilde{a}_1, \ldots, \tilde{a}_{n-1}\}$ of equivalence classes such that $b \subseteq \bigcup\{\tilde{a}_k : k < n\}$ which follows from Lemma 2.10. We set $\psi(b) = \bigcup\{\psi_{\tilde{a}_k}(b \cap \tilde{a}_k) : k < n\}$. Now, because $I(B) = I_{\mathrm{rk}(B)}(B)$ is a maximal ideal of B, if $\mathrm{rk}^B(b) = \mathrm{rk}^B(1^B)$, then $\mathrm{rk}^B(-b) < \mathrm{rk}^B(1^B)$, and we set $\psi(b) = -\psi(-b)$. The fact that ψ is a one-to-one homomorphism from B into $B(\lambda)$ is a consequence of the following obvious result.

Claim 2.15 *Let B' and B'' be two atomic algebras and ψ be a one-to-one function from $\mathrm{At}(B')$ onto $\mathrm{At}(B'')$. We suppose that, for each $b \in B'$, there is a unique element of B'', denoted by $\underline{\psi}(b)$, such that for every $a \in \mathrm{At}(B')$ we have $a \leq b$ if and only if $\psi(a) \leq \underline{\psi}(b)$. Then $\underline{\psi}$ is a one-to-one homomorphism from B' into B'', extending ψ.* $\qquad\Box$

That finishes the proof of (i) $\Rightarrow$ (iii).

2.6 Part C: *Proof of (ii) $\Rightarrow$ (vi): If B is embeddable in an ordinal algebra, then B has a well-founded generating tree which is a complete set of generalized atoms.*

Before proving this, let us establish some facts concerning the relation between lengths of a, b, $a \cup b$, $a \cap b$, $a - b$, $b - a$, etc., for members a and b of an interval algebra.

Claim 2.16 *Let $B(C)$ be an interval algebra and a, $b \in B(C)$ (we recall that the symmetric difference of a and b is $a \Delta b \overset{\text{def}}{=} (a - b) \cup (b - a)$). Then:*

(1) $\ell(a \cup b) + \ell(a \cap b) + \ell(a - b) + \ell(b - a) = \ell(a) + \ell(b) + \ell(a \Delta b)$;

(2) $\ell(a \cup b) + \ell(a \cap b) \leq \ell(a) + \ell(b)$;

(3) $\ell(a) + \ell(b) \geq \ell(a - b) + \ell(b - a)$;

(4) *if* $\ell(a - b) > \ell(a)$, *then* $\ell(b - a) < \ell(b)$;

(5) *if* $\ell(a - b) \geq \ell(a)$, *then* $\ell(b - a) \leq \ell(b)$.

Proof. Let us remark that (3) implies (4) and (5). First, we will show that: $(1) \Rightarrow (2) \Rightarrow (3)$. Clearly, for every c, $d \in B(C)$, $\ell(c \cup d) \leq \ell(c) + \ell(d)$, and hence $\ell(a \Delta b) \leq \ell(a - b) + \ell(b - a)$. The last inequality and (1) imply (2). Let us show (3). By (2), we have

$$\ell((a \cup b) \cap (-(a \cap b))) + \ell((a \cup b) \cup (-(a \cap b))) \leq \ell(a \cup b) + \ell(-(a \cap b)).$$

From this inequality, and the facts that $\ell(-(a \cap b)) - 1 \leq \ell(a \cap b)$, $\ell((a \cup b) \cup (-(a \cap b))) \geq 1$ and $a \Delta b = (a \cup b) \cap (-(a \cap b))$, it follows that $\ell(a \Delta b) \leq \ell(a \cup b) + \ell(a \cap b)$. Again, this last inequality and (1) implies (3). A direct proof of (2) and (3) can be found in Pouzet and Rival [47].

The present proof of (1) is suitable for the proof of Claim 2.17. Let $a = \bigcup\{[a_{2i}, a_{2i+1}) : i < \ell(a)\}$ and $b = \bigcup\{[b_{2i}, b_{2i+1}) : i < \ell(b)\}$, and let $\{c_0, \ldots, c_{q-1}\} \overset{\text{def}}{=} \{a_0, \ldots, a_{2\ell(a)-1}\} \cup \{b_0, \ldots, b_{2\ell(b)-1}\}$ with $c_0 < c_1 < \cdots < c_{q-1}$. For every $e \in E \overset{\text{def}}{=} \{a \cup b, a \cap b, a - b, b - a, a, b, a \Delta b\}$ and $i < q$, let $e^{(i)} = e \cap [0, c_i)$, and for $i > 0$, $\Delta(e, i) = \ell(e^{(i)}) - \ell(e^{(i-1)})$. Clearly, for every $e \in E$, $\ell(e^{(0)}) = 0$ and $\ell(e) = \sum_{i=1}^{q-1} \Delta(e, i)$. Let E_1 and E_2 denote respectively the elements appearing on the left side and on the right side of the identity (1), i.e. $E_1 = \{a \cup b, a \cap b, a - b, b - a\}$ and $E_2 = \{a, b, a \Delta b\}$. For $j = 1, 2$ and $i < q$, let $L(j, i) = \sum\{\ell(e^{(i)}) : e \in E_j\}$. Note that $L(1, 0) = L(2, 0) = 0$, and $L(j, i + 1) = L(j, i) + \sum\{\Delta(e, i + 1) : e \in E_j\}$, $j = 1, 2$. Let $\Delta^*(j, i) = \sum\{\Delta(e, i) : e \in E_j\}$. Let $\Delta^*(j, i) = \sum\{\Delta(e, i) : e \in E_j\}$. We will show that for every $i < q$, $\Delta^*(1, i) = \Delta^*(2, i)$. For then $L(1, i) = L(2, i)$ by induction on i; the case $i = q - 1$ gives the desired conclusion. Let $D = \{a \cap b, a - b, b - a, -(a \cup b)\}$. Note that for all $i < q - 1$, we have $[c_i, c_{i+1}) \subseteq a$ or $[c_i, c_{i+1}) \subseteq -a$, and similarly for b. Hence this is also true for any $e \in E$ in place of a. Hence for every $I < q$ there is a unique $d_i \in D$ such that $[c_i, c_{i+1}) \subseteq d_i$. We denote $d_{-1} = -(a \cup b)$. Also note the following rule, easily verified, for determining $\ell(e^{(i)})$ in terms of $\ell(e^{(i-1)})$:

$$\ell(e^{(i)}) = \begin{cases} \ell(e^{(i-1)}) & \text{if } [c_{i-2}, c_{i-1}) \subseteq e \text{ and } [c_{i-i}, c_i) \subseteq e \text{;} \\ \ell(e^{(i-1)}) & \text{if } [c_{i-2}, c_{i-1}) \subseteq e \text{ and } [c_{i-i}, c_i) \not\subseteq e \text{;} \\ \ell(e^{(i-1)}) + 1 & \text{if } [c_{i-2}, c_{i-1}) \not\subseteq e \text{ and } [c_{i-i}, c_i) \subseteq e \text{;} \\ \ell(e^{(i-1)}) & \text{if } [c_{i-2}, c_{i-1}) \not\subseteq e \text{ and } [c_{i-i}, c_i) \not\subseteq e \text{.} \end{cases}$$

It follows from this rule that $\Delta^*(j, i)$ is completely determined by the pair $\langle d_{i-1}, d_i \rangle$. On the basis of these considerations it is easy to check that $\Delta^*(1, i) = \Delta^*(2, i)$ for all i, by considering the 16 possibilities for $\langle d_{i-1}, d_i \rangle$:

($\bullet$) $\Delta^*(1, i) = \Delta^*(2, i) = 0$ if and only if $\langle d_{i-1}, d_i \rangle = \langle e, e \rangle$ for $e \in \{a \cap b, a - b, b - a\}$, or $\langle d_{i-1}, d_i \rangle = \langle e, -(a \cup b) \rangle$ for $e \in \{a \cap b, a - b, b - a\}$.

($\bullet$) $\Delta^*(1, i) = \Delta^*(2, i) = 1$ if and only if $\langle d_{i-1}, d_i \rangle = \langle e', e'' \rangle$ with distinct $e', e'' \in \{a \cap b, a - b, b - a\}$.

(•) $\Delta^\star(1,i) = \Delta^\star(2,i) = 2$ if and only if $\langle d_{i-1}, d_i \rangle = \langle -(a \cup b), e \rangle$ with $e \in \{a \cap b, \ a-b, \ b-a\}$.
□

Now, let C be a well-ordered chain, B a subalgebra of $B(C)$. For $a = \sum \{[a_{2i}, a_{2i+1}) : i < \ell(a)\}$ in $B(C)$, we set $f(a) = \langle a_0, \ldots, a_{2\ell(a)-1} \rangle$. Let a and b be such that $\ell(a) = \ell(b) \overset{\text{def}}{=} m$. We set $a \leq^\ell b$ whenever $f(a) \leq f(b)$ for the lexicographic ordering on $(C^+)^{2m}$, i.e. $a \leq^\ell b$ if and only if $a = b$ or, for the first $i < 2m$ such that $a_i \neq b_i$, $a_i < b_i$. So $\leq^\ell$ is a *linear order* on $\{c \in B(C) : \ell(c) = m\}$. Note that there is no relation between $a \leq b$ (i.e. $a \subseteq b$) in B and $a \leq^\ell b$.

Claim 2.17 (a) *If* $\ell(a-b) = \ell(a)$, $\ell(b-a) = \ell(b)$ *and* $a <^\ell a-b$, *then* $b-a <^\ell b$.
(b) *If* $\ell(a \cap b) = \ell(a)$, $\ell(a \cup b) = \ell(b)$ *and* $a <^\ell a \cap b$, *then* $a \cup b <^\ell b$.

Proof. (a) We use the notations appearing in the proof of Claim 2.16, in particular $a = \bigcup \{[a_{2i}, a_{2i+1}) : i < \ell(a)\}$, $b = \bigcup \{[b_{2i}, b_{2i+1}) : i < \ell(b)\}$,

$$\{c_0, \ldots, c_{q-1}\} \overset{\text{def}}{=} \{a_0, \ldots, a_{2\ell(a)-1}\} \cup \{b_0, \ldots, b_{2\ell(b)-1}\}$$

with $c_0 < c_1 < \cdots < c_{q-1}$, and for every $e \in E \overset{\text{def}}{=} \{a, b, a \cap b, a \cup b, a-b, b-a, -(a \cup b)\}$, and $i < q$, $e^{(i)} = e \cap [0^C, c_i)$. Assume $\ell(a-b) = \ell(a)$, $\ell(b-a) = \ell(b)$ and $a <^\ell a-b$. Because $a \neq a-b$, there is $r < q$ such that $[c_r, c_{r+1}) \subseteq a \cap b$. Let p be the first integer such that $[c_p, c_{p+1}) \subseteq a \cap b$. By the definition, we have $a \cap b \cap [0^C, c_p) = \emptyset$, and thus: $a^{(p)} - b^{(p)} = a^{(p)}$ and $b^{(p)} - a^{(p)} = b^{(p)}$. Let us consider the element $[c_{p-1}, c_p)$ (if $p > 0$). We distinguish four cases.

CASE 1: $p > 0$ AND $[c_{p-1}, c_p) \subseteq a \cap b$.
This is impossible by the choice of p.
CASE 2: $p > 0$ AND $[c_{p-1}, c_p) \subseteq a - b$.
This is impossible, since $c_p < c_{p+1}$ implies $a - b <^\ell a$.
CASE 3: $p > 0$ AND $[c_{p-1}, c_p) \subseteq b - a$.
This implies $b - a <^\ell a$, which is our conclusion.
CASE 4: $p = 0$ OR: $p \geq 1$ AND $[c_{p-1}, c_p) \subseteq -(a \cup b)$.
Let $\underline{C}$ be the well-founded chain obtained from C by adding a new immediate predecessor $\underline{c}$ to c_p, i.e. $\underline{c} \notin C$, $\underline{c} < c_p$ and if $x \in C$ verifies $x < c_p$, then $x < \underline{c}$. So, if $p > 0$, then $c_{p-1} < \underline{c} < c_p$; and for instance, if $p = 0$ and $c_p = 0^C$, then $\underline{c} = 0^{\underline{C}}$. Let a', $b' \in B(\underline{C})$ be defined as follows: $a' = a \cup [\underline{c}, c_p)$ and $b' = b$. Now $\ell(a') = \ell(a)$, $\ell(b') = \ell(b)$, $\ell(a'-b') = \ell(a-b)+1$ and $\ell(b'-a') = \ell(b-a)$. Consequently $\ell(a')+\ell(b') < \ell(a'-b')+\ell(b'-a')$, which is contradictory to Claim 2.16(3).

(b) Assume $\ell(a \cap b) = \ell(a)$, $\ell(a \cup b) = \ell(b)$ and $a <^\ell a \cap b$. Let p be the first integer such that $[c_p, c_{p+1}) \subseteq a - b$. By the definition, we have $(a - b) \cap [0^C, c_p) = \emptyset$ and thus $a^{(p)} \cup b^{(p)} = b^{(p)}$ and $a^{(p)} \cap b^{(p)} = a^{(p)}$. We will conclude using similar arguments as in part (a), via four cases.
CASE 1: $p > 0$ AND $[c_{p-1}, c_p) \subseteq a - b$.
This is impossible by the choice of p.
CASE 2: $p > 0$ AND $[c_{p-1}, c_p) \subseteq a \cap b$.
This is impossible, since in this case, $a \cap b <^\ell a$.
CASE 3: $p > 0$ AND $[c_{p-1}, c_p) \subseteq -(a \cup b)$.

We have $(a \cup b)^{(p-1)} = b^{(p-1)}$, and because $c_p < c_{p+1}$, we have $a \cup b <^\ell b$.

CASE 4: $p = 0$ OR: $p \geq 1$ AND $[c_{p-1}, c_p) \subseteq b - a$.

Suppose $p = 0$. We have $[c_0, c_1) \subseteq a - b$, and thus $a \cup b <^\ell b$. Now, suppose that $p > 0$ and $[c_{p-1}, c_p) \subseteq b - a$. As in Case 4 of (a), let $\underline{C}$ be the well-founded chain obtained from C by adding a new immediate predecessor $\underline{c}$ to c_p. So $c_{p-1} < \underline{c} < c_p$. Let $a' = a \cup [\underline{c}, c_p)$, $b' = b \cup \{\underline{c}\}$. Then $\ell(a') = \ell(a)$, $\ell(b') = \ell(b)$, $\ell(a' \cap b') = \ell(a \cap b) + 1$ and $\ell(b' \cup a') = \ell(b \cup a)$. Consequently $\ell(a' \cap b') + \ell(a' \cup b') > \ell(a') + \ell(b')$, which is contradictory to Claim 2.16(2).□

End of the proof of (ii) $\Rightarrow$ (vi). Let C be a well-ordered chain , B a subalgebra of $B(C)$. Let α be given and $\underline{a}$ be an atom of $D_\alpha(B)$. Let $U_{\underline{a}} \stackrel{\text{def}}{=} \{t \in B : t/I_\alpha(B) = \underline{a}\}$. Let $m(\underline{a}) \stackrel{\text{def}}{=} \min\{\ell(t) : t \in U_{\underline{a}}\}$, and $U_{\underline{a}, m(\underline{a})} \stackrel{\text{def}}{=} \{t \in U_{\underline{a}} : \ell(t) = m(\underline{a})\}$. For $t \in U_{\underline{a}, m(\underline{a})}$, we have $t = \bigcup\{[t_{2i}, t_{2i+1}) : i < m(\underline{a})\}$ and we set $f(t) = \langle t_0, \ldots, t_{2m(\underline{a})-1} \rangle$, which is an element of $(C^+)^{2m(\underline{a})}$. Let $\mu(\underline{a}) = \min(U_{\underline{a}, m(\underline{a})})$, under the order relation $\leq^\ell$, i.e. $\mu(\underline{a})$ is the unique element of $U_{\underline{a}, m(\underline{a})}$ such that $f(\mu(\underline{a}))$ is the smallest element of $f(U_{\underline{a}, m(\underline{a})}) \subseteq (C^+)^{2m(\underline{a})}$, under the lexicographic order relation. Let

$$G \stackrel{\text{def}}{=} \{\mu(\underline{a}) : \underline{a} \in \bigcup\{\text{At}(D_\alpha(B)) : \alpha \leq \text{rk}(B)\}\}.$$

We claim that G *satisfies* (vi). It is sufficient to show that if $a, b \in G$ are such that $\text{rk}^B(a) \leq \text{rk}^B(b)$, with $a \neq b$, then $a \subseteq b$ or $a \cap b = 0$. By contradiction, assume that $a \not\subseteq b$ and $a \cap b \neq 0$.

Then $a - b \neq 0$ and $a - b \neq a$. Note that $\text{rk}^B(b - a) = \text{rk}^B(b)$: for a contradiction, suppose that $\text{rk}^B(b - a) < \text{rk}^B(b)$. Then $\text{rk}^B(b) = \text{rk}^B(a \cap b) = \text{rk}^B(a)$, and thus $a = b$.

CASE 1: $\text{rk}^B(a - b) = \text{rk}^B(b)$.

Then, by the choice of a, we have $\ell(a) \leq \ell(a - b)$. First suppose that $\ell(a - b) = \ell(a)$ and thus $a <^\ell a - b$. By Claim 2.16(3), we have $\ell(b - a) \leq \ell(b)$. From the choice of b and the fact that $\text{rk}^B(b - a) = \text{rk}^B(b)$ it follows that $\ell(b - a) = \ell(b)$ and thus $b <^\ell b - a$. Because $a <^\ell a - b$, we obtain a contradiction with Claim 2.17(a). Now, suppose that $\ell(a - b) > \ell(a)$. Then by Claim 2.16(4), $\ell(b - a) < \ell(b)$. Because $\text{rk}^B(b - a) = \text{rk}^B(b)$, we obtain a contradiction with the choice of b.

CASE 2: $\text{rk}^B(a - b) < \text{rk}^B(b)$.

Then $\text{rk}^B(a \cap b) = \text{rk}^B(a)$ and thus $\ell(a \cap b) \geq \ell(a)$. We have $\text{rk}^B(a) < \text{rk}^B(b)$: otherwise $\text{rk}^B(a) = \text{rk}^B(a \cap b) = \text{rk}^B(b)$ and thus $a = b$. First, suppose that $\ell(a \cap b) = \ell(a)$. Then $a <^\ell a \cap b$. By Claim 2.16(2), $\ell(a \cup b) \leq \ell(b)$ and then $\ell(a \cup b) = \ell(b)$. Hence $b <^\ell a \cup b$. We obtain a contradiction with Claim 2.17(b). Now, suppose that $\ell(a \cap b) > \ell(a)$. By Claim 2.16(2), $\ell(a \cup b) < \ell(b)$, and we obtain a contradiction with the choice of b.

This finishes the proof of (ii) $\Rightarrow$ (vi).   □

Acknowledgement. The author thanks J. Donald Monk for helpful comments and information, Norbert Sauer, Robert Woodrow and the referees for many remarks and comments during the preparation of this work.

References

[1] U. Avraham and R. Bonnet (1992): Every superatomic subalgebra of an interval algebra is embeddable in an ordinal algebra, *Proc. Amer. Math. Soc.* **115** (3), 585–592.

[2] Z. Balogh, A. Dow, D. H. Fremlin and P.J. Nyikos (1988): Countable tightness and proper forcing, *Amer. Math. Bull.* **19**, 295–298.

[3] J. Baumgartner and S. Shelah (1987): Remarks on superatomic Boolean algebras, *Annals of Pure and Applied Logic* **33**(2), 109–129.

[4] J. Baumgartner and M. Weese (1987): Partition algebras for almost-disjoint families, *Trans. Amer. Math. Soc.* **274**, 619–630.

[5] M. Bekkali (1991): *On superatomic Boolean algebras*, Ph. D. Dissertation, University of Colorado, Boulder, viii + 122 pp.

[6] M. Bekkali (1991): Pseudo-tree algebras, submitted to *Order*.

[7] M. Bekkali (1991): Superatomic tree algebras, submitted to *Algebra Univ.*.

[8] M. Bekkali, R. Bonnet and M. Rubin (1993): Compact interval spaces in which all closed subset are homeomorphic to a clopen ones [I], *Order* **9**(1), 69–95.

[9] M. Bekkali, R. Bonnet and M. Rubin (1993): Compact interval spaces in which all closed subset are homeomorphic to a clopen ones [II], *Order* **9**(2), 177–200.

[10] R. Bonnet (1977): Sur le type d'isomorphie d'algèbres de Boole dispersées, in *Colloq. Intern. de Logique*, 107–122, Paris, C.N.R.S.

[11] R. Bonnet (1984): On homomorphism types of superatomic interval algebras, in *Models and Sets*, 67–81, Lectures Notes in Math. (Springer) p. 1103.

[12] R. Bonnet and H. Si-Kaddour (1984): Comparison of Boolean algebras, *Order* **4**(3), 273–283.

[13] R. Bonnet and S. Shelah (1989): On HCO spaces. An uncountable compact T_2 space, different of $\aleph_1 + 1$, which is homeomorphic to every of its uncountable closed subspace, (Sh 354) to appear *Israel J. Math.*

[14] R. Bonnet, M. Rubin and H. Si-Kaddour (1992): On Boolean algebras with well-founded set of generators, (2nd version), submitted to *Trans. Amer. Math. Soc.*

[15] R. Bonnet and M. Rubin (1991): On Boolean algebras which are isomorphic to each of it uncountable subalgebra, submitted to *Canadian J. Math.*

[16] R. Bonnet and M. Rubin (1992): On poset Boolean algebras, *in preparation*.

[17] G. Brenner (1982): Tree algebras, Ph. D., Univ. of Colorado.

[18] G. Brenner (1983): A simple construction of rigid and weakly homogeneous Boolean algebras, answering a question of Rubin, *Proc. Amer. Math. Soc.* **87**, 601–606.

[19] G.W. Day (1967): Superatomic Boolean algebras, *Pacific J. Math.* **23**, 479–489.

[20] A. Dow and S. Watson (1990): Skula spaces, *Comment. Math. Univ. Carolina.* **31**(1), 27–31.

[21] A. Dow and P. Simon (1988): Thin-tall Boolean algebras and their automorphism groups, *preprint*.

[22] J. Dugundji (1978): *Topology* (Tweltfh edition), Allyn and Bacon.

[23] B. Dushnik and E. W. Miller (1940): Concerning similarity transformations of linear ordered sets, *Bull. Amer. Math. Soc.* **46**, 322–326.

[24] R. Fraïssé (1986): *Theory of relations*, Studies in Math. Logic, North-Holland.

[25] S. Goncharov (1973): The constructivizability of superatomic Boolean algebras (Russian), *Alg. i. Log.* **12**, 17–22.

[26] P. Halmos (1963): *Lectures on Boolean Algebras*, Van Nostrand Math. Studies, **1**, Van Nostrand Company, New York.

[27] L. Henkin, J.D. Monk and A. Tarski (1971): *Cylindric Algebras. Part I. With An Introductory Chapter; General Theory of Algebras*, vi+508 pp, North-Holland.

[28] T. Jech (1978): *Set Theory*, Pure and Applied Mathematics, Academic Press.

[29] I. Juhász and W. Weiss (1978): On thin-tall scattered spaces, *Coll. Math.* **90**, 64–68.

[30] W. Just (1985): Two consistency results concerning thin-tall Boolean algebras, *Alg. Universalis* **20**, 135–142.

[31] J.L. Kelley (1955): *General Topology*, Van Nostrand Company.

[32] S. Koppelberg (1989): Special classes of Boolean algebras, *Handbook on Boolean Algebras*, volume 1, Part I, chapter 6, 239–284, editor J.D. Monk and R. Bonnet, North Holland.

[33] S. Koppelberg and J. D. Monk (1989): Pseudo-tree Boolean algebras, to appear in *Order*.

[34] K. Kunen (1980): *Set Theory*, Studies in Math. Logic, North-Holland.

[35] K. Kuratowski and A. Mostowski (1976): *Set Theory*, Studies in Math. Logic, Vol. 86, North-Holland.

[36] R. La Grange (1977): Concerning the cardinal sequence of a Boolean algebra, *Algebra Univ.* **7**, 307–312.

[37] A. Marcja (1982): An algebraic approach to superstability, *Boll. Un. Mat. Ital.* **1**, 71–76.

[38] J. Martinez (1991): A consistency result on thin-tall superatomic Boolean algebras, to appear in *Proc. Amer. Math. Soc.*

[39] R. Mayer and R.S. Pierce (1960): Boolean algebras with ordered basis, *Pacific Journal of Math.* **10**, 925–942.

[40] S. Mazurkiewicz and W. Sierpinski (1920): Contribution à la topologie des ensembles dénombrables, *Fund. Math.* **1**, 17–27.

[41] B. Molzan (1982): The theory of superatomic Boolean algebras in the logic with the binary Ramsey quantifier, *Z. Math. Logik, Grdl. der Math.* **28**, 365–376.

[42] T. Mori (1983): Results on a compact space which has certain properties and their applications to superatomic Boolean algebras, *Rep. Sci. Engrg. Saga Univ. Math* **11**, 1–8.

[43] A. Mostowski and A. Tarski (1939): Boolesche Ringe mitgeordneter Basis, *Fund. Math.* **32**, 69–86.

[44] A. Ostaszewski (1989): On countably compact, perfectly normal spaces, *J. London Math. Soc.* **14**(2), 505–516.

[45] R.S. Pierce (1989): Countable Boolean algebras, *Handbook on Boolean Algebras*, volume 3, Part II, chapter 21, 775–876, editor J.D. Monk and R. Bonnet, North Holland.

[46] M. Pouzet: *Ordered sets*, Edit. I. Rival . Nato Advanced Study Institutes Series. Series C. Volume **83**, 1982, 847.

[47] M. Pouzet and I. Rival (1984): Every countable lattice is a retract of a direct product of chains, *Algebra Universalis* **18**, 295–307.

[48] J. Roitman (1985): Height and weight of superatomic Boolean algebras, *Proc. Amer. Math. Soc.* **94**, no 1, 9–14.

[49] J. Roitman (1985): A very thin-thick superatomic Boolean algebra, *Alg. Universalis* **21**, no 2-3, 137–142.

[50] J. Roitman (1985): Thin-tall spaces with restricted homeomorphism, *preprint*.

[51] J. Roitman (1989): Superatomic Boolean algebras, *Handbook on Boolean Algebras*, volume 3, Part II, chapter 19, 719–740, editor J.D. Monk and R. Bonnet, North Holland.

[52] J. Roitman (1989): A space homeomorphic to each of its uncountable closed subspaces under CH, *preprint*.

[53] M. Rajagapolan (1976): A chain compact space which is not strongly scattered, *Israel J. Math* **23**, 117–125.

[54] J. Rosenstein (1982): *Linear Orderings*, Academic Press, New York.

[55] M. Rubin (1983): A Boolean algebra with few subalgebras, interval Boolean algebras and retractiveness, *Trans. Amer. Math. Soc.* **278**(1), 65–89.

[56] M. Rubin (1974): Theories of linear order, *Israel J. Math.* **17**, 392–443.

[57] M. Rubin and S. Shelah (1988): On the cardinality of superatomic subalgebras of an interval algebra (*manuscript*).

[58] S. Shelah (1983): Uncountable constructions for Boolean algebras, e.c. groups and Banach spaces, *Israel J. Math.* **51**, 273–297.

[59] S. Shelah (1992): Factor = quotient, uncountable Boolean algebras, number of endomorphisms and width (Sh. 397), *Mathematica Japonica* **37**(2), 385–400.

[60] S. Shelah and J. Steprans (1990): Homogeneous almost disjoint families, *preprint*.

[61] H. Si-Kaddour (1989): *Sur les Algèbres de Booles Superatomiques*, Diplôme de Doctorat, Université Claude-Bernard (Lyon 1).

[62] R. Sikorski (1964): *Boolean Algebras*, Ergebnisse Mathematik **25**, Springer-Verlag, Berlin.

[63] P. Simon and M. Weese (1985): Non-isomorphic thin-tall superatomic Boolean algebras, *Comment. Math. Univ. Carolina* **26**, 241–252.

[64] R. Tegárski (1968): Derivatives of Cartesian products and dispersed spaces, *Coll. Math.* **19**, 59–66.

[65] M. Weese (1976): The isomorphism problem of superatomic Boolean algebras, *Z. Math. Logik Grundlagen Math.* **22**, 439–440.

[66] M. Weese (1982): On the classification of scattered spaces, in: *Proc. of the Conference of Topology and Measure*, III, 347–356.

[67] M. Weese (1986): On the classification of superatomic Boolean algebras, *Open Days in Model Theory and Set Theory (Proc. Conf. Jadwissin)*, University of Leeds.

[68] M. Weese (1991): Subalgebra-rigid and homomorphism-rigid Boolean algebras (*manuscript*).

Representing Rings Using Continuous Functions

ANDREW B. CARSON

Mathematics Department,

University of Saskatchewan,

Saskatoon, Saskatchewan,

Canada S7N 0W0.

Abstract

Classic results due to Stone, and Arens and Kaplansky, showed that many rings can be represented using continuous (or continuous and twisted) functions, from some Boolean space to a field. This paper summarizes these early results and discusses recent efforts to characterize all rings having a similar representation, involving (possibly twisted) functions from some Boolean space (or from dense open subsets of it), into an indecomposable ring. Recent results were obtained using sheaves, ring localizations, and model theory.

1 Introduction (Historical Background)

We begin with a warning to analysts: In this talk functions always have topologically discrete range, and so are continuous iff they are locally constant. From this viewpoint many classical theorems can be construed as saying that certain rings are function rings. For example, if R is a commutative[1] semi-simple artinian ring, then the Wedderburn-Artin theorem asserts that $R \cong \oplus \sum_{0 \leq i \leq n} F_i$, for suitable $n \geq 0$ and fields F_i. If in addition R is an algebraic algebra over some field L with algebraic closure F, then $F_i \subseteq F$, for each $i \geq 0$, so that viewing $X := \{0, \ldots, n\}$ as a discretely topologized space, we have $R \cong \mathcal{C}(X, F; \{F_x : x \in X\})$, where we are using the following notation:

Definition 1.1 Let X be a Boolean space, F be a discretely topologized indecomposable ring, and let $\{F_x : x \in X\}$ be a family of subrings of F. Then $\mathcal{C}(X, F; \{F_x : x \in X\}) = \{f : f$ is a continuous function from X to F and $f(x) \in F_x$, for all $x \in X\}$. This notation is shortened to $\mathcal{C}(X, F)$, when $F_x = F$, for all $x \in X$. If $\mathcal{M}$ is a class of indecomposable rings, call R an $\mathcal{M}$-*function ring* iff it has the form $R \cong \mathcal{C}(Y, F; \{F_y : y \in Y\})$, for some suitable $\{F\} \cup \{F_y : y \in Y\} \subseteq \mathcal{M}$.

To my knowledge, Marshall Stone was the first to extend these results to *certain* non-artinian rings. Specifically (cf. [20]) he showed that any Boolean ring R has the form $\mathcal{C}(\mathbf{X}(R), Z_2)$, where $\mathbf{X}(R)$ denotes the space of all maximal ideals of R, with the Zariski topology. (In fact $\mathbf{X}(R)$ is also a Boolean space: that is to say it is compact, Hausdorff,

---

[1]To simplify our exposition, we shall restrict ourselves to commutative rings with 1, throughout this paper.

N.W. Sauer et al. (eds.), Finite and Infinite Combinatorics in Sets and Logic, 63–79.

© 1993 *Kluwer Academic Publishers. Printed in the Netherlands.*

and totally disconnected.) I have heard anecdotally that Stone and his friends were excited the most when they realized that R was the ring of *all* continuous functions $X \xrightarrow{f} F$, not just the ring of *some* of them. (Originally they had only viewed R as being a subdirect product of $\prod_{x \in \mathbf{X}(R)} Z_2$.) McCoy and Montgomery extended this to show that any p-ring R has the form $C(\mathbf{X}(R), Z_p)$, where p is a prime. (R is an m-ring iff $R \models (\forall r)(r^m = r)$.) In the late 1940's, Richard Arens and Irving Kaplansky tried to extend these results to any algebra over Z_p that satisfied $(\forall r)(r^{p^n} = r)$, where $n > 1$. Arguing in the tradition of Stone, they let $F_x = R/x$, for each $x \in \mathbf{X}(R)$, and obtained a subdirect product embedding $\Phi : R \longrightarrow \prod(F_x : x \in \mathbf{X}(R))$. They now let F be the algebraic closure of Z_p, and observed that $F_x \subseteq F$, for each $x \in \mathbf{X}(R)$, since R and thus $F_x \cong R/x$ was algebraic over Z_p. Since each $F_x \subseteq F$, and

$$C(\mathbf{X}(R), F; \{F_x : x \in \mathbf{X}(R)\}) \subseteq \prod(F_x : x \in \mathbf{X}(R)) \subseteq F^{\mathbf{X}(R)},$$

they hoped to extend Stone's results by establishing that

$$R \cong \Phi(R) = C(\mathbf{X}(R), F; \{F_x : x \in \mathbf{X}(R)\}).$$

However they now faced an obstacle unknown to Stone: for each $x \in \mathbf{X}(R)$, there are many embeddings $f_x : F_x \longrightarrow F$, insofar as F (unlike Z_p) has non-trivial automorphisms. To accomplish their goal they would have to patch these embeddings together (i.e. make them vary "continuously" with x) in such a way that $\Phi(R) \subseteq C(\mathbf{X}(R), F; \{F_x : x \in \mathbf{X}(R)\})$, where Φ is the map $R \longrightarrow \prod(F_x : x \in \mathbf{X}(R)) \subseteq F^{\mathbf{X}(R)}$ that they induce. When successful in this "patching" they obtained $\Phi(R) \subseteq C(\mathbf{X}(R), F; \{F_x : x \in \mathbf{X}(R)\})$, and easily continued to conclude $\Phi(R) = C(\mathbf{X}(R), F; \{F_x : x \in \mathbf{X}(R)\})$. Much to their surprise, however, they only succeeded in this "patching" in certain cases, of which the following were the most important: (i) $F_x = F$, for each $x \in \mathbf{X}(R)$, or (ii) R is countable. To represent R in other cases, they showed that $S = F \otimes_L R$ fell into the first case, and so obtained $R \subseteq S = C(\mathbf{X}(S), F)$. This fact enabled them to represent R as the ring of certain *twisted* functions $\mathbf{X}(S) \longrightarrow F$ (*not* $\mathbf{X}(R) \longrightarrow F$.) Their result in case (ii) followed from this. With Kelley's help they found a p^n ring R such that R did not have the form $C(\mathbf{X}(R), F; \{F_x : x \in \mathbf{X}(R)\})$. Thus the use of twisted functions was unavoidable, when representing some uncountable p^n-rings. Their results extended beyond p^n-rings, but they still required that R have only countably many idempotents and that R/x be a finite, normal, and solvable extension of L, in order to represent R using functions defined on $\mathbf{X}(R)$.

The next major advance occurred in 1966, when R. S. Pierce showed that *all* of these rings could be represented using sheaf sections defined on $\mathbf{X}(R)$, rather than twisted functions defined on $\mathbf{X}(R) \otimes_L F$ (cf. [19]).

2   The Pierce Sheaf of a Ring

In 1966 R.S. Pierce showed that any commutative ring R was isomorphic to the ring $\Gamma(\mathbf{X}(R), \mathbf{k}(R))$, of all global sections of some sheaf $\mathbf{k}(R)$ of indecomposable rings over the Stone space $\mathbf{X}(R)$,[2] of the Boolean ring $\mathbf{B}(R)$ of all central idempotents of R. (The addition

---

[2]This differs slightly from our use of $\mathbf{X}(R)$ in § 1 as there it denoted the maximal spectrum of R. However the two usages coincide in all of the rings that we have considered until now.

on $\mathbf{B}(R)$ must be redefined in terms of the operations on R by $e \oplus f = e + f - 2ef$. $\mathbf{X}(R)$ is called the *Boolean spectrum* of R.) The stalks of $\mathbf{k}(R)$ were given by $\mathbf{k}_x(R) = R/(Rx)$, for each $x \in \mathbf{X}(R)$. In fact Pierce's representation works for *all* rings R, but the stalks of $\mathbf{k}(R)$ need not be indecomposable when R is non-commutative. Pierce was most interested in $\mathbf{k}(R)$ when R was commutative and (von Neumann) regular.[3] In that case all of its stalks were commutative, regular, and indecomposable, and hence were easily seen to be fields. Even those p^n-rings that *did not* have the form $\mathcal{C}(\mathbf{X}(R), F; \{F_x : x \in \mathbf{X}(R)\})$ *did* have the form $\Gamma(\mathbf{X}(R), \mathbf{k}(R))$, despite the fact that each stalk $\mathbf{k}_x(R)$ of $\mathbf{k}(R)$ was exactly the same as the ring F_x from the subdirect product representation, in § 1. In essence, the Pierce sheaf $\mathbf{k}(R)$ was the disjoint union $\cup\{\mathbf{k}_x(R) : x \in \mathbf{X}(R)\}$ (which equalled the disjoint union $\cup\{F_x : x \in \mathbf{X}(R)\}$ when R was as in § 1.); $\mathbf{k}(R)$ itself was given the exact topology required to ensure that the natural embedding $\Phi : R \longrightarrow \prod_{x \in \mathbf{X}(R)} \mathbf{k}_x(R)$ was actually an isomorphism $\Phi : R \cong \Gamma(\mathbf{X}(R), \mathbf{k}(R)) \subseteq \prod_{x \in \mathbf{X}(R)} \mathbf{k}_x(R)$.

We shall not dwell upon the precise definition of a sheaf. However the most important property follows: if σ and $\tau \in \Gamma(\mathbf{X}(R), \mathbf{k}(R))$ and $x \in \mathbf{X}(R)$, then there exists a neighbourhood $N_x \subseteq \mathbf{X}(R)$ of x such that $\sigma|_{N_x} = \tau|_{N_x}$. This is hauntingly reminiscent of the fact that continuous functions are locally constant, and is actually the "hallmark" of a sheaf. The transition from continuous functions $\mathbf{X}(R) \overset{f}{\to} F$ to continuous sections $\mathbf{X}(R) \overset{\sigma}{\to} \mathbf{k}(R)$ allows the range[4] of the sections to change from point to point, freeing us from the sometimes impossible task of embedding each one in some "universal" indecomposable F, in such a way that the embeddings vary continuously with x.[5]

Example 2.1 To see the full significance of the topology on $\mathbf{k}(R)$, consider the following two rings, both of which are 2^2-rings. (In these examples GF(2^2) denotes the four element field):

(1) Let R be the ring obtained by Arens and Kaplansky in [1], with Kelley's help. It is a 2^2-ring having a point $\infty \in \mathbf{X}(R)$ such that:

   (a) $\mathbf{k}_\infty(R) \cong Z_2$,

   (b) $\mathbf{k}_x(R) \cong \mathrm{GF}(2^2)$, when $x \in \mathbf{X}(R) - \{\infty\}$.

Let $F = \mathrm{GF}(2^2)$, $F_\infty = Z_2$, and $F_x = F$, when $x \in \mathbf{X}(R) - \{\infty\}$. Although the range of the sections in $\Gamma(\mathbf{X}(R), \mathbf{k}(R))$ "varies" from point to point, (a) and (b) suggest that by picking suitable embeddings $F_x \cong \mathbf{k}_x(R) \longrightarrow F$, we could obtain an isomorphism $R \cong \mathcal{C}(\mathbf{X}(R), F; \{F_x : x \in \mathbf{X}(R)\})$. Arens and Kaplansky showed that R has no such representation.

(2) Let $S = \mathcal{C}(\mathbf{X}(R), F; \{F_x : x \in \mathbf{X}(R)\})$.

---

[3] A ring is *von Neumann regular* iff it satisfies $(\forall r)(\exists s)(rsr = r)$. Trivially any m-ring is regular; nontrivially it also is commutative. In addition, the following are equivalent for an algebraic algebra R: (i) R is semi-simple with respect to its Jacobson radical, (ii) R has no non-zero nilpotent elements, and (iii) R is von Neumann regular.

[4] Intuitively, the range of $\sigma \in \Gamma(\mathbf{X}(R), \mathbf{k}(R))$ at x is $\mathbf{k}_x(R)$.

[5] In fact the construction is so general that it even works when the different stalks vary in characteristic, and so could not be all embedded in the same F. This, however, does not concern us here.

It follows that $\mathbf{X}(R) = \mathbf{X}(S)$ and $\mathbf{k}_x(R) = \mathbf{k}_x(S)$, for all $x \in \mathbf{X}(R)$, so that $\mathbf{k}(R) = \mathbf{k}(S)$, *as sets*. However they must bear different topologies, as otherwise we would have

$$R \cong \Gamma(\mathbf{X}(R), \mathbf{k}(R)) = \Gamma(\mathbf{X}(S), \mathbf{k}(S)) \cong S = \mathcal{C}(\mathbf{X}(R), F; \{F_x : x \in \mathbf{X}(R)\}). \quad \square$$

Thus the *topology* of $\mathbf{k}(T)$ (not just the algebraic structure of its *stalks*) is crucial in determining the full algebraic structure of a ring T. The following result links this topology to the question "Which rings are function rings?"[6]

Definition 2.2 Suppose that X is a Boolean space and that F is an indecomposable ring. The sheaf $\mathbf{k}$ over X is the *simple F-sheaf* iff (i) $\mathbf{k} = X \times F$ as a topological space, where F is discrete and $X \times F$ bears the product topology, and (ii) $\mathbf{k}_x = \{x\} \times F$, for all $x \in X$. The sheaf $\mathbf{k}$ over X is *F-subsimple* iff there is an embedding $\mathbf{k} \subseteq X \times F$. If $\mathcal{M}$ is a class of indecomposable rings, then $\mathbf{k}$ is $\mathcal{M}$-simple (subsimple) iff there exists $F \in \mathcal{M}$ such that $\mathbf{k}$ is F-simple (subsimple). (Heuristically, the subsimple sheaves are those whose topology is Cartesian.)

Theorem 2.3 *Suppose that R and F are rings and that F is indecomposable. Then $\mathbf{k}(R)$ is F-subsimple iff R has the form $R \cong \mathcal{C}(\mathbf{X}(R), F; \{F_x : x \in \mathbf{X}(R)\})$, for suitably chosen $F_x \subseteq F$ such that $\mathbf{k}_x(R) \cong F_x$.* $\square$

We exploit these insights in the next section.

3   Which Rings Are Function Rings?

The significance of a ring being an algebra over a field is given in this observation from [3]:

Proposition 3.1 *A ring R is an algebra over some field L iff there exists an embedding of the simple sheaf $\mathbf{X}(R) \times L$ into $\mathbf{k}(R)$ which, for notational convenience, we denote by $\mathbf{X}(R) \times L \subseteq \mathbf{k}(R)$.* $\square$

In view of this the following question seemed reasonable: *if F is the algebraic closure of L, when can we also get an embedding $\mathbf{k}(R) \subseteq \mathbf{X}(R) \times F$, i.e., when is $\mathbf{k}(R)$ F-subsimple?* As noted in Theorem 2.3, R must be a function ring, whenever the answer is "yes". The following partial answer was obtained in [3] (1973), using this technical lemma:

Lemma 3.2 *Let $\mathbf{k}$ be a sheaf of fields over a Boolean space X and let $S = \Gamma(X, \mathbf{k}) \cong \Gamma(\mathbf{X}(S), \mathbf{k}(S))$. Suppose that $f(Z) = s_0 + s_1 Z + \ldots + s_{n-1} Z^{n-1} + Z^n \in S[Z]$. Note that $f_x(Z) := s_0(x) + \ldots + s_{n-1}(x) Z^{n-1} + Z^n \in \mathbf{k}_x[Z]$. Routinely (because $\mathbf{k}$ is a sheaf) $U := \{x \in \mathbf{X}(R) : f_x(Z) \text{ has a root in } \mathbf{k}_x\}$ is an open subset of X.*

(1) *Suppose that there exists a continuous section $\sigma : U \longrightarrow \mathbf{k}$, such that $\sigma(u)$ is a root of $f_u(Z)$, for all $u \in U$. Then there exists a sheaf $K \supseteq \mathbf{k}$ of fields over X such that $f_x(Z)$ has a root in K_x, for all $x \in X$. Moreover*

[6]Special cases of this linkage are used by Pierce in [19] to show that certain m-rings are function rings.

(2) *if* $\mathbf{B}(S)$ *is countable or is complete as a Boolean algebra, then a section* σ *exists, as described above.* □

Combining this with Zorn's lemma yields:

Proposition 3.3 *Retain the notation from the above lemma. Suppose that* $\mathbf{B}(S)$ *is countable or complete as a Boolean algebra. Then there exists a sheaf* $K \supseteq \mathbf{k}$ *of fields over* X *such that, for each* $x \in X$, *the stalk* K_x *is the algebraic closure of* $\mathbf{k}_x$. □

Results from [3] show that this *algebraic closure* need not be *topologically* unique, although of course it is *set theoretically* unique. Fortunately, where L is still a field, the algebraic closure of the simple sheaf $X \times L$ *does* turn out to be unique and is $X \times F$, where F is the algebraic closure of L. This leads to:

Theorem 3.4 *Suppose that* R *is a semi-simple algebraic algebra over a field* L *whose algebraic closure is* F. *Suppose in addition that* $\mathbf{B}(R)$ *is either countable or complete. Then:*

(1) *there is a sheaf embedding* $\mathbf{k}(R) \subseteq \mathbf{X}(R) \times F$, *so that*

(2) $R \cong C(\mathbf{X}(R), F; \{F_x : x \in \mathbf{X}(R)\})$, *for a suitably chosen family* $\{F_x : x \in \mathbf{X}(R)\}$ *of subfields of* F. *Moreover* $F_x \cong \mathbf{k}_x(R)$, *for all* $x \in \mathbf{X}(R)$.

Proof:
(1) By previous results, $X \times L \subseteq \mathbf{k}(R)$ and $\mathbf{k}(R)$ has an algebraic closure K. Hence $X \times L \subseteq \mathbf{k}(R) \subseteq K$ so that, clearly, K is also the algebraic closure of $X \times L$. Thus $\mathbf{k}(R) \subseteq K \cong X \times F$.
(2) This follows from (1) and preceding results. □

The existence (in Lemma 3.2) of the section $\sigma : U \longrightarrow \mathbf{k}$ picking roots of the polynomials $f_u(Z)$ in each stalk $\mathbf{k}_u$ over a point $u \in U$, is crucial in representing R (from Theorem 3.4) as a function ring. That such a section does not always exist follows as the ring R from Example 2.1, due to Arens and Kaplansky, is a semi-simple algebraic algebra over Z_2, yet is not a function ring. Heuristically, this means that the (analogue) of the axiom of choice may fail in sheaves. Van den Dries exploited this analogy by observing that *defined* choice functions *always* exist, even when the axiom of choice fails. Instead of considering sheaves of fields, he considered sheaves of ordered fields over Boolean spaces. In this context the "choice" section σ *does* always exist as it can be defined to pick the *largest* root of $f_u(Z)$ in $\mathbf{k}_u$, for each $u \in U$. Such sheaves arise as the Pierce sheaf of regular f-rings (cf. [22]). ((A partially ordered regular ring R is an f-ring iff it satisfies $(\forall r)(\exists e \in \mathbf{B}(R))(\mathbf{r}e \geq 0 \wedge \mathbf{r}(1 - \mathbf{e}) \leq 0)$. If we call it an algebra over some ordered field L, we mean that it is an algebra in the usual sense, and that the natural embedding $L \longrightarrow R$ is order preserving.) Hence he obtained:

Theorem 3.5 *Suppose that* R *is a regular* f-*ring that is an algebraic algebra over some ordered field* L. *Then:*

(1) *there exists a sheaf* $K \supseteq \mathbf{k}(R)$ *of real closed fields such that (for each* $x \in \mathbf{X}(R)$) *the stalk* K_x *is the real closure of* $\mathbf{k}_x(R)$. *Consequently*

68 CARSON

(2) R has the form $\mathcal{C}(\mathbf{X}(R), F; \{F_x : x \in \mathbf{X}(R)\})$, for some family $\{F_x : x \in \mathbf{X}(R)\}$ of subfields of the real closure F of L. (L is an ordered field as $L \subset R$.). Moreover

(3) if T is a formally real[7] regular ring such that $\mathbf{B}(T)$ is complete as a Boolean algebra, then it is an f-ring with respect to some order. $\square$

Subsequent work (cf. [2], [4], [10], [11], or [17]) yielded the following model theoretic result:

Theorem 3.6 *Suppose that R is an algebraically closed (real closed) regular ring (f-ring) that is an algebra over some (ordered) field L. Then $R \equiv \mathcal{C}(\mathbf{X}(R), F)$, where F is the algebraic (real) closure of L.* $\square$

However results from [8] show that some hypothesis such as "R is algebraically closed (or real closed)" is required in the last theorem to ensure that R is elementarily equivalent to a function ring.

Recently attempts have been made to characterize *all* rings that are function rings. (cf. [5], [6], and [7].) Specifically:

Goal: Given an elementary class $\mathcal{M}$ of indecomposable rings, we seek *all* of those rings that are $\mathcal{M}$-function rings. To simplify our task, we assume that $\mathcal{M}^\forall$ has the amalgamation property.[8]

We observe that if $R = \mathcal{C}(\mathbf{X}(R), F; \{F_x : x \in \mathbf{X}(R)\})$ (for some suitable $\{F\} \cup \{F_x : x \in \mathbf{X}(R)\} \subseteq \mathcal{M}$) then, as continuous functions are locally constant:

(1) elements of R have clopen (closed and open) support, when viewed as functions $\mathbf{X}(R) \longrightarrow F$.

As $\mathbf{B}(R)$ is just the collection[9] of all characteristic functions of the clopen subsets of $\mathbf{X}(R)$, it follows that (1) is equivalent to:

(2) $R \models (\forall \mathbf{r})(\exists \mathbf{e} \in \mathbf{B}(R))(\mathbf{re} = \mathbf{e} \wedge (\forall \mathbf{f} \in \mathbf{B}(R))[\mathbf{rf} = 0 \Rightarrow \mathbf{ef} = 0])$. Moreover:

(3) if $n \geq 1$ and $\bar{r} = (r_1, \ldots, r_n) \in R$, then there exists $m_{\bar{r}} \geq 1$ and a partition $\{C_1, \ldots, C_{m_{\bar{r}}}\}$ of $\mathbf{X}(R)$ into clopen sets such that $r_i|_{C_j}$ is constant, whenever $1 \leq i \leq n$ and $1 \leq j \leq m_{\bar{r}}$.

Since $\mathbf{k}_x(R) \cong F_x$, whenever $x \in \mathbf{X}(R)$, it follows that:

(4) retaining the above notation, let $\bar{z} = (z_1, \ldots, z_n)$ and suppose that x and y belong to the same element of the above partition. Then the map $t(\bar{r}(x)) \to t(\bar{r}(y))$ (where $t(\bar{z})$ varies over the logical terms in $\bar{z}$) is an isomorphism between the subrings of $\mathbf{k}_x(R)$ and $\mathbf{k}_y(R)$ generated by $\{\bar{r}(x)\}$ and $\{\bar{r}(y)\}$ respectively.

---

[7] A regular ring T is formally real iff $-e$ can not be written as a sum of squares, whenever $e \in \mathbf{B}(T) - \{0\}$.

[8] $\mathcal{M}^\forall$ denotes the class consisting of all substructures of models from $\mathcal{M}$. To ensure that each substructure is also a subring, we include the unary operation symbol "$-$" in our language.

[9] This follows from the indecomposability of all members of $\mathcal{M}$. Another consequence is this: If T has the form $\mathcal{C}(Y, G; \{G_y : y \in Y\})$, where Y is a Boolean space and G is indecomposable, then $\mathbf{X}(T) = Y$.

Letting $e_1, \ldots, e_{m_{\bar{r}}} \in \mathbf{B}(R)$ be the characteristic functions of $C_1, \ldots, C_{m_{\bar{r}}}$ respectively, we record the fact that (4) can be expressed in terms forcing over $\mathbf{B}(R) - \{0\}$ by

(5) $e_i \Vdash t(\bar{r}) = 0$ or $e_i \Vdash t(\bar{r}) \neq 0$, whenever $1 \leq i \leq m_{\bar{r}}$ and $t(\bar{z})$ is a term in $\bar{z}$.

(This forcing relation is defined on any atomic formula with parameters $\bar{r} \in R$ by

$$e \Vdash t(\bar{r}) = 0 \text{ iff } t(r_1 e, \ldots, r_n e) = 0, \text{ for all } e \in \mathbf{B}(R) - \{0\}.$$

It is extended inductively to arbitrary formulae in the manner that is usual for forcing relations. For example, $e \Vdash \neg \rho(\bar{r})$ iff there is no $f \in \mathbf{B}(R) - \{0\}$ such that $f < e$ and $f \Vdash \rho(\bar{r})$. In this context, "$<$" is the standard order on a Boolean algebra defined by $f < e$ iff $f = fe$.) It follows that (5) is actually an elementary formula involving the parameters e_i and $\bar{r} \in R$. We have now established:

(6) for each $n \geq 1$ and $\bar{r} = (r_1, \ldots, r_n) \in R$, there exists $m_{\bar{r}} \geq 1$ such that whenever $\mathcal{F}$ is a finite collection of formulae in $z_1, \ldots, z_n$, each of which is either atomic or the negation of an atomic formula, then R satisfies

$$(*(\mathcal{F})) \ (\exists e_1 \in \mathbf{B}(R)) \cdots (\exists e_{m_{\bar{r}}} \in \mathbf{B}(R)) \Big[(e_1 + \ldots + e_{m_{\bar{r}}} = 1) \wedge \bigwedge_{i \neq j} (e_i e_j = 0) \wedge$$

$$\bigwedge_{\lambda \in \mathcal{F}} \bigwedge_{1 \leq i \leq m_{\bar{r}}} (e_i \Vdash \lambda(\bar{r}) \vee e_i \Vdash \neg \lambda(\bar{r})) \Big].$$

Moreover, $*(\mathcal{F})$ is an elementary formula, for each finite set $\mathcal{F}$ as described above.

Finally we note that:

(7) for each $n \geq 1$ and $\{x_1, \cdots, x_n\} \subseteq \mathbf{X}(R)$, there exists $G \in \mathcal{M}$ and embeddings $\mathbf{k}_x(R) \longrightarrow G$.

This condition can also be expressed in terms of forcing (cf. [5], pages 63–66), and so turns out to be preserved under elementary equivalence. Hoping to catch the abstract essence of a function ring we make:

Definition 3.7 Let $\mathcal{M}$ be an elementary class of indecomposable rings. A ring R is an ω-abstract $\mathcal{M}$-function ring iff it satisfies (2), (6), and (7) from the preceding discussion.

The appearance of "ω" in this definition is justified by the next result, from [6].

Theorem 3.8 *Suppose that $\mathcal{M}$ is an elementary class of indecomposable rings such that $\mathcal{M}^\forall$ has the amalgamation property. Then:*

(A) *any countable ω-abstract $\mathcal{M}$-function ring R is an $\mathcal{M}$-function ring.*

(B) *Now suppose that $\mathcal{M}$ is the class of fields. Then there exists an ω-abstract $\mathcal{M}$-function ring S that is not an $\{F\}$-function ring, for any indecomposable ring F. Moreover, S can be chosen such that it is a von Neumann regular algebra over the rationals and (both of the following are possible)*

(a) $\mathbf{B}(S)$ *is countable, or*

(b) $\mathbf{B}(S)$ *is complete.*

Remark and Warning:

(1) The ω-abstract function rings *do* generalize the semi-simple algebraic algebras insofar as both are function rings in the countable case.

(2) However this generalization fails in the uncountable case, insofar as any semi-simple algebraic algebra satisfying (a) or (b) from Theorem 3.8 (B) must be a function ring, in contrast to the situation for ω-abstract function rings. Classically, a countable semi-simple algebraic algebra R was shown to be a function ring only because each open subset of $\mathbf{X}(R)$ was a union of disjoint clopen sets. However this condition does not suffice to guarantee that an ω-abstract function ring S be a function ring; (B) shows that S itself must be countable, before we can draw that conclusion.

SKETCH OF PROOF: (A) Let $R = \{r_1, \ldots, r_n, \ldots\}$. For each $n \geq 1$ let R_n be the ring generated by $\{r_1, \ldots, r_n\} \cup \mathbf{B}(R)$, and let $m_n \geq 1$ be associated with $r_1, \ldots, r_n$ as in (6) from the definition of an ω-abstract $\mathcal{M}$-function ring. Note that $\mathbf{B}(R_n) = \mathbf{B}(R)$ so that $\mathbf{X}(R_n) = \mathbf{X}(R)$, for all $n \geq 1$. The ring $F \in \mathcal{M}$ such that $R \cong \mathcal{C}(\mathbf{X}(R), F; \{F_x : x \in \mathbf{X}(R)\})$ (for suitable rings $F_x \subseteq F$) is obtained as follows.

(1) For each $n \geq 1$ and finite collection $\mathcal{F}$ of suitable formulae, obtain the elements $e_1, \ldots, e_{m_n} \in \mathbf{B}(R)$ whose existence is guaranteed by (6) from Definition 3.3. Let $C_{n,1}, \ldots, C_{n,m_n}$ be the associated partition of $\mathbf{X}(R)$.

(2) (This is the tricky part)

  (i) Show that $\{C_{n,1}, \ldots, C_{n,m_n}\}$ *can* be chosen in such a way that it is independent of $\mathcal{F}$, and

  (ii) $\{C_{n,1}, \ldots, C_{n,m_n}\}$ is a refinement of $\{C_{k,1}, \ldots, C_{k,m_k}\}$, whenever $k \leq n$.

(3) Let $\{c_{r,x} : r \in R \text{ and } x \in \mathbf{X}(R)\}$ be a new set of constant symbols.

(4) Write down axioms $\mathcal{T}$ guaranteeing that

  (i) $\mathcal{M} \subseteq \text{models}(\mathcal{T})$,

  (ii) $|\{c_{r_i,x} : x \in \mathbf{X}(R)\}| \leq m_i$, for all $i \geq 1$,

  (iii) $\{c_{r_n,x} = c_{r_n,y} : x \text{ and } y \in C_{n,j}\} \subseteq \mathcal{T}$, whenever $1 \leq j \leq m_n$ and $n \geq 1$.

(5) Use logical compactness and our hypothesis on R and $\mathcal{M}^{\forall}$ to show that $\mathcal{T}$ is consistent.

(6) Let F be a model of $\mathcal{T}$ and let $c_{r,x} \in F$ be the interpretation of $\mathbf{c}_{r,x}$.

(7) For each $r \in R$ let $\hat{r} : \mathbf{X}(R) \longrightarrow F$ be defined by $\hat{r}(x) = c_{r,x} \in F$.

Note that $\mathcal{T}$ has been defined so as to ensure that the map $r \to \hat{r}$ is an embedding $\hat{} : R \subseteq \mathcal{C}(\mathbf{X}(R), F)$. Hence (cf. Theorem 2.3) there is a family $\{F_x : x \in \mathbf{X}(R)\}$ of subrings of F such that $R \cong \mathcal{C}(\mathbf{X}(R), F; \{F_x : x \in \mathbf{X}(R)\})$.

(B) See [6]. $\square$

Note that $|\mathrm{range}(r_n)|$ may be dependent on the indexing of R used. Moreover, in our proof, $|\mathrm{range}(r_n)| \le m_n$ and $m_1 \le m_2 \le \cdots$. Thus we should not be surprised by the fact that some uncountable ω-abstract function rings are not function rings. The amalgamation property for $\mathcal{M}^\vee$ cannot be omitted from Theorem 3.8. (cf. [5, Example 5.33].)

Unfortunately a non-algebraic algebra may fail to be a function ring. Nonetheless it often is "almost" a function ring.

Example 3.9 Let $R = L^N$, where L is the field of rational numbers. Then $\mathbf{X}(R)$ is the Stone-Čech compactification of the natural numbers and one can show that $\mathbf{X}(R) \times L \subseteq \mathbf{k}(R)$. However there does not exist a ring F such that $\mathbf{k}(R) \subseteq \mathbf{X}(R) \times F$, as the element $r \in R$ given by $r(n) = n$ *must* correspond to a section $\hat{r} \in \Gamma(\mathbf{X}(R), \mathbf{k}(R))$ such that $\hat{r}(n) = n \in \mathbf{k}_n(R) \cong L$. As F is discrete and $\mathbf{X}(R)$ is compact, thus any continuous function $\mathbf{X}(R) \longrightarrow F$ must have finite range. Thus R is not a function ring. Note, however, that $\mathbf{X}(R) \times N$ *can* be embedded in $\mathbf{k}(R)$ as a dense subsheaf.

We will explore this theme further in the next section.

4 Rings That Are Almost A Function Ring

Many classic results in functional analysis represent certain operator algebras A using continuous (in the analyst's sense) functions defined not on $\mathbf{X}(A)$, but rather on dense open subsets of $\mathbf{X}(A)$. In this context two functions are identified iff they agree on the intersection of their domains. This trend, started by Stone, was continued by many others, including Kadison in the 1950's. In 1985, T.J. Jech developed some very nice structure results for *abstract operator algebras*. In essence, an algebraically closed complex algebra A (with involution *) is such an algebra iff it is von Neumann regular and its "real part" (i.e. $\{a \in A : a^* = a\}$) is real closed, archimedean, and locally bounded, over the reals. He showed that:

Theorem 4.1 *Any abstract operator algebra A is isomorphic to the ring of all complex valued continuous functions defined on dense open subsets of $\mathbf{X}(A)$, where the complex numbers have the usual topology.* $\square$

In [5, chapter 5], a characterization was given of *all* rings R having an elementary extension S that can be represented as the ring of certain specified functions defined on dense open subsets of $\mathbf{X}(S)$. In addition, several cases were described in which R itself has such a representation. The full details of this fact and the methods used to accomplish it are rather lengthy, but can be summarized as follows:

Definition 4.2 Let $\mathcal{M}$ be an elementary class of indecomposable rings. A sheaf K of rings over a Boolean space X is *(topologically) $\mathcal{M}$-semi-simple* iff there exists a subsheaf $k \subseteq K$ such that:

  (i) k is topologically a dense subspace of K,

  (ii) there exists $F \in \mathcal{M}$ such that there is an embedding $k \subseteq X \times F$ of sheaves, and

  (iii) $\Gamma(X, k)$ is *torsion free*, where this concept is defined following Proposition 4.3.

Clearly, a topologically semi-simple sheaf *almost* bears the topology of a Cartesian product. The linkage between ring representations and topologically semi-simple Pierce sheaves is given in:

Proposition 4.3 *A ring R can be represented as the ring of certain specified functions from dense open subsets of $\mathbf{X}(R)$ to some fixed $F \in \mathcal{M}$ iff $k(R)$ is topologically $\mathcal{M}$-semi-simple.* $\square$

We shall not here fully describe what we mean by "certain specified functions". However this situation can be re-expressed entirely in terms of ring theory and ring localizations.

SKETCH OF A DEFINITION: For any ring T let $\mathbf{Q}(T)$ be the ring theoretic localization of T with respect to the filter of T-ideals that are generated by dense ideals in $\mathbf{B}(T)$. For any T-module M define $\mathrm{tor}_T(M) = \{m \in M : Dm = 0,$ for some dense ideal D in $\mathbf{B}(T)\}$. Call M (tor_T)-torsion free if $\mathrm{tor}_T(M) = 0$. Say that an embedding $A \subseteq B$ of T-modules is (tor_T)-essential iff (whenever $b \in B - \{0\}$) there exists a dense ideal D in $\mathbf{B}(T)$ such that $(Db) \cap A \neq 0$. (Alternatively, say that B is an essential extension of A.)

In fact we have defined the correct concept of torsion to accompany this localization. Hence the standard map $T \longrightarrow \mathbf{Q}(T)$ given by general localization theory is an embedding iff T is torsion free; the embedding $T \subseteq \mathbf{Q}(T)$ is essential, when T is torsion free and we identify $T \subseteq \mathbf{Q}(T)$.

To see how all of this helps us to represent rings using functions, let S be a ring such that $\mathbf{k}(S)$ is $\mathcal{M}$-semi-simple and choose $F \in \mathcal{M}$ and a dense subsheaf $k \subseteq \mathbf{k}(S)$ such that $k \subseteq \mathbf{X}(S) \times F$. Let $R = \Gamma(\mathbf{X}(S), k)$. We now have $R \subseteq S$ and $R \subseteq \mathcal{C}(\mathbf{X}(R), F)$, so that $R \cong \mathcal{C}(\mathbf{X}(R), F; \{F_x : x \in \mathbf{X}(R)\})$, for some subrings $\{F_x : x \in \mathbf{X}(R)\}$ of F, satisfying $F_x \cong \mathbf{k}_x(R)$, for all $x \in \mathbf{X}(R)$. Moreover the embedding $R \subseteq S$ is $(\mathrm{tor}_R(R))$-essential, so that standard results about abstract ring localizations apply to give the following commutative diagram:

$$
\begin{array}{ccccc}
R & \cong & \mathcal{C}(\mathbf{X}(R), F; \{F_x : x \in \mathbf{X}(R)\}) & \subseteq & S \\
\downarrow & & & & \downarrow \\
\mathbf{Q}(R) & & \cong & & \mathbf{Q}(S)
\end{array}
$$

Thus, when $\mathbf{k}(S)$ is $\mathcal{M}$-semi-simple, S is closely related to a function ring R having the same localization as S. Moreover the elementary properties of this localization are completely (and independently) determined by those of S and of the function ring R.

As noted before, any countable semi-simple ring that is an algebraic algebra over some field is von Neumann regular *and* is a function ring. By Example 3.9 this result does not apply to non-algebraic von Neumann regular algebras. Nonetheless, such rings are "almost" function rings, in the sense that we have just described. In fact such rings are surprisingly common. Here are some results from [5, chapter 5]. To state these, let $\mathcal{F}$ denote the class of all fields and let $\mathcal{R}$ denote the class of all formally real fields.

Theorem 4.4 *Suppose that R is a von Neumann regular ring that is an algebra over some field L. Then $\mathbf{k}(R)$ is (topologically) $\mathcal{F}$-semi-simple.* $\square$

In particular any f-ring is $\mathcal{F}$-semi-simple. (We recall that if R is a regular f-ring then the stalks of $\mathbf{k}(R)$ are ordered fields.) The importance of $\mathcal{F}$ in Theorem 4.4 is emphasized by the following surprising contrast:

Theorem 4.5 *There exists a real closed regular f-algebra R over the field of real numbers, such that $\mathbf{k}(R)$ is not $\mathcal{R}$-semi-simple.* $\square$

Despite this counterexample, R *does* have an elementary extension $S \succ R$ such that $\mathbf{k}(S)$ is $\mathcal{R}$-semi-simple. In fact this holds far more generally.

Theorem 4.6 *Let $\mathcal{M}$ be any elementary class of indecomposable rings and suppose that R is a ring.*

(1) *Then there exists $S \succ R$ such that $\mathbf{k}(S)$ is $\mathcal{M}$-semi-simple iff*

    (*) *for any finite collection $\{x_1, \ldots, x_n\} \subseteq \mathbf{X}(R)$, there exists $F \in \mathcal{M}$ such that there are embeddings $\mathbf{k}_{x_i}(R) \subseteq F$, whenever $1 \leq i \leq n$.*

(2) *If (*) holds, $\mathcal{M}^\forall$ has the amalgamation property, and R is countable, then $\mathbf{k}(R)$ itself is $\mathcal{M}$-semi-simple.*

(3) *If (*) holds and $\mathbf{B}(R)$ is an atomic Boolean algebra, then $\mathbf{k}(R)$ is $\mathcal{M}$-semi-simple.* $\square$

By a variant of Theorem 4.5, the hypothesis "$\mathcal{M}^\forall$ has the amalgamation property" can not be omitted from Theorem 4.6(2). Curiously, the non-commutative analogue of Theorem 4.6(2) is false. As noted earlier, the class of rings such that (*) holds is elementary.

5 Representing Rings Using Twisted Functions

Observe that if Y is compact and F is discrete, then any continuous function $Y \longrightarrow F$ must have finite range. Heuristically there are two reasons that a ring R might not be a function ring, even though $\mathbf{k}(R)$ is topologically semi-simple:

(1) R might contain an element r that could never be represented as a function (continuous or otherwise) with finite range defined on $\mathbf{X}(R)$, or

(2) R can be represented using certain functions with finite range defined on $\mathbf{X}(R)$, but it is not possible to choose all of these functions to be continuous.

The first reason trivially applies to $R := \mathbb{Q}^{\mathbb{N}}$, where $\mathbb{Q}$ is the rational number field and $\mathbb{N}$ is the set of natural numbers. (Let $r \in R$ be the function $r(n) = n$, and suppose that there is

an embedding $\hat{\ } : R \longrightarrow F$, for some suitable F. It is not hard to show that $\mathbb{N} \subseteq \mathrm{range}(\hat{r}).$)
The second reason applies to the 2^2-ring R from Example 2.1, due to Arens and Kaplansky.
Nonetheless Arens and Kaplansky showed that it could be represented as the ring of certain
twisted functions defined on the Stone space of some Boolean algebra $A \supseteq \mathbf{B}(R)$. In this
section we shall see that such representations are *always* possible, when the second reason
prevents a ring from being represented using continuous functions defined on its Boolean
spectrum. This twisting can be viewed as a description of the topology on $\mathbf{k}(R)$.

We shall assume, throughout this section, that $\mathcal{M}$ is an elementary class of indecomposable rings such that any automorphism of some subring K of a model M of $\mathcal{M}$ can be
extended to an automorphism of M. (This will always be possible if $\mathcal{M}$ has the amalgamation property and $\mathcal{M}$ can be axiomatized using universal sentences (cf. [9, p. 153]).) In this
case it follows that $\mathcal{M}$ must be closed under substructures, so that we now require "$-$" to
be in our language, to ensure that any element of $\mathcal{M}$ actually is a ring. By similarly adding
a unary symbol "$^{-1}$", we can so axiomatize both the classes of all fields and all division
rings, of given fixed characteristic.) We make this assumption only to simplify the present
exposition; without it, the following definition of twisting would need to be formulated in
a more complicated and less traditional fashion.

Most of the results in this section are from [7].

Definition 5.1 Suppose that X is a Boolean space and that $F \in \mathcal{M}$.

(1) A *twisting* $\mathcal{T}$ on (X, F) is a pair $(\sim, \Phi)$, where $\sim$ is an equivalence relation on
   X and $\Phi = \{\Phi_{x,y} : x \sim y\}$ is a family of automorphisms of F.

(2) If $\{F_x : x \in X\}$ is a family of subrings of F and $\mathcal{T} = (\sim, \Phi)$ is a twisting on (Y, F), then $\mathcal{C}(Y, F; \{F_y : y \in Y\}; \mathcal{T}) := \{f \in \mathcal{C}(Y, F; \{F_y : y \in Y\}) :
   \Phi_{y,x}(f(x)) = f(y), \text{ for all } x \text{ and } y \in Y \text{ such that } x \sim y\}$.

(3) The above twisting $\mathcal{T}$ is *G-symmetric* iff G is a group of homeomorphisms of Y,
   whose orbits are just the equivalence classes of $\sim$.

(4) A ring R is a *twisted function ring* iff it has the form $R \cong \mathcal{C}(Y, F; \{F_y : y \in
   Y\}; \mathcal{T})$, as above.

Arens and Kaplansky showed that, whenever R was a semi-simple algebraic algebra over
some field with algebraic closure F, then $R \otimes_L F$ was a function ring defined on its Boolean
spectrum Y. They used this fact to represent R as a ring of certain twisted functions
defined on Y. We shall see that twisted function rings arise far more generally. In fact, for
any ring R such that there is a *nice* embedding $R \longrightarrow \mathcal{C}(Y, F)$, where $F \in \mathcal{M}$ and Y is a
suitable Boolean space, then R has the form $\mathcal{C}(Y, F; \{F_y : y \in Y\}; \mathcal{T})$, for some twisting $\mathcal{T}$.

Definition 5.2 Suppose that $^- : R \longrightarrow S = \mathcal{C}(\mathbf{X}(S), F; \{F_x : x \in \mathbf{X}(S)\})$ is an embedding, where Y is a Boolean space, and that $\{F\} \cup \{F_x : x \in \mathbf{X}(S)\} \subseteq \mathcal{M}$. Note[10] that
$\mathbf{X}(\mathcal{C}(Y, F)) = Y$ and let $\pi : Y \longrightarrow \mathbf{X}(R)$ be the continuous onto map given by Stone's

---

[10]This follows from the fact that, as F is indecomposable, $\mathbf{B}(\mathcal{C}(Y, F))$ is just the set of characteristic
functions of clopen subsets of Y.

theorem. Let $\hat{} : R \cong \Gamma(\mathbf{X}(R), \mathbf{k}(R))$ be the standard isomorphism. Call $\bar{}$ a *local monomorphism* iff the map $\mathbf{k}_{\pi(y)}(R) \longrightarrow F$ given by $\hat{r}(\pi(y)) \to \bar{r}(y)$ (where r varies over R) is a well defined monomorphism, for all $y \in Y$.

We shall see that R is a twisted function ring whenever there is a local monomorphism $R \longrightarrow \mathcal{C}(Y, F)$, for suitable Y and F. In general monomorphisms $R \longrightarrow \mathcal{C}(Y, F)$ need not be local.

Example 5.3 Let R be the polynomial ring $F[X_1, X_2]/(X_1 X_2)$, where $F = Z_2$, and let S be the classical quotient ring of R. Then R is indecomposable, so that $\mathbf{X}(R)$ consists of a single point p, and $\mathbf{k}_p(R) \cong R$. However $S \cong F(X_1) \oplus F(X_2)$, so that $\mathbf{X}(S)$ consists of two points q_1 and q_2, and $S \cong \mathcal{C}(\{q_1, q_2, \}, F(X'))$, for any indeterminate X'. The inclusion $R \subseteq S$ is not a local monomorphism, as there is no embedding $F[X_1, X_2]/(X_1 X_2) \longrightarrow F(X')$.

Fortunately, local embeddings can easily be obtained using:

Lemma 5.4 *Suppose that* $\bar{} : R \longrightarrow S$ *is a monomorphism, elements of* $\Gamma(\mathbf{X}(R), \mathbf{k}(R))$ *have clopen support, and S has the form $S = \mathcal{C}(\mathbf{X}(S), F; \{F_x : x \in \mathbf{X}(S)\})$, as in Definition 3.7. Then $\bar{}$ is a local monomorphism in each of the following cases:*

(1) $\mathbf{k}_x(R)$ *is a simple ring, for all* $x \in \mathbf{X}(R)$,

(2) R *is von Neumann regular,*

(3) $\mathbf{B}(R)$ *is a dense Boolean subalgebra of* $\mathbf{B}(S)$, *and*

(4) *there exists a ring* $T \supseteq S$ *such that* $R \prec T$ *and* $\mathbf{B}(T) = \mathbf{B}(S)$. $\square$

Theorem 5.5 *Suppose that R is a ring and that there is a local embedding $\bar{} : R \subseteq \mathcal{C}(Y, F)$, for some Boolean space Y and $F \in \mathcal{M}$. Let $\pi : Y \longrightarrow \mathbf{X}(R)$ be the continuous onto map given by Stone's theorem and let $\sim$ be the equivalence relation defined on Y by $y \sim z$ iff $\pi(y) = \pi(z)$.*

(1) *There exists a twisting $(\sim, \Phi)$ on (Y, F) such that R has the form*

$$R \cong \mathcal{C}(Y, F; \{F_y : y \in Y\}; \mathcal{T}).$$

(2) *If $\mathbf{B}(\mathcal{C}(Y, F))$ is $| \mathbf{B}(R) |^+$-saturated, then R has the form*

$$R \cong \mathcal{C}(Y, F; \{F_y : y \in Y\}; \mathcal{T}'),$$

for some symmetric twisting $\mathcal{T}'$.

SKETCH OF PROOF OF (1): For each $y \in Y$ let $F_y = \{\bar{r}(y) : r \in R\}$. Our hypothesis guarantees that $\mathbf{k}_{\pi(y)}(R) \cong F_y$. Define $\Phi_{z,y}$ to be the composite of

$$F_y \cong \mathbf{k}_{\pi(y)}(R) = \mathbf{k}_{\pi(z)}(R) \cong F_z,$$

whenever $y \sim z$, and then extend it to an automorphism of F. That $R \subseteq \mathcal{C}(Y, F; \{F_y : y \in Y\}; \mathcal{T})$ follows trivially from our definitions. In fact $R = \mathcal{C}(Y, F; \{F_y : y \in Y\}; \mathcal{T})$, although establishing this requires a bit more work. $\square$

As an application we have:

Corollary 5.6 *Suppose that R is a regular algebraic algebra over some field L with algebraic closure F. Let Y be the Stone space of the Boolean completion of $\mathbf{B}(R)$. Then there is a family $\{F_y : y \in Y\}$ of subfields of F and a twisting $\mathcal{T}$ on (Y, F) such that*

$$R \cong \mathcal{C}(Y, F; \{F_y : y \in Y\}; \mathcal{T}).$$

Proof: Let S be the subring of $\mathbf{Q}(R)$ generated by $R \cup \mathbf{B}(\mathbf{Q}(R))$. Show that S is also an algebraic algebra over L and that $\mathbf{B}(S)$ is the Boolean completion of $\mathbf{B}(R)$. By Theorem 3.4, S has the form $S \cong \mathcal{C}(\mathbf{X}(S), F; \{F_x : x \in \mathbf{X}(S)\})$. The result now follows from Theorem 5.5.   □

The space Y in the last corollary is closely related to X insofar as $\mathbf{B}(S)$ is an essential extension of $\mathbf{B}(R)$. Thus each atom in $\mathbf{B}(R)$ is also an atom in $\mathbf{B}(S)$, so that standard results on the elementary theory of Boolean algebras yield $\mathbf{B}(R) \subseteq_\forall \mathbf{B}(S)$. We shall see that if $R \cong S$, for some ring S having the form $\mathcal{C}(Y, F; \{F_y : y \in Y\}; \mathcal{T})$, then $R \cong T$ for some T having the form $T = \mathcal{C}(Z, F; \{F_y : y \in Z\}; \mathcal{S})$, such that $\mathbf{B}(R) \prec \mathbf{B}(T)$ and $\mathcal{S}$ is symmetric. In other words, although such a ring R can not always be represented using continuous functions defined on its Boolean spectrum $\mathbf{X}(R)$, it can be represented as the ring of certain twisted functions defined on a Boolean space $Z = \mathbf{X}(T)$ which, as $\mathbf{B}(R) \prec \mathbf{B}(T)$, we view as being a "non-standard" version of its Boolean spectrum. Moreover, this can be accomplished using a symmetric twisting.

Definition 5.7 A ring R is an $\mathcal{M}$-*non-standard function ring* iff it has the form $R \cong T = \mathcal{C}(Z, F; \{F_y : y \in Z\}; \mathcal{S})$, for some suitable $\{F\} \cup \{F_z : z \in Z\} \subseteq \mathcal{M}$, and symmetric twisting $\mathcal{S}$, where $\mathbf{B}(R) \prec \mathbf{B}(T)$. (More precisely, $\mathcal{S}$ is G-symmetric, for some group G of homeomorphisms of $\mathbf{X}(T)$.)

At the beginning of this section we claimed that, if R is a ring such that $\mathbf{k}(R)$ is topologically semi-simple, then R might fail to be a function ring because its elements could never be represented as continuous functions having finite range. The correctness of this claim is borne out by the following result:

Theorem 5.8 *Suppose that R is a ring and that $\mathcal{M}$ is as specified for this section. By F^Y we mean $\prod_{y \in Y}(F)$, even if Y also happens to be a topological space.*
  Then the following are equivalent:

(1) *there exist embeddings $\mathbf{k}_x(R) \subseteq F$, for some fixed $F \in \mathcal{M}$, such that range(r^*) is finite, for all $r \in R$, where * is the map*

$$^* : R \cong \Gamma(\mathbf{X}(R), \mathbf{k}(R)) \subseteq \prod_{x \in \mathbf{X}(R)} (\mathbf{k}_x(R)) \subseteq F^{\mathbf{X}(R)}$$

   induced by these embeddings;

(2) *there exists a local monomorphism $R \subseteq S = \mathcal{C}(Y, F)$, for some Boolean space Y and $F \in \mathcal{M}$;*

(3) *R is a twisted function ring;*

(4) *R is a non-standard function ring.* □

As an immediate application we have:

Corollary 5.9 *Suppose that R is an algebraic algebra over some field. Then R is a non-standard function ring.* □

Proof: This follows from Theorems 3.4 and 5.8. It also follows more directly from the preceding theorem as R is algebraic over L, and so any polynomial with coefficients from L has at most finitely many roots in F, where F is the algebraic closure of L. □

The representation of R given in the last corollary differs from that obtained by Arens and Kaplansky insofar as the domain of the twisted functions used to represent R is some "non-standard version" of $\mathbf{X}(R)$, not $\mathbf{X}(R \otimes_L F)$.

Theorem 5.8 easily implies these two surprising facts: (1) any twisted function ring can be re-represented as a non-standard function ring, and (2) if a ring can be represented using possibly discontinuous functions with finite range, defined on its Boolean spectrum, then it can be re-represented as the ring of all continuous suitably twisted functions defined on some non-standard version of its Boolean spectrum. This in turn can be construed as determining the topology on $\mathbf{k}(R)$.

Remark: Z. Chatzidakis and P. Pappas have recently established:

Theorem(cf. [23] and [24]) *Let G be a torsion abelian group, K be a field whose characteristic is not the order of any element in G, R be the group ring K[G], and F be the algebraic closure of L. Then R is a von Neumann regular algebraic algebra over K. Moreover:*

(1) *if R is countable, then it has the form* $C(\mathbf{X}(R), F; \{F_x : x \in \mathbf{X}(R)\})$, *and*

(2) *if K is either the real or rational field, $n \geq 3$, and $G = (Z/n)^{(\aleph_1)}$, then R does not have the form* $C(\mathbf{X}(R), F; \{F_x : x \in \mathbf{X}(R)\})$. □

The first part of this theorem follows from representations which we have discussed in Section 3. I consider the second part of this theorem to be particularly nice as (to my knowledge) it provides the first naturally occurring examples of rings that can be represented using twisted functions, but can not be represented using ordinary functions. In contrast the elegant example due to Arens and Kaplansky (which we discussed in Example 2.1) was constructed *using* twisted functions. Undoubtedly the Chatzidakis-Pappas work shows that important function rings of various kinds abound in "real life."

References

[1] R. Arens and I. Kaplansky, Topological representations of algebras, *Trans. Amer. Math. Soc.* **63** (1949), pages 457–481.

[2] S. Burris and H. Werner, Sheaf constructions and their elementary properties, *Trans. Amer. Math. Soc.* **248** (1979), pages 269–309.

[3]   A. B. Carson, Representation of semi-simple algebraic algebras *J. Alg.* **24** (1973), pages
      245–257.

[4]   A. B. Carson, The model completion of the theory of commutative regular rings,
      J. Alg. **27** (1973), pages 136–146.

[5]   A. B. Carson, *Model Completions, Ring Representations and the Topology of the Pierce
      Sheaf*, Pitman Research Notes in Mathematics **209** (1989), Longman Scientific and
      Technical, Essex and John Wiley, New York.

[6]   A. B. Carson, A characterization of function rings with Boolean domain, *Bull. Amer.
      Math. Soc.* (to appear).

[7]   A. B. Carson, A characterization of rings of twisted functions, preprint.

[8]   A. B. Carson, Rings that are not elementarily equivalent to a function ring, *Com. Alg.*
      18 (1990), pages 4225–4234.

[9]   G. Cherlin , Model theoretic algebra–selected topics, *Lecture Notes in Math.* (Springer-
      Verlag) **521**, 1976.

[10] S. D. Comer, Elementary properties of structures of sections, *Bol. Soc. Mat. Mexicana*
      2 (1978), pages 78–85.

[11] D. P. Ellerman, Sheaves of structures and generalized ultraproducts, *Ann. Math. Logic*
      7 (1974), pages 163–195.

[12] J. S. Golan, *Localization of Noncommutative Rings*, Pure and Applied Math Series
      (Marcel Dekker, New York) **30**, 1974.

[13] J. S. Golan, *Torsion theories*, Pitman Monographs in Pure and Applied Mathematics
      (Longman Scientifical and Technical, Essex) **29** (1986).

[14] T. J. Jech, Abstract theory of abelian operators: an application of forcing, *Trans.
      Amer. Math. Soc.* **289** (1985), pages 133–162.

[15] R. V. Kadison, A representation theory for commutative topological algebras, *Mem.
      Amer. Math. Soc.* **7** (1951).

[16] L. Lipshitz and D. Saracino, The model companion of the theory of commutative
      regular rings without nilpotent elements, *Proc. Amer. Math. Soc.* **37** (1973), pages
      381–387.

[17] A. MacIntyre, Model completeness for sheaves of structures, *Fundamenta Mathematica*
      LXXXI (1973), pages 73–89.

[18] N. H. McCoy and D. Montgomery, A representation of generalized Boolean rings, *Duke
      Math. Jour.* **3** (1937), pages 455–459.

[19] R. S. Pierce, Modules over commutative regular rings, *Mem. Amer. Math. Soc.* **70**
      (1967).

[20] M. H. Stone, The theory of representations for Boolean algebras, *Trans. Amer. Math. Soc.* **40** (1936), pages 37–111.

[21] M. H. Stone , A general theory of spectra, *Proc. Nat. Acad. Sci. U. S. A.* **26** (1940), pages 280–283.

[22] L. van den Dries, Artin-Schreier theory for commutative regular rings, *Ann. Math. Logic* **12** (1977), pages 113–150.

[23] Z. Chatzidakis and P. Pappas, Topological representations of abelian group rings, preprint.

[24] Z. Chatzidakis and P. Pappas, Von Neumann regular group rings not representable as rings of continuous functions, preprint.

Homogeneous Directed Graphs

GREGORY L. CHERLIN
Mathematics Department
Rutgers University
Hill Center, Busch Campus
New Brunswick, NJ 08903

Abstract

The classification of the (countable) homogeneous directed graphs is complete. It has been known for almost two decades that there are $2^{\aleph_0}$ homogeneous directed graphs. We use methods devised by Lachlan and Woodrow, and applied by them to the case of homogeneous undirected graphs, and by Lachlan to the case of tournaments. Ours is the first case in which an uncountable collection of structures is handled by these methods. The methods used will first be sketched in the context of tournaments, then in our case. They combine a rather artificial inductive procedure with a timely use of Ramsey's theorem introduced by Lachlan in his work on tournaments. We also describe some examples, including a rather strange one uncovered by the classification procedure, and some related problems, for the most part open. We devote considerable attention to a decision problem connected with the classification which remains open. This problem has been studied by Brenda Latka, and is equivalent to a recognition problem for well quasi-ordered families of tournaments defined by finitely many constraints.

1 Introduction

1.1. HOMOGENEITY

1.1.1. *Definition.* A structure Γ is said to be *homogeneous* if any isomorphism between finitely generated substructures of Γ is induced by an automorphism. (Since we will be interested in combinatorial structures like directed graphs, and there are no functions present, we may say "finite" rather than "finitely generated" in our context.)

The condition of homogeneity is obviously very strong – it is more or less the strongest transitivity hypothesis one could place on the automorphism group of a countable structure. Accordingly the question of providing a complete classification of all the homogeneous structures of various kinds arises.

1.1.2. *The finite case.* As far as finite homogeneous structures are concerned, there is a completely general classification theory due to Lachlan which provides a classification by finitely many natural invariants in all cases in which the structures under consideration are equipped with finitely many relations and no functions. An exposition of this theory with

N.W. Sauer et al. (eds.), Finite and Infinite Combinatorics in Sets and Logic, 81–95.

substantial simplifications is found in [12]. Lachlan's theory relies on the classification of the finite simple groups in what appears to be an essential way, though in the most common cases, where all relations are binary, this can be avoided [19]. It should be said that if one is interested in generating completely explicit solutions to classification problems for finite homogeneous structures this theory does not provide a very practical method in the present state of knowledge, and more limited but efficient approaches have been used in special cases [24, 8, 10, 14, 17].

In Lachlan's classification certain infinite examples are also captured as limits of finite structures, and Lachlan has shown that his classification covers exactly the homogeneous structures which are stable in Shelah's sense. In particular there are only countably many homogeneous stable structures for finite relational languages. (Here we are counting the isomorphism types of *countable* structures.)

1.2. CLASSIFICATION

1.2.1. *The classification problem.* When we consider the problem of classifying homogeneous structures in general, we run immediately into the following problem: there are $2^{\aleph_0}$ homogeneous directed graphs [11, 22]. What then can we mean by a classification of an uncountable family of structures? As far as the particular case of directed graphs is concerned, I think the description of homogeneous directed graphs given below is a convincing classification. In the general case I find Lachlan's proposed criterion convincing. This makes use of Fraïssé's theory of amalgamation classes. This theory will be reviewed in more detail below, but for our present purpose it suffices to say that each (countable) homogeneous structure is determined up to isomorphism by the isomorphism types of its finite substructures. Accordingly if $\mathcal{K}$ is a class of homogeneous structures then it is natural to study the entailment relation $\mathcal{A} \Longrightarrow_\mathcal{K} \mathcal{B}$ between two finite sets $\mathcal{A}, \mathcal{B}$ of finite structures defined by the condition:

> Any homogeneous structure embedding all structures in $\mathcal{A}$ also embeds some structure in $\mathcal{B}$. [Embedding signifies: embedding as an *induced* substructure.]

(The most striking case is that in which $\mathcal{B}$ contains only one structure.) Typically $\mathcal{K}$ will be taken to consist of all the countable homogeneous structures in some natural and fixed category, and we will drop the subscript $\mathcal{K}$ from the notation.

We will call a collection $\mathcal{K}$ of homogeneous structures *classifiable* if the relation $\Longrightarrow_\mathcal{K}$ is decidable. There is no direct relation between the countability of $\mathcal{K}$ and its classifiability in this sense. In particular the class of homogeneous directed graphs is classifiable in this sense, but uncountable.

1.2.2. *Classification technique.* There is no general theory of homogeneous relational structures beyond Lachlan's theory for the stable case, but Lachlan and Woodrow have developed techniques which apply in some quite special cases. These techniques depend heavily on Fraïssé's theory of amalgamation classes. They are best understood in the context of homogeneous tournaments, which I reworked in [4], incorporating some technical improvements I found while working on the case of directed graphs. The classification of homogeneous tournaments is now quite short, but in spite of technical improvements the classification of the homogeneous directed graphs is quite long and will be presented in the last four chapters of an AMS Memoir which should be in final form shortly.

1.3. CONTENTS

What I propose to do here is to review the Fraïssé theory, the ideas developed by Lachlan and Woodrow, and the application to the classification of homogeneous tournaments, and to sketch briefly the adaptation of these methods to the case of directed graphs. Finally, there are a number of problems about homogeneous directed graphs which are not settled by the classification, but which can be reformulated in purely combinatorial terms as a result of the classification, and hence are now ripe for attack. One such problem is the determination of all the reducts of each homogeneous directed graph up to interdefinability. This problem was solved in the undirected case by Simon Thomas in [24] using the Nešetřil–Rödl theorem as the main technical device, and some directed cases have been handled by Thomas' student J. Bennett (doctoral dissertation, in preparation).

1.3.1. *A decision problem.* Another quite curious problem is the determination of the "finitely constrained" classes of directed graphs which contain only countably many homogeneous members. Typical examples of such classes would be the class of *tournaments* and the class of *partial orders*, both of which were given explicit classifications prior to the treatment of the general case – in fact, these two classifications are invoked at appropriate points in the treatment of the general case. This problem will lead us directly into the study of well quasi-ordered families of tournaments and the general theory of well quasi-orderings. At the end I will summarize the results of my student B. Latka on this problem.

1.3.2. *Pseudofinite structures.* Hrushovski's article describes the extension of Lachlan's theory in a quite distinct direction, remaining in the context of finite structures and weakening the homogeneity condition. The resulting theory is much closer in spirit to Lachlan's original theory (and considerably less anecdotal in spirit).

2 Amalgamation classes

2.1. FRAISSE'S THEORY

2.1.1. *Fraïssé's correspondence.* It is fairly clear by a back-and-forth argument that a countable homogeneous structure is determined up to isomorphism by the isomorphism types of its finite substructures. Thus there is a bijective correspondence between homogeneous structures Γ and certain classes $\mathcal{A}$ of finite structures. Fraïssé observed that the corresponding classes $\mathcal{A}$ can be characterized intrinsically by the obvious properties of closure under isomorphism and downward, together with joint embedding (any two members of $\mathcal{A}$ embed jointly in a third), and one further property – the amalgamation property – which captures the essence of the homogeneity condition.

2.1.2. *Definition.* The class $\mathcal{A}$ has the *amalgamation property* if for any structures $A_0, A_1, A_2 \in \mathcal{A}$ and any embeddings $f_i : A_0 \to A_i$ $(i = 1, 2)$ there is a structure $A \in \mathcal{A}$ and there are embeddings $g_i : A_i \to A$ with $g_2 f_2 = g_1 f_1$.

For example, a dense linear ordering without endpoints is the homogeneous tournament associated with the class of finite transitive tournaments.

2.1.3. *Application: Many models.* Fraïssé's correspondence can be used to good effect both to generate new examples of homogeneous structures, and to prove classification theorems. For example, following Henson [11], we will now use it to produce $2^{\aleph_0}$ countable homogeneous directed graphs. (Our directed graphs have at most one oriented edge between any pair of vertices, and no loops.)

2.1.4. *Construction.* Let $\mathcal{T}$ be a class of tournaments, and let $\mathcal{A}(\mathcal{T})$ be the class of directed graphs H containing no tournaments other than those which embed in some element of $\mathcal{T}$. Then $\mathcal{A}(\mathcal{T})$ is an amalgamation class, and the corresponding countable homogeneous structure will be denoted $\Gamma(\mathcal{T})$. Thus a finite tournament embeds in $\Gamma(\mathcal{T})$ if and only if it embeds in some element of $\mathcal{T}$.

Now let T_n be the tournament obtained from a linear ordering (transitive tournament) of length n by reversing the edges between successive vertices, as well as the edge from the first to last element. The collection $(T_n : n \geq 6)$ forms an *antichain*, or in other words: there is no embedding of T_m in T_n for $m \neq n$. To see this, examine the edges in T_n which belong to two distinct oriented 3-cycles in T_n; call these edges 2-edges. If we consider the 2-edges momentarily without their orientation, we find that T_n contains a unique circuit of 2-edges, of length $n - 2$. Hence these tournaments form an antichain. Accordingly, if $\mathcal{T} = (T_n : n \in X)$ with X a set of integers greater than 5, then:

$$X = \{n : T_n \text{ embeds in } \Gamma(\mathcal{T})\}.$$

Thus we have produced $2^{\aleph_0}$ homogeneous directed graphs $\Gamma(\mathcal{T})$.

Looking ahead, we should point out that according to our classification of the homogeneous directed graphs, if one omits the graphs of the form $\Gamma(\mathcal{T})$ described above – which may be manufactured with total freedom – then it turns out that only countably many other countable homogeneous directed graphs exist.

2.1.5. *Application: Classification.* An early use of Fraïssé's correspondence to limit the possibilities, rather than to produce new examples, is found in Woodrow [26]. Here triangle-free homogeneous graphs are classified, and the triangle-freeness is used heavily to control the outcome of various amalgamation problems. I found it useful to proceed in an analogous fashion in the case of homogeneous directed graphs omitting I_n for some n, where I_n is an independent set of n vertices, that is, an edgeless directed graph on n vertices. The idea here is to study the amalgamation class "generated" by a finite set of graphs. More exactly, one studies the entailment relation $\implies$ introduced above. (This can easily be redefined without mentioning homogeneity in terms of the consequences of a series of attempted amalgamations). In general the set of "consequences" of a finite set $\mathcal{A}$ of finite structures is not itself an amalgamation class, but on the other hand many interesting amalgamation classes can be usefully characterized in this manner as the set of consequences of a suitable finite set $\mathcal{A}$.

2.2. THE METHODS OF LACHLAN AND WOODROW

In [19] a new idea enters the picture which makes more refined use of amalgamation classes. Consider the generic undirected graph Γ_n omitting the complete graph K_n, by which we mean the homogeneous graph associated to the amalgamation class $\mathcal{A}(K_{n-1})$ of all graphs which do not embed K_n. This is the "typical" homogeneous graph, and the main step in the classification of all homogeneous graphs is the characterization of the class $\mathcal{A}(K_{n-1})$ as the smallest amalgamation class containing K_{n-1} and two other small graphs which we will call A and B here. I am now going to deform the idea of [19] rather badly, in a way that seems to me to preserve the essential idea.

If $\mathcal{A}$ is an amalgamation class containing K_{n-1}, A, and B, let $\mathcal{A}^*$ be the class of graphs $H \in \mathcal{A}$ such that for any extension H^+ of H by a single additional vertex, if H^+ does not contain K_n then H^+ is again in $\mathcal{A}$. Consider the following statement:

if $\mathcal{A}$ is an amalgamation class containing K_{n-1}, A, and B, then so is $\mathcal{A}^*$.   (*)

Now it is fairly easy to see that (*) is equivalent to the result that we are in any case aiming at, namely that any such amalgamation class $\mathcal{A}$ contains all graphs that do not contain K_n. In the first place, the assertion (*) is a special case of this result. Conversely, it is not hard to see that given (*), our main result follows by "undergraduate induction" (I refer to the kind of argument which normally involves confusion on the part of the author as to which value of k is under consideration). Indeed, we may prove by induction on k that any graph of order k omitting K_n is in any amalgamation class $\mathcal{A}$ containing K_{n-1}, A, and B. For the base step we take $k = 1$, or even $k = 0$. For the inductive step, if $|H| = k + 1$ we remove a vertex, getting H_0 of order k. By induction H_0 is in $\mathcal{A}^*$, hence $H = H_0^+$ is in $\mathcal{A}$, and the proof is complete.

There is very little difficulty in checking that K_{n-1}, A, B satisfy the requirement for membership in $\mathcal{A}^*$, but unfortunately there is no obvious way to prove that the class $\mathcal{A}^*$ is again an amalgamation class. One responds to this difficulty by changing the definition of $\mathcal{A}^*$ so that with the new definition it is obviously an amalgamation class; one then must work harder to verify that certain specific graphs also satisfy the new definition. The final inductive argument remains exactly as above.

I will not give any more of the details relating to the case of undirected graphs. Instead I prefer to pass directly to the case of tournaments, which was treated by Lachlan in [17], because in this case a second idea is introduced on top of this "cheap induction" argument using associated amalgamation classes.

3 Homogeneous tournaments

3.1. THE CLASSIFICATION

There are five homogeneous tournaments, namely I_1 with one vertex, C_3 the oriented 3-cycle, $\mathbb{Q}$ the rational order, $\mathbb{Q}^*$ the dense local order, and T^∞ the random tournament. We will describe the last two in more detail.

3.1.1. *Notation.* If T is a tournament, we say that a vertex a *dominates* a vertex b if the edge between them is oriented toward b, and we write $'a$ and a' for the sets of vertices dominating or dominated by a, respectively. A tournament is called a *local order* if for each vertex a, the induced tournaments on a' and $'a$ are linear orders. The class of finite local orders is an amalgamation class and the corresponding homogeneous tournament is called the dense local order. It is rather easy to check that the first four tournaments on our list are exactly the homogeneous local orders.

The last tournament on our list, the random tournament, is the homogeneous tournament corresponding to the amalgamation class of all finite tournaments. The classification of the homogeneous tournaments amounts to the following:

3.1.2. *Theorem.* A homogeneous tournament which is not a local order embeds all finite tournaments.

We may rephrase this at once in terms of amalgamation classes. Let $[I_1, C_3]$ denote the tournament consisting of one vertex dominating an oriented 3-cycle. The local orders are the tournaments omitting $[I_1, C_3]$ and its dual $[C_3, I_1]$. Taking this duality into account, the classification theorem becomes:

3.1.3. *Theorem.* If $\mathcal{A}$ is an amalgamation class of tournaments and $[I_1, C_3] \in \mathcal{A}$, then all finite tournaments are in $\mathcal{A}$.

3.2. THE PROOF

The proof can be carried out for the most part using relatively general principles. The outstanding item of specific information required is the following:

3.2.1. *Lemma.* If $\mathcal{A}$ is an amalgamation class of tournaments and $[I_1, C_3]$ is in $\mathcal{A}$, then $[C_3, I_1]$ is in $\mathcal{A}$.

This can of course be proved by exhibiting some amalgamation diagrams. One can also argue as follows. With some effort it can be checked that a tournament which omits $[C_3, I_1]$ has the form $[L, S]$ where L is a linear order, S is a local order embedding C_3, and as the notation suggests, every vertex in L dominates every vertex in S. If such a tournament is homogeneous it clearly reduces either to L or to S, and hence omits $[I_1, C_3]$ as well.

Now we may prepare our cheap induction argument as follows.

3.2.2.   *Definition.*   Let $\mathcal{A}$ be an amalgamation class of tournaments. Then $\mathcal{A}^* =:$

$$\{H \in \mathcal{A} : \text{ any linear extension of } H \text{ lies in } \mathcal{A}\},$$

where a linear extension of H is a tournament $H \cup L$ with the induced subtournament on L linear (i.e., transitive).

What we need is then:

3.2.3.   *Proposition.*   Let $\mathcal{A}$ be an amalgamation class of tournaments containing $[I_1, C_3]$. Then:

(1) $\mathcal{A}^*$ is an amalgamation class.

(2) $\mathcal{A}^*$ contains $[I_1, C_3]$.

If this proposition is granted, our rather silly inductive argument succeeds. Furthermore, the first claim is purely formal. In brief, if A_0, A_1, A_2, f_1, f_2 are the data for an amalgamation problem in $\mathcal{A}^*$ and we list the (finitely many) possible solutions B_i to this problem, then the assumption that no B_i is in $\mathcal{A}^*$ quickly produces a contradiction, as each B_i would then have a linear extension $B_i \cup L_i$ lying outside A, and after gluing the L_i together into one long linear order L, we would find that $A_0 \cup L, A_1 \cup L, A_2 \cup L$ with the natural embeddings define an amalgamation problem in $\mathcal{A}$ which has no solution in $\mathcal{A}$!

This very formal analysis has reduced the classification problem to the following:

3.2.4.   *Lemma.*   If $\mathcal{A}$ is an amalgamation class of tournaments containing $[I_1, C_3]$ and H is a finite linear extension of $[I_1, C_3]$, then $H \in \mathcal{A}$.

The major innovation in [17] comes at this point. Using Ramsey's theorem, Lachlan reduces the preceding lemma to one of the following form (I use a variant of his original version). We will use the notation $L[C_3]$ for the composition of a linear tournament L with the tournament C_3, that is a series of copies of C_3 with each copy dominating later ones.

3.2.5.   *Lemma.*   If $\mathcal{A}$ is an amalgamation class of tournaments containing $[I_1, C_3]$ and H is an extension of a tournament of the form $L[C_3]$ with L finite and linear by a *single* vertex, then $H \in \mathcal{A}$.

The main point here is that $L[C_3]$ contains a potentially large number of disjoint copies of $[I_1, C_3]$. The connection between the two versions of the lemma is provided by Ramsey's theorem and an explicit amalgamation argument. In addition to [17], where the argument is presented with all details, a rather detailed sketch of the argument is given in [4].

3.3. THE FINAL STAGE OF THE PROOF.

It should be said that a third idea comes into play at this point in the argument. The foregoing lemma is quite concrete in its content, apart from our total lack of understanding as to how a certain vertex of H sits over the others. It turns out that this lemma can easily be proved by induction if we set it up properly. In fact the summary up to this point seems to cover the main ideas adequately, but this final point is rather important in practice and is really the main issue that requires attention when one is working out these proofs. Accordingly we will go into a little more detail concerning the proof of the last lemma.

In the first place, having gotten so much mileage out of the "amalgamation class" viewpoint, we find that it is now useful to shift our viewpoint back from amalgamation classes to homogeneous tournaments. So we translate the last lemma back to the context of homogeneity, getting:

3.3.1.   *Lemma.* Let Γ be a homogeneous tournament embedding $[I_1, C_3]$. Let H be an extension of $L[C_3]$ by a single vertex v, with L finite and linear. Then H embeds in Γ.

In this form the proof proceeds by induction on the length of L. Let C be the first copy of C_3 in $L[C_3]$, and let p be the type of the vertex v over C, that is, p consists of a specification, for each vertex of C, as to whether v is to dominate or be dominated by that vertex. Let ^pC be the set of vertices in $\Gamma - C$ which have this same type p over C, and let C' be the set of vertices in Γ dominated by all vertices of C. (The possibility $^pC = C'$ turns out to be illusory after one makes the modifications alluded to below.) We consider the structure $\mathbb{T} = (^pC, C')$ consisting of two tournaments together with some oriented edges connecting them. A structure of this type will be called a 2-tournament. To embed H into Γ, it is sufficient to embed the 2-tournament $\mathbb{H} = (\{v\}, L_0[C_3])$ into $\mathbb{T}$, where L_0 is L with its first vertex removed, and with v relating to $L_0[C_3]$ in $\mathbb{H}$ as it does in H.

3.3.2.   *A shift of category.*   Unfortunately as we began with tournaments and we are now working in a new category, the category of 2-tournaments, even though L_0 is shorter than L the induction fails. However if we start over, and rephrase our lemma in terms of homogeneous 2-tournaments (thereby strengthening it a little), then the indicated induction succeeds. To carry this out involves writing down (and proving) some specific properties of the 2-tournament $\mathbb{T}$. In [4] I introduce the notion of an *ample* 2-tournament at this point and carry out the appropriate inductive proof in a couple of easy pages.

4   Some homogeneous directed graphs

It is always more pleasant if a classification project turns up a few new examples along the way. I gave a catalog of the known homogeneous directed graphs in 1983, including one slightly odd one that turned up in the process of making that list. Much later (Summer 1988) I found one more, which turned out to be the last. This last example may perhaps be called the dense local partial order, by analogy with the dense local order.

4.1. EXAMPLES

4.1.1. *The dense local order.* The dense local order may be realized by partitioning the dense linear order $\mathbb{Q}$ into two dense sets Q_0, Q_1 and reversing the edges between Q_0, Q_1. This may also be described as follows: identify the two possible orientations of an edge with the integers modulo 2, and shift the orientations between Q_i, Q_j by $j - i$. There is another homogeneous directed graph that can be manufactured in a very similar way by partitioning $\mathbb{Q}$ into three dense sets Q_i labelled by the integers modulo 3, and identifying the two possible orientations of an edge with the numbers ± 1, while 0 represents the absence of an edge: if we shift the edges between Q_i and Q_j by $j - i$ we get a directed graph that may be pictured as a circle along which edges point up to 1/3 of the way around in the positive (say, counterclockwise) direction.

4.1.2. *The dense local partial order.* A similar approach starting with the generic partial order $\mathcal{P}$ produces another homogeneous directed graph. We partition $\mathcal{P}$ into three dense subsets P_i indexed by the integers modulo 3, we identify the three binary relations holding in $\mathcal{P}$ with the integers modulo 3 as above, and we shift the relations between P_i and P_j by $j - i$. The resulting homogeneous directed graph is rather hard to visualize and I would not claim to grasp it.

In any case, this is the most subtle example of a homogeneous directed graph, and it is not too far removed from a linear order.

5   The classification argument

As I have said, the classification of all homogeneous directed graphs involves a particularly long and detailed analysis. I will sketch the argument briefly here.

After disposing of the finite and imprimitive cases, it seems necessary to divide the primitive infinite case in two, according as our homogeneous directed graph Γ does or does not embed I_∞, an infinite set of independent vertices.

5.1. THE FIRST CASE

If Γ does not contain I_∞, then in fact it omits I_n for some n. If $n = 2$ we are of course dealing with tournaments. For any finite value of n the argument simply generalizes the argument for tournaments, and indeed this more general argument was the direct source for the presentation of the case of tournaments in [4]. One simplifying factor in the case of directed graphs omitting I_n is that any application of Ramsey's theorem to a sufficiently large set of vertices inevitably produces a long linear order, and for this reason the argument stays fairly close to the model argument given for tournaments.

5.2. THE SECOND CASE

The study of homogeneous directed graphs embedding I_∞ seems harder. In spirit the situation is rather similar to the classification of homogeneous undirected graphs by Lachlan and Woodrow, but they found a trick exploiting the symmetry of the relations in an essential way. I use an argument which is much closer to the analysis of homogeneous tournaments. This argument can also be used for undirected graphs, but is harder than the one used by Lachlan and Woodrow in that case. I will go into this case in substantially more detail.

5.2.1. *Formulation in terms of amalgamation classes.* Let Γ be a homogeneous directed graph embedding I_∞, and let $\mathcal{T}$ be the set of tournaments embedding in Γ. Let A be the directed graph consisting of the disjoint union of an isolated vertex and an oriented path of length 2 (order 3), and define $\mathcal{A}(\mathcal{T})$ as the set:

$$\{A\} \cup \{I_n : \text{ all } n\} \cup \mathcal{T}.$$

Let $\Gamma(\mathcal{T})$ be the homogeneous directed graph associated with the amalgamation class $\mathcal{A}(\mathcal{T})$. Our problem is to show that if Γ embeds A as well, then $\Gamma \simeq \Gamma(\mathcal{T})$; if Γ instead omits A then we easily fall back into one of a small number of known exceptional cases. In terms of amalgamation classes our goal has become:

5.2.2. *Theorem.* Let $\mathcal{A}$ be an amalgamation class of finite directed graphs with $\mathcal{A}(\mathcal{T}) \subseteq \mathcal{A}$. Then $\mathcal{A}$ contains every finite directed graph H such that every subtournament of H embeds in a tournament in $\mathcal{T}$.

This will be proved for finite sets $\mathcal{T}$, from which the general case follows instantly. The advantage of working with finite sets is that we may proceed by induction on the size n of the largest element of $\mathcal{T}$, as well as on the number of nonisomorphic tournaments of this maximal size in $\mathcal{T}$.

5.2.3. *The class $\mathcal{A}^*$.* Just as in the case of tournaments, we will introduce an auxiliary amalgamation class $\mathcal{A}^*$, but there is a small difference at this point. In our earlier treatment we made use of linear tournaments, but now we may make a similar definition using either linear tournaments or independent sets of vertices I_n. We will have to consider both possibilities, since we at some point will have to invoke Ramsey's theorem, and we simply cannot predict in advance which of the two configurations will be produced. In spite of this ambiguity, much of the proof goes as in the case of tournaments until we arrive at the relatively concrete stage corresponding to our final lemma on 2-tournaments, itself proved by an inductive argument. At this stage we require a similar lemma on homogeneous 2-digraphs, which we will now state explicitly.

5.2.4. *Ample 2-directed graphs.* A 2-digraph $\mathbb{H} = (\Gamma_1, \Gamma_2)$ will be called *ample* if it embeds the specific directed graph A introduced above in its second component Γ_2, and if every possible configuration of the form $(\{v\}, I_n)$ embeds in $\mathbb{H}$. With the class of finite tournaments $\mathcal{T}$ and the 2-directed graph $\mathbb{H}$ fixed, a configuration $(\{v\}, H)$ will be called *restricted* if for every subtournament T of H we have: T embeds in some element of $\mathcal{T}$, and $(\{v\}, T)$ embeds in $\mathbb{H}$.

5.2.5. *Lemma $\mathcal{T}$.* Let $\mathcal{T}$ be a finite set of finite tournaments and let **H** be an ample homogeneous 2-digraph. Then any restricted configuration $(\{v\}, H)$ embeds in **H**.

Now the point is that both this lemma and the main result must be proved simultaneously by induction over $\mathcal{T}$. The order of steps is as follows. First Lemma $\mathcal{T}$ is proved for configurations in which H is a disjoint union of tournaments and oriented paths of length 2. This provides the key step to complete the main result for $\mathcal{T}$. Then using the main result for $\mathcal{T}$, the general case of Lemma $\mathcal{T}$ follows, and we are ready to proceed to a new $\mathcal{T}$.

6 Decision problems

6.1. CLASSIFICATION PROBLEMS

With the classification of the directed graphs in hand we can consider various special cases. In general it is natural to consider *finitely constrained* classes of directed graphs, in which we study all the directed graphs which embed all directed graphs in one finite set $\mathcal{A}$ and omit all directed graphs in a second finite set $\mathcal{B}$. Typically $\mathcal{A}$ is empty, though one might for example put I_2 in $\mathcal{A}$. In particular the classification of homogeneous partial orders [23] falls into this framework.

6.1.1. *Three problems.* Let $\mathcal{K}$ be a finitely constrained class of directed graphs. Our classification result for directed graphs specializes directly to $\mathcal{K}$. Accordingly one may well expect the following natural problems to become trivial:

$$\text{Are there any homogeneous directed graphs in } \mathcal{K}? \qquad (A)$$

$$\text{Are there infinitely many homogeneous directed graphs in } \mathcal{K}? \qquad (B)$$

$$\text{Are there uncountably many homogeneous directed graphs in } \mathcal{K}? \qquad (C)$$

We also have Lachlan's original question: is the entailment relation $\Longrightarrow$ decidable for finite structures in $\mathcal{K}$? Our classification immediately yields a positive answer to this question, as any real classification must, and problem (A) is a special case of this.

6.1.2. *Problem B.* Problem (B) is a little less trivial. If some negative constraint H ($H \in \mathcal{B}$) is linear, then all but finitely many of the homogeneous directed graphs of the form $\Gamma(\mathcal{T})$ are excluded by this constraint. This immediately reduces problem (B) to a manageable special case. If on the other hand the negative constraints $H \in \mathcal{B}$ all involve C_3, then (B) is settled: there are indeed infinitely many homogeneous structures in $\mathcal{K}$, generated by linear tournaments of different sizes. So in either case (B) is settled quickly.

6.1.3. *Problem C.* Problem (C) is open, but can be rephrased in purely combinatorial terms. In the first place, a direct application of the classification theorem reduces the problem to the very special case in which there are no positive constraints, and the set $\mathcal{B}$ of negative constraints consists exclusively of tournaments. In this case we consider the class $\mathcal{T}_\mathcal{B}$ of all finite tournaments which contain no member of $\mathcal{B}$. If $\mathcal{T}_\mathcal{B}$ contains no infinite antichain (with respect to embeddability), then $\mathcal{T}_\mathcal{B}$ is said to be well quasi-ordered, and $\mathcal{B}$ is said to be *tight* – that is, $\mathcal{B}$ is a tight constraint. Our special case of problem (C) is equivalent to the problem of determining whether $\mathcal{T}_\mathcal{B}$ is well quasi-ordered.

It is certainly not out of the question that such a decision problem could turn out to be undecidable, either as stated or perhaps in a somewhat more general formulation for slightly richer combinatorial structures. However there is no obvious encoding to show this.

6.2.   RESULTS ON PROBLEM (C)

The known results in this area, due to my student B. Latka, all concern the classes $\mathcal{T}_{\{B\}}$ determined by just one negative constraint. She has shown that the problem is decidable in this case, and has also given a very simple decision procedure – the latter being much harder to obtain, as it involves the explicit solution of a number of critical cases. The soft approach to decidability does not seem to work in the context of more than one constraint.

Latka shows that all sufficiently large nonlinear tournaments constitute loose constraints, which certainly yields the decidability result. She then gives an explicit (and very short) list of the exceptions; this latter information would seem to be a prerequisite for a useful analysis of the case of two constraints, in our present state of knowledge.

6.2.1.   *Soft arguments.* To get Latka's first, comparatively soft result, one first exhibits two infinite antichains of finite tournaments which serve to show that a number of specific tournaments are loose. Let $[L_2, C_3]$ and $[C_3, L_2]$ be respectively tournaments in which two vertices dominate or are dominated by a 3-cycle C_3, and let $C_3[L_1, L_1, L_4]$ be a tournament in which one vertex of C_3 has been replaced by a linear tournament of length 4. One of Latka's antichains, a modification of a linear order in the spirit of Henson's example, is made up of tournaments omitting $C_3[L_1, L_1, L_4]$, while the other is made up of tournaments omitting $[L_2, C_3]$ and $[C_3, L_2]$. On the other hand every sufficiently large nonlinear tournament contains one of these three tournaments, by the pigeon-hole principle.

6.2.2.   *Hard arguments.* A closer look at the information afforded by Latka's two antichains reveals that every nonlinear tournament on at least seven vertices constitutes a loose constraint, and that among non-linear tournaments for which the problem is not settled by these antichains, there are only two maximal ones, of order 5 and 6 respectively. Let N_5 be the tournament obtained from the linear tournament of order 5 by reversing the arrows representing the successor relation, and let $C^+ = [L_1, C_3[L_1, L_1, L_3]]$ be the tournament in which one vertex dominates the tournament derived from C_3 by replacing one vertex by a linear tournament of order 3. Latka completes the analysis of the case of a single constraint by showing that tournaments omitting N_5 or C^+ admit a structure theorem of a type that allows us to apply Kruskal's Tree Theorem to conclude that the class is well quasi-ordered. In the case of N_5 the full Kruskal theorem is needed, basically because the class is closed under composition, while in the case of C^+ Kruskal's theorem for trees of bounded height – which is essentially the same as Higman's lemma for finite words in a well quasi-ordered alphabet – is adequate.

Thus in a certain precise sense, this completes the classification of all singly constrained classes of finite tournaments which admit a structure theorem.

7 Whither?

7.1. FINITE STRUCTURES

I have mentioned Hrushovski's work extending Lachlan's theory, discussed in Hrushovski's article in this volume.

7.2. 0-1 LAWS

Another interesting line is the study of 0-1 laws for probabilities in spaces consisting of the labelled structures in a finitely constrained amalgamation class. The very striking results of Kolaitis, Prömel, and Rothschild on the case of finite undirected graphs omitting a fixed K_{n+1} have not been generalized to the directed case, and perhaps unexpectedly involve new difficulties. Consider the case $n = 2$ (treated earlier in [5]): a random finite undirected graph with no triangles is bipartite, with probability approaching 1 as the number of vertices goes to infinity. The directed version of this result would apply to directed graphs omitting one of the two tournaments L_3, C_3 of order 3. For directed graphs omitting L_3, a similar result may hold, but not with the same proof. For directed graphs omitting C_3 the analogous statement is clearly false: there are more graphs of a given size obtained from a linear tournament by omitting some edges (as in the Albert–Frieze model for random partial orders) than there are bipartite graphs. So all of this remains quite mysterious.

7.3. THE FINITE MODEL PROBLEM

A somewhat related problem is the finite model problem. We know that the random (or generic) undirected graph has the finite model property: any first order property of this graph is true of some (actually, most) finite graphs. The analogous question for the generic graph omitting K_{n+1} is open, even for $n = 2$. By the 0-1 law, most finite triangle-free graphs contain no cycle of length 5, and are thus very different from the generic infinite triangle-free graph; but it is possible that this graph has (a few) good finite approximations. The most successful exploration of this case that I know of is due to Michael Albert (unpublished), but is still in its early stages.

7.4. OTHER CLASSIFICATION PROBLEMS

In principle the methods used to classify homogeneous graphs apply to any finite binary language. In practice these methods are quite cumbersome and appear to be near the point of collapsing under their own weight. At the same time the actual classification is strikingly simple: one essentially has the directed graphs corresponding to the most obvious amalgamation classes ($\Gamma(\mathcal{T})$ and Γ_n) and a handful of exceptions, finite, imprimitive, or linked to partial orders. It remains to be seen whether our classification techniques break down at the point where radically new examples appear.

References

[1]   Albert, M. and Frieze, A. (1989), Random graph orders, *Order* **6**, 19–30.

[2]   Cherlin, G., Homogeneous directed graphs, submitted to *AMS Memoirs*.

[3]   Cherlin, G. and Lachlan, A. (1986), Finitely homogeneous relational structures, *Trans. Amer. Math. Soc.* **296**, 815–850. MR 88f:03023

[4]   Cherlin, G. (1988), Homogeneous tournaments revisited, *Geometria Dedicatae* **26**, 231–240. MR 89k:05039

[5]   Erdős, P., Kleitman, D. and Rothschild, B. (1976), Asymptotic enumeration of K_n-free graphs, in *Colloquio Internazionale sulle Teorie Combinatorie*, vol. 2, 19–27. MR 57 #2984.

[6]   Fagin, R. (1976), Probabilities on Finite Models, *J. Symb. Logic* **41**, 50–58. MR 57 #16042.

[7]   Fraïssé, R. (1953), Sur l'extension aux relations de quelques propriétés connues des ordres, *C. R. Acad. Sci. Paris* **237**, 508–510. MR 57 #16042.

[8]   Gardiner, A. (1976), Homogeneous graphs, *J. Combin. Theory Ser. B* **20**, 94–102. MR 52 #7316.

[9]   Glebskiĭ, Yu., Kogan, D., Liogonkiĭ, M. and Talanov, V. (1969), Volume and fraction of satisfiability of formulas of the lower predicate calculus (Russian), *Kibernetika* **2** (1969), 17–27. MR 46 #42.

[10] Gol'fand, Ya. and Klin, M. (1978), On k-homogeneous graphs, in *Algoritmicheskie Issledovaniya v Kombinatorike*, (Russian), ed. I.A. Fradzhev, *Proceedings of an International Colloquium on Combinatorics and Graph Theory* (July 1976), 76–85, Nauka, Moscow. MR 80d:05043.

[11] Henson, C. W. (1972), Countable homogeneous relational structures and $\aleph_0$-categorical theories, *J. Symbolic Logic* **37**, 494–500. MR 48 #94.

[12] Knight, J. and Lachlan, A. (1987), Shrinking, stretching, and codes for homogeneous structures, in *Classification Theory, Chicago, 1985*, ed. J. Baldwin, LNM 1292, Springer, 1987. MR 90k:03033.

[13] Kolaitis, P., Prömel, H. and Rothschild, B. (1987), K_{l+1}-free graphs: Asymptotic structure and a 0-1 law, *Trans. Amer. Math. Soc.* **303**, 637–671. MR 88j:05016.

[14] Lachlan, A. (1982), Finite homogeneous simple digraphs, in *Logic Colloquium 1981*, 189–208, ed. J. Stern, North-Holland, New York. MR 85h:05049.

[15] Lachlan, A. (1984), On countable stable structures which are homogeneous for a finite relational language, *Israel J. Math* **49**, 69–153. MR 87h:03047a.

[16] Lachlan, A. (1985), Verification of the characterization of stable homogeneous 3-graphs, *Research Report* **85-10**, Simon Fraser University.

[17] Lachlan, A. (1984), Countable homogeneous tournaments, *Trans. Amer. Math. Soc.* **284**, 431–461. MR 85i:05118.

[18] Lachlan, A. (1987), Homogeneous structures, in *Proceedings of the ICM 1986*, 314–321, ed. A. Gleason, AMS, Providence.

[19] Lachlan, A. and Shelah, S. (1984), Stable structures homogeneous for a finite binary language, *Israel J. Math* **49**, 150–180. MR 87h:03047b.

[20] Lachlan, A. and Woodrow, R. (1980), Countable ultrahomogeneous undirected graphs, *Trans. Amer. Math. Soc.* **262**, 51–94. MR 82c:05083.

[21] Latka, B. (1991), Finitely Constrained Classes of Homogeneous Directed Graphs and Well Quasi-ordered Classes of Tournaments, doctoral dissertation, Rutgers University, 1991.

[22] Peretyatkin, M. (1973), On complete theories with a finite number of denumerable models, *Algebra i Logika* **12**, 550–576 and 618. MR 50 #6827.

[23] Schmerl, J. (1979), Countable homogeneous partially ordered sets, *Alg. Univ.* **9**, 317–321. MR 81g:06001.

[24] Sheehan, J. (1974), Smoothly embeddable subgraphs, *J. London Math. Society* **9**, 212–218. MR 51 #229.

[25] Thomas, S. (1991), Reducts of the random graph, *J. Symbolic Logic* **56**, 176–181.

[26] Woodrow, R. (1979), There are four countable ultrahomogeneous graphs without triangles, *J. Combinatorial Theory Ser. B* **27**, 168–179.

Ordinal Partition Behavior of Finite Powers of Cardinals

PAUL ERDŐS

Mathematical Institute
Hungarian Academy of Sciences
Budapest, Réaltanoda u. 13-15, H-1053 Hungary

A. HAJNAL*

Mathematical Institute
Hungarian Academy of Sciences
Budapest, Réaltanoda u. 13-15, H-1053 Hungary

JEAN A. LARSON

Department of Mathematics
University of Florida
Gainesville, Florida

Abstract

In the notation of Erdős and Rado, the expression $\alpha \to (\beta, p)^2$ means that for any graph on α either there is an independent subset of type β or there is a complete subgraph of size p. We discuss results for this relation where α and β are both finite powers of some cardinal. In particular, assume that λ is either a regular cardinal or a strong limit cardinal and that k and ℓ are positive integers. Then $\lambda^{1+k\ell} \to (\lambda^{1+k}, \ell+1)^2$. On the other hand, $\lambda^{k\ell} \not\to (\lambda^{1+k}, 2^{\ell-1}+1)^2$ holds provided $k \geq 4$. We prove that the positive result is sharp if λ is a successor cardinal of the form $\lambda = \theta^+ = 2^\theta$, while the negative result is sharp if the cofinality of λ is a weakly compact cardinal.

0   Introduction and Terminology

At the conference in honor of Eric Milner's Third Coming of Age, we considered the following problem: is it true that for every ordinal α, there is an ordinal β of the same cardinality such that $\beta \to (\alpha, 3)^2$?

For countable ordinals the answer to this question is "yes" by a theorem of Erdős and Milner (see [7] or [19]), but a little thought shows that under the Continuum Hypothesis (CH) the answer is "no" for $\omega_1{}^\omega$; we give a proof in Section 2. This result suggests looking at $\alpha = \lambda^n$ for cardinals λ and finite n. We prove for all cardinals λ which are either regular or strong limit cardinals, and for all positive integers k and ℓ, that $\lambda^{k\ell+1} \to (\lambda^{k+1}, \ell+1)^2$.

---

*Research supported by Hungarian Science Foundation OTKA grant 1908
Mathematics Subject Classification 03E10

N.W. Sauer et al. (eds.), Finite and Infinite Combinatorics in Sets and Logic, 97–115.
© 1993 *Kluwer Academic Publishers. Printed in the Netherlands.*

Thus we have a positive answer for all $\alpha = \lambda^m$ under the Generalized Continuum Hypothesis (GCH).

We prove that the result is sharp for finite powers of successor cardinals $\lambda = \theta^+ = 2^\theta$ using a well-known partition which shows $\lambda^2 \not\to (\lambda^2, 3)^2$.

For λ a weakly compact cardinal or a strong limit cardinal whose cofinality is weakly compact, we prove a stronger positive theorem: for positive integers $k \geq 4$ and ℓ, $\lambda^{k\ell+1} \to (\lambda^{k+1}, 2^\ell)^2$. To show this theorem is sharp, we lift counter-examples for finite powers of ω to all cardinals.

We start the paper in Section 1 with our strongest counter-examples. Section 2 contains the rest of our counter-examples, which are generalized from results for countable ordinals via pinning. In Section 3 we use the jumping around technique to prove the weak positive theorem which is true for all regular cardinals. Then we use canonization and modify the proof to extend it to singular strong limit cardinals, using some of the notation from the section on pinning. Canonization plays an even stronger role in Sections 4 and 5 where we generalize some of the known positive results for ω to weakly compact cardinals and singular strong limit cardinals whose cofinality is weakly compact. We conclude in Section 6 with a brief discussion of directions for further research.

1   Counter-examples for successor cardinals

The starting point for our investigation is a family of examples by A. Hajnal and J. Baumgartner.

Theorem 1.1 [10],[1] *For all cardinals θ, if $2^\theta = \theta^+ = \lambda$, then*

$$\lambda^2 \not\to (\lambda^2, 3)^2.$$

The proof is an elegant recursive construction of a graph on $[\lambda]^2_<$ with the following properties:

(i) all edges go "down", that is, if $\alpha < \beta$ and there is an edge between (α, γ) and (β, δ), then $\gamma > \delta$; and

(ii) there is at most one edge from a point in the αth column to any point in the later βth column, that is, if $\alpha < \beta$ then for any $\gamma > \alpha$, there is at most one $\delta > \beta$ so that (α, γ) and (β, δ) are joined by an edge.

Our proof proceeds by induction. The following lemma contains the basic idea for stepping up.

Lemma 1.2 *Suppose that θ is a cardinal and $2^\theta = \theta^+ = \lambda$. If $\lambda^{k\ell} \not\to (\lambda^{k+1}, \ell+1)^2$ for some positive integers k and ℓ, then $\lambda^{k\ell+k} \not\to (\lambda^{k+1}, \ell+2)^2$.*

Proof. Represent $\lambda^{k\ell+k}$ as $\lambda^k \times \lambda^{k\ell}$ under the lexicographic ordering. Let $f : [\lambda^{k\ell}]^2 \to \{0, 1\}$ be the graph whose existence witnesses the induction hypothesis, $\lambda^{k\ell} \not\to (\lambda^{k+1}, \ell+1)^2$. Let $g : [\lambda^2]^2 \to \{0, 1\}$ be the graph of Hajnal which witnesses $\lambda^2 \not\to (\lambda^2, 3)^2$ and has the two extra properties listed above.

Let $e_0 : \lambda^k \to \lambda$ and $e_1 : \lambda^{k\ell} \to \lambda$ be one-to-one enumerations. Define $h : \lambda \times \lambda^{k\ell} \to \{0,1\}$ by

$$h((\alpha,\gamma),(\beta,\delta)) = \begin{cases} g(\gamma,\delta) & \text{if } \alpha = \beta, \\ f((e_0(\alpha),e_1(\gamma)),(e_0(\beta),e_1(\delta))) & \text{otherwise.} \end{cases}$$

We think of $\lambda^{k\ell+1}$ as consisting of λ^k clumps each of size $\lambda^{k\ell}$. Within a clump, the graph matches the graph from the induction hypothesis. Edges between clumps come from the Hajnal graph. The following series of claims completes the proof.

Claim 1 *Any subset of a clump of size λ^{k+1} has an edge.*

Claim 2 *Any subset which meets λ many clumps in a set of size λ has an edge.*

Claim 3 *If a complete set of at least three elements meets two clumps, then one of the clumps (the later one) has only one point.*

Claim 4 *The graph has no complete set of size $\ell + 2$.*
Just use the induction hypothesis applied to the clumps and Claim 3.

Claim 5 *The graph has no independent set of size λ^{k+1}.*

Suppose that X is a subset of size λ^{k+1}. If it meets λ many clumps in a set of size λ, then it has an edge by Claim 2. Otherwise, it meets fewer than λ many clumps in a set of size λ. Since λ is regular, then the set of points of X from clumps which X meets in a set of size less than λ has type at most $\lambda^k < \lambda^{k+1}$. Furthermore λ^{k+1} has the property that it is not the sum of fewer than λ smaller subsets, so some clump contains a subset Y of X of size λ^{k+1}. Thus in this case the set X has an edge by Claim 1. $\square$

Theorem 1.3 *Suppose that θ is a cardinal and $2^\theta = \theta^+ = \lambda$. Then $\lambda^{k\ell} \not\to (\lambda^{k+1}, \ell + 1)^2$ for all positive integers k and ℓ.*

Proof. The proof is by induction on ℓ. For $\ell = 1$, use the empty graph. Continue to step up the proof by Lemma 1.2. $\square$

2 Counter-examples from pinning

In this section we discuss a modification of the pinning argument first introduced by Specker [18] to lift results from one ordinal to another. While the main thrust of this section is the transfer of negative results for the finite power of the cofinality of a singular cardinal to the corresponding finite power of the singular cardinal, we start with a simple example outside this paradigm.

Theorem 2.1 *Assume CH. If $\omega_1{}^\omega \le \beta < \omega_2$, then $\beta \not\to (\omega_1{}^\omega, 3)^2$.*

Proof. Under the Continuum Hypothesis (CH), we know that $\omega_1{}^\omega \not\to (\omega_1{}^\omega, 3)^2$ by a familiar argument: pin the ordinal $\omega_1{}^\omega$ onto $\omega_1 \cdot \omega$ and use the result of Erdős and Hajnal [4] that $\omega_1 \cdot \omega \not\to (\omega_1 \cdot \omega, 3)^2$. The pinning is especially simple: we write $\omega_1{}^\omega$ as the increasing sum of sets of size $\omega_1{}^n$ for $n < \omega$, and combine one-to-one mappings of the nth set in the summand onto the nth copy of ω_1 in $\omega_1 \cdot \omega$. Next suppose that β is an ordinal with $\omega_1 \cdot \omega \leq \beta < \omega_2$. We take a Milner–Rado paradoxical [14] decomposition of β, that is, a decomposition into countably many sets each of order type less than $\omega_1{}^\omega$. Let $\pi : \beta \to \omega_1 \cdot \omega$ be a combination of one-to-one mappings as before, each on some cell of the Milner–Rado paradoxical decomposition. Any subset A of β of type at least $\omega_1{}^\omega$ must meet infinitely many of the cells of the paradoxical decomposition in an uncountable set. Thus for any such set A, $\pi(A)$ has type at least $\omega_1 \cdot \omega$. Therefore we can lift the counter-example from $\omega_1 \cdot \omega$ to prove $\beta \not\to (\omega_1{}^\omega, 3)^2$.   □

Now we extend the notion of pinning to reflect the generality we use in the above proof.

Definition 2.2 We say that the *type γ subsets of α can be pinned to the type δ subsets of β*, in symbols $[\alpha]^\gamma \to [\beta]^\delta$, if there is a mapping $\pi : \alpha \to \beta$ so that the image $\pi(X)$ of every subset X of α of type γ has type at least δ. In this case the function π is called a *pinning map*.

E. Specker [18] first developed this notion to lift the example which shows that $\omega^3 \not\to (\omega^3, 3)$ to other finite powers of ω. Below give the straightforward generalization of the lemma relating pinning and partition relations to the broadened definition of pinning.

Proposition 2.3 *For ordinals α, β, γ and δ and any positive integer k, if $[\alpha]^\gamma \to [\beta]^\delta$ and $\beta \not\to (\delta, k)^2$, then $\alpha \not\to (\gamma, k)^2$.*

The main result of this section is that an ordinal which is a finite power of a singular cardinal can be pinned in a strong way onto the corresponding power of its cofinality.

Theorem 2.4 *For any singular cardinal λ of cofinality θ and any positive integers $k \leq n$,*

$$[\lambda^n]^{\lambda^k} \to [\theta^n]^{\theta^k}.$$

Before we give the proof, we state the corollary of Proposition 2.3 and Theorem 2.4 that led us to consider this pinning problem.

Corollary 2.5 *For any singular cardinal λ of cofinality θ and any positive integers $k \leq n$ and m, if $\theta^n \not\to (\theta^k, m)^2$, then $\lambda^n \not\to (\lambda^k, m)^2$.*

Proof of Theorem 2.4: To start the proof, let $\langle \mu_\alpha \mid \alpha < \theta \rangle$ be a continuous sequence of cardinals cofinal in λ starting with $\mu_0 = 0$ and with the property that successors in the sequence are regular cardinals. Define $p : \lambda \to \theta$ by setting $p(\beta)$ equal to the supremum of all the α for which μ_α is less than or equal to β. Since the sequence is continuous, this definition assigns to each element of λ an element of θ. For notational convenience, for each $\alpha < \theta$, let $A(\alpha)$ be the half-open interval $[\mu_\alpha, \mu_{\alpha+1})$. Note that λ is the union of the sets $A(\alpha)$ and that p is constantly α on $A(\alpha)$. Thus p is a pinning map which shows that $\lambda \to \theta$. Let $W_\ell(\lambda)$ be the collection of all increasing sequences from λ on which p is one-to-one. One can show by a simple induction on ℓ that $W_\ell(\lambda)$ has order type λ^ℓ under the lexicographic order. For each positive integer ℓ, define $\pi_\ell : W_\ell(\lambda) \to [\theta]_<^\ell$ by component-wise application of p: $\pi(\alpha_0, \alpha_1, \ldots, \alpha_{m-1}) = (p(\alpha_0), p(\alpha_1), \ldots, p(\alpha_{m-1}))$. We shall refer to all of these maps by the single name π. From here the proof proceeds by a series of claims.

Claim 1 *For any positive integer ℓ, if X, a subset of $W_\ell(\lambda)$, has cardinality λ, then $\pi(X)$ has cardinality at least θ.*

Proof. Suppose by way of contradiction that X is a subset of $W_\ell(\lambda)$ of cardinality λ so that $\pi(X)$ has cardinality less than θ. Then for some $\alpha < \theta$ and for all x in X, the image of x under π, namely $(p(x_0), p(x_1), \ldots, p(x_{\ell-1}))$, has the property that all of the $p(x_i)$ are less than μ_α. However in this case, X is a subset of the increasing sequences of length ℓ from $[0, \mu_{\alpha+1})$. Since the latter set has cardinality $\mu_{\alpha+1}^\ell = \mu_{\alpha+1}$, we have reached the contradiction that X has cardinality less than λ.

Before we proceed to the next claim, we introduce some more notation. For any positive integer $\ell > 1$ and any subsets Y of $W_\ell(\lambda)$ and A of λ, denote by ^-Y the set of all elements z of $W_{\ell-1}(\lambda)$ for which there is some b with $(b)^{\smallfrown}z$ in Y and denote by Y_A the set of all elements x of $W_\ell(\lambda)$ whose least element is in A.

Claim 2 *For any positive integers $\ell > 1$ and j, any subset Y of $W_\ell(\lambda)$ and any bounded subset A of λ, if Y_A has type at least λ^j, then ^-Y_A has type at least λ^j.*

Proof. Suppose by way of contradiction that ^-Y_A has type at most $\lambda^{j-1} \cdot \delta$ for some $\delta < \lambda$. Let μ be a regular cardinal which is an upper bound for A. Then Y_A can be regarded as the sum over a in A of $Y_{\{a\}}$, which can be thought of as a subset of $\{a\} \times {}^-Y_A$. Thus Y_A has type at most $\lambda^{j-1} \cdot \delta \cdot \mu$ which is less than λ^j. This contradiction proves the claim.

Claim 3 *For all positive integers $k \le n$ and all subsets X of $W_n(\lambda)$, if X has type at least λ^k then $\pi(X)$ has type at least θ^k.*

Proof. The proof is a double induction on k and on n. For $k = 1$, the statement is true for all n by Claim 1.

For the induction step, assume the statement is true for some fixed k and for all $n \ge k$. Next we prove by induction on $n \ge k+1$ that the statement is true for $k+1$.

Let n be a positive integer greater than or equal to $k+1$ and if n is strictly greater than $k+1$, assume that the statement is true for $k+1$ and all m with $k+1 \le m < n$. Suppose that Y is a subset of $W_n(\lambda)$ of type at least λ^{k+1}.

For the first case, assume that the set L of all $\alpha < \theta$ so that type $Y_{A(\alpha)}$ has type at least λ^k is unbounded in θ. Then by Claim 2, for α in L, the set $^-Y_{A(\alpha)}$ has type at least λ^k. By the induction hypothesis, for each α in L, the image $\pi(^-Y_{A(\alpha)})$ has type at least θ^k. Since $\pi(Y_{A(\alpha)})$ is simply $\{\alpha\} \times \pi(^-Y_{A(\alpha)})$, it follows in this case that $\pi(Y)$ has type at least θ^{k+1}.

For the second case, assume that the set L is bounded by ν in θ. Then Y is the union over α with $\mu_\alpha < \nu$ of the sets $Y_{A(\alpha)}$. Since λ^{k+1} has cofinality θ and Y is the sum of fewer than θ sets each of type less than λ^{k+1}, it follows that for some η with $\mu_\eta < \nu$, the set $Y_{A(\eta)}$ has type at least λ^{k+1}. If $n = k+1$, then $Y_{A(\eta)}$ has type at most $\lambda^k \cdot \mu_{\eta+1}$ which is strictly less than λ^{k+1}. The contradiction shows that if $n = k+1$, this case cannot happen and the basis set of the induction on n is completed. Thus we may assume that n is strictly greater than $k+1$ and that the statement is true for all m with $k+1 \le m < n$. We apply the induction hypothesis to $^-Y_{A(\eta)}$, and observe that $\pi(^-Y_{A(\eta)})$ has type at least θ^{k+1}. Since

$\pi(Y_{A(\eta)})$ is simply $\{\eta\} \times \pi(^-Y_{A(\eta)})$, it follows that $\pi(Y)$ has type at least θ^{k+1}. Thus the claim is true for n and $k+1$.

Hence by induction on n, the claim is true for all n greater than or equal to $k+1$. Therefore, by induction on k, the claim is true.

Claim 3 completes the proof of Theorem 2.4, since it proves that π is the required pinning map.   □

We conclude this section with a discussion of the general negative results for finite powers of cardinals that can be gotten from pinning.

We start the discussion with a list of results about finite powers of ω and in particular those of the form $\omega^n \nrightarrow (\omega^3, m)^2$, which was studied by many people. E. Specker [18] produced an interesting example which shows $\omega^3 \nrightarrow (\omega^3, 3)^2$. E.C. Milner [12] showed that the example is unique in a sense that we discuss in §5. Others involved in the investigation were Galvin, Hajnal (see [3]), Haddad and Sabbagh (see [9]) and Milner (see [12], [13]). The culmination of this work is the following theorem of Nosal, a proof of which appears in [19] starting on page 164.

Theorem 2.6 (Nosal [15,16]) *For all positive integers t, $\omega^{t+2} \to (\omega^3, 2^t)^2$ and $\omega^{t+2} \nrightarrow (\omega^3, 2^t + 1)^2$.*

The canonical counter-example showing $\omega^3 \nrightarrow (\omega^3, 3)^2$ generalizes immediately to all regular cardinals and hence by Corollary 2.5 to all cardinals. We can also use Corollary 2.5 to lift Theorem 1.3 to cardinals whose cofinality is a successor cardinal. For simplicity in our discussion, at this point we will only lift the case $k = 2$ of Theorem 1.3 which gives a target of θ^3.

Theorem 2.7 *Assume that λ is a cardinal and ℓ and t are positive integers. Then*

(1) $\lambda^{t+2} \nrightarrow (\lambda^3, 2^t + 1)^2$;

(2) *if the cofinality of λ is a successor cardinal $\theta = \mu^+ = 2^\mu$, then $\lambda^{\ell \cdot 2} \nrightarrow (\lambda^3, \ell+1)^2$.*

Since $t + 2 \le \ell \cdot 2$ for $\ell = \lceil (t+2)/2 \rceil$, the second conclusion is stronger when it applies except when $\ell = 1$ and $t = 0$, in which case the two are identical.

We continue our discussion with a result of Nosal about finite powers of ω, of the form $\omega^n \to (\omega^m, p)^2$, for $m \ge 5$.

Theorem 2.8 (Nosal [17]) *For all positive integers m and n such that $5 \le m \le n$, $\omega^n \to (\omega^m, 2^{\lfloor (n-1)/(m-1) \rfloor})^2$ and $\omega^n \nrightarrow (\omega^m, 2^{\lfloor (n-1)/(m-1) \rfloor} + 1)^2$.*

Nosal's counter-examples are canonical and generalize immediately to all regular cardinals. So we again use Corollary 2.5 on them and on the part of Theorem 1.3 that applies to the situation.

Theorem 2.9 *Assume that λ is a cardinal and k and ℓ are positive integers with $k \geq 4$. Then*

(1) $\lambda^{k\ell} \nrightarrow (\lambda^{k+1}, 2^{\ell-1} + 1)^2$;

(2) *if the cofinality of λ is a successor cardinal $\theta = \mu^+ = 2^\mu$, then*

$$\lambda^{k\ell} \nrightarrow (\lambda^{k+1}, \ell + 1)^2.$$

For $\ell = 1, 2$, the results match, but for larger values of ℓ, the part with the more restrictive hypothesis gives the better result.

The remaining case to consider has us looking at results of the form $\omega^n \to (\omega^4, p)^2$. In her thesis, Nosal [16] proved $\omega^6 \nrightarrow (\omega^4, 3)^2$. Thus three is the least positive integer p for which $\omega^6 \nrightarrow (\omega^4, p)^2$. In his thesis, Darby [2] determined that for $n = 7, 8$, the least positive integer p for which $\omega^n \nrightarrow (\omega^4, p)^2$ is $p = 5$, and for $n = 10$ the least positive integer p is $p = 9$. He also showed $\omega^9 \to (\omega^4, 5)^2$. More generally, he proved that if $\omega^n \nrightarrow (\omega^4, p)^2$ then $\omega^{n+2} \nrightarrow (\omega^4, 2p - 1)^2$. A corollary to this result is the following theorem which represents the best general negative result currently known for this situation.

Theorem 2.10 (Darby [2]) *For all positive integers t, $\omega^{2t+4} \nrightarrow (\omega^4, 2^t + 1)^2$.*

Unfortunately, in the comparable positive doubling result by Erdős and Milner, the initial exponent n goes up by threes where for the above result the initial exponent n goes up by twos. Thus more work remains to be done even for finite powers of ω. The results again lift immediately to regular cardinals. The theorem below is the application of Corollary 2.5 on Darby's result lifted to regular cardinals and on the part of Theorem 1.3 that applies to the situation.

Theorem 2.11 *Assume that λ is a cardinal and that t and ℓ are positive integers. Then*

(1) $\lambda^{2t+4} \nrightarrow (\lambda^4, 2^t + 1)^2$;

(2) *if the cofinality of λ is a successor cardinal $\theta = \mu^+ = 2^\mu$, then $\lambda^{3\ell} \nrightarrow (\lambda^4, \ell+1)^2$.*

If we let $n = 6s + 6$, then $n = 2(3s + 1) + 4 = 3(2s + 2)$, so the first part of the theorem gives $\lambda^n \nrightarrow (\lambda^4, 2^{3s+1} + 1)^2$, while the second part gives $\lambda^n \nrightarrow (\lambda^4, 2s + 3)^2$.

In summary, these results give general bounds on the size p of the finite set one can get in a partition relation $\lambda^n \to (\lambda^m, p)^2$, when λ is a cardinal and n and m are finite positive integers with $m > 2$.

3   General positive results

In this section, under the assumption of the Generalized Continuum Hypothesis (GCH), we prove a general positive partition relation for finite ordinal powers of cardinal numbers, $\lambda^{k\ell+1} \to (\lambda^{k+1}, \ell + 1)^2$. We start by using the technique of jumping around to prove it for regular cardinals, and for this result we do not need the GCH.

Theorem 3.1 *Suppose that λ is a regular cardinal. Then $\lambda^{k\ell+1} \to (\lambda^{k+1}, \ell+1)^2$ for all positive integers k and ℓ.*

Proof. The proof is by induction on ℓ. For $\ell = 1$, the result is true by the definition of the partition relation. For the induction step, assume the result is true for m and prove it for $\ell = m + 1$.

Case 1 There is a vertex of valence at least λ^{km+1}.

Let x_0 be the vertex of valence at least λ^{km+1} and let W be the set of vertices to which x_0 is joined. By the induction hypothesis, either W has an independent set of type λ^{k+1} and we are immediately done, or it has a complete set of size $m + 1 = \ell$ to which we adjoin x_0 to get the required complete set.

Case 2 Every vertex has valence less than λ^{km+1}.

Partition the underlying set into the sum of λ^k many sets A_ξ of type λ^{km+1}. Enumerate λ^k in order type λ as $\{\xi(\alpha) \mid \alpha < \lambda\}$ so that every element of λ^k occurs λ many times. We plan to "jump around" to choose λ many points from each set A_ξ by recursion. Suppose that $\{a_{\xi(\alpha)} \mid \alpha < \beta\}$ have been chosen with $a_{\xi(\alpha)}$ from $A_{\xi(\alpha)}$. Since every vertex has valence less than λ^{km+1} and fewer than λ points have been chosen, there is some point $a_{\xi(\beta)}$ in $A_{\xi(\beta)}$ which is not joined to any of the points chosen so far. At the end of the recursion we have chosen λ many point from each A_ξ. Therefore the set X consisting of all these points has type $\lambda \cdot \lambda^k = \lambda^{k+1}$ and is the independent set required to prove the partition relation. $\square$

Our next goal is to use the technique of "canonization" to reduce questions about partitions of finite powers of a singular strong limit cardinal to questions about partitions for the corresponding power of the cofinality. Before we start, we introduce some terminology that will be useful in the discussion. Erdős, Hajnal and Rado [6] have defined the notion of a canonical sequence of subsets of a singular cardinal for a coloring of the r-tuples. Since we will be discussing more than one notion of canonicity, we will qualify this notion with the word "singularly".

Definition 3.2 Let $\mathcal{A} = \langle A_\xi \mid \xi < \theta \rangle$ be a sequence of pairwise disjoint subsets of a set W and write A for the union of the sequence and let $f : [W]^r \to \tau$ be a coloring. We say that $\mathcal{A}$ is *(singularly) canonical* with respect to f if $f(u) = f(v)$ whenever u and v are in $[A]^r$ and $|u \cap A_\xi| = |v \cap A_\xi|$ for every $\xi < \theta$.

We shall further assume without loss of generality that whenever $\mathcal{A}$ is a (singularly) canonical sequence of subsets of λ for a coloring f then also the following conditions are satisfied:

(1) the sequence $\mathcal{A}$ is $\ll$-increasing where for sets X and Y we write $X \ll Y$ to mean that every element of X is less than every element of Y;

(2) for all $\xi < \theta$, the size of set A_ξ is a regular cardinal; and

(3) the sequence is increasing in the cardinality of the sets.

If a sequence fails to satisfy these extra conditions, then we can thin it further to satisfy these demands as well.

Rather than quote the full-blown General Canonization Lemma of Erdős, Hajnal and Rado [6] and [5], we give only the corollary that we need (see Corollary 28.2 of [5]).

Theorem 3.3 (Singular Cardinals Canonization) *Let λ be a singular strong limit cardinal of cofinality θ and let $\langle \lambda_\xi \mid \xi < \theta \rangle$ be an increasing sequence of cardinals cofinal in λ. Let r be a positive integer, $\tau < \lambda$ a cardinal and $f : [\lambda]^r \to \tau$ a coloring. Then there is a sequence $\mathcal{B} = \langle B_\xi \mid \theta \rangle$ of pairwise disjoint subsets of λ with $B_\xi \subseteq \lambda_\xi$ so that $\bigcup \mathcal{B}$ has cardinality λ and $\mathcal{B}$ is (singularly) canonical with respect to f.*

The proof of the main theorem will be further delayed while we make some more preparations. We would like to use the Canonization Lemma to make a coloring of the pairs from the finite power of a singular cardinal λ^n look as much as possible like one inherited from a coloring of the power of its cofinality θ^n. In order to state the lemma nicely we use the terminology *similar* defined below from the study of canonical partitions of finite powers of regular cardinals (a coloring of the pairs from θ^n is *canonical* if similar pairs receive the same color).

Definition 3.4 Suppose that θ is a regular cardinal. Two pairs $\{\vec{a}, \vec{b}\}_<$ and $\{\vec{c}, \vec{d}\}_<$ from $[\theta]^n_<$ are said to be *similar*, written $\{\vec{a}, \vec{b}\} \sim \{\vec{c}, \vec{d}\}$, if for all i and j less than n,

$$\vec{a}(i) < \vec{b}(j) \iff \vec{c}(i) < \vec{d}(j) \quad \text{and} \quad \vec{a}(i) > \vec{b}(j) \iff \vec{c}(i) > \vec{d}(i).$$

It is not difficult to see that $\sim$ is an equivalence relation with only finitely many equivalence classes, which we will refer to as *similarity types*. Hajnal and later Galvin independently proved the existence and prevalence of canonical partitions of finite powers of ω by showing that for any complete graph on ω^n colored with m colors, there is an infinite set N so that on sequences from N the coloring is canonical. See Theorem 7.2.7 of [19] for a proof of this result. It extends in the obvious way to finite powers of weakly compact cardinals.

Definition 3.5 Suppose that λ is a singular cardinal of cofinality θ and that n is a positive integer. Let $\mathcal{A} = \langle A_\xi \mid \xi < \theta \rangle$ be a $\ll$-increasing sequence of pairwise disjoint subsets of λ so that $A = \bigcup \mathcal{A}$ has cardinality λ and let $f : [W]^2 \to \tau$ be a coloring where $W = [\lambda]^n_<$. We say that $\mathcal{A}$ is $\sim$-*canonical* for f if $f(x, y) = f(u, v)$ whenever $\{x, y\}_<$ and $\{u, v\}_<$ are similar pairs from A with $|x \cap A_\xi| = |u \cap A_\xi|$ and $|y \cap A_\xi| = |v \cap A_\xi|$ for all $\xi < \theta$.

Lemma 3.6 (Canonization for λ^n) *Suppose that λ is a singular strong limit cardinal of cofinality θ, that n is a positive integer and that f is a coloring of the pairs of $W = [\lambda]^n_<$ with τ colors. Then there is a $\ll$-increasing sequence $\mathcal{B} = \langle B_\xi \mid \xi < \theta \rangle$ of pairwise disjoint subsets of λ with $B = \bigcup \mathcal{B}$ of cardinality λ which is $\sim$-canonical for f.*

Proof. The similarity type of a pair $\{x, y\}_<$ from $[\lambda]^n_<$ can be matched with the pair of functions (u, v) so that, for $x \cup y$ enumerated in increasing order as $\langle z_j \mid j < |x \cup y| \rangle$, the sequence x is $(z_{u(0)}, z_{u(1)}, \ldots, z_{u(n-1)})$ and the sequence y is $(z_{v(0)}, z_{v(1)}, \ldots, z_{v(n-1)})$. For each such pair of functions (u, v), define $g_{(u,v)} : [\lambda]^{2n}_< \to 2$ by

$$g_{(u,v)}(a_0, a_1, \ldots, a_{2n-1})(u, v) = f((a_{u(0)}, a_{u(1)}, \ldots, a_{u(n-1)}), (a_{v(0)}, a_{v(1)}, \ldots, a_{v(n-1)})).$$

Let $\mathcal{U}$ be the set of all pairs of increasing functions from n into $2n$. Code the family of functions $g_{(u,v)}$ for (u, v) in U into a single function g by the usual trick of replacing the original range 2 by a new range consisting of the functions from U into 2, and defining g so that on the (u, v) coordinate, it gives the value of $g_{(u,v)}$.

Let $\langle \lambda_\xi \mid \xi < \theta \rangle$ be an increasing sequence of cardinals cofinal in λ. By the Canonization Lemma, there is a sequence $\mathcal{B} = \langle B_\xi \mid \theta \rangle$ of pairwise disjoint subsets of λ with $B_\xi \subseteq \lambda_\xi$ so that $\bigcup \mathcal{B}$ has cardinality λ and $\mathcal{B}$ is canonical with respect to g.

Suppose that $\{x, y\}_<$ and $\{s, t\}_<$ are similar pairs from B with $|x \cap A_\xi| = |u \cap A_\xi|$ and $|y \cap A_\xi| = |v \cap A_\xi|$ for all $\xi < \theta$. Let (u, v) be the pair from $\mathcal{U}$ matched with $\{x, y\}_<$. Since $\{s, t\}_<$ is similar it is also matched with (u, v). Let w be chosen from $\bigcup \mathcal{B}$ of length $2n - |x \cup y|$ so that $x \cup y \ll w$ and $s \cup t \ll w$. Notice that $f(x, y) = g_{(u,v)}(x \cup y \cup w)$ and $f(s, t) = g_{(u,v)}(s \cup t \cup w)$. Since for every $\xi < \theta$, $|x \cap A_\xi| = |u \cap A_\xi|$ and $|y \cap A_\xi| = |v \cap A_\xi|$, and since the two pairs are similar, we see that $|(x \cup y \cup w) \cap B_\xi| = |(s \cup t \cup w) \cap B_\xi|$ for every $\xi < \theta$. Since $\mathcal{B}$ is canonical for g, hence for $g_{(u,v)}$, it follows that $g_{(u,v)}(x \cup y \cup w) = g_{(u,v)}(s \cup t \cup w)$, so $f(x, y) = f(s, t)$ as desired. Therefore $\mathcal{B}$ is $\sim$-canonical for f. $\square$

We are now ready to prove the main theorem of this section. We state it without the assumption of GCH, but note that all cardinals fall into one of the two classifications below under that assumption.

Theorem 3.7 *Suppose that λ is either a regular cardinal or a strong limit cardinal and further suppose that k and ℓ are positive integers. Then $\lambda^{k\ell+1} \to (\lambda^{k+1}, \ell + 1)^2$.*

Proof. For regular cardinals, see Theorem 3.1. Suppose that λ is a singular strong limit cardinal of cofinality θ.

We plan to work by induction, so notice that if $\ell = 1$ then for all k, the theorem is trivially true.

Suppose that the theorem is true for some fixed k and a particular m. We must show it is also true for k and $\ell = m + 1$. Let $W = [\lambda]^n_<$ be the set of increasing sequences from λ of length $n = k\ell + 1 = km + k + 1$ and suppose $f : [W]^2 \to 2$ is any coloring.

By Lemma 3.6 (Canonization for λ^n), there is a $\ll$-increasing sequence $\mathcal{B} = \langle B_\xi \mid \xi < \theta \rangle$ of pairwise disjoint subsets of λ with $B = \bigcup \mathcal{B}$ of cardinality λ which is $\sim$-canonical for f.

Let D be the collection of all sequences of length n whose elements are chosen from $<$, $>$ and $=$. Let $\equiv$ be the sequence of all $=$'s. For each element $\diamond$ of D, define colorings $g_\diamond : [\theta]^n_< \to 2$ for $\diamond \neq \equiv$ and $h_\diamond : [[\theta]^n_<]^2 \to 2$ by $g_\diamond(\vec{\eta}) = f(x, y)$ and $h_\diamond(\vec{\eta}, \vec{\zeta}) = f(x, z)$ where $x = (x_0, x_1, \ldots, x_{n-1})$, $y = (y_0, y_1, \ldots, y_{n-1})$ and $z = (z_0, z_1, \ldots, z_{n-1})$ are chosen so that $x_i \diamond_i y_i$ are both in $B_{\vec{\eta}(i)}$ and z_i is in $B_{\vec{\zeta}(i)}$ for all $i < n$, so that for all $\eta < \zeta$, whenever $\vec{\eta}(i) = \vec{\zeta}(j)$, then $x_i \diamond_i z_j$. Since $\mathcal{B}$ is canonical and the definition of $g_\diamond$ for $\diamond \neq \equiv$ is based on pairs that meet each $B_{\vec{\xi}}$ in sets of the same size, $g_\diamond$ is well-defined. Define $g_\equiv$ on $[\theta]^n$ to be the constantly 0 function. Since $\mathcal{B}$ is canonical for λ^n and any two pairs $\{x, z\}$ and $\{u, w\}$ selected according to the above criteria are similar and have both pairs x, u and z, w meeting each B_ξ in the same size set, the coloring $h_\diamond$ is also well-defined.

The argument at this point breaks into cases.

Case 1 There is some $\diamond = (\diamond_0, \diamond_1, \ldots, \diamond_{n-1})$ in D and some σ in $[\theta]^n_<$ so that $g_\diamond(\sigma) = 1$.

Suppose that p is a positive integer. For each $i < n$, choose $u_0(i), u_1(i), u_2(i), \ldots, u_{p-1}(i)$ from $B_{\sigma(i)}$ so that $u_0(i) \diamond_i u_1(i) \diamond_i u_2(i) \diamond_i \ldots \diamond_i u_{p-1}(i)$. For each $j < p$ let

$$\vec{u}_j = (u_j(0), u_j(1), \ldots, u_j(n-1)).$$

Then for $i < j < p$, $f(\vec{u}_i, \vec{u}_j) = g_\diamond(\sigma) = 1$. Thus the graph has a complete set $P \subseteq W$ of size p. Since p was arbitrary, in this case we are done.

Case 2 There are an element $\vec{\eta}$ in $[\theta]^n_<$, an element $\diamond$ of D and a set $S \subseteq [\theta]^n_<$ of type at least θ^{km+1} so that for all $\vec{\zeta}$ in S, $h_\diamond(\vec{\eta}, \vec{\zeta}) = 1$.

Since θ^{km+1} is indecomposable, there is a subset T of S of type at least θ^{km+1} so that either $\vec{\eta} < \vec{\zeta}$ for all $\vec{\zeta} \in T$ or $\vec{\zeta} < \vec{\eta}$ for all $\vec{\zeta} \in T$. Since there are only finitely many similarity types of pairs of sequences of length n, there is a subset U of T of type at least θ^{km+1} so that for all $\vec{\zeta}$ and $\vec{\zeta}'$ in Z', the pairs $\{\vec{\eta}, \vec{\zeta}\}$ and $\{\vec{\eta}, \vec{\zeta}'\}$ are similar. Since θ^{km+1} cannot be split into fewer than θ subsets of type less than θ^{km+1}, and there are fewer than θ many finite sequences of elements less than or equal to $\vec{\eta}(n-1) = \max \vec{\eta}$, there are $q < n$, a sequence $\vec{\sigma}$ of length q and a subset V of U of type at least θ^{km+1}, so that for all $\vec{\zeta}$ from V regarded as an increasing function from n into θ, the restriction of $\vec{\zeta}$ is $\vec{\zeta}|_q = \vec{\sigma}$. Notice that since $\vec{\sigma}$ and $\vec{\eta}$ are both increasing, for any i there is at most one j and for any j there is at most one i with $\vec{\eta}(i) = \vec{\sigma}(j)$. We would like to define a sequence x and a set W' of type λ^{km+1} so that each y in W' meets each B_ξ in at most one point and has the further property that the set or increasing sequence $\vec{\zeta} = \pi(y) = \{\xi \mid y \cap B_\xi \neq \emptyset\}$ is in V and $\pi(x) = \eta$. Furthermore, we want to guarantee that for y in W', we have $f(x,y) = h_\diamond(\pi(x), \pi(y))$. To that end, we construct x with $\pi(x) = \eta$ and W' so that any two elements of W' agree on their first q elements with a sequence s chosen so that $\pi(s) = \sigma$ and x and s fit the pattern decreed by $\diamond$. In particular, choose $x = (x_0, x_1, \ldots, x_{n-1})$ and $s = (s_0, s_1, \ldots, s_{q-1})$ so that $x_i \in B_{\vec{\eta}(i)}$ for all $i < n$ and $s_i \in B_{\vec{\sigma}(i)}$ for all $i < q$, and whenever $\vec{\eta}(i) = \vec{\sigma}(j)$, then also $x_i \diamond_i s_j$ in the case that $\vec{\eta} < \vec{\zeta}$ for all ζ in V and $s_j \diamond_j x_i$ in the case that $\vec{\zeta} < \vec{\eta}$ for all ζ in V.

Let W' be the set of all increasing sequences y from B of length n which have s as an initial segment, meet each B_ξ in at most one point and have the further property that the set or increasing sequence $\vec{\zeta} = \pi(y) = \{\xi \mid y \cap B_\xi \neq \emptyset\}$ is in V. Since V has type at least θ^{km+1} and all sequences in V have σ as an initial segment, W' has type at least λ^{km+1}. By the definition of $h_\diamond$, for any y in W', we have $f(x,y) = h_\diamond(\eta, \pi(y)) = 1$.

By the induction hypothesis, either W' has an independent set of type λ^{k+1} and we are immediately done, or it has a complete set of size $m + 1 = \ell$ to which we adjoin x to get the required complete set.

Case 3 For every σ in $[\theta]^n_<$ and every $\diamond$ in D, $g_\diamond(\sigma) = 0$ and for every element $\vec{\eta}$ in $[\theta]^n_<$ and every $\diamond$ in D, the set $S(x, \diamond) \subseteq [\theta]^n_<$ of all $\vec{\zeta}$ for which $h_\diamond(\vec{\xi}, \vec{\zeta}) = 1$ has type less than θ^{km+1}.

Modify the "jumping around" argument of Case 2 of Theorem 3.1 to get a set $S \subseteq [\theta]^n_<$ of type at least θ^{k+1} so that $h_\diamond(\vec{\eta}, \vec{\zeta}) = 0$ for all of the finitely many functions $h_\diamond$ for $\diamond$ in D and all the pairs $\vec{\eta}, \vec{\zeta}$ from S. Let W' be the set of all sequences y from B of length n so that the set or increasing sequence $\vec{\zeta} = \pi(y) = \{\xi \mid y \cap B_\xi \neq \emptyset\}$ in S. Since for any pair $\{x,y\}_<$ from W', there is a sequence $\diamond$ in D so that either $\pi(x) = \pi(y)$ and $f(x,y) = g_\diamond(\pi(x), \pi(y)) = 0$ or $\pi(x) \neq \pi(y)$ and $f(x,y) = h_\diamond(\pi(x), \pi(y)) = 0$, it follows that W' is an independent set. Since W' has type at least λ^{k+1}, it is the set required to prove the theorem.

These three cases complete the proof of the inductive step of the theorem, which thereby follows. $\square$

4   A Positive Result for Cardinals of Weakly Compact Cofinality

In this section we use the canonization of the previous section and modify the proof of a partition relation for finite powers of ω to deduce positive results for finite powers of cardinals of weakly compact cardinality. Some open questions remain for the finite powers of such cardinals.

Our techniques are similar to those of Baumgartner [1] in his proof that if λ is a singular strong limit cardinal whose cofinality is weakly compact, then $\lambda^2 \to (\lambda^2, 3)^2$.

The result for ω that we would like to lift is one of Erdős and Milner (see [7] or Milner's thesis): for all finite ℓ and all countable μ, $\omega^{1+\mu\ell} \to (\omega^{1+\mu}, 2^\ell)^2$.

If λ is itself weakly compact and $\mu = k$ is finite, then one can lift the result immediately to $\lambda^{k\ell+1}$ by using the fact that $\lambda \to (\lambda)_n^m$ in each place where Ramsey's Theorem is used for ω. Thus the interesting case of the next theorem is the singular cardinals case.

Theorem 4.1 *Assume that λ is either a weakly compact cardinal or a singular strong limit cardinal of weakly compact cofinality and further assume that k and ℓ are positive integers. Then $\lambda^{k\ell+1} \to (\lambda^{k+1}, 2^\ell)^2$.*

Proof. Since the proof for weakly compact cardinals is routine and we will basically repeat it in the singular cardinals case, we will omit it. Thus assume that λ is a singular strong limit cardinal whose cofinality, θ, is a weakly compact cardinal (we include the possibility that $\theta = \omega$). The proof is by induction on ℓ.

The basis case with $\ell = 1$ is trivially true.

To start the induction step, assume that $\ell = m + 1$ and that $\lambda^{km+1} \to (\omega^{k+1}, 2^m)^2$. For notational convenience, let $n = k\ell + 1$ and let $W = [\lambda]^n$. Suppose that $f : [W]^2 \to 2$ is a graph. Use Lemma 3.6 to get $\mathcal{B} = \{ B_\xi \mid \xi < \theta \}$ which is $\sim$-canonical for f. Furthermore, assume that $\mathcal{B}$ has been thinned if necessary to satisfy the constraints listed in the previous section. Let $B = \bigcup \mathcal{B}$, and let D be the set of all sequences of length n whose elements are chosen from $=$, $<$ and $>$. As before, let $\equiv$ be the sequence of all $=$'s. Define $g_\equiv$ on $[\theta]^n$ to be the constantly 0 function. For each element $\diamond$ of D, define colorings $g_\diamond : [\theta]^n_< \to 2$ for $\diamond \neq \equiv$ and $h_\diamond : [[\theta]^n_<]^2 \to 2$ by $g_\diamond(\vec\eta) = f(x,y)$ and $h_\diamond(\vec\eta, \vec\zeta) = f(x,z)$ where $x = (x_0, x_1, \ldots, x_{n-1})$, $y = (y_0, y_1, \ldots, y_{n-1})$ and $z = (z_0, z_1, \ldots, z_{n-1})$ are chosen so that $x_i \diamond_i y_i$ are both in $B_{\vec\eta(i)}$ and z_i is in $B_{\vec\zeta(i)}$ for all $i < n$, so that for all $\eta < \zeta$, whenever $\vec\eta(i) = \vec\zeta(j)$, then $x_i \diamond_i z_j$. As we saw in the previous section, since $\mathcal{B}$ is canonical for λ^n, $g_\diamond$ and $h_\diamond$ are well-defined.

Since θ is weakly compact (or $\theta = \omega$), we can find a set $L \subseteq \theta$ of cardinality θ so that all the functions $g_\diamond$ are canonical on $[L]^n_<$ and all the functions $h_\diamond$ are canonical on $[[L]^n]^2_<$.

At this point the proof breaks into three cases depending on the homogeneous values of the functions $g_\diamond$ and $h_\diamond$.

Case 1 There is some $\diamond = (\diamond_0, \diamond_1, \ldots, \diamond_{n-1})$ in D and some σ in $[L]^n_<$ so that $g_\diamond(\sigma) = 1$.

Use the argument of Case 1 of Theorem 3.7 to show that f has arbitrarily large complete sets.

Case 2 For all $\diamond$ in D, $g_\diamond$ is constantly 0 on $[L]^n_<$ but there are $\diamond$ in D and a pair $\{\sigma, \tau\} \subseteq [L]^n$ with $h_\diamond(\sigma, \tau) = 1$ so that for some $p, q \leq k$ the following conditions hold:

(1) $\sigma(n-1) < \tau(q)$;

(2) if $q = 0$, then $p = 0$;

(3) if $q > 0$, then $\tau(q - 1) < \sigma(p)$; and

(4) if $p > 0$, then $\sigma(p - 1) \leq \tau(q - 1)$.

In this case we will show that f has a complete set of size 2^ℓ. For notational convenience, let us call p and q satisfying the above conditions *tail parameters* for σ, τ. We will refer to such parameters in the next case.

In the current case, there are two subcases that differ only slightly, depending on whether $\sigma < \tau$ or $\tau < \sigma$. One of these must occur since σ and τ are different. For simplicity we will only treat the subcase in which $\sigma < \tau$, and leave the other to the reader.

Let $s = (s_0, s_1, \ldots, s_{p-1})$ and $t = (t_0, t_1, \ldots, t_{q-1})$ be chosen so that $s_i \in B_{\sigma(i)}$, $t_j \in B_{\tau(j)}$ and if $\sigma(i) = \tau(j)$, then $s_i \diamond_i t_j$. (Note to the reader: in the case $\tau < \sigma$, this requirement becomes $t_j \diamond_j s_i$.) Let S be the set of all those x in $[B]^n_<$ which have s as an initial segment and which meet each B_ξ in at most one point. Since $p \leq k$, S has order type at least λ^{km+1}. Apply the induction hypothesis to S. If we get an independent set of type λ^{k+1}, then we are done. Assume instead that we get a complete set $U \subseteq S$ of size 2^m.

Let ξ_0 be greater than every ξ with $B_\xi \cap (\bigcup U) \neq \emptyset$. Since the set of ξ's for which B_ξ meets $\bigcup U$ is finite, there is always such a ξ_0. Let T be the set of all those y in $[B]^n_<$ which have t as an initial segment, which meet each B_ξ in at most one point and satisfy the condition that if $y \cap B_\xi \neq \emptyset$ for $\xi < \xi_0$, then ξ is one of $\tau(0), \tau(1), \ldots, \tau(q-1)$. Since $q \leq k$, T has order type at least λ^{km+1}. Apply the induction hypothesis to T. As before, if we get an independent set of type λ^{k+1}, then we are done. Assume instead that we get a complete set $V \subseteq T$ of size 2^m.

We plan to show that $S \cup T$ is the desired complete set. It certainly has size $2^m + 2^m = 2^\ell$. Since S and T are complete, we need only show that *cross pairs* are joined by edges. For any z in $S \cup T$, let $\pi(z) = \{\xi \mid z \cap B_\xi \neq \emptyset\}$. Let $x \in S$ and $y \in T$ be arbitrary. Note that $\{\pi(x), \pi(y)\}$ is similar to $\{\sigma, \tau\}$ by construction. Since s is an initial segment of x and t is an initial segment of y, $f(x, y) = h_\diamond(\pi(x), \pi(y))$ by the definition of $h_\diamond$. Since $h_\diamond$ is canonical on $[L]^n_<$, it follows that $f(x, y) = h_\diamond(\sigma, \tau) = 1$. Thus $S \cup T$ is the complete set of size 2^ℓ required to prove this case.

Case 3 For all $\diamond$ in D, $g_\diamond$ is constantly 0 on $[L]^n_<$ and $h_\diamond$ is constantly 0 on all pairs $\{\sigma, \tau\}$ for which there are tail parameters $p_{\sigma,\tau}, q_{\sigma,\tau} \leq k$.

In the two previous cases we found a complete set of size 2^ℓ if the induction hypothesis did not give us a large independent set. In this case we build an independent set of size λ^{k+1}.

As a first step we will build a set $M \subseteq [L]^n_{\not<}$ of type θ^{k+1} so that all pairs from M admit tail parameters less than or equal to k.

Since the cardinality of L is regular, we can write L as the increasing sum of θ many sets $\mathcal{L} = \{ L_\alpha \mid \alpha < \theta \}$ each of size n. Since $[\theta]^{\leq n}$ has cardinality θ and $\theta \cdot \theta = \theta$, we can partition $\mathcal{L} = \bigcup \{ M_{\vec{\mu}} \mid \vec{\mu} \in [\theta]^{\leq n} \}$ into the disjoint union of θ many sets each of size θ indexed by finite sequences from θ of length at most n. Let M be the set of all increasing sequences $(\xi_0, \xi_1, \ldots, \xi_{k-1}) * \vec{p}$ where ξ_0 is the least element of some member of $M_\emptyset$, ξ_{i+1} is the least element of some member of $M_{(\xi_0, \xi_1, \ldots, \xi_i)}$ for all $i < k - 1$, and $\vec{p}$ is the first $n - k = km + 1$ elements of $M_{(\xi_0, \xi_1, \ldots, \xi_{k-1})}$. The set M has order type θ^{k+1} by construction. Suppose that σ and τ are in M. Notice their intersection is an initial segment of both σ and τ. We call this property *polite*. Furthermore either $(\sigma(k), \ldots, \sigma(n-1)) \ll (\tau(k), \ldots, \tau(n-1))$ or $(\tau(k), \ldots, \tau(n-1)) \ll (\sigma(k), \ldots, \sigma(n-1))$. These two facts guarantee that $\{\sigma, \tau\}$ admits tail parameters.

Let Z be the set of all sequences z in $[B]^n_{\not<}$ for which $\pi(z) = \{ \xi \mid z \cap B_\xi \neq \emptyset \}$ is in M. Since M has type θ^{k+1}, it follows that Z has type λ^{k+1}. For any x, y in Z, since $\{\pi(x), \pi(y)\} \subseteq M$ admit tail parameters, $f(x, y) = h_\diamond(\pi(x), \pi(y)) = 0$ for some suitable choice of $\diamond$. Thus Z is the desired large independent set.

Case 3 concludes the induction step of the theorem. In each case we got either a complete set of size 2^ℓ or an independent set of size λ^{k+1}. Therefore by induction the theorem follows.

5 Another Positive Result for Cardinals of Weakly Compact Cofinality

The second result for ω that we would like to lift is the following result (see page 164 of [19]): for all positive integers t, $\omega^{2+t} \to (\omega^3, 2^t)^2$.

If λ is itself weakly compact, then one can lift this result immediately to λ^{2+t} by using the fact that $\lambda \to (\lambda)^m_n$ in each place where Ramsey's Theorem is used for ω. Thus the interesting case of the next theorem is the singular cardinals case.

Theorem 5.1 *Assume that λ is either a weakly compact cardinal or a singular strong limit cardinal of weakly compact cofinality and further assume that t is a positive integer. Then $\lambda^{2+t} \to (\lambda^3, 2^t)^2$.*

Milner [12] has shown that the canonical counter-example for $\omega^3 \not\to (\lambda^3, 3)^2$ is unique in the following sense. Given any counter-example $f : [[\omega]^3_{\not<}]^2 \to 2$, we can find an infinite set $L \subseteq \omega$ so that f is canonical on pairs from $[L]^3_{\not<}$. By a procedure like that outlined at the end of the last proof, we can define a set $M \subseteq [L]^3_{\not<}$ so that all pairs are polite. Milner has shown that any pair $a = (a_0, a_1, a_2) < (b_0, b_1, b_2) = b$ from M which does satisfy the inequality

$$(*): \quad a_0 < a_1 < b_0 < a_2 < b_1 < b_2$$

is joined in the graph while any pair that does not satisfy (∗) cannot be joined in the graph, since one can build an arbitrarily large set of triples from L so that any pair from it is similar to $\{a, b\}$, which would contradict the fact that the graph is a counter-example.

We would like to extend this description to weakly compact cardinals and larger finite powers. First, for notational convenience, given a sequence $\sigma = (\sigma_0, \sigma_1, \ldots, \sigma_{n-1})$, let us write $\sigma[i, j)$ for the subsequence $(\sigma_i, \sigma_{i+1}, \ldots, \sigma_{j-1})$. Recall $A \ll B$ means $\max A < \min B$.

The next definition describes the kind of pairs we will need to consider in looking at sets of type θ^3, where θ is regular.

Definition 5.2 For integers $0 \leq i < j < k < n$, a pair $\{\sigma, \tau\} \subseteq [\theta]^n_{\lessgtr}$ *fits* (i, j, k)-*blocks* if σ, τ are polite, $\sigma[0, i) = \tau[0, i)$, and for any two sequences from the following list, either they are equal or they are $\ll$-comparable:

$$\sigma[i, j), \quad \sigma[j, k), \quad \sigma[k, n), \quad \tau[i, j), \quad \tau[j, k), \quad \tau[k, n).$$

Lemma 5.3 *Suppose that $n \geq 3$ is an integer, θ is a regular cardinal and $h : [[\theta]^n_{\lessgtr}]^2 \to 2$ is a canonical graph. Further suppose there is some pair $\{\sigma, \tau\}$ with $h(\sigma, \tau) = 1$ and some $0 \leq i < j < k < n$ so that $\{\sigma, \tau\}$ fits (i, j, k)-blocks and σ, τ do not satisfy the following inequality:*

$$(∗∗): \quad \sigma[0, i) = \tau[0, i) < \sigma[i, j) < \sigma[j, k) < \tau[i, j) < \sigma[k, n) < \tau[j, k) < \tau[k, n).$$

Then h has arbitrarily large complete sets in which any pair is similar to $\{\sigma, \tau\}$.

Proof. Given a graph on $[\theta]^n_{\lessgtr}$ and a pair $\{\sigma, \tau\}$, define a canonical graph on $[\omega]^3_{\lessgtr}$ by joining only pairs similar to $\{(\sigma(i), \sigma(j), \sigma(k)), (\tau(i), \tau(j), \tau(k))\}$. Let p be an arbitrary positive integer. Use Milner's result to get a complete set $P \subseteq [\omega]^3_{\lessgtr}$ of size p. By construction, any pair from P must be similar to $\{(\sigma(i), \sigma(j), \sigma(k)), (\tau(i), \tau(j), \tau(k))\}$. Blow up the points of $\bigcup P$ to subsequences and use these subsequences along with $\sigma[0, i)$ to build a complete set $Q \subseteq [\theta]^n_{\lessgtr}$ of size p. □

Now we extend to canonical partitions of singular cardinals of weakly compact cofinality.

Lemma 5.4 *Suppose that $n \geq 3$ is a positive integer, λ is a singular cardinal of weakly compact cofinality and $f : [[\lambda]^n_{\lessgtr}]^2 \to 2$ is a graph for which $B = \{B_\xi \mid \xi < \theta\}$ is $\sim$-canonical. Further suppose there is some pair $\{s, t\} \subseteq [\bigcup B]^n_{\lessgtr}$ with $f(s, t) = 1$ and some $0 \leq i < j < k < n$ so that $\{s, t\}$ fits (i, j, k)-blocks, both s and t meet each B_ξ in at most one point and together s, t fail to satisfy the inequality (∗∗) of Lemma 5.3. Then f has arbitrarily large complete sets in which any pair is similar to $\{s, t\}$.*

Proof. Let θ be the cofinality of λ. Let $\sigma = \pi(s)$ and $\tau = \pi(t)$ be defined as before. We saw in Case 1 of the proof of Theorem 3.7 that if $\sigma = \tau$, then there are arbitrarily large complete sets. So suppose that $\sigma \neq \tau$. Let $\diamond = (\diamond_0, \diamond_1, \ldots, \diamond_{n-1})$ be a sequence of $=$, $<$ and $>$'s of length n so that if $\sigma < \tau$, then for all $i < n$ $\diamond_i$ is chosen so that if $s(i)$ and $t(j)$ are both in B_ξ then $s(i) \diamond_i t(j)$, and if $\tau < \sigma$, then for all $i < n$ $\diamond_i$ is chosen so that if $s(j)$ and $t(i)$ are both in B_ξ then $t(i) \diamond_i s(j)$. Define $h_\diamond$ on $[\theta]^n$ by $h_\diamond(\vec{\eta}, \vec{\zeta}) = 1$ if and only if $\{\vec{\eta}, \vec{\zeta}\}$ is similar to $\{\sigma, \tau\}$.

By the previous lemma, there are arbitrarily large complete sets $P \subseteq [\theta]^n_{<}$ so that every pair from P is similar to $\{\sigma, \tau\}$. In particular, note that since s, t is polite, all the sequences in P have a common initial segment and are otherwise disjoint. For each u in P build a sequence $\rho(u)$ from $[\lambda]^n_{<}$ by choosing $\rho(u)(i)$ from $B_{u(i)}$ and using $\diamond$ on the common initial segment so that the set Q of all the sequences $\rho(u)$ has the property that any pair from it is similar to $\{s, t\}$. Since $\mathcal{B}$ is canonical for f, Q is a complete set. By this method we can construct arbitrarily large complete sets for f.   □

We are now ready to prove Theorem 5.1.

Proof of Theorem 5.1. The proof starts the same way as the induction step of the proof of Theorem 4.1. Namely we start with $n = 2 + t$ and with a graph $f : [[\lambda]^n]^2 \to 2$, find a family $\mathcal{B}$ which is $\sim$-canonical for f, use it to define families of functions $g_\diamond$ on $[\tau]^n$ and $h_\diamond$ on $[[\theta]^n]^2$ and find $L \subseteq \theta$ which makes all of these canonical. Let $B = \bigcup \mathcal{B}$.

Claim 1 Suppose that there is some $\diamond = (\diamond_0, \diamond_1, \ldots, \diamond_{n-1})$ in D and some σ in $[L]^n_{<}$ so that $g_\diamond(\sigma) = 1$. Then f has arbitrarily large complete sets.

Use the argument of Case 1 of Theorem 3.7.

Claim 2 Suppose there are a pair $\{\sigma, \tau\} \subseteq [L]^n_{<}$, a choice of $0 \le i < j < k < n$ so that $\{\sigma, \tau\}$ fits (i, j, k)-blocks, and some $h_\diamond$ so that $h_\diamond(\sigma, \tau) = 1$ and σ, τ fail to satisfy the inequality (∗∗) of Lemma 5.3. Then f has arbitrarily large complete sets.

The function $h_\diamond$ can be lifted back to a function $H_\diamond$ on $[B]^n$ which joins only those pairs s, t, which fit the definition of $h_\diamond$. That is, $H_\diamond$ joins only those s, t in for which $\pi(s)$ and $\pi(t)$ are distinct subsets of L, $h_\diamond(\pi(s), \pi(t)) = 1$ and if $\pi(s) < \pi(t)$, then whenever $\pi(s)(i) = \pi(t)(j)$ is true, also $s(i) \diamond_i t(j)$ is true, and if $\pi(t) < \pi(s)$, then whenever $\pi(s)(i) = \pi(t)(j)$ is true, also $t(j) \diamond_j s(i)$ is true. All these pairs are also joined by f. Thus Lemma 5.4 gives the conclusion that f has arbitrarily large complete sets.

Let $\equiv$ be the constantly $=$ sequence.

Claim 3 Suppose $h_\equiv(\sigma, \tau) = 1$ for every pair $\{\sigma, \tau\} \subseteq [L]^n_{<}$ which fits (i, j, k)-blocks for some $0 \le i < j < k < n$ and also satisfies the inequality (∗∗) of Lemma 5.3. Then there is a complete set of size 2^t.

By a theorem of Nosal (see [15] and page 162 of [19]), modified to the weakly compact case, there is a complete set $Y \subseteq [L]^n$ of size 2^t for $h_\equiv$ in which any pair from Y is polite in the terminology of Section 4, fits (i, j, k)-blocks for some suitable choice of $0 \le i < j < k < n$ and also satisfies the inequality (∗∗) of Lemma 5.3. Let X be the set of all sequences x in $[B]^n_{<}$ with $\pi(x)$ in Y so that if $x_i \in B_\xi$, then $x_i = \min B_\xi$. Since Y has size 2^t and π is a one-to-one mapping between Y and X, it follows that X has size 2^t. Notice that for any pair x, y from X, the pair $\pi(x)$, $\pi(y)$ is similar to x, y. Thus by construction, $f(x, y) = h_\equiv(\pi(x), \pi(y))$ for all x, y in X. In other words, X is the complete set required to prove the claim.

Claim 4 Suppose that none of the hypotheses of Claims 1 to 3 hold. Then there is an independent set of type λ^3.

Since the hypothesis of Claim 1 does not hold, $g_\diamond$ is constantly 0 on $[L]^n_{<}$ for all $\diamond$ in D.

Since the hypothesis of Claim 2 does not hold, $h_\circ$ has value 0 on all pairs ρ, χ from $[L]^n$ which fit (p, q, r)-blocks for some $0 \leq p < q < r < n$ but fail to satisfy the inequality $(**)$ of Lemma 5.3. Moreover, since the hypothesis of Claim 3 does not hold, there is a pair $\{\sigma, \tau\} \subseteq [L]^n_<$ which fits (i, j, k)-blocks for some $0 \leq i < j < k < n$, which satisfies the inequality $(**)$ of Lemma 5.3 and also has $h_\equiv(\sigma, \tau) = 0$.

In this case we build a set $M \subseteq [\theta]^n_<$ of type θ^3 so that every pair from M fits (i, j, k)-blocks as follows. Construct $\mathcal{M} = \{ M_{\vec{v}} \mid \vec{v} \in [\theta]^{<n} \}$ as in Case 3 of Theorem 4.1 so that each set $M_{\vec{v}}$ has size n. Let M be the set of all increasing sequences $\vec{u} * \vec{v} * \vec{w} * \vec{z}$ where $\vec{u}$ is the first i elements of the least member of $M_\emptyset$, $\vec{v}$ is the first $j - i$ elements of some member of $M_{\vec{u}}$, $\vec{w}$ is the first $k - j$ elements of some member of $M_{\vec{u}*\vec{v}}$, and $\vec{z}$ is the first $n - k$ elements of some member of $M_{\vec{u}*\vec{v}*\vec{w}}$. This construction guarantees that every pair from M fits (i, j, k)-blocks.

Let Z be the set of all sequences z in $[B]^n_<$ for which $\pi(z) = \{\xi \mid z \cap B_\xi \neq \emptyset\}$ is in M. Then Z has type λ^3 since M has type θ^3 and all sequences in Z have a common initial segment of length i. Suppose $\{x, y\}$ is an arbitrary pair from Z. If $\pi(x) = \pi(y)$, then $f(x, y) = g_\circ(\pi(x)) = 0$ for some $\diamond$ since the hypothesis of Claim 1 is not satisfied. Otherwise $\pi(x) \neq \pi(y)$ and $\{\pi(x), \pi(y)\}$ fits (i, j, k)-blocks. If the pair satisfies $(**)$, then $f(x, y) = h_\equiv(\pi(x), \pi(y)) = 0$ since $\{\pi(x), \pi(y)\}$ must be similar to $\{\sigma, \tau\}$. If it fails to satisfy $(**)$, then $f(x, y) = h_\circ(\pi(x), \pi(y)) = 0$ for some $\diamond$ since the hypothesis of Claim 2 is not satisfied. Since $\{x, y\}$ was arbitrary, we have shown that Z is the required independent set.

The four claims we considered constitute four cases. In the first three, we were able to construct a complete set of size at least 2^t. In the last case, to which we default when none of the other claims is pertinent, we were able to construct an independent set of type λ^3 as required to finish the proof of Theorem 5.1.   □

6   Directions for Further Research

Since Baumgartner, Hajnal and Todorčević treat elsewhere in this volume those partition relations of the form $\lambda \to (\alpha, \beta)^2$ where λ is a cardinal and α and β are ordinals, we omit a discussion of them here.

Instead we turn to the next interesting question about finite powers of cardinals: for which cardinals λ does $\lambda^2 \to (\lambda^2, 3)^2$? Baumgartner [1] has solved this problem completely in the constructible universe, showing $\lambda^2 \to (\lambda^2, 3)^2$ if and only if the cofinality of λ is weakly compact. The proof is via the theorem of Jensen [11] which says that cofinality of λ is weakly compact if and only if the λ-Souslin hypothesis holds. Baumgartner builds a counter-example from a Souslin tree.

Such counter-examples can be used to show that for appropriate λ, for every n there is some p so that $\lambda^n \not\to (\lambda^2, p)^2$, but that is beyond the scope of the present paper. It remains open to determine the function $f(\lambda, p)$ defined in L to be the least power n so that the above relation holds. The first interesting case would be for the first inaccessible cardinal.

If we move out of the Constructible Universe, then there are many more questions. Is it consistent that $\omega_1^2 \to (\omega_1^2, 3)^2$? Baumgartner stated this problem among others in his paper in 1975 [1]. Since that time, he has shown using MA that it is consistent to assume $\omega_1 \cdot \omega \to (\omega_1 \cdot \omega, 3)^2$ and $\omega_1 \cdot \omega^2 \to (\omega_1 \cdot \omega^2, 3)^2$.

References

[1] J.E. Baumgartner, Partition relations for uncountable ordinals, *Israel J. Math.* **21** (1975), 296–307.

[2] C. Darby, Countable Ramsey games and partition relations, Ph.D. Dissertation, University of Colorado, 1990.

[3] P. Erdős and A. Hajnal, Unsolved problems in set theory, in *Axiomatic Set Theory (Proc. Sympos. Pure Math., Univ. of Calif., Los Angeles, Ca., 1967)*, D.S. Scott, eds., AMS, Providence, RI, **13 (I)**, 1971, 17–48.

[4] P. Erdős and A. Hajnal, Ordinary partition relations for ordinal numbers, *Periodica Math. Hung.* **1** (1971), 171–185.

[5] P. Erdős, A. Hajnal, A. Máté and R. Rado, *Combinatorial Set Theory: Partition Relations for Cardinals*, North Holland, Amsterdam, 1984.

[6] P. Erdős, A. Hajnal and R. Rado, Partition relations for cardinal numbers, *Acta Math. Acad. Sci. Hungar.* **16** (1965), 93–196.

[7] P. Erdős and E.C. Milner, A theorem in the partition calculus, *Canad. Math. Bull.* **15** (1972), 501–505.

[8] Corrigendum, *ibid*, **17** (1974), 305.

[9] L. Haddad and G. Sabbagh, Nouveaux résultats sur les nombres de Ramsey generalisés, *C.R. Acad. Sci. Paris Sér. A-B* **268** (1969), A1516–A1518.

[10] A. Hajnal, A negative partition relation, *Proc. Nat. Acad. Sci. USA* **68** (1971), 142–144.

[11] R. Jensen, The fine structure of the constructible hierarchy, *Ann. Math. Logic* **4** (1972), 229–308.

[12] E.C. Milner, A finite algorithm for the partition calculus, in *Proc. of the Twenty-Fifth Summer Meeting of the Canadian Mathematical Congress*, Lakehead Univ., Thunder Bay, Ont., W.R. Eames et al., eds., (1981), 117–128.

[13] E.C. Milner, Partition relations for ordinal numbers, *Canad. J. Math.* **21** (1969), 317–334.

[14] E.C. Milner and R. Rado, The pigeon-hole principle for ordinal numbers, *Proc. London Math. Soc. (3)* **15** (1965), 750–768.

[15] E. Nosal, On a partition relation for ordinal numbers, *J. London Math. Soc. Ser. 2* **8** (1974), 306–310.

[16] E. Nosal, On arrow relations $\omega^n \to (k, \omega^m)^2$, Ph. D. Dissertation, University of Calgary, 1974.

[17] E. Nosal, Partition relations for denumerable ordinals, *J. Comb. Th. (B)* **27** (1979), 190–197.

[18] E. Specker, Teilmengen von Mengen mit Relationen, *Comment. Math. Helv.* **31** (1957), 302–314.

[19] N. Williams, *Combinatorial Set Theory*, North Holland, Amsterdam, 1977.

Some Subdirect Products of Finite Nilpotent Groups

DAVID M. EVANS

School of Mathematics,
University of East Anglia,
Norwich NR4 7TJ, England

1 Introduction

The starting point for this paper is the following intriguing lemma of E. Hrushovski:

Lemma 1.1 ([2] Lemma 4.11) *Let I be an indexing set and $(G_i : i \in I)$ a sequence of groups. Suppose there exists a subgroup H of the direct product $G = \Pi_{i \in I} G_i$ with the following two properties:*

> (i) *for all $i, j \in I$ the restriction to H of the natural projection from G to $G_i \times G_j$ is onto, i.e. $\{(f(i), f(j)) : f \in H\} = G_i \times G_j$;*
>
> (ii) *for all $i \in I$ there exists a finite subset $J(i)$ of I containing i such that if $f \in H$ and $f(j) = 1$ for all $j \in J(i) \setminus \{i\}$ then $f(i) = 1$.*

Then each G_i is nilpotent of class at most $|J(i)| - 2$.

This result arises in the context of a "trivial" proof of the nilpotency of certain groups (the groups $\mathrm{Aut}(M_{i,0}/M_{i-1})$ defined in the proof of 4.8 of [2]) arising in the analysis of disintegrated totally categorical structures. (The reader should be aware that there are difficulties with this "trivial" proof. The lemma shows only that the groups $\mathrm{Aut}(M_{i,0}/M_{i-1})$ are nilpotent-by-abelian [1].) In this context there is an additional constraint on the groups G_i (and the subgroup H):

> (iii) the indexing set I is the set of n-subsets from a set Ω (for some $n \in \mathbf{N}$) and $\mathrm{Sym}(\Omega)$ acts on G so that it permutes the coordinates in the same way as $\mathrm{Sym}(\Omega)$ acts on the n-subsets from Ω. Moreover, H is stabilised by this action of $\mathrm{Sym}(\Omega)$.

Hrushovski raises the question as to whether there exist finite *non-abelian* groups G_i and a subgroup H of their direct product satisfying (i), (ii), and (iii). In this paper we construct such groups (the additional constraint (iii) enables us to give a very straightforward proof via dimension counting that condition (ii) holds in our groups (see Lemma 3.4)). We can take Ω to be infinite, and the G_i can have any nilpotency class less than $n/2$ (cf. Theorem 3.5). We do not know whether this class can be increased (or indeed whether it is bounded in terms of n).

The groups H which we construct fit into the original model theoretic context, but the results here are disappointingly negative. Let M be the canonical structure associated with

N.W. Sauer et al. (eds.), Finite and Infinite Combinatorics in Sets and Logic, 117–124.
© 1993 *Kluwer Academic Publishers. Printed in the Netherlands.*

the semidirect product $H]\mathrm{Sym}(\Omega)$. So M is of course disintegrated, totally categorical and of Morley rank n. If we now compute the groups $\mathrm{Aut}(M_{i,0}/M_{i-1})$ for $2 \leq i < n$, then we find that these are all *abelian*: the nilpotency built into the automorphism group of M is absorbed by imaginary elements of rank less than n. So we have not settled the question in 4.9 of [2] as to whether the groups $\mathrm{Aut}(M_{i,0}/M_{i-1})$ can ever be non-abelian. We omit the details of these computations.

2   Constructing the finite groups

The aim of this section is to construct, for each $n \in \mathbb{N}$, a finite 2-group which is nilpotent of class n with generators $g_1, \ldots, g_n$ such that $\mathrm{Sym}(n)$ acts as a group of automorphisms of the group by permuting these generators. Quotients of this group will be used as the groups G_i.

We construct the nilpotent groups via a Lie algebra.

Notation: For $n \in \mathbb{N}$ we denote by $\underline{n}$ the set $\{1, 2, \ldots, n\}$. If $k \leq n$ then $[\underline{n}]^k$ is the set of k-sets from $\underline{n}$. F is the field with two elements. If X is a set, then FX denotes the F-vector space with basis X. From now on, fix $n \geq 3$.

Construction: For k with $3 \leq k \leq n$ let $w \in [\underline{n}]^k$. Define $R(w)$ to be the subspace of $F[w]^2$ given by

$$R(w) = \langle \sum \{w' : w' \in [w'']^2\} : w'' \in [w]^3 \rangle.$$

Let $X_k = \bigoplus_{w \in [\underline{n}]^k} (F[w]^2/R(w))$. Define $X_1 = F\underline{n}$ and $X_2 = F[\underline{n}]^2$. Set $V_n = X_1 \oplus X_2 \oplus \ldots \oplus X_n$. We now define an operation $[,] : V_n \times V_n \to V_n$. $[,]$ is to be (skew) symmetric, trace zero (i.e. $[a, a] = 0$ for all $a \in V_n$), and bilinear. We set $[X_i, X_j] = \{0\}$ if $i, j \geq 2$. For $w' + R(w) \in X_k$ with $k \geq 3$ and $w' \in [w]^2$ and $a \in \underline{n}$ let

$$[a, w' + R(w)] = \begin{cases} 0 & \text{if } a \in w, \\ w' + R(w \cup \{a\}) & \text{if } a \notin w. \end{cases}$$

For $a, b \in \underline{n}$ with $a \neq b$

$$[a, b] = \{a, b\}.$$

For $a, b, c \in \underline{n}$ and $b \neq c$

$$[a, \{b, c\}] = \begin{cases} 0 & \text{if } a \in \{b, c\}, \\ \{b, c\} + R(\{a, b, c\}) & \text{otherwise.} \end{cases}$$

Lemma 2.1 *The partial operation $[,]$ defined above extends in a unique fashion to a (skew) symmetric, trace zero bilinear operation on V_n. With this operation, V_n is a Lie algebra.*

Proof. The uniqueness in the first statement is clear, because the operation has already been defined on pairs from a spanning set for V_n. Existence is either obvious, or via the following formal argument. [Notation: $\mathcal{P}(X)$ denotes the power set of X, for a set X.]

Let

$$V^* = X_1 \oplus X_2 \oplus \left(\bigoplus \{\{w\} \times F[w]^2 : w \in \mathcal{P}(\underline{n}), |w| \geq 3\} \right)$$

and regard V_n as the quotient of this by

$$W^* = \{0\} \oplus \{0\} \oplus (\bigoplus\{\{w\} \times R(w) : w \in \mathcal{P}(\underline{n}), |w| \geq 3\}).$$

Define $[[,]] : V^* \times V^* \to V^*$ to be the unique skew symmetric, trace zero bilinear operation satisfying $[[x, y]] = 0$ unless one of x, y has a component in X_1;

$$[[k, (w, \{a, b\})]] = \begin{cases} (w \cup \{k\}, \{a, b\}) & \text{if } k \notin w, \\ 0 & \text{otherwise,} \end{cases}$$

where $k \in \underline{n}$, $\{a, b\} \in [w]^2$ and $|w| \geq 3$;

$$[[k, \{a, b\}]] = \begin{cases} (\{a, b, k\}, \{a, b\}) & \text{if } k \notin \{a, b\}, \\ 0 & \text{otherwise.} \end{cases}$$

$[[k, l]] = \{k, l\}$ if $k \neq l$, where $k, l \in \underline{n}$.

Note that $[[W^*, W^*]] = \{0\}$. Furthermore, if we (temporarily) regard $F[w]^2$ as a subset of $F[w \cup \{k\}]^2$ then $R(w) \subseteq R(w \cup \{k\})$. So $[[X_1, W^*]] \subseteq W^*$. It follows that $[[,]]$ induces a well-defined skew symmetric bilinear operation on V^*/W^*, and this gives us the desired operation on V_n.

To check that V_n is a Lie algebra, it only remains to verify the Jacobi identity:

$$[v, [u, w]] + [w, [v, u]] + [u, [w, v]] = 0 \quad \text{for all } u, v, w \in V_n.$$

It is sufficient to do this when u, v, w come from a spanning set for V_n, and by definition of $[,]$ we may assume that at least two of u, v, w are in X_1.

The cases to consider are these:

$u, v, w \in \underline{n}$, ALL DISTINCT: the left hand side of the equation becomes

$$(\{u, w\} + R(\{u, v, w\})) + (\{v, u\} + R(\{u, v, w\})) + (\{v, w\} + R(\{u, v, w\})$$

which is indeed zero in V_n;

$u \neq v \in X_1$, $w = w' + R(x)$: if either u or v is in x then the identity clearly holds, otherwise, the left hand side of the equation is $(w' + R(x \cup \{v, u\})) + 0 + (w' + R(x \cup \{v, u\}))$, which is zero;

$u = v \in X_1$: clearly zero.

So the Jacobi identity holds and we have a Lie algebra.

Notice that $[V_n, V_n] = [X_1, V_n] = \bigoplus_{i \geq 2} X_i$, $[V_n, \bigoplus_{i \geq j} X_i] = \bigoplus_{i \geq j+1} X_i$ and $[V_n, X_n] = \{0\}$. So V_n is nilpotent of class $n + 1$.

As usual, the adjoint map adj: $V_n \to gl(V_n)$ (where $gl(V_n)$ is the Lie algebra of endomorphisms of V_n) given by $\text{adj}(v)(w) = [v, w]$ is a homomorphism of Lie algebras. Moreover, by the nilpotency of V_n, $(\text{adj}(v))^n = 0$ for all $v \in V_n$, and so if we let $g_v : V_n \to V_n$ be given by $g_v = 1 + \text{adj}(v)$, then $g_v \in GL(V_n)$.

Let G_n be the subgroup of $GL(V_n)$ generated by $g_1, \ldots, g_n$.

Lemma 2.2 *The following results hold:*

(i) *each g_i is an involution;*

(ii) *if $\{j,k\} \in [w]^2$ then*

$$[g_i, g_{\{j,k\}+R(w)}] = \begin{cases} 1 & \text{if } i \in w, \\ g_{\{j,k\}+R(w\cup\{i\})} & \text{otherwise;} \end{cases}$$

(iii) *G_n is nilpotent of class n.*

Proof. Notice that if $i_1, \ldots, i_r \in \underline{n}$ and two of the i_s are equal, then $\text{adj}(i_1)\ldots\text{adj}(i_r)$ is the zero transformation. Thus $(1 + adj(i))^2 = 1$ whence (i) is proved.

Also, $\text{adj}(\{j,k\} + R(w))|\bigoplus_{i\geq 2} X_i = 0$ and

$$\text{adj}(\{j,k\} + R(w))(i) = \begin{cases} 0 & \text{if } i \in w, \\ \{j,k\} + R(w \cup \{i\}) & \text{otherwise.} \end{cases}$$

So $(g_{\{j,k\}+R(w)})^2 = 1$. The commutator in (ii) is therefore

$$((1 + \text{adj}(i))(1 + \text{adj}(g_{\{j,k\}+R(w)})))^2$$
$$= (1 + \text{adj}(i) + \text{adj}(g_{\{j,k\}+R(w)}) + \text{adj}(i)\text{adj}(g_{\{j,k\}+R(w)}))^2$$
$$= 1 + \text{adj}([i, g_{\{j,k\}+R(w)}]),$$

as required for (ii). (Notice that we used here the fact that $\text{adj}(v)\text{adj}(w) + \text{adj}(w)\text{adj}(v) = \text{adj}([v,w])$.)

For the last part, let $G_n^{(i)} = \langle g_{w'+R(w)} : w' \in [w]^2 \text{ and } |w| \geq i \rangle$. Then $G_n^{(i+1)} \leq G_n^{(i)}$ and by (ii) $G_n^{(i)}$ is normal in G_n. Clearly $G_n^{(n)} = 1$. Again by (ii), we have $[G_n, G_n^{(i)}] = G_n^{(i+1)}$. Furthermore $[\text{adj}(1), [\text{adj}(2), [\ldots, \text{adj}(n-1)]\ldots]] \neq 0$, so $G_n^{(n-1)} \neq 1$, and we have (iii).

Notation: $G_n^{(i)} = \langle g_{w'+R(w)} : w' \in [w]^2 \text{ and } |w| \geq i \rangle$.

Lemma 2.3 *$G_n^{(i)}/G_n^{(i+1)}$ is isomorphic to X_i for $i = 1, \ldots, n-1$.*

Proof. Each of these factor groups is an elementary abelian 2-group. The only question is its size, i.e., whether or not there is some "collapse" amongst the generators.

Notice first that if $v_1, v_2 \in \bigoplus_{i\geq 2} X_i$ then $\text{adj}(v_1)\text{adj}(v_2) = 0$, so $g_{v_1}g_{v_2} = g_{v_1+v_2}$. Thus to demonstrate the lemma for $i \geq 2$ we need only show that if

$$\prod_{w\in[\underline{n}]^i} g_{v(w)+R(w)} \in G_n^{i+1} \qquad\qquad (*)$$

(where $v(w) \in F[w]^2$) then $v(w) \in R(w)$. We prove this for the case $w = \{1, \ldots, i\}$, without loss of generality.

Commutating $(*)$ with $i+1$ gives

$$\prod_{w\in[\underline{n}]^i, i+1\notin w} g_{v(w)+R(w)} \in G_n^{i+2}.$$

Repeating this successively with $i + 2, \ldots, n - 1$ gives

$$\prod_{w \in [\underline{n}]^i, i+1, \ldots, n-1 \not\subseteq w} g_{v(w)+R(w)} = 1.$$

Applying the expression on the left to $n \in X_1$ gives the equality in V_n

$$n + (v(w) + R(\underline{n})) = n.$$

Thus $v(w) \in R(\underline{n})$. By Lemma 2.4 below, $R(\underline{n}) \cap F[\underline{i}]^2 = R(\underline{i})$. Thus $v(w) \in R(w)$, as required.

Finally, we want to show that G_n^1/G_n^2 is elementary abelian of order 2^n. Suppose $g_1^{a_1}, \ldots, g_n^{a_n} \in G_n^2$ for some $a_1, \ldots, a_n \in \{0, 1\}$. Then working in G_n^2/G_n^3 we have $[g_1, g_1^{a_1} \cdots g_n^{a_n}]G_n^3 = G_n^3$. It follows (by standard commutator identities) that

$$[g_1, g_1^{a_1} \cdots g_{n-1}^{a_{n-1}}]g_{\{1,n\}}^{a_n}G_n^3 = G_n^3.$$

Continuing in this way, we get $g_{\{1,2\}}^{a_2} \cdots g_{\{1,n\}}^{a_n} \in G_n^3$ and so by what we have already proved, $a_2 = \cdots = a_n = 0$. But clearly then also $a_1 = 0$, as required.

Lemma 2.4 *Let $w \in [\underline{n}]^k$. For $a \in \underline{n} \setminus w$, consider $F[w]^2$ as a subset of $F[w \cup \{a\}]^2$. Then $R(w) = F[w]^2 \cap R(w \cup \{a\})$.*

Proof. Suppose $v = \sum\{\sum[w']^2 : w' \in A\} \in F[w]^2$ where $A \subseteq [w \cup \{a\}]^3$. We may assume that $a \in w'$ for each $w' \in A$. If $\{b, c\} \in \text{supp}(v)$ then $\{a, b, c\} \in A$ and there must be an even number of $w' \in A$ with $\{a, b\} \subseteq w'$. Construct a graph Γ with vertex set w and edge set $\{w' \setminus \{a\} : w' \in A\}$. Then by the above, every vertex in Γ has even valency. To show that $v \in R(w)$ it is enough to show that by adding triangles to Γ (in a Boolean fashion) we can reduce Γ to the null graph (i.e. no edges) on w. But adding a triangle to Γ preserves the property that all valencies are even. So we can isolate a vertex by adding an appropriate number of triangles at that vertex, and then proceed inductively.

Notation. For any group K let $Z_0(K) = 1$ and let $Z_{i+1}(K)$ be the subgroup of K which contains $Z_i(K)$ such that $Z(K/Z_i(K)) = Z_{i+1}(K)/Z_i(K)$.

Lemma 2.5 *For $i = 0, 1, \ldots, n - 1$ we have $Z_i(G_n) = G_n^{n-i}$.*

Proof. From Lemma 2.3, any $g \in G_n$ can be written in a unique way as

$$g = g_1^{a_1} \cdots g_n^{a_n} \prod_{i=2,\ldots,n} \left(\prod_{w \in [\underline{n}]^i} g_{v(w)+R(w)} \right). \qquad (**)$$

Now

$$g^{g_j} = g_1^{a_1} g_{\{1,j\}}^{a_1} \cdots g_n^{a_n} g_{\{n,j\}}^{a_n} \cdot \prod_{i=2,\ldots,n} \left(\prod_{w \in [\underline{n}]^i} g_{v(w)+R(w)} \right) \cdot \prod_{i=2,\ldots,n} \left(\prod_{w \in [\underline{n}]^i, j \not\subseteq w} g_{v(w)+R(w \cup \{j\})} \right).$$

Collecting this expression to the right, and comparing the result with the normal form in $(**)$ gives that if $g \in Z(G_n)$, then $a_i = 0$ for all $i \in \underline{n}$, and $v(w) \in R(w)$ for all $w \in [\underline{n}]^i$ with $2 \le i \le n - 1$. This proves the lemma for the case $i = 1$. The rest of the lemma is proved inductively in a similar fashion (using $(**)$).

Lemma 2.6 *There is an embedding* $\phi : \mathrm{Sym}(\underline{n}) \to \mathrm{Aut}(G_n)$ *with* $\phi(\sigma)g_i = g_{\sigma i}$.

Proof. There is clearly an embedding $\psi : \mathrm{Sym}(\underline{n}) \to \mathrm{Aut}(V_n)$ given by extending the map $\psi(\sigma)i = \sigma i$ in an obvious fashion. Now $im\psi$ normalises G_n and

$$\psi(\sigma)g_i\psi(\sigma)^{-1}j = \begin{cases} \{\sigma i, j\} + j & \text{if } \sigma i \neq j, \\ 0 + j & \text{otherwise.} \end{cases}$$

So $\psi(\sigma)g_i\psi(\sigma)^{-1} = g_{\sigma i}$.

3   Constructing a subdirect product

In this section Ω will be a (possibly infinite) set and $n \in \mathbb{N}$ such that $3 \leq n < |\Omega|/2$.

For each $w \in [\Omega]^n$ let G_w be a group with generators $\{g_{i,w} : i \in w\}$ which is isomorphic to the group G_n constructed in Section 2 (of course, we mean that the isomorphism should be induced by the map $g_i \mapsto g_{\phi i, w}$ where ϕ is some (any) bijection between $\underline{n}$ and w). Let $1 \leq k \leq n$.

Let $G_w^k = G_w/Z_{k-1}(G_w)$ and $G^k = \prod_{w \in [\Omega]^n} G_w^k$. Write $G = G^1$. There is an action of $\mathrm{Sym}(\Omega)$ (the symmetric group on Ω) on G given by $(f^\sigma)(w) = \sigma f(\sigma^{-1}w)$ for $\sigma \in \mathrm{Sym}(\Omega)$, $f \in G$, and $w \in [\Omega]^n$, where the action of σ on $\cup G_w$ is given by $\sigma g_{i,w} = g_{\sigma i, \sigma w}$ extended multiplicatively in a unique fashion. This clearly induces an action on G^k.

Construction: For $a, b \in \Omega$ with $a \neq b$ define $h_{a,b} \in G$ by

$$h_{a,b}(w) = \begin{cases} 1 & \text{if } a \notin w \text{ or } b \in w, \\ g_{a,w} & \text{if } a \in w \text{ and } b \notin w. \end{cases}$$

Let H be the subgroup of G generated by $\{h_{a,b} : a, b \in \Omega\}$. We denote by H^k the image of H in G^k. The image of $h_{a,b}$ in H^k will still be denoted by $h_{a,b}$.

Lemma 3.1 *The following results hold:*

(i) H^k *is* $\mathrm{Sym}(\Omega)$ *invariant;*

(ii) *if* $w, w' \in [\Omega]^n$ *are distinct then* $\{(h(w), h(w')) : h \in H^k\} = G_w^k \times G_{w'}^k$.

Proof.
(i) If $\sigma \in \mathrm{Sym}(\Omega)$ then

$$^\sigma h_{a,b}(w) = \sigma h_{a,b}(\sigma^{-1}w) = \left\{ \begin{array}{ll} 1 & \text{if } a \notin \sigma^{-1}w \text{ or } b \in \sigma^{-1}w, \\ g_{\sigma a, w} & \text{if } a \in \sigma^{-1}w \text{ and } b \notin \sigma^{-1}w. \end{array} \right\} = h_{\sigma a, \sigma b}(w).$$

(ii) Let $b \in w' \backslash w$ and $a \in w$. Then $h_{a,b}(w) = g_{a,w}$ and $h_{a,b}(w') = 1$.

For $x \subseteq \Omega$ and $K \leq G^k$ let $K(X) = \{f|[X]^n : f \in K\}$. We define the *support* of $f \in G^k(X)$ to be $\{w \in [X]^n : f(w) \neq 1\}$ and the *weight* of f to be the cardinality of its support.

Our ultimate aim is to show that if X is large enough then $H^k(X)$ has no elements of weight 1, provided that k is at least $n/2$.

Lemma 3.2 *The following results hold:*

(i) $Z(H^k(X)) = H^k(X) \cap Z(G^k(X)) = H^k(X) \cap (\prod_{w \in [X]^n} Z(G_w^k))$.

(ii) *if* $f \in H^k(X)$ *has weight 1, then there exists* $g \in Z(H^k(X))$ *of weight 1.*

Proof.

(i) Obvious.

(ii) Say $f(w) \neq 1$ with $w \in [X]^n$. By 3.1(i) $\{h(w) : h \in H^k(X)\} = G_w^k$ and so as G_w^k is nilpotent, by commutating f with an appropriate selection of elements of $H^k(X)$ we obtain an element of $Z(H^k(X))$ of weight 1.

For $A, B \subseteq \Omega$ with $2 \leq |A| \leq n$ and $v \in [A]^2$ let

$$h_{v,A,B}(w) = \begin{cases} g_{(v+R(A)),w} & \text{if } A \subseteq w \text{ and } B \cap w = \emptyset, \\ 1 & \text{otherwise.} \end{cases}$$

(We are adopting the notational convenience that $R(A) = 0$ if $|A| = 2$.) Then

$$[h_{a,b}, h_{a',b'}] = h_{\{a,a'\},\{a,a'\},\{b,b'\}}$$

if $a \neq a'$. More generally

$$[h_{a,b}, h_{v,A,B}] = \begin{cases} h_{v,A \cup \{a\}, B \cup \{b\}} & \text{if } a \notin A, \\ 1 & \text{otherwise.} \end{cases}$$

Lemma 3.3 $Z(H^k(X)) = \langle h_{v,A,B} | [X]^n : A \in [X]^{n-k}, v \in [A]^2, B \subseteq \Omega, |B| \leq n - k \rangle$.

Proof. It is clear from the commutation rule above that the group on the right is contained in $Z(H^k(X))$.

Suppose $g \in Z(H^k(X))$. Using the commutation rule, we can write

$$g = \prod_{i=1,\ldots,m} \left(\prod_{j=1,\ldots,m(i)} h_{v(i,j),A(i),B(i,j)} | [X]^n) \right),$$

where $|B(i,j)| \leq n - k$. Suppose there is some i with $|A(i)| < n - k$. Let $w \in [X]^n$ be such that $A(i) \subseteq w$. Then $g(w) \in Z_k(G_w)$ and so by Lemma 2.5 $\sum\{v(i,j) : B(i,j) \cap w = \emptyset\} \in R(A(i))$. Thus $\prod_{j=1,\ldots,m(i)} h_{v(i,j),A(i),B(i,j)} | [X]^n = 1$. This is what is required.

Lemma 3.4 *If* $k > n/2$ *and* $m = |X|$ *is sufficiently large, then* $Z(H^k(X))$ *has no element of weight 1.*

Proof. By Lemma 3.3, $Z(H^k(X))$ is an elementary abelian 2-group, and its dimension as a vector space over the field of two elements is at most $\binom{m}{n-k}^2 \binom{n-k}{2}$ (because in choosing a generator $h_{v,A,B} | [X]^n$ for $Z(H^k(X))$ there are $\binom{m}{n-k}$ choices for A, then $\binom{n-k}{2}$ choices for v, and (if m is large enough in comparison with $n - k$) at most a further $\binom{m}{n-k}$ choices for B). This is $O(m^{2(n-k)})$ (where $O(\ )$ is the usual order notation). However, if $Z(H^k(X))$ contains a vector of weight 1, because $Z(H^k(X))$ is $\text{Sym}(X)$-invariant, it contains a vector whose support is $\{w\}$ for any $w \in [X]^n$. So the dimension of $Z(H^k(X))$ is at least $\binom{m}{n}$. But this is $O(m^n)$. However, by choice of k, $2(n - k) < n$. Thus for large enough m we have a contradiction.

We can summarise what we have just proved as:

Theorem 3.5 *There exist non-abelian groups* $(g_i : i \in I)$ *and a subgroup* H *of their direct product which satisfy hypotheses* (i) *and* (ii) *of Lemma* 1.1. *The groups can be taken to also satisfy* (iii), *and have any nilpotency class less than* $n/2$.

Proof: Let Ω be an infinite set. Let $n \geq 3$ and $k > n/2$. In the notation of Lemma 1.1, let $I = [\Omega]^n$, let G_w be the group G_w^k defined at the beginning of this section, let $G = \prod_{w \in [\Omega]^n} G_w^k$, and let H be the group H^k introduced above. We claim that the groups G_w and H satisfy the conditions (i), (ii), and (iii) of the introduction: condition (i) is from 3.1(ii); condition (ii) is from 3.2(ii) and 3.4; condition (iii) is contained in the definition of G^k and 3.1(i); and the nilpotency class of G_w^k comes from 2.2(iii).

References

[1] David M. Evans and Ehud Hrushovski, On the automorphism groups of finite covers, to appear in *Annals Pure and Appl. Logic.*

[2] Ehud Hrushovski, Totally categorical structures, *Trans. Amer. Math. Soc.* **313** (1) 131–159.

Three Remarks on End-Faithfulness

GEŇA HAHN*

Université de Montréal

JOZEF ŠIRÁŇ†

Comenius University Bratislava

Abstract

We introduce the notion of a lexicographic tree and prove that its existence is equivalent to the existence of a normal tree in an infinite graph. Further, we show that for any free set of ends in a connected countable graph and for any choice of pairwise disjoint rays representing this set (one ray per end), there exists an end-faithful spanning tree in the graph which contains the chosen rays.

1 Introduction

The concept of an end in an (infinite) graph appeared first in Halin's 1964 paper [3]. Results of that paper include the fact that every countable connected graph has an end-faithful spanning tree, that is, a tree which represents the end structure of the graph. In 1967 Jung [4] showed that such a spanning tree can be chosen so that it has the property of depth-first-search trees, namely that no edge **not** in the tree connects different branches (trees with this property are called normal). Other proofs of Halin's basic result have since appeared; see, for example, the papers by Diestel [2] and Polat [6]. Halin's conjecture that every connected graph has an end-faithful spanning tree was only recently disproved, independently, by Thomassen [9] and by Seymour with Thomas [7].

Our aim is to explore other natural extensions of the ideas involved in this direction. We introduce the notion of a *lexicographic* tree and show that such a tree exists in a graph if and only if a normal one does and that it also is end-faithful. Then we prove that any countable set of rays representing a free subset of the ends of a graph can be included in an end-faithful free spanning tree. Finally, we show that under some conditions a prescribed set of ends can be "omitted" from a spanning tree in the sense that a spanning tree can be constructed in which all other ends are represented by rays while the prescribed ones are "rayless".

All our graphs will be simple, connected and infinite. For undefined graph-theoretic terms see [1]. A *ray* P in G is a one-way infinite path $p_0 p_1 p_2 \ldots$. For any $i < \omega$, an *i-tail* of P is a one-way infinite subpath $p_i p_{i+1} \ldots$ of P; most of the time the i is irrelevant and

---

*Partially supported by a grant from the NSERC

†Work done while visiting the Université de Montréal and supported by the NSERC research grant of the first author

N.W. Sauer et al. (eds.), Finite and Infinite Combinatorics in Sets and Logic, 125–133.
© 1993 *Kluwer Academic Publishers. Printed in the Netherlands.*

we speak simply of a *tail* of P. An *initial segment* of P can also be defined for each i and the index i can be safely ignored in most cases: it is $p_0 p_1 \ldots p_i$. Let P and Q be two rays in G. We say that P and Q are *equivalent*, $P \sim Q$, if there is a ray R which intersects both P and Q infinitely often. It is easy to see that this is equivalent to saying that for each finite set $U \subset V(G)$, all but finitely many tails of P and Q lie in the same component of $G - U$. This defines an equivalence relation on the set of rays in G; the equivalence classes of this relation are called *ends* of G. We will denote ends by Greek letters $\alpha, \beta, \ldots$ and the set of ends by $\mathcal{V}(G)$. By definition, any two distinct ends $\alpha, \beta \in \mathcal{V}(G)$ can be *separated* by a finite set of vertices in the sense that there exists a finite set $U \subset V(G)$ such that for all rays $P \in \alpha$, $Q \in \beta$ the tails of P and Q belong to different components of $G - U$. Let H be a subgraph of G. Observe that if two rays are equivalent in H, they are also equivalent in G. Thus for each end α of H there is a unique end $f(\alpha)$ of G such that $\alpha \subseteq f(\alpha)$. The mapping $f : \mathcal{V}(H) \to \mathcal{V}(G)$ is a natural one. It need be neither one-to-one nor onto. In case it is bijective, we say that H is *end-faithful* in G. This paper is interested in end-faithful trees and spanning trees.

2   Lexicographic trees

Let $\rho : V(G) \to \mu$ be a well-ordering of $V(G)$ by an ordinal μ. The pair (G, ρ) will be called a *labeled graph*; we write simply G when the ordering ρ is clear from the context. This allows us to look at rays as sequences of ordinals less than μ and to define a natural lexicographic order $\prec$ on the set of rays. That is, if P and Q are distinct rays with $P = (u_i)$, $Q = (v_i)$, $i < \omega$, then $P \prec Q$ if $\rho(u_i) \leq \rho(v_i)$ for the least i such that $u_i \neq v_i$.

Lemma 1 *Let S be a set of rays of a labeled graph (G, ρ). Then S has an infimum in (G, ρ), that is, there is a ray R in G such that $R \prec P$ for each $P \in S$ and if $Q \prec P$ for all $P \in S$ then $Q \prec R$.*

Proof. Throughout the proof the word "least" means "minimum in the well ordering given by ρ". Let v_0 be the least initial vertex among all the initial vertices of rays in S and let S_0 be the set of rays of S beginning at v_0. We construct a decreasing sequence of sets $S_0 \supset S_1 \supset \ldots$ of rays. If v_i and S_i have been constructed, let v_{i+1} be the least among the unique neighbours of v_i on the rays in S_i which are not v_{i-1}. Let S_{i+1} be the set of rays whose initial segment is $v_0, \ldots, v_{i+1}$. Let $R = (v_i)_{i<\omega}$. We claim that R is the infimum of S.

Clearly R is a ray. Note that it might be outside of S. By construction, $R \prec P$ for any $P \in S$. If Q is also a ray smaller than every ray in S, then in no coordinate can it be bigger than R lest it be bigger than some ray in S. $\square$

Corollary 1 *Every end in (G, ρ) has an infimum.* $\square$

It should be emphasized that different orderings usually lead to different infima. Also, it is important to note that the infimum of an end is not necessarily *in* the end. For an example see Figure 1. For obvious reasons we will call an infimum of an end α which lies in α the *minimum* of α. Labeled graphs in which every end has a minimum have the following nice property.

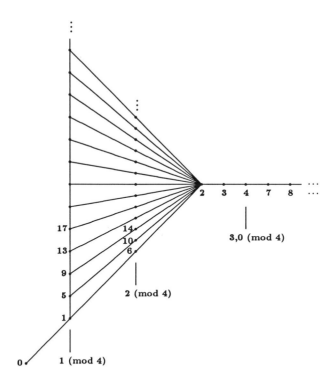

Figure 1

Proposition 1 *Let (G, ρ) be a labeled graph and suppose that every end α has a minimum $P_\alpha \in \alpha$. Then the graph T obtained as the union $\bigcup_{\alpha \in \mathcal{V}(G)} P_\alpha$ is a tree end-faithful in G.*

Proof. Since G is connected, each P_α begins at v_0 with $\rho(v_0)$ minimum. Hence T is connected. Also, every pair of rays shares a common initial segment. If no two rays meet at a point outside of the common initial segment, there cannot be any cycles. Suppose P and Q so meet and let v be the first point of intersection beyond the initial segment. More precisely, suppose $P = v_0, \ldots, v_i, v_{i+1}, \ldots, v_k, \ldots$ and $Q = v_0, \ldots, v_i, v'_{i+1}, \ldots, v'_l, \ldots$, with $v_k = v'_l = v$ and $\{v_{i+1}, \ldots, v_{k-1}\} \cap \{v'_{i+1}, \ldots, v'_{l-1}\} = \emptyset$. Now without loss of generality $\rho(v_{i+1}) < \rho(v'_{i+1})$ and so the ray $Q' = v_0, \ldots, v_i, v_{i+1}, \ldots, v_k, v'_{l+1}, \ldots$ is smaller than Q and belongs to the same end as Q, contradicting the minimality of Q. Thus T is a tree.

It remains to show that T is end-faithful. Clearly every end α of G is represented by P_α in T. Assume that T contains a ray Q which is equivalent to some P_α in G but not equivalent in T. Without loss of generality we may assume that Q begins at v_0 and let $v_0, \ldots, v$ be the longest common initial segment of P and Q and let $P = v_0, \ldots, v, u_1, \ldots$ and $Q = v_0, \ldots, v, w_1, \ldots$, with $u_i \neq w_j$ for all i, j. The edge vw_1 lies on some ray P_β for some $\beta \in \mathcal{V}(G)$. Now P_β is not equivalent to Q for otherwise $P_\beta \sim P_\alpha$ which is a contradiction. Thus P_β shares with Q only a finite segment and we may put $P_\beta = v_0, \ldots, v, w_1, \ldots, w_k, z_1, \ldots$ with $z_i \neq w_j$ for all i, j. Let $M = \{v_0, \ldots, v, u_1, w_1, \ldots, w_k\}$.

Since $Q \not\sim P_\beta$ in G, there is a finite set X of vertices which separates them. But $Q \sim P_\alpha$ in G, so $M \cup X$ cannot separate P_α from Q. Therefore there is a path B in G from some vertex $u_i \in P_\alpha$ to some vertex $w_j \in Q$, $u_i, w_j \notin M \cup X$, not containing any vertices of P_α and Q other than the endpoints and avoiding $M \cup X$. By the minimality of P_α, $u_1 < w_1$. But then the ray $v_0, \ldots, v, u_1, \ldots, u_i, B, w_j, w_{j-1}, \ldots, w_k, z_1, z_2, \ldots$ is a ray in β smaller than P_β, a contradiction.   □

We will call an end-faithful tree obtained as in Proposition 1 a *lexicographic* tree in (G, ρ). Its shape and even existence depend on the labeling of the vertices of G by μ. The lexicographic tree has another nice property. Following Jung [4], a *normal* tree (T, u) in a graph G is a rooted tree with a root u such that any pair of vertices $v, w \in V(T)$ adjacent in G lies on a path in T beginning at u.

Lemma 2 *Let v_0 be the least vertex of (G, ρ). If (G, ρ) has a lexicographic tree T then (T, v_0) is normal.*

Proof. If not, there are rays P and Q in T and an edge $uv \notin E(T)$ with both $u \in V(P)$ and $v \in V(Q)$ further from v_0 than the last vertex of the common initial segment of P and Q. This contradicts the minimality of either P or Q in a manner similar to that in the proof of Proposition 1.   □

Interestingly enough, the converse also holds.

Lemma 3 *Let (T, u) be a normal tree in G. Then there is an ordering ρ of $V(G)$ such that T is a lexicographic tree in (G, ρ).*

Proof. Define a new ordering ρ on $V(G)$ as follows. Denote by $d_T(x, y)$ the distance in T between x and y. For each $i < \omega$ let $V_i = \{v \in V(T) : d_T(v, u) = i\}$ and let $V_\omega = V(G) \backslash V(T)$. Let $\mu \geq |V(G)|$ be an ordinal and let, for each $i \leq \omega$, $\rho_i : V_i \to \mu$ be a well-ordering of V_i. For each $v \in V_i$ let $\rho(v) = (i, \rho_i(v))$; the well ordering of $V(G)$ is then given by $\rho(w) < \rho(v)$ if and only if $i < j$, or $i = j$ and $\rho_i(w) < \rho_i(v)$, for $w \in V_i$, $v \in V_j$. Note that ρ has the property that if v and w are vertices in T such that w is a vertex internal to the unique u–v path in T, then $\rho(v) > \rho(w)$.

We claim that each end has a minimum under the lexicographic ordering $\prec$ induced by ρ. Let α be an end. Jung's results [4] tell us that T contains a ray P_α from α and, moreover, every ray $Q \in \alpha$ intersects α infinitely often. We claim that under $\prec$ P_α is a minimum in α. To see this, let Q be any other ray in α and assume, without loss of generality, that Q begins at u (which is the least element under $\leq$). Let $u, v_1, \ldots, v_k$ be the longest common initial segment of Q and P_α. Let w be the vertex on Q following v_k; since Q begins at u, the edge $v_k w$ cannot belong to T, by normality. If w is a vertex in V_ω then clearly $P_\alpha \prec Q$. Otherwise w lies in T and, by the normality of T, v_k is an internal vertex of the unique u–w path on T. Thus, $\rho(v_k) < \rho(w)$ and so if v follows v_k on P_α then $\rho(v) \leq \rho(w)$. This shows that $P_\alpha \prec Q$.   □

We conclude that the existence of a normal tree in a graph G is equivalent to the existence of a lexicographic tree in some labeling of G. Thus the necessary and sufficient condition given by Jung [4] for the existence of a normal tree also applies to the existence of a lexicographic one.

Remark 1 We note that if G is countable, then the existence of an end-faithful tree guarantees the existence of an end-faithful spanning tree. This is proven in [4]. It is important to realize that countability is essential for the existence of an end-faithful *spanning* tree, as shown by the examples of uncountable graphs without such trees constructed by Thomassen [9] and Seymour with Thomas [7].

3 Trees with prescribed rays

We will call a subset $\mathcal{V}$ of ends of G *free* if for every end $\alpha \in \mathcal{V}$ there is a finite set $U \subset V(G)$ such that for every other end $\beta \in \mathcal{V}$ the set U separates α from β. A routine argument shows that a free set of ends is necessarily countable. The aim of this section is to show that any prescribed set of disjoint rays "representing" a free set of ends can be extended to an end-faithful spanning tree. We do not assume here that the whole set of ends of G is countable; under such an assumption a stronger result can be proved, see [8]. Our argument uses ideas of Halin's original proof [3].

Let G be countable and let $\mathcal{V}$ be a countable subset of ends of G. For each $\alpha \in \mathcal{V}$ choose a ray $Q_\alpha \in \alpha$. It is easy to see that there are tails P_α of Q_α which are pairwise disjoint. Call the set of such P_α a $\mathcal{V}$-*representing set*.

Theorem 1 *Let G be countable and let $\mathcal{V}$ be a free set of ends in G. Let $\mathcal{P}$ be a $\mathcal{V}$-representing set. Then there is an end-faithful spanning tree of G containing $\mathcal{P}$.*

Proof. Let $V(\mathcal{P}) = \cup_{P \in \mathcal{P}} V(P)$ and put $W = V(G) \backslash V(\mathcal{P})$. The set $M = W \cup \mathcal{P}$ of the vertices in W and the rays in $\mathcal{P}$ is countable. Let $g : M \to \omega$ be an enumeration. We will define a nested sequence of trees T_i, $i < \omega$ in G with the property that each T_i contains only a finite number of elements of M. Put $T_0 = \{g^{-1}(0)\}$; this is either a vertex of W or a ray in $\mathcal{P}$. Suppose we have T_j for $j < i$. Let $H_i = G - V(T_{i-1})$. Let $b_i \in M$, b_i in H_i, be such that $g(b_i)$ is minimal; note that b_i may be either a ray or a vertex. Let G_i be the component of H_i containing b_i. Let S_i be the set of vertices of T_{i-1} adjacent to some vertex of G_i. For each $P \in \mathcal{P}$ let v_P be its initial vertex. Then let $S_i' \subseteq S_i$ be the set of vertices v that lie on rays $P \in \mathcal{P}$ such that $d_P(v, v_P) \geq i$ (where $d_P(v, v_P)$ is the distance in P between the two vertices). By construction, S_i contains at least one element; S_i', however, could be empty. We choose a vertex $u_i \in V(T_{i-1})$ in two different ways, depending on S_i'.

When $S_i' \neq \emptyset$, choose $u_i \in S_i'$ with the property that $u_i \in V(T_j) \setminus V(T_{j-1})$ with j as large as possible. The vertices chosen in this way are of the *first kind*. If $S_i' = \emptyset$, take $u_i \in S_i$ with $u_i \in V(T_j) \setminus V(T_{j-1})$ with j as large as possible. The vertices chosen in this manner will be of the *second kind*.

Let $v_i \in V(G_i)$ be a vertex adjacent to u_i. There is a path Q_i from b_i to v_i such that Q_i intersects each ray of $\mathcal{P}$ at most once. Let $\mathcal{P}_i$ be the set of rays in $\mathcal{P}$ intersected by Q_i. Define T_i as $T_{i-1} \cup \{u_i v_i\} \cup Q_i \cup \mathcal{P}_i$. Note that this is a tree containing only a finite number of elements of M.

Put $T = \cup_{i < \omega} T_i$. Clearly, T is a spanning tree of G. It remains to show that T is end-faithful.

Let R be a ray in G not equivalent to any ray in $\mathcal{P}$. Then R meets no T_i infinitely often, for otherwise R meets some ray in T_i infinitely often and is thus equivalent to a ray in $\mathcal{P}$. It then follows that for each $i < \omega$, a tail of R belongs to a single component B of H_i. Since at some stage $k_i \geq i$ of the construction the component B will be reached (i.e., $b_{k_i} \in B$), we will have $B = G_{k_i}$, a component of H_{k_1}. Thus we can construct a nested sequence $G_{k_1} \supset G_{k_2} \supset \ldots$ of connected subgraphs each of which contains a tail of R. The choice of the u_i and the fact that $R \not\sim P$ for any $P \in \mathcal{P}$ guarantee that $V(T_{k_i})$ contains at most a finite number of vertices u_{k_j} of the first kind. Further, $V(T_{k_i})$ contains only a finite number of vertices u_{k_j} of the second kind, by their choice as those having the maximal index l among those in $V(T_{l+1}) \backslash V(T_l)$. Thus for each k_i there is a $j > i$ such that each edge $u_{k_s}, v_{k_s}, s \geq j$, lies in G_{k_i}. From this it follows that there is a ray P_R in T with the following properties.

- P_R contains an infinite number of edges of the form $u_{k_l} v_{k_l}$.

- For each k_i, a tail of P_R lies in G_{k_i}.

It is immediate from the second property that $R \sim P_R$ in G. Thus , for every ray R in G there is a ray in T equivalent to it in G.

Now let Q and R be disjoint rays in T. We have to show that $Q \not\sim R$ in G. Assume the contrary. Then Q and R cannot both belong to $\mathcal{P}$. Consider first the case when, say, $Q \in \mathcal{P}$, $R \notin \mathcal{P}$. Let r be the least such that $Q \in T_r$ and let $P_0, \ldots, P_l \in \mathcal{P}$ be the remaining rays in T_r. Obviously, $R \not\sim P_i$ for any $i \leq l$. Therefore, there is a finite set S separating R from all these P_i. Let $n \geq r$ be the least such that $S \subset V(T_n)$. It follows from this that for each $j > n$ no edge of G joins a vertex of H_j to a vertex in $V(P_i) \backslash S$, for any $i \leq l$. On the other hand, as we already saw earlier, tails of R lie in a nested sequence $G_{k_1} \supset G_{k_2} \supset \ldots$. Moreover, since $R \sim Q$, there are infinitely many vertices on Q joined to an infinite number of vertices in G_{k_i} for each $k_i \geq n$.

Obviously, R can be written in the form $\cup_{j < \omega} R_{m_j}$ where each R_{m_j} is a finite subpath of the subgraph added at the m_j-th step of our construction to connect b_{m_j} with u_{m_j} (and, hence, to extend T_{m_j-1} to T_{m_j}). Note that by the rules of our construction the sets of indices $\{m_j : j < \omega\}$ and $\{k_i : i < \omega\}$ have an infinite intersection. Now we have two possibilities.

1. Only a finite number of the R_{m_j} contain edges which also belong to rays in $\mathcal{P}$. Then take a sufficiently large $m_j > n$ such that R_t does not contain any edge of a ray in $\mathcal{P}$ for every $t \geq m_j$. At the m_{j+1}-st step of the construction we then have $S'_{m_{j+1}} \neq \emptyset$ because of the fact that $R \sim Q$ (see the arguments above with the finite separating set S). Thus a $b_{m_{j+1}} - Q$ path should have been chosen at the m_{j+1}-st step, a contradiction.

2. An infinite number of R_{m_j} contain edges which also belong to rays in $\mathcal{P}$. Then, necessarily, the ray R intersects an infinite subset $\mathcal{P}' \subset \mathcal{P}$. But this, together with the assumption that $Q \sim R$, implies that Q cannot be separated from all members of $\mathcal{P}'$ by means of a single finite set. This contradicts our assumption that $\mathcal{V}$ is a free subset of ends.

It remains to consider the case where neither Q nor R belongs to $\mathcal{P}$. From the fact that $Q \sim R$ we derive a contradiction in a manner analogous to the preceding case; we leave the details to the reader.   □

4 $\mathcal{V}$-faithful spanning trees

There is another direction that can be taken when generalizing Halin's results. In the previous section we prescribed a set of rays to be included in an end-faithful spanning tree. Here we consider trees avoiding a given free set of ends.

Let G be a graph and let P be a ray in G. Say that P is *dominated* in G if there is a vertex u and infinitely many paths from u to P pairwise intersecting in exactly u. It is easy to see that for any end α, if $P \in \alpha$ is dominated then every ray in α is. Thus we can say that α is dominated. Let $\mathcal{V} \subseteq \mathcal{V}(G)$. Say a subgraph H is $\mathcal{V}$-*faithful* if the natural mapping $f : \mathcal{V}(H) \to \mathcal{V}(G)$ is a bijection between $\mathcal{V}(H)$ and $\mathcal{V}$.

Let $\mathcal{V} \subseteq \mathcal{V}(G)$ and assume there is a $\mathcal{V}$-faithful spanning tree T in G. Then, as is shown in [8], every ray in $\mathcal{V}(G) \setminus \mathcal{V}$ is dominated. We are interested in conditions under which the converse holds. That is, if $\mathcal{D} = \mathcal{V}(G) \setminus \mathcal{V}$ is the set of all dominated rays in G, when does there exist a $\mathcal{V}$-faithful spanning tree in G? It was proved in [8] that such a tree exists if $\mathcal{V}$ is countable. Here we give another sufficient condition.

Theorem 2 *Let G be a countable graph and let $\mathcal{D} \subseteq \mathcal{V}(G)$ be the set of all dominated ends in G. Let also $\mathcal{V} = \mathcal{V}(G) \setminus \mathcal{D}$. Assume that $\mathcal{D}$ is a free set of ends. Then G contains a $\mathcal{V}$-faithful spanning tree.*

Proof. We already know that $\mathcal{D}$ is countable. We will adapt the proof of Theorem 1. Let $g : V(G) \to \omega$ be a labeling of the graph. For each end $\alpha \in \mathcal{D}$ choose a vertex u_α which dominates it. Note that it may happen that $u_\alpha = u_\beta$ for $\alpha \neq \beta$. Let $D = \{u_\alpha : \alpha \in \mathcal{D}\}$. We will, again, define a nested sequence of finite trees T_i, $i < \omega$. Let $T_0 = \{g^{-1}(0)\}$ and suppose that T_j has been defined for all $j < i$. Let $H_i = G - V(T_{i-1})$. Let $b_i \in V(H_i)$ be such that $g(b_i)$ is minimal. Let G_i be the component of H_i containing b_i. Let S_i be the set of vertices of T_{i-1} adjacent to some vertex of G_i. Put $S_i' = S_i \cap D$. The vertices u_i will be defined as follows (this is different from the proof of Theorem 1).

When $S_i' \neq \emptyset$, choose $u_i \in S_i'$ with $g(u_i)$ maximum. The vertices chosen in this way are of the *first kind*. If $S_i' = \emptyset$, take $u_i \in S_i$ with $u_i \in V(T_j) \setminus V(T_{j-1})$ with j as large as possible. The vertices chosen in this manner will be of the *second kind*.

Take a vertex $v_i \in V(G_i)$ adjacent to u_i. Since G_i is connected, there is a path Q_i from b_i to v_i in G_i. Define T_i as $T_{i-1} \cup \{u_i v_i\} \cup Q_i$. Obviously T_i is a tree.

Put $T = \cup_{i<\omega} T_i$. Clearly T is a spanning tree of G. We now show that T is $\mathcal{V}$-faithful. Let R be a ray not in $\mathcal{D}$. Then, for each $i < \omega$, a tail of R is contained in a component B of H_i; after some steps we will have $B = G_{k_i}$ for some $k_i \geq i$. As in the proof of Theorem 1, this produces a decreasing sequence $G_{k_1} \supset G_{k_2} \supset \ldots$ of connected graphs containing tails of R. Suppose now that, for some i, there are infinitely many $j > i$ such that u_{k_j} is a vertex of the first kind belonging to $V(T_i)$. Since T_i is finite, this would imply that R is dominated by a vertex of T_i, a contradiction. Thus, for each i, there are at most finitely many vertices u_{k_j}, $j > i$ of the first kind in T_i. By the rules of our construction the same holds for the vertices of the second kind. Therefore, for each i, there is an $s > i$ such that for every $j \geq s$ each edge $u_{k_j} v_{k_j}$ lies in G_{k_i}. From this it follows that there is a ray P_R in

T with the following properties (as in Theorem 1).

- P_R contains an infinite number of edges of the form $u_{k_l} v_{k_l}$.
- For each k_i, a tail of P_R lies in G_{k_i}.

Clearly $R \sim P_R$ in G and, therefore, for every ray R not in $\mathcal{D}$ there is a ray in T equivalent to it in G.

Next, we have to prove that T contains no ray equivalent to a ray in $\mathcal{D}$. Assume the contrary and let P be such a ray in T. Consider first the case when P contains an infinite number of vertices of D, say u_{α_i}, $i < \omega$. Now P itself is dominated by, say, u_β. But then it is not difficult to see that the end β cannot be separated from all the α_i by a single finite set. This contradicts the assumption that $\mathcal{D}$ is free. On the other hand, if P contains only a finite number of vertices of D — and we may clearly suppose that the number is zero — then a contradiction can be obtained in the manner of the proof of Theorem 2 of [8] (the idea is that the strategy of building T forces us to use an infinite number of $P - u_\beta$ paths instead of the ray P in T).

It remains to show that no two disjoint rays in T are equivalent in G. Thus, let Q and R be disjoint rays in T and assume that $Q \sim R$ in G. As we already know, neither is dominated in G. Since we are only interested in tails of rays, we may, without loss of generality, assume that only three cases occur.

1. Neither Q nor R contains a vertex of D.
2. The ray Q contains infinitely many vertices of D while R contains none.
3. Both Q and R contain an infinite number of vertices of D.

In each of these cases the rules of construction force us, as before, to use an infinite number of disjoint $Q - R$ paths (which exist, since $Q \sim R$ in G) instead of building the parallel rays Q, R in T. We omit the repetitive details. $\square$

Remark 2 It is interesting to note that Polat [5] gives an example of a countable graph G with uncountably many dominated ends and an uncountable set $\mathcal{V}$ of ends not dominated and such that G has no $\mathcal{V}$-faithful spanning tree.

It is an open problem whether our results hold without the assumption that the (necessarily countable) subset of prescribed ends is free.

References

[1] J.A. Bondy and U.S.R. Murty, *Graph theory with applications*, Macmillan (1976).

[2] R. Diestel, Ends of graphs: recent results and open problems, to appear in *Discrete Math.*

[3] R. Halin, Über unendliche Wege in Graphen, *Math. Ann.* **157** (1964) 125–137.

[4] H.A. Jung, Wurzelbäume und unendliche Wege in Graphen, *Math. Nachrichten* **41** (1969) 1–22.

[5] N. Polat, Multiterminal spanning trees of infinite graphs, preprint.

[6] N. Polat, Topological aspects of infinite graphs, *Cycles and Rays*, Hahn, Sabidussi, Woodrow, eds., NATO ASI Ser. C, Vol. 301 (1990) 197–220.

[7] P. Seymour, R. Thomas, An end-faithful spanning tree counterexample, in print.

[8] J. Širáň, End-faithful forests and spanning trees in infinite graphs, to appear in *Discrete Math.*

[9] C. Thomassen, Infinite graphs with no end-preserving spanning trees, preprint.

True Embedding Partition Relations

A. HAJNAL*
*Mathematical Institute of the
Hungarian Academy of Sciences
Budapest, Hungary*

Abstract

This paper is primarily a survey of recent results of the author and others on transfinite generalizations of Nešetřil–Rödl type Ramsey theory. In the second half of the paper large universal graphs are studied in detail.

0 Introduction.

This paper is a detailed version of the author's talk given at the Banff Conference. It is intended to be a survey of a certain branch of infinitary combinatorics — or rather set theory — since quite a few of the recent results are consistency proofs. We will be talking about infinitary Ramsey theory. Transfinite generalizations of Ramsey's theorem were first studied by Erdős and Rado; see e.g. [3]. Let us first recall a special instance of the ordinary partition relation introduced by Erdős and Rado. The set of subsets of a set X of cardinal μ is denoted by $[X]^\mu$.

Let κ be a cardinal, $1 \leq r < \omega$, γ an ordinal. A map $f : [\kappa]^r \to \gamma$ is said to be an *r-partition* of κ with γ-colors. A subset $X \subset \kappa$ is *homogeneous* with respect to f if the elements of $[X]^r$ are monochromatic, i.e. there is a $\nu < \gamma$ such that

$$f(A) = \nu \quad \text{for} \quad A \in [X]^r .$$

The partition relation $\kappa \to (\lambda)^r_\gamma$ is said to hold if for every r-partition f of κ with γ colors there exists a subset $X \subset \kappa$ of size λ homogeneous for f; $\kappa \nrightarrow (\lambda)^r_\gamma$ denotes the negation of this statement. Ramsey's celebrated theorem can be stated in this notation as

$$\omega \to (\omega)^k_\gamma \quad \text{for} \quad 1 \leq k < \omega,\ 1 \leq \gamma < \omega,$$

and the simplest often quoted finite result is

$$6 \to (3)^2_2 .$$

The following formulation of this simple remark is closer to the heart of combinatorialists.

Assume we color the edges of a complete graph of six vertices with two colors, say red and blue. Then there exists either a red triangle or a blue triangle. Or to twist this a

*Research supported by Hungarian National Science Foundation OTKA grant 1908.

N.W. Sauer et al. (eds.), Finite and Infinite Combinatorics in Sets and Logic, 135–151.
© 1993 *Kluwer Academic Publishers. Printed in the Netherlands.*

little further, if we color the edges of a graph $\mathcal{G}$ with two colors arbitrarily there will be a monochromatic triangle, provided $K_6 \subset \mathcal{G}$, i.e. $\mathcal{G}$ contains a complete graph of size six. We started to speculate with Erdős in [3] whether there are other graphs having this property, and we stated the problem whether there is a $\mathcal{G}$ not containing a K_4 having the above property, which we could write down as

$$\mathcal{G} \to (K_3)_2^2.$$

The existence of such a graph $\mathcal{G}$ was proved by Folkman [8]. His proof was surprisingly hard. Despite much effort of other people, and prizes offered by Paul Erdős, we do not know as of today if such a graph $\mathcal{G}$ having at most 10^6 vertices exists.

The next important idea toward generalization is due to Deuber [1]. To illuminate this let us look again at Folkman's result.

There is a graph $\mathcal{G}$ not containing a K_4 such that $\mathcal{G} \to (K_3)_2^2$. What other graphs $\mathcal{H}$ can we expect to occur monochromatically in every two coloring of some $\mathcal{G}$ not containing a K_4? Once the question is asked the answer seems obvious; we can expect every $\mathcal{H}$ which does not contain a K_4. However there is one more step in the process of generalization we have to understand.

We can require the existence of a monochromatic subgraph $\mathcal{H}'$ isomorphic to $\mathcal{H}$, in which case as a direct generalization of the Erdős–Rado partition relation we can write

$$\mathcal{G} \to (\mathcal{H})_\gamma^2,$$

the exponent 2 referring to edge colorings. However we can require, more strongly, the existence of an $\mathcal{H}'$ isomorphic to $\mathcal{H}$ such that $\mathcal{H}'$ is an *induced* subgraph of $\mathcal{G}$ and is monochromatic. In this case we will have to use a different notation:

$$\mathcal{G} \rightarrowtail (\mathcal{H})_\gamma^2.$$

Once again to emphasize the difference:

Clearly, by Ramsey's theorem,

$$K_n \to (C_5)_2^2$$

if n is large enough, but for

$$\mathcal{G} \rightarrowtail (C_5)_2^2$$

we need a more sophisticated $\mathcal{G}$.

We will give the formal definition of these "true embedding" partition relations in the next section. For finite graphs or hypergraphs the theory of the true embedding partition relations was extensively studied by Nešetřil and Rödl; see e.g. [13]–[16]. As our aim in this paper is to study or rather to discuss infinitary generalizations, we will not discuss finitary results, with one exception in §5. Also we will not discuss Nešetřil–Rödl type generalizations well-known to combinatorialists, where say instead of coloring of edges certain subgraphs are colored, and we also disregard natural and (sometimes easy) generalizations for r-uniform hypergraphs with $r \geq 3$ or general relational structures.

We will focus on listing the main results for the true embedding partition relations in case of vertex and edge colorings, and we will try to isolate the simplest unsolved problems.

We will give quite a few details on set-ideals defined in large universal (homogeneous) graphs which play an important role in obtaining positive partition relations. Finally using these methods I will try to outline the proof of an old Erdős–Hajnal–Pósa [6] theorem stated at the beginning of section 4 and the proof of a recent theorem of mine [9] stated at the end of section 4.

1   The definitions of the "true embedding" partition relations and a further possible generalization.

Our set theoretical notation will be standard. As for graphs, we must make a few remarks.

We consider a graph $\mathcal{G} = \langle V, G \rangle$ an ordered pair, $G \subset [V]^2 = \{e : e \subset V \wedge |e| = 2\}$, i.e. we consider undirected graphs without loops and multiple edges. The vertex set V will usually be a cardinal κ.

$\mathcal{H} = \langle W, H \rangle$ is an *induced subgraph* of $\mathcal{G} = \langle V, G \rangle$ if $W \subset V$ and $H = G \cap [W]^2$. For $W \subset V$, $\mathcal{G}[W]$ is the subgraph of $\mathcal{G}$ induced by W: $\mathcal{G}[W] = \langle W, G \cap [W]^2 \rangle$.

$\overline{\mathcal{G}}$, the complement of $\mathcal{G}$, is the graph $\langle V, [V]^2 \setminus G \rangle$.

For τ a cardinal, K_τ is (the isomorphism type) of the complete graph of size τ. $\mathcal{N}_\tau$ is the class of graphs not containing a K_τ as a subgraph. For $3 \leq n < \omega$, C_n is the circuit of length n.

Definition 1.
1.1. $\mathcal{G} \to (\mathcal{H})^1_\gamma$ means the following statement.

For every $f : V[\mathcal{G}] \to \gamma$ there is a subgraph $\mathcal{H}' \subset \mathcal{G}$ such that $\mathcal{H}'$ is isomorphic to $\mathcal{H}$ and the vertex set of $\mathcal{H}'$ is monochromatic, i.e. f is constant on $V(\mathcal{H}')$.

1.2. $\mathcal{G} \to (\mathcal{H})^2_\gamma$ means the following statement.

For every $f : G \to \gamma$ there is a subgraph $\mathcal{H}' \subset \mathcal{G}$ such that $\mathcal{H}'$ is isomorphic to $\mathcal{H}$ and the edge set of $\mathcal{H}$ is monochromatic, i.e. f is constant on H'.

1.3 $\mathcal{G} \rightarrowtail (\mathcal{H})^1_\gamma$ means the following statement.

For every $f : V[\mathcal{G}] \to \gamma$ there is an induced subgraph $\mathcal{H}'$ of $\mathcal{G}$ such that $\mathcal{H}'$ is isomorphic to $\mathcal{H}$ and f is constant on $V(\mathcal{H}')$.

1.4 $\mathcal{G} \rightarrowtail (\mathcal{H})^2_\gamma$ means the following statement.

For every $f : G \to \gamma$ there is an induced subgraph $\mathcal{H}'$ of $\mathcal{G}$ such that $\mathcal{H}'$ is isomorphic to $\mathcal{H}$ and f is constant on H'.

As usual we denote by $\not\to$ and $\not\rightarrowtail$ the negation of the corresponding statements. It was only lately discovered, at least in set theory, that instead of 1.4 we can investigate a more general statement and that it is often quite convenient to do so.

1.5 $\mathcal{G} \rightarrowtail (\mathcal{H})^2_{\gamma, \delta}$ means the following statement.

For every $f : G \to \gamma$ and for every $g : ([V]^2 - G) \to \delta$ there is an induced subgraph $\mathcal{H}'$ of $\mathcal{G}$ isomorphic to $\mathcal{H}$ such that f is constant on H' and g is constant on $\overline{H'}$ (the edge set of the complement of $\mathcal{H}'$).

Note that

$$\mathcal{G} \rightarrowtail (\mathcal{H})^2_\gamma \qquad \text{is equivalent to} \qquad \mathcal{G} \rightarrowtail (\mathcal{H})^2_{\gamma, 1}$$

and some of the old results extend to this more general relation without any difficulty.

Throughout this paper we will be interested in existence theorems starting with a prefix $\forall \mathcal{H} \, \forall \gamma, \delta \, \exists \mathcal{G}$ followed by some of the above partition relations.

However, as we have already explained in the introduction, for the first two relations the simple existence is given by the classical Erdős–Rado result that for every λ and γ there is a κ with

$$\kappa \to (\lambda)^i_\gamma \qquad i = 1, 2 \, .$$

So in general, the existence problem for the relations is formulated as follows. Let $\mathcal{E}_0, \mathcal{E}_1$ be two classes of graphs. Is it true that $\forall \mathcal{H} \in \mathcal{E}_0 \, \forall \gamma, \delta \, \exists \mathcal{G} \in \mathcal{E}_1 \ldots$, where $\ldots$ is replaced by one of the above partition relations. In this paper we restrict our attention to infinitary problems where $\mathcal{E}_0 = \mathcal{E}_1 = \mathcal{N}_\tau$ for some cardinal τ, possibly finite, but we will be always interested in how small a $\mathcal{G}$ can we find for given $\mathcal{H}, \gamma, \tau$ and possibly δ.

As is well-known, for every cardinal λ we can find special graphs $\mathcal{G}[\lambda]$ containing all graphs of size λ^+ as induced subgraphs. Clearly if for some $\mathcal{H}$ and γ there is a suitable $\mathcal{G}$ then there is one among the $\mathcal{G}[\lambda]$. That is why we need a detailed description of these special graphs. This will be done in the next chapter.

2   Special universal (homogeneous) graphs.

Let $\mathcal{G} = \langle \kappa, G \rangle$ be a graph. With some abuse of notation it will be convenient to denote the graph by $\mathcal{G}_0$ and the complement of it by $\mathcal{G}_1$, i.e. $G_0 = G$, $G_1 = [\kappa]^2 \setminus G$.

We will denote by $\mathcal{F}n(\kappa, \lambda)$ the set

$$\{\varepsilon \ : \ \varepsilon \text{ is a function} \wedge \text{Dom}(\varepsilon) \subset \kappa \wedge R(\varepsilon) \subset 2 \wedge |\varepsilon| < \lambda\},$$

i.e. the set of 0-1 valued partial functions defined on a subset of κ of size less than λ.

For an arbitrary $\varepsilon \in \mathcal{F}n(\kappa, \lambda)$, using the graph $\mathcal{G}$ we can define a subset $\mathcal{G}_\varepsilon$ of κ as follows:

$$\mathcal{G}_\varepsilon = \{\alpha \in \kappa \setminus \text{Dom}(\varepsilon) \ : \ \forall \beta \in \text{Dom}(\varepsilon) \, \{\alpha, \beta\} \in G_i \Leftrightarrow \varepsilon(\beta) = i\} \, .$$

In particular, if Dom $(\varepsilon) = \{\beta\}$ and $\varepsilon(\beta) = i$, then $\mathcal{G}_\varepsilon$ is the set of vertices adjacent to β if $i = 0$ and the set of vertices $\alpha \neq \beta$ not adjacent to β if $i = 1$.

We denote these sets by $\mathcal{G}_i(\beta)$ as well for $i < 2$ respectively. We need the following facts.

2.1.   There is a (up to isomorphism) unique graph $\mathcal{U} = \langle \omega, \mathcal{U} \rangle$ such that $|\mathcal{U}_\varepsilon| = \omega$ for every $\varepsilon \in \mathcal{F}n(\omega, \omega)$. Every countable graph $\mathcal{G}$ is isomorphic to an induced subgraph of $\mathcal{U}$. This $\mathcal{U}$ is the well-known homogeneous countable graph.

2.2.   Let $3 \leq n < \omega$. There is a (up to isomorphism) unique homogeneous graph $\mathcal{U}_n = \langle \omega, U_n \rangle$ such that $\mathcal{U}_n \in \mathcal{N}_n$ (it is K_n-free) and every countable $\mathcal{G} \in \mathcal{N}_n$ is isomorphic to an induced subgraph of $\mathcal{U}_n$. $\mathcal{U}_n$ satisfies the following condition: for all $\varepsilon \in \mathcal{F}n(\omega, \omega)$, $|(\mathcal{U}_n)_\varepsilon| = \omega$ provided $K_{n-1} \not\subset \mathcal{U}_n[\{\alpha \in \text{Dom}(\varepsilon) : \ \varepsilon(\alpha) = 0\}]$.

If we call the $(\mathcal{U}_n)_\varepsilon$ for $\varepsilon \in \mathcal{F}n(\omega, \omega)$ briefly *types* (what they really are in model theory), the above condition means that every permissible type is infinite, where a type is permissible if making it non-empty does not force a K_n to be included in $\mathcal{U}_n$.

2.3.   For every $\lambda \geq \omega$ there is a graph $\mathcal{G}[\lambda] = \mathcal{G}$ with vertex set $\kappa = 2^\lambda$ such that

$$|\mathcal{G}_\varepsilon| = \kappa \quad \text{for all} \quad \varepsilon \in \mathcal{F}n(\kappa, \lambda^+) \, .$$

This graph is universal for all graphs of size λ^+. Unfortunately $\mathcal{G}[\lambda]$ is unique (up to isomorphism) only if we assume $2^\lambda = \lambda^+$. We do not need this and in what follows $\mathcal{G}[\lambda]$ is any fixed graph satisfying 2.3.

For logicians, $\mathcal{U}, \mathcal{U}_n, \mathcal{G}[\lambda]$ are saturated models; the rest of us should construct them by (transfinite) induction.

3 Results and problems on vertex partitions.

The existence problem for the relation 1.3 and for $\mathcal{N}_\tau$ classes is solved by the following:

Theorem *Let* $3 \leq \tau$ *and* $\kappa = \mathrm{cf}(\kappa) \geq \omega$. *For every* $\mathcal{H} \in \mathcal{N}_\tau$ *with* $|\mathcal{H}| \leq \kappa$ *there is a* $\mathcal{G} \in \mathcal{N}_\tau$, $|\mathcal{G}| \leq 2^\kappa$, *such that*

$$\mathcal{G} \rightarrowtail (\mathcal{H})^1_\kappa .$$

This was proved in Hajnal–Komjáth [10] and earlier for $\tau < \omega$ in Komjáth–Rödl [11]. The main difficulty in the proof is the case $\tau = \omega$ since there are no universal graphs for graphs in $\mathcal{N}_\omega$ with properties analogous to $\mathcal{U}_n$. ($\mathcal{G} \in \mathcal{N}_\omega$ is not a first order statement.) We also proved with Komjáth that it is consistent that $2^\kappa > \kappa^+$ and the theorem is true with $|\mathcal{G}| \leq \kappa^+$.

However in case $\tau < \omega$ and $\kappa = \omega$ one can do better.

Theorem (El-Zahar, Sauer [2]) *For* $3 \leq n < \omega$, $\gamma < \omega$,

$$\mathcal{U}_n \rightarrowtail (\mathcal{U}_n)^1_\gamma .$$

This is a very nice theorem with a very complicated proof. Earlier it was proved for $n = 3$ by Komjáth and Rödl in [11].

Note that this theorem can be interpreted as follows.

Let $I = \{X \subset \omega \; : \; \mathcal{U}_n$ *is not isomorphic to a subgraph of* $\mathcal{U}_n[X]\}$. *Then* I *is an ideal in* $\mathcal{P}(\omega)$.

The first problem I want to state is

Problem 1 Find a good characterization of the El-Zahar–Sauer ideal I.

Rather than explaining what I mean by a good characterization, I will state an easy result.

Let $\mathcal{U}_n^* = \langle \omega_1, U_n^* \rangle$ be a graph on the vertex set ω_1 satisfying the following conditions:

(i) for every $\varepsilon \in \mathcal{F}n(\omega_1, \omega)$, $|(\mathcal{U}_n^*)_\varepsilon| = \omega_1$ provided ε is a permissible type, i.e.

$$K_{n-1} \not\subset \mathcal{U}_n^*[\{\alpha < \omega_1 \; : \; \varepsilon(\alpha) = 0\}];$$

(ii) for every $\alpha < \omega_1$, $|\mathcal{U}_n^*(\alpha) \cap \alpha| < \omega$.

It is obvious that such a graph $\mathcal{U}_n^*$ exists.

Let $I^* = \{X \subset \omega_1 \; : \; \exists\{\varepsilon_\xi \; : \; \xi < \omega_1\} \subset \mathcal{F}n(\omega_1, \omega)$ such that the ε_ξ have pairwise disjoint domains, they are permissible and $|(\mathcal{U}_n^*)_{\varepsilon_\xi} \cap X| < \omega_1$ for $\xi < \omega_1\}$.

Now it is easy to see the following facts:

(A) I^* is an ideal, and

(B) $\mathcal{U}_n$ is isomorphic to an induced subgraph of $\mathcal{U}_n^*[X]$ for $X \notin I^*$.

Note that (A) follows from (ii) using a set-mapping argument and using the following fact. Assume $\varepsilon, \varepsilon'$ are permissible types with disjoint domains with no edges between the domains; then ε and ε' can be amalgamated, i.e. $\varepsilon \cup \varepsilon'$ is a permissible type as well.

4 Results and problems on edge partitions, no restrictions on $\mathcal{H}$ and $\mathcal{G}$.

As we have already mentioned, the existence problem for the relation 1.4 was asked by Deuber. For finite graphs this was settled by three sets of authors.

Theorem (Deuber [1]; Erdős, Hajnal, Pósa [6]; Nešetřil, Rödl) $\forall |\mathcal{H}| < \omega \quad \forall \gamma, \delta < \omega$ $\exists \mathcal{G} |\mathcal{G}| < \omega$ *such that*

$$\mathcal{G} \rightarrowtail (\mathcal{H})^2_{\gamma,\delta}$$

(though all authors claimed only $\mathcal{G} \rightarrowtail (\mathcal{H})^2_\gamma$).

With Erdős and Pósa we started to investigate infinitary generalizations and proved

Theorem (Erdős–Hajnal–Pósa [6]) $\forall |\mathcal{H}| \leq \omega \quad \forall \gamma < \omega \quad \exists \mathcal{G} |\mathcal{G}| \leq 2^\omega$

$$\mathcal{G} \rightarrowtail (\mathcal{H})^2_\gamma,$$

or rather

$$\mathcal{G} \rightarrowtail (\mathcal{H})^2_{\gamma,\gamma}.$$

(We also proved that a countable graph will not do in general, because $\mathcal{U} \not\rightarrowtail (K_{\omega,\omega})^2_2$.)

This happened in 1973. We realized that our proof will generalize neither for uncountably many colors nor for uncountable $\mathcal{H}$. The real reason behind this became clear only a decade later. We proved with Komjáth:

Theorem (Hajnal–Komjáth [10]) *It is consistent with ZFC that there is a graph $\mathcal{H}$ of size ω_1 such that for all $\mathcal{G}$,*

$$\mathcal{G} \not\rightarrowtail (\mathcal{H})^2_2.$$

The proof of this is technically not complicated, one can get it just adding one Cohen-real, but just because of this the effect is quite strong. It gives also that we can not get a suitable $\mathcal{G}$ from large cardinal axioms either.

Since in [10], we made a technical mistake in writing down the proof of a more general theorem implying this, I will outline a correct proof here.

Let $P = \{p : \text{Dom}(p) \in \omega \wedge R(p) \subset 2\}$ ordered by reverse inclusion be the partial order adding one Cohen real to V.

Let $r : \omega \to 2$ be the generic function, $p_n = r|n$ for $n \in \omega$.

Define $\mathcal{H} = \langle \omega_1, H \rangle$ in V^P as follows. First define in V a sequence of injections $f_\alpha : \alpha \to \omega$ for $\alpha < \omega_1$, and for $\beta < \alpha < \omega_1$ put $\{\alpha, \beta\} \in H \Leftrightarrow r(f_\alpha(\beta)) = 0$. This is a graph invented by Shelah.

Let $\mathcal{G} = \langle \lambda, G \rangle$ be a graph in V^P, $\dot{G}$ a name for G. Define a mapping $f : G \to 2$ in V^P as follows. For $\{\alpha, \beta\} \in G$ let $n(\alpha, \beta) = \min\{n : p_n \Vdash \{\alpha, \beta\} \in \dot{G}\}$, and $f(\alpha, \beta) = r(n(\alpha, \beta))$. Assume indirectly that in V^P there is an $F : \omega_1 \to \lambda$ and a $j < 2$ such that F maps ω_1 onto some $C \subset \lambda$, F is an isomorphism between $\mathcal{H}$ and $\mathcal{G}[C]$, and for $\{\alpha, \beta\} \in H$ $f(\{F(\alpha), F(\beta)\}) = j$.

Let n_0 be such that p_{n_0} forces all the above facts.

Since $|P| \leq \omega$ there is an $\bar{F} \in V$, $|\bar{F}| = \omega_1$ with $\bar{F} \subset F$.

Let $A = \text{Dom}(\bar{F})$, $B = \text{Ran}(\bar{F})$. First note that $[A]^2 \cap H \notin V$, since if $\alpha_0, \alpha_1, \ldots,$ $\alpha_k, \ldots, \alpha$ are the first $\omega + 1$ elements of A, then $D = \{\alpha_k : k < \omega\} \in V$, $f''_\alpha D \in V$, while $D \cap \mathcal{H}(\alpha) \notin V$. It follows that $[B]^2 \cap G \notin V$, hence there is an $n_1 \geq n_0$ with $n(\alpha, \beta) = n_1$ for some $\alpha \neq \beta \in B$. Indeed otherwise $e \in G \Leftrightarrow \exists n < n_0 \ p_n \models e \in \bar{G}$ for $e \in [B]^2$. Now

$$p_{n_1} \Vdash f(\alpha, \beta) = j,$$
$$p_{n_1} \Vdash r(n_1) = j, \quad \text{a contradiction.}$$

Soon afterwards it was proved that for target graphs $\mathcal{H}$ of size ω_1 this is really independent.

Theorem (Shelah [18]) *It is consistent with ZFC that $\forall \mathcal{H} \ \forall \gamma \ \exists \mathcal{G} \quad \mathcal{G} \rightarrowtail (\mathcal{H})^2_\gamma$.*

In Shelah's model, obtained by a class forcing, GCH is violated cofinally. So it is still possible to prove our theorem with Komjáth from GCH but I do not think that this is the case.

The real problem is what happens to countable graphs.

Problem 2 *Can $\forall \gamma \ \exists \mathcal{G} \ \mathcal{G} \rightarrowtail (\mathcal{U})^2_{\gamma, \gamma}$ be proved in ZFC?*

The simplest unsolved case is $\gamma = \omega$.

I have proved recently the following:

Theorem (Hajnal [9]) $\forall \mathcal{H} |\mathcal{H}| < \omega \ \forall \gamma \ \exists \mathcal{G} \ |\mathcal{G}| \leq \exp_{|\mathcal{H}|+5}(\gamma)$

$$\mathcal{G} \rightarrowtail (\mathcal{H})^2_{\gamma, \gamma} .$$

I will give some details of the rather complicated proof of this in the closing chapters. Here $\exp_{k+1}(\gamma) = 2^{\exp_k(\gamma)}$ is the iterated exponentiation. It would be nice to find a smaller bound for the size of $\mathcal{G}$, it could be as small as $(2^{|\gamma|})^+$.

For later purposes, let me quote another theorem from [10] and a related problem.

Theorem *It is consistent that there is a triangle free graph $\mathcal{H}$ of size ω_1 such that for every $\mathcal{G}$ with $\mathcal{G} \to (\mathcal{H})^2_\omega$*

$$K_\omega \subset \mathcal{G} \qquad \text{holds.}$$

Problem 3 *Can $\mathcal{G} \to (\mathcal{H})^2_\omega$ be replaced by $\mathcal{G} \to (\mathcal{H})^2_2$ in the above result and can $K_\omega \subset \mathcal{G}$ be improved to $K_{\omega_1} \subset \mathcal{G}$?*

To conclude this chapter let me remark that our trick with Komjáth can not work for forcing a counterexample to Problem 2. This is shown by the following fact.

Let V denote, as usual, a model of ZFC, and P a partial order. Assume P forces a counterexample to Problem 2, i.e. $V^P \models \forall \mathcal{G}\ \mathcal{G} \not\rightarrow (\mathcal{U})^2_{\gamma,\gamma}$. Let $\delta = \gamma^{|P|}$. Then a weaker negative relation must have been true in the ground model V,

$$V \models \forall \mathcal{G}\ \mathcal{G} \not\rightarrow (\mathcal{U})^2_{\delta,\delta}$$

and maybe we started with Shelah's model. Note that the proof of the above remark is based on the fact that the $\mathcal{U}$ of V^P is the same as the $\mathcal{U}$ of V.

5   Results and problems on edge partitions for the $\mathcal{N}_\tau$ classes.

In this case we only know essentially one new consistency result of Shelah, a strengthening of his result stated towards the end of this section as a theorem of Komjáth and Shelah. However to be able to formulate unsolved problems, I will state this result in stages. First let me recall a very old Erdős–Hajnal problem from [3]. See also [4] and [5].

Problem 4 *Does there exist a $\mathcal{G} \in \mathcal{N}_4$, $|\mathcal{G}| = (2^\omega)^+$ such that $\mathcal{G} \rightarrow (K_3)^2_\omega$ holds?*

As far as I remember this is a \$250 problem. Note that $\mathcal{G} \rightarrow (K_3)^2_\omega$ is equivalent to $\mathcal{G} \rightarrowtail (K_3)^2_\omega$ since a complete subgraph of $\mathcal{G}$ is trivially an induced subgraph as well. Now Shelah proved the following.

Theorem *The following is consistent for every cardinal $\tau \geq 3$. For every γ there is a $\mathcal{G} \in \mathcal{N}_\tau$ such that for all $\sigma < \tau$*

$$\mathcal{G} \rightarrow (K_\sigma)^2_\gamma.$$

In this case however there are no consistency theorems from the other direction, some instances of this result are easy theorems of ZFC and maybe all of them can be proved. Let me quote some old results from [3].

Theorem *Assume GCH. Then*

(i) *there is a $\mathcal{G} \in \mathcal{N}_\omega$, $|\mathcal{G}| = \omega_4$ such that for every $\ell < \omega$ $\mathcal{G} \rightarrow (K_\ell)^2_\omega$;*

(ii) *there is a $\mathcal{G} \in \mathcal{N}_{\omega_2}$, $|\mathcal{G}| = \omega_2$ such that $\mathcal{G} \rightarrow (K_{\omega_1})^2_\omega$.*

Problem 5 Can one prove (i) with $|\mathcal{G}| = \omega_2$?

It is especially easy to see (ii). Let $\mathcal{G}$ be a graph on ω_2 establishing $\omega_2 \not\rightarrow (\omega_2)^2_2$. Then $\mathcal{G} \rightarrow (K_{\omega_1})^2_\omega$ because otherwise, by the Erdős–Rado theorem $\omega_2 \rightarrow ((\omega_1)_\omega, \omega_2)^2$ we get $K_{\omega_2} \subset \overline{\mathcal{G}}$, a contradiction. So the really interesting simplest instances of this Shelah result are $\tau \leq \omega$, $\gamma = \omega$.

Again the case $\tau = \omega$ is more difficult. A proof for this consistency is given in Komjáth–Shelah [12] where they also prove a theorem which seemed at that time a full generalization.

Theorem (Komjáth–Shelah [12]) *The following is consistent for every $\tau \geq 3$:*

$$\forall \gamma \forall \mathcal{H} \in \mathcal{N}_\tau \exists \mathcal{G} \in \mathcal{N}_\tau\ \mathcal{G} \rightarrowtail (\mathcal{H})^2_\gamma.$$

Here we know, by the result of Hajnal–Komjáth [9], that this can not be proved in ZFC.

This was improved by Shelah during this conference; $\mathcal{G} \rightarrowtail (\mathcal{H})^2_\gamma$ can be replaced by $\mathcal{G} \rightarrowtail (\mathcal{H})^2_{\gamma,\gamma}$ which is much stronger, as we have already explained. It has a corollary for finite graphs.

Corollary (Nešetřil–Rödl [17]) $\forall 3 \leq \tau < \omega \; \forall \gamma < \omega \; \forall \mathcal{H}(|\mathcal{H}| < \omega \wedge \mathcal{H} \in \mathcal{N}_\tau) \Rightarrow$
$\exists \mathcal{G} \in \mathcal{N}_\tau |\mathcal{G}| < \omega \; \mathcal{G} \rightarrowtail (\mathcal{H})^2_{\gamma,\gamma}$.

This follows by compactness and absoluteness from Shelah's result too.

Finally I want to state a problem which is hard to formulate precisely. We know from Nešetřil and Rödl that for $n \geq 3$ $\mathcal{U}_n \rightarrowtail (\mathcal{H})^2_2$ holds for $\mathcal{H} \in \mathcal{N}_n \wedge |\mathcal{H}| < \omega$.

Problem 6 *Find a direct "set theoretical" proof of the above statement or just a proof of $\mathcal{U}_4 \rightarrow (K_3)^2_2$ (which follows from Folkman's result already mentioned).*

6 Ideals in $\mathcal{G}[\lambda]$. An outline of the proof of the Erdős–Hajnal–Pósa theorem.

For a fixed $\lambda \geq \omega$ we are going to consider the graph $\mathcal{G}[\lambda]$ on the vertex set $\kappa = 2^\lambda$ defined in 2.3. We remark that we could formulate the definitions so that everything we say should work for the homogeneous graph $\mathcal{U}$ and for $\mathcal{G}[\lambda]$ at the same time, but we do not want to complicate notation.

We will briefly denote $\mathcal{G}[\lambda]$ by $\mathcal{G}$. G_0 is the edge set of $\mathcal{G}$, G_1 is the edge set of the complement. $\mathcal{P}(\kappa)$ denotes the power set of κ. $I \subset \mathcal{P}(\kappa)$ is an ideal if $\kappa \notin I$, $A \subset B \in I \Rightarrow A \in I$ and $A, B \in I \Rightarrow A \cup B \in I$. I is a ϱ-complete ideal for some $\varrho \geq \omega$ if $I' \subset I \wedge |I'| < \varrho$ implies that $\cup I' \in I$. A good example of a λ^+-complete ideal in $\mathcal{P}(\kappa)$ is $I = [\kappa]^{<\kappa}$. We can consider sets in the ideal to be small, sets not in the ideal are not small. At our first approach we want to extend $[\kappa]^{<\kappa}$ to an ideal I more strongly related to $\mathcal{G}$. We want an I with the following property:

> If $X \subset \kappa$ is not small, then the degree of almost all $\alpha < \kappa$ with respect to X should not be small, both in $\mathcal{G}$ and in the complement of $\mathcal{G}$.

This means really that $|\{\alpha < \kappa \; : \; \mathcal{G}_i(\alpha) \cap X \in I\}| < \kappa$ should hold for $i < 2$ provided $X \notin I$.

To achieve this we need a general procedure to extend ideals $I \subset \mathcal{P}(\kappa)$.

Let $\mathcal{F}$ briefly denote the set $\mathcal{F}n(\kappa, \lambda^+)$. There is a natural partial order on $\mathcal{F}$. For $\varepsilon, \varepsilon' \in \mathcal{F}$ we write $\varepsilon' \overset{\leq}{=} \varepsilon$ if $\varepsilon' \supset \varepsilon$, i.e. ε' is an extension of ε. Note that $\varepsilon' \overset{\leq}{=} \varepsilon \Rightarrow \mathcal{G}_{\varepsilon'} \subset \mathcal{G}_\varepsilon$. A subset $D \subset \mathcal{F}$ is said to be *dense* if for all $\varepsilon \in \mathcal{F}$ there is $\varepsilon' \in D$ with $\varepsilon' \overset{\leq}{=} \varepsilon$. D is *open* if $\varepsilon \in D$ and $\varepsilon' \overset{\leq}{=} \varepsilon$ imply $\varepsilon' \in D$. It is a well known fact that $\mathcal{F}$ has the λ^+-Baire property, i.e.

6.1.   the intersection of at most λ dense-open subsets of $\mathcal{F}$ is dense-open.

Assume now that $I \subseteq \mathcal{P}(\kappa)$ is a λ^+-complete ideal, $[\kappa]^{<\kappa} \subset I$, and $X \subset \kappa$. Let us denote by $D(X, I)$ the set $\{\varepsilon \in \mathcal{F} \; : \; \mathcal{G}_\varepsilon \cap X \in I\}$. Clearly $D(X, I)$ is open.

6.2. Definition. $X \in \tilde{I} \Leftrightarrow D(X, I)$ is dense.

We can now prove the following.

Lemma Assume $I \subseteq \mathcal{P}(\kappa)$ is a λ^+-complete ideal and $\kappa \notin \tilde{I}$. Then $I \subset \tilde{I}$, $\tilde{I}$ is a λ^+-complete ideal and $\tilde{\tilde{I}} = \tilde{I}$.

Before we give the proof let us remark that if $I = [\kappa]^{<\kappa}$ then $\kappa \notin \tilde{I}$ since $|\mathcal{G}_\varepsilon| = \kappa$ for every $\varepsilon \in \mathcal{F}$.

Proof of the Lemma $I \subset \tilde{I}$ is obvious. Assume $\{A_\xi : \xi < \lambda\} \subset \tilde{I}$ and let $A = \bigcup_{\xi < \lambda} A_\xi$. $D(A_\xi, I)$ is dense-open by the assumption. Let $D = \bigcap \{D(A_\xi, I) : \xi < \lambda\}$. D is dense-open by 6.1. We claim $D \subset D(A, I)$. Assume $\varepsilon \in D$. Then $\mathcal{G}_\varepsilon \cap A_\xi \in I$ for $\xi < \lambda$ and $\mathcal{G}_\varepsilon \cap A = \bigcup_{\xi < \lambda} (\mathcal{G}_\varepsilon \cap A_\xi) \in I$ since I is λ^+-complete. Hence $\varepsilon \in D(A, I)$. It follows that $D(A, I)$ is dense-open, $A \in \tilde{I}$.

To finish the proof assume now that $A \in \tilde{\tilde{I}}$, i.e. $D(A, \tilde{I})$ is dense-open. Let $\varepsilon \in \mathcal{F}$ be arbitrary. There is an extension $\varepsilon' \stackrel{\supseteq}{=} \varepsilon$ with $\varepsilon' \in D(A, \tilde{I})$. This means $\mathcal{G}_{\varepsilon'} \cap A \in \tilde{I}$. By the definition of $\tilde{I}$ there is an extension $\varepsilon'' \stackrel{\supseteq}{=} \varepsilon'$ with $\mathcal{G}_{\varepsilon''} \cap \mathcal{G}_{\varepsilon'} \cap A \in I$. $\mathcal{G}_{\varepsilon''} \cap \mathcal{G}_{\varepsilon'} = \mathcal{G}_{\varepsilon''}$, hence $\varepsilon'' \in D(A, I)$. $D(A, I)$ is dense-open, $A \in \tilde{I}$. □

I gave a detailed proof, because in both papers [6] and [9] a clumsier ideal is used.

6.3. Definition. I is a *good* ideal if $[\kappa]^{<\kappa} \subset I$, I is λ^+-complete and $\tilde{I} = I$.

As a corollary of the lemma, we know

6.4. There is a good ideal I, since $\widetilde{[\kappa]^{<\kappa}} = I$ is a good ideal.

For the proof of the Erdős–Hajnal–Pósa theorem we only need 6.4. Let me first list some properties of good ideals. The detailed proofs can be left to the reader.

6.5. If I is a good ideal and $A \notin I$ then there is no sequence $\{\varepsilon_\nu : \nu < \lambda^+\} \subset \mathcal{F}$ with pairwise disjoint domains such that $\mathcal{G}_{\varepsilon_\nu} \cap A \in I$.

This is true since $\{\varepsilon' \in \mathcal{F} : \exists \nu < \lambda^+ \; \varepsilon' \stackrel{\supseteq}{=} \varepsilon_\nu\}$ is dense-open. Formerly we used 6.5 as a definition of a good ideal. The two definitions are generally not equivalent.

6.6. If I is a good ideal, $A \notin I$ and $|\mathcal{H}| \leq \lambda^+$ then $\mathcal{H}$ is isomorphic to an induced subgraph of $\mathcal{G}[A]$.

This is true because if we try to embed $\mathcal{H}$ into $\mathcal{G}[A]$ by transfinite induction λ^+-times and we get stuck at each attempt, we get a sequence $\{\varepsilon_\nu : \nu < \lambda^+\} \subset \mathcal{F}$ contradicting 6.5.

As a corollary we get

6.7. For every $|\mathcal{H}| \leq \lambda^+$,

$$\mathcal{G} \rightarrowtail (\mathcal{H})^1_\lambda.$$

But the most important corollary of 6.5 is

6.8. Assume I is a good ideal, $A \notin I$. Then

$$|\{\alpha < \lambda \ : \ \mathcal{G}_i(\alpha) \cap A \in I\}| \leq \lambda \qquad \text{for} \qquad i < 2 \,.$$

Proof. Define ε_α^i by $\text{Dom}(\varepsilon_\alpha^i) = \{\alpha\}$ and $\varepsilon_\alpha^i(\alpha) = i$. If $T = \{\alpha < \lambda \ : \ \mathcal{G}_i(\alpha) \cap A \in I\}$ and $|T| \geq \lambda^+$ then $\{\varepsilon_\alpha^i \ : \ \alpha \in T\}$ contradicts 6.5. $\square$

Now I will try to outline the proof of the Erdős–Hajnal–Pósa theorem $\mathcal{G}[\omega] \gg (\mathcal{U})_{\gamma,\gamma}^2$ for $\gamma < \omega$. We write $\lambda = \omega$ and let I be a good ideal.

Let $f \ : \ [\kappa]^2 \to \gamma$ be a 2–partition of κ ($= 2^\omega$) with γ colors. (Note that if "$\gamma = \delta$" we can use the same function for coloring the edges of G and the edges of $\overline{G}$.)

For $i < 2$, $n < \gamma$ define

$$\mathcal{G}_{i,n}(\alpha) = \{\beta < \kappa \ : \ \{\alpha,\beta\} \in G_i(\alpha) \ \wedge \ f(\{\alpha,\beta\}) = n\},$$

the set of points adjacent to α in $\mathcal{G}_i$, for which $\{\alpha,\beta\}$ gets the color n. Clearly

6.9 $\mathcal{G}_i(\alpha) = \bigcup\limits_{n<\gamma} \mathcal{G}_{i,n}(\alpha)$ for $i < 2$ and $\alpha < \kappa$.

We now state the

Main Lemma *Assume A, B are two disjoint sets with $A, B \notin I$. Then for $i < 2$ there are $A' \subset A$, $B' \subset B$, $A', B' \notin I$ and two colors $n_i < \gamma$ such that for every $A'' \subset A'$, $B'' \subset B'$, $A'', B'' \notin I$, there is an $\alpha \in A''$ with $\mathcal{G}_i(\alpha) \cap B'' \notin I$.*

Indeed assuming indirectly that this is false for some $i < 2$, for every $n < \gamma$, choosing successively a counterexample for each color $n < \gamma$, we would get subsets $A^* \subset A$, $B^* \subset B$, $A^*, B^* \notin I$ such that $\mathcal{G}_{i,n}(\alpha) \cap B^* \in I$ for every $\alpha \in A^*$ and $n < \gamma$. But, by 6.8, $\mathcal{G}_i(\alpha) \cap B^* \notin I$ for some $\alpha \in A^*$ and this contradicts 6.9.

It should now be mentioned that the ideal I of 6.4 is trivially nowhere κ-saturated, i.e. for each $A \notin I$ we can find κ pairwise disjoint subsets of A not in I.

Using the main lemma and the above remark we can choose subsets A_k, B_k for $k < \omega$ in such a way that $A_k, B_k \notin I$, $A_k \cap B_k = \emptyset$, $B_k \supset A_{k+1} \cup A_{k+2} \cup \ldots$ and A_k, B_k satisfy the main lemma with $n_{k,i}$ for $i < 2$. By thinning out, using $\gamma < \omega$, we may really assume $n_{k,i} = n_i$ for $k < \omega$ and $i < 2$. Note that the sets A_k are pairwise disjoint. We plan to pick $\alpha_k \in A_k$ in such a way that $k \mapsto \alpha_k$ is a true embedding of $\mathcal{U} = \langle \omega, U \rangle$ into $\mathcal{G}$, and for $\ell < k < \omega$,

$$\{\ell, k\} \in U \Rightarrow f(\{\alpha_\ell, \alpha_k\}) = n_0,$$
$$\{\ell, k\} \notin U \Rightarrow f(\{\alpha_\ell, \alpha_k\}) = n_1 \,.$$

We will choose α_k by induction on k. Using the fact that A_0, B_0 satisfy the main lemma, we can choose $\alpha_0 \in A_0$ in such a way that for $1 \leq k < \omega$,

$$A_k' = \mathcal{G}_{0,n_0}(\alpha_0) \cap A_k \notin I \quad \text{for} \quad \{0, k\} \in U \qquad \text{and}$$
$$A_k' = \mathcal{G}_{1,n_1}(\alpha_0) \cap A_k \notin I \quad \text{for} \quad \{0, k\} \notin U \,.$$

It should be clear that we can pick $\alpha_1 \in A_1'$ and obtain sets A_k'' for $2 \leq k < \omega$ analogously. Continuing this procedure, the resulting sequence $\alpha_0, \alpha_1, \ldots, \alpha_k, \ldots$ will satisfy our requirements.

7   Product ideals in $\mathcal{G}[\lambda]$.   An outline of the proof of the result of [9].

To generalize the previous proof for infinite γ, at first glance, seems to be quite hopeless. In the proof of the main lemma we will end up with the empty set instead of A^* and B^* and we can only thin out if we get a much longer sequence $A_\alpha \; : \; \alpha < \varphi$ of disjoint sets not in the ideal. In the meantime the Hajnal–Komjáth independence proof tells us that if we get an uncountable sequence of the A_α it "can not be too homogeneous".

Still there is a way out of these difficulties. The first thing is to realize that the main lemma is really a statement about the product ideal $I \times I$ in $\mathcal{P}(A \times B)$. The second idea is that we will have to fix a sequence $A_\alpha \; : \; \alpha < \varphi,\ A_\alpha \notin I$, in advance and try to manipulate ideals in $A = \prod_{\alpha < \varphi} A_\alpha$. First let us choose the parameters. We prepare the embedding of an $\mathcal{H}$ with $|\mathcal{H}| = k < \omega$ into some $\mathcal{G}[\lambda]$ defined on $\kappa = 2^\lambda$ where $f \; : \; [\kappa]^2 \to \gamma$ is a given 2-partition of κ with γ colors.

7.1. We choose $\varphi = \exp_k(\gamma)^+$, $\lambda = \exp_3(\varphi)$, $\kappa = 2^\lambda$ and we choose a good ideal I_0 in $\mathcal{G}[\lambda]$ and pick $A_\alpha \; : \; \alpha < \varphi$ pairwise disjoint sets, $A_\alpha \notin I_0$.

We are going to work in the product

$$A = \prod_{\alpha < \varphi} A_\alpha \; .$$

First we try to lift to the product the small theory of ideals we developed in the previous section. This will go smoothly.

Let P be the set of boxes, i.e.

$$P = \{ B \subset A \; : \; B = \prod_{\alpha < \varphi} B_\alpha \ \wedge \ \forall \alpha < \varphi \ B_\alpha \subset A_\alpha \}.$$

B, C, D will run over the elements of P.

Our starting point is

7.2. Assume I is a λ^+-complete ideal in $\mathcal{P}(\kappa)$, $A_\alpha \notin I$. Then

$$T[I] = \{ B \subset A \; : \; \exists \alpha < \varphi \ B_\alpha \in I \}$$

λ^+-generates a λ^+-complete ideal in $\mathcal{P}(A)$, which we will denote by $J[I]$. $J[I]$ consists of those subsets of A which can be covered with the union of $\leq \lambda$ elements of $T[I]$. It is obvious that $J[I]$ is indeed a λ^+-complete ideal in $\mathcal{P}(A)$.

Now we need a partial order analogous to $\mathcal{F}$ defined in §6.

7.3. We simply write $\mathcal{F}^* = {}^\varphi\mathcal{F}$, i.e. we write

$$\mathcal{F}^* = \{ \underline{\varepsilon} \; : \; \underline{\varepsilon} = \langle \varepsilon_\alpha \; : \; \alpha < \varphi \rangle \ \wedge \ \forall \alpha < \varphi \ \varepsilon_\alpha \in \mathcal{F} \}$$

and we define a partial order $\stackrel{\leq}{=}$ on $\mathcal{F}^*$ by $\underline{\varepsilon} \stackrel{\leq}{=} \underline{\varepsilon}' \Leftrightarrow \forall \alpha < \varphi \ \varepsilon_\alpha \stackrel{\leq}{=} \varepsilon'_\alpha$. It is clear that $\langle \mathcal{F}^*, \stackrel{\leq}{=} \rangle$ still has the λ^+-Baire property, i.e.

7.4. the intersection of $\leq \lambda$ dense-open subsets of $\mathcal{F}^*$ is dense-open in $\mathcal{F}^*$.

For $\underline{\varepsilon} \in \mathcal{F}^*$ define $\mathcal{G}_{\underline{\varepsilon}} = \prod_{\alpha < \varphi} \mathcal{G}_{\varepsilon_\alpha} \cap A_\alpha$.

Note that ε_α can be an arbitrary element of $\mathcal{F}$; we do *not* assume $\mathrm{Dom}(\varepsilon_\alpha) \subset A_\alpha$.
Let now J be an ideal in $\mathcal{P}(A)$ (in what follows we will briefly say an A-ideal) and $X \subset A$.
Just like in 6.2 we define $D^*(X, J) = \{\underline{\varepsilon} \in \mathcal{F}^* \ : \ \mathcal{G}_{\underline{\varepsilon}} \cap X \in J\}$ and

7.5. $X \in \tilde{J} \Leftrightarrow D^*(X, J)$ is dense in $\mathcal{F}^*$.
We now need a little extra lemma for product ideals.

7.6. If I is a good ideal on κ, $A_\alpha \notin I$ for $\alpha < \varphi$, then $A \notin \tilde{J}[I]$.
Having seen this we can proceed to prove the analogue of the lemma given in §6.

Lemma *Assume J is a λ^+-complete A-ideal and $A \notin \tilde{J}$. Then $J \subset \tilde{J}$, $\tilde{J}$ is a λ^+- complete ideal and $\tilde{\tilde{J}} = \tilde{J}$.*

This can be seen with the same proof as the lemma of §6.

7.7. Definition. J is a *good A-ideal* if it is λ^+- complete, $\tilde{J} = J$ and $J[I_0] \subset J$.

Now, just like in §6.

7.8. There is a good A-ideal, since by the lemma, by 7.2 and by 7.6 $\tilde{J}[I_0]$ is a good A-ideal.
It is now very easy to get the analogue of 6.5.

7.9. If J is a good A-ideal and $X \notin J$, then there is no sequence $\langle \underline{\varepsilon}^\nu \ : \ \nu < \lambda^+ \rangle$ such that $G_{\underline{\varepsilon}^\nu} \cap X \in J$ for $\nu < \lambda^+$ and $\langle \varepsilon_\alpha^\nu \ : \ \nu < \lambda^+ \rangle$ have pairwise disjoint domains for every fixed $\alpha < \varphi$.
The next thing to understand is what happens to the all important property 6.8 of the good ideals I, what replaces the sets $\mathcal{G}_i(\alpha)$, $i < 2$. To find this is one of the main ideas of the proof. Let $B \in P$, $\alpha < \varphi$, be given. Let us choose two disjoint sets Δ_i, $i < 2$, $\alpha \notin \Delta_0 \cup \Delta_1 \subset \varphi$. For $\beta \in \Delta_0$ we want "to go with edges into A_β", for $\beta \in \Delta_1$ we intend to use the non-edges.
We will call $\langle \alpha, \Delta_0, \Delta_1 \rangle$ a *triple*. For any such triple and $x \in B_\alpha$ define

$$\mathcal{G}^B_{\alpha, \Delta_0, \Delta_1}(x) = C$$

where

$$C_\beta = \begin{cases} \mathcal{G}_i(x) \cap B_\beta & \text{for } \beta \in \Delta_i, \ i < 2, \\ B_\beta & \text{otherwise .} \end{cases}$$

So $\mathcal{G}^B_{\alpha, \Delta_0, \Delta_1}(x)$ is a subset of the product carrying all the information about $\mathcal{G}_i(x)$'s we need. Using 7.9 we can prove the claim corresponding to 6.8.

7.10. Assume J is a good A-ideal, $B \notin J$, and $\langle \alpha, \Delta_0, \Delta_1 \rangle$ is a triple. Then

$$|\{x \in B_\alpha \ : \ \mathcal{G}^B_{\alpha, \Delta_0, \Delta_1}(x) \notin J\}| \leq \lambda.$$

We now have to find an equality corresponding to the simple equality 6.9.

If $\mathcal{G}^B_{\alpha,\Delta_0,\Delta_1}(x) = C$ and $\beta \in \Delta_i$, then $C_\beta = \mathcal{G}_i(x) \cap B_\beta = \bigcup_{\nu<\gamma} \mathcal{G}_{i,\nu}(x) \cap B_\beta$,

where just like in the previous proof,

$$\mathcal{G}_{i,\nu}(x) = \{y < \kappa \; : \; y \in \mathcal{G}_i(x) \wedge f(\{x,y\}) = \nu\}.$$

For $\Delta_0, \Delta_1 \subset \varphi$ let $\mathcal{N}_{\Delta_0,\Delta_1} = \{\langle g_0, g_1\rangle \; : \; g_i \in {}^{\Delta_i}\gamma; \; i < 2\}$.

We will call a $\langle g_0, g_1\rangle \in \mathcal{N}_{\Delta_0,\Delta_1}$ a Δ_0, Δ_1-pattern.

Let us define $\mathcal{G}^B_{g_0,g_1}(x) = D$ where

$$D_\beta = \mathcal{G}_{i,g_i(\beta)}(x) \cap B_\beta \qquad \text{for} \quad \beta \in \Delta_i, \quad i < 2$$

and $D_\beta = B_\beta$ otherwise.

With all these notations, using simply the distributive law we get the analogue of 6.9.

7.11. Assume J is a good A-ideal, $B \notin J$, $\langle\alpha, \Delta_0, \Delta_1\rangle$ is a triple, $x \in B_\alpha$. Then

$$\mathcal{G}^B_{\alpha,\Delta_0,\Delta_1}(x) = \bigcup\{\mathcal{G}^B_{g_0,g_1}(x) \; : \; \langle g_0, g_1\rangle \in \mathcal{N}_{\Delta_0,\Delta_1}\},$$

and considering that $|\mathcal{N}_{\Delta_0,\Delta_1}| \leq |\gamma|^\varphi < \lambda$ if $\mathcal{G}^B_{\alpha,\Delta_0,\Delta_1}(x) \notin J$ (which, by 7.10 happens most of the time), then for some pattern $\langle g_0, g_1\rangle \in \mathcal{N}_{\Delta_0,\Delta_1}$, $\mathcal{G}^B_{g_0,g_1}(x) \notin J$.

We are now in a position to state the

Main Lemma of this chapter.

Assume J is a good A-ideal, $B \subset A$, $B \in P$, $B \notin J$. Assume further that for each triple $\langle\alpha, \Delta_0, \Delta_1\rangle$ a set $S^0_{\alpha,\Delta_0,\Delta_1} \subset \mathcal{N}_{\Delta_0,\Delta_1}$ is given in such a way that the patterns $\langle g_0, g_1\rangle \notin S^0_{\alpha,\Delta_0,\Delta_1}$ are totally lost, i.e

$$\forall x \in B_\alpha \qquad \mathcal{G}^B_{g_0,g_1}(x) \notin J.$$

(Note that if we choose $S^0_{\alpha,\Delta_0,\Delta_1} = \mathcal{N}_{\Delta_0,\Delta_1}$ then this condition holds vacuously, and indeed we will make this choice in the first application of the main lemma.) Then there are $C \subset B$, $J \subset J'$ and $\emptyset \neq S_{\alpha,\Delta_0,\Delta_1} \subset S^0_{\alpha,\Delta_0,\Delta_1}$ such that $C \notin J'$, J' is a good A-ideal, and

(1) the patterns outside $S_{\alpha,\Delta_0,\Delta_1}$ are totally lost i.e.

$$\forall\langle\alpha, \Delta_0, \Delta_1\rangle \; \forall\langle g_0, g_1\rangle \notin S_{\alpha,\Delta_0,\Delta_1} \; \forall x \in C_\alpha \qquad \mathcal{G}^C_{g_0,g_1}(x) \in J,$$

(2) the patterns in $S_{\alpha,\Delta_0,\Delta_1}$ are not lost for subsets not in J', i.e.

$$\forall\langle\alpha, \Delta_0, \Delta_1\rangle \; \forall\langle g_0, g_1\rangle \in S_{\alpha,\Delta_0,\Delta_1} \; \forall D \subset C \wedge D \notin J' \; \exists x \in D_\alpha \qquad \mathcal{G}^D_{g_0,g_1}(x) \notin J.$$

Note that if in (2) we could say $\mathcal{G}^D_{g_0,g_1}(x) \notin J'$ one could hope for a positive result, arrowing $\mathcal{U}$ with countably many colors, but I was unable to do that.

The proof of this is based on a tree argument or ramification system argument. Assume that at a certain stage indicated by an upper suffix ψ, we have a box B^ψ and for $\alpha, \Delta_0, \Delta_1$ we have sets $S^\psi_{\alpha,\Delta_0,\Delta_1} \subset S^0_{\alpha,\Delta_0,\Delta_1}$ in such a way that (1) holds with B^ψ and $S^\psi_{\alpha,\Delta_0,\Delta_1}$.

Let us look at the ideal J_0 λ^+-generated by the sets of the form:

$$\{D \; : \; \exists\langle\alpha,\Delta_0,\Delta_1\rangle \exists\langle g_0,g_1\rangle \in S^\psi_{\alpha,\Delta_0,\Delta_1} \forall x \in D_\alpha \; \mathcal{G}^D_{g_0,g_1}(x) \in J\} \cup J \;.$$

If $B^\psi \notin \tilde{J}_0$ then $J' = \tilde{J}_0$, $C = B^\psi$ satisfy the requirements of the main lemma. Otherwise we get a partition of B^ψ to at most λ boxes and for each smaller box at least one more pattern is lost. Now standard ramification technique gives us a decreasing sequence of length $(2^{2^\varphi})^+ \leq \lambda$ of boxes such that at every stage, for some triple $\langle\alpha,\Delta_0,\Delta_1\rangle$ we lose a pattern $\langle g_0,g_1\rangle$.

Considering that the number of triples is at most 2^φ, an easy application of the Erdős–Rado theorem gives us that we must have lost $(2^\varphi)^+$ patterns for the same triple $\langle\alpha,\Delta_0,\Delta_1\rangle$. This however contradicts $|\mathcal{N}_{\Delta_0,\Delta_1}| \leq 2^\varphi$.

Now let J_0 be a fixed good A-ideal, $B^0 = A \notin J_0$, $S^0_{\alpha,\Delta_0,\Delta_1} = N_{\Delta_0,\Delta_1}$. Applying the main lemma we get a good A-ideal $J_1 \supset J_0$, $B^1 \subset B^0$, $B^1 \notin J_1$, and $S^1_{\alpha,\Delta_0,\Delta_1} \subset S^0_{\alpha,\Delta_0,\Delta_1}$. Repeating this $k-1$ times we get sequences $B^0 \supset \ldots \supset B^{k-1}$, $J_0 \subset \ldots \subset J_{k-1}$,

$$S^0_{\alpha,\Delta_0,\Delta_1} \supset \ldots \supset S^{k-1}_{\alpha,\Delta_0,\Delta_1} \neq \emptyset,$$

satisfying the requirements of the main lemma.

Write $\langle\alpha,\Delta_0,\Delta_1\rangle \sim \langle\alpha',\Delta'_0,\Delta'_1\rangle$ if there is a monotone map sending $\{\alpha\} \cup \Delta_0 \cup \Delta_1$ onto $\{\alpha'\} \cup \Delta'_0 \cup \Delta'_1$ and this map sends α to α', Δ_i to Δ'_i for $i < 2$.

By the choice of $\varphi = \exp_k(\gamma)^+$ and by the Erdős–Rado theorem there is a $\Gamma \subset \varphi$, $\mathrm{typ}\Gamma = |\gamma|^+$, such that whenever $|\{\alpha\} \cup \Delta_0 \cup \Delta_1| \leq k$ and $\langle\alpha,\Delta_0,\Delta_1\rangle \sim \langle\alpha',\Delta'_0,\Delta'_1\rangle$ and $\{\alpha\} \cup \Delta_0 \cup \Delta_1$, $\{\alpha'_1\} \cup \Delta'_0 \cup \Delta'_1 \subset \Gamma$ then the monotone map sends $S^{k-1}_{\alpha,\Delta_0,\Delta_1}$ onto $S^{k-1}_{\alpha',\Delta'_0,\Delta'_1}$.

Let $\alpha_0 = \min\Gamma$ and split Γ into the union of two disjoint sets Γ_0,Γ_1 of type $|\gamma|^+$.

We know that $S^{k-1}_{\alpha_0,\Gamma_0,\Gamma_1} \neq \emptyset$. This really follows from 7.11.

Let $\langle g_0,g_1\rangle \in S^{k-1}_{\alpha_0,\Gamma_0,\Gamma_1}$. There are $\Gamma'_i \subset \Gamma_i$, $|\Gamma'_i| = |\gamma|^+$ and $\nu_i < \gamma$ for $i < 2$ such that $g_i(\beta) = \nu_i$ for $\beta \in \Gamma'_i$. Let $\bar{\nu}_i$ denote the constant ν_i function for $i < 2$.

Using the fact that for every $\langle\alpha,\Delta_0,\Delta_1\rangle$ with $\{\alpha\} \cup \Delta_0 \cup \Delta_1 \subset \Gamma$ there are $\Delta'_i \subset \Gamma'_i$ with $\langle\alpha,\Delta_0,\Delta_1\rangle \sim \langle\alpha_0,\Delta'_0,\Delta'_1\rangle$ it follows that for every $\{\alpha\} \cup \Delta_0 \cup \Delta_1 \subset \Gamma$, $|\{\alpha\} \cup \Delta_0 \cup \Delta_1| \leq k$,

$$\langle\bar{\nu}_0|\Delta_0,\bar{\nu}_1|\Delta_1\rangle \in S^{k-1}_{\alpha,\Delta_0,\Delta_1} \subset S^j_{\alpha,\Delta_0,\Delta_1} \qquad \text{for} \quad j < k \;.$$

Let $\alpha_0,\ldots,\alpha_{k-1}$ be the first k elements of Γ and let $\mathcal{H} = \langle k, H\rangle$ be given. We claim that we can pick elements $x_n \in B^{k-1}_{\alpha_n}$ in such a way that $n \mapsto x_n$ is a true embedding of $\mathcal{H}$ into $\mathcal{G}$ and for $n < m < k$,

$$\{n,m\} \in H_i \Rightarrow f(\{x_n,x_m\}) = \nu_i \qquad \text{for} \quad i < 2 \;.$$

We can define x_n by induction on n.

Let $\Delta_i = \{1 \leq m < k \; : \; \{0,m\} \in H_i\}$ for $i < 2$.

Let $x_0 \in B^{k-1}_{\alpha_0,\Delta_0,\Delta_1}$ with

$$\mathcal{G}^{B^{k-1}}_{\bar{\nu}_0|\Delta_0,\bar{\nu}_1|\Delta_1}(x_0) \notin J_{k-2} \;.$$

Clearly we can continue this procedure applying the same reasoning for

$$\mathcal{G}^{B^{k-1}}_{\bar{\nu}_0|\Delta_0,\bar{\nu}_1|\Delta_1}(x_0) \qquad \text{and the ordinals} \qquad \alpha_1,\ldots,\alpha_{k-1} \;.$$

The sequence $\{x_n \; : \; n < k\}$ clearly satisfies all our requirements. $\qquad\square$

References

[1] W. Deuber, Partitionstheoreme für Graphen, *Math. Helv.* **50** (1975), 311–320.

[2] M. El-Zahar, N. Sauer, The indivisibility of the homogeneous K_n-free graphs, *Journal of Combinatorial Theory (B)* **47** (1989), 162–170.

[3] P. Erdős, A. Hajnal, On decompositions of graphs, *Acta Math. Acad. Sci. Hung.* **18** (1967), 359–377.

[4] P. Erdős, A. Hajnal, Unsolved problems in set theory, *Proc. Symp. Pure Math.* **13** (1971), part I, 17–48.

[5] P. Erdős, A. Hajnal, Unsolved and solved problems in set theory, *Proc. Symp. Pure Math.* **25** (1974), 269–287.

[6] P. Erdős, A. Hajnal, L. Pósa, Strong embedding of graphs into colored graphs, *Infinite and Finite Sets* (Keszthely, 1973), Coll Math. Soc. J. Bolyai **10**, 585–595.

[7] P. Erdős, R. Rado, A partition calculus in set theory, *Bull. Amer. Math. Soc.* **62** (1956), 427–489.

[8] J. Folkman, Graphs with monochromatic complete subgraphs in every edge coloring, *SIAM J. Appl. Math.* **18** (1970), 115–124.

[9] A. Hajnal, Embedding finite graphs into graphs colored with infinitely many colors, *Israel Journal of Mathematics* **73** (1991), 309–319.

[10] A. Hajnal, P. Komjáth, Embedding graphs into colored graphs, *Trans. Amer. Math. Soc.* **307** (1988), 395–409, and A. Hajnal, P. Komjáth, Corrigendum to "Embedding graphs into colored graphs", *Trans. Amer. Math. Soc.* **322** (1992), 475.

[11] P. Komjáth, V. Rödl, Coloring of universal graphs, *Graphs and Combinatorics* **2** (1986), 55–60.

[12] P. Komjáth, S. Shelah, A consistent partition theorem for infinite graphs, *Acta Math. Hung.*, to appear.

[13] J. Nešetřil, V. Rödl, Partitions of vertices, *Comment. Math. Univ. Carolin.* **17** (1976), 85–95.

[14] J. Nešetřil, V. Rödl, A Ramsey graph without triangles exists for any graph without triangles, *Infinite and Finite sets* (Keszthely, 1973), *Coll. Math. Soc. J. Bolyai* **10** 1127–1132.

[15] J. Nešetřil, V. Rödl, The Ramsey property of graphs with forbidden subgraphs, *Journal of Combinatorial Theory* (B) **20** (1976), 243–249.

[16] J. Nešetřil, V. Rödl, Partitions of finite relational and set systems, *Journal of Combinatorial Theory* (A) **22** (1977), 289–312.

[17] J. Nešetřil, V. Rödl, On Ramsey graphs without cycles of short odd length, *Comment. Math. Univ. Carolin.* **20.3** (1979), 565–582.

[18] S. Shelah, Consistency of positive partition theorems for graphs and models, *Set Theory and Applications* (J. Steprans, S. Watson, eds), Springer Lect. Notes **1401**, 167–193.

Lattices Related to Separation in Graphs

R. HALIN

Universität Hamburg
Mathematisches Seminar
Bundesstrasse 55, D-2000 Hamburg 13
Germany

Abstract

This article provides a brief survey of the literature on lattices generated by cuts (i.e. minimal separating sets) or primitive sets in graphs.

1 Introduction

If a certain vertex a (or a set of vertices A) in a graph G is fixed as its "origin" or "root" and a certain set of configurations in G is considered, then for a pair C, D of these configurations it is natural to call C a predecessor of D (with respect to the origin) if each path from the origin to D meets C. If the set of configurations is chosen appropriately, by this predecessor relation a lattice structure may be induced, and by this idea possibly problems of connectivity and separation in graphs become accessible to lattice theoretical tools. This has been done in several ways, and it is the purpose of this expository article to give a brief survey of the present state of these investigations. A review of the literature on the topic will be provided, with a more detailed exposition of the investigations not yet available in English, in particular the fundamental results of Escalante. Further some recent results of the author on cardinality problems for cuts in infinite graphs will be outlined. Some typical proofs will be carried through, and several open problems will be presented.

In the References the literature is collected which is concerned with graphs and their lattices of separation (as far as it is known to the author). Under "Further References" other articles or books are listed which are used in this paper.

2 Terminology and notation

In this note graphs are assumed to be undirected and not to contain loops or multiple edges; they may be infinite. In general we use the standard notation and terminology of graph theory, in so far as it can be supposed to be understood without extra explanation.

Let G be a graph. If H is a subgraph of G or a subset of $V(G)$, ∂H denotes the set of vertices of $G - H$ which are adjacent to at least one vertex of H. For $a \in V(G)$ then $\partial a \, (= \partial\{a\})$ is the set of neighbours of a; $|\partial a|$ is the degree of a.

If $C \subseteq V(G)$, $a \in V(G) - C$, then C_a will denote the (connected) component of $G - C$ which contains a. By convention, the term ∂C_a has to be read as $\partial(C_a)$. For vertices $a, b \in V(G) - C$ obviously $C_a \neq C_b$ means that C separates a and b in G.

153

N.W. Sauer et al. (eds.), Finite and Infinite Combinatorics in Sets and Logic, 153–167.
© 1993 *Kluwer Academic Publishers. Printed in the Netherlands.*

C is called an a,b-cut if $C_a \neq C_b$, but no proper subset of C also separates a,b (in G).

If S separates a,b, then $\partial(\partial S_a)_b$ is an a,b-cut contained in S; that S is an a,b-cut is equivalent with $S = \partial(\partial S_a)_b$ (or, by symmetry, also with $S = \partial(\partial S_b)_a$).

By $a \| b$ we indicate that $a \neq b$ are non-adjacent vertices. A graph G together with a vertex pair (a,b) with $a \| b$ will be called *double-rooted*. By $\mathcal{C}_G(a,b)$ we denote the set of all a,b-cuts for that double-rooted graph; it is non-empty, as $S = V(G) - \{a,b\}$ separates a,b.

The Menger number (or local connectivity number) $\kappa_G(a,b)$ is equal to $\min |C|$ for all $C \in \mathcal{C}_G(a,b)$; by Menger's theorem it is also the maximal number of internally disjoint a,b-paths in G. (Here an a,b-path may be defined as a connected subgraph of G containing a,b which is minimal with respect to inclusion.) If P is an a,b-path and $x,y \in V(P)$, by P_{xy} we denote the subpath of P which connects x,y. The *internal* vertices of the a,b-path P are the vertices of P different from a and b.

$C \subseteq V(G)$ is called a *cut* of G if there are $a \| b$ such that $C \in \mathcal{C}(a,b)$, or, equivalently, if there are components $H \neq J$ of $G - C$ with $\partial H = \partial J = C$.

A *minimal a,b-cut* is an a,b-cut of cardinality $\kappa_G(a,b)$; let $\mathcal{C}_G^*(a,b)$ denote the set of minimal a,b-cuts in G.

If $a \neq b$ are vertices, an a,b-*edge cut* is a set of edges which separates a,b such that no proper subset also has this separation property. $\overline{\mathcal{C}}_G(a,b)$ will denote the set of a,b-edge cuts in G, and $\overline{\mathcal{C}}_G^*(a,b)$ the set of elements of $\overline{\mathcal{C}}_G(a,b)$ of smallest cardinality.

In a lattice we have the two operations supremum (or join) and infimum (or meet); we denote them by sup and inf, respectively. We also write

$$a \sqcup b, \quad a \sqcap b$$

for the sup and inf of two elements a,b of the lattice in question.

A subset S of a complete lattice L will be called *relatively complete* in L if for any *non-empty* family of elements of S its inf and sup (with respect to L) again belongs to S. Then (if $S \neq \emptyset$) it forms a sublattice of L which is complete and faithful in L with regard to forming infima and suprema of non-empty families; however its zero- (unit-) element may be different from the zero- (unit-) element of L.

For any $M \subseteq L$ there exists the relatively complete hull in L, consisting of all infima and suprema of non-empty families of elements from M. The empty set $\emptyset$ is also considered as a sublattice of L and to be relatively complete.

A lattice L fulfills the *Jordan-Dedekind chain condition* if for each pair a,b of comparable elements of L all maximal chains between a and b have the same finite cardinality.

A lattice L (of finite length) is called *upper semi-modular* if for any $x,y \in L$ holds: if x and y are upper neighbours of ("cover") $x \sqcap y$, then $x \sqcup y$ is an upper neighbour of x and of y.

Dually *lower semi-modular* lattices are defined.

$L \simeq L'$ indicates that the lattices L, L' are isomorphic.

$\mathbb{N}$ denotes the set of natural numbers $1,2,3,\ldots$, and by ω the cardinality of $\mathbb{N}$ is denoted. If S is a set by 2^S the set of its subsets is denoted, and $|S|$ denotes its cardinality.

3   Cut-lattices

Let G with a,b be a double-rooted graph. For $C, D \in \mathcal{C}_G(a,b)$ put $C \trianglelefteq D$, and call C a *predecessor* of D, if $C_a \subseteq D_a$ holds. It is easy to verify that the following statements are equivalent:

   (i) $C \trianglelefteq D$;

   (ii) each path from a to D meets C;

   (iii) each path from b to C meets D;

   (iv) each a, b-path (if oriented from a to b) first meets C, then D.

   Clearly $\trianglelefteq$ is a reflexive and transitive relation in $\mathcal{C}_G(a, b)$. It is also antisymmetric. For assume $C \trianglelefteq D$, $D \trianglelefteq C$; then $C_a \subseteq D_a$ and $D_a \subseteq C_a$, hence $C_a = D_a$, and by $C = \partial C_a$, $D = \partial D_a$ we find $C = D$.

   So $\trianglelefteq$ is a partial order. Clearly the predecessor relation with respect to $\mathcal{C}_G(b, a)$ is the dual of that in $\mathcal{C}_G(a, b)$. Moreover we have:

Theorem 1. $\mathcal{C}_G(a, b)$ *is a complete lattice with respect to $\trianglelefteq$; it is called the cut-lattice of G with respect to a, b. For a non-empty family $C_i (i \in I)$ of a, b-cuts we have*

   (1) $\inf_{i \in I} C_i = \partial(\bigcup_{i \in I} C_i)_a$,

   (2) $\sup_{i \in I} C_i = \partial(\bigcup_{i \in I} C_i)_b$.

Especially we see that the inf and sup of a family of cuts C_i are contained in the union of these cuts. The zero- and the unit-element of the lattice $\mathcal{C}_G(a, b)$ equals $\partial(\partial a)_b$ and $\partial(\partial b)_a$, respectively.

   Proof. By symmetry it suffices to show (2). Let S denote the right-hand side of (2); it consists of all vertices of $\cup C_i$ from which there is an edge into the component H of $G - \cup C_i$ which contains b. Clearly S separates a, b. S is an a, b-cut: Otherwise there would be an $s \in S$ such that also $S - \{s\}$ separates a, b. There is a $j \in I$ with $s \in C_j$. Let P be an a, b-path with $V(P) \cap C_j = \{s\}$. There must be an $s' \in S$ on P before s (otherwise $S - \{s\}$ would not separate a, b). Let x be a neighbour of s' in H. By $P_{as'}$, the edge $[s', x]$ and an x, b-path in H we would find an a, b-path avoiding C_j, which is impossible. Hence S must be an a, b-cut.

   Each a, s-path, for an $s \in S$, meets C_i for every $i \in I$ (otherwise we find, in a similar way, an a, b-path avoiding C_i). Therefore $C_i \trianglelefteq S$ for all $i \in I$.

   If we have $C_i \trianglelefteq T$ for a $T \in \mathcal{C}_G(a, b)$ and all $i \in I$, then let P be an arbitrary a, b-path. P has a last vertex s in common with S. If s is in C_i then there must be a vertex t in T following s on P (possibly $s = t$), by $C_i \trianglelefteq T$. We conclude $S \trianglelefteq T$, and (2) follows. $\square$

   In what follows lattice-theoretical concepts and notation (as sup, inf, $\sqcup$, $\sqcap$, etc.) if used in relation to a (double rooted) graph G will always refer to the predecessor relation $\trianglelefteq$, as in the last theorem.

Theorem 2. *If $a \| b$ and $\kappa_G(a, b) = n < \infty$, then $\mathcal{C}_G^*(a, b)$ is a finite distributive sublattice of $\mathcal{C}_G(a, b)$.*

   Proof. Let $P_1, \ldots, P_n$ be n internally disjoint a, b-paths. If $C \in \mathcal{C}_G^*(a, b)$ then for each $i \in \{1, \ldots, n\}$ exactly one vertex is contained in $P_i \cap C$; so we see that $\mathcal{C}_G^*(a, b)$ is finite.

Let $C, D \in \mathcal{C}_G^*(a, b)$ and c_i, d_i denote the vertex in $C \cap P_i, D \cap P_i$ respectively. If we walk on P_i from a to b, one of the vertices c_i, d_i occurs later (perhaps $c_i = d_i$); denote this vertex by s_i and put $S := \{s_1, \ldots, s_n\}$. S is the sup of C and D. This will be clear if we know that S separates a, b. But otherwise there is a path Q connecting a vertex x of a P_i which lies properly before s_i with a vertex y of a P_j which lies properly behind s_j and having nothing else in common with $H := P_1 \cup \ldots \cup P_n$. But then if s_i is in C, say, C would not separate a, b, which is a contradiction.

We have still to show the distributivity. Obviously $\mathcal{C}_G^*(a, b)$ is a sublattice of $\mathcal{C}_H(a, b) = \mathcal{C}_H^*(a, b)$, which is the direct product (or cardinal product in the sense of Birkhoff [17]) of the chains $\mathcal{C}_{P_i}(a, b)$, and this is a distributive lattice.   □

If $\kappa_G(a, b)$ is infinite, $\mathcal{C}_G^*(a, b)$ can be non-distributive. Namely if $\mathcal{C}_G(a, b)$ is non-distributive, then by adding more than $|V(G)|$ new vertices and joining each of these to a and b by edges we get a graph J so that $\mathcal{C}_J(a, b) = \mathcal{C}_J^*(a, b)$ is isomorphic to $\mathcal{C}_G(a, b)$.

For the position of the elements of $\mathcal{C}_G^*(a, b)$ in $\mathcal{C}_G(a, b)$ in the case of a finite graph G the following result of B. Meyer [12] is of interest. Here the lattice intervals $[X, Y]$ are formed with respect to $\mathcal{C}_G(a, b)$.

Theorem 3. *If G is finite and $a \| b$, then for any $C, D \in \mathcal{C}_G^*(a, b)$*

$$[C \sqcap D, C \sqcup D] \simeq [C \sqcap D, C] \times [C \sqcap D, D].$$

In particular, if $S \in [C \sqcap D, C \sqcup D]$, then

$$[C \sqcap D, C] \simeq [S \sqcap D, S \sqcup C].$$ □

In the set of all a, b-edge cuts of a graph G the predecessor relation $\trianglelefteq$ can be defined quite analogously as for the a, b-cuts. Escalante [1] proved the analogues of Theorems 1 and 2 for edge cuts; in fact the case of edge cuts can be reduced to that of cuts in a certain derived graph which essentially is the line graph of the given graph (see [1] and [10]). By $\lambda_G(a, b)$ we denote the line connectivity of a, b in G (i.e. the maximal number of edge disjoint a, b-paths). Then we have

Theorem 4. *$\overline{\mathcal{C}}_G(a, b)$ is a complete lattice with respect to $\trianglelefteq$, the predecessor relation for edge cuts. If $\lambda_G(a, b) < \infty$ then $\overline{\mathcal{C}_G^*}(a, b)$ is a distributive sublattice of $\overline{\mathcal{C}}_G(a, b)$.*   □

Theorem 5. *If G is finite then $\overline{\mathcal{C}}_G(a, b)$ fulfills the Jordan-Dedekind chain condition.*

The latter result was first obtained by Escalante; he tragically died before he could publish the proof. A proof is given by B. Meyer [11].

Meyer [11, section 3] also considers the case that for the double root a, b distinct ends (i.e. equivalence classes of one-way infinite paths, considered as improper vertices) are chosen. In the case of a locally finite graph he bases his studies on an extension of Menger's theorem [19]; the analogue of Theorem 1 is proved, and the lattice of minimal a, b-cuts (if a, b are ends) may be infinite, but it must be fully distributive (i.e. the distributive laws hold also for infinite families) and compactly generated. (Here the lattices in question must be completed by adding a zero- and a unit-element.) Meyer conjectures that then the lattice of all cuts is compactly generated, where the cuts of finite order are compact elements.

4   The representation problem

In section 3 in various ways lattices were associated with connectivity properties in graphs. It is a natural question to ask which lattices can occur (be represented) as a cut lattice of one of the described types. Here particularly Escalante gave answers in some relevant cases in his doctoral thesis [1].

Theorem 6. (Escalante [1]). *To every complete lattice $(L, \leq)$ there can be found a graph G with vertices $a \| b$ such that the cut lattice $C_G(a, b)$ is isomorphic to L.*

Proof. For $x \in L$ let $\bar{x}$ denote the set of all $y \in L$ with $x \not\leq y$. Let L' be a disjoint copy of L, and let $x \mapsto x'$ be a bijection of L onto L'. For $x' \in L'$, $y \in L$ we draw the edge $[x', y]$ if and only if $y \in \bar{x}$; so we get a bipartite graph (with bipartition L', L).

Further we choose two new vertices $a \neq b$ and draw all edges $[a, x']$, $x' \in L'$, and $[x, b]$, $x \in L$. The arising graph (with vertex set $L \cup L' \cup \{a, b\}$) will be denoted by G. We assert that the lattice of a, b-cuts in G is isomorphic to L.

For $x \in L$ put $C_x := \bar{x} \cup \{y' \in L' | \bar{y} \not\subseteq \bar{x}\}$. Then $x' \notin C_x$, $C_x \cap L = \bar{x}$ and clearly C_x is an a, b-cut. Further one easily checks

$$x \leq y \Longleftrightarrow \bar{x} \subseteq \bar{y} \Longleftrightarrow C_x \trianglelefteq C_y$$

for any $x, y \in L$. Hence $x \mapsto C_x$ is a monomorphism of L into $C_G(a, b)$. To complete the proof we only have to show that this mapping is surjective.

Let $C \in C_G(a, b)$ be given. Let Y be the set of all $y \in L$ with $\bar{y} \subseteq C \cap L$, and let Z be the union of all the sets $\bar{y}$ with $y \in Y$. Clearly $Z \subseteq C \cap L$.

If there were a vertex $c \in (C \cap L) - Z$, then each neighbour x' of c in L' has a neighbour d in $L - C$ (otherwise $c \in Z$); hence all these neighbours x' of c in L' must belong to C, and $C - c$ would separate a, b too, which is a contradiction.

So we find

$$Z = C \cap L.$$

Let x be the sup of Y in $(L, \leq)$. Then $\bar{x} \supseteq \bar{y}$ for all $y \in Y$, hence $\bar{x} \supseteq C \cap L$. If there were a vertex u in $\bar{x} - C$, then, for each $y \in Y$, y' and u would be non-adjacent (since all the neighbours of y' in L lie in C). Therefore $y \leq u$ for all $y \in Y$, hence $x \leq u$, and this contradicts $u \in \bar{x}$. Thus we have

$$C \cap L = \bar{x} = C_x \cap L,$$

and this obviously implies $C = C_x$. □

For any complete lattice L let $\gamma(L)$ denote the smallest cardinal γ such that there exists a graph G with $|V(G)| = \gamma$ which represents L in the sense of Theorem 6. Then the last proof yields

$$\gamma(L) \leq 2 \cdot |L| + 2.$$

Problem: Can this bound for $\gamma(L)$ be improved, especially if L is infinite? For example, is there a countable graph representing the real compact interval $[0,1]$? Perhaps $\gamma(L) = \omega$ for every complete lattice L of cardinality 2^ω; analogous questions also for greater infinite cardinals seem to be quite challenging. Also lower bounds for $\gamma(L)$ (L finite or infinite) seem to be of interest.

Further let $\delta(L)$ be the smallest value of $\sup|C|$, $C \in \mathcal{C}_G(a,b)$, for all representations $\mathcal{C}_G(a,b)$ of L. What bounds can be given for $\delta(L)$? We shall show $\delta(L) = \gamma(L)$ for all infinite L in Section 7.

Another question is: How many different orders (cardinalities) of cuts must at least occur in any finite representation of a finite L? If this minimal number is 1, then L must be distributive (by Theorem 2). Is also the converse true? This means: Can every finite distributive lattice L be represented in a finite double-rooted graph G so that $\mathcal{C}_G(a,b) = \mathcal{C}_G^*(a,b)$?

A further important result of Escalante is the following.

Theorem 7. (Escalante [1]). *To every finite distributive lattice $(L, \leq)$ there exists a finite graph G with vertices $a\|b$ such that $\mathcal{C}_G^*(a,b)$ is isomorphic to L.*

Proof. Let Q be the set of sup-irreducible elements of L; thus $q \in Q$ if and only if q has exactly one lower neighbour with respect to $\leq$. Further let $\overline{L}$ be the set of all initial subsets of Q; this means: $A \subseteq Q$ is in $\overline{L}$ if and only if from $q \in A$, $p \in Q$ and $p \leq q$ it follows $p \in A$.

For $x \in L$ let $\overline{x} := \{q \in Q | q \leq x\}$. Then $\overline{x} \in \overline{L}$ for every $x \in L$. Especially, if 0 and 1 denote the zero- and unit- element of L, then $\overline{0} = \emptyset$, $\overline{1} = Q$. By a well-known decomposition theorem for lattices (see Hermes [20], p. 113 or Birkhoff [17], p. 142)

$$x \mapsto \overline{x} \quad (\text{for } x \in L)$$

is an isomorphism φ of $(L, \leq)$ onto $(\overline{L}, \subseteq)$.

Now let Q' be a disjoint copy of Q and let $x \mapsto x'$ be a bijection of Q onto Q'; for $T \subseteq Q$ let T' denote the image of T under this mapping. For $p \in Q$, $q' \in Q'$ draw the edge $[p, q']$ if and only if $p \leq q$.

Further add two new vertices $a \neq b$ and connect a with all members of Q' and b with all members of Q by edges. In this way a graph G with $V(G) = Q \cup Q' \cup \{a, b\}$ is defined.

Obviously $[q', q]$ is in $E(G)$ for every $q \in Q$, and $Q, Q' \in \mathcal{C}_G^*(a,b)$; we see $\kappa_G(a,b) = |Q|$.
For $F \subseteq Q$ let

$$F^* := F \cup (Q' - F').$$

Clearly each $C \in \mathcal{C}_G^*(a,b)$ equals F^* for $F = Q \cap C$.
If $F \subseteq Q$, then by construction

$$\begin{aligned}
F^* \in \mathcal{C}_G^*(a,b) &\iff \partial F' \cap (Q - F) = \emptyset \\
&\iff \forall_{q \in F} \forall_{p \in Q} ([p, q'] \in E(G) \Rightarrow p \in F) \\
&\iff \forall_{q \in F} \forall_{p \in Q} (p \leq q \Rightarrow p \in F) \\
&\iff F \in \overline{L}.
\end{aligned}$$

We see, by the above isomorphism φ, that $C_G^*(a, b)$ consists of the sets $\bar{x}^*$ with $x \in L$; and because of

$$\bar{x}^* \trianglelefteq \bar{y}^* \iff \bar{x} \subseteq \bar{y} \iff x \leq y$$

(for any $x, y \in L$) the theorem follows.   □

By Theorem 6 and 7 the following problem is suggested. Let (L, L^*) be an ordered pair consisting of a finite lattice L and a distributive sublattice L^* of L. We call the pair (L, L^*) *graph-representable*, briefly: *representable*, if there is a graph G with double root a, b such that an isomorphism φ of $C_G(a, b)$ onto L exists which induces an isomorphism of $C_G^*(a, b)$ onto L^*. Question: *Which pairs are representable?*
    E. Dahlhaus (private communication; see [8], p. 258) was the first to give an example of a non-representable pair (L, L^*); it is reproduced in Fig. 1.

Fig. 1

By Theorem 3 of Meyer a strong restriction of representable pairs is established; obviously it also covers Dahlhaus' example. Meyer [11, pp. 17–20] gave a class of representable pairs; but the problem of characterizing all representable pairs is far from being solved.
    It is easy to carry over Theorems 1, 2, 6, 7 to directed graphs (Escalante [1], p. 217). Many more difficulties arise if one tries to prove the analogs of Theorems 6 and 7 for edge cuts.

Theorem 8. (Escalante–Gallai [2]) *To every finite distributive lattice L^* there is a finite graph G with vertices $a \neq b$ such that $\overline{C}_G^*(a, b)$, the lattice of minimal a, b-edge cuts in G, is isomorphic to L^*.*   □

Escalante's Theorem 5 gives a strong restriction on the possible lattices of edge cuts $\overline{C}_G(a, b)$. But the Jordan–Dedekind chain condition is by far not sufficient for the representability of a finite lattice as the lattice of edge cuts of a graph: Meyer [11, p. 53] pointed out that the distributive lattice of Fig. 2 cannot appear as a $\overline{C}_G(a, b)$. If $G = K_{2,3}$ and $a \neq b$ are vertices of degree 2 in $K_{2,3}$, then $\overline{C}_G(a, b)$ is neither upper nor lower semimodular (its Hasse diagram is a 6-gon), as Escalante [1, p. 219] pointed out.

Fig. 2

The characterization of the lattices representable in the form $\overline{C}_G(a, b)$ remains one of the major open problems in this area.

Further problems which may be of interest arise if we restrict ourselves, in the above representation problems, to certain classes of graphs or to certain classes of lattices. For instance the lattices $C_G(a, b)$ for triangulated (or chordal) graphs G must be chains. (This follows easily from the fact that the triangulated graphs are just those graphs in which every cut induces a complete subgraph.) Which other graphs G also have the property that $C_G(a, b)$ for each pair $a\|b$ is a chain? Conversely: Can every complete chain be represented in a triangulated graph (or interval graph)?

What can be said, for example, about the (edge) cut-lattices of finite (or also infinite) planar graphs? Edge cuts seem to be particularly interesting here, with regard to the rôle they play in the duality problems.

One may ask further which lattices are representable in a vertex-transitive graph, or the like. These are only a few examples from a great variety of open problems.

5   Primitive sets

In this section we briefly consider another way of associating lattices with the connectivity structure of graphs. At first glance it seems to be closely related to the cut lattices, but in fact it leads to quite different problems.

Let $G = (V, E)$ be a graph. For $A, T \subseteq V$ and $i = 1, 2, 3$ we define the *connection graph*

$$G(A \xrightarrow{i} T)$$

of type i from A to T in G as follows:

$G(A \xrightarrow{1} T)$ denotes the subgraph of G induced by all vertices x which can be reached from A by paths $\subseteq G$ which have no internal vertex in common with T (Halin [7]);

$G(A \overset{2}{\rightarrow} T)$ is the subgraph of G formed by all those edges e (together with their end vertices) such that at least one end vertex of e can be reached by a path $\subseteq G$ starting in A and not meeting T (Sabidussi [15]);

$G(A \overset{3}{\rightarrow} T)$ is the union of all the components of $G - T$ which have non-empty intersection with A (Pym and Perfect [14], Polat [13]).

Let $A \neq \emptyset$ be fixed, and for any $T, T' \subseteq V$ put

$$T \overset{i}{\sim} T'$$

if $G(A \overset{i}{\rightarrow} T) = G(A \overset{i}{\rightarrow} T')$.

Clearly each $\overset{i}{\sim}$ ($i = 1, 2, 3$) is an equivalence relation in 2^V. It can be shown that in each of the corresponding equivalence classes there is a unique smallest element (with respect to inclusion); each such smallest representative is called an A-*primitive set* (of vertices) *of type i in G*.

For $T, U \subseteq V$ put $T \trianglelefteq_A U$ if each path in G from A to U meets T. Then we have

Theorem 9. (Halin [8], Sabidussi [15], Polat [13], Pym and Perfect [14]). *With respect to $\trianglelefteq_A$ the A-primitive sets of type i in G form a complete lattice $L_i(G; A)$, $i = 1, 2, 3$. We have $L_i(G; A) \subseteq L_{i+1}(G; A)$ for $i = 1, 2$, and $L_i(G; A)$ is faithful in $L_{i+1}(G; A)$ with respect to forming infima. (In general the faithfulness does not hold for suprema.)*

Clearly for $A = \{a\}$ and any vertex $b \| a$ the predecessor relation in $\mathcal{C}_G(a, b)$ coincides with the restriction of $\trianglelefteq_a$. M. Fuchs [3] showed that $\mathcal{C}_G(a, b)$ is a subset of $L_1(G; a)$ which is faithful with respect to infima; he also gave the following simple example (Fig. 3) showing that in general $\mathcal{C}_G(a, b)$ is not a sublattice of $L_1(G; a)$.

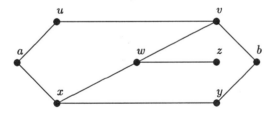

Fig. 3

Let $T = \{v, x\}$, $S = \{u, w, y\}$. The sup of S, T with respect to $L_1(G; a)$ is $\{v, w, y\}$, whereas in $\mathcal{C}_G(a, b)$ it is $\{v, y\}$.

A lattice L will be called L_i-*representable* if there is a graph G and a subset $A \neq \emptyset$ of $V(G)$ such that $L_i(G; A)$ is isomorphic to L.

Using Escalante's method (see the proof of Theorem 6) Fuchs [3] and independently Hager [6] showed

Theorem 10. *Every complete lattice is L_1-representable.*

Fuchs [3] also characterised the lattices which can be represented in the form $L_1(T; a)$ where T is a tree with root a.

The L_3-representable lattices were characterized by Polat [13]. His theorem is very deep, but very complicated and lengthy even to formulate. It implies that each L_3-representable lattice must be upper semimodular; whereas Hager [6] pointed out by an example that it is not necessarily lower-semimodular.

The logical implications between the various kinds of representability were exhibited by Hager [6]:

Theorem 11.

(a) *L_1-representability does not imply L_2-representability.* (This is shown by the lattice in Fig. 4.)

(b) *L_3-representability implies L_2-representability.*

(c) *L_2-representability does not imply L_3-representability.*

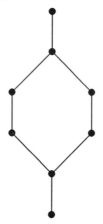

Fig. 4

The characterisation of the L_2-representable lattices remains an open problem.

Without giving any details we mention that the articles by Hager [4], [5], Polat [13], Sabidussi [15], [16] to a large extent are concerned with establishing morphisms of graphs and lattices in such a way that the relations between graphs (in which a "root" A is distinguished) and their various lattices of primitive sets result in functors of the corresponding categories. (The morphisms of graphs are certain contractions.)

6 Lattices generated by finite cuts

Let us return to the study of the cut lattices $\mathcal{C}_G(a,b)$. In recent articles [9],[10] the present author considered questions of cardinality related to these lattices. As basic problems we may ask for example: If a cardinal δ is given, can an upper bound for the cardinality of the set of a,b-cuts of order $\leq \delta$ be given? What can be said about the sublattice generated by these cuts? What follows about the structure of the underlying graph G if many a,b-cuts of order $\leq \delta$ exist?

If δ is a cardinal, by $\mathcal{C}_G(a,b;<\delta)$, $\mathcal{C}_G(a,b;\leq\delta)$ we denote the set of a,b-cuts of cardinality $<\delta$, or $\leq\delta$, respectively. In this section $\mathcal{C}_G(a,b;<\omega)$ will be studied, whereas in the next section we shall be concerned with $\mathcal{C}_G(a,b;\leq\delta)$ for infinite δ.

If $\mathcal{M} \subseteq \mathcal{C}_G(a,b)$, by $\mathcal{L}(\mathcal{M})$ we denote the sublattice (with respect to $\trianglelefteq$) of $\mathcal{C}_G(a,b)$ generated by $\mathcal{M}$, and $\mathcal{L}^c(\mathcal{M})$ will denote the relatively complete sublattice generated by $\mathcal{M}$ in $\mathcal{C}_G(a,b)$. $\mathcal{L}^c(\mathcal{M})$ consists of the infima and suprema of all non-empty families of elements $\in \mathcal{M}$. If $\mathcal{M}$ is empty than also $\mathcal{L}(\mathcal{M}) = \mathcal{L}^c(\mathcal{M}) = \emptyset$, by convention.

By Theorem 1 we know that the infima and suprema of any family C_i ($i \in I; I \neq \emptyset$) of a,b-cuts are subsets of $\bigcup_{i\in I} C_i$. Hence $\mathcal{C}_G(a,b;<\delta)$ and $\mathcal{C}_G(a,b;\leq\delta)$ are sublattices of $\mathcal{C}_G(a,b)$ for every infinite δ. In particular the finite a,b-cuts form a sublattice of $\mathcal{C}_G(a,b)$; and we also see that for any finite $\mathcal{M} \subseteq \mathcal{C}_G(a,b;<\omega)$ the lattice $\mathcal{L}(\mathcal{M})$ is again finite. By Theorem 1 we do not, however, get an estimate of the number of finite a,b-cuts. Basic for this section is the following observation (Halin [9], [10]):

Theorem 12. *For any $n \in \mathbf{N}$ and any double-rooted graph G we have*

$$|\mathcal{C}_G(a,b;\leq n)| < \infty.$$

Proof. By induction on n. For $n = 1$ the cuts to be considered are just the articulation vertices separating a and b; they all must lie on any a,b-path in G. If $n \geq 2$ and our assertion is true for $n-1$, then let P be any a,b-path. If there were infinitely many C_i in $\mathcal{C}_G(a,b;\leq n)$ then there would be infinitely many of them which share the same vertex x of P; the corresponding sets $C_i - \{x\}$ would form infinitely many distinct a,b-cuts of order $\leq n-1$ in $G - \{x\}$, which contradicts the induction hypothesis. $\square$

So we have the alternative that $\mathcal{C}_G(a,b;<\omega)$ is either finite or countable. It is the latter alternative which interests us in the first place.

If $C_1 \triangleleft C_2 \triangleleft C_3 \triangleleft \ldots$ is a strictly increasing sequence of finite a,b-cuts, then $\sup_{i\in\mathbf{N}} C_i = S$ must be infinite. This follows easily from Theorem 1. (Namely if $s \in S$ then $s \in C_i$ for some $i \in \mathbf{N}$, and then $s \in C_i$ for all $n > i$. If S were finite we would find $S = C_n$ for sufficiently large n, which is impossible.)

In [10] the following is proved.

Theorem 13. *If $\mathcal{M}$ is an infinite set of finite a,b-cuts, then there is a strictly monotonic (decreasing or increasing) sequence of finite a,b-cuts in $\mathcal{L}(\mathcal{M})$. Hence, by the preceding remark, $\mathcal{L}(\mathcal{M})$ cannot be a relatively complete sublattice of $\mathcal{C}_G(a,b)$.*

Remark: Simple examples [9] show that such a monotonic sequence cannot always be found in $\mathcal{M}$ itself, so that really $\mathcal{L}(\mathcal{M})$ is needed.

In particular we see from Theorem 13 that, if $C_G(a, b; < \omega)$ is infinite, this sublattice of $C_G(a, b)$ is never faithful with respect to forming infima and suprema of countable families. Then $\mathcal{L}^c(C_G(a, b; < \omega))$ contains $C_G(a, b; < \omega)$ properly. What can be said about its cardinality? Clearly this cardinality cannot exceed 2^ω. Indeed this cardinal can be attained. An example is given in [10].

What configurations must be present if G contains infinitely many finite a, b-cuts? It is then, by Theorem 13, no restriction of generality to suppose that a strictly increasing sequence of such cuts exists. In [10] the following result was proved:

Theorem 14. *Let $C_1 \lhd C_2 \lhd C_3 \lhd \ldots$ be a strictly increasing sequence of finite a, b-cuts in a graph G and put $\sup_{n \in \mathbf{N}} C_n = S$. Further let T denote the set of vertices appearing in C_n for infinitely many n. Then there is a one-way infinite path R in G starting at a and disjoint from T, and there are infinitely many disjoint paths P_n in G where each P_n connects a vertex r_n of R with a vertex s_n of S and has nothing but r_n, s_n in common with R and T. In particular, G must contain a subdivision of one of the graphs in Fig. 5 and 6 (where in Fig. 6 b and b' may coincide). Here the circles left white inside indicate vertices from S.*

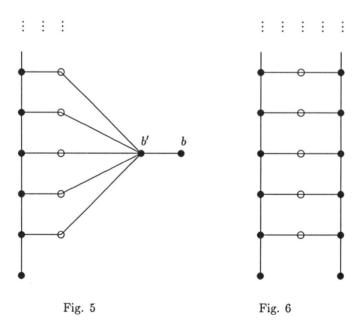

Fig. 5   Fig. 6

The graphs of Fig. 5 and 6 (and all their subdivisions) themselves satisfy the hypotheses of Theorem 14. Therefore they may be considered, in a certain sense, as the minimal configurations, or prototypes, of all the graphs G with a pair $a \| b$ such that $C_G(a, b; < \omega)$ is infinite, and Theorem 14 is best possible.

At this point we put the following problem: Can the real compact interval $[0, 1]$ be represented in the form $\mathcal{L}^c(C_G(a, b; < \omega))$ with a countable graph G such that the finite a, b-cuts correspond to the rationals?

7 Infinite a, b-cuts

We now study the a, b-cuts of a graph G whose cardinalities do not exceed a given infinite cardinal δ. The graph P_δ consisting of δ internally disjoint a, b-paths of length 3 shows that for a graph G of cardinality δ we may have $|\mathcal{C}_G(a, b; \leq \delta)| = 2^\delta$. The bound 2^δ however cannot be exceeded for any set of the form $\mathcal{C}_G(a, b; \leq \delta)$. This is a consequence of the following:

Theorem 15. (Halin [10]). *Let δ be an infinite cardinal. Then for any graph G and any pair $a\|b$ of vertices of G there exists a graph A with the following properties:*

1) $|V(A)| \leq \delta$;

2) $a, b \in V(A) \subseteq V(G)$;

3) *two vertices $x \neq y$ of A are adjacent in A if and only if they either are adjacent in G or fulfill $\kappa_G(x, y) > \delta$;*

4) $\mathcal{C}_A(a, b) = \mathcal{C}_G(a, b; \leq \delta)$, *and this equality also holds for the lattice structures;*

5) $\mathcal{C}_A(a, b)$ *is a relatively complete sublattice of $\mathcal{C}_G(a, b)$.*

As corollaries we immediately have:

Theorem 15'. *If δ is an infinite cardinal, then $\mathcal{C}_G(a, b; \leq \delta)$ is a complete lattice of cardinality $\leq 2^\delta$, and it is a relatively complete sublattice of $\mathcal{C}_G(a, b)$.*

Theorem 15''. *For every infinite complete lattice L we have $\gamma(L) = \delta(L)$.*

Proof. Clearly $\gamma(L) \geq \delta(L)$, and, by Theorems 12, 13, $\delta(L)$ must be infinite. Let $\mathcal{C}_G(a, b)$ be a representation of L with $|C| \leq \delta(L)$ for all $C \in \mathcal{C}_G(a, b)$. Then by Theorem 15 we find a graph A with $|V(A)| \leq \delta(L)$ and also representing L; hence $\gamma(L) \leq \delta(L)$. $\square$

Just to give a flavour of the ideas which are involved in this area of infinite graph theory we shall present a proof of the partial statement of Theorem 15' that $|\mathcal{C}_G(a, b; \leq \delta)| \leq 2^\delta$. For a full proof of Theorem 15 the reader is referred to [10, section 4].

For each vertex pair $x\|y$ of G with $\kappa_G(x, y) \leq \delta$ we choose an $f(x, y) \subseteq V(G)$ which separates x, y in G and has cardinality $\leq \delta$. If $J \subseteq G$, then J_f is defined as the subgraph (of G) induced by J and all the $f(x, y)$ with $x, y \in V(J)$, $x\|y$ and $\kappa_G(x, y) \leq \delta$. Now let $H_0 = \{a, b\}$ and define the sequence $H_0 \subseteq H_1 \subseteq H_2 \subseteq \ldots$ inductively by

$$H_{n+1} := (H_n)_f.$$

Clearly $H := \bigcup_{n \in \mathbb{N}} H_n$ is an induced subgraph of G. By induction we find $|V(H_n)| \leq \delta$ for every n, and therefore we also have $|V(H)| \leq \delta$.

Now it suffices to show that every a, b-cut in G of cardinality $\leq \delta$ is contained in H. Otherwise there is a $C \in \mathcal{C}_G(a, b; \leq \delta)$ which contains a vertex c not in H. Let P be an a, b-path having only c in common with C. There are vertices x in C_a and y in C_b lying on P such that P_{xy} contains c and has only x, y in common with H. C also separates x, y, because of $C_a = C_x \neq C_b = C_y$; hence $\kappa_G(x, y) \leq \delta$. There is an $n \in \mathbb{N}$ with $x, y \in V(H_n)$;

then $f(x,y) \subseteq H_{n+1} \subseteq H$. But P_{xy} shows that x,y cannot be separated (in G) by a subset of $V(H)$. By this contradiction the proof is complete.   □

If δ is an uncountable limit cardinal one may ask analogous questions as were treated in Section 6 for ω. If the generalized continuum hypothesis is assumed it follows from Theorem 15$'$ that $\mathcal{C}_G(a,b;<\delta)$ has cardinality $\leq \delta$. The problem whether $\mathcal{L}^c(\mathcal{C}_G(a,b;<\delta)$ can have cardinality 2^δ (as in the case ω) is left open.

With regard to Theorem 14 it is worth mentioning that for the graph P_δ there are chains of length 2^δ in $\mathcal{C}_G(a,b;\leq\delta)$ (for $\delta \geq \omega$) and nevertheless P_δ does not even contain a finite path of length 6.

It is not difficult to extend the results of sections 6 and 7 to edge cuts (see [10, section 5]).

References

[1]   F. Escalante, Schnittverbände in Graphen, *Abh. Math. Sem. Univ. Hamburg* **38** (1972), 199–220.

[2]   F. Escalante and T. Gallai, Note über Kantenschnittverbände in Graphen, *Acta Math. Acad. Sci. Hungar.* **25** (1975), 93–98.

[3]   M. Fuchs, Verbände A-primitiver Mengen und Schnittverbände in Graphen, Diploma Thesis, Hamburg, 1978.

[4]   M. Hager, Primitive Mengen in Graphen, Doctoral Thesis, Hamburg, 1980.

[5]   M. Hager, On Halin-lattices in graphs, *Discrete Math.* **46** (1983), 235–246.

[6]   M. Hager, Primitive sets in graphs, *Europ. J. Combin.* **3** (1982), 29–34.

[7]   R. Halin, Über trennende Eckenmengen in Graphen und den Mengerschen, *Satz. Math. Ann.* **157** (1964), 31–41.

[8]   R. Halin, *Graphentheorie*, 2nd Edition, Berlin–Darmstadt, 1989, Chap. 11, §1.

[9]   R. Halin, Some finiteness results concerning separation in graphs, to appear in *Discrete Math.*

[10] R. Halin, Lattices of cuts in graphs, submitted.

[11] B. Meyer, Algebraische Aspekte des Trennens und Verbindens in Graphen, Doctoral Thesis, Hamburg, 1983.

[12] B. Meyer, On the lattices of cutsets in finite graphs, *Europ. J. Combin.* **3** (1982), 153–157.

[13] N. Polat, Treillis de séparation des graphes, *Canad. J. Math.* **28** (1976), 725–752.

[14] J.S. Pym and H. Perfect, Submodular functions and independence structures, *J. Math. Anal. Appl.* **30** (1970), 1–31.

[15] G. Sabidussi, Weak separation lattices of graphs, *Canad. J. Math.* **28** (1976), 691–724.

[16] G. Sabidussi, On maps related to Halin separation lattices, *Europ. J. Combin.* **6** (1985), 257–264.

FURTHER REFERENCES

[17] G. Birkhoff, *Lattice Theory*, New York, 1948.

[18] R. Diestel, *Graph Decompositions – a Study in Infinite Graph Theory*, Oxford University Press, Oxford, 1990.

[19] R. Halin, A note on Menger's theorem for infinite locally finite graphs, *Abh. Math. Sem. Univ. Hamburg* **40** (1974), 111–114.

[20] H. Hermes, *Verbandstheorie*, Springer, Berlin, 1955.

Ramsey Numbers for Sets of Five Vertex Graphs With Fixed Number of Edges

HEIKO HARBORTH

Technische Universität Braunschweig, Germany

From time to time I like to hunt for small Ramsey numbers. The classical Ramsey number $r = r(G, H)$ is the smallest r, such that every 2–coloring of the edges of the complete graph K_r contains a graph G with all edges of color 1, or a graph H with all edges of color 2. In [1] it was proposed to ask for the special Ramsey numbers for sets of graphs having fixed numbers of vertices and edges, that means to determine the smallest $r = r_{m,n}(s, t)$, such that every 2–coloring of the edges of K_r contains any graph with m vertices and s edges of color 1, or any graph with n vertices and t edges of color 2. Since $1 \le s \le \binom{m}{2}$ and $1 \le t \le \binom{n}{2}$ an $\binom{m}{2}$ by $\binom{n}{2}$ rectangular array of Ramsey numbers has to be determined for graphs with m and n vertices.

All values $r_{3,n}(s, t)$ for $n \le 7$ are given in [4]. The table for $r_{4,5}(s, t)$ is complete up to $r_{4,5}(6, 10) = r(K_4, K_5)$ and can be found in [1] or [6]. Here the table for $r_{5,5}(s, t) = r_5(s, t)$ is discussed.

Since $r_n(s, t) = r_n(t, s)$ the values below the main diagonal follow by symmetry. It is easy to see that

$$r_n(s, t) = n \quad \text{if and only if} \quad s + t \le 1 + \binom{n}{2},$$

so that only the numbers below the secondary diagonal and not below the main diagonal remain in question (see Table 1). The Ramsey numbers on the first and second parallel line to the secondary diagonal are determined for $r_n(s, t)$ in [6]. On the main diagonal $r_5(7, 7) = 10$ and $r_5(8, 8) = 14$ are proved in [5], and $r_5(9, 9) = 22$ can be found in [2]. For the bounds of $r_5(9, 10) = r(K_5 - e, K_5)$ and $r_5(10, 10) = r(K_5)$ see [3,7].

Theorem 1. $r_5(4, 10) = 9$.

Proof. If a coloring of K_9 contains a $c1K_3$ (graph K_3 of color 1) then any other $c1e$ (edge of color 1) determines a $c1(K_5 - 6e)$ (graph with 5 vertices and 4 edges of color 1). Thus one vertex of this $c1K_3$ together with 4 other vertices determine a $c2K_5$. Then the only connected graphs with at most 3 $c1e$'s are K_1, K_2, P_3, P_4, and $K_{1,3}$ (P_i is a path with i vertices). Any collection of disjoint copies of these graphs in K_9 leaves 5 independent vertices, that is, a $c2K_5$, and ≤ 9 is proved. Two $c1P_4$'s in K_8 prove > 8 (see Figure 1).

N.W. Sauer et al. (eds.), Finite and Infinite Combinatorics in Sets and Logic, 169–174.
© 1993 *Kluwer Academic Publishers. Printed in the Netherlands.*

$s \backslash t$	1	2	3	4	5	6	7	8	9	10
1	5	5	5	5	5	5	5	5	5	5
2		5	5	5	5	5	5	5	5	6
3			5	5	5	5	5	5	7	9
4				5	5	5	5	6	7	9
5					5	5	7	8	10	13
6						6	7	9	11	14
7							10	11	13	
8								14		
9									22	30–34
10										43–52

Table 1. $r_5(s,t)$.

Theorem 2. $r_5(5,9) = 10$.

   Proof. If $c2d \geq 7$ for one vertex in K_{10} then $r_{5,4}(5,5) = 7$ guarantees a $c1(K_5 - 5e)$ or a $c2(K_5 - e)$. If $c1d \geq 4$ for one vertex v then the 4 endvertices determine, together with v, a $c1(K_5 - 5e)$, or they determine a $c2K_4$. Any further vertex of K_{10} has 3 $c2e$'s to this $c2K_4$, and a $c2(K_5 - e)$ occurs, or 2 $c1e$'s to this $c2K_4$, and together with v a $c2(K_5 - 5e)$ is determined. — It remains that all vertices of K_{10} have $c1d = 3$. Then all 15 $c1e$'s determine a $c1C_3, c1C_4$, or $c1C_5$ (C_i is the cycle graph with i vertices) which imply a $c1(K_5 - 5e)$. — The proof for > 9 follows from 3 $c1K_3$'s in K_9 (see Figure 1).

Theorem 3. $r_5(6,8) = 9$.

   Proof. If $c2d \geq 6$ for a vertex in K_9 then $r_{5,4}(6,4) = 6$ implies a $c1(K_5 - 4e)$ or a $c2(K_5 - 2e)$. — Since not all vertices of K_9 can have $c1d = 3$ it remains to assume a vertex v with $c1d \geq 4$. Without a $c1(K_5 - 4e)$ the 4 endvertices determine at most one $c1e$, and then exactly one $c1e$, say (v_1, v_2), occurs since any of the remaining vertices of K_9 has exactly 2 $c1e$'s and 2 $c2e$'s to the 4 endvertices if a $c1(K_5 - 4e)$ and a $c2(K_5 - 2e)$ is avoided. Because of $c2d \leq 5$ one $c1e$ connects v_1 to a remaining vertex w, and v, w, v_1, v_2, together with another endvertex gives $c1(K_5 - 4e)$. — The proof for > 8 follows from the cube graph of color 1 (see Figure 1).

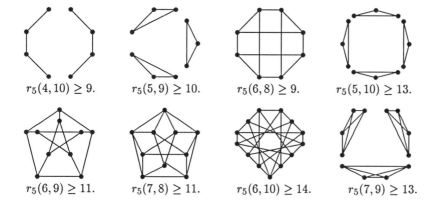

$r_5(4,10) \geq 9.$	$r_5(5,9) \geq 10.$	$r_5(6,8) \geq 9.$	$r_5(5,10) \geq 13.$
$r_5(6,9) \geq 11.$	$r_5(7,8) \geq 11.$	$r_5(6,10) \geq 14.$	$r_5(7,9) \geq 13.$

Figure 1.

Theorem 4. $r_5(5,10) = 13.$

Proof. For $c2d \geq 10 = r_{5,4}(5,6)$ a $cl(K_5 - 5e)$ or a $c2K_5$ occurs in K_{12}. — If $cld \geq 4$ for one vertex then the 4 endvertices determine a $c2K_4$, or a $cl(K_5 - 5e)$ exists. Because of $c2d \leq 9$ any endvertex is connected by 2 cle's to 2 other vertices. These are connected by $c2e$'s to one another and to the 3 other endvertices if a $cl(K_5 - 5e)$ is avoided, and then a $c2K_5$ is determined. — It remains that $cld = 3$ for all vertices of K_{13}. However, this is impossible since twice the number of cle's cannot equal $3 \cdot 13$. — Four disjoint clK_3's in K_{12} prove > 12 (see Figure 1).

Theorem 5. $r_5(6,9) = 11.$

Proof. For $c2d \geq 7 = r_{5,4}(6,5)$ and for $cld \geq 5 = r_{4,5}(2,9)$ a $cl(K_5 - 4e)$ or a $c2(K_5 - e)$ exists in K_{11}. — It remains $cld = 4$ for all vertices of K_{11}. Let S be the set of 4 vertices with cle's to a vertex v, and T be the set of the remaining 6 vertices of K_{11}. Without a $cl(K_5 - 4e)$ at most one cle is in S. Both vertices of this cle have 2 cle's to 4 different vertices of T, or a $cl(K_5 - 4e)$ occurs. Any cle between these 4 vertices in T determines 2 clK_3's with a common vertex or a clC_5 with a cle as diagonal, that is, a $cl(K_5 - 4e)$, or there is a $c2K_4$ which together with v gives a $c2K_5$. If no cle is in S then every vertex has a triple of cle's to T such that every vertex of T has a pair of cle's to S. Every triple determines a $c2K_3$, or a $cl(K_5 - 4e)$ occurs. Thus there are $4\binom{3}{2}$ $c2e$'s, that is, $\binom{6}{2} - 12 = 3$ cle's in T, whereas there should be 6 cle's in T because of $cld = 4$. — The Petersen graph of color 1 in K_{10} proves > 10 (see Figure 1).

Theorem 6. $r_5(7,8) = 11.$

Proof. For $c2d \geq 7 = r_{5,4}(7,4)$ and for $cld \geq 6 = r_{4,5}(3,8)$ a $cl(K_5 - 3e)$ or a $c2(K_5 - 2e)$ exists in K_{11}. Since $cld = 5$ (= odd) for all vertices of K_{11} is impossible one vertex v with $c2d = 6$ remains to be discussed. Let S be the set of 6 endvertices.
Avoiding a $cl(K_5 - 3e)$ and a $c2(K_4 - 2e)$ in S the edges of S are colored as follows.

Because of $r(C_4, C_4) = 6$ a $c1C_4$ can be assumed. Its diagonals are $c2e$'s since at least 3 $c2e$'s from each of the remaining vertices of S to the $c1C_4$ determine a $c2C_4$. Then each of the remaining 2 vertices of S has exactly 2 $c1e$'s and 2 $c2e$'s to the $c1C_4$. All possibilities lead to 2 colorings of the edges of S, where the graph of color 2 consists (1) of 2 disjoint $c2K_3$'s, or (2) of a $c2C_6$.

In case (1) any of the remaining vertices of K_{11} has at least 2 $c1e$'s to each of the 2 $c2K_3$'s so that a $c1(K_5 - 2e)$ is guaranteed. In case (2) each of the 4 remaining vertices has exactly 2 $c1e$'s to S, and these are possible only to a pair of opposite vertices of the $c2C_6$. Thus 2 of the 4 remaining vertices, one of these pairs, and v determine a $c1(K_5 - 3e)$.

The corresponding $c1$–graph in Figure 1 proves > 10. It contains 5 edge disjoint $c1K_3$, so that no $c1(K_4 - e)$ occurs. However, every $K_5 - 3e$ contains a $K_4 - e$. In the corresponding $c2$–graph every vertex has $c2d = 5$. The 5 endvertices determine a $c2C_5$ or a $c2P_5$, so that 4 of the endvertices determine at most 3 $c2e$'s. Since every $K_5 - 2e$ contains a vertex of degree 4 no $c2(K_5 - 2e)$ exists in this coloring of K_{10}.

Theorem 7. $r_5(6, 10) = 14$.

Proof. If $c2d \geq 10 = r_{5,4}(6,6)$ then a $c1(K_5 - 4e)$ or a $c2K_5$ exists in K_{14}. From $r(K_3, K_4) = 14$ a $c1K_3$ can be assumed if a $c2K_5$ is avoided. Since $c1d \geq 4$ for all vertices at least 2 $c1e$'s are incident to each of the vertices of the $c1K_3$. The 6 endvertices are disjoint and connected by $c2e$'s only, or $c1(K_5 - 4e)$ occurs, and thus a $c2K_6$ is determined.

The proof of > 13 follows from the $c1$–graph in Figure 1. No 5 independent vertices occur since there exist 2 vertices at distance 2 on the 13–gon, and among the 4 vertices nonadjacent to these vertices only 2 are independent. Thus no $c2K_5$ exists in the corresponding $c2$–graph.

Any $K_5 - 4e$ contains a K_3 or a C_4. Since none of both types of diagonals in the 13–gon is side of a triangle or of a C_4 also a $c1(K_5 - 4e)$ does not occur.

Theorem 8. $r_5(7, 9) = 13$.

Proof. If $c2d \geq 10 = r_{5,4}(7,5)$, or if $c1d \geq 7 = r_{4,5}(3,9)$ then a $c1(K_5 - 3e)$ or a $c2(K_5 - e)$ exists in K_{13}. Since not all degrees can be odd it suffices to discuss that one vertex occurs with (1) $c1d = 6$ or (2) $c2d = 8$.

Figure 2. Color 1 of all K_8 without $c1(K_5 - 3e)$ and $c2(K_4 - e)$.

(1) Let v be a vertex which is connected to the set S of 6 vertices v_i by 6 $c1e$'s. Without $c1(K_5 - 3e)$ and $c2(K_5 - e)$ there are only two possibilities to color the edges within S: (a) (v_1, v_2), (v_3, v_4), and (v_5, v_6) or (b) (v_1, v_2), (v_2, v_3), (v_4, v_5) and (v_5, v_6) are $c1e$'s, and all other edges are of color 2.

(a) Without $c1(K_5 - 3e)$ and $c2(K_5 - e)$ any of the 6 remaining vertices w_i can only have a pair of $c1e$'s to both vertices of a $c1e$ in S, or a triple of $c1e$'s each to one vertex of the 3 $c1e$'s in S. Since 2 pairs of $c1e$'s from w_i and w_j to the same $c1e$ in S determine a $c1(K_5 - 3e)$ at most 3 pairs of $c1e$'s are possible. Since 2 triples of $c1e$'s from w_i and w_j to S which are connected to more than one common vertex of S, force a $c1(K_5 - 3e)$ or a $c2(K_5 - e)$ there exist at most 4 such triples. A triple and a pair of $c1e$'s from w_i and w_j to S, however, determine a $c1(K_5 - 3e)$ or a $c2(K_5 - e)$.

(b) From each of the remaining 6 vertices w_i only exactly 2 $c1e$'s are possible, which can only be incident to endvertices of different of the 2 $c1P_3$'s in S. Since there are 4 possibilities for 2 of the w_i's the pairs of $c1e$'s are connected to the same pair of vertices of S, and a $c2(K_5 - e)$ can be found.

(2) At first all colorings of K_8 without $c1(K_5 - 3e)$ and $c2(K_4 - e)$ are constructed. If $c1d \geq 5$ in K_8 then without $c1(K_5 - 3e)$ at most 3 $c1e$'s are possible between the endvertices, and in all possibilities a $c1(K_5 - 3e)$ or a $c2(K_4 - e)$ follows. If $c2d \geq 5$ in K_8 then any $c2P_3$ implies a $c2(K_4 - e)$, and at most 2 $c1e$'s guarantee a $c1(K_5 - 2e)$. Thus only degrees 3 and 4 remain to be discussed.

(i) All 8 vertices have $c2d = 3$: Let v be connected by 3 $c1e$'s to v_1, v_2, v_3. — Let one edge, say (v_1, v_2), be a $c2e$. Then $c2e$'s (v_1, w_1) and (v_2, w_2) exist, and v_1 is connected by $c1e$'s to w_2, w_3, w_4. Without $c1(K_5 - 3e)$ in (v, v_1, w_2, w_3, w_4) the vertices w_2, w_3, w_4 determine a $c2K_3$. Since (v_1, w_3), (v_1, w_4), (v_2, w_3), (v_2, w_4), (v_1, v_3), and (v_2, v_3) are already 6 $c1e$'s the vertices v_3, w_3, w_4 determine another $c2K_3$. Both together form a $c2(K_4 - e)$. — Now it can be assumed that no $c2K_3$ occurs. Then exactly 3 $c2e$'s, and thus also 3 $c1e$'s, are within the vertices w_1, w_2, w_3, w_4 which together with v imply a $c1(K_5 - 3e)$.

(ii) One vertex v with $c2d = 4$ exists: The 4 endvertices v_i can only determine (α) none, (β) one, and (γ) two disjoint $c2e$'s.

(α) Without $c1(K_5 - 3e)$ the remaining 3 vertices w_i have only $c2e$'s to the 4 vertices v_i which form a $c1K_4$. Then without a $c2(K_4 - e)$ the 3 vertices w_i and v determine another $c1K_4$, and a coloring of K_8 is complete (its $c1$–graph is the first in Figure 2).

(β) Without $c1(K_5 - 3e)$ at most one $c1e$ connects each w_i to the vertices v_i. Without $c2(K_4 - e)$ at least one $c1e$ connects each w_i to the one $c2e$, say (v_1, v_2). Thus all 3 vertices w_i are connected by $c2e$'s to v_3 and v_4. If one edge (w_i, w_j) is a $c2e$ then with v_3 and v_4 a $c2(K_4 - e)$ is given, and otherwise w_1, w_2, w_3 together with v and v_1 or v_2 determine a $c1(K_5 - 3e)$.

(γ) Now for all vertices v with $c2d = 4$ it can be assumed that the vertices v_i determine a $c1C_4$, say (v_1, v_2, v_3, v_4), where the 2 diagonals (v_1, v_3) and (v_2, v_4) are $c2e$'s. Each w_i has at most 2 $c1e$'s to the $c1C_4$, or a $c1(K_5 - 3e)$ is obtained, and at least 2 $c1e$'s or a $c2(K_4 - e)$ occurs. No pair of $c2e$'s from w_i to a diagonal of the $c1C_4$ is possible since otherwise w_i, the corresponding pair of $c2e$'s, and v determine a $c2(K_4 - e)$. The pairs of $c1e$'s from w_i and w_j cannot go to the same $c1e$ of the $c1C_4$ since otherwise with v a $c1(K_5 - 3e)$ exists. Thus the 3 pairs of $c1e$'s are possible only to 3 consecutive edges of the $c1C_4$, say (w_1, v_1), (w_1, v_2), (w_2, v_2), (w_2, v_3), (w_3, v_3), and (w_3, v_4) are $c1e$'s, and the other edges to the vertices v_i are $c2e$'s. Then v, v_1, v_2, w_1, w_2 determine a $c1(K_5 - 3e)$ or (w_1, w_2) is a $c2e$. By symmetry also (w_2, w_3) is $c2e$. If v_1, w_1, w_2, w_3 is no $c2(K_4 - e)$ then (w_1, w_3) is a $c1e$. — The coloring of K_8 is complete, and the $c1$–graph which is selfcomplementary is the second graph in Figure 2.

Now a vertex v in K_{13} with $c2d = 8$ is considered. If in the set S of the endvertices v_i the edges are colored as in the first graph in Figure 2 then each of the remaining 4 vertices w_i is adjacent to all v_i only in color 2, or a $c1(K_5 - 3e)$ occurs. To avoid a $c2(K_5 - e)$ all edges determined by v and the 4 vertices w_i are $c1e$'s, and this gives a $c1K_5$.

For the second graph in Figure 2 any $c2$–diagonal of a $c1C_4$ must have one $c1e$ to one w_i, or a $c2(K_5 - e)$ is determined since a $c2P_3$ exists for the vertices w_i, or a $c1(K_5 - 3e)$ occurs. Let v_1, v_2, v_3, v_4 be the consecutive vertices of a $c1C_4$ in S, and v_1 is connected by $c1e$'s to v_5 and v_8, v_2 to v_5 and v_6, v_3 to v_6 and v_7, v_4 to v_7 and v_8. If (v_1, w_i) is a $c1e$ then (v_5, w_i), and by symmetry (v_8, w_i), are $c2e$'s since without a $c1(K_5 - 3e)$ the edge (v_5, w_i) as $c1e$ forces (v_2, w_i), (v_7, w_i), and (v_8, w_i) to be $c2e$'s, and with v a $c2(K_5 - e)$ occurs. Then (v_3, w_i) is a $c1e$ if a $c2(K_5 - e)$ is avoided. By symmetry (v_6, w_i) and (v_7, w_i) are $c2e$'s, and without $c2(K_5 - e)$'s the edges (v_2, w_i) and (v_4, w_i) are $c1e$'s. Then w_i, v_1, v_2, v_3, v_4 determine a $c1(K_5 - 2e)$, and the proof of ≤ 13 is complete. — Three disjoint $c1K_4$'s in K_{12} prove > 12 (see Figure 1).

The missing three entries in Table 1 besides the two difficult numbers $r(K_5 - e, K_5)$ and $r(K_5)$ may be hunted later. In general it could be of interest to determine the values $r_n(s, \binom{n}{2} - s + i)$ for $i \geq 4$ (see [6] for $i = 2, 3$), that means all entries on the next parallel lines to the secondary diagonal.

References

[1] R. Bolze and H. Harborth, The Ramsey number $r(K_4 - x, K_5)$, In: *The Theory and Applications of Graphs* (Kalamazoo, Mich., 1980), Wiley, New York, 1981, 109–116.

[2] C. Clapham, G. Exoo, H. Harborth, I. Mengersen and J. Sheehan, The Ramsey number of $K_5 - e$, *J. Graph Theory* **13** (1989), 7–15.

[3] G. Exoo, A lower bound for $r(K_5 - e, K_5)$, *Utilitas Math.* **38** (1990), 187–188.

[4] U. Grenda and H. Harborth, The Ramsey number $r(K_3, K_7 - e)$, *J. Combinatorics Information Syst. Sci.* **7** (1982), 166–169.

[5] H. Harborth and I. Mengersen, Eine Ramsey–Zahl für fünf Knoten und acht Kanten, *Elem. Math.* **39** (1984), 6–9.

[6] H. Harborth and I. Mengersen, Ramsey numbers for graph sets with fixed numbers of edges, *Colloquia Math. Soc. János Bolyai* (to appear).

[7] S.P. Radziszowski, Small Ramsey numbers, Manuscript, May 1991.

Finite Structures with Few Types

EHUD HRUSHOVSKI*

MIT

Abstract

I will report on joint work with G. Cherlin on the quasi-finite axiomatizability of smoothly approximable structures, and on finite structures with few types. Let L be a finite language, k an integer, and $C(L, k)$ be the class of finite L-structures with at most k 5-types. The large members of $C(L, k)$ with no nontrivial 0-definable equivalence relation are known to be bi-interpretable with (finite unions of) certain classical geometries. (Work of Cherlin–Lachlan, Kantor–Liebeck–Macpherson.) A dimension of an L-structure M is the dimension of a geometry interpretable in M, such that every automorphism of the geometry lifts to an automorphism of M (with some other conditions.) $C(L, k)$ can be naturally (and effectively) divided into a finite number of families $C_\phi(L, k)$. Within each C_ϕ, a finite number of dimensions is identified; every first order statement is equivalent to a Boolean combination of statements asserting that a given dimension is finite (and fixed). In particular, the isomorphism type of a structure in F_i is determined by its dimensions. These dimensions can be varied essentially independently. This generalizes Lachlan's theory of shrinking and stretching homogeneous structures for a finite relational language. The proof involves methods of stability theory (geometries, orthogonality, modularity, stable groups) applied in this unstable context.

In this talk all infinities will be assumed to be nonstandard finite numbers. Theories will have the finite model property: every sentence has a finite model. We will consider this condition in conjunction with a restriction on the number of types, and derive a structure theory analogous to superstability.

One condition restricting the number of types is $\aleph_0$-categoricity: for each n, M has finitely many n-types. The strength of this (together with the finite model property) is unknown. Our condition is stronger. We consider the expansion M^* of M obtained by interpreting naturally the language L^*, the set of all nonstandard formulas of L with a finite number of variables. We say that M is *quasi-finite* if M^* is $\aleph_0$-categorical. As it turns out, it suffices to require that M^* have finitely many 5-types. The difference between M and M^* will be explained; we will see that the extra structure of M^* arises from the addition of dimension-quantifiers to the logic.

A central idea here, as in superstability, is that the structure is controlled by an assembly of *geometries* (or regular types). In stability theory, one usually starts with a global condition on the theory (say, having a small number of models in some power) and attempts to say something about the geometries; it is an achievement to show that they exist at all. In the present endeavor the starting point is a perfect knowledge of the geometries; the goal is

*Supported by the NSF and by a SLOAN FELLOWSHIP.

N.W. Sauer et al. (eds.), Finite and Infinite Combinatorics in Sets and Logic, 175–187.
© 1993 *Kluwer Academic Publishers. Printed in the Netherlands.*

to obtain a global theory (again, finding the number of models of a given nonstandard size might be an example.)

Our knowledge of the geometries is due to classification of the finite simple groups, entering through [6] and [12]. We begin by describing these. We will limit the description to two varieties, polar and quadratic; the others (symplectic, orthogonal, and unitary) provide extra color but are in substance the same.

A polar geometry consists of a pair (V, V^*) of vector spaces over a finite field F, together with a nondegenerate bilinear map $V \times V^* \to F$. Thus V^* can be identified with the space of homomorphisms $V \to F$, and vice versa. A polar geometry is determined by the field and a single dimension, $\dim_F(V) = \dim_F(V^*)$. The first order theory of (V, V^*) has $\aleph_0$ completions, specifying whether the dimension is finite or infinite; the latter is $\aleph_0$-categorical. The polar geometry considered in this form in fact has elimination of quantifiers. The infinite one also has *a geometric form of elimination of imaginaries*: every imaginary element e satisfies $\mathrm{acl}(e) = \mathrm{acl}(B)$ (where $\mathrm{acl}(X)$ is the algebraic closure of X) for some set B of real elements. This property, while often considered in the past for its own sake, will play a surprisingly critical role in the way structures can be coordinatized by such a geometry.

A symplectic space can be viewed as a polar geometry (V, V^*) together with an isomorphism $f : V \to V^*$ satisfying $(v, f(v)) = 0$; usually however one assumes f is the identity. Suppose $G = GF(2)$. One can find a map $q : V \to F$ such that $q(u+v) - q(u) - q(v) = (u, v)$. If q_1, q_2 are such maps, then $q_1 - q_2$ is linear; so the space Q of such maps forms an affine space over V^*. The structure (V, Q) thus described is the quadratic geometry over $GF(2)$. It too admits elimination of quantifiers and, in the infinite case, weak elimination of imaginaries.

Technically, in a geometry, the algebraic closure of a point should consist of the point alone; to achieve this we identify scalar multiples, getting projective versions of the above structures. This done, algebraic closure defines a combinatorial geometry; the exchange law holds, any two algebraically independent sets have the same cardinality, giving a well-defined notion of dimension. It is still true that the isomorphism type of a geometry of a given type is determined by its dimension. (This requires splitting the orthogonal case into two types.)

Both geometries come in a number of variants; for example it is sometimes important to consider a polar geometry in which V, V^* cannot be distinguished. For the purpose of this talk however one may think of a *geometry* as being one of the structures described above.

We say that one structure M is *embedded* in another, N, if the universe of M is 0-definable in N, and the 0-definable relations of M are precisely the relations on M that are 0-definable in N. Our theory will show that a quasi-finite structure is to a large extent determined by the geometries embedded in it. Outside the stable context, however, the notion of embedding is weak. For example, the generic bipartite graph is the union of two embedded degenerate geometries (sets with no structure); so is the generic bipartite (2,2)-hypergraph omitting a (4,4)-clique; the difference between them is in no way reflected in the geometries. The appropriate notion is that of a *stably embedded* structure: every relation of M definable with parameters in N, is definable with parameters in M. It is an easy exercise to determine the bipartite structures N in which the two parts are stably embedded pure sets; up to renaming of the relations, N is either the 2-sorted pure set (of some pair of cardinals), or has a 0-definable bijection between the two sorts (and no additional structure.) This is an example of the *theory of orthogonality*.

Definition A (rank 1) *geometry* is a structure J satisfying (in every elementary extension):

(i) $\mathrm{acl}(a) = \{a\}$;

(ii) if $a \in \mathrm{acl}(B,c) - \mathrm{acl}(B)$ then $c \in \mathrm{acl}(B,a)$;

(iii) if $e \in J^{\mathrm{eq}}$ then $\mathrm{acl}(e) = \mathrm{acl}(B)$ for some $B \subseteq J$;

(iv) for any 0-definable $J' \subseteq J$, if $a, a' \in J$ and $\mathrm{tp}(a/J') = \mathrm{tp}(a'/J')$ then $a = a'$.

We shall say that a substructure is *fully embedded* if it is embedded, and stably embedded.

Lemma 1 *Let J_1, J_2 be geometries. Assume J_1, J_2 are fully embedded in a structure M. Then either:*

a) J_1, J_2 *are orthogonal: every 0-definable subset of M^n is a Boolean combination of sets $D_1 \times D_2$, D_i a 0-definable subset of J_i^n; or*

b) *there exists a 0-definable bijection between J_1 and J_2. In either case, the induced structure on $J_1 \cup J_2$ is completely determined.*

Proof. Suppose (a) fails.

CLAIM 1: There exist $a_i \in J_i$ with $a_2 \in \mathrm{acl}(a_1)$.

PROOF: Let $D \subseteq J_1^{n_1} \times J_2^{n_2}$ be a counterexample to (a). For $b \in D^{n_2}$, let $D(b) = \{x \in J_1^{n_1} : (x,b) \in D\}$. Then for some b, $D(b)$ is not $\mathrm{acl}(\emptyset)$-definable in J_1. Since J_1 is stably embedded, $D(b)$ is definable with parameters in D_1^{eq}; if canonical parameters c_1 are chosen, they are definable over b. By condition (iii) of the definition of a geometry, there exists $a_1 \in J_1 \cap \mathrm{acl}(c_1)$. Let $\phi(x,y) \in \mathrm{tp}(d,b)$ be such that $\phi(x,b)$ has finitely many solutions, and $C(a_1) = \{b' \in J_2^{n_2} : \phi(a_1, b')\}$. Let c_2 be the canonical parameter of $C(a_1)$. Clearly $c_2 \in \mathrm{dcl}(a_1)$ (where $\mathrm{dcl}(X)$ is the definable closure of X), and $a_1 \in \mathrm{acl}(c_2)$. Thus $c_2 \notin \mathrm{acl}(\emptyset)$. Pick $a_2 \in J_2 \cap \mathrm{acl}(c_2)$ (using (iii) for J_2.). Again let e be a canonical parameter for the locus of a_1 over a_2. If $a_1' \in \mathrm{acl}(e) \cap J_1$, then $a_1' \in \mathrm{acl}(a_2) \subseteq \mathrm{acl}(a_1)$, so $a_1' = a_1$. Thus $\mathrm{acl}(e) = \mathrm{acl}(a_1)$. So $a_2 \in \mathrm{acl}(a_1)$, proving the claim.

CLAIM 2: $\mathrm{dcl}(a_1) = \mathrm{dcl}(a_2)$.

PROOF: If a_1' is a conjugate of a_1 over a_2, then $\mathrm{acl}(a_1') = \mathrm{acl}(a_2) = \mathrm{acl}(a_1)$, so $a_1' = a_1$. Hence there exist nonempty, 0-definable $D_i \subseteq J_i$ and a 0-definable bijection $F : D_1 \to D_2$. Observe that such a bijection is unique; if f, g are two such, then $f^{-1}g$ is a 0-definable permutation of D_2, and must be the identity.

CLAIM 3: For any $a_1 \in J_1$ there exists a unique $a_2 \in J_2$ with $\mathrm{acl}(a_1) = \mathrm{acl}(a_2)$.

PROOF: Let a_1 be any element of J_1. Using (iv), find $\phi(x,y)$ such that with $C(a_1) = \{b' \in D_1 : \phi(a, b')\}$, a_1 is determined by $C(a_1)$. Let c_2 be a canonical parameter for $F[C(a_1)]$. From this point the argument of Claim 1 applies. Similarly, for any $a_2 \in J_2$ there exists a unique $a_1 \in J_1$ with $\mathrm{acl}(a_1) = \mathrm{acl}(a_2)$, and so $\mathrm{dcl}(a_1) = \mathrm{dcl}(a_2)$. This finishes the proof.

It follows from Lemma 1 that non-orthogonality is an equivalence relation on stably embedded, embedded geometries. The interaction of a finite number of such geometries is thus completely understood given the following:

Lemma 2 *Let $J_1, \ldots, J_n$ be fully embedded geometries, orthogonal in pairs. Then $J_1 \cup \ldots \cup J_n$ is stably embedded (as a model-theoretic disjoint union of structures.)*

Proof. Suppose not, and choose a counterexample with n minimal. In a manner similar to Lemma 1, one can find $a_i \in J_i$ such that for each i, $a_i \in \mathrm{acl}(\{a_j : j \neq i\})$. Let $R = \{(a_1, \ldots, a_n) \in \Pi_i J_i : \text{for all } i, a_i \in \mathrm{acl}(\{a_j : j \neq i\}).\}$. One shows that for each i and each choice of a_j $(j \neq i)$ there exists a unique a_i with $(a_1, \ldots, a_n) \in R$. From this one can conclude there exists an Abelian group A and a regular action of A on J_i, for each i, such that R is an orbit of the diagonal action of A on $\Pi_i J_i$. But then an element of A can be coded by a pair of elements of any J_i; so there exists a nontrivial relation between J_1^2 and J_2^2; contradicting the orthogonality in pairs.

As the proof of the lemma suggests, the orthogonality theory in this form would fail if condition (iv) were relaxed; there exist *affine geometries* that can interact in groups without a pairwise interaction. They should probably be viewed not as geometries in their own right but rather as principal homogeneous spaces for groups defined over a single geometry. In a more general theory, such spaces can have rank higher than one. In the pseudo-finite context, irreducible groups defined over the basic geometries are Abelian of rank 1; and the only affine spaces that need be considered are those over the underlying Abelian group of the basic geometries (in their linear, rather than projective version.) We will return to these spaces later.

Definition Let N be a structure with finitely many 1-types. N is *coordinatized by Lie geometries* if for each $a \in N$ there exist $a_0, \ldots, a_n \in \mathrm{dcl}(a)$, $a_n = a$, such that for each i one of the following holds:

(a) $a_i \in \mathrm{acl}(a_{i-1})$;

(b) there exists $i' < i$, an $a_{i'}$-definable, fully embedded projective Lie geometry $J^i = J_a^i$ such that $a_i \in J^i$;

(c) there exist i', J_a^i as in (b) and $i'' < i$, $i' < i''$, and an $a_{i''}$-definable affine geometry (V^i, A^i) with projectivization J^i, such that $a_i \in A^i$.

Definition M is *Lie-coordinatizable* if there exists a finite cover N of M such that N is coordinatized by linear Lie geometries. This means that M is fully embedded in N, and there exists a 0-definable $f : N \to M$ in N with finite kernel.

By virtue of the orthogonality theory, if M is Lie, one can identify a canonical version of each geometry. By definition a fully embedded geometry J defined over a is *canonical* if for any conjugate $a' \neq a$ of a, $J_a, J_{a'}$ are embedded over a, a' and are orthogonal. It can be shown that for every fully embedded J_b there exists $a \in \mathrm{dcl}(b)$ and a canonical J_a such that J_b is the localization of J_a by b (in the appropriate sense.)

The existence of a canonical geometry gives a reasonable notion of dimension. By a _dimension form_ we will mean a partial type J in two variables, determining a complete type P in its first variable, and such that if $a \in P$ then $J_a = \{y : (x,y) \in J\}$ is a canonical geometry. M has _true J-dimension_ if $\dim J_a^M$ does not depend on the choice of $a \in P^M$. Note that in the quadratic case, the dimension is an even integer, or ∞. Two dimension forms J, J' are equivalent if there exists a 0-definable bijection between J and J' and between P and P' preserving the projection $J \to P$ and $J' \to P'$.

Theorem 1 _Let M be Lie-coordinatizable._

(a) _M admits a finite language._

Assume a finite language is chosen for M.

(b) _There are finitely many dimension forms for M, up to equivalence, for which M has true infinite dimension._

Choose a representative set C of such forms.

(c) _There exists a sentence $\phi = \phi_M \in \mathrm{Th}(M)$ with the following property. Call a model N of ϕ true if it has true J-dimension for each $J \in P$. In this case let $d(N) = (\dim_J N : J \in P)$._

(c1) _Every true model of ϕ is determined up to isomorphism by $d(N)$._

(c2) _For any sufficiently large reasonable choice of d, there exists a true model N of ϕ with $d(N) = d$. Denote it by $M(d)$._

(c3) _There exists a homogeneous embedding of $M(d)$ into M. It is unique up to an isomorphism of M._

In (c2) "reasonable" choice is just one in which the parity of the dimension is appropriate to the type of geometry. In the orthogonal case, we let the dimension include the information $(\pm)$ determining the signature of the space, if it is finite. The words "sufficiently large" can actually be deleted if one is dealing with a Lie-coordinatized structure; in general a coordinatizable structure is bi-interpretable with a coordinatized one, and the dimension must be large enough to make the bi-interpretation work.

DESCRIPTION OF PROOF: We pass to the bi-interpretable Lie-coordinatized structure N. By extending the orthogonality theory, and developing some understanding of the interaction of affines, one shows that (b) holds, and that homogeneous substructures of N of prescribed dimensions exist, and they are determined by their dimensions, up to isomorphism, and in the finite case up to an automorphism of N. This is analogous to the "Zilber envelopes" of [5].

We view N as obtained from a smaller structure N_1 in the following way. There exists a complete type D of N_1, a type C of N, and a definable map $p : C \twoheadrightarrow D$ in N. N consists of N_1 together with C. Moreover for $d \in D$, $\pi^{-1}(d) = C_d$ is finite, or affine over a vector space V_d in N, or a basic geometry. By the definition of Lie coordinatization, N is obtained in finitely many steps from a single point by such a process.

If C_d is a (nonaffine) geometry and the corresponding canonical geometry lies in N_1, then $N \subseteq \mathrm{dcl}(N_1)$. If C_d is itself a canonical geometry, orthogonal to every geometry in N_1, then C is a "free" extension of N by copies of C_d, and the isomorphism type of N is completely determined. With some care, it may also be viewed as a reduct of a structure in which C essentially has the form $D \times J$ for some geometry J; so if the structure of N_1 is assumed known, then N also becomes known.

The affine case was the most difficult one in the stable case. Surprisingly, in the higher level of generality, a single affine cover can be understood in terms of a finite cover[1]. An affine geometry A with corresponding vector geometry V can be viewed as an exact sequence $0 \to V \to V' \to F \to 0$, where F is the underlying field, V' is a space containing V, and A is identified with the inverse image of $1 \in F$. The context in which we work allows us to take the dual of a group, as a new kind of imaginary sort. Dualizing each of the groups in the above sequence we get $0 \to F^* \to V'^* \to V^* \to 0$; V'^* is a finite cover of V^*. This construction can be carried out uniformly; for example it transforms a cover C of D (with each C_d affine over V) into a finite cover C^* of $D \times V^*$. One must of course show that the structure of the covers is sufficiently linear, that knowledge of C^* yields knowledge of C. This requires an analysis of the affine geometries (analyzing the imaginary sorts), as well as some understanding of the possible algebraic relations among different affines along a cover.

The remaining case is that of finite covers. This is dealt with using the Ahlbrandt–Ziegler method, as generalized in [10]. We have a sentence ϕ_1 such that every true model M_1 of ϕ_1 of dimension d is isomorphic to $N_1(d)$, a homogeneous substructure of N' determined uniquely by its dimensions. We let ϕ be a sentence stating that M is a finite cover of M_1, and describing the behavior of the cover over small subsets of M_1. The problem is to show that if M is a model of ϕ, and f is a homogeneous embedding of M_1 into N_1, then f can be lifted into a homogeneous embedding of M into N. One lifts f above increasingly larger initial sections of an appropriately chosen enumeration of M_1; it is shown by combinatorial methods that the type of an element of N over an analogous initial section of N_1 is determined by a bounded sub-type, and that this fact can be expressed with a bounded number of formulas; using definability properties of the initial sections, this is transferred to M and M_1.

It is shown in [11] how to use the power of [5] in order to derive Lachlan's theory of stable finitely homogeneous structures; this requires a number of ideas, and the result from [6] that the class in question is a first order class. Our analog is:

Theorem 2 *The following conditions are equivalent (for a model M of nonstandard finite size).*

(a) *M is Lie-coordinatizable.*

(b) *M is quasi-finite.*

(c) *M is 5-quasi-finite (i.e. M^* has finitely many 5-types).*

(d) *For some c, for all n, $s_n(M^*) \leq c^{n^2}$.*

(a)–(d) are also closely related to Lachlan's notion of smoothly approximable structures; see [12]. This notion is equivalent to (a)–(d) together with a certain technical condition on the quadratic geometries (pointed out to us by David Evans.) This equivalence can be used to show that "weakly smoothly approximable" is the same as "smoothly approximable".

---

[1] It seems that the reduction cannot, in general, be applied to iterated covers.

For us the most significant equivalence is of (a) and (c). For any integer k, and finite language L, let $C(L, k)$ be the class of finite L-structures with $\leq k$ 5-types. Let $TF_5(k)$ be the theory of $C(L, k)$. Then one sees immediately that a first-order theory has a 5-quasi-finite model iff it contains $TF_5(k)$ for some k. (In fact for $k = s_5(T)$, as we will see later.) This will allow us to apply Theorem 1 to study the structure of $C(L, k)$.

<u>DESCRIPTION OF PROOF:</u> (a)$\Rightarrow$(b),(f) is a consequence of the theory of envelopes mentioned earlier. The other nontrivial implication is (e)$\Rightarrow$(a). Since (e) is not easily seen to be preserved under addition of parameters, one first uses [12] to show that (e) implies that M^* is locally Lie.

Definition Let M be a nonstandardly finite structure. Let W be a collection of Lie geometries definable in M, each given with an element of M as a defining parameter. M is *locally Lie* if:

(i) M has finitely many sorts, and finitely many 1-types in each sort. We are also given a tree structure on M (of finite height); $\leq$ will refer to the tree ordering. If $A_b \in W$ is affine, then the corresponding linear geometry J_b is also in W (or part of a pair in W).

(ii) If $J_b \in W$ is projective or linear then J_b is embedded in (M^*, b).

(iii) If $a \in M$ then there exists $b < a$ and $J_b \in W$ with $a \in J_b$; or $a \in \mathrm{acl}(b)$.

We modify M by replacing each pure projective geometry in W by a polar pair, and each characteristic-2 symplectic geometry by a quadratic pair. We then show that each $J_b \in W$ is stably embedded in M^*. This shows that M^* is Lie-coordinatized. At this point we need:

Theorem 3 *The class of Lie-coordinatizable structures is closed under interpretations.* ꞞALSᴱ ·

The ideas of the proof of Theorem 3 will be discussed later. The proof that each geometry in W is coordinatized uses induction. One essentially reduces to two cases: two embedded geometries J_1, J_2, or a geometry J_1 and its affine A, forming the entire nonstandardly finite structure. Then assuming only that J_1, J_2 (respectively J_1) are embedded, one must show they are stably embedded. This is done by reversing in effect Lemma 1, using simple properties of the automorphism groups of the finite geometries (simplicity of the derived group, solvability of the outer automorphism group) to derive an orthogonality theory directly.

We are now in position to finitize our results. However first order logic is not quite appropriate as a language for describing $C(L, k)$. To see this consider the following example.

Example Let L be a language with two predicates V, S and three maps $\pi_i : S \to V$. Let $d = (d_1, d_2, d_3)$ be a triple of integers, with $d_2 \neq d_3$. Let $M_d = (V_d, S_d, \pi_d)$, where V_d is a d_1-dimensional vector space, S_d is a set, $\pi_d : S_d \to V_d^3$ is the interpretation of (π_1, π_2, π_3), and $\pi_d^{-1}(x, y, z)$ has d_2 elements if $x + y + z = 0$, d_3 otherwise. The vector space structure is a uniform feature of the M_d's, but cannot be discussed in a first-order fashion.

To remedy this, we will add *dimension quantifiers*. (In the orthogonal case, *Witt index quantifiers sQ* will also be needed.) We use comparative quantifiers; one can ask whether two sets have the same cardinality, but not what the cardinality is. In the most general form, the dimension quantifiers can be applied to an arbitrary pair of definable sets. Thus if $\phi_1(x,y)$, $\phi_2(x,y)$ are two formulas, we get a new formula $(dQx)(\phi_1,\phi_2)(y)$. A model for $L[dQ]$ includes an interpretation of all formulas satisfying the usual Tarski rules, and the following rule for the new quantifier: if $\{x : \phi_1(x,b)\}$ is finite, then

$$(dQx)(\phi_1,\phi_2)(b) \iff |\{x : \phi_1(x,b)\}| < |\{x : \phi_2(x,b)\}|.$$

This quantifier is far more powerful than what we need, but is easier to describe, and does not have more strength than the theory can handle. To demonstrate this we mention

Theorem 4 $TF_5(k)$ *is decidable, uniformly in* k. *In fact given* k *and a sentence* ϕ *of* $L(dQ,sQ)$, *one can effectively decide whether there exists a (finite) model of* $TF_5(k) \cup \{\phi\}$.

This is proved by means of a COMPLETENESS THEOREM for the language $L(dQ,sQ)$ and the class $C(L,k)$. There is a natural set of axioms for $C(L,k)$, and one shows that if a sentence ϕ of $L(dQ,sQ)$ does not follow from the axioms, then there exists a structure in $C(L,k)$ in which ϕ is false. It follows of course that the first-order part of $TF_5(k)$ is decidable, but there does not seem to be a natural completeness theorem for the first order part alone.

The idea of the completeness theorem is this: if a sentence ϕ does not follow from the axioms, get an infinite model M using the usual completeness theorem, and then apply Theorem 1 to shrink M to a finite model (taking some care to preserve the meaning of the dimension quantifiers.) We need to know however that our axioms imply that M is Lie-coordinatizable. It is easy to state that a particular set Φ of formulas gives a bi-interpretation of M with a structure N, and an analysis of N as a Lie-coordinatized structure. The problem is that one cannot quantify over all possible finite sets Φ of formulas. It is therefore necessary to have a BOUNDEDNESS THEOREM: There exists a finite set Φ of formulas of $L(dQ,sQ)$, such that for any L-structure M, if M is Lie-coordinatizable then the coordinatization is given by formulas in Φ. (More generally, for any finite k one can find a finite Φ such that if M is Lie, then every k-ary formula is equivalent to one in Φ.) This is again a consequence of Theorem 3. If the boundedness theorem were false, one would have Lie-coordinatized structures M_n in which a coordinatization is given by formulas of increasing minimal complexity; at the non-standard limit, we would get a Lie-coordinatized structure in which a coordinatization is not given by any standard formula, and so the reduct to the standard structure is not coordinatizable at all, in contradiction with Theorem 3.

In practice, we will use the dimension quantifiers in a restricted way: starting with the first order structure, we apply the dimension quantifiers only to the canonical geometries; this yields additional structure and new canonical geometries, to which we again apply the quantifiers; this process will terminate in a number of steps equal at most to the bound on the number of 2-types.

At this point the compactness theorem can be applied to Theorem 1, and we obtain:

Theorem 5 *Let L be a finite language, k an integer. Then $C(L, k)$ is the finite union of families $\mathcal{F}_1, \ldots, \mathcal{F}_n$.*

(a) *Each $\mathcal{F}_i$ is the class of finite models of some sentence ϕ_i of $L[dQ, sQ]$. The ϕ_i's are pairwise contradictory.*

(b) *With each $\mathcal{F}_i$ there is associated a Lie-coordinatizable structure M_i; $\mathcal{F}_i$ is the class of envelopes of M_i from Theorem 1. Hence each member N of $\mathcal{F}_i$ is determined by certain numerical dimensions $d(N)$. These dimensions vary freely in $\mathcal{F}_i$ above a certain minimum, except that distinct dimensions cannot be equal.*

(c) *If $d(N) \leq d(N')$ then there exists a homogeneous embedding of N into N', unique up to an automorphism of N'.*

(d) *Membership in $C(L, k)$ can be determined in polynomial time: indeed $N \in C(L, k)$ iff N satisfies some ϕ_i. Membership in $\mathcal{F}_i$ can also be tested in polynomial time, and $d(N)$ can be computed in polynomial time. In particular the isomorphism problem can be solved in polynomial time within $C(L, k)$.*

(e) *Conversely, given i and a dimension d, let $N_i(d)$ be the member of $\mathcal{F}_i$ of dimension d. Then the cardinality $n_i(d)$ of $N_i(d)$ is a polynomial in d^*, where d^* is a certain explicit exponential in d. $N_i(d)$ can be constructed in time polynomial in its size.*

Some explanatory remarks may be in order.

(b) The restriction of distinct dimensions may seem strange. It is largely a matter of choice. Consider the example above; if $d_2 = d_3$, there is no vector space structure at all in M_d, and so it does not seem to belong with the others.

(c) A map $f : A \to B$ of L-structures is a *homogeneous embedding* if the type of the image of a tuple from A determines the type of the tuple.

(d) The exponent of the polynomial time bound is recursive in k, but we have no further information on its growth. We note that the number of 5-types is not known to be computable in polynomial time, in general; this would be equivalent to showing that graph isomorphism can be solved in polynomial time. Note that the classes $C(\text{"graphs"}, k)$ in which isomorphism is poly-time are orthogonal to those considered in [14]; here $\text{Aut}(M)$ can be shown to have a bounded number of chief factors for $M \in C(L, k)$, whereas Luks (in effect) considered classes in which each factor is bounded. We view Theorem 5 as a generalization of Lachlan's theory of shrinking and stretching, which was of course its inspiration. Lachlan's results initially applied to prove similar results for finitely homogeneous structures of bounded rank (in a certain technical sense.) In [6] it was shown that the bounded rank condition is equivalent to stability. This extended the scope of the theory to apply to a substantial class of $\aleph_0$-categorical, $\aleph_0$-stable structures of disintegrated type. The method of proof, however, broke out of the homogeneous framework; it included in effect a recognition theorem for the Lie geometries mentioned above. This was complemented by a similar theorem for the elusive affine geometries, in [12], which opened the door to

the present paper. The extension of the theory all the way to the smoothly approximable
context was envisioned by Lachlan at an early date.

We refer the reader to [11] for an exposition of the shrinking and stretching theory itself.

One of the central problems raised in [11] was the question of effectiveness of stretching:
given a finite $N \in C(L, k)$, and dimension-forms $J_1, \ldots, J_d$, can one effectively tell whether
in fact there exists M with $\dim J(M)$ infinite for $J = J_1, \ldots, J_d$, such that N is an envelope
of M? We answer this affirmatively in the general context. (In order to apply this to the
homogeneous context, we need to prove also that one can tell in finite time whether the
stretching will turn out to be homogeneous. This can be done with the same methods.[2])

We now briefly describe the proof of Theorem 3. There is certainly no direct relation
between the Lie coordinatization of a structure N and that of a reduct M; the geometries
of M could well be high dimensional objects from the point of view of N, and may be
defined over geometries of different types. One is forced to derive global properties from
the existence of a coordinatization. The ones we have found are these.

Theorem 6 *M is Lie-coordinatizable if, and only if, the following conditions hold.*

(a) *M has FINITE RANK. We use here a simple theory of rank, in which*
$\mathrm{rk}(a/Bb) < \mathrm{rk}(B)$ *if $b \in \mathrm{acl}(Bb) - \mathrm{acl}(a)$.*

(b) *INDEPENDENCE THEOREM: Let $r_i(i = 1, ..., n)$ and $r_{ij}(1 \leq i < j \leq n)$
be types over $B = \mathrm{acl}^{\mathrm{eq}}(B)$. Suppose the variables of r_{ij} are x_i, x_j, the variable
of r_i is x_i, and r_{ij} contains r_i and r_j. Suppose also $r_{ij}(ac)$ implies that a, c
are independent. Then there exists a type r extending all these, and such that
$r(a, \ldots)$ implies that $a, \ldots$ are independent.*

(c) *MODULARITY: for any $a, b \in M^{\mathrm{eq}}$, a, b are independent over $\mathrm{acl}^{\mathrm{eq}}(a) \cap
\mathrm{acl}^{\mathrm{eq}}(b)$.*

(c') *MODULARITY IN GROUPS: If A is a definable Abelian group, then every
definable subset of A is a Boolean combination of cosets of definable subgroups
and of a fixed finite number of definable sets.*

(d) *$\aleph_0$-CATEGORICITY.*

(e) *$\mathrm{Th}(M)$ has the finite model property; and M does not interpret the random
bipartite graph.*

(f) *RANK/MEASURE PROPERTY: Let A be an Abelian group, D a definable
subset of full rank. Then $\liminf \mathrm{card} h(A)/\mathrm{card} h(D) > 0$, where h varies over
all definable homomorphisms of A onto a finite group.*

Some of these properties are trivially inherited by reducts; some require work and help
from the others. (b) is of course a standard property of independence in the stable context.
Many other properties, such as uniqueness, do not hold here. (Though failures of uniqueness
satisfy a chain condition; this may be related to (c').) (c) is a decisive property in the stable
context. In particular it entails (c') (and is equivalent to it for groups; see [9].) Here it

---

[2]There is also a proof along the lines used in [11] to solve the problem in the binary case. The difficulty
is to decide whether a finitely constrained class is an amalgamation class. For binary languages this is easy,
and for the same reason it is easy for classes with a canonical amalgamation. We observe that amalgamation
classes with a stable limit are close enough to having a canonical amalgamation to make the proof work.

seems (c′) must be derived separately. (e) appears at present to be a strange condition, whose main effect is to rule out the generic graph.

Given properties (a)–(f), one obtains a coordinatization into rank 1 sets using (a)–(c). It remains to recognize these sets as stably embedded Lie geometries (up to algebraic closure.) If the rank 1 set D has a trivial combinatorial geometry, (b) will imply that any definable relation on D must behave essentially like the generic graph; this is ruled out by (e). Thus D must be a stably embedded disintegrated strongly minimal set. (In practice the proof is more complicated, since (b) does not easily go down to the reduct.) If the geometry is nontrivial, the set can be replaced by a definable Abelian group A. (This requires some analysis of piecewise-definable groups; showing that if they have bounded rank, then they are unions of definable subgroups. Piecewise definable groups of unbounded rank play a large role in finite covers, but we have little information about them.) If there are no definable homomorphisms on A into a finite group, then by (c′) the group is stable, and stably embedded. Otherwise, again using the result on piecewise-definable groups, one gets a definable dual group A^* to A. If A, A^* are algebraically related, one shows they are definably isomorphic, and one obtains a symplectic, orthogonal or unitary structure. It remains to understand the finite base of definable sets in (c′), and classify (A, A^*) as a Lie geometry.

Problems

A. In 5(e), can $N(d)$ be constructed in polynomial time in d? Of course, there is insufficient time to construct the underlying set; but this set can be viewed as known, one can ask whether the basic relations on it can be recognized in polynomial time.

B. Generalize the theory to structures coordinatized also by stably embedded pseudo-finite fields. (they form a geometry, and the independence theorem holds.) Does every such structure have the finite model property? Can the reducts be characterized? Further generalize to "tier 2 geometries", with infinite-dimensional vector spaces over pseudo-finite fields allowed as geometries.

C. Using (B), develop a horizontal theory, in which the field can also be viewed as a varying parameter. The theory should encompass a sufficiently large class of structures to include all but finitely many elements of $C(L, k)$ for each L, k. (It may include all r-closed finite permutation groups with $\leq k$ orbits on pairs, with an appropriate choice of language for each such permutation structure. The choice of language was not very important in the context of few 5-types because of the boundedness theorem, but no such result will be available in the larger context.)

D. Given G acting on X, let $Cl_r(G)$ be the group of all permutations σ of X such that for each r-element subset Y of X, there exists τ in G with $\sigma|Y = \tau|Y$. For example $Cl_r(SL_n) = GL_n$ for $4 \leq r \ll n$. Then given k and r, we have developed a theory for the class of r-closed finite permutation groups with at most k orbits on 5-tuples. (Any model of $TF_5(k)$ has at most c^{n^2} n-types. Hence there exists a locally finite language L_k which can reasonably be considered as a uniform language for all finite structures with at most 5 k-types.) It can be shown that there exists an r depending on k such that $Cl_r(G)/G$ is solvable for all G in this class. Can the theory be extended to arbitrary permutation groups with boundedly many orbits on 5-types? A beginning of a positive answer would follow from:

D1. Given k,n there exists r such that every $G \in CG(k)$ contains a subgroup H with $Cl_n(G) = Cl_n(H)$ and $Cl_r(H) = H$.

D2. Let k be given. For some m, b the map sending G to $Cl_r(G)$ is m-to-one on $CG(k)$.

E.[3] Does PF_4 imply PF_5? At all events PF_3 does not. In fact, any theory in PF_0 is interpretable in one in PF_3. To get a continuum of theories in PF_3, let X be any set of positive integers. Let $F_p(X)$ be the p-element field, with an additional unary predicate P, whose interpretation is $\{n \bmod p : 0 < n < p, n \in X\}$. Let M_p be the model whose universe is the projective line over F_p, with the full structure induced from F_p. Take any ultraproduct of the M_p (p prime). Then fixing 3 points one can interpret a field of characteristic 0 with a distinguished subset; and the intersection of this subset with the standard integers is X.

References

[1] G. Ahlbrandt and M. Ziegler, Quasi-finitely axiomatizable totally categorical theories, *Annals of Pure and Applied Logic* **30** (1986) 63–82.

[2] P. Cameron, Finite permutation groups and finite simple groups, *Bull. London Math. Soc.* **13** (1981) 1–22.

[3] R.M. Carter, Simple groups of Lie type, *Wiley Classics Library*, London–New York (1989).

[4] P. Cameron and W.M. Kantor, 2-Transitive and antiflag transitive collineation groups of finite projective spaces, *J. Algebra* **60** (1979) 384–422.

[5] G. Cherlin, L. Harrington and A. Lachlan, $\aleph_0$-categorical, $\aleph_0$-stable structures, *Annals of Pure and Applied Logic* **18** (1980) 227–270.

[6] G. Cherlin and A.H. Lachlan, Stable finite homogeneous structures, *Trans. Amer. Math. Soc.* **296** (1986) 815–850.

[7] D. Evans and E. Hrushovski, On the automorphism groups of finite covers, to appear in *Annals of Pure and Applied Logic*.

[8] L. Harrington, Lachlan's homogeneous structures, talk given at ASL meeting, Orsay (1985).

[9] E. Hrushovski and A. Pillay, Weakly Normal Groups, *Logic Colloquium* **85**, Paris, North Holland (1986).

[10] E. Hrushovski, Totally categorical structures, *Trans. Amer. Math. Soc.* **313** #1 (1989) 131–159.

---

[3]The answer is affirmative. Dugald Macpherson informs me that the corresponding change from 5 to 4 can be made in [12], and this suffices for the non-primitive case.

[11] J. Knight and A.H. Lachlan, Shrinking, stretching and codes for homogeneous structures, in *Classification Theory*, ed. J. Baldwin, Lecture Notes in Mathematics **1292**, Chicago, Springer (1985).

[12] W.M. Kantor, M.W. Liebeck and H.D. Macpherson, $\aleph_0$-categorical structures smoothly approximable by finite substructures, *Proc. London Math. Soc.* (3) **59** (1989) 439–463.

[13] A.H. Lachlan, On countable stable structures which are homogeneous for a finite relational language, *Israel J. Math.* **49** (1984) 69–153.

[14] E. M. Luks, Isormorphism of graphs of bounded valence can be tested in polynomial time, *Proc. 21 IEEE Symp. on Foundations of Computer Science, Syracuse, 1980*, 42–49, IEEE, New York (1980).

[15] D. Macpherson, Interpreting groups in ω-categorical structures, *J. Symb. Logic* **56** (1991) 1317–1324.

Recognition Problem in Reconstruction for Decom Relations

PIERRE ILLE

C.N.R.S. - U.R.A. 225

Marseille (France)

Abstract

Let R be a binary and reflexive relation on a finite set E. A subset X of E is an *interval* of R if for $a, b \in X$ and $x \in E \setminus X$, we have: $R(a,x) = R(b,x)$ and $R(x,a) = R(x,b)$. So $\emptyset$, E and every singleton are intervals of R (called *trivial* intervals). The relation R is called *decomposable* if R has a non-trivial interval. On the other hand, we say that two relations R and R' on the same set E are *hypomorphic* when for each element x of E, the restrictions of R and R' to $E \setminus \{x\}$ are isomorphic. In this paper, we show that if R and R' are two hypomorphic relations on a set of cardinality ≥ 12 and if R is decomposable then R' is also decomposable.

1 Introduction

If E is a set then a *binary relation* with *base* E is a function from $E \times E$ into $\{+, -\}$. If X is a subset of E then R/X denotes the restriction of R to X. The relation R is *reflexive* when for every element x of E, we have: $R(x,x) = +$. The cardinality of E will be denoted by $\operatorname{card}(E)$.

In this paper, we only consider binary and reflexive relations with finite base. The following notion of interval has been introduced by R. Fraïssé [2]: a subset X of the base E of a relation R is an *interval* of R when for all elements a, b of X, and x of the complement $E \setminus X$ of X, we have: $R(a,x) = R(b,x)$ and $R(x,a) = R(x,b)$. For example, this notion is the classical notion of interval whenever R is a linear ordering $\leq$ on a set E ($R(x,y) = +$ means $x \leq y$). We say that a relation R with base E is *decomposable* if there is an interval X of R such that $2 \leq \operatorname{card}(X) < \operatorname{card}(E)$. Otherwise R is *indecomposable*. An *interval partition* P of a relation R on a set E is a partition of E, the elements of which are intervals of R. If P is an interval partition of R then we define the *quotient relation* R/P with base P in the following way: for given elements X and Y of P, $(R/P)(X,Y) = +$ when (i) $X = Y$ or (ii) $X \neq Y$ and if $x \in X$, $y \in Y$ then $R(x,y) = +$.

Before recalling the decomposability theorem, let us introduce some notation and definitions which are useful in the following sections as well.

Definition 1

(1) The class of decomposable relations is denoted by Δ.

(2) Δ_1 is the class of relations R such that there exists an interval partition $Q(R)$ of R with the following properties:

N.W. Sauer et al. (eds.), Finite and Infinite Combinatorics in Sets and Logic, 189–198.

(i) $3 \leq \text{card}[Q(R)] < \text{card}(E)$, where E is the base of R, and $R/Q(R)$ is indecomposable.

(ii) A proper subset A of E is an interval of R if and only if there is an element X of $Q(R)$ such that A is an interval of the restriction R/X.

(3) Δ_2 is the class of relations R such that there exists an interval partition $Q(R)$ of R with the following properties:

(i) $2 \leq \text{card}[Q(R)] < \text{card}(E)$, where E is the base of R, and $R/Q(R)$ is a linear ordering.

(ii) A subset A of E is an interval of R if and only if either there is an element X of $Q(R)$ such that A is an interval of the restriction R/X or there is an interval Q of $R/Q(R)$ such that $\text{card}(Q) \geq 2$ and A is the union of elements of Q.

(4) Δ_3 (resp. Δ_4) is the class of relations R such that there exists an interval partition $Q(R)$ of R with the following properties:

(i) $2 \leq \text{card}[Q(R)] < \text{card}(E)$, where E is the base of R, and $R/Q(R)$ is the complete graph on $Q(R)$ (i.e. for all elements X and Y of $Q(R)$, $(R/Q(R))(X,Y) = +$) (resp. $R/Q(R)$ is the empty graph on $Q(R)$ (i.e. if X and Y belong to $Q(R)$ then $(R/Q(R))(X,Y) = +$ if and only if $X = Y$)).

(ii) A subset A of E is an interval of R if and only if either there is an element X of $Q(R)$ such that A is an interval of the restriction R/X or there is a subset Q of $Q(R)$ such that $\text{card}(Q) \geq 2$ and A is the union of elements of Q.

Proposition 1 [3] (decomposability theorem) *The family* $\{\Delta_1, \ldots, \Delta_4\}$ *is a partition of* Δ.

Now we introduce some notions which are often used in reconstruction theory (see [1] and [7]).

Let R and R' be two relations on the same set E: we say that R and R' are *hypomorphic* when for each element x of R, the restrictions $R/(E\setminus\{x\})$ and $R'/(E\setminus\{x\})$ are isomorphic. A relation R is *reconstructible* if for each relation R' such that R and R' are hypomorphic, R and R' are isomorphic. A class Σ of relations is said to be *recognizable* if for every relation R in Σ and for each relation R' such that R and R' are hypomorphic, R' belongs to Σ.

P.K. Stockmeyer [6] has proved that generally relations are not reconstructible. More precisely, he has found for each integer n, two hypomorphic tournaments T and T' on a set E with $\text{card}(E) \geq n$ such that T and T' are not isomorphic (a relation R on a set E is a *tournament* when for all elements $x \neq y$ of E we have: $R(x,y) \neq R(y,x)$). However, the tournaments T and T' are indecomposable so that a natural question is the following: is a decomposable relation reconstructible? This problem is still open. Generally, before proving that the elements of a class Σ of relations are reconstructible, we show that Σ is recognizable. The purpose of this paper is to show that Δ is recognizable.

In Section 3, we show the following theorem.

Theorem 1 *Let R be an element of Δ_i $(i \in \{2,3,4\})$ with base E. If $\operatorname{card}(E) \geq 7$ then for every relation R' such that R and R' are hypomorphic, R' belongs to Δ_i.*

In Section 4, we prove the following theorem.

Theorem 2 *Let R be an element of Δ_1 with base E. If $\operatorname{card}(E) \geq 12$ then for every relation R' such that R and R' are hypomorphic, R' belongs to Δ_1.*

The proof of this theorem uses two lemmas which are proved in Sections 5 and 6.

2   Preliminaries

First we recall the following properties of intervals obtained in [2].

Proposition 2 *Let R be a relation with base E and let I be the set of intervals of R.*

(1) *The base E and the empty set $\emptyset$ belong to I, and for each element x of E, the singleton $\{x\}$ belongs to I. These intervals are called trivial intervals of R.*

(2) *If X and Y belong to I then the intersection $X \cap Y$ belongs to I.*

(3) *If X and Y belong to I and if $X \cap Y \neq \emptyset$ then the union $X \cup Y$ belongs to I.*

(4) *If X and Y are elements of I such that $X \setminus Y = \{x \in X / x \notin Y\}$ is not empty then $Y \setminus X$ belongs to I.*

(5) *If F is a subset of E and if X is an element of I then $X \cap F$ is an interval of R/F. Moreover, if F belongs to I then a subset X of F is an interval of R/F if and only if X belongs to I.*

In the sequel, we shall use some properties about indecomposable restrictions of an indecomposable relation which have been obtained by J.H. Schmerl [5] and myself [4].

Proposition 3 (J.H. Schmerl) *Let R be an indecomposable relation with base E.*

(1) *If $\operatorname{card}(E) \geq 5$ then there is a subset X of E such that $\operatorname{card}(X) = 3$ or 4 and R/X is indecomposable.*

(2) *Let X be a subset of E such that: $\operatorname{card}(X) \geq 3$, $\operatorname{card}(E \setminus X) \geq 2$ and R/X is indecomposable. Then there are distinct elements a and b of $E \setminus X$ such that $R/(X \cup \{a,b\})$ is indecomposable.*

(3) *If $\operatorname{card}(E) \geq 5$ then there is a subset X of E such that: $\operatorname{card}(X) \geq 3$, $\operatorname{card}(E \setminus X) = 1$ or 2 and R/X is indecomposable.*

Definition 2 *Let R be a relation with base E and let X be a subset of E.*

(1) *For every element u of X, we denote by $X(u)$ the set of elements x of $E \setminus X$ such that $\{u,x\}$ is an interval of $R/(X \cup \{x\})$.*

(2) X_2 is the set of elements x of $E \setminus X$ such that X is an interval of $R/(X \cup \{x\})$.

(3) X_1 is the set of elements x of $E \setminus X$ such that $R/(X \cup \{x\})$ is indecomposable.

Proposition 4 (P. Ille) *Let R be a relation with base E and let X be a subset of E such that $\operatorname{card}(X) \geq 3$ and R/X is indecomposable.*

(1) *The family $S(X) = \{X_1, X_2, X(u) \ (u \in X)\}$ is a partition of $E \setminus X$. Furthermore, if x belongs to X_2 (resp. $X(u)$, where $u \in X$) and if Y is a non-trivial interval of $R/(X \cup \{x\})$ then $Y = X$ (resp. $Y = \{u, x\}$).*

(2) *If a and b are distinct elements of X_1 then $R/(X \cup \{a, b\})$ is decomposable if and only if $\{a, b\}$ is an interval of $R/(X \cup \{a, b\})$.*

(3) *If a belongs to X_1 and b to X_2 (resp. $X(u)$, where $u \in X$) then $R/(X \cup \{a, b\})$ is decomposable if and only if $X \cup \{a\}$ (resp. $\{u, b\}$) is an interval of $R/(X \cup \{a, b\})$.*

(4) *If a belongs to X_2 and b to $X(u)$, where $u \in X$, then the following assertions are equivalent:*

  (i) *$R/(X \cup \{a, b\})$ is decomposable.*
  (ii) *$X \cup \{b\}$ is an interval of $R/(X \cup \{a, b\})$.*
  (iii) *$\{u, b\}$ is an interval of $R/(X \cup \{a, b\})$.*

(5) *If a belongs to $X(u)$ and b to $X(v)$, where u and v are distinct elements of X, then the following assertions are equivalent:*

  (i) *$R/(X \cup \{a, b\})$ is decomposable.*
  (ii) *$\{u, a\}$ is an interval of $R/(X \cup \{a, b\})$.*
  (iii) *$\{v, b\}$ is an interval of $R/(X \cup \{a, b\})$.*

Proposition 5 (P. Ille) *Let R be an indecomposable relation with base E.*

(1) *If $\operatorname{card}(E) \geq 6$ and if a is an element of E then there is a subset X of E satisfying: $\operatorname{card}(X) = 4$ or 5, a belongs to X and R/X is indecomposable.*

(2) *Let X be a subset of E such that: $\operatorname{card}(X) \geq 3$, $\operatorname{card}(E \setminus X) \geq 6$ and R/X is indecomposable. Then there is a subset Y of E satisfying: Y contains X, $\operatorname{card}(E \setminus Y) = 2$ and R/Y is indecomposable.*

(3) *If $\operatorname{card}(E) \geq 7$ then there is a subset X of E such that: $\operatorname{card}(E \setminus X) = 2$ and R/X is indecomposable.*

(4) *If $\operatorname{card}(E) \geq 11$ and if a is an element of E then there is a subset X of E such that a belongs to X, $\operatorname{card}(E \setminus X) = 2$ and R/X is indecomposable.*

We finish this section with some remarks about Proposition 1 (decomposability theorem).

Remark 1

(1) Proposition 1 can be extended to relations R on a set E with $\operatorname{card}(E) = 2$. In this case, R belongs to $\Delta_2 \cup \Delta_3 \cup \Delta_4$.

(2) Let R be an element of $\Delta_2 \cup \Delta_3 \cup \Delta_4$ and let X be an element of $Q(R)$. If R/X is decomposable or if $\operatorname{card}(X) = 2$ then R and R/X do not belong to the same class Δ_i $(i \in \{2, 3, 4\})$.

(3) Let R be an element of $\Delta_2 \cup \Delta_3 \cup \Delta_4$ with base E and let X be a subset of E such that R/X is decomposable or $\operatorname{card}(X) = 2$. If R and R/X do not belong to the same class Δ_i then there is an element of $Q(R)$ containing X.

3 Proof of Theorem 1

Since R and R' are hypomorphic, for each element x of E, there is an isomorphism, denoted by f_x, from $R/(E \setminus \{x\})$ onto $R'/(E \setminus \{x\})$.

Suppose that there is a subset X of E such that: R'/X is indecomposable, $\operatorname{card}(E \setminus X) = 2$ and X is not an interval of R'. Since X is not an interval of R' there is an element x of $E \setminus X$ such that X is not an interval of $R'/(X \cup \{x\})$. Let y be the element of $E \setminus (X \cup \{x\})$: since $R/f_y^{-1}(X)$ is indecomposable, there is an element Y of $Q(R)$ which contains $f_y^{-1}(X)$ and since $f_y^{-1}(X)$ is not an interval of $R/f_y^{-1}(X \cup \{x\})$, $f_y^{-1}(x)$ belongs to Y. Thus, there is an element z of $E \setminus X$ such that $Q(R) = \{E \setminus \{z\}, \{z\}\}$.

If X is a subset of E such that $E \setminus X = \{x\}$ and R'/X is indecomposable then $f_x^{-1}(X)$ is contained in an element of $Q(R)$ in such a way that $Q(R) = \{X, \{x\}\}$.

The relation R' is decomposable. Otherwise, by Propositions 3 (2) and 5 (1), for each element a of E there is a subset X of E such that: $a \in X$, $\operatorname{card}(E \setminus X) = 1$ or 2 and R'/X is indecomposable. By previous facts, there is an element b of $E \setminus X$ such that $Q(R) = \{E \setminus \{b\}, \{b\}\}$.

The relation R' belongs to $\Delta_2 \cup \Delta_3 \cup \Delta_4$. If $R' \in \Delta_1$, then consider an element X of $Q(R')$ such that $\operatorname{card}(X) \geq 2$. Let x be an element of X and let A be a subset of $E \setminus \{x\}$ such that for every element Y of $Q(R')$, $A \cap (Y \setminus \{x\})$ is a singleton. Since R'/Y is indecomposable, $f_x^{-1}(Y)$ is contained in an element of $Q(R)$ so that $Q(R) = \{E \setminus \{x\}, \{x\}\}$.

The relations R and R' belong to the same class Δ_i. Otherwise, there is an element X of $Q(R)$ such that $\operatorname{card}(X) \geq 2$ and for each element x of X, $Q(R') = \{E \setminus \{x\}, \{x\}\}$.

4 Proof of Theorem 2

Before proving Theorem 2, let us introduce some definitions and present some properties which are useful in what follows.

Definition 3. Let T be a relation on a set F.

(1) $I(T)$ is the relation with base F defined in the following way: for $a, b \in X$, $I(T)(a, b) = +$ when $T/(F \setminus \{a, b\})$ is indecomposable. Notice that $I(T)$ is not necessarily reflexive.

(2) For each element x of F, $V(T)(x)$ is the set of elements y of $F \setminus \{x\}$ such that $I(T)(x,y) = +$.

(3) $i(T)$ is the number of pairs $\{x,y\}$ such that $I(T)(x,y) = +$.

Lemma 1 *Let T be a relation on a set F with $\mathrm{card}(F) \geq 5$ and let x be an element of F. If $I(T)(x,x) = -$ and if $V(T)(x) \neq \emptyset$ then either there is an element y of $V(T)(x)$ such that $F \setminus \{x,y\}$ is an interval of $T/(F \setminus \{x\})$ and we have $V(T)(x) = \{y\}$ or there are distinct elements y and z of $V(T)(x)$ such that $\{y,z\}$ is an interval of $T/(F \setminus \{x\})$ and $V(T)(x) = \{y,z\}$.*

The proof of this lemma is a direct consequence of the following fact. If y belongs to $V(T)(x)$ and if $F \setminus \{x,y\}$ is denoted by X then y belongs to X_2 or $X(u)$ (see Definition 2), where u is an element of X.

Now we are ready to prove Theorem 2. For each element x of E, there is an isomorphism, denoted by f_x, from $R/(E \setminus \{x\})$ onto $R'/(E \setminus \{x\})$.

By Theorem 1, it is sufficient to show that R' is decomposable. Since R and R' are hypomorphic, for each element x of E, $I(R)(x,x) = I(R')(x,x)$, $\mathrm{card}[V(R)(x)] = \mathrm{card}[V(R')(x)]$ and so $i(R) = i(R')$.

Suppose that $\mathrm{card}[Q(R)] < \mathrm{card}(E) - 2$. By Proposition 5 (4), we can assume that for each element a of E, there is a subset X of E satisfying: $a \in X$, $\mathrm{card}(E \setminus X) = 2$ and R'/X is indecomposable. Either X is an interval of R' or there is an element x of $E \setminus X$ such that X is not an interval of $R'/(X \cup \{x\})$. Let us denote by b the unique element of $E \setminus (X \cup \{x\})$. Since $R/f_b^{-1}(X)$ is indecomposable, there is an element Y of $Q(R)$ containing $f_b^{-1}(X)$ and since $f_b^{-1}(X)$ is not an interval of $R/f_b^{-1}(X \cup \{x\})$, x belongs to Y so that $Q(R) = \{E \setminus \{b\}, \{b\}\}$. Thus we obtain the following contradiction: for each $a \in E$ there is $b \in E \setminus \{a\}$ such that $Q(R) = \{E \setminus \{b\}, \{b\}\}$.

Suppose that $\mathrm{card}[Q(R)] = \mathrm{card}(E) - 2$ and that there are distinct elements a, b and c of E such that $\{a,b,c\} \in Q(R)$. We obtain the following definition of $I(R)$: if $x,y \in E$ then $I(R)(x,y) = +$ if and only if $x \neq y$ and $\{a,b,c\}$ contains $\{x,y\}$. Necessarily, $I(R) = I(R')$. By Lemma 1, if u and v are distinct elements of $\{a,b,c\}$ then $\{u,v\}$ is an interval of $R'/(E \setminus \{w\})$ (where $\{a,b,c\} = \{u,v,w\}$) in such a way that $\{a,b,c\}$ is an interval of R'.

Suppose that $\mathrm{card}[Q(R)] = \mathrm{card}(E) - 2$ and that there are different elements a, b, c and d of E such that $\{a,b\}$ and $\{c,d\}$ belong to $Q(R)$. In this case, we have the following definition of $I(R)$: if $x,y \in E$ then $I(R)(x,y) = +$ if and only if $x \neq y$, $\{x,y\} \cap \{a,b\} \neq \emptyset$ and $\{x,y\} \cap \{c,d\} \neq \emptyset$. There are distinct elements u and v of $\{a,b,c,d\}$ with $I(R')(u,v) = -$ and if $w \in \{a,b,c,d\} \setminus \{u,v\}$ then $I(R')(u,w) = I(R')(v,w) = +$ so that by Lemma 1, $\{u,v\}$ is an interval of $R'/(E \setminus \{w\})$. Therefore, $\{u,v\}$ is an interval of R'.

Suppose that $\mathrm{card}[Q(R)] = \mathrm{card}(E) - 1$. There are distinct elements a and b of E such that $\{a,b\} \in Q(R)$. Let us denote by Γ the set of elements γ of $E \setminus \{a,b\}$ such that $I(R)(a,\gamma) = +$. The relation $I(R)$ has the following properties.

(i) Let x be an element of E: $I(R)(x,x) = +$ if and only if $x = a$ or b.

(ii) Let x and y be distinct elements of E such that $\{x,y\} \neq \{a,b\}$: $I(R)(x,y) = +$ if and only if $\{x,y\} \cap \{a,b\} \neq \emptyset$ and $\{x,y\} \cap \Gamma \neq \emptyset$.

Suppose that for $x \in \{a, b\}$ and $\gamma \in \Gamma$, we have: $I(R')(x, \gamma) = +$. If Γ has at least two elements γ and γ' then $\{a, b\}$ is an interval of $R'/(E \setminus \{\gamma\})$ and of $R'/(E \setminus \{\gamma'\})$ so that $\{a, b\}$ is an interval of R'. Hence, we can assume that $\mathrm{card}(\Gamma) \leq 1$. By Proposition 5 (4), we can also assume that for each element u of E, there are distinct elements $x(u)$ and $y(u)$ of $E \setminus \{u\}$ such that $I(R')(x(u), y(u)) = +$. In particular, for $u = a$ we have $\{x(a), y(a)\} \cap \Gamma \neq \emptyset$ so that Γ is a singleton $\{\gamma\}$. For $u = \gamma$ we have $\{x(\gamma), y(\gamma)\} = \{a, b\}$. Finally, we obtain the following definition of $I(R')$: for given x and y in E, $I(R')(x, y) = +$ if and only if either $x = y$ and $x \in \{a, b\}$ or $x \neq y$ and $\{x, y\} = \{a, b\}$, $\{a, \gamma\}$ or $\{b, \gamma\}$. Therefore, we can apply Lemma 2 (see Section 5).

Suppose that there is an element γ' of Γ such that $I(R')(a, \gamma') = -$. Since $\mathrm{card}[V(R')(a)] = \mathrm{card}[V(R)(a)] \geq \mathrm{card}(\Gamma)$, $I(R')(a, b) = +$ and for every element γ of $\Gamma \setminus \{\gamma'\}$, $I(R')(a, \gamma) = +$. Since $\mathrm{card}[V(R')(a)] = \mathrm{card}[V(R')(b)]$ and since $I(R')(a, b) = +$, there is an element γ of $\Gamma \setminus \{\gamma'\}$ such that $I(R')(b, \gamma) = -$. So, we obtain the following definition of $I(R')$.

(a) If x is an element of E, then $I(R')(x, x) = +$ if and only if $x = a$ or b.

(b) If x and y are distinct elements of $\{a, b, \gamma, \gamma'\}$, then $I(R')(x, y) = +$ if and only if $\{x, y\} = \{a, \gamma\}$, $\{\gamma, \gamma'\}$, $\{\gamma', b\}$, or $\{b, a\}$.

(c) If x belongs to $E \setminus \{a, b, \gamma, \gamma'\}$ then for every $y \in E$, we have: $I(R')(x, y) = +$ if and only if $x \in \Gamma \setminus \{\gamma, \gamma'\}$ and $y \in \{a, b\}$.

If $\Gamma \neq \{\gamma, \gamma'\}$ then $\{a, \gamma, \gamma'\}$ is an interval of $R'/(E \setminus \{b, \gamma'\})$. Since $I(R')(b, \gamma') = +$, $\Gamma = \{\gamma, \gamma'\}$ and we can use Lemma 3 (see Section 6).

5   Proof of Lemma 2

In this section, we prove the following result.

Lemma 2 *Let T be a relation on a set E with $\mathrm{card}(E) \geq 11$ and let a, b and γ be distinct elements of E. Suppose that for all elements x and y of E, we have: $I(T)(x, y) = +$ if and only if either $x = y$ and $x \in \{a, b\}$ or $x \neq y$ and $\{x, y\} = \{a, b\}$, $\{a, \gamma\}$ or $\{b, \gamma\}$. Then T is decomposable.*

Before proving this lemma, let us introduce some notation.

(a) X is the set $\{a, b, \gamma\}$.

(b) For $x \in E \setminus X$, $T(x)$ is the relation $T/(X \cup \{x\})$.

(c) Y is the set of elements y of $E \setminus X$ such that X is an interval of $T(y)$.

(d) Z is the complement $E \setminus (X \cup Y)$.

Before studying the relation $T(z)$, for $z \in Z$, let us make some remarks. First, if $Z = \emptyset$ then X is an interval of T. So we can assume that $Z \neq \emptyset$. Secondly, by Proposition 5 (2), we can assume that if U is a subset of E such that $3 \leq \mathrm{card}(U) \leq 5$ and $\mathrm{card}(U \cap X) \geq 2$, then T/U is decomposable. Finally, observe that since $I(T)(a, \gamma) = I(T)(b, \gamma) = +$, $\{a, b\}$

is an interval of $T/(E \setminus \{\gamma\})$. Thus, we can also assume that $\{a, b\}$ is not an interval of T/X.

Now consider an element z of Z. By a previous remark, $T(z)$ is decomposable so that by Proposition 1, we can define the interval partition $Q[T(z)]$ of $T(z)$ (see Definition 1). Since X is not an interval of $T(z)$, all non-trivial intervals of $T(z)$ contain z. Thus, we can suppose that $\mathrm{card}[Q(T(z))] = 2$ and $\{a\} \in Q[T\{z\}]$ or $\{b\} \in Q[T(z)]$. Let us denote by $Z(a)$ (resp. $Z(b)$) the set of all elements z of Z such that $\{a\}$ (resp. $\{b\}$) belongs to $Q[T(z)]$. We obtain the following properties.

(i) If $z \in Z(a)$ (resp. $z \in Z(b)$) then the non-trivial intervals of $T(z)$ are $\{z, b\}$, $\{z, b, \gamma\}$ (resp. $\{z, a\}$, $\{z, a, \gamma\}$).

(ii) If $z \in Z(a)$ (resp. $z \in Z(b)$) then $\{b, \gamma\}$ (resp. $\{a, \gamma\}$) is the unique non-trivial interval of $T(z)$.

(iii) $Z = Z(a)$ or $Z = Z(b)$.

In what follows, we suppose that $Z = Z(a)$, so that if $Y = \emptyset$ then $E \setminus \{a\}$ is an interval of T. Thus, we can also assume that $Y \neq \emptyset$. Now, we claim that if $y \in Y$ and if $z \in Z$ then $X \cup \{z\}$ is an interval of $T/(X \cup \{y, z\})$. Assuming this claim, we obtain that $E \setminus Y$ is an interval of T and the proof is complete.

So, consider an element y of Y and an element z of Z. By previous hypothesis, $T/(X \cup \{y, z\})$ has at least one non-trivial interval I and $T/\{b, \gamma, y, z\}$ has at least a non-trivial interval K. We can assume the following.

(i) z belongs to I. Otherwise, $I \cap (X \cup \{z\})$ is an interval of $T(z)$ which does not contain z so that there is an element x of X such that $I = \{x, y\}$. Since y belongs to Y, $X \setminus \{x\}$ is an interval of T/X and so $x = a$. It follows that $X \cup \{z\}$ is an interval of $T/(X \cup \{y, z\})$.

(ii) y belongs to I. Otherwise, I is an interval of $T(z)$ in such a way that I contains $\{b, z\}$ and so $X \cup \{z\}$ is an interval of $T/(X \cup \{y, z\})$.

Since $y \in Y$ and $z \in Z$, $I \neq \{y, z\}$ and there is an element x of X such that I contains $\{x, y, z\}$. Thus, $I \setminus \{y\}$ is a non-trivial interval of $T(z)$ so that $a \notin I$. Since $(X \cup \{z\}) \setminus \{a\}$ is an interval of $T(z)$, $(X \cup \{x, y\}) \setminus \{a\}$ is an interval of $T/(X \cup \{y, z\})$.

Now, we can assume the following about K.

(i) y belongs to K. Otherwise, K is an interval of $T(z)/\{b, z, \gamma\}$ and since $\mathrm{card}(K) \geq 2$, K contains $\{b, z\}$ so that $X \cup \{z\}$ is an interval of $T/(X \cup \{y, z\})$.

(ii) γ belongs to K. Otherwise, $K \cap \{b, z\}$ contains at least one element x and $\{a, \gamma\}$ is an interval of $T(z)$. Since $\{b, z\}$ is an interval of $T(z)$, $\{a, \gamma\}$ would be an interval of $T(z)$.

If $K \neq \{y, \gamma\}$ then there is an element x of $\{b, z\}$ such that $K = \{y, \gamma, x\}$. Thus, $K \cap \{z, b, \gamma\} = \{\gamma, x\}$ would be an interval of $T/\{z, b, \gamma\}$ and so of $T(z)$. It follows that $K = \{y, \gamma\}$ and since $\{b, z\}$ is an interval of $T/\{b, z, \gamma\}$, $X \cup \{z\}$ is an interval of $T/(X \cup \{y, z\})$.

6   Proof of Lemma 3

In this section, we prove the following result.

Lemma 3 *Let T be a relation on a set E with $\mathrm{card}(E) \geq 12$ and let a, b, γ and γ' be distinct elements of E. Suppose that for all elements x and y of E, we have: $I(T)(x,y) = +$ if and only if either $x = y$ and $x \in \{a, b\}$ or $x \neq y$ and $\{x, y\} = \{a, b\}$, $\{a, \gamma\}$, $\{b, \gamma'\}$ or $\{\gamma, \gamma'\}$. Then T is decomposable.*

Before proving this lemma, let us introduce some notation.

(a) X is the set $\{a, b, \gamma, \gamma'\}$.

(b) For $x \in E \setminus X$, $T(x)$ is the relation $T/(X \cup \{x\})$.

(c) Y is the set of elements y of $E \setminus X$ such that X is an interval of $T(y)$.

(d) Z is the set $E \setminus (X \cup Y)$.

We begin the proof of this lemma with some remarks. First, if $Z = \emptyset$ then X is an interval of T. So we can suppose that $Z \neq \emptyset$. Secondly, by Proposition 5 (2), we can assume that if U is a subset of E such that U contains $\{a, \gamma'\}$ or $\{b, \gamma\}$ and $3 \leq \mathrm{card}(U) \leq 6$ then T/U is decomposable. Lastly, since $I(T)(a, \gamma) = I(T)(\gamma, \gamma') = +$ and since $I(T)(b, \gamma') = I(T)(\gamma', \gamma) = +$, $\{a, \gamma'\}$ is an interval of $T/(E \setminus \{\gamma\})$ and $\{b, \gamma\}$ is an interval of $T/(E \setminus \{\gamma'\})$. Thus, we can assume that $\{a, \gamma'\}$ is not an interval of $T/\{a, \gamma, \gamma'\}$ and $\{b, \gamma\}$ is not an interval of $T/\{b, \gamma, \gamma'\}$.

Let z be an element of Z. Since X is not an interval of $T(z)$, all non-trivial intervals of $T(z)$ contain z. It follows that $\mathrm{card}[Q(T(z))] = 2$ or 3. Let us denote by Z_2 (resp. Z_3) the set of elements z of Z such that $\mathrm{card}[Q(T(z))] = 2$ (resp. $\mathrm{card}[Q(T(z))] = 3$). In the sequel, we study the relation $T(z)$, for $z \in Z$, distinguishing two cases.

CASE 1: LET z BE AN ELEMENT OF Z_2.

Let us denote by A the element of $Q[T(z)]$ which contains z. If γ (resp. γ') does not belong to A then $\{a, \gamma'\}$ is an interval of $T/\{a, \gamma, \gamma'\}$ (resp. $\{b, \gamma\}$ is an interval of $T/\{b, \gamma, \gamma'\}$). So, γ and γ' belong to A and $\{a\} \in Q[T(z)]$ or $\{b\} \in Q[(T(z)]$. Let us denote by $Z_2(a)$ (resp. $Z_2(b)$) the set of elements z of Z_2 such that $\{a\}$ (resp. $\{b\}$) belongs to $Q[T(z)]$. We obtain the following properties.

(i) If z belongs to $Z_2(a)$ (resp. $Z_2(b)$) then the non-trivial intervals of $T(z)$ are $\{z, b\}$, $\{z, b, \gamma'\}$ and $\{z, b, \gamma', \gamma\}$ (resp. $\{z, a\}$, $\{z, a, \gamma'\}$ and $\{z, a, \gamma', \gamma\}$).

(ii) If z belongs to $Z_2(a)$ (resp. $Z_2(b)$) then the non-trivial intervals of T/X are $\{b, \gamma'\}$ and $\{b, \gamma', \gamma\}$ (resp. $\{a, \gamma'\}$ and $\{a, \gamma', \gamma\}$).

(iii) $Z_2 = Z_2(a)$ or $Z_2 = Z_2(b)$.

In what follows, we assume that $Z_2 = Z_2(a)$.

CASE 2: LET z BE AN ELEMENT OF Z_3.

Let us denote by F the base of $T(z)$. We obtain the following: the non-trivial intervals of $T(z)$ are $F \setminus \{a\}$, $F \setminus \{b\}$ and $F \setminus \{a, b\}$.

It follows that $Z = Z_2(a)$ or $Z = Z_3$. As in the proof of Lemma 2, we show that if $y \in Y$ and $z \in Z$ then $X \cup \{z\}$ is an interval of $T/(X \cup \{y, z\})$. Thus, $E \setminus Y$ is an interval of T and we can assume that $Y = \emptyset$. But in this case, since $Z = Z_2(a)$ or $Z = Z_3$, $E \setminus \{a\}$ is an interval of T.

References

[1] J.A. Bondy and R.L. Hemminger, Graph reconstruction, a survey, *J. Graph Theory* **1** (1977), 227–268.

[2] R. Fraïssé, L'intervalle en theorie des relations, ses generalisations, filtre intervallaire et cloture d'une relation, in *Orders, Description and Roles* (M. Pouzet, D. Fichard eds.), North-Holland (1984), 313–342.

[3] M. Habib, Substitution des structures combinatoires, theorie et algorithmes, Doctoral Dissertation, Univ. P. and M. Curie, Paris (1981).

[4] P. Ille, Indecomposable relations, preprint (1990).

[5] J.H. Schmerl, Indecomposable partially ordered sets and other binary relational structures, preprint (1987).

[6] P.K. Stockmeyer, The falsity of the reconstruction conjecture for tournaments, *J. Graph Theory* **1** (1977), 19–25.

[7] S.M. Ulam, A collection of mathematical problems, in *Interscience Tracts in Pures and Appl. Math.* **8** (1960).

The Average Size of Nonsingular Sets in a Graph

WILFRIED IMRICH

Institut für Mathematik und Angewandte Geometrie,
Montanuniversität,
A-8700 Leoben, Austria.

NORBERT SAUER

Department of Mathematics,
University of Calgary,
Calgary, Alberta
T2N 1N4, Canada.

WOLFGANG WOESS

Dipartimento di Matematica,
Università di Milano,
Via C. Saldini, 50,
20133 Milano, Italy.

Abstract

A nonsingular set in a finite graph is defined as the vertex set of an induced subgraph which has no isolated point. If G is a graph without isolated points and with at least two vertices and B is a connected subgraph, then the average size of those nonsingular sets in G which contain B is at least $(|G| + |B|)/2$. This result is used to prove the following: if $\mathcal{F}$ is a family of sets which is closed with respect to union, and if none of the generating sets in $\mathcal{F}$ has more than two elements, then the average size of a set in $\mathcal{F}$ is at least half of the size of the largest set in $\mathcal{F}$.

1 Introduction and statement of results

Let G be a finite graph (unoriented, without loops or multiple edges). Subsets A of $V(G)$ are identified with induced subgraphs; $|A|$ denotes the number of vertices. We call a subset A of G *nonsingular*, if it has no isolated vertex. If B is a subgraph of G, then we define

$$\mathcal{F}(G, B) = \{A \subseteq G \mid B \subseteq A \text{ and } A \text{ is nonsingular}\}.$$

In this note, we are interested in the average size of an element in $\mathcal{F}(G, B)$. To this end, we introduce a generating function:

$$f(t|G, B) = \sum_{A \in \mathcal{F}(G,B)} t^{|A|}.$$

Then the average size in question is given by $f'(1|G, B)/f(1|G, B)$, where $f'(t|G, B)$ denotes the derivative with respect to t.

N.W. Sauer et al. (eds.), Finite and Infinite Combinatorics in Sets and Logic, 199–205.
© 1993 *Kluwer Academic Publishers. Printed in the Netherlands.*

Theorem 1. *If G is a nonsingular graph (with at least two vertices) and B is a connected subgraph (possibly void), then*

$$\frac{f'(1|G,B)}{f(1|G,B)} \geq \frac{|G|+|B|}{2}.$$

Observe that $\mathcal{F}(G,B)$ is a family of sets which is closed with respect to set union. Thus, our result can be related to the study of union closed families.

Let $\mathcal{F}$ denote an arbitrary finite family of finite sets which is closed with respect to set union. P. Frankl [1], [2] has conjectured the following.

Conjecture. *There is an element which occurs in at least half of the sets of $\mathcal{F}$.*

In other words, if V is the largest set in $\mathcal{F}$ and, for x in V, $\mathcal{F}_x = \{A \in \mathcal{F} \mid x \in A\}$, then one is looking for an x such that $|\mathcal{F}_x| \geq |\mathcal{F}|/2$. In discussions of the conjecture at the 1985 Banff meeting, R. Graham and several others observed the following.

Lemma 1. *If $\mathcal{F}$ contains a nonvoid set A with no more than two elements, then $|\mathcal{F}_x| \geq |\mathcal{F}|/2$ for some $x \in A$.*

As no proof of this observation seems to have appeared in print, we include a proof in this paper. At the same meeting in Banff, R. Graham claimed to have an example of a union closed family $\mathcal{F}$ with the following properties: there is an $A \in \mathcal{F}$ with three elements, such that $|\mathcal{F}_x| < |\mathcal{F}|/2$ for every $x \in A$, but the conjecture is true for this $\mathcal{F}$.

Consider the following statement.

The average size of a set in $\mathcal{F}$ is at least half of the size of the largest set in $\mathcal{F}$. (1)

It is easy to see that statement (1) implies the validity of the conjecture for $\mathcal{F}$: assume that

$$\sum_{A \in \mathcal{F}} |A|/|\mathcal{F}| \geq |V|/2,$$

Suppose that $|\mathcal{F}|/2 > |\mathcal{F}_x|$ for every x in V. Then

$$|V| \cdot |\mathcal{F}|/2 > \sum_{x \in V} |\mathcal{F}_x| = \sum_{A \in \mathcal{F}} |A|,$$

a contradiction.

Thus, in particular, the conjecture is true for $\mathcal{F}(G,B)$ as in Theorem 1. (However, we remark that Lemma 1 is needed in the proof of Theorem 1.) A family of *generators* $\mathcal{E}$ of $\mathcal{F}$ is a subfamily of $\mathcal{F}$ with the following properties:

(i) every set in $\mathcal{F}$ is a finite union of sets in $\mathcal{E}$,

(ii) $\mathcal{E}$ is minimal with respect to (i).

It is easy to see that $\mathcal{E}$ is unique and no set in $\mathcal{E}$ is empty. Theorem 1 can be applied to the case when no generator has more than two points.

Theorem 2. *If $|A| \leq 2$ for every generator A of $\mathcal{F}$, then the average size of a set in $\mathcal{F}$ is at least half of the size of the largest set in $\mathcal{F}$.*

Note that Theorem 2 says that, if $|A| \leq 2$ for every generator A of $\mathcal{F}$, then statement (1) is true. In general it is not true that the average size of a set in a finite union closed family $\mathcal{F}$ is at least half of the size of the largest set in $\mathcal{F}$. Actually, the ratio between the average size and the largest size of a set in $\mathcal{F}$ may be arbitrarily small. Here is an example: let F be a set with $|F| = 2^n$ and let $E \subset F$ with $|E| = 2n$, and define

$$\mathcal{F} = \{A \subset F : |A| \geq 2^n - 1 \text{ or } A \subseteq E\}.$$

In this case, the average size of a set in $\mathcal{F}$ is asymptotically equal to $n + 1$.

In a preliminary version of this note, we raised the question whether statement (1) was true for every union closed family $\mathcal{F}$ where all generators have the same size. This has been disproved recently by P. Wójcik [3].

2 Proofs

We start with Lemma 1, as discussed at Banff.

Proof of Lemma 1. If $A = \{x\}$ for some $x \in V$, where V is the largest set in $\mathcal{F}$, then $B \mapsto B \cup \{x\}$ is injective from $\mathcal{F} \setminus \mathcal{F}_x$ into $\mathcal{F}_x$, so $|\mathcal{F}_x| \geq |\mathcal{F}|/2$.

Now assume that $A = \{x, y\}$ with $|\mathcal{F}_x| \geq |\mathcal{F}_y|$. Let $a = |\{B \in \mathcal{F} \mid x \in B, y \notin B\}|$, $b = |\{B \in \mathcal{F} \mid y \in B, x \notin B\}|$, $c = |\{B \in \mathcal{F} \mid A \subseteq B\}|$ and $d = |\{B \in \mathcal{F} \mid A \cap B = \emptyset\}|$. Then $|\mathcal{F}| = a + b + c + d$, $|\mathcal{F}_x| = a + c$ and $|\mathcal{F}_y| = b + c$. Hence $a \geq b$, and (as in the first case) $c \geq d$, so again $|\mathcal{F}_x| \geq |\mathcal{F}|/2$. ∎

Now we turn to the study of $\mathcal{F}(G, B)$, where G is a graph. Note that

$$f(t|G, \emptyset) = f(t|G, \{x\}) + f(t|(G \setminus \{x\}), \emptyset). \tag{2}$$

In particular, if G is a graph without isolated vertices, then Lemma 1 implies that for some x in G,

$$f(1|G, \{x\}) \geq f(1|(G \setminus \{x\}), \emptyset). \tag{3}$$

Proof of Theorem 1. We first assume that G is connected.

We use induction on $n = |G|$ and, for each n, induction on $k = |G| - |B|$.

For $n = 2$ the verification is trivial.

Assume that $|G| = n$ and that the statement of the theorem is true for every graph $|G'|$ with $|G'| = m$, $2 \leq m < n$. First, let $k = 0$. Then $B = G$, and $f(t|G, B) = t^{|G|}$. We have

$$f'(1|G, B) = |G| = \frac{|G| + |B|}{2} f(1|G, B),$$

so the result follows. Now let $k = |G| - |B|$ be positive and assume also that the statement is true for all B' such that $0 \leq |G| - |B'| < k$.

CASE 1. B IS NONVOID.

As G and B are connected and $G \setminus B \neq \emptyset$, there is an x in $G \setminus B$ such that

$$B' = B \cup \{x\}$$

is connected.

CASE 2. $B = \emptyset$.

By (3), there is a vertex x such that $f(1|G, \{x\}) \geq f(1|(G \setminus \{x\}), \emptyset)$. Set

$$B' = \{x\}.$$

In both cases,

$$f(t|G, B) = f(t|(G \setminus \{x\}), B) + f(t|G, B').$$ (4)

We intend to differentiate and to obtain inequalities for the right-hand side.

The graph $G \setminus \{x\}$ may be disconnected, and it may have isolated points. Let $C = C(x, B)$ be the set of those isolated points of $G \setminus \{x\}$ which are not in B. Then, for every y in C, the only neighbour of y in G is x. In particular,

$$G' = G \setminus C$$

is connected. No point of C can be an element of a nonsingular set of $G \setminus \{x\}$. Hence

$$f(t|(G \setminus \{x\}), B) = f(t|(G' \setminus \{x\}), B).$$ (5)

Now write $G' \setminus \{x\}$ as the union of its connected components: $G' \setminus \{x\} = G_1 \cup G_2 \cup ... \cup G_s$, listed so that $B \subseteq G_1$ and $|G_i| \geq 2$ for $i = 2, ..., s$. A set A is in $\mathcal{F}(G' \setminus \{x\}, B)$ if and only if $A = A_1 \cup A_2 \cup ... \cup A_s$, where $A_1 \in \mathcal{F}(G_1, B)$ and $A_i \in \mathcal{F}(G_i, \emptyset)$ for $i = 2, ..., s$. This and (5) yield

$$f(t|G \setminus \{x\}, B) = f(t|G_1, B) \cdot f(t|G_2, \emptyset) \cdot ... \cdot f(t|G_s, \emptyset).$$

Note that we may have $|G_1| = 1$ only if $|B| = 1$. In this case, $f(t|G_1, B) = f'(t|G_1, B) = 0$. Otherwise, we may apply the induction hypothesis (for n) to (G_1, B). In both cases

$$f'(t|G_1, B) \geq \frac{|G_1| + |B|}{2} f(t|G_1, B).$$

The induction hypothesis also applies to $(G_i, \emptyset)$, giving

$$f'(t|G_i, \emptyset) \geq \frac{|G_i|}{2} f(t|G_i, \emptyset), \quad i = 2, ..., s.$$

This, (5) and the product rule for differentiation give

$$f'(1|(G \setminus \{x\}), B) \geq \frac{|G| - |C| - 1 + |B|}{2} f(1|(G \setminus \{x\}), B).$$ (6)

On the other hand, B' is a connected subgraph of G'. For any $U \subseteq C$, considering those nonsingular sets in $\mathcal{F}(G, B')$ which contain U and do not contain $C \setminus U$, we get

$$f(t|G, B') = \sum_{U \subseteq C} f(t|G' \cup U, B' \cup U).$$

Differentiating and applying the induction hypothesis (for k, if $U = C$, and for n otherwise) to each of the summands, we obtain

$$f'(1|G, B') \geq \sum_{U \subseteq C} \frac{|G| - |C| + |B| + 1 + 2|U|}{2} f(1|G' \cup U, B' \cup U).$$ (7)

Combining inequalities (6) and (7), we obtain from (4)

$$f'(1|G, B) \geq \frac{|G| + |B|}{2} f(1|G, B) + \frac{\Phi}{2},$$ (8)

where

$$\Phi = \sum_{U \subseteq C} (|U| - |C \setminus U| + 1) f(1|G' \cup U, B' \cup U) - (|C| + 1) f(1|(G \setminus \{x\}), B).$$

The proof will be complete if we can show that $\Phi \geq 0$. We observe that if $U \subseteq C$ is nonvoid, then $A \in \mathcal{F}(G' \cup U, B' \cup U)$ if and only if $A \cup (C \setminus U) \in \mathcal{F}(G, B' \cup C)$. Hence, $f(1|(G' \cup U), B' \cup U) = f(1|G, B' \cup C)$ whenever $\emptyset \neq U \subseteq C$. (If B is nonvoid, this also holds if $U = \emptyset$.) Consequently,

$$\Phi = \left(\sum_{U \subseteq C} |U| - |C \setminus U| + 1 \right) f(1|G, B' \cup C)$$
$$+ (|C| - 1) \left(f(1|G, B' \cup C) - f(1|G', B') \right) - (|C| + 1) f(1|(G \setminus \{x\}), B)$$
$$= 2^{|C|} f(1|G, B' \cup C) - (|C| + 1) f(1|(G \setminus \{x\}), B)$$
$$+ (|C| - 1) \left(f(1|G, B' \cup C) - f(1|G', B') \right).$$

If $A \in \mathcal{F}(G', B')$ then $x \in A$, and $A \cup C$ is nonsingular: $A \in \mathcal{F}(G, B' \cup C)$. Thus, $f(1|G, B' \cup C) \geq f(1|G', B')$, and

$$\Phi \geq (|C| + 1) \left(f(1|G, B' \cup C) - f(1|(G \setminus \{x\}), B) \right),$$

since $2^{|C|} \geq |C| + 1$. There are two cases to consider.

CASE 1. B IS NONVOID.
   If $A \in \mathcal{F}(G \setminus \{x\}, B) = \mathcal{F}(G' \setminus \{x\}, B)$, then $A \cup \{x\}$ is nonsingular, as x has a neighbour in $B \subseteq A$. In other words, $A \cup \{x\} \cup C$ is nonsingular (even if $C = \emptyset$), and is an element of $\mathcal{F}(G, B' \cup C)$. Thus, $f(1|(G \setminus \{x\}), B) \leq f(1|G, B' \cup C)$, so $\Phi \geq 0$.

CASE 2. $B = \emptyset$.
   We have to consider two subcases.

(a) C IS NONVOID. In this case, $\{x\} \cup C$ is nonsingular, and if $A \in \mathcal{F}(G \setminus \{x\}, \emptyset) = \mathcal{F}(G' \setminus \{x\}, \emptyset)$ then

$$A \cup \{x\} \cup C = A \cup B' \cup C \in \mathcal{F}(G, B' \cup C).$$

Again, $f(1|(G \setminus \{x\}), B) \leq f(1|G, B' \cup C)$, and $\Phi \geq 0$.

(b) $\underline{C = \emptyset}$. Then

$$f(1|G, B' \cup C) - f(1|(G \setminus \{x\}), B) = f(1|G, \{x\}) - f(1|(G \setminus \{x\}), |(G \setminus \{x\}), \emptyset),$$

which is nonnegative by the choice of x, and $\Phi \geq 0$.

In all three cases, $\Phi \geq 0$, and by (8), the proof is completed (for connected G).
Now suppose in general that G is nonsingular. Then we can use the formula

$$f(t|G, B) = f(t|G_1, B) \cdot f(t|G_2, \emptyset) \cdot \ldots \cdot f(t|G_s, \emptyset),$$

where $G_1, ..., G_s$ are the connected components of G and $B \subseteq G_1$. Differentiation and application of the result for connected graphs yield the theorem. ∎

There may be several ways to derive Theorem 2. Our proof uses the full strength of Theorem 1.

Proof of Theorem 2. We may assume that $\emptyset \in \mathcal{F}$ (adding the empty set to $\mathcal{F}$ makes the average size smaller). If all generators of $\mathcal{F}$ are singletons, then Theorem 1 is trivially true. Hence, we assume that $\mathcal{E}$, the set of all generators of $\mathcal{F}$, contains two-element sets and maybe also singletons. Observe that there may be points x and y such that $\{x\}$ and $\{x, y\}$ are generators. In this case, $\{y\}$ cannot be a generator.

Let V be the largest set in $\mathcal{F}$, and set $B = \{x \mid \{x\} \in \mathcal{E}\}$. Choose a new set $\bar{B} = \{\bar{x} \mid x \in B\}$ in one-to-one correspondence with B, disjoint from V. We define a graph G with vertex set $V \cup \bar{B}$; the edges are   (i) all $\{x, y\} \in \mathcal{E}$,   (ii) all $\{x, \bar{x}\}$, where $x \in B$, and   (iii) all $\{\bar{x}, \bar{y}\}$, where $\bar{x}, \bar{y} \in \bar{B}$, $\bar{x} \neq \bar{y}$.

Thus, $\bar{B}$ induces a complete subgraph of G which is connected to the rest of G by the edges $\{x, \bar{x}\}$, $x \in B$. (If B is empty, then so is $\bar{B}$.) Now, G has no isolated point, and we may apply Theorem 1 to $(G, \bar{B})$:

$$f'(1|G, \bar{B}) \geq \frac{|G| + |B|}{2} f(1|G, \bar{B}).$$

By our construction, we obviously have $A \in \mathcal{F}$ if and only if $A \cup \bar{B} \in \mathcal{F}(G, \bar{B})$. Hence, $f(1|G, \bar{B}) = |\mathcal{F}|$, and

$$f'(1|G, \bar{B}) = \sum_{A \in \mathcal{F}} |A \cup \bar{B}| = |B| \cdot |\mathcal{F}| + \sum_{A \in \mathcal{F}} |A|.$$

We obtain

$$\sum_{A \in \mathcal{F}} |A| \geq \frac{|G| + |B|}{2} |\mathcal{F}| - |B| \cdot |\mathcal{F}| = \frac{|V|}{2} |\mathcal{F}|. \qquad ∎$$

References

[1] I. Rival, Ed., *Graphs and Order. Problem Session 1*, p 525. Reidel, Dordrecht - Boston, 1985.

[2] R.P. Stanley, *Enumerative Combinatorics, Vol. I.* Exercise 39a, p 161. Woodsworth & Brooks / Cole Math. Series, Monterey, Calif., 1986.

[3] P. Wójcik, *Density of union-closed families.* Preprint, Adam Mickiewicz University, Poznań, Poland.

Some Canonical Partition Ordinals

JEAN A. LARSON*

University of Florida,

Gainesville, Florida

Abstract

In the notation of Erdős and Rado, the expression $\alpha \rightarrow (\alpha, m)^2$ means that for any graph on α either there is an independent subset of type α or there is a complete subgraph of size m. Baumgartner called such ordinals *partition ordinals*. The following two theorems show that κ^ω and $\kappa^\omega \lambda^\omega$ are partition ordinals whenever κ and λ are Ramsey cardinals and $\lambda < \kappa$.

Theorem. For all positive integers m and all Ramsey cardinals κ,

$$\kappa^\omega \rightarrow (\kappa^\omega, m)^2.$$

Theorem. For all positive integers m and all Ramsey cardinals κ and λ with $\lambda < \kappa$,

$$\kappa^\omega \lambda^\omega \rightarrow (\kappa^\omega \lambda^\omega, m)^2.$$

0   Introduction and Notation

J. Baumgartner [1] has labeled *partition ordinals* those ordinals α with the property that $\alpha \rightarrow (\alpha, m)^2$ for all $m < \omega$ where the arrow notation of Erdős and Rado [5] indicates that for any graph on the pairs from a set order isomorphic to α, either there is a subset order isomorphic to α no pair of which is joined by an edge (an *independent set*) or there is a subset of size n all of whose pairs are joined by edges (a *complete* subgraph of size n). One can phrase this statement in terms of partitions by letting Δ_1 be the set of edges of the graph and Δ_0 be the pairs which are not edges. Then the arrow relation holds if every partition of the pairs $[\alpha]^2 = \Delta_0 \cup \Delta_1$ has the property that either there is a homogeneous set X of type α for the Δ_0 side of the partition (all of whose pairs are in Δ_0) or for every $n < \omega$ there is a homogeneous set Y_n of size n for the Δ_1 side of the partition (all of whose pairs are in Δ_1).

The known countable partition ordinals are ω, ω^2 and ω^ω. The fact that ω is a partition ordinal is a consequence of Ramsey's Theorem. E. Specker [14] proved that ω^2 is a partition ordinal. C.C. Chang [2] proved that $\omega^\omega \rightarrow (\omega^\omega, 3)^2$ and E.C. Milner generalized the proof to show that ω^ω is a partition ordinal.

A simple complete bipartite graph gives a counter-example for the partition relation for decomposable ordinals. Thus the first interesting counter-example is for ω^3. These results

---

*Research was partially supported by National Science Foundation Grant DMS-8600931.

N.W. Sauer et al. (eds.), Finite and Infinite Combinatorics in Sets and Logic, 207–224.

and their proofs are the starting point for generalizations to certain ordinals made up from weakly compact and Ramsey cardinals that are the main results of this paper.

The nice counter-example for ω^3 is what is called a "canonical partition". Galvin and Hajnal [3] used Ramsey's Theorem to prove the existence of canonical partition relations for finite sequences of arbitrary fixed length with elements from ω. Here is a generalization of canonical partition to arbitrary sets of finite sequences with elements from a well ordered set.

Definition 0.1 Suppose that A is a well ordered set. Two pairs $\{\vec{u}, \vec{v}\}$ and $\{\vec{x}, \vec{y}\}$ of sequences from A are *similar* if and only if

(1) $\text{length}(\vec{u}) = \text{length}(\vec{x})$ and $\text{length}(\vec{v}) = \text{length}(\vec{y})$;

(2) for all $i < \text{length}(\vec{u})$, for all $j < \text{length}(\vec{v})$,

$$u_i \leq x_j \Leftrightarrow v_i \leq y_j \quad \text{and} \quad u_i \geq x_j \Leftrightarrow v_i \geq y_j.$$

A partition Δ of the pairs $[W]^2$ from a set W of finite sequences from A is *canonical* if and only if whenever two pairs from W are similar, then also they are in the same cell of the partition.

For a first impression of canonical partitions, consider the canonical counter-example for ω^3. First represent ω^3 as the set $W(3)$ of increasing triples from ω under the lexicographic ordering. Next define the canonical partition $\Delta = \Delta_0 \cup \Delta_1$ by $\{\vec{u}, \vec{v}\}$ is in Δ_1 if and only if it is similar to $\{(0, 1, 3), (2, 4, 5)\}$. The reader may check that every subset of $W(3)$ of type ω^3 must contain such a pair, so there is no full size set homogeneous for the Δ_0 side of the partition. Furthermore no set of three elements has all three pairs in Δ_1, so there is no homogeneous set for the other side either.

This example generalizes very nicely to λ^3 for any regular cardinal λ. Since λ^3 is order isomorphic to increasing triples from λ, the same description applied to the appropriate set gives rise to a canonical partition which shows that λ^3 is not a partition ordinal.

Specker [14] was the first to prove that ω^3 is not a partition ordinal, but his proof is not the transparent one given above. E.C. Milner [10] has shown that this partition is the only canonical counter-example for ω^3.

Haddad and Sabbagh [6] have a simple proof that ω^2 is a partition ordinal (E. Specker [14] first proved it), that starts by using Ramsey's Theorem for ω to reduce any partition of the increasing pairs from ω, $W(2)$, to a canonical partition of the increasing pairs from an infinite subset of ω. Then they show that the desired partition property holds for canonical partitions.

J. Larson [7] has taken a similar approach to simplifying the proof of Chang [2] that ω^ω is a partition ordinal. (See also [15] for an exposition of this proof.)

Thus these countable ordinals might be called canonical partition ordinals. The proof of Haddad and Sabbagh for ω^2 lifts to κ^2 for a weakly compact cardinal κ. Thus there are larger canonical partition ordinals.

J. Baumgartner [8] has shown that if α is a partition ordinal and λ is a weakly compact cardinal with $\alpha < \lambda$, then the product $\lambda\alpha$ is also a partition ordinal. Larson [8] has generalized the argument to show that in such a case $\lambda^2\alpha$ is also a partition ordinal. These results have some but not all of the flavor of canonical partition relations, since the role of α in the result has nothing canonical about it. However these results suggest that products of weakly compact cardinals to the first and second power are canonical partition ordinals, since the proof applied recursively would give subsets on which a partition would be canonical. To make things more interesting, one would like to include more complicated ordinals in analogy to ω^ω. The appropriate analog is the ω power of a Ramsey cardinal. Thus the main result of the paper is that κ^ω and $\kappa^\omega \lambda^\omega$ are partition ordinals when $\lambda < \kappa$ are Ramsey cardinals.

Theorem 0.2 For all positive integers m and all Ramsey cardinals κ,

$$\kappa^\omega \to (\kappa^\omega, m)^2.$$

Theorem 0.3 For all positive integers m and all Ramsey cardinals κ and λ with $\lambda < \kappa$,

$$\kappa^\omega \lambda^\omega \to (\kappa^\omega \lambda^\omega, m)^2.$$

The basic idea of the proof is to take an arbitrary partition of the ordinal at hand and to make it as canonical as possible on a restricted set. Unfortunately this process does not proceed as far as one could have hoped for, but it does go far enough to prove the above theorems.

One can use the notion of pinning to derive a corollary from the second theorem. An ordinal α can be *pinned to* an ordinal β, written $\alpha \to \beta$, if and only if there is a function $f : \alpha \to \beta$ such that every subset of α of order type α is carried by f to a subset of β of order type β. We refer to f as a *pinning map*. The basic fact about pinning maps in given below. For a proof see [1].

Proposition 0.4 *If $\alpha \to \beta$ and $\alpha \to (\alpha, m)^2$, then $\beta \to (\beta, m)^2$.*

The corollary to Theorem 0.3 is that $\kappa^\omega \omega$ is a partition ordinal when κ is a Ramsey cardinal. The proof gives the appearance of requiring that κ be larger than some other Ramsey cardinal, but that hypothesis is not necessary. In my thesis under the direction of J. Baumgartner I constructed a direct proof of this result as well as giving a proof that κ^ω is a partition ordinal. The proofs of those results given here are simpler. The proof that $\kappa^\omega \lambda^\omega$ is also a partition ordinal is entirely new.

Theorem 0.5 *For all positive integers m and all Ramsey cardinals κ,*

$$\kappa^\omega \to (\kappa^\omega, m)^2.$$

Proof. We shall only prove it in the special case that $\lambda < \kappa$ is another Ramsey cardinal. For in this case, we may pin $\kappa^\omega \lambda^\omega$ onto $\kappa^\omega \omega$. Represent $\kappa^\omega \lambda^\omega$ as products of finite sequences from λ with finite sequences from κ, $[\lambda]^{<\omega} \times [\kappa]^{<\omega}$, ordered lexicographically. Represent $\kappa^\omega \omega$ as the product $\omega \times [\kappa]^{<\omega}$, ordered lexicographically. Define $f : [\lambda]^{<\omega} \times [\kappa]^{<\omega} \to \omega \times [\kappa]^{<\omega}$ by $f(\vec{\alpha}, \vec{\beta}) = (|\vec{\alpha}|, \vec{\beta})$. Then f is a pinning map, so the corollary follows from Theorem 0.3. $\square$

One might ask whether $\kappa^{\omega+1} = \kappa^\omega \kappa$ is also a partition ordinal, but Larson [9] has shown that if κ is uncountable then $\kappa^{\omega+1} \not\to (\kappa^{\omega+1}, 3)^2$. As a corollary to that result, one sees that if κ^α is a partition ordinal and κ is uncountable and $\alpha \geq \omega + 1$, then α must be indecomposable. Thus one might ask if there are partition ordinals κ^α with $\alpha > \omega$.

The next section of the paper discusses basic facts about partitions of sequences. Section 2 contains the proof of Theorem 0.2. The beginning of the section shows how to make any partition of κ^ω as canonical as possible and the end contains discussions of how to get homogeneous sets. Section 3 treats the problem of making a partition of $\kappa^\omega \lambda^\omega$ as canonical as possible. The situation is not as satisfactory as in the previous section. Section 4 contains the proof of Theorem 0.3 with its discussions of how to get homogeneous sets.

1   Form and Shape

The ordinal λ^n is represented by the set $[\lambda]^n_<$ of increasing sequences of ordinals less than λ of length n, ordered lexicographically, and the ordinal λ^ω is represented by $[\lambda]^\omega_<$, the union of the sets $[\lambda]^n_<$, ordered first by length and then lexicographically. To simplify notation, the expression $\vec{\alpha}$ will represent a finite sequence of ordinals listed in increasing order. For the special case of finite sequences of natural numbers, the alternate notation s is often used. By an abuse of notation, sets of ordinals, increasing sequences of ordinals and finite increasing functions whose domain is a natural number will all be identified, and notation appropriate for one formulation will be used in any of the situations. In particular, the notation $\vec{\alpha}$ will be regarded as a function with domain n. Furthermore, the notation $\vec{\alpha}[i, j)$ will refer to the finite set or increasing sequence consisting of the image of the half-open interval from i to j under the function $\vec{\alpha}$. Unless otherwise noted, sequences denoted by letters, such as $\vec{\alpha}, \vec{\beta}$, are assumed to be listed in increasing order both alphabetically and as sequences.

An important concept in the study of partitions of finite powers of cardinals is that of *canonical partition*, which we introduced in the previous section. Below is another notion that has been used in conjunction with canonical partitions.

Definition 1.1 A pair $\{\vec{\alpha}, \vec{\beta}\}_<$ from $[\lambda]^{<\omega}$ is said to be *clear* if either $\vec{\alpha}$ and $\vec{\beta}$ are disjoint or else they have the same length and the points in common to the two sequences form an initial segment of each of them. A graph is *clear* if only clear pairs are joined.

Hajnal and later Galvin independently proved the existence and prevalence of canonical partitions on finite powers of ω by showing that for any multicolored graph, there is an infinite set $N \subseteq \omega$ so that on sequences from N the graph is canonical. Milner remarked in [10] that one could then define by recursion a subset of type ω^n on which the graph was also clear. These ideas generalize very naturally to $[\lambda]^n$ for weakly compact cardinals λ, and with somewhat more work to $[\lambda]^\omega_<$ for Ramsey cardinals.

Lemma 1.2 (see Theorem 7.2.7 and Lemma 7.3.4 of [15]) *If λ is a Ramsey cardinal, then for any graph on $[\lambda]^\omega_<$ colored with m colors, there is a full cardinality subset L of λ so that the graph is canonical for pairs from $[L]^\omega_<$, the set of finite increasing sequences from L.*

Not all similarity types of pairs have the same chance of appearing in large subsets of a set $[\lambda]^n_{\lessgtr}$; some can be avoided altogether. Notice that the similarity type of a clear pair from $W(n)$ can be described in terms of two increasing sequences of integers up to n. First define a partial order $\ll$ on finite sequences from ω by $\mathbf{a} \ll \mathbf{b}$ if and only if the maximum of $\mathbf{a}$ is less than the minimum of $\mathbf{b}$.

The notions of *form* and *shape* were introduced to describe a collection of similarity types large enough so that given an infinite set $H \subseteq \omega$, one can construct a set $X \subseteq [\omega]^{<\omega}$ of type ω^ω so that all pairs of different lengths had one of the prescribed similarity types.

Now modify these definitions of *form* $2k$ and *form* $2k + 1$ from [15] which applied to sequences of different lengths from W to apply to clear sequences.

Definition 1.3 For $\vec{\alpha}$ and $\vec{\beta}$ from $[\lambda]^{<\omega}$ and some integer ℓ, say the pair $\{\vec{\alpha}, \vec{\beta}\}_<$ has *form* ℓ if there are increasing sequences of integers $\mathbf{s} = (s_0, s_1, \ldots, s_p)$ and $\mathbf{t} = (t_0, t_1, \ldots, t_q)$ with $p = \lceil (\ell+1)/2 \rceil$, $q = \lfloor (\ell+1)/2 \rfloor$, s_p equal to the length of $\vec{\alpha}$ and t_q equal to the length of $\vec{\beta}$, so that

$$\vec{\beta}[0, t_0) \ll \vec{\alpha}[s_0, s_1) \ll \vec{\beta}[t_0, t_1) \ll \vec{\alpha}[s_1, s_2) \ll \ldots \ll \vec{\beta}[t_{p-2}, t_{p-1}) \ll \vec{\alpha}[s_{p-1}, s_p),$$

so that if $|\vec{\alpha}| = |\vec{\beta}|$, then $\vec{\alpha}[0, s_0) = \vec{\beta}[0, t_0)$, while if $|\vec{\alpha}| < |\vec{\beta}|$, then $\vec{\alpha}[0, s_0) \ll \vec{\beta}[0, t_0)$, and if p and q are equal, then one has the additional inequality $\vec{\alpha}[s_{p-1}, s_p) \ll \vec{\beta}[t_{p-1}, t_p)$. The pair of sequences $(\mathbf{s}, \mathbf{t})$ is called the *shape* of the pair $\{\vec{\alpha}, \vec{\beta}\}_<$. Denote the set of all such pairs by $R(\mathbf{s}; \mathbf{t})$.

Lemma 1.4 *The clear, canonical multicolored graphs on $[\lambda]^n$ colored with m colors are those whose edges of color ℓ are the union of some of the sets $R(\mathbf{s}; \mathbf{t})$, where $\max \mathbf{s} = \max \mathbf{t} = n$, for each of the m colors.*

The notation $R(\mathbf{s}; \mathbf{t})$ is modeled on that of E. Nosal, who uses these sets in [13] to build a collection of examples to show $\omega^{th+h} \nrightarrow (\omega^{1+h}, 2^t + 1)^2$, for h at least four and t positive from such sets. She also has the best results for a target set of ω^3 in [12].

P. Erdős and E.C. Milner have shown that $\omega^{1+\mu m} \rightarrow (\omega^\mu, 2^m)^2$. The result dates back to 1959; proofs occur in [4], [11] and [15]. These proofs generalize to finite powers of weakly compact cardinals. For use later on, we prove a result for finite powers of weakly compact cardinals similar to this one but with a stronger conclusion. To that end, *monotonicity* is introduced.

Definition 1.5 A subset of $[\lambda]^n_{\lessgtr}$ is *monotonically increasing* if for all increasing triples $\vec{\alpha} < \vec{\beta} < \vec{\gamma}$ from it, the pair $\{\vec{\alpha}, \vec{\beta}\}_<$ is similar to the pair $\{\vec{\alpha}, \vec{\gamma}\}_<$. It is *monotonically decreasing* if, for such triples, $\{\vec{\alpha}, \vec{\gamma}\}_<$ is similar to the pair $\{\vec{\beta}, \vec{\gamma}\}_<$. Write $\lambda^n \nearrow (\lambda^k, m)^2$ and $\lambda^n \searrow (\lambda^k, m)^2$ to indicate that for any graph on $[\lambda]^n_{\lessgtr}$ either there is an independent set of type λ^k or a homogeneous set of size m which is monotonically increasing, respectively decreasing.

Lemma 1.6 *If $\lambda = \omega$ or λ is weakly compact, then $\lambda^{km} \nearrow (\lambda^k, m)^2$ and $\lambda^{km} \searrow (\lambda^k, m)^2$ for all positive integers k and $m > 1$.*

Proof. The proof proceeds by induction on m. For $m = 2$ the lemma is trivial, since any pair is by default both monotonically increasing and monotonically decreasing.

Next suppose the lemma is true for m and let G be a graph on $[\lambda]^{km+m}$. For the sake of concreteness, assume that G has no independent set of type λ^k. Find a set $L \subseteq \lambda$ of cardinality λ so that for pairs of sequences from L, the graph is canonical. By recursion one can construct a set X_0 of sequences from L of type λ^k so that any pair from X_0 has form ℓ for some $\ell < 2k+2$ and shape $\mathbf{s}, \mathbf{t}$ for some pair with $\mathbf{s} \cup \mathbf{t} \subseteq \{0, 1, \ldots, k-2, k-1, km+m\}$. By assumption this set is not independent, so suppose that $\{\vec{\beta}, \vec{\gamma}\}_<$ is a joined pair from the set. For monotonically increasing sets, consider the collection U of all sequences from $[\lambda]^n_<$ which agree with $\vec{\gamma}$ up to its last block and continue with values greater than $\vec{\beta} \cup \vec{\gamma}$. This set has type at least $\lambda^k m$. Thus by the induction hypothesis and the assumption above, U has a monotonically increasing set Y of size m. Adjoin to Y either $\vec{\beta}$ or a variant $\vec{\alpha}$ gotten by replacing the last block of $\vec{\beta}$ with one whose elements are larger than all the elements of sequences in Y. The set Y with the newly adjoined point is the required monotonic increasing set. For monotonically decreasing sets reverse the roles of $\vec{\beta}$ and $\vec{\gamma}$. $\square$

The construction of homogeneous sets in the above lemma can be modified so that each pair has a specified form and shape. Let $\mathbf{s}^-$ denote a sequence obtained from $\mathbf{s}$ by the deletion of its maximum element. Call a pair of sequences $(\mathbf{s}, \mathbf{t})$ a *shape* if it is the shape of some pair $\{\vec{\alpha}, \vec{\beta}\}_<$.

Remark 1.7 For any subset $L \subseteq \lambda$ of cardinality λ, and any collection of shapes $(\mathbf{s}_0, \mathbf{t}_0)$, $(\mathbf{s}_1, \mathbf{t}_1), \ldots, (\mathbf{s}_{m-2}, \mathbf{t}_{m-2})$, and for any ℓ, if $\max \mathbf{s}_i = \max \mathbf{t}_i = \ell$, for all $i < m-1$, and $\mathbf{s}_0^- \cup \mathbf{t}_0^- \ll \mathbf{s}_1^- \cup \mathbf{t}_1^- \ll \ldots \ll \mathbf{s}_{m-2}^- \cup \mathbf{t}_{m-2}^-$, then there are a monotonic increasing subset $U = \{\mathbf{u}_i \mid i < m\}_<$ and a monotonic decreasing subset $V = \{\mathbf{v}_i \mid i < m\}_<$ of $[L]^\ell$ so that for all $i < j < m$, the pair $\{\mathbf{u}_i, \mathbf{u}_j\}_<$ has shape $(\mathbf{s}_i, \mathbf{t}_i)$ and the pair $\{\mathbf{v}_j, \mathbf{v}_i\}_<$ has shape $(\mathbf{s}_{m-j-1}, \mathbf{t}_{m-j-1})$.

The next lemma gives a similar result for partitions of pairs of sequences of different lengths.

Lemma 1.8 *For any subset $L \subseteq \lambda$ of cardinality λ, and any infinite set $M \subseteq \omega$ and any increasing sequence of integers $-1 \le \ell_0 \le \ell_1 \le \ldots \le \ell_{m-1}$ all even or all odd, there are an increasing sequence $U = \{\mathbf{u}_i \mid i < m\}$ and a decreasing sequence $V = \{\mathbf{v}_i \mid i < m\}$ of elements of $[L]^{<\omega}$ of different lengths so that for $i < j < m$, both $\{\mathbf{u}_i, \mathbf{u}_j\}_<$ and $\{\mathbf{v}_j, \mathbf{v}_i\}_<$ have form ℓ_i, and have shapes $(\mathbf{s}_{ij}, \mathbf{t}_{ij})$ and $(\mathbf{z}_{ij}, \mathbf{w}_{ij})$ respectively from M with $\mathbf{s}_{ij} \ll \mathbf{t}_{ij}$ and $\mathbf{z}_{ij} \ll \mathbf{w}_{ij}$.*

Proof. For $i < m$, let $q_i = \lfloor (\ell_i + 1)/2 \rfloor$. Choose a $\ll$-increasing sequence $S_0, S_1, \ldots, S_{m-1}$ of subsets of M so that S_i is a set of size $2 + q_i$ which starts above $f(\ell_i)$. We plan to define $\mathbf{u}_i$ as the concatenation of subsequences $\mathbf{u}_i(j)$ for $j \le q_i$ together with $\mathbf{u}_i(\omega)$ if ℓ_i is even. We plan to define $\mathbf{v}_i$ as a similar concatenation of subsequences.

The subsequences $\mathbf{u}_i(0)$ and $\mathbf{v}_i(0)$ are to have length the minimum of S_i. For $j > 0$, the subsequences $\mathbf{u}_i(j)$ and $\mathbf{v}_i(j)$ have length the difference between the jth and $(j-1)$st terms of S_i, where we use the $(q_i + 1)$st term in the choice of length for $\mathbf{u}_i(\omega)$ and $\mathbf{v}_i(\omega)$ as needed.

Having specified the lengths of the subsequences, we can choose them minimal from L so that the subsequences are ordered by $\ll$ and satisfy the following inequalities:

$$\mathbf{u}_i(r) \ll \mathbf{u}_j(t) \iff r < t \quad \text{or} \quad r = t < \omega \ \& \ i < j \quad (\text{or} \quad r = t = \omega \ \& \ j < i).$$

Any pair of such sequences $\{\,\mathbf{u}_i, \mathbf{u}_j\,\}_<$ has form ℓ_i and shape $(\mathbf{s}_i, \mathbf{t}_i)$ where $\mathbf{s}_{ij}$ is a subset of S_i and $\mathbf{t}_{ij}$ is a subset of S_j.   □

2   One infinite power

In this section we show that for any Ramsey cardinal λ the ordinal λ^ω is a partition ordinal. First we restrict our attention to certain pairs which we designate as *acceptable*. We have seen that any graph can be made canonical when restricted to sequences from $[L]^{<\omega}$ for a suitably chosen set L. One way to show that λ^ω is a partition ordinal is to use a canonical graph on $[L]^{<\omega}$ to define a canonical graph on $[\omega]^{<\omega}$, and use the fact that ω^ω is a partition ordinal to produce the desired set. On the one hand, an independent set of type ω^ω can be used to build an independent set of type λ^ω after analysis either of the proof for ω^ω or a delicate inquiry into what sets of type ω^ω look like. On the other hand, a homogeneous subset of $[\omega]^{<\omega}$ of size m can be easily transformed into a homogeneous subset of $[\lambda]^{<\omega}$ of the same size. More difficulties arise in generalizing the proof to the product of two infinite powers of Ramsey cardinals since the corresponding statement about countable ordinals that one would like to use is simply not true. Consequently the proof for one infinite power is presented from scratch with an eye toward what is needed for two.

Definition 2.1 Suppose that L is a subset of λ, M is a subset of ω and $f : \omega \to \omega$ is an increasing function with $f(n) \geq n$. A pair $\{\,\vec{\alpha}, \vec{\beta}\,\}_<$ is a (L, M, f)-*acceptable pair* if the following conditions are satisfied:

(1) $\vec{\alpha} \cup \vec{\beta} \subseteq L$;

(2) for some positive integer k and sequences $\mathbf{s}$, $\mathbf{t}$ from M, the pair $\vec{\alpha}, \vec{\beta}$ has form k and shape $(\mathbf{s}, \mathbf{t})$ where $f(k) < \min(\mathbf{s} \cup \mathbf{t})$;

(3) if $|\vec{\alpha}| < |\vec{\beta}|$, then $\mathbf{s} \ll \mathbf{t}$.

The following easy lemma follows from the definition.

Lemma 2.2 *If a pair* $\{\,\vec{\alpha}, \vec{\beta}\,\}_<$ *is a* (L, M, f)-*acceptable pair and if* $L \subseteq L'$, $M \subseteq M'$ *and* $n \leq f'(n) \leq f(n)$ *for all* n, *then the pair* $\{\,\vec{\alpha}, \vec{\beta}\,\}_<$ *is also* (L', M', f')-*acceptable.*

For convenience in construction, introduce the notion of a *pearl* $\vec{\delta}$ of a well-ordered set enumerated as $D = \{\,d_\xi \mid \xi < \mu\,\}$ to mean any sequence of consecutive elements of D beginning with a limit point of D.

Lemma 2.3 *For any subset* $L \subseteq \lambda$ *of cardinality* λ, *any subset* $M \subseteq \omega$ *of cardinality* ω *and any increasing function* $f : \omega \to \omega$ *with* $f(k) \geq k$ *for all* $k < \omega$, *there is a subset* U *of* $[L]^{<\omega}$ *of type* λ^ω *under the lexicographic ordering so that every pair from X is* (L, M, f)-*acceptable.*

Proof. The first order of business is to define a collection of sequences from M so that all shapes in the proposed set are subsequences of them. Define by recursion on ω, sequences $\mathbf{s}_i$ of length $i + 1$ from M so that the sequences are ordered under $\ll$ in the same order as their indices are ordered in ω and so that $f(2 + 2i) < \min \mathbf{s}_i$.

Next partition L into many full size subsets. Let L be the disjoint union of the sets $L_{\vec{\alpha}}$ each of size λ where $\vec{\alpha}$ is an increasing (possibly empty) sequence from L. Let $\{\,\mathbf{a}_j \mid j < \omega\,\}$ be a $\ll$-increasing sequence of increasing sequences from $L_\emptyset$ with $|\mathbf{a}_j| = \mathbf{s}_j(0)$ and let σ be an element of L greater than every element in all the sets $\mathbf{a}_j$.

Let $U(j)$ be the set of all those increasing sequences $\vec{\alpha} \subseteq L$ of length $\mathbf{s}_j(j)$ with $\mathbf{a}_j$ as an initial segment so that for all i with $0 < i < j$, $\vec{\alpha}[\mathbf{s}_j(i), \mathbf{s}_j(i{+}1))$ is a pearl of $L_{\vec{\alpha}[0,\mathbf{s}_j(i))} \setminus \sigma$. By construction $U(j)$ has order type λ^j, and any pair from $U(j)$ has form k for some $k < 2j+2$ and shape $(\mathbf{s}, \mathbf{t})$ for some $\mathbf{s} \subseteq \mathbf{s}_j$ and $\mathbf{t} \subseteq \mathbf{s}_j$.

Let U be the union over all j of the sets $U(j)$. Notice that U has order type λ^ω. Furthermore, any pair $\{\vec{\alpha}, \vec{\beta}\}_<$ of elements from $U(j)$ and $U(k)$, respectively, will have $j < k$, will be of form ℓ for some $\ell < 2j+2$ and have shape $(\mathbf{s}, \mathbf{t})$ for some $\mathbf{s} \subseteq \mathbf{s}_j$ and $\mathbf{t} \subseteq \mathbf{s}_k$. Thus U is the desired (L, M, f)-acceptable set. $\qquad\square$

Clearly Ramsey's Theorem should be brought to bear on the problem so that graphs can be made canonical in more senses. We shall use the following lemma which appears in [15] for use in the proof that ω^ω is a partition ordinal.

Lemma 2.4 (see Lemma 7.3.1 of [15]) *Let $S \in [\omega]^\omega$. Suppose for each $\mathbf{a}$ from $[\omega]^{<\omega}$, there are given an integer $m(\mathbf{a})$ and a partition $\Delta(\mathbf{a})$ of $[S]^{m(\mathbf{a})}$ into two parts. Then there is an infinite subset H of S such that for each $\mathbf{a}$ from $[H]^{<\omega}$, the set $\{\, h \in H \mid h > \max(\mathbf{a})\,\}$ is homogeneous for $\Delta(\mathbf{a})$.*

Definition 2.5 Suppose that $\{\vec{\alpha}, \vec{\beta}\}_<$ is a pair of sequences from λ of form k and shape $(\mathbf{s}, \mathbf{t})$ and $\{\vec{\gamma}, \vec{\delta}\}_<$ is a pair of sequences from λ of form ℓ and shape $(\mathbf{u}, \mathbf{v})$. Call the two pairs *block-similar* and denote this relationship by $\{\vec{\alpha}, \vec{\beta}\}_< \sim_b \{\vec{\gamma}, \vec{\delta}\}_<$ if $\{\mathbf{s}, \mathbf{t}\}_< \sim \{\mathbf{u}, \mathbf{v}\}_<$, $\mathbf{s} \leq \mathbf{t}$ if and only if $\mathbf{u} \leq \mathbf{v}$ and $\mathbf{s} \geq \mathbf{t}$ if and only if $\mathbf{u} \geq \mathbf{v}$. A partition Δ of the pairs of finite sequences from $[\lambda]^{<\omega}$ is *block-canonical* on a set A if and only if whenever two pairs from A are block-similar, then also they are in the same cell of the partition.

In order to use either Ramsey's theorem or the Ramsey property of a cardinal λ on pairs of sequences, introduce some more notation. Call a shape $(\mathbf{s}, \mathbf{t})$ *minimal* if $\mathbf{s} \cup \mathbf{t} = \{0, 1, \ldots, r-1\} = r$ for some natural number $r < \omega$.

Lemma 2.6 (Super canonicity for one cardinal power) *Assume λ is a Ramsey cardinal. For any graph G on $[\lambda]^{<\omega}$ regarded as a partition of the edges into two cells, there are a subset L of λ of cardinality λ, an infinite subset M of ω and an increasing function $f : \omega \to \omega$ with $k \leq f(k)$ for all k, so that the graph is block-canonical for (L, M, f)-acceptable pairs.*

Proof. Let S be the collection of all minimal shapes and let $\pi : S \to \omega$ be a one-to-one function with the additional property that $|\mathbf{s} \cup \mathbf{t}| \leq \pi(\mathbf{s}, \mathbf{t})$ for all shapes $(\mathbf{s}, \mathbf{t})$ in S. For each $\sigma = (\mathbf{s}, \mathbf{t})$ in S and each $\mathbf{z}$ in $[\omega]^{\pi(\sigma)}$, let $\{\vec{\alpha}(\sigma, \mathbf{z}), \vec{\beta}(\sigma, \mathbf{z})\}_<$ be a pair of finite sequences from λ of shape $(\mathbf{u}, \mathbf{v})$ where $\{\mathbf{u}, \mathbf{v}\} \sim \{\mathbf{s}, \mathbf{t}\}$, $\mathbf{u} \cup \mathbf{v}$ is an initial segment of $\mathbf{z}$ and furthermore $\mathbf{s} \leq \mathbf{t}$ if and only if $\mathbf{u} \leq \mathbf{v}$, and $\mathbf{s} \geq \mathbf{t}$ if and only if $\mathbf{u} \geq \mathbf{v}$.

Define partitions $\Delta(\sigma, \mathbf{z})$ of $[\lambda]^r$ for $r = 2+2\max \mathbf{z}$ by $\Delta(\sigma, \mathbf{z})(A) = 1$ if and only if the pair $\{\vec{\gamma}, \vec{\delta}\}_<$ is joined by an edge where $\{\vec{\gamma}, \vec{\delta}\}_<$ is the unique pair similar to $\{\vec{\alpha}(\sigma, \mathbf{z}), \vec{\beta}(\sigma, \mathbf{z})\}_<$ whose union is an initial segment of A.

Use the fact that λ is a Ramsey cardinal to get a set L of λ many indiscernibles for the partitions $\Delta(\sigma, \mathbf{z})$.

Next define partitions $\Delta(\pi(\sigma))$ of $[\omega]^{\pi(\sigma)}$ by $\Delta(\pi(\sigma))(\mathbf{z}) = \Delta(\sigma, \mathbf{z})(A)$ where A is a subset of L of size $r = 2 + 2 \max \mathbf{z}$. By Lemma 2.4, there is an infinite subset M of ω such that for each $\mathbf{a}$ from $[M]^{<\omega}$, the set $\{ h \in H \mid h > \pi(\sigma) \}$ is homogeneous for $\Delta(\pi(\sigma))$. Finally define $f : \omega \to \omega$ so that $f(k)$ is greater than k and greater than $\pi(\sigma)$ for any minimal shape σ associated with pairs of form k.

The two uniformizations and the definition of f guarantee that the graph is canonical on (L, M, f)-acceptable pairs. □

Theorem 2.7 *If λ is a Ramsey cardinal, then λ^ω is a partition ordinal and $\lambda^\omega \to (\lambda^\omega, m)^2$ for all $m < \omega$.*

Proof. Let G be a graph on pairs of finite sequences from λ. Use Lemma 2.7 to find a subset L of λ of cardinality λ, an infinite subset M of ω and an increasing function $f : \omega \to \omega$ with $f(k) \geq k$ for all $k < \omega$ so that the graph is block-canonical on all (L, M, f)-acceptable pairs.

If there is no (L, M, f)-acceptable pair that is joined by an edge in G, then use Lemma 2.3 to get a subset U of $[L]^{<\omega}$ of type λ^ω under the lexicographic ordering so that every pair from U is (L, M, f)-acceptable and U is the set required to prove the theorem.

Otherwise, there is an (L, M, f)-acceptable pair $\{ \vec{\alpha}, \vec{\beta} \}_<$ of form k and shape $\sigma = (\mathbf{s}, \mathbf{t})$ which is joined in the graph G. Let m be an arbitrary positive integer.

For the first case, assume that $\max \mathbf{s} \neq \max \mathbf{t}$. Since the pair is (L, M, f)-acceptable, it follows that $\mathbf{s} \ll \mathbf{t}$. Use Lemma 1.8 to build a set Y from $[L]^{<\omega}$ of size m so that any pair from Y has form k and shape from M above $f(k)$. Then every pair from Y is (L, M, f)-acceptable and is block-similar to $\{ \vec{\alpha}, \vec{\beta} \}_<$. Since the graph is block-canonical, every pair from Y is joined.

For the second case, assume that $\max \mathbf{s} = \max \mathbf{t}$. Let $r = |\mathbf{s} \cup \mathbf{t}|$. Let $(\mathbf{s}_0, \mathbf{t}_0), (\mathbf{s}_1, \mathbf{t}_1)$, ..., $(\mathbf{s}_{m-2}, \mathbf{t}_{m-2})$ be a collection of shapes from M similar to $(\mathbf{s}, \mathbf{t})$ that pairwise satisfy the same inequalities so that for some ℓ, $\max \mathbf{s}_i = \max \mathbf{t}_i = \ell$, for all $i < m - 1$, and $\mathbf{s}_0^- \cup \mathbf{t}_0^- \ll \mathbf{s}_1^- \cup \mathbf{t}_1^- \ll \ldots \ll \mathbf{s}_{m-2}^- \cup \mathbf{t}_{m-2}^-$.

By Remark 1.7 there is a set $U = \{ \vec{\alpha}_i \mid i < m \}$ of size m listed in increasing order of finite sequences from L so that $\{ \vec{\alpha}_i, \vec{\alpha}_j \}_<$ has shape $(\mathbf{s}_i, \mathbf{t}_i)$. Thus all these pairs are (L, M, f)-acceptable and block-similar to $\{ \vec{\alpha}, \vec{\beta} \}_<$. Since G is canonical for such pairs, U is a complete subgraph of size m.

In either case we constructed a complete subgraph of size m. Since m was arbitrary, we can always construct the set required for the finite alternative of the partition relation. Thus the theorem follows. □

3   Canonical Partitions for Two Cardinal Powers

In this section and the next, we generalize the approach from §2 to the case of two cardinal powers.

Definition 3.1 A pair $\{(\vec{\alpha}, \vec{\gamma}), (\vec{\beta}, \vec{\delta})\}_<$ is a (K, L, M, f)-*acceptable pair* if the following conditions are satisfied:

(1) $\{\vec{\alpha}, \vec{\beta}\}_<$ is (L, M, f)-acceptable of form k and shape $(\mathbf{s}, \mathbf{t})$ for some positive integer k and sequences $\mathbf{s}$, $\mathbf{t}$;

(2) $\{\vec{\gamma}, \vec{\delta}\}_<$ is (K, M, f)-acceptable of form ℓ and shape $(\mathbf{u}, \mathbf{v})$ for some positive integer ℓ and sequences $\mathbf{u}$, $\mathbf{v}$;

(3) $\mathbf{s} \ll \mathbf{u}$, $\mathbf{t} \ll \mathbf{v}$, $f(|\vec{\alpha}|) < \ell < \min(\mathbf{u} \cup \mathbf{v})$ and if $|\vec{\beta}| < |\vec{\gamma}|$, then $f(|\vec{\beta}|) < \ell$;

(4) if $|\vec{\alpha}| < |\vec{\beta}|$, then $|\vec{\gamma}| \neq |\vec{\delta}|$;

(5) if $|\vec{\gamma}| \neq |\vec{\delta}|$, then $\mathbf{t}$, $\mathbf{u}$ and $\mathbf{v}$ are linearly ordered by $\ll$.

The following easy lemma follows from the definition.

Lemma 3.2 *If a pair* $\{(\vec{\alpha}, \vec{\gamma}), (\vec{\beta}, \vec{\delta})\}_<$ *is a* (K, L, M, f)-*acceptable pair and* $K \subseteq K'$, $L \subseteq L'$, $M \subseteq M'$ *and* $n \leq f'(n) \leq f(n)$ *for all* $n < \omega$, *then the pair is also* (K', L', M', f')-*acceptable.*

Lemma 3.3 *For any subset* $K \subseteq \kappa$ *of cardinality* κ, *any subset* $L \subseteq \lambda$ *of cardinality* λ, *any subset* $M \subseteq \omega$ *of cardinality* ω, *and any increasing function* $f : \omega \to \omega$ *with* $n \leq f(n)$, *there is a subset* X *of* $[L]^{<\omega} \times [K]^{<\omega}$ *of type* $\kappa^\omega \lambda^\omega$ *under the lexicographic ordering so that every pair from* X *is* (K, L, M, f)-*acceptable.*

Proof. The first order of business is to define a collection of sequences from M so that all shapes in the proposed set are subsequences of them. To that end, order the union $I = \omega \cup [\omega]^2_<$ by the reverse lexicographic order $<_r$, where i is identified with $(-1, i)$. Under $<_r$, the set I is ordered in type ω. Define by recursion on I, sequences $\mathbf{s}_i$, $\mathbf{s}_{ij}$ from M and a function $g : I \to \omega$ so that the sequences are ordered under $\ll$ in the same order as their indices are ordered in I under $<_r$, $|\mathbf{s}_i| = i + 1$, $2i + 2 < \min \mathbf{s}_i$, $|\mathbf{s}_{ij}| = 3g(i, j)$, $4g(i, j) + 2 < \min \mathbf{s}_{ij}$ and $g(i, j) = \max\{1 + f(\max(\mathbf{s}_w)) \mid w <_r (i, j)\}$.

Next partition K and L into many full size subsets. Let L be the disjoint union of the sets $L_{\vec{\alpha}}$ each of size λ where $\vec{\alpha}$ is an increasing (possibly empty) sequence from L. Similarly let K be the disjoint union of the sets $K_{\vec{\gamma}}$ each of size κ where $\vec{\gamma}$ is an increasing (possibly empty) sequence from K. Let $\{\mathbf{a}_j \mid j < \omega\}$ be a $\ll$-increasing sequence of increasing sequences from $L_\emptyset$ with $|\mathbf{a}_j| = \mathbf{s}_j(0)$ and let σ be an element of L greater than every element in all the sets $\mathbf{a}_j$.

Let $U(j)$ be the set of all those increasing sequences $\vec{\alpha} \subseteq L$ of length $\mathbf{s}_j(j)$ with $\mathbf{a}_j$ as an initial segment so that for all i with $0 < i < j$, $\vec{\alpha}[\mathbf{s}_j(i), \mathbf{s}_j(i+1))$ is a pearl of $L_{\vec{\alpha}[0, \mathbf{s}_j(i))} \setminus \sigma$. By construction $U(j)$ has order type λ^j, and any pair from $U(j)$ has form k for some $k < 2j + 2$ and shape $(\mathbf{s}, \mathbf{t})$ for some $\mathbf{s} \subseteq \mathbf{s}_j$ and $\mathbf{t} \subseteq \mathbf{s}_j$.

Let U be the union over all j of the sets $U(j)$. Notice that U has order type λ^ω. Furthermore, any pair $\{\vec{\alpha}, \vec{\beta}\}_<$ of elements from $U(j)$ and $U(k)$, respectively, will have $j < k$, be of form ℓ for some $\ell < 2j + 2$ and have shape $(\mathbf{s}, \mathbf{t})$ for some $\mathbf{s} \subseteq \mathbf{s}_j$ and $\mathbf{t} \subseteq \mathbf{s}_k$.

Let U^+ be the set of all triples $(i, g(j, k), \vec{\alpha})$ where $\vec{\alpha}$ is either $\emptyset$ or in $U(j)$ and $i < g(j, k) < \omega$. Order U^+ lexicographically by $<_\ell$ in order type $\lambda^\omega \cdot \omega^2$. Define by recursion a set of increasing sequences $\mathbf{b}(i, g(j, k), \vec{\alpha})$ from $K_{\emptyset}$ well-ordered by $\ll$ for $j < k < \omega$, $i < g(j, k)$ and $\vec{\alpha}$ in $U(j)$ of length $|\mathbf{b}(i, g(j, k), \vec{\alpha})| = \mathbf{s}_{jk}(i) - \mathbf{s}_{jk}(i - 1)$ where $\mathbf{s}_{jk}(-1) = 0$ so that any pair is ordered by $\ll$ in the same order as their indices are ordered by $<_{\ell}$. For each $\vec{\alpha} \in U(j)$ and each $k > j$, let $\mathbf{b}(\vec{\alpha}, g(j, k))$ be the concatenation of the sequences $\mathbf{b}(i, g(j, k), \emptyset)$ for $i < g(j, k)$ and $\mathbf{b}(i, g(j, k), \vec{\alpha})$ for $g(j, k) \leq i < 2g(j, k)$. Let τ be an element of K greater than every element in all the sets $\mathbf{b}(i, g(j, k), \vec{\alpha})$.

Suppose that $\vec{\alpha}$ is an element of $U(j)$ and that $k > j$. Let $X(\vec{\alpha}, k)$ be the collection of all pairs $(\vec{\alpha}, \vec{\beta})$ with $\vec{\beta}$ a sequence from K of length $\max \mathbf{s}_{jk}$ with $\mathbf{b}(\vec{\alpha}, g(j, k))$ as an initial segment so that for all i with $2g(j, k) \leq i < 3g(j, k)$, $\vec{\beta}[\mathbf{s}_{jk}(i), \mathbf{s}_{jk}(i+1))$ is a pearl of $K_{\vec{\beta}(0,t)} \backslash \tau$ for $t = \mathbf{s}_{jk}(i)$. Since for each fixed sequence length, there are κ many pearls of that length and they are well-ordered under $\ll$, $X(\vec{\alpha}, k)$ has order type $\kappa^{g(j,k)}$. For completeness, let $X(\vec{\alpha}, k) = \emptyset$ for $k \leq j$ when $|\vec{\alpha}| = \mathbf{s}_j(j)$. Finally let X be the union of all the sets $X(\vec{\alpha}, k)$ for $\vec{\alpha}$ in U and k in ω. The reader may check that X is a set of type $\kappa^{\omega} \lambda^{\omega}$ satisfying all the requirements of the definition of (K, L, M, f)-acceptable. □

Definition 3.4 The *outline* of a $(\kappa, \lambda, \omega, id)$-acceptable pair $\{(\vec{\alpha}, \vec{\gamma}), (\vec{\beta}, \vec{\delta})\}_{<}$ is the pair $(\star, \diamond)$ assigned according to the following table:

(1) $(=, =<)$ if $|\vec{\alpha}| = |\vec{\beta}|$, $|\vec{\gamma}| = |\vec{\delta}|$ and $\vec{\gamma} < \vec{\delta}$;

(2) $(=, =>)$ if $|\vec{\alpha}| = |\vec{\beta}|$, $|\vec{\gamma}| = |\vec{\delta}|$ and $\vec{\gamma} > \vec{\delta}$;

(3) $(=, <)$ if $|\vec{\alpha}| = |\vec{\beta}|$, $|\vec{\gamma}| < |\vec{\delta}|$;

(4) $(=, >)$ if $|\vec{\alpha}| = |\vec{\beta}|$, $|\vec{\gamma}| > |\vec{\delta}|$;

(5) $(<, <)$ if $|\vec{\alpha}| < |\vec{\gamma}| < |\vec{\beta}| < |\vec{\delta}|$;

(6) $(<, >)$ if $|\vec{\alpha}| < |\vec{\gamma}| < |\vec{\delta}| < |\vec{\beta}|$;

(7) $(<, <<)$ if $|\vec{\alpha}| < |\vec{\beta}| < |\vec{\gamma}| < |\vec{\delta}|$.

Definition 3.5 Suppose that $\{(\vec{\alpha}, \vec{\gamma}), (\vec{\beta}, \vec{\delta})\}_{<}$ and $\{(\vec{\eta}, \vec{\zeta}), (\vec{\xi}, \vec{\theta})\}_{<}$ are pairs which are $(\kappa, \lambda, \omega, id)$-acceptable of outline other than $(<, <<)$. Call the two pairs *doubly similar* and denote this relationship by

$$\{(\vec{\alpha}, \vec{\gamma}), (\vec{\beta}, \vec{\delta})\}_{<} \sim_d \{(\vec{\eta}, \vec{\zeta}), (\vec{\xi}, \vec{\theta})\}_{<}$$

if $\{\vec{\alpha}, \vec{\beta}\}_{<} \sim \{\vec{\eta}, \vec{\xi}\}_{<}$, $\{\vec{\gamma}, \vec{\delta}\}_{<} \sim_b \{\vec{\zeta}, \vec{\theta}\}_{<}$, $\vec{\gamma} < \vec{\delta}$ if and only if $\vec{\zeta} < \vec{\theta}$, and $|\vec{\gamma}| < |\vec{\beta}|$ if and only if $|\vec{\zeta}| < |\vec{\xi}|$. For pairs of outline $(<, <<)$ we weaken the condition $\{\vec{\alpha}, \vec{\beta}\}_{<} \sim \{\vec{\eta}, \vec{\xi}\}_{<}$ to require that for $\sigma = (\mathbf{s}, \mathbf{t})$ the shape of $\{\vec{\alpha}, \vec{\beta}\}_{<}$ and $\tau = (\mathbf{u}, \mathbf{v})$ the shape of $\{\vec{\eta}, \vec{\xi}\}_{<}$ we have $\mathbf{s} = \mathbf{u}$ and $|\mathbf{t}| = |\mathbf{v}|$.

Notice that the definition of doubly similar is not quite what one would hope for where in each component the sequences are block-similar. Instead we have required the first components satisfy the stricter requirement that they be similar, that is, that they have the same shape, except in the case of outline $(<, <<)$. We would like to make canonical partitions a bit more uniform than simply requiring that doubly similar pairs fall in the same cell, and to that end we introduce some more terminology.

We call a shape $\sigma = (\mathbf{s}, \mathbf{t})$ *persistent* if $\max \mathbf{s} = \max \mathbf{t}$ or $\mathbf{s} \ll \mathbf{t}$ or $\mathbf{t} \ll \mathbf{s}$. Before we proceed to the canonization lemma for two cardinal powers, we also introduce the notion of a shape *recurring*.

Definition 3.6 We say a shape σ *occurs with ℓ* for (K, L, M, f)-acceptable pairs in one cell of a partition in outline $(\star, \diamond) \neq (<, <<)$ if there is some (K, L, M, f)-acceptable pair $\{ (\vec{\alpha}, \vec{\gamma}), (\vec{\beta}, \vec{\delta}) \}_<$ of outline $(\star, \diamond)$ in that cell so that $(\vec{\alpha}, \vec{\beta})$ has shape σ and $(\vec{\gamma}, \vec{\delta})$ has form ℓ. We say $\sigma = (\mathbf{s}, \mathbf{t})$ *occurs with ℓ* in outline $(\star, \diamond)$ if in the witnessing pair, $\{ (\vec{\alpha}, \vec{\gamma}), (\vec{\beta}, \vec{\delta}) \}_<$, $(\vec{\alpha}, \vec{\beta})$ has shape $\sigma' = (\mathbf{s}', \mathbf{t}')$, where $\mathbf{s}' = \mathbf{s}$ and $(\vec{\beta}, \vec{\delta})$ has form ℓ. We say a shape *recurs* in outline $(\star, \diamond)$ for (K, L, M, f)-acceptable pairs in a cell of a partition if there are infinitely many ℓ for which it occurs in outline $(\star, \diamond)$ with ℓ in that cell of the partition.

With these definitions in hand we can now describe canonical subsets of $[\lambda]^{<\omega} \times [\kappa]^{<\omega}$.

Definition 3.7 A partition Δ of the pairs from $[\lambda]^{<\omega} \times [\kappa]^{<\omega}$ is *doubly canonical* on a set A for (K, L, M, f)-acceptable pairs if and only if whenever two (K, L, M, f)-acceptable pairs from A are doubly similar, then also they are in the same cell of the partition, and whenever a persistent shape occurs in outline $(\star, \diamond)$ for some (K, L, M, f)-acceptable pair among the edges, then every similar shape from M chosen from points above $f(k)$ recurs in outline $(\star, \diamond)$ for (K, L, M, f)-acceptable pairs among the edges.

Lemma 3.8 (Super canonicity for two cardinal powers) *Assume $\lambda < \kappa$ are Ramsey cardinals. For any graph G on $[\lambda]^{<\omega} \times [\kappa]^{<\omega}$ regarded as a partition of the edges into two cells, there are a subset K of κ of cardinality κ, a subset L of λ of cardinality λ, an infinite subset M of ω and an increasing function $f : \omega \to \omega$ with $k \leq f(k)$ for all k, so that the graph is doubly canonical for (K, L, M, f)-acceptable pairs.*

Proof. Let T be the collection of all shapes. Let $\rho : T \to \omega$ be a one-to-one function with the additional property that $|\mathbf{u} \cup \mathbf{v}|$ is less than or equal to $\rho(\mathbf{u}, \mathbf{v})$ for all shapes $(\mathbf{u}, \mathbf{v})$ in S. For each shape $\tau = (\mathbf{u}, \mathbf{v})$ in T let $\{ \vec{\gamma}(\tau), \vec{\delta}(\tau) \}_<$ be a pair $[\kappa]^{<\omega}$ so that $\{ \vec{\gamma}(\tau), \vec{\delta}(\tau) \}_<$ has shape $(\mathbf{u}', \mathbf{v}')$ where $\{ \mathbf{u}', \mathbf{v}' \} \sim \{ \mathbf{u}, \mathbf{v} \}$, $\mathbf{u}' \cup \mathbf{v}'$ is an initial segment of $\mathbf{z}$ and furthermore $\mathbf{u} \leq \mathbf{v}$ if and only if $\mathbf{u}' \leq \mathbf{v}'$, and $\mathbf{u} \geq \mathbf{v}$ if and only if $\mathbf{u}' \geq \mathbf{v}'$.

For $\vec{\alpha} < \vec{\beta}$ from $[\lambda]^{<\omega}$, define partitions $\Delta(\vec{\alpha}, \vec{\beta}, \tau, \Delta)$ of $[\kappa]^r$ for $r = 2 + 2 \max \cup \tau$ and for Δ one of $<$ and $>$ by $\Delta(\vec{\alpha}, \vec{\beta}, \tau, <)(A) = 1$ if and only if the pair $\{ (\vec{\alpha}, \vec{\gamma}), (\vec{\beta}, \vec{\delta}) \}_<$ is joined by an edge and $\Delta(\vec{\alpha}, \vec{\beta}, \tau, >)(A) = 1$ if and only if the pair $\{ (\vec{\alpha}, \vec{\delta}), (\vec{\beta}, \vec{\gamma}) \}_<$ is joined by an edge where $\{ \vec{\gamma}, \vec{\delta} \}_<$ is the unique pair similar to $\{ \vec{\alpha}(\tau), \vec{\beta}(\tau) \}_<$ whose union is an initial segment of A.

Use the fact that κ is a Ramsey cardinal to get a set K of κ many indiscernibles for the partitions $\Delta(\vec{\alpha}, \vec{\beta}, \tau, \Delta)$.

Next define partitions $\Delta(\sigma, \tau, \Delta)$ of $[\lambda]^r$ for $r = 2 + 2 \max \cup \sigma$ by $\Delta(\sigma, \tau, \Delta)(A) = \Delta(\vec{\alpha}, \vec{\beta}, \tau, \Delta)(B)$ where B is a subset of K of the appropriate size, and $\{ \vec{\alpha}, \vec{\beta} \}_<$ is the unique pair similar to $\{ \vec{\alpha}(\sigma), \vec{\beta}(\sigma) \}_<$ whose union is an initial segment of A.

Use the fact that λ is a Ramsey cardinal to get a set L of λ many indiscernibles for the partitions $\Delta(\sigma, \tau, \Delta)$.

For a minimal persistent shape $\sigma = (\mathbf{s}, \mathbf{t})$ and a sequence $\mathbf{z}$, let shape$(\sigma, \mathbf{z})$ be the shape $(\mathbf{s}', \mathbf{t}')$ similar to σ where $\mathbf{s}' \cup \mathbf{t}'$ is an initial segment of $\mathbf{z}$.

For $\diamond$ one of $<$ and $>$, define partitions $\Theta(\sigma, \mathbf{z}, \tau, \diamond)(\mathbf{w})$ where σ and τ are persistent minimal shapes and if the elements of τ have the same max, then so do the elements of σ as follows: $\Theta(\sigma, \mathbf{z}, \tau, \diamond)(\mathbf{w}) = \Delta(\text{shape}(\sigma, \mathbf{z}), \text{shape}(\tau, \mathbf{w}), \Delta)$. In addition, for persistent minimal shapes $\sigma = (\mathbf{s}, \mathbf{t})$ and $\tau = (\mathbf{u}, \mathbf{v})$ satisfying $\mathbf{s} \ll \mathbf{t}$ and $\mathbf{u} \ll \mathbf{v}$, define a partition, $\Theta(\sigma, \mathbf{z}, \tau, <<)(\mathbf{w})$ by $\Theta(\sigma, \mathbf{z}, \tau, <<)(\mathbf{w}) = \Delta((\mathbf{z}, \mathbf{t}'), (\mathbf{u}', \mathbf{v}'), <)$ where $(\mathbf{z}, \mathbf{t}')$ is similar to σ, $(\mathbf{u}', \mathbf{v}')$ is similar to τ and where $\mathbf{z} \ll \mathbf{u}' \ll \mathbf{t}' \ll \mathbf{v}'$ and $\mathbf{u}' \cup \mathbf{t}' \cup \mathbf{v}'$ is an initial segment of $\mathbf{w}$.

By Lemma 2.4, there is an infinite subset H of ω such that for each ℓ from ω, the set $\{h \in H \mid h > \ell\}$ is homogeneous for the partitions $\Theta(\sigma, \mathbf{z}, \tau, \diamond)$ where τ is a minimal persistent shape for pairs of form ℓ.

For persistent minimal shapes σ, and for $\diamond$ one of $<, >$ and $<<$, define partitions $\Theta(\sigma, \diamond)(\mathbf{z})$ by $\Theta(\sigma, \diamond)(\mathbf{z}) = 1$ if and only if for infinitely many ℓ, $\Theta(\sigma, \mathbf{z}, \tau, \diamond)(\mathbf{w}) = 1$ where $\mathbf{w}$ is chosen from a suitable tail of H and $\tau = (\mathbf{u}, \mathbf{v})$ is the minimal shape of pairs of form ℓ with $\mathbf{u} \ll \mathbf{v}$. If $\diamond$ is $<<$, then the previous definition only makes sense if $\sigma = (\mathbf{s}, \mathbf{t})$ where $\mathbf{s} \ll \mathbf{t}$. For persistent minimal shapes $\sigma = (\mathbf{s}, \mathbf{t})$ with $\max \mathbf{s} = \max \mathbf{t}$, and for $\diamond$ one of $<$ and $>$, define partitions $\Theta(\sigma, = \diamond)(\mathbf{z})$ by $\Theta(\pi, = \diamond)(\mathbf{z}) = 1$ if and only if for infinitely many ℓ, $\Theta(\sigma, \mathbf{z}, \tau, \diamond)(\mathbf{w}) = 1$ where $\mathbf{w}$ is chosen from a suitable tail of H and $\tau = (\mathbf{u}, \mathbf{v})$ is some minimal shape of pairs of form ℓ with $\max \mathbf{u} = \max \mathbf{v}$. Again by Lemma 2.4, there is an infinite subset M of ω such that for each k from ω, the set $\{m \in M \mid m > k\}$ is homogeneous for the partitions $\Theta(\sigma, \diamond)$ where σ is a minimal shape of form k.

Now define an increasing function f from ω to ω by recursion on j. Let $f(0) = 1$. Suppose that $f(j-1)$ has been defined. Choose $f(j)$ subject to the following constraints. First, $f(j)$ is greater than $f(j-1)$ and hence greater than j. Second, if $\Theta(\sigma, \diamond) = 0$ for some σ of form k for some $k \leq j$, then $f(j)$ is greater than all ℓ for which is there is a persistent minimal shape τ of form ℓ so that $\Theta(\sigma, \mathbf{z}, \tau, \diamond)(\mathbf{w}) = 1$ where $\mathbf{z}$ is a sequence of elements less than or equal to j and $\mathbf{w}$ is chosen from a suitable tail of M.

The four uniformizations and the definition of f guarantee that the graph is uniform doubly similar (K, L, M, f)-acceptable pairs. Furthermore suppose some (K, L, M, f)-acceptable pair $\{(\vec{\alpha}, \vec{\gamma}), (\vec{\beta}, \vec{\delta})\}_<$ is joined by an edge, the shape of $\{\vec{\alpha}, \vec{\gamma}\}_<$ is σ' and the outline of the pair is $(\star, \diamond)$. Let σ be the minimal shape similar to σ and let k be the form associated with the shape. Notice that σ is persistent. Since M is homogeneous for $\Theta(\sigma, \diamond)$ above k, then any shape from M above $f(k)$ which is similar to σ' recurs for (K, L, M, f)-acceptable pairs of outline $(\star, \diamond)$. Therefore the graph is doubly canonical for (K, L, M, f)-acceptable pairs. $\qquad \square$

4   Some homogeneous sets

We start this section with another lemma on constructing sets of sequences whose pairs have the desired form and shape and will conclude it with the proof of Theorem 0.3, here relabeled as Theorem 4.5.

Lemma 4.1 (Masterful Mingling) *For any collection of integers ℓ_{ij} all even or all odd satisfying the constraint that $\ell_{ij} > 2i + 2$, there is an increasing sequence $U = \{\mathbf{u}_i \mid i < m\}$ from $[L]^{<\omega}$ so that all pairs from U and V are (L, M, f)-acceptable pairs of sequences of different lengths and for $i < j < m$, the pair $\{\mathbf{u}_i, \mathbf{u}_j\}_<$ has form ℓ_{ij}.*

Proof. First suppose that all the ℓ_{ij} are odd and let ℓ be the largest among them.

On a global level, we plan to define each $\mathbf{u}_i$ as the concatenation of sequences $\mathbf{u}_i(p,q)$ where $1 \le p < q \le m$ and one of p, q is i, together with $\mathbf{u}_i(\omega, i)$ and $\mathbf{u}_i(0,i)$. The subsequences $\mathbf{u}_i(p,q)$ and $\mathbf{u}_j(r,s)$ will be defined in such a way that for $(p,q) \ne (r,s)$, they are ordered by $\ll$ and $\mathbf{u}_i(p,q) \ll \mathbf{u}_j(r,s)$ if and only if $p > r$ or $p = r$ and $q < s$.

Suppose that we have successfully defined these sequences and for fixed $i < j$ that $\mathbf{u}_i\hat{\,}$ and $\mathbf{u}_j\hat{\,}$ are obtained from $\mathbf{u}_i$ and $\mathbf{u}_j$ by the omission of $\mathbf{u}_i(i,j)$ and $\mathbf{u}_j(i,j)$. With a bit of work, one computes that the pair $\{\mathbf{u}_i\hat{\,}, \mathbf{u}_j\hat{\,}\}_<$ has form $2i - 1$. Let k_{ij} be the integer so that $\ell_{ij} = 2i - 1 + 2k_{ij}$.

On a local level, we plan to define $\mathbf{u}_i(i,j)$ and $\mathbf{u}_j(i,j)$ as the concatenations of $\mathbf{u}_i(i,j,k)$ and $\mathbf{u}_j(i,j,k)$ respectively for $1 \le k \le k_{ij}$, so that all the sequences are ordered by $\ll$ and $\mathbf{u}_p(i,j,q) \ll \mathbf{u}_r(i,j,s)$ if and only if $q < s$ or $q = s$ and $p < r$. Then $\mathbf{u}_i\hat{\,}$ and $\mathbf{u}_j\hat{\,}$ get expanded to $\mathbf{u}_i$ and $\mathbf{u}_j$ by the addition of k_{ij} extra blocks each, so $\{\mathbf{u}_i, \mathbf{u}_j\}_<$ has form $2i - 1 + 2k_{ij} = \ell_{ij}$ as desired.

Consideration of both the global and the local level of construction shows that we plan to define $\mathbf{u}_i$ as the concatenation of a total of

$$K_i = 1 + \sum_{j>i} k_{ij} + \sum_{j<i} k_{ji} + 1$$

many subsequences. The two levels of construction completely specify the order under $\ll$ of the $\sum_i K_i$ many subsequences. So we can complete the definitions by recursion using elements from L once we determine the lengths of the subsequences. To that end, take any $\ll$-increasing sequence S_i of sets of length K_i from M above $f(\ell)$, make the length of $\mathbf{u}_i(\omega, i)$ the least element of S_i and make the length of the tth subsequence the difference between the tth and $(t-1)$st terms of S_i. Thus if all the ℓ_{ij} are odd, then the desired set U can be constructed.

Finally suppose all the ℓ_{ij} are even. Let k_{ij} be the integer so that $\ell_{ij} = 2i + 2k_{ij}$. The construction of the first case can be augmented by tacking on to each $\mathbf{u}_i$ a final sequence $\mathbf{u}_i(0,i)$ so that $\mathbf{u}_m(0,m) \ll \mathbf{u}_{m-1}(0,m-1) \ll \ldots \ll \mathbf{u}_1(0,1)$. In this case the sets S_i must be chosen one larger as well, but otherwise the construction proceeds as above. The set U so defined has the property that $\{\mathbf{u}_i, \mathbf{u}_j\}_<$ has form $2i - 1 + 2k_{ij} + 1 = \ell_{ij}$. Thus in either case the lemma follows. ∎

The next lemma says that if there is an edge between a pair of one of the outlines $(=,=<)$, $(=,=>)$, $(=,<)$ or $(=,>)$, then the graph contains arbitrarily large finite complete subgraphs.

Lemma 4.2 *Suppose a graph is doubly canonical for (K, L, M, f)-acceptable pairs and a minimal shape $\sigma = (\mathbf{s}, \mathbf{t})$ with $\max \mathbf{s} = \max \mathbf{t}$ recurs for (K, L, M, f)-acceptable pairs. Then the graph contains arbitrarily large finite complete subgraphs.*

Proof. Let m be a positive integer. We must show that the graph contains a complete subgraph of size m.

Define sequences $\mathbf{s}_i$ and $\mathbf{t}_i$ from $M \setminus f(k+1)$ for $i < m - 1$ all with the same maximum element n so that $\mathbf{s}_i, \mathbf{t}_i$ is similar to $\mathbf{s}, \mathbf{t}$ and so that if $i < j$ then $\mathbf{s}_i^- \cup \mathbf{t}_i^- \ll \mathbf{s}_j^- \cup \mathbf{t}_j^-$. By Remark 1.7 there is a monotonic increasing subset $\{\mathbf{u}_0, \mathbf{u}_1, \ldots, \mathbf{u}_{m-1}\}_<$ of $[L]^n$ so that for all $i < j < m$, the pair $\mathbf{u}_i, \mathbf{u}_j$ has shape $(\mathbf{s}_i, \mathbf{t}_i)$, which is similar to $(\mathbf{s}, \mathbf{t})$.

Since σ recurs, we may choose ℓ so that ℓ is greater than $f(n)$ and so that σ occurs with ℓ. Let $\{(\vec{\alpha}, \vec{\gamma}), (\vec{\beta}, \vec{\delta})\}_<$ witness that σ occurs with ℓ.

For the first case, suppose that $|\vec{\gamma}| = |\vec{\delta}|$ and $\tau = (\mathbf{z}, \mathbf{w})$ is a minimal shape similar to the shape of the pair $\{\vec{\gamma}, \vec{\delta}\}$. Define sequences $\mathbf{z}_i$ and $\mathbf{w}_i$ from $M \setminus f(\ell+1)$ for $i < m - 1$ all with the same maximum element p so that $\mathbf{z}_i, \mathbf{w}_i$ is similar to $\mathbf{z}, \mathbf{w}$ and so that if $i < j$ then $\mathbf{z}_i^- \cup \mathbf{w}_i^- \ll \mathbf{z}_j^- \cup \mathbf{w}_j^-$. If $\vec{\gamma} < \vec{\delta}$, then use Remark 1.7 to get a monotonic increasing subset $\{\mathbf{v}_0, \mathbf{v}_1, \ldots, \mathbf{v}_{m-1}\}_<$ of $[K]^p$ so that for all $i < j < m$, the pair $\mathbf{v}_i, \mathbf{v}_j$ has shape $(\mathbf{z}_i, \mathbf{w}_i)$, which is similar to $(\mathbf{z}, \mathbf{w})$. On the other hand, if $\vec{\gamma} > \vec{\delta}$, then use Remark 1.7 to get a monotonic decreasing subset $\{\mathbf{v}_0, \mathbf{v}_1, \ldots, \mathbf{v}_{m-1}\}_<$ of $[K]^p$ so that for all $i < j < m$, the pair $\mathbf{v}_i, \mathbf{v}_j$ has shape $(\mathbf{z}_{m-j-1}, \mathbf{w}_{m-j-1})$, which is similar to $(\mathbf{z}, \mathbf{w})$. Note that in either case, for all $i < j < m$, the pair $\{(\mathbf{u}_i, \mathbf{v}_i), (\mathbf{u}_j, \mathbf{v}_j)\}_<$ is (K, L, M, f)-acceptable and doubly similar to $\{(\vec{\alpha}, \vec{\gamma}), (\vec{\beta}, \vec{\delta})\}_<$ and thus is joined by an edge in the graph. Therefore in the first case, the lemma holds.

For the second case, suppose that $|\vec{\gamma}| \neq |\vec{\delta}|$ and that $\tau = (\mathbf{z}, \mathbf{w})$ is a minimal shape similar to the shape of the pair $\{\vec{\gamma}, \vec{\delta}\}$. By assumption $\{(\vec{\alpha}, \vec{\gamma}), (\vec{\beta}, \vec{\delta})\}_<$ is (K, L, M, f)-acceptable, so it follows that $\{\vec{\gamma}, \vec{\delta}\}$ is (K, M, f)-acceptable. By Lemma 1.8 there is a set Y listed in increasing order as $\mathbf{y}_i$ so that each pair from it is (K, M, f)-acceptable and block-similar to $\{\vec{\gamma}, \vec{\delta}\}$ and thus has form ℓ as desired. If $\vec{\gamma} < \vec{\delta}$, then for all $i < j < m$, the pair $\{(\mathbf{u}_i, \mathbf{y}_i), (\mathbf{u}_j, \mathbf{y}_j)\}_<$ is (K, L, M, f)-acceptable and doubly similar to $\{(\vec{\alpha}, \vec{\gamma}), (\vec{\beta}, \vec{\delta})\}_<$ and thus is joined by an edge in the graph. On the other hand, in $\vec{\gamma} > \vec{\delta}$, then for all $i < j < m$, the pair $\{(\mathbf{u}_i, \mathbf{y}_{m-i}), (\mathbf{u}_j, \mathbf{y}_{m-j})\}_<$ is (K, L, M, f)-acceptable and doubly similar to $\{(\vec{\alpha}, \vec{\gamma}), (\vec{\beta}, \vec{\delta})\}_<$ and thus is joined by an edge in the graph. Therefore in the second case, the lemma also holds. $\square$

Lemma 4.3 *Suppose a graph is doubly canonical for (K, L, M, f)-acceptable pairs and some shape $\sigma = (\mathbf{s}, \mathbf{t})$ recurs in outline $(<, \diamond)$ for $\diamond$ one of $<$ and $>$. Then the graph contains arbitrarily large finite complete subgraphs.*

Proof. Let m be a positive integer. We must show that the graph contains a complete subgraph of size m. Let R be a number so large that for any partition of the pairs from R into two classes there is a homogeneous set of size m.

Let k be the form of the shape σ. Choose R sequences from M above $f(k)$ of length $\lceil (k+1)/2 \rceil$. Use Lemma 1.8 and these sequences to build a monotonic increasing set $U' = \{\mathbf{u}_i' \mid i < R\}$ so that any pair from U is (L, M, f)-acceptable of form k and shape similar to σ. Divide the pairs from U' into two classes, EVEN and ODD, by putting those pairs into EVEN whose shape occurs with infinitely many even integers in outline $(<, \diamond)$. Let $U = \{\mathbf{u}_i \mid i < m\}$ be a homogeneous subset of U' say for the EVEN side of the partition.

For the first case, assume that $\diamond$ is $<$. Since the shape of $\{\mathbf{u}_i, \mathbf{u}_j\}_<$ occurs infinitely often for even integers, we can find an even integer ℓ_{ij} greater than $f(|\mathbf{u}_m|)$ and greater than $2i + 2$ so that it occurs with ℓ_{ij}. By Lemma 4.1 (Masterful Mingling) there is a set $Y = \{\mathbf{y}_i \mid i < m\}$ all of whose pairs are (K, M', f)-acceptable for $M' = M \setminus f(|\mathbf{u}_m|)$ so that $\{\mathbf{y}_i, \mathbf{y}_j\}_<$ has form ℓ_{ij}. In this case the set X of all pairs $(\mathbf{u}_i, \mathbf{y}_i)$ is a complete subgraph of size m since any pair from it fits outline $(<, <)$.

For the second case, assume $\diamond$ is $>$. Let $V = \{\mathbf{v}_i \mid i < m\}$ be U listed in decreasing order. Find even integers k_{ij} greater than $f(|\mathbf{u}_m|)$ and greater than $2i + 2$ so that the shape of $\{\mathbf{v}_j, \mathbf{v}_i\}_<$ occurs with k_{ij}. Use Lemma 4.1 again to get a set $Z = \{\mathbf{z}_i \mid i < m\}_<$. In this case the set X of pairs $(\mathbf{v}_i, \mathbf{z}_i)$ has the property that any pair from it fits the outline $(<, >)$, so it is the complete subgraph of size m required for this case.

We assumed above that the homogeneous set came from the even side of the partition. By interchanging the use of even and odd, the proof above adapts to the other case as well. □

Lemma 4.4 *Suppose a graph is doubly canonical for (K, L, M, f)-acceptable pairs and some shape $\sigma = (\mathbf{s}, \mathbf{t})$ of form k recurs in outline $(<, <<)$ for (K, L, M, f)-acceptable pairs. Then the graph contains arbitrarily large finite complete subgraphs.*

Proof. Let m be a positive integer. We must show that the graph contains a complete subgraph of size m.

Let $p = |\mathbf{s}|$, $q = |\mathbf{t}|$. By recursion define a sequence of triples (Z_i, r_i, W_i), for $i < 2m$ as follows. Let Z_0 be any set of p elements of M above $f(k)$. Suppose that Z_i has been defined. Let Z_i' be any sequence of q elements of M above Z_i. Since (Z_i, Z_i') is similar to σ, it recurs in outline $(<, <<)$. Let r_i be an integer greater than $f(\max Z_i)$ and at least four that occurs with (Z_i, Z_i'). Let W_i be a sequence of r_i elements of M above $f(r_i)$ and let Z_{i+1} be a sequence of p elements above W_i. Continue as above until all the required sequences and numbers are chosen.

Since we have chosen $2m$ triples, either at least m of the numbers r_i are even or at least m of them are odd. Select a consistent set of indices and rename the triples $(S_0, \ell_0, T_0), (S_1, \ell_1, T_1), \ldots, (S_{m-1}, \ell_{m-1}, T_{m-1})$. For each $i < m$, let $\{(\vec{\alpha}_i, \vec{\gamma}_i), (\vec{\beta}_i, \vec{\delta}_i)\}_<$ be a pair which witnesses that (S_i, S_{i+1}^*) occurs with ℓ_i where S_{i+1}^* is S_{i+1} if k is odd and is obtained from S_{i+1} by the deleting of the penultimate point otherwise.

Use Lemma 1.8 together with the sequences $S_0, S_1, \ldots, S_{m-1}$ to produce an increasing set $U = \{\mathbf{u}_i \mid i < m\}_<$ so that any pair $\{\mathbf{u}_i, \mathbf{u}_j\}_<$ has form k and shape $(\mathbf{s}_{ij}, \mathbf{t}_{ij})$ where $\mathbf{s}_{ij} = S_i$.

Use Lemma 1.8 applied to $\ell_0, \ell_1, \ldots, \ell_{m-1}$ together with the sequences $T_0, T_1, \ldots, T_{m-1}$ to produce an increasing set $V = \{\mathbf{v}_i \mid i < m\}_<$ so that any pair $\{\mathbf{v}_i, \mathbf{v}_j\}$ has form ℓ_i. Let X be the set of pairs $x_i = (\mathbf{u}_i, \mathbf{v}_i)$. For any $i < j < m$, the pair $\{x_i, x_j\}_<$ is doubly similar to $\{(\vec{\alpha}_i, \vec{\gamma}_i), (\vec{\beta}_i, \vec{\delta}_i)\}_<$, so it must be joined by an edge. Thus X is the complete set of size m required to prove the lemma.   □

Theorem 4.5 *If $\lambda < \kappa$ are Ramsey cardinals, then $\kappa^\omega \lambda^\omega$ is a partition ordinal and $\kappa^\omega \lambda^\omega \to (\kappa^\omega \lambda^\omega, m)^2$ for all $m < \omega$.*

Proof. Let G be a graph on pairs $(\vec{\alpha}, \vec{\gamma})$ from $[\lambda]^{<\omega} \times [\kappa]^{<\omega}$. Use Lemma 3.5 to find a subset K of κ of cardinality κ, a subset L of λ of cardinality λ, an infinite subset M of ω and an increasing function $f : \omega \to \omega$ with $f(k) \geq k$ for all $k < \omega$ so that the graph is doubly canonical on all (K, L, M, f)-acceptable pairs.

If there is no (K, L, M, f)-acceptable pair that is joined by an edge in G, then use Lemma 3.3 to get a set X of type $\kappa^\omega \lambda^\omega$ under the lexicographic ordering so that every pair from X is (K, L, M, f)-acceptable and X is the set required to prove the theorem.

Otherwise, there is a (K, L, M, f)-acceptable pair $\{\vec{\alpha}, \vec{\beta}\}_<$ of form k and shape $\sigma = (\mathbf{s}, \mathbf{t})$ which is joined in the graph G. Notice that for some $(\diamond, \star)$, $\{\vec{\alpha}, \vec{\beta}\}_<$ witnesses that the shape σ occurs in outline $(\diamond, \star)$ among the edges of the graph for (K, L, M, f)-acceptable pairs. Since the graph is doubly canonical on (K, L, M, f)-acceptable pairs, it follows that σ recurs in outline $(\diamond, \star)$. Thus by one of Lemmas 4.3, 4.4 and 4.5, there are arbitrarily large complete sets in the graph as required for the other alternative of the theorem. $\Box$

References

[1] J.E. Baumgartner, Remarks on partition ordinals, *Set Theory and its Applications*, eds. J. Steprans and S. Watson, Lecture Notes in Mathematics 1401 (1989), pp. 3–17.

[2] C.C. Chang, A partition theorem for the complete graph on ω^ω, *J. Combinatorial Theory Ser. A* **54** (1973), pp. 396–452.

[3] P. Erdős and A. Hajnal, Unsolved Problems in set theory, *Axiomatic Set Theory* (Proc. Sympos. Pure Math., Univ. of Calif., Los Angeles, Ca., 1967), D.S. Scott, eds., AMS, Providence, RI **13** (I) (1971), pp. 17–48.

[4] P. Erdős and E.C. Milner, A theorem in the partition calculus, *Canad. Math. Bull.* **15** (1972), pp. 501–505.

[4a] *ibid* Corrigendum, **17** (1974), page 305.

[5] P. Erdős and R. Rado, A partition calculus in set theory, *Bull. Amer. Math. Soc.* **62** (1956), pp. 427–489.

[6] L. Haddad and G. Sabbagh, Nouveaux résultats sur les nombres de Ramsey generalisés, *C.R. Acad. Sci. Paris Sér. A-B* **268** (1969), pp. A1516–A1518.

[7] J.A. Larson, A short proof of a partition theorem for the ordinal ω^ω, *Ann. Math. Logic* **6** (1973), pp. 129–145.

[8] J.A. Larson, Partition theorems for certain ordinal products, *Infinite and Finite sets, Coll. Math. Soc. J. Bolyai* **10**, Keszthely, Hungary (1973), pp. 1017–1024.

[9] J.A. Larson, A counter-example in the partition calculus for an uncountable ordinal, *Israel J. Math.* **36** (1980), pp. 287–299.

[10] E.C. Milner, A finite algorithm for the partition calculus, *Proc. of the Twenty-Fifth Summer Meeting of the Canadian Mathematical Congress*, Lakehead Univ., Thunder Bay, Ont., W.R. Eames et al., eds. (1981), pp. 117–128.

[11] E.C. Milner, Partition relations for ordinal numbers, *Canad. J. Math.* **21** (1969), pp. 317–334.

[12] E. Nosal, On a partition relation for ordinal numbers, *J. London Math. Soc. Ser. 2* **8** (1974), pp. 306–310.

[13] E. Nosal, Partition relation for denumerable ordinals, *J. Comb. Th. (B)* **27** (1979), pp. 190–197.

[14] E. Specker, Teilmengen von Mengen mit Relationen, *Comment. Math. Helv.* **31** (1957), pp. 302–314.

[15] N. Williams, *Combinatorial Set Theory*, North Holland, Amsterdam (1977).

The Group of Automorphisms of a Relational Saturated Structure

D. LASCAR

C.N.R.S., Université Paris 7,
Equipe de logique Mathématique,
Tour 45-55, 5ᵉ étage,
2, Place Jussieu, 75251, Paris CEDEX 05

Abstract

In this paper, we will be concerned mainly with the automorphism group of a saturated structure. Since a complete theory has, up to isomorphism, at most one saturated model in a given cardinality, these groups may be considered as an invariant attached to the theory. Moreover, they are equal for two theories which are bi-interpretable. I will be especially interested in the facts that enable us to answer the following question: what information about a complete theory T can be derived from the group of automorphisms of one of its saturated models?

1 Introduction

The notation will be standard. If $\mathcal{M}$ is a structure, the group of automorphisms of $\mathcal{M}$ will be written $\mathrm{Aut}(\mathcal{M})$. If A is a subset of $\mathcal{M}$, $\mathrm{Aut}_A(\mathcal{M})$ is the subgroup of $\mathrm{Aut}(\mathcal{M})$ of all automorphisms leaving A pointwise fixed. In Section 3, I will introduce several topologies on $\mathrm{Aut}(\mathcal{M})$, all of them compatible with the group structure. Up to that point, $\mathrm{Aut}(\mathcal{M})$ will be endowed with the pointwise convergence topology, that is, the one for which a basis of open sets is

$$\{f \cdot \mathrm{Aut}_A(\mathcal{M}); f \in \mathrm{Aut}(\mathcal{M}), A \text{ finite}, A \subseteq \mathcal{M}\}.$$

It turns out, that, if $\mathcal{M}$ is an uncountable saturated structure, then there is a smallest element in the set of all normal subgroups of small index: it is the subgroup of strong automorphisms. This group will be defined and studied in the second section. In the third section, I will define appropriate topologies on $\mathrm{Aut}(\mathcal{M})$, for saturated $\mathcal{M}$, and show how to reconstruct one topological group $\mathrm{Aut}(\mathcal{M})$ from another group $\mathrm{Aut}(\mathcal{N})$, where $\mathcal{M}$ and $\mathcal{N}$ are saturated models of the same theory. In the fourth section, I will define the small index property and some of its variants: they provide a way to recover the topological structure of $\mathrm{Aut}(\mathcal{M})$ from the abstract group structure. Lastly, in the fifth section, I will rapidly present some of the tools which are used to prove the theorems mentioned in this paper.

N.W. Sauer et al. (eds.), Finite and Infinite Combinatorics in Sets and Logic, 225–236.
© 1993 *Kluwer Academic Publishers. Printed in the Netherlands.*

2 Strong automorphisms

Let $\mathcal{M}$ be a saturated structure. When studying the automorphism group of $\mathcal{M}$, it is natural to look for normal subgroups of this group. The group of strong automorphisms, whose definition follows, will play an important role in our discussion:

Definition 2.1 The *group of strong automorphisms* is the subgroup of $\mathrm{Aut}(\mathcal{M})$ generated by the set

$$\{g \in \mathrm{Aut}(\mathcal{M}); \text{ there exists } M_0 \prec M \text{ such that } g \text{ is the identity on } M_0\}.$$

This group will be denoted $\mathrm{Aut}f(\mathcal{M})$.

If A is a subset of $\mathcal{M}$, with $\mathrm{card}(A) < \mathrm{card}(\mathcal{M})$, we can relativise this notion, and define the *group of A-strong automorphisms* as the group generated by the set

$$\{g \in \mathrm{Aut}(\mathcal{M}); \text{ there exists } M_0, A \subseteq M_0 \prec M, \text{ such that } g \text{ is the identity on } M_0\}$$

and this group will be denoted by $\mathrm{Aut}f_A(\mathcal{M})$.

Strong automorphisms are very important in the context of stability theory. In that case, there is an alternative definition. Let us define another normal subgroup of $\mathrm{Aut}(\mathcal{M})$:

$H(\mathcal{M}) = \{f \in \mathrm{Aut}(\mathcal{M}); \text{for all } n \in \mathbb{N}, \text{ for all equivalence relations } R \text{ on } M^n$ definable without parameters and having a finite number of classes, and for all $a \in M^n$, we have $M \models R(a, f(a))\}$.

and more generally

$H_A(\mathcal{M}) = \{f \in \mathrm{Aut}(\mathcal{M}); \text{for all } n \in \mathbb{N}, \text{ for all equivalence relations } R \text{ on } M^n$ definable with parameters in A and having a finite number of classes, and for all $a \in M^n$, we have $M \models R(a, f(a))\}$.

It is easy to see that an automorphism which leaves pointwise fixed an elementary submodel belongs to $H(\mathcal{M})$. So

$$\mathrm{Aut}f_A(\mathcal{M}) \subseteq H_A(\mathcal{M})$$

It is a consequence of the finite equivalence relation theorem (see [15], p. 96) that, if T is stable, the converse holds:

Proposition 2.2 *If T stable, then for every saturated model $\mathcal{M}$ of T and for every subset A of $\mathcal{M}$ of smaller cardinality, $\mathrm{Aut}f_A(\mathcal{M}) = H_A(\mathcal{M})$.*

This proposition remains true for many theories. For example, for the theory of dense linear orderings, with or without endpoints, or for the theory of the random graph, then it is true, just because $\mathrm{Aut}f_A(\mathcal{M}) = H_A(\mathcal{M}) = \mathrm{Aut}_A(\mathcal{M})$. Here is an example, due to Poizat, where the conclusion of Proposition 2.2 does not hold:

Example 2.3 We describe a model of T: the universe of the model is the disjoint union of $\mathbf{R}$ and the unit circle C identified with $\mathbf{R}/2\pi$. In the language, there are function and relation symbols defining: the additive group structure on $\mathbf{R}$, the action of $\mathbf{R}$ on C, and a ternary relation $R(x,y,z)$ whose interpretation is: x and y belong to C, z belongs to $\mathbf{R}$, and the length of the shortest arc from x to y is less than z. In any model $\mathcal{M}$ of T, the formula

$$\bigwedge_{n\in\omega} R(x,y,n^{-1})$$

defines an equivalence relation E on C. It is not very difficult to see that E has exactly $2^{\aleph_0}$ classes, and that $H(\mathcal{M}) = \mathrm{Aut}(\mathcal{M})$, so $H(\mathcal{M})$ is transitive on C, and that any automorphism in $\mathrm{Aut}f(\mathcal{M})$ leaves fixed every class modulo E.

We are now going to list some facts about these groups of strong automorphisms; $\mathcal{M}$ will denote a saturated structure and A a subset of $\mathcal{M}$ of smaller cardinality. The proofs can be found in [8].

Fact 2.4 $\mathrm{Aut}f_A(\mathcal{M})$ is the subgroup of $\mathrm{Aut}_A(\mathcal{M})$ generated by the set

$$X = \{g : \exists M_1 \succ M \text{ and } M_0, \text{ such that } A \subseteq M_0 \prec M_1 \text{ and } g \text{ can be extended to} \\ \text{an automorphism of } \mathrm{Aut}_{\mathcal{M}_0}(\mathcal{M}_1)\}.$$

It can be proved that if the theory of $\mathcal{M}$ is stable, then $X = \mathrm{Aut}f_A(\mathcal{M})$. Here is an example, found in [18], where

$$X = \{g : \exists M_1 \succ M_0, A \subseteq M_0 \prec M_1 \text{ such that } g \text{ can be extended to an automor-} \\ \text{phism of } \mathrm{Aut}_{\mathcal{M}_0}(\mathcal{M}_1)\}$$

is not a group.

Example 2.5 The countable universal cyclic order. As the base set, we take a countable subset of the unit circle in $\mathbf{C}$, and the language contains only one ternary relation R, whose interpretation is defined by: $R(x,y,z)$ is true if and only if y lies on the anticlockwise path from x to z (we will say that y is between x and z). We also demand that, given any distinct x and z, there exists a third point y such that $R(x,y,z)$. The structure $\mathcal{M}$ that we get is $\aleph_0$-categorical and admits elimination of quantifiers. It is easy to see that if $g \in \mathrm{Aut}(\mathcal{M})$, and if g leaves a point, say a, fixed, then $g \in X_0$, because it is possible to extend $\mathcal{M}$ elementarily by adding between a and all the other points of $\mathcal{M}$, a model $\mathcal{M}_0$ of $Th(\mathcal{M})$, and we can extend g by leaving fixed all the points in $\mathcal{M}_0$. From this, we deduce that every automorphism is strong: given $g \in \mathrm{Aut}(\mathcal{M})$, choose any point a. If $g(a) = a$, we are done; if not, choose a point b, which does not lie between a and $g(a)$. There exists $h \in \mathrm{Aut}(\mathcal{M})$ such that $h(a) = g(a)$ and $h(b) = b$. Then h and $h^{-1} \circ g$ are in X_0 and $g \in \mathrm{Aut}f(\mathcal{M})$.

Now, if $g \in \mathrm{Aut}(\mathcal{M})$ leaves a point a fixed, then we can see that, for all $b \in M$ and for all $n \in \mathbf{N}$, either $g(b), g^2(b), \ldots, g^n(b)$ lie between b and $g^{n+1}(b)$, or they all lie between $g^{n+1}(b)$ and b. The same is true, of course, if g can be extended to an automorphism of an elementary extension of $\mathcal{M}$ which leaves a point fixed, that is if $g \in X_0$. But it is easy to construct an automorphism which does not fulfil this condition.

Fact 2.6 Let $\mathcal{M}_0$ be an elementary submodel of $\mathcal{M}$ of smaller cardinality containing A. Then $\mathrm{Aut}f_A(\mathcal{M})$ is the normal subgroup of $\mathrm{Aut}_A(\mathcal{M})$ generated by $\mathrm{Aut}_{\mathcal{M}_0}(\mathcal{M})$.

In fact, $\mathrm{Aut}_A(\mathcal{M})$ is generated, as a group, by two subgroups of the form $\mathrm{Aut}_{\mathcal{M}_0}(\mathcal{M})$.

Fact 2.7 Let $\mathcal{M}_0$ be an elementary submodel of $\mathcal{M}$ of smaller cardinality containing A. Then there exists $\mathcal{M}_1, A \subseteq \mathcal{M}_1 \prec \mathcal{M}$, such that $\mathrm{Aut}f_A(\mathcal{M})$ is the subgroup of $\mathrm{Aut}(\mathcal{M})$ generated by $\mathrm{Aut}_{\mathcal{M}_0}(\mathcal{M}) \cup \mathrm{Aut}_{\mathcal{M}_1}(\mathcal{M})$.

Fact 2.8 The cardinality of the group

$$\Gamma_A(\mathcal{M}) = \mathrm{Aut}_A(\mathcal{M})/\mathrm{Aut}f_A(\mathcal{M})$$

is not bigger than $sup(2^{\aleph_0}, 2^{\mathrm{card}(A)})$. If $\mathcal{M}$ and $\mathcal{M}'$ are two saturated models of the same complete theory, both containing A and of bigger cardinality than A, then there is a natural isomorphism from $\Gamma_A(\mathcal{M})$ onto $\Gamma_A(\mathcal{M}')$.

Thus, in the case where A is the empty set, $\Gamma(\mathcal{M})$ does not depends on the particular saturated model $\mathcal{M}$ of T, and will be denoted $\Gamma(T)$. We have seen examples where this group is trivial (the dense linear ordering or the random graph), and the theories of algebraically closed fields show that it can be of cardinality $2^{\aleph_0}$. Concerning $\aleph_0$-categorical theories, there are two results worth mentioning.

Fact 2.9 (Cherlin–Hrushovski, see [9]) There is an $\aleph_0$-categorical theory T such that $\Gamma(T) = (\mathbf{Z}/2\mathbf{Z})^\omega$.

In fact, if G is any separable profinite group, there is an $\aleph_0$-categorical theory T such that $\Gamma(T)$ is isomorphic to G.

Fact 2.10 (Hrushovski, unpublished) If T is $\aleph_0$-categorical and $\aleph_0$-stable, then $\Gamma(T)$ is finite.

This is a corollary of a theorem of Hrushovski [6], stating that if T is an $\aleph_0$-categorical, ω-stable theory expressed in a language L, then there is a finite sublanguage $L_0 \subseteq L$ such that every symbol in L is equivalent, modulo T, to a formula of L_0 (consider the structure whose universe is $\mathcal{M}$, the countable model of T, and whose automorphism group is $\mathrm{Aut}f(\mathcal{M})$).

The next theorem expresses the fact that, if $\mathcal{M}$ is saturated and uncountable, then $\mathrm{Aut}f(\mathcal{M})$ is the least subgroup of small index.

Theorem 2.11 Let $\mathcal{M}$ be a saturated structure of cardinality $\lambda^+ = 2^\lambda$, and G a normal subgroup of $\mathrm{Aut}(\mathcal{M})$ whose index is at most λ^+. Then $\mathrm{Aut}f(\mathcal{M}) \subseteq G$ (so, by fact 2.8, the index of G is not bigger than $2^{\aleph_0}$.

For the countable case, the situation is not as clear. Let us try a conjecture:

Conjecture 2.12 Let $\mathcal{M}$ be a saturated countable structure, and G a normal subgroup of $\mathrm{Aut}(\mathcal{M})$ of countable index. Then $\mathrm{Aut}f(\mathcal{M}) \subseteq G$.

This conjecture has been proved in the case where $\mathcal{M}$ is ω-stable and such that $\Gamma_A(\mathcal{M})$ is finite whenever A is finite. In the general case, I am not sure that this is the right conjecture. Maybe the condition that $\mathcal{M}$ be saturated is not strong enough, and what is needed is something generalising the $\aleph_\varepsilon$-saturation.

3 The reconstruction problem

One of the main motivations in studying automorphism groups is the following question: assume that we know what is the automorphism group $\mathrm{Aut}(\mathcal{M})$, then what do we know about $\mathcal{M}$ itself, or about the theory of $\mathcal{M}$? We have already said, in the very first paragraph of this paper, that we certainly cannot get more than the theory up to bi-interpretability. Let us throw some light on the question by an answer to it, in a simple case (see [5]):

Theorem 3.1 *Assume that T and T' are $\aleph_0$-categorical theories and that $\mathcal{M}$ and $\mathcal{N}$ are the countable models of T and T' respectively. Then the two following conditions are equivalent:*

1°) *there exists a bicontinuous isomorphism from $\mathrm{Aut}(\mathcal{M})$ onto $\mathrm{Aut}(\mathcal{N})$.*

2°) *T and T' are bi-interpretable.*

In other words, if we know the topological group $\mathrm{Aut}(\mathcal{M})$, then we know everything about T. There are two kinds of possible extensions of this theorem: first, we may assume that $\mathrm{Aut}(\mathcal{M})$ and $\mathrm{Aut}(\mathcal{N})$ are isomorphic groups, not taking into account the topology. Second, we may ask what happens if T and T' are no longer necessarily $\aleph_0$-categorical. It is clear that we cannot hope for much if we take any model of T, and it seems reasonable to think that it is the automorphism group of the saturated models which conveys information about the theory. That is why we will restrict our attention to $\mathrm{Aut}(\mathcal{M})$ for saturated structures $\mathcal{M}$. But other options are possible. First one may study $\mathrm{Aut}(\mathcal{M})$ for a countable recursively saturated structure $\mathcal{M}$ (it is the option that has been chosen by R. Kaye, R. Kossak and H. Kotlarski; it seems particularly appropriate for models of Peano Arithmetic ; see [7]). The advantage is that any countable theory has a countable recursively saturated model, and that its automorphism group is very rich, and seems to give a lot of information about $\mathcal{M}$. It is still possible, always with countable $\mathcal{M}$, to try to recover the theory of $\mathcal{M}$ in $\mathcal{L}_{\omega_1,\omega}$.

The first question that I want to discuss is the following: assume that $\mathcal{M}$ is countably saturated and that we know the topological group $\mathrm{Aut}(\mathcal{M})$. We will show that in fact we know a lot about T, the theory of $\mathcal{M}$: we may reconstruct the category of all models of T which are countably saturated, with elementary maps as morphisms. This is done in two steps:

1°) Reconstruction of the semigroup of elementary endomorphisms of $\mathcal{M}$. This is the Cauchy completion of $\mathrm{Aut}(\mathcal{M})$ with the appropriate uniform topological structure. Define a sequence of automorphisms $(g_n; n \in \omega)$ to be Cauchy if, for all open subgroups G of $\mathrm{Aut}(\mathcal{M})$, there exists an integer n such that, for all $m, p > n$, $g_m^{-1} \circ g_p \in G$. Two Cauchy sequences $(g_n; n \in \omega)$ and $(h_n; n \in \omega)$ are equivalent if, for all open subgroups G of $\mathrm{Aut}(\mathcal{M})$, there exists an integer n such that, for all $m > n$, $h_m^{-1} \circ g_m \in G$. On the equivalence classes of Cauchy sequences, it is possible to define a multiplication by

$$\text{class of } (g_n; n \in \omega) \times \text{class of } (h_n; n \in \omega) = \text{class of } (g_n \circ h_n; n \in \omega)$$

and this defines a semigroup isomorphic to the semigroup of elementary endomorphisms of $\mathcal{M}$, which we shall denote by $\text{End}(\mathcal{M})$.

2°) Reconstruction of $\mathcal{C}$, the category of countably saturated models. We first reconstruct the category $\mathcal{C}_1$ of countably saturated models: it is a category equivalent to $\text{End}(\mathcal{M})$. Then $\mathcal{C}_1$ is a full subcategory of $\mathcal{C}$ and $\mathcal{C}$ is just the completion of $\mathcal{C}_1$ under inductive limits (this is just the translation of the fact that a countably saturated model is the union of its countable elementary submodels which are saturated). Thus, we get $\mathcal{C}$ by taking the completion of $\mathcal{C}_1$ under inductive limits. There is a formal way to do this: $\mathcal{C}$ is the category of the ind-objects of $\text{End}(\mathcal{M})$; (see [14], Exposé 1).

It follows that, if we know the topological group $\text{Aut}(\mathcal{M})$, where $\mathcal{M}$ is a countable saturated structure, we can answer the following questions: is T stable? Is T superstable? Is T $\aleph_1$-categorical? What is the number of countably saturated models of T in some uncountable cardinality? We can also reconstruct the automorphism group of a saturated model of T of higher cardinality. So, if T and T' are two theories having countable saturated models $\mathcal{M}$ and $\mathcal{M}'$, if $\text{Aut}(\mathcal{M})$ and $\text{Aut}(\mathcal{M}')$ are topologically isomorphic, and if $\mathcal{N}$ and $\mathcal{N}'$ are saturated models of T and T' respectively of the same cardinality, then $\text{Aut}(\mathcal{N})$ and $\text{Aut}(\mathcal{N}')$ are also topologically isomorphic.

Let us now examine the situation for uncountable saturated structures. Let $\mathcal{M}$ be a saturated structure of cardinality λ, and we will suppose that λ is regular. For each infinite cardinal μ less than or equal to λ, we define another topology on $\text{Aut}(\mathcal{M})$, that we will call $\mathcal{T}_\mu$. A basis of open sets for this topology is

$$\{h \cdot \text{Aut}_A(\mathcal{M}) : h \in \text{Aut}(\mathcal{M}) \text{ and } A \text{ is a subset of } M \text{ of cardinality strictly less than } \mu\}.$$

It is easy to see that $\text{Aut}(\mathcal{M})$, endowed with $\mathcal{T}_\mu$, is a topological group. The topology $\mathcal{T}_\omega$ is just the pointwise convergence topology.

Assume now that we are given the group $\text{Aut}(\mathcal{M})$ endowed with the topology $\mathcal{T}_\lambda$, where $\mathcal{M}$ is a saturated model $\mathcal{M}$ of cardinality λ. I want to explain how to reconstruct the topological group $(\text{Aut}(\mathcal{N}), \mathcal{T}_\mu)$, where $\mathcal{N}$ is the saturated structure elementarily equivalent to $\mathcal{M}$ of regular cardinality μ. If $\lambda < \mu$, then the technique used for countable $\mathcal{M}$ can be generalized. With the Cauchy sequences of length λ, we reconstruct the semigroup of endomorphisms of $\mathcal{M}$, and then the category of all λ-saturated models (we have to take directed unions of cofinality at least λ).

Going down is a little more difficult. We will assume that, for some $\kappa > \lambda$, there is a saturated model $\mathcal{N}$ of T of cardinality κ. This is a very weak hypothesis (in fact, we can work in a generic extension of the universe which does not change anything below λ and in which there is such a cardinal). So, from what has already been said, we can reconstruct $\text{Aut}(\mathcal{N})$, and inside this group we can define the family

$$\mathcal{G} = \{\text{Aut}_{\mathcal{M}_0}(\mathcal{N}) : \mathcal{M}_0 \prec \mathcal{N}, \mathcal{M}_0 \text{ saturated of cardinality } \lambda\},$$

and consequently, we also know the topology $\mathcal{T}_\lambda$ on $\text{Aut}(\mathcal{N})$. We need the following fact:

Fact 3.2 Let $\mathcal{M}_0 \prec \mathcal{N}$ be a saturated model of cardinality λ, ν an ordinal less than or equal to λ and $(a_i : i < \nu)$ a sequence of elements of $\mathcal{M}_0$, $\nu \leq \lambda$. Set $A_i = \{a_j : j < i\}$. There exists an increasing sequence $(\mathcal{N}_i : i < \nu)$ of submodels of $\mathcal{N}$, all of them being saturated and of cardinality λ, such that, for all $i < \nu$, $A_i \subseteq \mathcal{N}_i$ and $\text{Aut}f_{A_i}(\mathcal{N})$ is equal to the group generated by $\text{Aut}_{\mathcal{M}_0}(\mathcal{N}) \cup \text{Aut}_{\mathcal{N}_i}(\mathcal{N})$.

The proof (by induction on ν) is rather technical. It is a generalisation of Theorem 6.16 of [8].

Let H be a subgroup of $\mathrm{Aut}(\mathcal{N})$. We will say that H has the *finite support property* if:

1°) H is open for $\mathcal{T}_\lambda$;

2°) whenever ν is a limit ordinal and $(G_i : i < \nu)$ is a decreasing sequence of elements of $\mathcal{G}$ such that $\bigcap_{i<\nu} G_i \subseteq H$, then there exists $i < \nu$ such that $G_i \subseteq H$. We have the two following facts:

Fact 3.3 If A is finite, then $\mathrm{Aut}f_A(\mathcal{N})$ has the finite support property. This is because $\mathrm{Aut}_{\mathcal{M}_0}(\mathcal{N}) \subseteq \mathrm{Aut}f_A(\mathcal{N})$ if and only if $A \subseteq \mathcal{M}_0$.

Fact 3.4 If H has the finite support property, then there exists A finite such that

$$\mathrm{Aut}f_A(\mathcal{N}) \subseteq H.$$

OUTLINE OF THE PROOF: We know that there is a saturated model $\mathcal{M}_0$ of cardinality λ such that $\mathrm{Aut}_{\mathcal{M}_0}(\mathcal{N}) \subseteq H$. Let ν be the least cardinal number such that there exists a subset A of $\mathcal{M}_0$ of cardinality ν and such that $\mathrm{Aut}f_A(\mathcal{N}) \subseteq H$. We assume that ν is infinite and deduce a contradiction. Enumerate $A = (a_i : i < \nu)$. Let $(N_i : i < \nu)$ be a sequence of submodels, like in Fact 3.2. Because H has the finite support property, there exists an ordinal $i < \nu$ such that $\mathrm{Aut}_{N_i}(\mathcal{N}) \subseteq H$, and, calling $B = (a_j : j < i)$, $\mathrm{Aut}_B(\mathcal{N}) \subseteq H$, contradicting the minimality of ν.

So, if μ is uncountable, we can recover the topology $\mathcal{T}_\mu$ on $\mathrm{Aut}(\mathcal{N})$: the open subgroups are just those which are included in the intersection of less than μ subgroups having the finite support property. For μ countable, I do not think it is possible, but we may recover the topology $\mathcal{T}_a$ whose definition follows: if B is any structure, a basis of open sets for the topology $\mathcal{T}_a$ on $\mathrm{Aut}(B)$ is

$$\{g \cdot \mathrm{Aut}f_A(B); g \in \mathrm{Aut}(B), A \text{ is a finite subset of } B\}.$$

Assume now that μ is a regular cardinal less than κ and that there is a saturated structure elementarily equivalent to $\mathcal{M}$ of cardinality μ. Then the set

$$\{\mathrm{Aut}_{\mathcal{M}_0}(\mathcal{N}) : \mathcal{M}_0 \prec N \text{ and } \mathcal{M}_0 \text{ is saturated of cardinality } \mu\}$$

is the unique subset X of subgroups of $\mathrm{Aut}(\mathcal{N})$ such that:

a) the elements of X are open for T_{μ^+};

b) any two elements in X are conjugate;

c) any subgroup which is open for T_{μ^+} includes an element of X;

d) X is closed by decreasing intersection of length μ.

So $\mathrm{Aut}(\mathcal{M}_0)$ is isomorphic (as a group) to $N(G)/G$, where G is any element of X and $N(G)$ denotes the normalizer of G (note that if $G = \mathrm{Aut}_{\mathcal{M}_0}(\mathcal{N})$, then $N(G)$ is the subgroup of automorphisms leaving $\mathcal{M}_0$ setwise fixed). If, moreover, μ is uncountable then we can also recover the $\mathcal{T}_\mu$ topology on $\mathrm{Aut}(\mathcal{M}_0)$ since we know it on $\mathrm{Aut}(\mathcal{N})$. If μ is countable, then it is the topology $\mathcal{T}_a$ that we get. But unfortunately, in this case, the other direction (recovering $\mathcal{T}_\lambda$ from $\mathcal{T}_a$) is not so clear. The natural thing to try is to prove, once again, that the semigroup of endomorphisms of $\mathcal{M}$ (the countable saturated model) is the Cauchy completion of $(\mathrm{Aut}(\mathcal{M}), \mathcal{T}_a)$. This is not in general true. It is true, for example, if $\mathcal{M}$ is stable and $\aleph_\varepsilon$-saturated (compare with the last remark of Section 2). And, indeed, in that case, we can reconstruct $\mathcal{T}_\lambda$ from $\mathcal{T}_a$.

If $\mathcal{M}$ is a saturated model of cardinality λ, $\mathcal{T}^*$ will be $\mathcal{T}_\lambda$ if λ is uncountable or $\mathcal{T}_a$ if $\lambda = \aleph_0$. With this notation, we get the following theorem:

Theorem 3.5 *Let λ_1, λ_2 be regular cardinals, λ_1 uncountable, T and T' be two theories, $\mathcal{M}_1, \mathcal{M}_2$ be saturated models of T' of cardinality λ_1 and λ_2 respectively, $\mathcal{N}_1, \mathcal{N}_2$ be saturated models of T' of cardinality λ_1 and λ_2 respectively, and assume that $(\mathrm{Aut}(\mathcal{M}_1), \mathcal{T}^*)$ is isomorphic to $(\mathrm{Aut}(\mathcal{N}_1), \mathcal{T}^*)$. Then $(\mathrm{Aut}(\mathcal{M}_2), \mathcal{T}^*)$ is isomorphic to $(\mathrm{Aut}(\mathcal{N}_2), T^*)$.*

4 The small index property

The question that I want to examine now is the following: is it possible to recover the topological structure of $\mathrm{Aut}(\mathcal{M})$ from the group structure? The main idea to answer this question is contained in the following definition (see [5]):

Definition 4.1 Let $\mathcal{M}$ be a countable structure. We will say that $\mathrm{Aut}(\mathcal{M})$ (or sometimes $\mathcal{M}$) has the *small index property* if every subgroup H of $\mathrm{Aut}(\mathcal{M})$ of index smaller than $2^{\aleph_0}$, is open for $\mathcal{T}_\omega$.

It is clear that the open (for $\mathcal{T}_\omega$) subgroups of a countable structure have countable index. If $\mathcal{M}$ has the small index property, then the converse is true, and this allows us to reconstruct T_ω from the group structure. The small index property has been proved for quite a lot of $\aleph_0$-categorical theories: for the theory of the infinite set in the language containing only equality ([13] or [1]), the theory of dense linear orderings ([17]), the theory of vector spaces over a finite field ([2]). We will see later some general theorems in this direction. The only way that I know to construct an example of a saturated countable structure which does not have the small index property makes use of the group $\Gamma(X)$. Let us begin with a remark. There are $\aleph_0$ subgroups of the form $\mathrm{Aut}_A(\mathcal{M})$ with $A \subseteq M$ and A finite, and the same number of right cosets modulo these subgroups. A subgroup G which contains one of those groups is a union of cosets, so there are at most $2^{\aleph_0}$ such subgroups. Now, assume that we have succeeded in constructing a theory T and $2^{2^{\aleph_0}}$ subgroups of $\Gamma(T)$ of countable index. By lifting them to $\mathrm{Aut}(\mathcal{M})$, we will get $2^{2^{\aleph_0}}$ subgroups of $\mathrm{Aut}(\mathcal{M})$ of countable index, and it is impossible that all of them contain a subgroup of the form $\mathrm{Aut}_A(\mathcal{M})$.

Thus, the example in Fact 2.9 is also an example where the small index property fails: there are $2^{2^{\aleph_0}}$ subgroups of $(\mathbf{Z}/2\mathbf{Z})^\omega$ of index 2 (one for each ultrafilter on ω). So we get an $\aleph_0$-categorical theory for which the small index property fails, even with a normal subgroup of finite index. It should be mentioned that Evans and Hewitt in [4] have constructed an example proving that it is definitely impossible to reconstruct an $\aleph_0$-categorical theory from its pure automorphism group. In [9], it is shown that the axiom of choice is needed to construct a counterexample to the small index property. As far as I know, the following conjecture still stands open:

Conjecture 4.2 *Let $\mathcal{M}$ be a countable saturated structure, and G a subgroup of* $\mathrm{Aut}(\mathcal{M})$ *of index less than* $2^{\aleph_0}$. *Then G is open for* $(\mathrm{Aut}(\mathcal{M}), \mathcal{T}^*)$.

Note that Fact 2.10 shows that, if $\mathcal{M}$ is ω-stable and $\aleph_0$-categorical, then the topologies $\mathcal{T}^*$ and $\mathcal{T}_\omega$ coincide, so Conjecture 4.2 gives the small index property in this case.

Now, I want to show how to recover the $\mathcal{T}^*$-topology on $\mathrm{Aut}(\mathcal{M})$ when $\mathcal{M}$ is a saturated structure of regular uncountable cardinality λ, assuming that every subgroup of index less than or equal to λ is open. This fact no longer characterises the open subgroups, since the converse is not necessarily true (it will be true if we assume that for any $\mu < \lambda$, $2^\mu < \lambda$; see [16] for more about that). But certainly, the index of a subgroup of the form $\mathrm{Aut}_A(\mathcal{M})$, for a finite A, is not bigger than λ. Thus the open subgroups are exactly those which contain the intersection of strictly less than λ subgroups, each of them being of index at most λ.

The countable case is more difficult, and I do not know how to do it in general. It can be done in some cases, for example if $\mathcal{M}$ is ω-stable.

Here is a list of the known results about the small index property and its variants. The cases of the infinite set, of the dense linear ordering or the vector spaces have already been mentioned. The two following theorems has been proved using Ehrenfeucht–Mostowski models:

Theorem 4.3 [9]: *If $\mathcal{M}$ is a saturated ω-stable structure of cardinality $\lambda^+ = 2^\lambda$, then every subgroup of* $\mathrm{Aut}(\mathcal{M})$ *of index at most λ^+ is open in* $(\mathrm{Aut}(\mathcal{M}), \mathcal{T}^*)$.

Theorem 4.4 *If $\mathcal{M}$ is a saturated, countable, strongly minimal set, then every subgroup of* $\mathrm{Aut}(\mathcal{M})$ *of countable index is open in* $(\mathrm{Aut}(\mathcal{M}), \mathcal{T}^*)$.

Recently, S. Shelah devised a new technique to prove the small index property for the countable random graph (I want to thank I. Hodkinson for showing me his notes on this). With this technique, it can be proved that the small index property is true for the countable model of any ω-categorical, ω-stable theory. Lastly, during the conference itself, Shelah proved that if $\mathcal{M}$ is any uncountable saturated model of cardinality $\lambda = \lambda^{<\lambda}$, then every subgroup of $\mathrm{Aut}(\mathcal{M})$ of index at most λ is open in $(\mathrm{Aut}(\mathcal{M}), \mathcal{T}^*)$.

Of course, the small index property allows us to solve the reconstruction problem, but there are others ways: an example is the work of M. Rubin [12] on the reconstruction problem for various graphs and partial orderings.

5   The technical tools

In the investigation of the automorphism groups, there are some important tools that I want to present in this section. The first one is certainly the topological structure. Let us begin with the topology $\mathcal{T}_\omega$ on $\mathrm{Aut}(\mathcal{M})$ where $\mathcal{M}$ is countable.

The topological space that we get is a Polish space: it is clear that it is Hausdorff, separable (there is a countable basis of open sets). Assume without lost of generality that the base set of $\mathcal{M}$ is $\mathbf{N}$. If f and g are distinct automorphisms, define

$$d(f, g) = 2^{-(\min(n: f(n) \neq g(n) \text{ or } f^{-1}(n) \neq g^{-1}(n)))}$$

The distance d defines the topology $\mathcal{T}_\omega$, and turns $\mathrm{Aut}(\mathcal{M})$ into a complete metric space.

We will assume that the reader is familiar with all the nice properties shared by Polish spaces. I will give an example of a theorem that we get for free from these properties:

Theorem 5.1 *If G is a subgroup of* Aut($\mathcal{M}$) *of index less than* $2^{\aleph_0}$ *and is a Baire set, then G is open.*

(A Baire set is a set which is equal to an open set modulo a meager set. For closed subgroups, this theorem has been proved by D.Evans [3].)

We know that G is equal to an open set O, modulo a meager set M. But O cannot be empty, otherwise G would be meager, and Hodkinson has proved that a meager subgroup has index $2^{\aleph_0}$ (notice that if you are willing to assume CH, then Baire's theorem directly gives you this result). By performing a translation, we may assume that O is an open subgroup of Aut($\mathcal{M}$), and since $G \cap O$ is a comeager subgroup of O, it cannot have a coset disjoint to itself (by Baire's theorem), so $O \subseteq G$ and G is open.

Let us shift now to the topologies $\mathcal{T}^*$ on Aut($\mathcal{M}$) for larger $\mathcal{M}$. Of course, we no longer have a Polish space, but the Baire theorem is still true:

Theorem 5.2 *If $\mathcal{M}$ is a structure of cardinality λ, then, in $\mathcal{T}^*$, the intersection of λ dense open subsets is dense.*

We will say that $X \in$ Aut($\mathcal{M}$) is λ-comeager if it contains the intersection of λ dense open sets (for $\mathcal{T}^*$, of course). A λ-meager set is a set whose complement is λ-comeager. Up to now, in this section, we have taken very little advantage of the fact that we are dealing with automorphism groups. The following definition appears in [18]:

Definition 5.3 Let $\mathcal{M}$ be a countable saturated structure and $g \in$ Aut($\mathcal{M}$). Then g is said to be *generic* if the conjugacy class of g is comeager in Aut($\mathcal{M}$) (in T_ω).

In [18], J. Truss proves that several kinds of countable structures (for example the infinite set without any relation, the dense linear ordering, the random graph) admit generic automorphisms. But it can be seen that there are no generic automorphisms as soon as Aut$f(\mathcal{M})$ is a proper subgroup of Aut($\mathcal{M}$). The following notion seems more adapted to our aim:

Definition 5.4 Let $\mathcal{M}$ be a saturated structure of cardinality λ and $g \in$ Aut($\mathcal{M}$). Then g is said to be *locally generic* if the conjugacy class of g is λ-comeager in Aut$f(\mathcal{M})$.

The situation, here again, is simpler if we restrict ourself to the uncountable case. We will introduce a class of theories for which locally generic automorphisms do exist:

Definition 5.5 Let T be a theory, and $\mathcal{C}$ its monster model. If $\mathcal{M}_0 \prec \mathcal{M}_1$, $\mathcal{M}_0 \prec \mathcal{M}_2$ are three elementary submodels of $\mathcal{C}$, we say that $\mathcal{M}_1$ and $\mathcal{M}_2$ are *Aut-free above* $\mathcal{M}_0$ if, whenever $g_1 \in$ Aut($\mathcal{M}_1$) and $g_2 \in$ Aut($\mathcal{M}_2$) are such that $g_1 \upharpoonright \mathcal{M}_0 = g_2 \upharpoonright \mathcal{M}_0 \in$ Aut($\mathcal{M}_0$), then $g_1 \cup g_2$ is an elementary map. We say that T has the *amalgamation property for automorphisms* if, whenever $\mathcal{M}_0 \prec \mathcal{M}_1$, $\mathcal{M}_0 \prec \mathcal{M}_2$ are three elementary submodels of $\mathcal{C}$, there exists $\mathcal{N}_2$, $\mathcal{M}_0 \prec \mathcal{N}_2 \prec \mathcal{C}$, and g, an elementary map from $\mathcal{M}_2$ onto $\mathcal{N}_2$ leaving $\mathcal{M}_0$ pointwise fixed, such that $\mathcal{M}_1$ and $\mathcal{N}_2$ are Aut-free above $\mathcal{M}_0$.

A structure has the amalgamation property for automorphisms if its complete theory has it. Every stable theory has this amalgamation property (if $\mathcal{M}_1$ and $\mathcal{M}_2$ are independent over $\mathcal{M}_0$, then they are Aut-free); dense linear orderings, the random graph and many other structures also have this property. See [9] for an example of a theory which fails to have it. The notion of a generic automorphism is an important tool in the proofs of the small index property. In the countable case, the technique of Shelah, that we have been speaking about, needs "generic sequences of automorphisms", that is, finite sequences $(g_1, g_2, \ldots, g_n)$ of automorphisms which are such that the set

$$\{(\alpha g_1 \alpha^{-1}, \alpha g_2 \alpha^{-1}, \ldots, \alpha g_n \alpha^{-1}) : \alpha \in \mathrm{Aut}(\mathcal{M})\}$$

is comeager on a nonempty open subset of the product of n copies of $\mathrm{Aut}(\mathcal{M})$. Such sequences can be proved to exist for all integers n provided, first, that the theory has the amalgamation property for automorphisms, and second, that the following property holds: if $f_1, f_2, \ldots, f_n$ are finite partial elementary maps from finite subsets of $\mathcal{M}$ onto finite subsets of $\mathcal{M}$, then there exists a finite A in $\mathcal{M}$ containing all the domains and the ranges of the f_i's, and elementary permutations $g_1, g_2, \ldots, g_n$ of A which extend the f_i's. As an example, $\aleph_0$-stable and $\aleph_0$-categorical theories satisfy both these properties. In the uncountable case, the notion of generic sequences can be generalised in an obvious way, even for sequences of infinite length. Roughly speaking, what is needed to prove the corresponding version of the small index property is the existence of generic sequences of length μ, for all μ strictly less than the cardinality of $\mathcal{M}$. And it is a good surprise to discover that no hypothesis at all (except a little bit of G.C.H.) is needed to prove this existence.

References

[1] J.D. Dixon, P.M. Neumann, S. Thomas, Subgroups of small index in infinite symmetric groups, *Bull. London Math. Soc.* **18** (1986) 580–586.

[2] D.M. Evans, Subgroups of small index in general linear groups, *Bull. London Math. Soc.* **18** (1986) 587–590.

[3] D. Evans, A note on automorphism groups of countably infinite structures, *Arch. Math. (Basel)* **49** (1987) 479–483.

[4] D.M. Evans, P.R. Hewitt, Counterexample to a conjecture on relative categoricity, preprint, University of East Anglia, 1988.

[5] W. Hodges, Categoricity and permutation groups, *Logic colloquium '87* (eds. Ebbinghaus et al., North Holland, Amsterdam, 1989) 53–72 .

[6] E. Hrushovski, Totally categorical structures, *Trans. Amer. Math. Soc.* **313**, 1 (May 1989) 131–159.

[7] R. Kaye, R. Kossak, H. Kotlarski, Automorphisms of recursively saturated models of arithmetic, Institute of Mathematics, Polish Academy of Sciences, Preprint 479, Oct. 1990.

[8] D. Lascar, On the category of models of a complete theory, *J. Symbolic Logic* **47** (1982) 249–266.

[9] D. Lascar, Autour de la propriété du petit indice, *Proc. London Math. Soc. (3)* **62** (1991) 25–53.

[10] D. Lascar, Le groupe d'automorphisme d'un ensemble fortement minimal, to appear in *J. Symbolic Logic*.

[11] D. Lascar, Les beaux automorphismes, to appear in *Archive for Math. Logic*.

[12] M. Rubin, On the reconstruction of $\aleph_0$-categorical structure from their automorphism group, Preprint, University of Beer Sheva, 1987.

[13] S.W. Semmes, Endomorphisms of infinite symmetric group, *Abstract Amer. Math. Society* **2** (1981) 426.

[14] Seminaire de geométrie algébrique du Bois Marie, 63–64, SGA4, LNM269, Springer Verlag, Berlin, Heidelberg, New York, 1972.

[15] S. Shelah, Classification theory and the number of non isomorphic models, *Studies in Logic*, North Holland, Amsterdam, 1978.

[16] S. Shelah, S. Thomas, Subgroups of small index in infinite symmetric group II, *J. Symbolic Logic* **54** (1989) 1, 95–99.

[17] J.K. Truss, Infinite permutation groups; subgroups of small index, *J. Algebra* **120** (1989) 494–515.

[18] J.K. Truss, Generic automorphisms of homogeneous structures, preprint, University of Leeds.

On Canonical Ramsey Numbers for Coloring Three-Element Sets

HANNO LEFMANN[1,2]

Fakultät für Mathematik, SFB
343, Universität Bielefeld,
W 4800 - Bielefeld 1, Germany

VOJTĚCH RÖDL

Department of Mathematics and Computer Science,
Emory University,
Atlanta, GA,
USA 30322

Abstract

In this note we will study the growth of canonical Ramsey numbers $er(3,l)$ for arbitrary colorings of three-element sets, where $er(3,l)$ is the least positive integer n such that for every coloring of the three-element subsets of the set $\{1, 2, \ldots, n\}$ there always exists an l-element subset colored canonically according to the theorem of Erdős and Rado. In particular, it will be shown that there exist positive constants $c, c' > 0$ such that

$$2^{2^{cl^2}} \leq er(3,l) \leq 2^{2^{2^{c'l^5}}}$$

holds for every integer $l \geq 4$.

1   Introduction

In 1930 Ramsey proved his famous theorem about colorings of finite sets:

Theorem 1 [9] *Let k, l, m with $l \geq k$ be positive integers. Then there exists a least positive integer $n = r_m(k,l)$ such that for every coloring $\Delta : [\{1, 2, \ldots, n\}]^k \longrightarrow \{1, 2, \ldots, m\}$ of the k-element subsets of the set $\{1, 2, \ldots, n\}$ with m colors there exists an l-element subset $X \subseteq \{1, 2, \ldots, n\}$ such that $\Delta(S) = \Delta(T)$ for all k-element subsets $S, T \subset [X]^k$, that is, the set $[X]^k$ is colored monochromatically.*

Due to stimulation by P. Erdős, much interest was drawn towards the study of the growth of the numbers $r_m(k,l)$, c.f. [2] and [6]. For colorings of two-element sets it is known that

---

[1] Part of this work has been done during the first author's visit at Emory University, Department of Mathematics and Computer Science, Atlanta, Georgia 30322, USA and Georgia Institute of Technology, School of Mathematics, Atlanta, Georgia 30332, USA

[2] Research partially supported by DFG - Deutsche Forschungsgemeinschaft

N.W. Sauer et al. (eds.), Finite and Infinite Combinatorics in Sets and Logic, 237–247.

Theorem 2 [5] [7] *There exist positive constants c, c' such that for all positive integers l, m with $l \geq 3$ and $m \geq 2$,*

$$2^{cml} \leq r_m(2, l) \leq 2^{c'ml \log m} . \tag{1}$$

The lower bound can be found in [7], the upper in [5].

In order to state the corresponding results for colorings of arbitrary k-element sets, it is convenient to work with the following notation. For positive integers k, n let $p_k(n)$ be a tower function defined by $p_1(n) = 2^n$ and for positive integers $k \geq 2$ let $p_k(n) = 2^{p_{k-1}(n)}$; hence $p_k(n)$ describes a tower of k twos and on the top an n.

For the general Ramsey numbers it is known that

Theorem 3 [3] *Let k, l, m be positive integers. Then there exist positive constants $c_k, c_k^m > 0$ such that*

$$
\begin{align}
r_2(k, l) &\geq p_{k-2}(c_k l^2) && \text{if } l \geq l_0(k) \tag{2} \\
r_4(k, l) &\geq p_{k-1}(c_k l) && \text{if } l \geq l_0(k) \tag{3} \\
r_m(k, l) &\leq p_{k-1}(c_k^m l) . && \tag{4}
\end{align}
$$

In Ramsey's theorem the number of colors is fixed. If one drops this assumption one cannot expect any more to get monochromatic subsets, as one can color the k-element subsets in a one-to-one way for example. However, as has been shown by Erdős and Rado [ER 50], it turns out that one can get always canonically colored subsets. From now on let the underlying set $\{1, 2, \ldots, n\}$ be totally ordered by the usual order of the set of positive integers.

Theorem 4 [4] *Let k, l be positive integers with $k < l$. Then there exists a least positive integer $n = er(k, l)$ such that for every coloring $\Delta : [\{1, 2, \ldots, n\}]^k \longrightarrow \omega$ of the k-element subsets of the set $\{1, 2, \ldots, n\}$ with an arbitrary number of colors there always exists a (possibly empty) subset $I \subseteq \{1, 2, \ldots, k\}$ and an l-element subset $X \subset \{1, 2, \ldots, n\}$ which is colored canonically, i.e.:*

for all k-element subsets $\{x_1, x_2, \ldots, x_k\}_<, \{y_1, y_2, \ldots, y_k\}_< \in [X]^k$ with $x_1 < x_2 < \ldots < x_k$ and $y_1 < y_2 < \ldots < y_k$,

$$\Delta(\{x_1, x_2, \ldots, x_k\}_<) = \Delta(\{y_1, y_2, \ldots, y_k\}_<) \quad \Longleftrightarrow \quad x_i = y_i \text{ for all } i \in I .$$

Thus the theorem of Erdős and Rado, also called the Canonical Ramsey Theorem, says that for positive integers n, n large enough, and for arbitrary colorings of the k-element subsets of an n-element set, one can always find an l-element subset such that the set of its k-element subsets is colored according to one of 2^k canonical cases. For coloring singletons, i.e. $k = 1$, there are two possibilities for the l-element set X: either all elements are colored the same, or they are colored pairwise differently. For general k, these canonical colorings are determined by subsets of the index set $\{1, 2, \ldots, k\}$. For example, if I is equal to the empty set, then $[X]^k$ is colored monochromatically. If $I = \{1\}$, then two k-element subsets of X are colored the same if and only if they have the same minimum. Moreover, $I = \{1, 2, \ldots, k\}$ describes just the one-to-one situation. Notice that one cannot omit any of these 2^k cases without violating the theorem: this can be easily seen by considering the colorings which for a given subset $I \subseteq \{1, 2, \ldots, k\}$ associate to every k-element subset $\{x_1, x_2, \ldots, x_k\}_<$ its I-subset, namely the set $\{x_i \mid i \in I\}$.

Not much is known about the growth of the Erdős-Rado numbers $er(k, l)$. For coloring singletons it is well known that $er(1, l) = (l - 1)^2 + 1$, c.f. also [6]. For $k = 2$, coloring two-element sets, the following is known.

Theorem 5 [8] *There exist positive constants $c, c' > 0$ such that for every positive integer l,*

$$2^{cl^2} \leq er(2, l) \leq 2^{2^{c'l^3}} .$$

In this paper we will investigate the growth of the Erdős-Rado numbers for colorings of three-element sets. The only upper bound known for $er(3, l)$ follows from the original proof of Erdős and Rado [4] and is of the order $p_5(cl)$ for some positive constant $c > 0$. Here we will give new bounds for the numbers $er(3, l)$, a lower bound being double-exponential and an upper bound being three-times exponential in l:

Theorem 6 *There exist positive constants $c, c' > 0$ such that for all positive integers $l \geq 4$,*

$$p_2(cl^2) \leq er(3, l) \leq p_3(c'l^5) .$$

2 Proofs

In this section we will prove Theorem 6. Let us consider at first the lower bound for $er(3, l)$. The following observation exhibits some connection of these numbers to the classical Ramsey numbers:

Lemma 1 *For all positive integers l,*

$$er(3, l) \geq r_{l-3}(3, l) . \tag{5}$$

Proof. In order to see this put $n = r_{l-3}(3, l)$ and choose a coloring

$$\Delta : [\{1, 2, \ldots, n - 1\}]^3 \longrightarrow \{1, 2, \ldots, l - 3\}$$

in such a way that for no subset X with $|X| = l$ the set $[X]^3$ is colored monochromatically with respect to Δ. But then, as there are only $l - 3$ colors available, no set X with $|X| = l$ can be colored canonically, as this would require in any nonmonochromatic case at least $l - 2$ colors. By Theorem 3 we have $r_{l-3}(3, l) \geq p_2(cl)$, hence $er(3, l) \geq p_2(cl)$ for $l \geq l_0$. In order to get a lower bound of the order $p_2(cl^2)$ we will give an improvement of the lower bound for the numbers $r_{l-3}(3, l)$.

Here it is convenient to work with the so-called arrow notation. For positive integers k, l, m, n let

$$n \nrightarrow (l)_m^k$$

denote the following statement: for every set X with $|X| = n$ there exists a coloring $\Delta : [X]^k \longrightarrow \{1, 2, \ldots, m\}$ such that for every subset $S \subseteq X$ with $|S| = l$ the restriction $\Delta|_{[S]^k}$ is not a constant coloring.

With this notation the lower bound for $r_m(2, l)$ given in Theorem 2 can be restated as

$$n \not\rightarrow \left(\frac{c \log n}{m}\right)_m^2 . \tag{6}$$

Once good lower bounds for Ramsey numbers $r.(k', .)$ are established, one obtains lower bounds for higher order numbers $r.(k, .)$ for $k > k'$ by the negative stepping up lemma, which we state here for the special case $k = 3$:

Lemma 2 [2] *Let l, m, n be positive integers. Assume that $n \not\rightarrow (l)_m^2$; then it is valid that*

$$2^n \not\rightarrow (l+1)_{2m}^3 .$$

It follows that

$$2^n \not\rightarrow \left(\frac{c' \log n}{m}\right)_{2m}^3$$

by combining (6) and Lemma 2, and hence $r_{l-3}(3, l) \geq p_2(c'l^2)$ for some positive constant $c' > 0$. This implies $er(3, l) \geq p_2(cl^2)$ for some positive constant $c > 0$.

Now we will proceed with the proof of the upper bound. Put $n = r_{68}(4, l^5) \leq p_3(cl^5)$ and let $\Delta \colon [\{1, 2, \ldots, n\}]^3 \longrightarrow \omega$ be an arbitrary coloring. This coloring induces another coloring $\Delta_0 \colon [\{1, 2, \ldots, n\}]^4 \longrightarrow C$, where $|C| \leq 68$, of the 4-element subsets as follows: for elements $x_1, x_2, x_3, x_4 \in \{1, 2, \ldots, n\}$ with $x_1 < x_2 < x_3 < x_4$ define $\Delta_0(\{x_1, x_2, x_3, x_4\}_<)$ to be

mono	if $\Delta(\{x_1, x_2, x_3\}) = \Delta(\{x_2, x_3, x_4\})$
min	if $\Delta(\{x_1, x_2, x_3\}) = \Delta(\{x_1, x_3, x_4\}) \neq \Delta(\{x_2, x_3, x_4\})$
max	if $\Delta(\{x_2, x_3, x_4\}) = \Delta(\{x_1, x_2, x_4\}) \neq \Delta(\{x_1, x_2, x_3\})$
middle	if $\Delta(\{x_1, x_2, x_3\}) = \Delta(\{x_1, x_2, x_4\})$ and $\Delta(\{x_1, x_3, x_4\}) = \Delta(\{x_2, x_3, x_4\})$ but $\Delta(\{x_1, x_2, x_4\}) \neq \Delta(\{x_1, x_3, x_4\})$
min-middle	if $\Delta(S) = \Delta(T)$ if and only if $\{S, T\} = \{\{x_1, x_2, x_3\}, \{x_1, x_2, x_4\}\}$ for all $S, T \in [\{x_1, x_2, x_3, x_4\}]^3$ with $S \neq T$
middle-max	if $\Delta(S) = \Delta(T)$ if and only if $\{S, T\} = \{\{x_1, x_3, x_4\}, \{x_2, x_3, x_4\}\}$ for all $S, T \in [\{x_1, x_2, x_3, x_4\}]^3$ with $S \neq T$
min-max	if $\Delta(S) = \Delta(T)$ if and only if $\{S, T\} = \{\{x_1, x_2, x_4\}, \{x_1, x_3, x_4\}\}$ for all $S, T \in [\{x_1, x_2, x_3, x_4\}]^3$ with $S \neq T$
one-to-one	if $\Delta(S) = \Delta(T)$ implies $S = T$ for all $S, T \in [\{x_1, x_2, x_3, x_4\}]^3$
P	if none of the above are valid, and P is the maximal subset of the set of all pairs $\{(i, j) \mid 1 \leq i < j \leq 4\}$ such that $\Delta(\{x_1, x_2, x_3, x_4\} \setminus \{x_i\}) = \Delta(\{x_1, x_2, x_3, x_4\} \setminus \{x_j\})$ for all $(i, j) \in P$.

The reason for such a choice of the coloring Δ_0 is that the first eight colors should describe the corresponding canonical coloring patterns. Concerning the other colors P, we will see later that for every five-element subset $X \subset \{1, 2, \ldots, n\}$ the set $[X]^4$ can not be monochromatic in a color P with respect to the coloring Δ_0. Indeed, we will see later that we do not need 68 colors to have a well defined coloring Δ_0. It will suffice to use 11 colors, as for the choice of P there are essentially four possibilities.

By choice of n, $n = r_{68}(4, l^5)$, there exists a subset $X \subseteq \{1, 2, \ldots, n\}$ of size $m = l^5$ monochromatic with respect to Δ_0 in some color C. Let $X = \{x_1, x_2, \ldots, x_m\}$ with $x_1 < x_2 < \ldots < x_m$. According to the value of C we distinguish cases:

$C = \mathbf{mono}$: We will show that the set $[X]^3$ is colored monochromatically w.r.t. Δ. We know that for every four elements $y_1 < y_2 < y_3 < y_4$ in X,

$$\Delta(\{y_1, y_2, y_3\}) = \Delta(\{y_2, y_3, y_4\}) . \tag{7}$$

Let $2 \leq i < j < k \leq m$. By (7) it follows that

$$\Delta(\{x_i, x_j, x_k\}) = \Delta(\{x_{i-1}, x_i, x_j\}) = \Delta(\{x_{i-2}, x_{i-1}, x_i\}) = \Delta(\{x_{i-1}, x_i, x_{i+1}\}),$$

hence by induction we have

$$\Delta(\{x_i, x_j, x_k\}) = \Delta(\{x_1, x_2, x_3\}),$$

which implies that the set $[X]^3$ is colored monochromatically with respect to Δ.

$C = \mathbf{min}$: We will show that $[X]^3$ is colored according to a minimum coloring w.r.t. Δ. By definition of Δ_0 we know that for every four elements $y_1 < y_2 < y_3 < y_4$ in X,

$$\Delta(\{y_1, y_2, y_3\}) = \Delta(\{y_1, y_3, y_4\}) \tag{8}$$

and

$$\Delta(\{y_1, y_2, y_3\}) \neq \Delta(\{y_2, y_3, y_4\}) . \tag{9}$$

Let $1 \leq i < j < k < m$ be integers. Then we conclude by (8) that

$$\Delta(\{x_i, x_j, x_k\}) = \Delta(\{x_i, x_k, x_{k+1}\}) = \Delta(\{x_i, x_{i+2}, x_k\}) = \Delta(\{x_i, x_{i+1}, x_{i+2}\}), \tag{10}$$

hence two three-element subsets of X which have the same minimum are colored the same w.r.t. Δ.

Now let $i < j \leq m - 2$ be positive integers. We will show that the sets $\{x_i, x_{i+1}, x_{i+2}\}$ and $\{x_j, x_{j+1}, x_{j+2}\}$ are colored differently w.r.t. Δ. Notice that by (8),

$$\Delta(\{x_i, x_{i+1}, x_{i+2}\}) = \Delta(\{x_i, x_{i+2}, x_{j+1}\}) = \Delta(\{x_i, x_{j+1}, x_{j+2}\}),$$

whereas (9) applied to the elements $x_i, x_j, x_{j+1}, x_{j+2}$ implies

$$\Delta(\{x_i, x_j, x_{j+1}\}) \neq \Delta(\{x_j, x_{j+1}, x_{j+2}\}),$$

hence

$$\Delta(\{x_i, x_{i+1}, x_{i+2}\}) \neq \Delta(\{x_j, x_{j+1}, x_{j+2}\})$$

for all $1 \leq i < j \leq m - 2$, which with (10) gives: for all $S, T \in [X]^3$,

$$\Delta(S) = \Delta(T) \qquad \text{if and only if} \qquad \min S = \min T .$$

$C = \mathbf{max}$: This case corresponds to the maximum coloring, i.e. for all $S, T \in [X]^3$,

$$\Delta(S) = \Delta(T) \qquad \text{if and only if} \qquad \max S = \max T.$$

By symmetry the proof is similar to the case $C = \min$.

$C = \mathbf{middle}$: By definition of the coloring Δ_0 we know that for all elements $y_1 < y_2 < y_3 < y_4$ in X,

$$\Delta(\{y_1, y_2, y_3\}) = \Delta(\{y_1, y_2, y_4\}), \tag{11}$$
$$\Delta(\{y_1, y_3, y_4\}) = \Delta(\{y_2, y_3, y_4\}), \tag{12}$$
$$\Delta(\{y_1, y_2, y_4\}) \neq \Delta(\{y_1, y_3, y_4\}). \tag{13}$$

Hence we have, by (12) and (11) for positive integers $i < j < k \leq m$,

$$\Delta(\{x_i, x_j, x_k\}) = \Delta(\{x_{j-1}, x_j, x_k\}) = \Delta(\{x_{j-1}, x_j, x_{j+1}\}),$$

which means that two three-element subsets of X are colored the same w.r.t. Δ if they have the same middle element. Now, for positive integers $1 < i < j < m$ we will show that the sets $\{x_{i-1}, x_i, x_{i+1}\}$ and $\{x_{j-1}, x_j, x_{j+1}\}$ are colored differently w.r.t. Δ. By (11) and (12) we have

$$\Delta(\{x_{i-1}, x_i, x_{i+1}\}) = \Delta(\{x_{i-1}, x_i, x_{j+1}\}),$$
$$\Delta(\{x_{j-1}, x_j, x_{j+1}\}) = \Delta(\{x_{i-1}, x_j, x_{j+1}\}).$$

On the other hand, (13) applied to the elements $x_{i-1}, x_i, x_j, x_{j+1}$ gives

$$\Delta(\{x_{i-1}, x_i, x_{j+1}\}) \neq \Delta(\{x_{i-1}, x_j, x_{j+1}\})$$

and hence

$$\Delta(\{x_{i-1}, x_i, x_{i+1}\}) \neq \Delta(\{x_{j-1}, x_j, x_{j+1}\}).$$

Therefore, for all $\{y_1, y_2, y_3\}_<, \{z_1, z_2, z_3\}_< \in [X]^3$ with $y_1 < y_2 < y_3$ and $z_1 < z_2 < z_3$,

$$\Delta(\{y_1, y_2, y_3\}) = \Delta(\{z_1, z_2, z_3\}) \quad \text{if and only if} \quad y_2 = z_2. \tag{14}$$

$C = \mathbf{min\text{-}middle}$: By definition of the coloring Δ_0 we know that for all elements $y_1 < y_2 < y_3 < y_4$ in X, and for all three-element subsets $S, T \in [\{y_1, y_2, y_3, y_4\}]^3$ with $S \neq T$,

$$\Delta(S) = \Delta(T) \quad \text{if and only if} \quad \{S, T\} = \{\{y_1, y_2, y_3\}, \{y_1, y_2, y_4\}\}. \tag{15}$$

Hence the color of a three-element set $\{x_i, x_j, x_k\}_<$ depends only on the first two elements $\{x_i, x_j\}$. For $X_0 = X \setminus \{x_m\}$ induce another coloring $\Delta' : [X_0]^2 \longrightarrow \omega$ by

$$\Delta'(\{x_i, x_j\}) = \Delta(\{x_i, x_j, x_m\})$$

for $i < j$. The coloring Δ' is well-defined and has the property that for all two-element subsets $S, T \in [X_0]^2$ with $|S \cap T| = 1$ it follows that $\Delta'(S) \neq \Delta'(T)$. Otherwise the sets $S \cup \{x_m\}$ and $T \cup \{x_m\}$ would be colored the same with respect to Δ, contradicting (15).

We will need the following lemma, which has been proven for $k = 2$ by Babai [1].

Lemma 3 *Let k, n with $k < n$ be positive integers with $n \geq n_0$. Let X be a set with $|X| = n$. Let $\Delta : [X]^k \longrightarrow \omega$ be a coloring such that for all $S, T \in [X]^k$, $|S \cap T| = k - 1$ implies $\Delta(S) \neq \Delta(T)$. Then there exists a one-to-one colored subset $Y \subseteq X$ which satisfies*

$$|Y| \geq \left(\frac{1}{2} - o(1) \right) \cdot [(k!)^2 \cdot n]^{\frac{1}{2k-1}} . \tag{16}$$

Proof. We call an unordered pair $\{B, B'\}$ of k-element subsets $B, B' \in [X]^k$ a j-pair if $|B \cap B'| = j$. A j-pair $\{B, B'\}$ is called bad if $\Delta(B) = \Delta(B')$.

Set $l = \frac{1}{2} \cdot \lfloor [(k!)^2 \cdot n]^{\frac{1}{2k-1}} \rfloor$ and set $m = 2l$. For an m-element subset $M \in [X]^m$ denote by $B_j(M)$ the number of bad j-pairs $\{B, B'\}, B, B' \in [X]^k$. Set $B(M) = \sum_{j=0}^{k-2} B_j(M)$.

It is sufficient to prove the following.

Claim 1. *There exists an m-element subset $M_0 \in [X]^m$ with*

$$B(M_0) \leq (1 + o(1)) \cdot \frac{1}{2} \cdot \left(\frac{m^k}{k!} \right)^2 \cdot \frac{1}{n} .$$

Lemma 3 follows from Claim 1 easily. As

$$(1 + o(1)) \cdot \frac{1}{2} \cdot \left(\frac{m^k}{k!} \right)^2 \cdot \frac{1}{n} \leq \left(\frac{1}{2} + o(1) \right) \cdot m,$$

due to the choice of l, we may destroy each bad pair of M_0 by deleting one vertex. This leaves us with a subset $Y \subseteq M_0$ with at least $(\frac{1}{2} - o(1)) \cdot m \geq (1 - o(1)) \cdot l$ vertices and without any bad pair, i.e. $[Y]^k$ is colored one-to-one.

Proof of Claim 1. For $j = 0, 1, \ldots, k-2$ fix a j-element subset $A \subset X$. Let $r_1, r_2, \ldots, r_t$ be the sizes of the color classes of k-sets containing A. We have clearly

$$r_1 + r_2 + \cdots + r_t = \binom{n-j}{k-j} .$$

On the other hand, as by assumption two different k-element subsets of X, which are colored the same, do not intersect in $k - 1$ elements, we have that

$$r_i \leq \frac{\binom{n-j}{k-j-1}}{k-j} \leq \frac{n^{k-j-1}}{(k-j)!}$$

for every $i = 1, 2, \ldots, t$. This means that the number of bad j-pairs containing a fixed j-set A is bounded from above by

$$\sum_{i=1}^{t} \binom{r_i}{2} \leq \frac{\binom{n-j}{k-j} \cdot (k-j)}{\binom{n-j}{k-j-1}} \cdot \binom{\frac{\binom{n-j}{k-j-1}}{k-j}}{2} \leq \frac{1}{2} \cdot \frac{n^{2k-2j-1}}{[(k-j)!]^2} .$$

It follows that

$$\sum_{M \in [X]^m} B_j(M) \leq \binom{n}{j} \cdot \frac{1}{2} \cdot \frac{n^{2k-2j-1}}{[(k-j)!]^2} \cdot \binom{n - (2k - j)}{m - (2k - j)}$$

$$\leq \frac{1}{2} \cdot \frac{n^{2k-j-1}}{j! \cdot [(k-j)!]^2} \cdot \binom{n - (2k - j)}{m - (2k - j)}$$

and thus the expectation $E(B_j(M))$ of $B_j(M)$, if the set M is chosen uniformly at random among all m-element subsets of X, is

$$
\begin{aligned}
E(B_j(M)) &\leq \frac{1}{2} \cdot \frac{n^{2k-j-1}}{j! \cdot [(k-j)!]^2} \cdot \frac{\binom{n-(2k-j)}{m-(2k-j)}}{\binom{n}{m}} \\
&\leq \frac{1}{2} \cdot \frac{n^{2k-j-1}}{j! \cdot [(k-j)!]^2} \cdot \left(\frac{m}{n}\right)^{2k-j} \\
&\leq \frac{1}{2} \cdot \frac{1}{j! \cdot [(k-j)!]^2} \cdot \frac{m^{2k-j}}{n}.
\end{aligned}
$$

Set $B = B_0 + B_1 + \cdots + B_{k-2}$. Then

$$
E(B(M)) \leq \sum_{j=0}^{k-2} \frac{1}{2} \cdot \frac{1}{j! \cdot [(k-j)!]^2} \cdot \frac{m^{2k-j}}{n} \leq \left(\frac{1}{2} + o(1)\right) \cdot \frac{m^{2k}}{(k!)^2} \cdot \frac{1}{n}
$$

and thus there exists a subset $M_0 \in [X]^m$ which satisfies the Claim. $\square$

The proof of Claim 1 finishes the proof of Lemma 3. $\square$

The coloring Δ' satisfies the assumptions of Lemma 3 for $k = 2$. Thus there exists a subset $X_1 \subseteq X_0$ with $|X_1| \geq \left(\frac{1}{2} - o(1)\right) \cdot (4m)^{1/3} \geq l$, which is colored one-to-one with respect to Δ'. This implies immediately with (15) that for all $\{y_1, y_2, y_3\}_<, \{z_1, z_2, z_3\}_< \in [X_1]^3$ with $y_1 < y_2 < y_3$ and $z_1 < z_2 < z_3$,

$$
\Delta(\{y_1, y_2, y_3\}) = \Delta(\{z_1, z_2, z_3\}) \quad \text{if and only if} \quad y_1 = z_1 \text{ and } y_2 = z_2.
$$

$C = \textbf{middle-max}$: By symmetry this case is similar to the case "min-middle". The proof shows the existence of a subset $X_1 \subseteq X$ with $|X_1| \geq l$ such that for all $\{y_1, y_2, y_3\}_<$, $\{z_1, z_2, z_3\}_< \in [X_1]^3$ with $y_1 < y_2 < y_3$ and $z_1 < z_2 < z_3$,

$$
\Delta(\{y_1, y_2, y_3\}) = \Delta(\{z_1, z_2, z_3\}) \quad \text{if and only if} \quad y_2 = z_2 \text{ and } y_3 = z_3. \tag{17}
$$

$C = \textbf{min-max}$: Hence for each four-element subset $y_1 < y_2 < y_3 < y_4$ of X we have: for all three-element subsets $S, T \in [\{y_1, y_2, y_3, y_4\}]^3$ with $S \neq T$,

$$
\Delta(S) = \Delta(T) \quad \text{if and only if} \quad \{S, T\} = \{\{y_1, y_2, y_4\}, \{y_1, y_3, y_4\}\}. \tag{18}
$$

Thus the color of a three-element subset of X depends only on its minimum and maximum. Put $X_1 = \{x_{2i} \mid 1 \leq i \leq \lfloor m/2 \rfloor\}$ and define another coloring $\Delta_1 : [X_1]^2 \longrightarrow \omega$ by

$$
\Delta_1(\{x_{2i}, x_{2j}\}_<) = \Delta(\{x_{2i}, x_{2i+1}, x_{2j}\})
$$

for all integers $1 \leq i < j \leq \lfloor m/2 \rfloor$. Observe that by (18) this coloring Δ_1 has the property that for all $S, T \in [X_1]^2$ with $\Delta_1(S) = \Delta_1(T)$ and $|S \cap T| = 1$,

$$
\text{either } \min S = \max T \text{ or } \max S = \min T. \tag{19}
$$

We need the following variant of Lemma 3:

Lemma 4 *Let X be a totally ordered set with $|X| = n$, where $n \geq n_0$. Let $\Delta \colon [X]^2 \longrightarrow \omega$ be a coloring with the property: for every three-element subset $\{y_1, y_2, y_3\} \subset X_1$ with $\Delta(\{y_1, y_2\}) = \Delta(\{y_2, y_3\})$, either $y_1 < y_2 < y_3$ or $y_3 < y_2 < y_1$. Then there exists a one-to-one colored subset $Y \subseteq X$ satisfying*

$$|Y| \geq \frac{1}{2} \cdot n^{1/3} . \tag{20}$$

Proof. We will follow a similar approach as in the proof of Lemma 3. Again, we call an unordered pair $\{\{x_1, x_2\}_<, \{x_3, x_4\}_<\}$ of edges bad if $\Delta(\{x_1, x_2\}) = \Delta(\{x_3, x_4\})$. According to our assumption on Δ, this may happen either if $x_2 = x_3$ or $x_4 = x_1$, i.e. if the edges are consecutive in which case we refer to a bad 1-pair, or if all vertices x_1, x_2, x_3, x_4 are distinct, in which case we talk about bad 0-pairs.

Let $r_1, r_2, \ldots, r_t$ be the sizes of the color classes. Clearly,

$$\sum_{i=1}^{t} r_i = \binom{n}{2} .$$

Each color class is a disjoint union of increasing paths and thus contains at most $n - 1$ edges, i.e. $r_i \leq n - 1$ for $i = 1, 2, \ldots, t$. Therefore we have at most

$$\sum_{i=1}^{t} (r_i - 1) \leq \binom{n}{2} - \frac{n}{2} < \frac{n^2}{2}$$

bad 1-pairs and less than

$$\sum_{i=1}^{t} \binom{r_i}{2} \leq \frac{n}{2} \cdot \binom{n-1}{2} \leq \frac{n^3}{4}$$

bad 0-pairs. Consider a random subset M of vertices, where each vertex is chosen independently of the others with probability $p = n^{-2/3}$. Let $B(M)$ be the random variable, which counts the number of bad pairs in a set M. Then

$$E(B(M)) < p^3 \cdot \frac{n^2}{2} + p^4 \cdot \frac{n^3}{4}$$

and thus by Markov's inequality for any $\varepsilon > 0$

$$\mathrm{Prob}\left(B(M) \geq (1 + \varepsilon) \cdot \left(p^3 \cdot \frac{n^2}{2} + p^4 \cdot \frac{n^3}{4} \right) \right) < \frac{1}{1 + \varepsilon} .$$

Set $\varepsilon = 1/8$; as

$$\frac{p^3 \cdot n^2}{2} + \frac{p^4 \cdot n^3}{4} < \left(\frac{1}{4} + \varepsilon \right) \cdot pn$$

for $n \geq n_0 > 64$ we infer that

$$\mathrm{Prob}\left(B(M) \leq (\frac{1}{2} + \varepsilon) \cdot pn \right) \geq \frac{\varepsilon}{1 + \varepsilon} .$$

On the other hand, by Chernoff's inequality

$$\mathrm{Prob}\left(|M| \geq (1 - \varepsilon) \cdot pn \right) = 1 - o(1)$$

and thus if $n \geq n_1$, there exists a subset $M_0 \subseteq X$ such that

$$|M_0| \geq \frac{7}{8} \cdot pn \quad \text{and} \quad B(M_0) \leq \frac{3}{8} \cdot pn.$$

After deleting one vertex from each bad pair we obtain a one-to-one colored subset $Y \subseteq M_0$ with

$$|Y| \geq \frac{7}{8} \cdot pn - \frac{3}{8} \cdot pn \geq \frac{1}{2} \cdot pn \geq \frac{1}{2} \cdot n^{1/3}$$

which finishes the proof of Lemma 4. $\square$

By (19) the coloring Δ_1 satisfies the assumptions of Lemma 4. Thus there exists a subset $X_2 \subseteq X_1$ with $|X_2| \geq \frac{1}{2} \cdot \lfloor m/2 \rfloor)^{1/3} \geq l$, which is colored one-to-one w.r.t. Δ_1. This implies with (18) that for all $\{y_1, y_2, y_3\}, \{z_1, z_2, z_3\} \in [X_2]^3$ with $y_1 < y_2 < y_3$ and $z_1 < z_2 < z_3$,

$$\Delta(\{y_1, y_2, y_3\}) = \Delta(\{z_1, z_2, z_3\}) \quad \text{if and only if} \quad y_1 = z_1 \text{ and } y_3 = z_3.$$

$C = \textbf{one-to-one}$: By assumption we know that all three-element subsets of a four-element subset of X are colored pairwise differently. This means that for all sets $S, T \in [X]^3$ with $|S \cap T| = 2$ we have $\Delta(S) \neq \Delta(T)$. Now apply Lemma 3 with $k = 3$ to the restriction $\Delta|_{[X]^3}$. This gives a subset $X_1 \subseteq X$ with $|X_1| \geq \left(\frac{1}{2} - o(1)\right) \cdot (36m)^{1/5} \geq l$ which is colored one-to-one w.r.t. Δ.

$C = P$: Suppose that $[X]^4$ with $|X| \geq 5$ is colored with respect to Δ_0 in color P. So for all elements $y_1 < y_2 < y_3 < y_4$ in X and for all pairs $(i, j) \in P$ with $i < j$,

$$\Delta(\{y_1, y_2, y_3, y_4\} \setminus \{y_i\}) = \Delta(\{y_1, y_2, y_3, y_4\} \setminus \{y_j\}), \tag{21}$$

where P is maximal and the pattern of the coloring Δ on these four elements could not be covered by the cases handled before.

We will show that in this case $|X| \leq 4$. Notice that P is nonempty, as otherwise we are in the "one-to-one" case. Moreover, we have $(1, 4) \notin P$, as otherwise we are in the "monochromatic" situation. Then as $(1, 4) \notin P$ transitivity implies that $|P| < 4$. Hence, we do have $1 \leq |P| \leq 3$. According to the value of $|P|$ we distinguish three cases.

Assume first that $|P| = 3$. Consider a graph with vertex set $V = \{1, 2, 3, 4\}$ and edge set E, where $\{i, j\}_< \in E$ if and only if $(i, j) \in P$. If the three edges determine a complete graph on three vertices, not including both vertices 1 and 4, this results in the "min" - resp. "max" case. Otherwise, the three edges give a connected graph on V. But then transitivity implies that $(1, 4) \in P$, a contradiction.

Second, suppose that $|P| = 1$, i.e., in each four-element subset exactly two three-element subsets determined by some pair $(i, j) \in P$ are colored the same. Then as $(1, 4) \notin P$ there are at most five possible choices for (i, j), namely $(1, 2), (1, 3), (2, 3), (2, 4), (3, 4)$. As $(1, 2)$ yields the "middle-max" case, $(2, 3)$ the "min-max" case and $(3, 4)$ the "middle-min" case, we are therefore left with two possible pairs, $(1, 3)$ and $(2, 4)$. Consider five elements $x_1 < x_2 < x_3 < x_4 < x_5$ and assume first the case $(i, j) = (1, 3)$. Then it follows by (21) that

$$\Delta(\{x_3, x_4, x_5\}) = \Delta(\{x_2, x_3, x_5\}),$$
$$\Delta(\{x_3, x_4, x_5\}) = \Delta(\{x_1, x_3, x_5\}),$$
$$\Delta(\{x_2, x_3, x_5\}) = \Delta(\{x_1, x_2, x_5\}),$$

thus $\Delta(\{x_1, x_3, x_5\}) = \Delta(\{x_1, x_2, x_5\})$, but this would imply $(2, 3) \in P$, which is not possible. Similarly, for the case $(i, j) = (2, 4)$ we infer

$$\Delta(\{x_1, x_2, x_3\}) = \Delta(\{x_1, x_3, x_4\}) = \Delta(\{x_1, x_3, x_5\}) = \Delta(\{x_1, x_4, x_5\}),$$

but $\Delta(\{x_1, x_3, x_5\}) = \Delta(\{x_1, x_4, x_5\})$ would give again $(2, 3) \in P$.

Suppose next that $|P| = 2$. We know that $(1, 4) \notin P$ and by transitivity it follows that there are only two possibilities for P, namely either $P = \{(1, 2), (3, 4)\}$ or $P = \{(1, 3), (2, 4)\}$. The case $P = \{(1, 2), (3, 4)\}$ yields the "middle" case, hence cannot occur. Concerning the second possibility, i.e. $P = \{(1, 3), (2, 4)\}$ we inferred above (when $|P| = 1$) that $(1, 3) \in P$ implies $(2, 3) \in P$. Therefore P cannot have exactly two elements.

This shows that $[X]^4$ cannot be colored monochromatically by some color P w.r.t. Δ_0. This finishes the proof of Theorem 6.

3   Final Comments

We have seen in Lemma 1 that $er(3, l) \geq r_{l-3}(3, l)$. It would be interesting to know how far the numbers $r_{l-3}(3, l)$ and $er(3, l)$ are apart from each other. In particular, it is not clear whether there exists a positive integer k such that $er(3, l)$ can be bounded from above by $p_2(c \cdot l^k)$ for some positive constant $c > 0$.

References

[1] L. Babai, An anti-Ramsey Theorem, *Graphs and Combinatorics* **1**, 1985, 23–28.

[2] P. Erdős, A. Hajnal, A. Maté and R. Rado, *Combinatorial Set Theory: Partition Relations for Cardinals*, Académiai Kiadó, Budapest, 1984.

[3] P. Erdős, A. Hajnal and R. Rado, Partition relations for cardinal numbers, *Acta Math. Acad. Sci. Hung.* **16**, 1965, 93–196.

[4] P. Erdős and R. Rado, A combinatorial theorem, *J. London Math. Soc.* **25**, 1950, 249–255.

[5] P. Erdős and E. Szemerédi, On a Ramsey type theorem, *Periodica Mathematica Hungarica* **2**, 1972, 295–299.

[6] R. L. Graham, B. L. Rothschild and J. Spencer, *Ramsey theory*, 2nd edition, John Wiley, New York, 1989.

[7] H. Lefmann, A note on Ramsey numbers, *Studia Scientiarum Mathematicarum Hungarica* **22**, 1987, 445–446.

[8] H. Lefmann and V. Rödl, On canonical Ramsey numbers for complete graphs versus paths, 1990, to appear in *J Combin. Theory Ser B*.

[9] F. P. Ramsey, On a problem of formal logic, *Proc. London Math. Soc. (2)* **30**, 1930, 264–286.

Large Subgroups of Infinite Symmetric Groups

DUGALD MACPHERSON

School of Mathematical Sciences
Queen Mary and Westfield College
Mile End Road
London E1 4NS
England

Abstract

This is a survey of work by many people on various notions of largeness for subgroups of infinite symmetric groups. The primary concern is with maximal subgroups of infinite symmetric groups, of which several new examples are given. Subgroups of small index in symmetric groups (and in other permutation groups) are considered, as are questions about covering symmetric groups with families of subgroups. Examples are given of maximal subgroups of other large permutation groups, such as $GL(\kappa, F)$ (where κ is an infinite cardinal) and $\text{Aut}(\mathbb{Q}, \leq)$. The paper concludes with a discussion of other notions of largeness in symmetric groups, such as oligomorphic, Jordan, maximal closed, and various strong transitivity conditions on infinite subsets.

1 Introduction

This paper is a survey of results, mostly recent, on maximal subgroups of infinite symmetric groups. For the bulk of the paper, 'large' will mean 'maximal', where of course a *maximal subgroup* of a group G is a maximal *proper* subgroup of G. However in §2, 'large' will mean 'of small index', and in §7 other notions of large, mostly symmetry conditions, will be discussed. Most of the results we mention have already appeared in the literature, so we omit many proofs.

We shall adopt some of the notation of [47]. We fix a set Ω of infinite cardinality κ, and let $S := \text{Sym}(\Omega)$, the full symmetric group on Ω. If $G \leq S$ and $\alpha \in \Omega$ then G_α is the stabiliser of α, and if $\Gamma \subseteq \Omega$ then $G_{\{\Gamma\}}, G_{(\Gamma)}$ denote respectively the setwise and pointwise stabilisers of Γ in G; also, G^Γ denotes the permutation group induced on Γ by $G_{\{\Gamma\}}$. If $g \in S$ then $\text{Supp}(g) := \{\alpha \in \Omega : \alpha g \neq \alpha\}$, $\text{Fix}(g) := \Omega \backslash \text{Supp}(g)$, and $\deg(g) := |\text{Supp}(g)|$. If Γ is an infinite set and $\Delta \subseteq \Gamma$ we say that Δ is a *moiety* of Γ if $|\Delta| = |\Gamma| = |\Gamma \backslash \Delta|$.

The normal subgroups of S were first classified by Baer in [6] in the general case (and earlier by J. Schreier and S. Ulam in the countable case), and are described in [65] (where a proof of the classification is given). For every infinite cardinal $\lambda \leq \kappa$ let $B_\lambda := \{g \in S : \deg(g) < \lambda\}$. Also let $\text{Alt}(\Omega)$ be the group of all even permutations in B_ω. (Note that the isomorphism types of these groups vary with κ.) Then $\text{Alt}(\Omega)$ and the B_λ are all normal in S, and indeed are the only proper non-trivial normal subgroups of S. Quotients of successive groups in this chain are all simple. These results have been extended by Bertram, Droste, Göbel, Moran, and others, culminating in Moran [53]. They have complete results on the

N.W. Sauer et al. (eds.), Finite and Infinite Combinatorics in Sets and Logic, 249–278.

following question: given $g, h \in S$ with $h \in S \backslash B_\kappa$, what is the smallest number n such that g can be written as a product of n conjugates of h? (In general the answer is 4, but for certain h this number can be reduced). It is also known that S has no non-trivial outer automorphisms. There is, I believe, no complete description of the outer automorphisms of the quotients $B_{\lambda+}/B_\lambda$ — there are some non-obvious examples.

One way of saying a subgroup of S is large is to say it has small index in S. Here, since $|S| = 2^\kappa$, 'small index' will usually mean 'of index less than 2^κ'. Subgroups of S of small index were first considered by Semmes [66]. They are also examined in [19] and [69]. In §2 we survey these results, and discuss generalisations to subgroups of small index in other permutation groups.

In §3 we consider some questions which are clearly related to large subgroups of S. If G is a group of uncountable cardinality, let $c(G)$ be the smallest cardinality of a well-ordered chain of proper subgroups of G with union G. Also, let $c_0(G)$ be the size of the smallest set of cosets of proper subgroups whose union is G, and let $c_1(G)$ be the size of the smallest set of proper subgroups of G with union G. We give in §3 some results from [47] on the sizes of these numbers for $G = S$. In particular, we sketch a proof that $c(S) > \kappa$. This has an application, also mentioned in §3, to classifying the supplements of the normal subgroups of S.

We turn in §§4 and 5 to our main topic, maximal subgroups of S. In §4 we describe some of the known examples and discuss the corresponding permutation representations on the coset spaces. We also give a few examples of maximal subgroups. Then in §5 we turn to the question: is every proper subgroup of S contained in a maximal subgroup of S? In general, by a result of Baumgartner, Shelah and Thomas [9], the answer is no, but there are positive results in restricted cases.

Many of these questions are natural with other familiar permutation groups playing the role of S. In particular, if κ is an infinite cardinal and F is a field, many of the above results go through for $GL(\kappa, F)$, the group of all non-singular linear transformations of a κ-dimensional vector space over F. Some of these results are discussed in §6. We also give there some examples of maximal subgroups of the group of all order-automorphisms of the rationals.

In §7 we turn to other notions of large. These are notions based on S as a permutation group rather than as an abstract group. I mention various notions, such as 'oligomorphic' (dense in the automorphism group of an ω-categorical structure), Jordan, maximal closed, plus other strong transitivity conditions.

I conclude this section with a little general background. First, recall that S carries a natural topology; there is a basis of open sets consisting of the collection of all cosets of the pointwise stabilisers in S of finite subsets of Ω. Under this topology, S is a topological group. A subgroup of S is closed in this topology if and only if it is the full automorphism group of some first-order structure of Ω (and this is what we mean when we refer to a *closed* subgroup of S). If $\kappa = \omega$ then the topology has a natural metric: identify Ω with ω, and for $f, g \in S$, put $d(f, g) = 1/2^n$ if $f(x) = g(x)$ and $f^{-1}(x) = g^{-1}(x)$ for all $x < n$, but $f(n) \neq g(n)$ or $f^{-1}(n) \neq g^{-1}(n)$. With this metric S is a complete metric space, so we can apply methods of Baire category to S.

If $G \leq S$ and $\Gamma \subseteq \Omega$, then we say that Γ is *full* for G if $G^\Gamma = \text{Sym}(\Gamma)$. Following [19], if $\Gamma \subseteq \Omega$ we often identify $S_{(\Omega \backslash \Gamma)}$ with $\text{Sym}(\Gamma)$ (but note that the assertion that $\text{Sym}(\Gamma) \leq G$ is stronger than that Γ is full for G). I list next a few elementary lemmas that we use repeatedly.

Lemma 1.1 ([19]) *Let* $\Gamma_1, \Gamma_2 \subseteq \Omega$ *and suppose* $|\Gamma_1 \cap \Gamma_2| = \mathrm{Min}\{|\Gamma_1|, |\Gamma_2|\}$. *Then* $\mathrm{Sym}(\Gamma_1 \cup \Gamma_2) = \langle \mathrm{Sym}(\Gamma_1), \mathrm{Sym}(\Gamma_2) \rangle$.

Lemma 1.2 ([55; Note 3(iii) of §4]) *If* $G \leq S$ *and all moieties of* Ω *are full for* G *then* $G = S$.

Lemma 1.3 ([46; Lemma 2.3]) *If* $G \leq S$ *and* Γ_1, Γ_2 *are subsets of* Ω *that are full for* G, *and if* $|\Gamma_1 \cap \Gamma_2| = \kappa$ *and* $\Gamma_1 \cup \Gamma_2 = \Omega$ *then* $G = S$.

Lemma 1.4 ([46; Lemma 2.4]) *If* $G \leq S$ *and some moiety of* Ω *is full for* G *then there is* $g \in S$ *such that* $\langle G, g \rangle = S$.

Also, recall that a permutation group G on a set X is *primitive* if there is no non-trivial G-invariant partition of X. It is easily seen that a subgroup of a group G is maximal in G if and only if the action of G by right multiplication on the right cosets of the subgroup is primitive (this is part of the motivation for investigating maximal subgroups of S).

Recall that a *filter* on Ω is a subset of $\mathcal{P}(\Omega)$ containing Ω but not $\emptyset$, and closed upwards and under finite intersections. An *ideal* on Ω is a subset of $\mathcal{P}(\Omega)$ containing $\emptyset$ but not Ω, and closed downwards and under finite unions. Note that there is a natural duality between filters and ideals (to obtain an ideal from a filter, replace each set in the filter by its complement). An *ultrafilter* is a maximal filter. If $\mathcal{S} \subseteq \mathcal{P}(\Omega)$ and $G \leq S$ then

$$G_{\{\mathcal{S}\}} := \{g \in G : \forall \Gamma \subseteq \Omega, \Gamma \in \mathcal{S} \Leftrightarrow \Gamma g \in \mathcal{S}\}.$$

If $\mathcal{F}$ is a filter then define $S_{(\mathcal{F})} := \{g \in S : \mathrm{Fix}(g) \in \mathcal{F}\}$. It is immediate that $S_{(\mathcal{F})} \trianglelefteq S_{\{\mathcal{F}\}}$.

If $\mathcal{M}$ is a relational structure and $g \in \mathrm{Sym}(M)$ then g is a *piecewise-automorphism* of $\mathcal{M}$ if there is $n < \omega$ and a partition $M = M_1 \dot\cup \ldots \dot\cup M_n$ such that $g|_{M_i}$ is an embedding for each i. As noted by Stoller [72], the set of all piecewise automorphisms of $\mathcal{M}$ forms a group. We denote it by $PA(\mathcal{M})$.

If $G \leq S$ and k is a positive integer we say that G is *k-homogeneous* (respectively *k-transitive*) if it is transitive on the set of unordered k-subsets of Ω (respectively on the set of ordered k-subsets). We say that G is *highly homogeneous* (respectively *highly transitive*) if it is k-homogeneous (respectively k-transitive) for all $k < \omega$.

For general model-theoretic background we refer to [16]. We adopt the convention that $\mathcal{M}$ will in general stand for a first-order structure with domain M.

I am grateful to many people for conversations and correspondence on the material of this paper. In particular, I mention Samson Adeleke, Paul Bankston, Peter Cameron, David Evans, Peter Neumann, Juris Steprans, Cheryl Praeger and Simon Thomas.

2 Subgroups of Small Index

In this section I sketch some of the main results on subgroups of small index of S. I also discuss briefly similar results for other groups, particularly automorphism groups of ω-categorical structures. There is a rapidly growing literature on this last topic, and the account here will be very cursory.

First, I remark that results on subgroups of small index of finite symmetric groups go back a long way. Clearly S_n has intransitive and imprimitive subgroups of quite small index; for example, if n is even then the stabiliser of a partition of $\{1, \ldots, n\}$ into two parts of equal size has index $\binom{n}{n/2}/2$ in S_n. However, a theorem of Bochert (see [12]) gives that any primitive subgroup of S_n, other than A_n, has index in S_n at least $((n+1)/2)!$. Using the classification of finite simple groups and the O'Nan–Scott Theorem, Cameron [12] has given much better bounds.

The following is the first result on subgroups of S of small index. It was first proved by Semmes [66] and then independently proved by Neumann. A short proof can be found in [19].

Theorem 2.1 *If $\kappa = \omega$ and $G \leq S$ with $|S : G| < 2^\omega$ then there is a finite $\Delta \subseteq \Omega$ such that $S_{(\Delta)} \leq G \leq S_{\{\Delta\}}$.*

The proof given in [19] runs roughly as follows. First, partition Ω into disjoint moieties $\Sigma_i (i < \omega)$ and consider the subgroup of G which fixes setwise each Σ_i; this will have small index in the corresponding subgroup of S. We find a moiety Σ of Ω such that $\mathrm{Sym}(\Sigma) \leq G$ (Σ will be one of the Σ_i). Now let $\{\Gamma_\lambda : \lambda < 2^\omega\}$ be a family of almost disjoint moieties of Σ (*almost disjoint* means that pairwise they have finite intersection). For each $\lambda < 2^\omega$ let x_λ be the involution in S interchanging Γ_λ with $\Omega \backslash \Sigma$ and fixing $\Sigma \backslash \Gamma_\lambda$ pointwise. By the pigeon-hole principle there are distinct $\mu, \nu < 2^\omega$ such that $x := x_\mu x_\nu^{-1} = x_\mu x_\nu \in G$. Now $\Sigma = \Omega \backslash \Gamma_\nu x_\nu$. Put $\Sigma' := \Omega \backslash \Gamma_\mu x_\nu$. Then $\Sigma x = \Sigma'$, so $\mathrm{Sym}(\Sigma') \leq G$. Now $\Sigma \cup \Sigma' = \Omega \backslash (\Gamma_\mu \cap \Gamma_\nu) x_\nu$, and $\Sigma \cap \Sigma' = \Omega \backslash (\Gamma_\mu \cup \Gamma_\nu) x_\nu$. Put $\Gamma := (\Gamma_\mu \cap \Gamma_\nu) x_\nu$. Then Γ is finite, $\Sigma \cup \Sigma' = \Omega \backslash \Gamma$ and $|\Sigma \cap \Sigma'| = \omega$, and by Lemma 1.1 we have that $G \geq S_{(\Gamma)}$. Now choose Δ to be a minimal such set Γ.

A similar proof yields the following (no longer assuming $\kappa = \omega$).

Theorem 2.2 *([18; Theorem 2]) If $G \leq S$ and $|S : G| \leq \kappa$, then there is $\Delta \subset \Omega$ with $|\Delta| < \kappa$ such that $S_{(\Delta)} \leq G$.*

To see that we could not expect here both inequalities of Theorem 2.1, consider $G := B_\omega \cdot S_{(\Delta)}$ where Δ is given in Theorem 2.2. A closely related result, also from [19], is the following.

Theorem 2.3 *Suppose $G \leq S$ and $|S : G| < 2^\kappa$. Put*

$$\mathcal{F} := \{\Gamma \subseteq \Omega : \exists \Delta \subseteq \Gamma \text{ with } |\Omega \backslash \Delta| = \kappa \text{ and } S_{(\Delta)} \leq G\}.$$

Then

(i) *$\mathcal{F}$ is a filter on Ω and contains a moiety of Ω;*

(ii) *$S_{(\mathcal{F})} \leq G \leq S_{\{\mathcal{F}\}}$.*

This theorem focusses attention on the index in S of the stabiliser of a filter, and this is tackled in [69]. The authors first note the following. Let $^\Omega\Omega$ denote the set of functions from Ω to Ω, and if $\mathcal{F}$ is a filter on Ω and $f, g \in {}^\Omega\Omega$ write $f \approx_\mathcal{F} g$ if $\{\alpha \in \Omega : f(\alpha) = g(\alpha)\} \in \mathcal{F}$.

Theorem 2.4 ([68; Theorem 4]) *Let $\mathcal{F}$ be a filter on Ω containing a moiety of Ω. Then*

$$|S : S_{\{\mathcal{F}\}}| = |\{f/\approx_\mathcal{F}: f \in {}^\Omega\Omega\}|.$$

The last two results, together with some calculations with filters, then yield

Theorem 2.5 ([68; Corollaries 2 and 3])

(i) *Assume GCH and that no inner model of ZFC contains a measurable cardinal. Let $G \leq S$. Then $|S : G| < 2^\kappa$ if and only if there is $\Delta \subseteq \Omega$ such that $|\Delta| < \mathrm{cf}(\kappa)$ and $S_{(\Delta)} \leq G$.*

(ii) *Suppose κ is measurable. Then there is a forcing extension in which κ is regular and in which there is a group $G \leq S$ such that $|S : G| < 2^\kappa$ and $S_{(\Delta)} \not\leq G$ for all $\Delta \subseteq \Omega$ with $|\Delta| < \kappa$.*

The paper [19] sparked off a lot of work on ω-categorical structures. Recall that a first-order structure $\mathcal{M}$ is ω-categorical if it is countably infinite and every countable structure which is elementary equivalent to $\mathcal{M}$ is isomorphic to $\mathcal{M}$. The following fundamental theorem is due independently to Engeler [21], Ryll-Nardzewski [64] and Svenonius [73]. We refer to Chang and Keisler [16] for any unfamiliar model-theoretic notions.

Theorem 2.6 *Let $\mathcal{M}$ be a countably infinite structure. Then the following are equivalent:*

(i) *$\mathcal{M}$ is ω-categorical;*

(ii) *for all $n < \omega$, each complete n-type of $\mathrm{Th}(\mathcal{M})$ is isolated;*

(iii) *for all $n < \omega$, $\mathrm{Th}(\mathcal{M})$ has just finitely many complete n-types over $\emptyset$;*

(iv) *$\mathrm{Aut}\, \mathcal{M}$ has finitely many orbits on M^n for each $n < \omega$.*

By the equivalence (i)$\Leftrightarrow$(iv), ω-categorical structures have very rich symmetry groups, so there are natural questions, first raised by Cherlin, about the extent to which an ω-categorical structure $\mathcal{M}$ is determined by the abstract group structure of its automorphism group (if we ignore questions about the choice of language, it is totally determined by the *permutation* group structure of $\mathrm{Aut}\, \mathcal{M}$). If $\kappa = \omega$ then the group S is the automorphism group of an ω-categorical structure, namely a set with no relations. Now Theorem 2.1 says that any transitive action of S on a countable set Σ is, roughly speaking, an action on the set of S-translates of a finite set $\Delta \subseteq \Omega$; for if $\alpha \in \Sigma$ then S_α has countable index in S, so we can apply Theorem 2.1. In particular, any countable structure with a transitive automorphism group isomorphic to S is ω-categorical, and interpretable in a pure set. The most general statement along these lines was worked out independently by David Evans and Daniel Lascar, and is given as Theorem 2.7 below. First, we remark that if $\mathcal{M}$ and $\mathcal{N}$ are ω-categorical structures, then $\mathcal{M}$ and $\mathcal{N}$ are biinterpretable if and only if $\mathrm{Aut}\, \mathcal{M}$ and $\mathrm{Aut}\, \mathcal{N}$ are isomorphic as topological groups (this is given in [3]; it follows easily from their definition of biinterpretability). Next, note that the proof of Lemma 2.10 of Rubin [62] gives that any continuous isomorphism between the automorphism groups of two ω-categorical structures is a homeomorphism. Also, note that if G is a closed permutation group on a countably infinite set then every open subgroup of G has countable index (this follows from the definition of the topology). Now say that a countable structure $\mathcal{M}$ has the *small index property* if, for every $G \leq$ Aut $\mathcal{M}$ with $|$Aut $\mathcal{M} : G| < 2^\omega$ there is finite $A \subset \mathcal{M}$ such that (Aut $\mathcal{M})_{(A)} \leq G$ (so this says that the subgroups of Aut $\mathcal{M}$ of index less than 2^ω are precisely the open subgroups). It is easily seen that if $\mathcal{M}$ and $\mathcal{N}$ are ω-categorical structures and $\phi :$ Aut $\mathcal{M} \to$ Aut $\mathcal{N}$ is an isomorphism of abstract groups, then if $\mathcal{M}$ has the small index property then ϕ is continuous. Putting these observations (also discussed in [31]) together we obtain the following theorem.

Theorem 2.7 *If $\mathcal{M}$ and $\mathcal{N}$ are ω-categorical structures whose automorphism groups are isomorphic as abstract groups, and if $\mathcal{M}$ has the small index property, then $\mathcal{M}$ and $\mathcal{N}$ are biinterpretable.*

Several ω-categorical structures other than a pure set have been shown to have the small index property. These include: infinite dimensional projective and affine spaces over a finite field (Evans [22]) possibly enriched by a non-degenerative symplectic, orthogonal or unitary form (Evans [24]); ($\mathbb{Q}$, $\leq$) and the countable atomless Boolean algebra (Truss [76]); all ω-categorical abelian groups ([26]); certain countable trees in the poset-theoretic sense ([20] but note that the proof given in [19; Theorem 4.1] has a small gap, as we do not really prove that $H_{(C)}^{T \setminus C} = \prod_{c \in C}$ Aut(T_C); this is easily filled in by an argument using almost disjoint sets).

On the basis of some of these examples I rashly conjectured in [44] that every ω-categorical structure has the small index property. Udi Hrushovski found a counterexample to this, described in [32]. Further interesting work on these topics can be found in: Rubin [62], where the general question of recovering an ω-categorical structure from its automorphism group is examined (and the small index property is proved for subgroups of countable index which are $\forall\exists$-definable in the abstract group language). Evans and Hewitt [25] (where two ω-categorical structures are exhibited whose automorphism groups are isomorphic as abstract groups but not as topological groups); Lascar [37], where an analogue of the small index property is proved for saturated models of size ω_1 of ω-stable theories; Lascar [38], where a version of the small index property is proved for countable saturated models of almost strongly minimal theories (this implies the results in [19] and [22], at least for subgroups of countable index); Möller [52], where subgroups of small index are examined for the automorphism groups of countable *graph-theoretic* trees (which are not ω-categorical).

3   Covering the Symmetric Group by Subgroups

The following question was asked some years ago by P. Deutsch, and again later by Thomas Forster: if $\kappa = \omega$, is S the union of a countable chain (ordered under inclusion) of proper subgroups? More generally, one can define the three invariants $c(G), c_0(G), c_1(G)$ of an uncountable group G as in the introduction. Note that $c(G)$ is the smallest size of *any* chain of proper subgroups of G with union G. Also, as noted in [47], we always have $c(G) \leq$ cf$(|G|)$ (if $|G| = \lambda$, write $G = \{g_\mu : \mu < \lambda\}$, choose cofinal $E \subseteq \lambda$ of size cf(λ), and for $\mu \in E$ put $G_\mu := \langle g_\nu : \nu < \mu \rangle$; then $G = \cup(G_\mu : \mu \in E)$, and the G_μ all have smaller cardinality than G). Also $c_0(G) \leq c_1(G) \leq c(G)$. In [47] the following results are proved on these numbers for the case $G = S$.

Theorem 3.1

(i) $\kappa < c(S) \leq \mathrm{cf}(2^\kappa)$,

(ii) $c_0(S) = \kappa$,

(iii) $c_1(S) \leq \kappa^\omega$, and if κ is uncountable and regular, $c_1(S) = \kappa$,

(iv) if $\kappa = \omega$ then $c_1(S) > \kappa$.

The most substantial of these are (iv) and (ii). I will give the proof that $c(S) > \kappa$, since the method of argument may have other applications (see §6).

So suppose that S is the union of a chain $(G_\lambda : \lambda < \kappa)$ of proper subgroups. Then clearly, for each $\lambda < \kappa$, there is no $g \in S$ such that $\langle G_\lambda, g \rangle = S$. Hence, by Lemma 1.4, for each $\lambda < \kappa$ there is no full moiety for G_λ. Let $\Omega := \dot\bigcup(\Omega_\lambda : \lambda < \kappa)$ be a partition of Ω into moieties, and for each $\lambda < \kappa$ choose $g_\lambda \in \mathrm{Sym}(\Omega_\lambda)\backslash G^{\Omega_\lambda}$. Let $g \in S$ be the unique permutation inducing g_λ on Ω_λ for each $\lambda < \kappa$. Then $g \in S\backslash\bigcup(G_\lambda : \lambda < \kappa)$, a contradiction.

Alan Mekler, and independently Simon Thomas, have pointed out that, for $\kappa = \omega$, it is consistent with ZFC that $c(S) < \mathrm{cf}(2^\kappa)$. Essentially, let L be the constructible universe, pick any cardinal μ of uncountable cofinality in L, and adjoin μ Cohen reals. In the resulting model of set theory cardinalities and cofinalities are unchanged, and S is a union of a λ-chain for all $\lambda \leq \mu$ such that $\mathrm{cf}(\lambda) > \omega$.

The proof that $c(S) > \kappa$ may have several applications, and I mention one. Scott [65] asked whether, for $\lambda \leq \kappa$, the normal subgroup B_λ has a complement in S. Rabinovich [58] proved that it does not, and later Semmes [67], using GCH, characterised the *supplements* of B_λ in S, that is, the subgroups G of S such that $B_\lambda \cdot G = S$. Later Neumann gave a characterisation of these supplements just in ZFC. The result is the following; its proof uses Theorem 3.1(i) above, and is given in [47].

Theorem 3.2 *If λ is a cardinal with $\omega \leq \lambda \leq \kappa$, and $G \leq S$, then $B_\lambda \cdot G = S$ if and only if there is $\Delta \subset \Omega$ with $|\Delta| < \lambda$ such that $G^{(\Omega\backslash\Delta)} = \mathrm{Sym}(\Omega\backslash\Delta)$.*

Using GCH, Semmes was able to strengthen the conclusion $G^{(\Omega\backslash\Delta)} = \mathrm{Sym}(\Omega\backslash\Delta)$ above to $\mathrm{Sym}(\Omega\backslash\Delta) \leq G$. In fact, GCH is not needed for this refinement, and a necessary and sufficient condition for the stronger conclusion to hold for all κ is the following: whenever μ, ν are cardinals with $\mu < \nu$, $2^\mu < 2^\nu$. In particular, the supplements of B_ω are precisely the groups G such that there is a finite $\Delta \subseteq \Omega$ with $S_{(\Delta)} \leq G \leq S_{\{\Delta\}}$. For $\kappa = \omega$ this was folklore. Note that for $\kappa = \omega$ such groups are precisely the subgroups of S of index less than 2^ω, and indeed this particular case of Theorem 3.2 can be deduced from Theorem 2.1.

4 Maximal Subgroups of S

In this section I give a number of examples of maximal subgroups of S. Most of these are taken from [47] and [10] but there are a few new examples. I have recently realised that several results on this topic, given without attribution in [47], were proved much earlier by F. Richman in [60].

By way of background, I begin by stating a theorem of Liebeck, Praeger and Saxl on maximal subgroups of finite symmetric groups. It is given in [40], and is based on the O'Nan–Scott Theorem ([41]; see also [12] and [5]). If n is a positive integer we denote by S_n and A_n the symmetric and alternating groups respectively of degree n.

Theorem 4.1 *Let n be a positive integer, let X be A_n or S_n, and let G be a maximal subgroup of X with $G \neq A_n$. Then G is of one of the following types (and groups in the list which are not maximal are known).*

(a) $G = (S_m \times S_k) \cap X$, *with $n = m + k$ and $m \neq k$ (the intransitive case);*

(b) $G = (S_m \, wr \, S_k) \cap X$, *with $n = mk, m > 1$ and $k > 1$ (the imprimitive case);*

(c) $G = AGL(k, p) \cap X$, *with $n = p^k$ and p prime (the affine case);*

(d) $G = (T^k \cdot (\mathrm{Aut}\, T \times S_k)) \cap X$, *with T a non-abelian simple group and $n = |T|^{k-1}$ (the diagonal case);*

(e) $G = (S_m \, wr \, S_k) \cap X$, *with $n = m^k$, $m \geq 5$ and $k > 1$ (the wreath case);*

(f) $T \trianglelefteq G \trianglelefteq \mathrm{Aut}T$, *with T a non-abelian simple group and $T \neq A_n$ (the almost simple case).*

It seems hopeless to try to prove an analogous theorem in the infinite case, but at the end of §7 we will discuss an attempt to handle maximal closed subgroups of S similarly.

Before listing some maximal subgroups of S, I first note the following lemma, proved in [60].

Lemma 4.2 *Let G be a highly transitive permutation group on an infinite set X and let A be a finite non-empty subset of X. Then $G_{\{A\}}$ is a maximal subgroup of G.*

Example 4.3 Let Γ be a finite subset of Ω. Then $S_{\{\Gamma\}}$ is a maximal subgroup of S. This is immediate from Lemma 4.2. Note that any maximal subgroup of S either contains B_ω or is a supplement for B_ω. Hence, by Theorem 3.2, any maximal subgroup of S not of the form above contains B_ω, so is highly transitive. This may be regarded as a rather uninformative infinite analogue of the O'Nan–Scott Theorem.

The next two families of examples (and some others) are given in Brazil, Covington, Penttila, Praeger, and Woods [10]. Example 4.4 can be found in Ball [8].

Example 4.4 Let Γ be a subset of Ω of size λ where $\omega \leq \lambda < \kappa$, and let $G := \{g \in S : |\Gamma g \Delta \Gamma| < \lambda\}$. Then G is maximal in S. For let $h \in S \backslash G$; by Lemma 1.1 it is easy to find $k \in \langle G, h \rangle$ such that $\Gamma \cap \Gamma k = \emptyset$. Now $\langle G, h \rangle$ contains $\langle \mathrm{Sym}(\Omega \backslash \Gamma), \mathrm{Sym}(\Omega \backslash \Gamma k) \rangle$, which by Lemma 1.1 is S. Note that if $\lambda > \omega$ then $G = B_\lambda \cdot S_{\{\Gamma\}}$, but, as pointed out by John McDermott, if $\lambda = \omega$ then $G > B_\lambda \cdot S_{\{\Gamma\}}$.

Example 4.5 Consider a partition of Ω into μ parts each of size λ, say $\Omega := \dot{\cup}(\Omega_\alpha : \alpha < \mu)$, where λ, μ are cardinals, λ infinite, with $\lambda \cdot \mu = \kappa$. For any $\Sigma \subseteq \Omega$, put $n(\Sigma) := |\{\alpha : |\Omega_\alpha \backslash \Sigma| = \lambda\}|$; then put $\mathcal{F} := \{\Sigma \subseteq \Omega : n(\Sigma) < \mu\}$. Then $\mathcal{F}$ is a filter, and contains each $\Omega \backslash \Omega_\alpha$. We may regard $S_{\{\mathcal{F}\}}$ as the almost stabiliser of the partition. Clearly $S_{\{\mathcal{F}\}}$ contains the product of B_λ with the wreath product preserving the partition. It is shown in [10] that the group $S_{\{\mathcal{F}\}}$ is always maximal in S. There is a (very elegant) proof in the case when both λ and μ are infinite. The case where μ is finite is also mentioned in [47], and earlier by Richman in [60]. There is also a maximal subgroup of S, constructed in [10], associated with partitions of Ω into parts each of a fixed finite size.

Example 4.6 Let $\mathcal{F}$ be an ultrafilter on Ω. Then $S_{(\mathcal{F})} = S_{\{\mathcal{F}\}}$ and is maximal in S. This was first proved by Richman in [60] and repeated in [47]. Richman's proof of maximality is the more informative, since he in fact proves the stronger result that S is highly transitive on the right cosets of $S_{\{\mathcal{F}\}}$. His argument uses an observation, proved for example in the proof of Theorem 6.4 of [47], that the stabiliser of an ultrafilter on Ω has two orbits on the set of moieties of Ω, one on those in $\mathcal{F}$, one on those in the dual ideal. Let O be an orbit of S on the set of ultrafilters on Ω, let k be a positive integer, and let $(\mathcal{F}_1, \ldots, \mathcal{F}_k), (\mathcal{G}_1, \ldots, \mathcal{G}_k)$ be two tuples of elements from O, each consisting of distinct elements. Then there are partitions into moieties $\Omega = \Gamma_1 \dot{\cup} \ldots \dot{\cup} \Gamma_k = \Delta_1 \dot{\cup} \ldots \dot{\cup} \Delta_k$ such that $\Gamma_i \in \mathcal{F}_i$ and $\Delta_i \in \mathcal{G}_i$ for each i. It is easily verified by the above observation on orbits that for each i there is $g_i \in S$ such that $\Gamma_i g_i = \Delta_i$ and $\mathcal{F}_i g_i = \mathcal{G}_i$. Now let g be the unique element of S inducing g_i on Γ_i for each i. Then $\mathcal{F}_i g = \mathcal{G}_i$ for each i, as required.

I have several further remarks on Example 4.6.

Remark 1. By a theorem of Pospisil (see Jech [33; p. 256]) there are 2^{2^κ} distinct ultrafilters on Ω. Hence, since $|S| = 2^\kappa$, S has 2^{2^κ} non-conjugate maximal subgroups. If the ultrafilter happens to be principal then the corresponding stabiliser is exactly the stabiliser of a point. Note that, as $S_{\{\mathcal{F}\}} = S_{(\mathcal{F})}$, there are strong restrictions on those subgroups of S which stabilise an ultrafilter. Note too that, by Lemma 4.2 and as pointed out in [60], if A is any finite set of ultrafilters on Ω which lie in the same S-orbit, then $S_{\{A\}}$ (given the obvious meaning) is maximal in S. In [60] it is in fact shown that such a group $S_{\{A\}}$ can be regarded as the almost stabiliser of a partition of Ω into parts of size $|A|$, in a sense of 'almost stabiliser' different to that in Example 4.5.

Remark 2. The proof of Example 4.6 shows that S has 2^{2^κ} inequivalent highly transitive actions. However, by the comment at the end of Example 4.3 and the fact that 2-transitive permutation groups are primitive and so point stabilisers in them are maximal subgroups, we have: any 2-transitive action of S other than the natural one has B_ω in its kernel, so is unfaithful.

Remark 3. For certain ultrafilters, the action of S on the corresponding orbit of ultrafilters has an even higher degree of transitivity. Let us say that a permutation group G on a set X is ω-*transitive*, if it is transitive on the set of ω-sequences of distinct elements of X, and ω-*homogeneous* if it is transitive on the set of unordered countably infinite subsets of X (in the case when X is countable, we require that each sequence enumerates a moiety of X, and that the subsets are moieties of X). I thank J. Steprans for communicating the following result to me.

Proposition 4.7 (Steprans) *Let $\kappa = \Omega$, and identify Ω with $\mathbb{N}$, so the set of ultrafilters on Ω is identified with its Stone-Čech compactification $\beta\mathbb{N}$, with the usual Stone topology (the principal ultrafilters are just the elements of $\mathbb{N}$). Then there are orbits O_1, O_2 of S on $\beta\mathbb{N} \setminus \mathbb{N}$ such that S is ω-transitive on O_1 but is not ω-homogeneous on O_2.*

Proof. Recall that a *weak p-point* in $\beta\mathbb{N} \setminus \mathbb{N}$ is an ultrafilter $\mathcal{F} \in \beta\mathbb{N} \setminus \mathbb{N}$ such that, for all countable $C \subseteq \beta\mathbb{N} \setminus \mathbb{N}$ with $\mathcal{F} \notin C$, $\mathcal{F}$ is not in the closure $\bar{C}$ of C. It is shown in van Mill [51], using just ZFC, that there are weak p-points in $\beta\mathbb{N} \setminus \mathbb{N}$. Now let $\mathcal{F}$ be a weak p-point in $\beta\mathbb{N} \setminus \mathbb{N}$, and let O_1 be the S-orbit of $\mathcal{F}$. Note that every countable subset of O_1 is closed. We must show that given $\{\mathcal{F}_i : i < \omega\} \subseteq O_1$, there is a partition $\mathbb{N} = \dot\bigcup(A_i : i < \omega)$ into moieties of $\mathbb{N}$ such that $A_i \in \mathcal{F}_i$ for all i (for then we can argue as in Example 4.6). So let $\{\mathcal{F}_i : i < \omega\} \subseteq O_1$. Since $\{\mathcal{F}_1, \mathcal{F}_2, \ldots\}$ is closed, there is a basic open set containing $\mathcal{F}_0$ but disjoint from $\{\mathcal{F}_1, \mathcal{F}_2, \ldots\}$ so there is $A_0 \in \mathcal{F}_0 \setminus \bigcup(\mathcal{F}_i : i > 0)$, with A_0 a moiety of $\mathbb{N}$. Now for each $i > 0$ let $\mathcal{F}_i' = \{B \cap (\Omega \setminus A_0) : B \in \mathcal{F}_i\}$. We find that $\{\mathcal{F}_i' : i > 0\}$ is a set of distinct ultrafilters on $\mathbb{N} \setminus A_0$ and no one of these ultrafilters is a limit point of the rest of them. Thus, an inductive argument yields $A_i \in \mathcal{F}_i$ for all i as required.

We choose O_2 to be an S-orbit on $\beta\mathbb{N} \setminus \mathbb{N}$ such that not every countable subset of O_2 is closed (by Steprans [personal communication] such an orbit exists in ZFC). Then O_2 has a countable non-discrete subset X say. Since O_2 is Hausdorff it has a countable discrete subset Y say. Clearly X and Y lie in different S-orbits, so S is not ω-homogeneous on O_2.

Note that, if p-points exist, we can choose O_1 to be an orbit of p-points (a *p-point* is an ultrafilter $\mathcal{F}$ on $\mathbb{N}$ such that given $\{A_i : i < \omega\} \subseteq \mathcal{F}$, there is $A \in \mathcal{F}$ such that $A \setminus A_i$ is finite for all $i < \omega$). Rudin [63] showed that, given the Continuum Hypothesis, there are 2^{2^ω} distinct p-points on $\mathbb{N}$. Shelah (see Wimmers [78]) has shown that it is consistent with ZFC that there are no p-points on $\mathbb{N}$.

Remark 4. In some sense, it seems that the action of S on an orbit of ultrafilters is a 'non-standard' version of the natural action. I have not been able to formalise this, but the impression is reinforced by the following lemma. Recall that an ultrafilter on Ω is *uniform* if it contains no subsets of Ω of size less than κ.

Lemma 4.8 *Let $\mathcal{F}$ be a uniform ultrafilter on Ω. Then $B_\kappa \trianglelefteq S_{\{\mathcal{F}\}}$, and $S_{\{\mathcal{F}\}} \setminus B_\kappa$ is simple.*

Proof. It is immediate that $B_\kappa \trianglelefteq S_{\{\mathcal{F}\}}$. To prove simplicity, choose $g \in S_{\{\mathcal{F}\}} \setminus B_\kappa$. Then as $g \in S_{(\mathcal{F})}$, $\mathrm{Fix}(g) \in \mathcal{F}$ and is a moiety of Ω. Let $h \in S_{\{\mathcal{F}\}}$. Again, $\mathrm{Fix}(h) \in \mathcal{F}$. Hence, by the transitivity of $S_{\{\mathcal{F}\}}$ on the set of moieties of Ω in $\mathcal{F}$, there is $k \in S_{\{\mathcal{F}\}}$ with $(\mathrm{Fix}(g))k \subseteq \mathrm{Fix}(h)$. Hence the normal closure of $\langle k^{-1}gk \rangle$ in $S_{\{\mathcal{F}\},(\mathrm{Fix}(h))}$ contains h, and the result follows.

Lemma 4.8 suggests that $S_{\{\mathcal{F}\}}$ is structurally quite similar to S. It would be interesting to know whether they are ever elementary equivalent as abstract groups. Note that if Ω is countable and $\mathcal{F}$ is a non-principal ultrafilter on Ω then $S_{\{\mathcal{F}\}}$ is not elementary equivalent to S. To see this, observe that an element g of S consisting of a single infinite cycle has abelian centraliser in S (namely $\langle g \rangle$), whilst no element h of $S_{\{\mathcal{F}\}}$ has abelian centraliser, since $S_{\{\mathcal{F}\}}$ induces the symmetric group on a set disjoint to $\mathrm{Supp}(h)$.

We consider next some examples of filters on Ω which are not ultrafilters but whose stabilisers are maximal in S. The following enables us to extend Richman's maximality result to filters which are not quite ultrafilters.

Proposition 4.9 *Let $\mathcal{F}$ be a filter on Ω which contains a moiety of Ω, and suppose that $S_{\{\mathcal{F}\}}$ has three orbits on the set of moieties of Ω. Then $S_{\{\mathcal{F}\}}$ is maximal in S.*

Proof. First note that $\mathcal{F}$ is not an ultrafilter. Let Π be the ideal on Ω dual to $\mathcal{F}$.

CLAIM 1: Let Γ be a moiety of Ω lying in $\mathcal{F}$. Then there is a moiety Δ of Γ with $\Delta \in \mathcal{F}$.

PROOF OF CLAIM 1: Suppose not, and let Σ_1, Σ_2 be moieties of $\Gamma, \Omega\backslash\Gamma$ respectively. Then it is easily seen that each of $\Gamma, \Omega\backslash\Gamma, \Sigma_1, \Sigma_2$ lie in different $S_{\{\mathcal{F}\}}$-orbits, a contradiction.

CLAIM 2: $S_{\{\mathcal{F}\}}$ is transitive on the set of moieties of Ω which lie in $\mathcal{F}$ (and similarly on those lying in Π).

PROOF OF CLAIM 2: Let Γ, Δ be moieties of Ω lying in Π. Then $\Gamma \cup \Delta \in \Pi$, and by Claim 1 there is $\Sigma \in \Pi$ such that both Γ, Δ are moieties of Σ. Since $S_{(\Omega\backslash\Sigma)} \leq S_{\{\mathcal{F}\}}$, there is $g \in S_{\{\mathcal{F}\}}$ with $\Gamma g = \Delta$, as required.

To prove the proposition, now choose $g \in S\backslash S_{\{\mathcal{F}\}}$.

CLAIM 3: There is a moiety Σ of Ω such that just one of $\Sigma, \Sigma g$ lies in $\mathcal{F}$.

PROOF OF CLAIM 3: Suppose not. Let $\Gamma \in \mathcal{F}$. Suppose first $|\Gamma| < \kappa$. Let Δ be a moiety of Ω lying in Π. Then $\Gamma \cup \Delta$ is a moiety of Ω, and as $\Gamma \cup \Delta \supseteq \Gamma$, $\Gamma \cup \Delta \in \mathcal{F}$, whence $(\Gamma \cup \Delta)g \in \mathcal{F}$. Also $\Omega\backslash\Delta g \in \mathcal{F}$, so, as $\Gamma g \supseteq (\Gamma \cup \Delta)g \cap (\Omega\backslash\Delta g)$, $\Gamma g \in \mathcal{F}$. Next, suppose $|\Omega\backslash\Gamma| < \kappa$. Choose any moiety $\Delta \in \mathcal{F}$. Then $\Gamma \cap \Delta \in \mathcal{F}$, so $(\Gamma \cap \Delta)g \in \mathcal{F}$, so $\Gamma g \in \mathcal{F}$. This contradicts that $g \notin S_{\{\mathcal{F}\}}$.

Now choose Σ as in Claim 3, with $\Sigma \in \mathcal{F}$. If $\Sigma g \in \Pi$ then every moiety of Ω which lies in $\Pi \cup \mathcal{F}$ is full for $\langle S_{\{\mathcal{F}\}}, g \rangle$, so by Lemma 1.3, $\langle S_{\{\mathcal{F}\}}, g \rangle = S$. If $\Sigma g \notin \Pi \cup \mathcal{F}$, then, as $S_{\{\mathcal{F}\}}$ is transitive on the set of moieties of Ω which do not lie in $\Pi \cup \mathcal{F}$, every moiety of Ω is full for $\langle S_{\{\mathcal{F}\}}, g \rangle$. Hence by Lemma 1.2, $\langle S_{\{\mathcal{F}\}}, g \rangle = S$.

The above, together with recent results of Denise Ramsay, yields new maximal subgroups of S.

Proposition 4.10 *There is a filter $\mathcal{F}$ on Ω, containing a moiety of Ω, such that $S_{\{\mathcal{F}\}}$ has precisely three orbits on moieties of Ω, so is maximal in S.*

Proof. Let $(R, \leq)$ be the following totally ordered set constructed in Ramsay [59].

$$R := \{(\alpha_1, \ldots, \alpha_{2n+1}) : 0 < n < \omega, \alpha_j < \kappa \text{ for } j = 1, \ldots, n, \text{ with } \alpha_{2n+1} \text{ a successor ordinal}\}.$$

We order R lexicographically. Note that by Lemma 1.12 of [59] every dense subset of R is isomorphic to R. Now let Π be the set of all subsets of R which do not have any subset order-isomorphic to R.

CLAIM 1: Π is an ideal on R.

PROOF OF CLAIM 1: We must show that if $\Gamma, \Delta \in \Pi$ then $\Gamma \cup \Delta \in \Pi$. If not, then we may suppose that $\Gamma \cup \Delta = R$. Now one of Γ, Δ, say Γ, is dense in some interval of R, and it follows from Theorem 2.12 of [59] that Γ is isomorphic to R, a contradiction.

We now identify Ω with R and $\mathcal{F}$ with the filter on R dual to Π. Let $PA(R, \leq)$ denote the group of piecewise automorphisms of R (as defined in the introduction).

CLAIM 2: $S_{\{\Pi\}} \geq PA(R, \leq)$.

PROOF OF CLAIM 2: Let $A \in \Pi$, and let $g \in PA(R, \leq)$ with corresponding partition $R = R_1 \dot{\cup} \ldots \dot{\cup} R_n$ (so g is order-preserving on each R_i). None of $A \cap R_1, \ldots, A \cap R_n$ embeds R, so none of $(A \cap R_1)g, \ldots, (A \cap R_n)g$ embeds R, so these all lie in Π. Hence, by Claim 1, $Ag = (A \cap R_1)g \cup \ldots \cup (A \cap R_n)g \in \Pi$, as required.

Next, note that if $A \in \Pi$ and A is a moiety of R then there is $B \in \Pi$ such that A is a moiety of B; for by Claim 1 $R \backslash A$ embeds a copy of R, which in turn embeds a copy A^* of A, and we may put $B := A \cup A^*$. It follows as in Claim 2 of Proposition 4.9 that $S_{\{\Pi\}}$ is transitive on the set of moieties of R which lie in Π, and on those which lie in $\mathcal{F}$. Let C, D be moieties of R not lying in Π or $\mathcal{F}$. By Lemma 3.3 of [59] there are piecewise automorphisms $g_1 : C \rightarrow D$, $g_2 : R \backslash C \rightarrow R \backslash D$. If we choose $g \in S$ to agree with g_1 on C and g_2 on $R \backslash C$ then $Cg = D$ and $g \in PA(R, \leq) \leq S_{\{\Pi\}}$, as required.

In the particular case when $\kappa = \aleph_0$, we have that $(R, \leq) \cong (\mathbb{Q}, \leq)$ (by Theorem 1.1(i) of [59]), and Π is the ideal of all *scattered* subsets of $\mathbb{Q}$ (that is, those which do not embed a copy of $\mathbb{Q}$). There are several other natural ideals on $\mathbb{Q}$ whose stabilisers in $\mathrm{Sym}(\mathbb{Q})$ are maximal in $\mathrm{Sym}(\mathbb{Q})$. Identify Ω with $\mathbb{Q}$ and give $\mathbb{Q}$ its natural ordering. For any subset $\mathcal{S} \subseteq \mathcal{P}(\Omega)$ such that Ω is not the union of finitely many elements of $\mathcal{S}$ let $\Pi_{\mathcal{S}}$ be the smallest ideal on Ω containing $\mathcal{S}$. Let $\mathcal{B}$ be the set of intervals (a, b) of $\mathbb{Q}$ where $a, b \in \mathbb{Q}$. Let $\mathcal{S}$ be the set of scattered subsets of $\mathbb{Q}$ and let $\mathcal{T}$ be the set of subsets A of $\mathbb{Q}$ such that, for any interval I of $\mathbb{Q}$, $A \cap I$ is not dense in I.

Proposition 4.11

(i) If Π is any of $\Pi_{\mathcal{S}}, \Pi_{\mathcal{T}}, \Pi_{\mathcal{B} \cup \mathcal{S}}, \Pi_{\mathcal{B} \cup \mathcal{T}}$, then $S_{\{\Pi\}}$ is maximal in S.

(ii) $S_{\{\Pi_{\mathcal{B}}\}}$ is not maximal in S.

Proof. The case $\Pi = \Pi_{\mathcal{S}}$ was handled in Proposition 4.10. We here handle $\Pi_{\mathcal{T}}$ and $\Pi_{\mathcal{B} \cup \mathcal{T}}$, and omit $\Pi_{\mathcal{B} \cup \mathcal{S}}$, which is similar. Note that $\mathrm{Aut}(\mathbb{Q}, \leq)$ fixes each ideal.

Suppose first that $\Pi = \Pi_{\mathcal{T}}$, and let $g \in S \backslash S_{\{\Pi\}}$. We may suppose there is $A \in \Pi$ with $Ag \notin \Pi$. Thus, there is a bounded open interval J of $\mathbb{Q}$ such that $Ag \cap J$ is dense in J. It follows (for example by expressing J as the union of two dense codense sets intersecting in a dense codense set, one of them contained in $Ag \cap J$) that there is $h \in \mathrm{Aut}(\mathbb{Q}, \leq)$ such that $(Ag \cap J) \cup (Ag \cap J)h = J$ and $(Ag \cap J) \cap (Ag \cap J)h$ is infinite. By Lemma 1.1 it follows that $\mathrm{Sym}(J) \leq \langle S_{\{\Pi\}}, g \rangle$. Next, choose $k \in S_{\{\Pi\}}$ such that $J \cap Jk$ is infinite and $J \cup Jk = \mathbb{Q}$ (for example, we may choose $a, b \in J$ with $a < b$, and choose k to induce order-isomorphisms $J \cap (-\infty, a) \rightarrow (-\infty, a), J \cap (b, \infty) \rightarrow (b, \infty)$ and $(\mathbb{Q} \backslash J) \cup (a, b) \rightarrow (a, b)$). Then Lemma 1.1, applied to the sets J, Jk, gives $\langle S_{\{\Pi\}}, g \rangle = S$.

Next, suppose $\Pi := \Pi_{\mathcal{B} \cup \mathcal{T}}$. Let $g \in S \backslash S_{\{\Pi\}}$. We may suppose there is $A \in \Pi$ with $Ag \notin \Pi$. It follows that, without loss of generality, there are $a_0 < b_0 < \ldots < a_i < b_i < \ldots$ in $\mathbb{Q}$ such that $(a_i : i < \omega)$ is unbounded above in $\mathbb{Q}$ and $Ag \cap (a_i, b_i)$ is dense in (a_i, b_i) for all $i < \omega$. We now find $h_1, h_2 \in \mathrm{Aut}(\mathbb{Q}, \leq)$ and an order anti-automorphism h_3 such that

(i) $Ag \cap Agh_1$ is infinite and $Ag \cup Agh_1 \supseteq \bigcup_{i<\omega}(a_i, b_i)$,

(ii) $(Ag \cup Agh_1)$ and $(Ag \cup Agh_1)h_2$ have infinite intersection, and their union contains $(0, \infty)$,

(iii) $(0, \infty), (0, \infty)h_3$ have infinite intersection and union $\mathbb{Q}$. Since $h_1, h_2, h_3 \in S_{\{\Pi\}}$ and since $\text{Sym}(A) \leq S_{\{\Pi\}}$, Lemma 1.1 yields that $\langle S_{\{\Pi\}}, g \rangle = S$.

(ii) Let $\Pi := \Pi_B$ and let J be the ideal of $\mathbb{Q}$ generated by B and all images under $\text{Aut}(\mathbb{Q}, \leq)$ of $\mathbb{Z}$. It is easy to see that $S_{\{\Pi\}} \leq S_{\{J\}}$. Now choose $g \in S$ to induce a bijection between $\mathbb{Z}$ and $(0, 1)$ and to be the identity elsewhere. Then $g \in S_{\{J\}} \backslash S_{\{\Pi\}}$ so $S_{\{\Pi\}} < S_{\{J\}} < S$.

Remark It is clear that there are many further ideals on $\mathbb{Q}$ whose stabilisers are maximal in S, constructed in the manner of Proposition 4.11. I know of no useful characterisation of such ideals. In [10] there are further results on ideals whose stabilisers are maximal in S.

All the examples considered in this section of maximal subgroups G of S have a common feature: namely that some moiety in Ω is full for G. However, as noted in [47], this does not always hold. First, note the following result (Lemma 6.9 from [47]) which is an immediate consequence of Zorn's Lemma (for clearly we may assume that A is finite).

Lemma 4.12 *If $G < S$ and there is $A \subseteq S$ with $|A| < c(S)$ such that $\langle G, A \rangle = S$, then G is contained in a maximal subgroup of S.*

Now let $G := PA(\omega, \leq)$, and identify Ω with ω. It is easily checked that $G \neq S$, and that G is transitive on the set of moieties of Ω. However, as shown in Lemma 6.10 of [47], there is $g \in S$ with $\langle G, g \rangle = S$. Hence by Lemma 4.13 G lies in a maximal subgroup H of S, and by Lemma 1.2 no moiety of Ω is full for H.

5 Containment in Maximal Subgroups of S

I turn here to the question: if $G < S$, is G necessarily contained in a maximal subgroup of S? By the following theorem of Baumgartner, Shelah and Thomas [9], it is consistent with ZFC that the answer in general is no.

Theorem 5.1 *Suppose that (F_κ) holds. There is a subgroup $G < S$ such that the set $\{H : G \leq H < S\}$, ordered under inclusion, is a chain whose order type is the initial ordinal of cardinality 2^κ.*

Here, (F_κ) is the following set-theoretic principle:

whenever $T < S/B_\kappa$ with $|T| < 2^\kappa$, there is an element of infinite order $\pi \in (S/B_\kappa) \backslash T$ such that $\langle T, \pi \rangle = T * \langle \pi \rangle$ (where $*$ denotes the free product).

Thomas shows in [9] that (F_κ) is consistent with but not provable from ZFC.

It is natural to look for large classes of subgroups G of S which are contained in a maximal subgroup of S. This is tackled in [48] and [9]. I sketch some of the results.

First, note that if G is an imprimitive subgroup of S, then, after first replacing G by a larger subgroup if necessary, we find that some moiety of Ω is full for G, and hence, by

Lemmas 1.4 and 4.12, that G is contained in a maximal subgroup of S. Next, observe that if $\mathcal{F}$ is a filter on Ω which contains a moiety then, by Lemmas 1.4 and 4.12 and the fact that $S_{(\mathcal{F})} \leq S_{\{\mathcal{F}\}}$, $S_{\{\mathcal{F}\}}$ is contained in a maximal subgroup of S. Thus, given $G < S$, we wish to find a G-invariant ideal on Ω which contains a moiety. The existence of such an ideal is clearly equivalent to the existence of a moiety Γ of Ω with the property that, for all $n < \omega$ and $g_1, \ldots, g_n \in G$, $\Omega \neq \Gamma g_1 \cup \ldots \cup \Gamma g_n$ (for given such a Γ we may take Π to be the ideal generated by the G-translates of Γ). Let $I(G)$ denote the set of all such moieties. I give below some of the results on $I(G)$.

Theorem 5.2 *If* $G < S$ *and* $|G| \leq \kappa$ *then* $I(G) \neq \emptyset$.

The proof of this, given in [48], is straightforward. One lists the elements of G and builds the moiety. In the case when $\kappa = \omega$, one can in fact use the Baire Category Theorem to obtain that the union of $I(G)$ with the set of finite and cofinite subsets of Ω is residual in $\mathcal{P}(\Omega)$. (Here we identify $\mathcal{P}(\Omega)$ with 2^ω and give it the product topology induced from the discrete topology on $\{0,1\}$.)

In [9], some possible strengthenings of Theorem 5.2 are investigated. Let

$$C(\kappa) := \min\{|G| : G \leq S, I(G) = \emptyset\},$$

and let

$$A(\kappa) := \min\{\lambda : \text{no family of almost disjoint subsets of } \Omega \text{ has size greater than } \lambda\}$$

Then the results of [9] include

Theorem 5.3

(i) $A(\kappa) \leq C(\kappa)$,

(ii) *for certain uncountable cardinals* κ *it is consistent with ZFC that there is* $G < S$ *with* $|G| < 2^\kappa$ *and* $I(G) = \emptyset$.

Note that by (i), for $\kappa = \omega$ Theorem 5.2 can be strengthened to: for every $G < S$ with $|G| < 2^\omega$, $I(G) \neq \emptyset$.

In [48] the following theorem is proved.

Theorem 5.4 *If* $\kappa = \omega$ *then every subgroup of* S *which is not highly transitive is contained in a maximal subgroup of* S.

By the remarks about imprimitive groups, together with the Ryll-Nardzewski Theorem (Theorem 2.6), this follows immediately from the following two results.

Theorem 5.5 *Let* $\kappa = \omega$ *and* $G < S$. *If there is* $n < \omega$ *such that* G *has infinitely many orbits on* Ω^n, *then* $I(G) \neq \emptyset$.

Theorem 5.6 *If* M *is an* ω-categorical structure and $G := \mathrm{Aut}\mathcal{M}$ *is primitive but not highly transitive, then* $I(G) \neq \emptyset$.

The proof of Theorem 5.6 is based on the Ehrenfeucht–Mostowski construction, together with a little model-theoretic stability theory. I sketch below an alternative proof of Theorem 5.6 which was suggested to me by Simon Thomas.

Sketch proof of Theorem 5.6. Identify M with Ω. First, call a subset Σ of Ω *definable* if it is a union of orbits of $G_{(\Delta)}$ for some finite $\Delta \subseteq \Omega$ (equivalently, if it is a definable subset of M in the structure $\mathcal{M}$). Let $\mathcal{B}$ be the Boolean algebra of all definable subsets of M, and let Π be the ideal of $\mathcal{B}$ consisting of all finite subsets of M. Following closely the argument of Evans [23] we distinguish two cases.

Case (i) $\mathcal{B}/\Pi$ has an atom.
In this case there is a definable subset Γ of M such that for every definable $\Delta \subseteq M$, $\Gamma \cap \Delta$ is finite or cofinite in Γ. We may suppose there are $g_1, \dots, g_n \in G$ such that $M = \Gamma g_1 \cup \dots \cup \Gamma g_n$. Now $\mathcal{M}$ has Morley rank one, and the argument at the end of the proof of Theorem 1.7 of [48] applies.

Case (ii) $\mathcal{B}/\Pi$ is atomless.
Now let $H := \operatorname{Aut} \mathcal{B}$. Since H permutes the atoms of $\mathcal{B}$, it has an induced action on M, so we may assume $G \leq H < S$. Furthermore, by Corollary 2.2 of [23], we may identify $\mathcal{B}$ with the Boolean algebra of subsets of $\mathbf{Q}$ generated by finite subsets of $\mathbf{Q}$ and intervals $\{(a + \pi, b + \pi) \cap \mathbf{Q} : a, b \in \mathbf{Q}\}$, and hence we identify M with $\mathbf{Q}$ and H with the automorphism group of the corresponding Boolean algebra. Thus, it suffices to show $\mathbf{Z} \in I(H)$. Suppose $\mathbf{Z} \notin I(H)$. Then there is an interval J of $\mathbf{Q}$ and $g \in H$ such that $K := \mathbf{Z} \, g \cap J$ is dense in J. There are $C_1, C_2, C_3 \in \mathcal{B}$ such that $C_1 \cap K, C_2 \cap K, C_3 \cap K$ are moieties of K and have pairwise infinite symmetric difference. The same does not hold for Kg^{-1} in place of K, which is a contradiction.

I remark that the above argument proves something slightly stronger than Theorem 5.6, since in Case (ii), up to conjugacy in S, we always embed G in the same maximal subgroup of S.

It is natural to try to generalise Theorem 5.4 to uncountable κ. To do this, it suffices to consider a first-order structure $\mathcal{M}$ with domain $M = \Omega$ such that if $G = \operatorname{Aut} \mathcal{M}$ then G is primitive but not highly transitive. Various analogues of Theorem 5.5 are possible, such as the following [47; Proposition 3.1].

Theorem 5.7 *Let $G < S$. Suppose there is $\lambda < \kappa$ with $\kappa^\lambda = \kappa$, and $\Sigma \subseteq \Omega$ such that $|\Sigma| = \lambda$ and $G_{(\Sigma)}$ has κ orbits on Ω. Then $I(G) \neq \emptyset$.*

As a partial analogue of Theorem 5.6 I give the following. I refer to [7] or [68] for background on stability theory.

Theorem 5.8 *Suppose that M is a stable structure of uncountable cardinality κ with an indiscernible subset I of size κ. If $G := \operatorname{Aut} \mathcal{M}$ is primitive but not highly transitive, then $I(G) \neq \emptyset$.*

Proof. We identify formulas $\phi(x, \bar{a})$ with their solution sets in M. For every formula $\phi(x, \bar{y})$ there is $n < \omega$ such that either for every $\bar{a} \in M^{1(\bar{y})}, |I \cap \phi(x, \bar{a})| < n$, or for every $\bar{a} \in M^{1(\bar{y})}, |I \backslash \phi(x, \bar{a})| < n$. We may suppose that $I \notin I(G)$, so there are $r < \omega$ and $g_1, \dots, g_r \in G$ such that $M = I g_1 \cup \dots \cup I g_r$. It follows easily that $\mathcal{M}$ has Morley rank one. Hence, by the Finite Equivalence Relation Theorem ([68]; see also [7; ch. IV; Example

1.25]), and primitivity, $\mathcal{M}$ is strongly minimal. Now since M is not an indiscernible set, it is easily checked that any independent (under model-theoretic algebraic closure) subset of M of size κ lies in $I(G)$. This yields the theorem.

Remark Under certain conditions models of stable theories are guaranteed to have moieties which are indiscernible sets. This holds if, for example, λ is sufficiently large, Th($\mathcal{M}$) is λ-stable, and $\mathcal{M}$ has size $\kappa := \lambda^+$ ([7; ch. V; Theorem 1.28]).

6   Related Results on Other Permutation Groups

It is natural to ask the questions considered in §§ 2–5 for permutation groups other than the symmetric groups. Possible groups to consider include the following: for any infinite cardinal κ and any field F the group $GL(\kappa, F)$ of all invertible linear transformations of a vector space of dimension κ over F; the corresponding infinite-dimensional symplectic, orthogonal or unitary groups (but note that in uncountable dimension these are not unique — see [15] and [29]); the group of all order-automorphisms of the rationals; the automorphism group of the random graph [14]; various Jordan groups associated with trees (see [1] and [55] for more on Jordan groups); normal subgroups of S and of some of the above groups. Very little seems to have been done on this, so this section consists mostly of open questions.

First, we consider $G := GL(\kappa, F)$, where κ is infinite and F is a field. There is a classification of the normal subgroups of G due to Rosenberg [61]. For each infinite cardinal $\lambda \leq \kappa$, let

$$G_\lambda := \{g \in G : \dim(V/\text{Fix}(g)) < \lambda\}.$$

Also let F^* denote the multiplicative group of F. Then the proper normal subgroups of G not contained in G_ω are precisely those of the form $G_\lambda \cdot Z$ where $\lambda \leq \kappa$ is infinite and $Z \leq F^*$. The remaining normal subgroups of G are also described in [61] (they were classified by Dieudonné in [18] and are analogous to the normal subgroups in the finite-dimensional case).

We now describe some maximal subgroups of G. Let V be a κ-dimensional vector space over F such that $G = GL(V)$, and let $\mathcal{S}(V)$ be the set of all subspaces of V. An *ideal* of $\mathcal{S}(V)$ is a subset of $\mathcal{S}(V)$ containing the trivial subspace, not containing V, and closed downwards and under finite sums. A *filter* on $\mathcal{S}(V)$ is a subset of $\mathcal{S}(V)$ containing V but not the trivial subspace, and closed upwards and under finite intersections. Note that there is no natural map taking ideals to filters. Note too that any filter is contained in a maximal filter, and any ideal in a maximal ideal. If λ is a cardinal with $\lambda \leq \kappa$, define the equivalence relation $=_\lambda$ on $\mathcal{S}(V)$ by the rule: $U =_\lambda W$ if $\dim((U + W)/U) + \dim((U + W)/W) < \lambda$. The following result is proved in [46].

Theorem 6.1

(i)   *The stabiliser in G of any maximal filter or ideal on $\mathcal{S}(V)$ is maximal in G.*

(ii) *If $U \in \mathcal{S}(V)$ with $\dim U = \lambda < \kappa$, then $\{g \in G : Ug =_\lambda U\}$ is a maximal subgroup of G.*

(iii) *If $V = V_1 \oplus V_2$ where $\dim V_1 = \dim V_2 = \kappa$, then $\{g \in G : V_1g =_\kappa V_1$ and $V_2g =_\kappa V_2$, or $V_1g =_\kappa V_2$ and $V_2g =_\kappa V_1\}$ is a maximal subgroup of G.*

Note that also the stabiliser in G of any finite-dimensional subspace of V is maximal in G. The above theorem suggests a number of open questions.

Problem 6.2 How many maximal ideals or filters does V have? (It is immediate, by applying Pospisil's Theorem to a basis, that V has at least 2^{2^κ} maximal ideals and at least 2^{2^κ} maximal filters - see Lemma 2.9 of [46]).

Problem 6.3 If $\mathcal{F}$ is a maximal filter on V, how transitive is G on the set of G-translates of $\mathcal{F}$? The same question arises for maximal ideals, and in both cases it is shown in [46] that the degree of transitivity is at least two.

Problem 6.4 Does the analogue of Theorem 6.1(iii) hold for decompositions into more than two parts? Is, in some appropriate sense, the almost stabiliser in V of a tensor decomposition of V, or of a symplectic, orthogonal or unitary form, ever maximal in G?

Problem 6.5 If $H \leq G$ and $H^U = GL(U)$ for every $U \leq V$ with $\dim U = \dim(V/U) = \kappa$, must we have $H = G$? (This is suggested by Lemma 1.2 above.)

I remark that, by an argument like the proof of Theorem 3.1(i), we have

Theorem 6.6 $c(G) > \kappa$.

This, together with remarks in §3, suggests the following problem.

Problem 6.7 Classify the supplements of the normal subgroups of G.

It is also natural to try to classify the maximal subgroups of the normal subgroups of G and S. I know of no results on maximal subgroups of, say G_ω (this probably deserves further attention). However, the following result on maximal subgroups of B_ω was proved by Ball [8].

Theorem 6.8

(i) *The maximal subgroups of* $\mathrm{Alt}(\Omega)$ *are precisely the stabilisers in* $\mathrm{Alt}(\Omega)$ *of proper non-empty subsets of* Ω, *and of partitions of* Ω *into finite non-empty pieces all of the same size.*

(ii) *The maximal subgroups of* B_ω *are precisely the stabilisers in* B_ω *of subsets and partitions as in* (i), *together with* $\mathrm{Alt}(\Omega)$.

This result follows immediately from the theorem of Wielandt [77] that any primitive subgroup of S which contains a non-identity element of B_ω contains $\mathrm{Alt}(\Omega)$. Adeleke and Neumann [2] have an analogous theorem for primitive subgroups of S containing an element of degree λ where λ is infinite but less than κ, but it does not seem to yield a classification of maximal subgroups of B_{λ^+}.

I conclude this section with a few examples of maximal subgroups of another familiar permutation group, $\mathrm{Aut}(\mathbb{Q}, \leq)$. Following Glass [27], if $(I, \leq)$ is a chain, then

$A(I) := \mathrm{Aut}(I, \leq)$, and $A(I)$ has normal subgroups

$$L(I) := \{g \in A(I) : (\exists x \in I)(\forall y > x)(yg = y)\},$$

$$R(I) := \{g \in A(I) : (\exists x \in I)(\forall y < x)(yg = y)\},$$

$$B(I) := L(I) \cap R(I).$$

In the case when $(I, \leq) = (\mathbb{Q}, \leq)$, by results of Higman [30], Holland [33] and Lloyd [43] (see also Glass [27; Theorem 2.3.2]) these are the only proper non-trivial normal subgroups. There are two obvious non-conjugate maximal subgroups of $A(\mathbb{Q})$, namely the stabilisers of a rational and of an irrational (maximality of these groups follows from the primitivity of the corresponding permutation representations). We obtain five further non-conjugate maximal subgroups.

Let R, G be unary relation symbols ('red', 'green' respectively) and let E be a binary relation symbol. We form structures $(I_j, \leq, R, G)$ (for $j = 1, \ldots, 4$) and $(J, \leq, E)$ as follows. First, let $\mathbb{Q} = A \dot\cup B$ be a partition of $\mathbb{Q}$ into two dense codense subsets. To obtain I_j from $\mathbb{Q}$ replace each point of A by a copy of $\mathbb{Q}$ in which each point satisfies R, and each element of B by a set, each point of it satisfying G, of order type 1, $1 + \mathbb{Q}$, $\mathbb{Q} + 1$, $1 + \mathbb{Q} + 1$, according as $j = 1, \ldots, 4$ respectively. Give the new set I_j the natural ordering $\leq$. Thus, I_j consists of two dense codense sets of intervals, the red intervals each having order type $\mathbb{Q}$, the green intervals having order type depending on j. To obtain J, replace each point of A by a copy of $\mathbb{Q}$ satisfying R, each point of B by a copy of $\mathbb{Q}$ satisfying G, and let E be the equivalence relation on J saying that two points have the same colour. Again give J the natural ordering $\leq$. Note that $(J, \leq)$ and each $(I_j, \leq)$ are order-isomorphic to $(\mathbb{Q}, \leq)$. Let $H_j := \mathrm{Aut}(I_j, \leq, R)$ and $M_j := L(I_j) \cdot H_j$ for $j = 1, \ldots, 4$. Also put $H := \mathrm{Aut}(J, \leq, E)$ and $M := L(J) \cdot H$.

Theorem 6.9 *The group M_j is maximal in $A(I_j)$ for $j = 1, \ldots, 4$ and M is maximal in $A(J)$.*

Proof. Let $j \in \{1, \ldots, 4\}$ and let $g \in A(I_j) \backslash M_j$. Put $K_j := \langle M_j, g \rangle$. Then one of the following holds.

(i) There is an increasing sequence $(a_i : i < \omega)$ of red points cofinal in I_j such that each interval $[a_{2i}, a_{2i+1}]$ is red and for each i just one of $a_{2i}g, a_{2i+1}g$ is red.

(ii) The same as (i) holds except that each interval $[a_{2i}, a_{2i+1}]$ is green (this case cannot occur if $j = 1$).

Suppose that (i) holds (the proof for (ii) being similar).

Next, let $(b_i : i < \omega)$ be an increasing sequence from I_j which is cofinal in I_j, and let $h \in A(I_j)$ be the identity on $\{x < b_0\} \dot\cup \dot\cup([b_{2i+1}, b_{2i+2}] : i < \omega)$ and have each interval (b_{2i}, b_{2i+1}) as an orbital of positive parity (in the sense of Glass [27]). It is easily seen that, since in $(a_{2i}g, a_{2i+1}g)$ both the maximal red and the maximal green intervals (when condensed to a point) are dense and codense, there is $k \in M_j$ such that $[b_{2i}, b_{2i+1}]k \subset (a_{2i}g, a_{2i+1}g)$ for all $i < \omega$, so $[b_{2i}, b_{2i+1}]kg^{-1} \subset (a_{2i}, a_{2i+1})$. By considering the subgroup of H which stabilises each maximal red interval, we obtain that there is $f \in H_j$ such that $h = kg^{-1}fgk^{-1}$, whence $h \in K_j$. By the classification of the normal subgroups of $A(\mathbb{Q})$ we have $L(I_j) \cdot \langle h \rangle^{A(I_j)} = A(I_j)$. Since our initial requirement on h depended only on its conjugacy class in $A(I_j)$, we have that $K_j = A(I_j)$, as required.

The proof that $L(J) \cdot M$ is a maximal subgroup of $A(J)$ is similar, so we omit details. Note that if $g \in A(J) \backslash M$, then there is an increasing sequence $(a_i : i < \omega)$ cofinal in J such that each interval $[a_{2i}, a_{2i+1}]$ lies in a single E-class, but for each i, $a_{2i}g, a_{2i+1}g$ lie in distinct E-classes.

Remark Note that the groups M_2, M_3 are conjugate in Aut $(A(\mathbb{Q}))$ when we identify I_2, I_3 with $\mathbb{Q}$, but not in $A(\mathbb{Q})$. I know of no maximal subgroups of $A(\mathbb{Q})$ other than ones similar to the above. Also, I have been unable to answer either of the following questions.

Question 6.10 *What is $c(A(\mathbb{Q}))$?*

Question 6.11 *What are the supplements of $L(\mathbb{Q})$ in $A(\mathbb{Q})$?*

It might also be interesting to investigate similar questions for other familiar infinite permutation groups, such as the automorphism group of the countable universal homogeneous graph of Rado (see [14] for information on this group). The only maximal subgroups I know for this group are stabilisers of certain finite sets (e.g. vertices, unordered edges, unordered non-edges).

7 Other Large Subgroups of S

In this section I survey work on several other notions of 'large' for subgroups of S. The account will be superficial, and its goal is to give an idea of what has been tackled.

One way of saying that a subgroup of S is large is to say that it is very transitive, i.e. that it has few orbits. This suggests the following definition (which we restrict to the case when $\kappa = \omega$). Following Cameron [14] we call a subgroup G of S *oligomorphic* if, for each positive integer n, G has finitely many orbits on Ω^n. It is an immediate consequence of the Ryll-Nardzewski Theorem (Theorem 2.6 above) that a subgroup of S is oligomorphic if and only if there is an ω-categorical structure $\mathcal{M}$ with domain Ω such that G is dense in Aut$\mathcal{M}$. Oligomorphic permutation groups have been studied in depth, particularly by Cameron, and I recommend [14] as a general account. Questions which have been considered include: the local and asymptotic behaviour of the integer sequences describing the number of orbits (of an oligomorphic permutation group) on the unordered (or ordered) k-subsets of Ω; the small index property (see §2) for closed oligomorphic permutation groups; embeddings of various groups into closed oligomorphic permutation groups (for example, by a theorem of Hodges [personal communication], any closed oligomorphic permutation group contains a dense free subgroup of rank 2^ω). I should also mention a theorem of Cameron [11] that any highly homogeneous but not highly transitive permutation group is dense in the automorphism group of one of: a dense linear order; the betweenness relation induced from a dense linear order; a dense circular order; the (quaternary) separation relation induced from a dense circular order. There has also been a great deal of recent model-theoretic work on ω-categorical structures.

A second and more technical way of saying that a subgroup G of S is large is to say that it is a Jordan group. This means that there is a subset Γ (the *Jordan* set) of Ω such that

$G_{(\Omega \setminus \Gamma)}$ is transitive on Γ, and none of the following degenerate cases happens:

(i) $|\Gamma| \leq 1$;

(ii) $\Gamma = \Omega$;

(iii) $|\Omega \setminus \Gamma| = k$ and G is $(k+1)$-transitive for some positive integer k.

Finite Jordan groups have been classified independently by Kantor [35] and Neumann [55]. Information about them was used in the independent classifications by Cherlin [17] and Mills of strictly minimal sets (Zil'ber [79] gave a classification which does not use Jordan groups). Infinite Jordan groups G on Ω have been examined in a series of papers by Adeleke and Neumann (see for example [1]) and a great deal of information is available about them. As a brief advertisement for them I note the following strengthening of a remark made after Example 4.3.

Theorem 7.1 *Let $\kappa = \omega$ and let G be a primitive but not highly transitive subgroup of S. Then $B_\omega \cdot G$ is not maximal in S.*

Proof. Suppose for a contradiction that $B_\omega \cdot G$ is maximal in S. By Theorems 5.5 and 5.6 there is a moiety Γ of Ω such that Ω is not the union of finitely many G-translates of Γ. Let Π be the ideal on Ω generated by the G-translates of Γ. Then $B_\omega \cdot G \geq S_{\{\Pi\}}$. Then $\mathrm{Sym}(\Gamma) \leq B_\omega \cdot G$, so as $|B_\omega \cdot G : G| \leq \omega$, by Theorem 2.1 there is $\Delta \subseteq \Gamma$ with $\Gamma \setminus \Delta$ finite such that $\mathrm{Sym}(\Delta) \leq G$. In particular Δ is a Jordan set for G, and Theorem 3 of [1] shows that this cannot happen.

A third form of largeness is to have a lot of transitivity on infinite subsets of Ω. If $G \leq S$ and λ is an infinite cardinal less than κ, define $f_G(\lambda)$ to be the number of orbits on the set of unordered λ-subsets of Ω. Also let $f_G(\kappa)$ be the number of G-orbits on moieties of Ω. I describe below several results on this function.

Theorem 7.2 *If $\kappa = \omega$ and $G < S$ is primitive but not highly transitive then $f_G(\kappa) = 2^\kappa$.*

This was proved in [45] and also by Evans in [23] using arguments close to the proof described above of Theorem 5.6. It also follows immediately from the main theorem of Litman and Shelah [42]. The obvious problem of finding a sensible generalisation for uncountable κ is still open.

The above result suggests the question: what cardinals between ω and 2^ω can arise as $f_G(\kappa)$ when $\kappa = \omega$ and $G < S$? Results on this include the following.

Theorem 7.3 (Läuchli and Neumann [39]) *Identify Ω with $\mathbb{Q}$ and let $G := PA(\mathbb{Q}, \leq)$. Then $f_G(\omega) = \omega_1$.*

The proof uses Laver's work on scattered chains. The following generalisation is due to Ramsay [59].

Theorem 7.4 *There is $G < S$ such that $f_G(\kappa) = \kappa^+$.*

Using Martin's Axiom, Shelah and Thomas [70] obtained the following strengthening of Theorem 7.3.

Theorem 7.5 *(MA) Let $\kappa = \omega$. For every cardinal λ with $\omega < \lambda < 2^\omega$, there is $G < S$ with $f_G(\omega) = \lambda$.*

Recall that a *Polish space* is a complete separable metric space. Given the various theorems on the number of classes of analytic and coanalytic equivalence relations on a Polish space, it is natural to put descriptive set-theoretic restrictions on G (where G inherits the metric on S described in §1). Thus one obtains the following easy result. Recall that a subset of a perfect Polish space [53; p. 17] is *analytic* if it is a continuous image of Baire space $^\omega\omega$ [53; IE6], and is *coanalytic* if it is the complement of an analytic set.

Proposition 7.6 *Let $\kappa = \omega$, give S the usual metric described in §1, and let $G \le S$ be an analytic set. Then either $f_G(\kappa) = 2^\omega$ or $f_G(\kappa) \le \omega_1$.*

Proof. Identify Ω with ω and $\mathcal{P}(\Omega)$ with 2^ω, and give $\mathcal{P}(\Omega)$ the product topology induced from the discrete topology on $\{0,1\}$. Also give $\mathcal{P}(\Omega) \times \mathcal{P}(\Omega)$ and $\mathcal{P}(\Omega) \times G$ the product topologies. Define a mapping $\phi : \mathcal{P}(\Omega) \times G \to \mathcal{P}(\Omega) \times \mathcal{P}(\Omega)$ by the rule: if $\Gamma \subseteq \Omega$ and $g \in G$, then $\phi(\Gamma, g) = (\Gamma, \Gamma g)$. It is easily checked that this map is continuous. Furthermore, the image of ϕ is precisely the equivalence relation which says that two subsets of Ω are in the same G-orbit. Since a continuous image of an analytic set is analytic, $\mathrm{Im}(\phi)$ is an analytic equivalence relation. The proposition now follows from the theorem of Burgess (Theorem 8.26 of Mansfield and Weitkamp [50]) that an analytic equivalence relation cannot have strictly between $\aleph_1$ and $2^{\aleph_0}$ equivalence classes. We remark that the group $PA(\mathbb{Q}, <)$ which arose in Theorem 7.3 is analytic; indeed, it is a countable union of closed sets.

It might also be interesting to impose topological restrictions on several of the groups considered below.

The notion of piecewise automorphism was introduced by Stoller [72] essentially to show that the function f is not monotonic increasing. Stoller pointed out that if $G := PA(\omega_1, \le)$ then $f_G(\omega_1) = 1$ (since any two uncountable subsets of ω_1 are order-isomorphic), whilst $f_G(\omega) = \omega_1$. This suggests questions along the following lines, raised by Peter Neumann in [56]: if $\kappa = \omega_2$, is there $G < S$ such that $f_G(\omega_1) = 1$ but $f_G(\omega) \neq 1$? Several examples of such groups have now been given. Shelah and Thomas [71] and Nyikos [57] have given examples under the assumptions $MA + 2^\omega > \omega_2$ and $MA + 2^\omega > \omega_1$ respectively. Furthermore Hajnal [28] has obtained the following result.

Theorem 7.7 *Assume that $\lambda > \omega$ and the set theoretic principle $\square_\lambda$ holds for λ. Put $\kappa = \lambda^+$. Then there is $G < S$ such that $f_G(\lambda) = 1$ but $f_G(\mu) > 1$ for all cardinals μ with $\omega < \mu < \lambda$.*

I mention here some other results of Neumann [56].

Theorem 7.8 *Let $G \leq S$.*

(i) *If λ is a cardinal such that $\omega \leq \lambda \leq \kappa$ and $f_G(\lambda) = 1$ then G is highly transitive on Ω.*

(ii) *If $f(\kappa)$ is finite but all the G-orbits on the set of moieties of Ω have length greater than κ, then G is highly transitive on Ω.*

(iii) *If for every moiety Σ of Ω there is $g \in G$ such that $\Sigma \cap \Sigma g = \emptyset$ then G is highly transitive on Ω.*

Part (i) above is in fact implied by a purely combinatorial result of Kierstead and Nyikos [36] which says essentially that any infinite uniform hypergraph with, for some infinite cardinal λ, exactly one isomorphism type of λ-element substructure, must be trivial (for if G is not highly transitive then for some positive integer k there is a non-trivial G-invariant k-uniform hypergraph on Ω). Theorem 7.8 is also related to some partial monotonicity results in [49] for the function which counts, for each infinite cardinal λ, the number of λ-element substructures, up to isomorphism, of a relational structure. For $\kappa = \omega$, part (iii) above would follow from Theorems 5.5 and 5.6. Finally, I mention a recent theorem of Thomas [75]. Its analogue for countable κ is easy.

Theorem 7.9 *Let κ be an uncountable cardinal such that $\kappa^\lambda > 2^\kappa$ for all $\lambda < \kappa$. Suppose that $G \leq S$ acts transitively on the set of moieties of Ω. Then for each moiety Γ of Ω, $G_{\{\Gamma\}}$ is transitive on Γ.*

I turn now to my last notion of largeness for subgroups of S. Assume for the rest of the paper that $\kappa = \omega$. If $G < S$ we say that G is *maximal closed* in S if $(G; \Omega)$ is closed and S has no proper closed subgroup properly containing G. A few families of maximal closed subgroups of S are known, but in most cases the proof of maximality is a substantial result. Some examples of maximal closed subgroups of S are given in [47]. They include the following:

(i) the stabiliser in S of any finite subset of Ω, or of any partition of Ω into parts of the same size;

(ii) the stabiliser of a countable dense separation relation on Ω (by Cameron's result on highly homogeneous groups mentioned earlier in the section);

(iii) certain Jordan groups (by the results of Adeleke and Neumann [1]);

(iv) the automorphism groups of certain ω-categorical, ω-stable structures (directly, or by quoting model-theoretic results from [17]);

(v) the automorphism group of the countable universal homogeneous triangle-free graph, or of (up to reversal) the countable universal homogeneous two-graph; this last object is described in [13; Ex. 6; p. 45] and both results are proved by Thomas in [74].

The following question was raised in [47].

Question 7.10 Is every closed subgroup of S contained in a maximal closed subgroup of S?

The examples suggest that maximal closed subgroups of S are the right analogue of maximal subgroups of finite symmetric groups. Hence it is natural to look for an analogue of the O'Nan–Scott Theorem [41] describing maximal closed subgroups of S, or possibly arbitrary primitive closed subgroups of S. Note that, since any imprimitive maximal closed subgroup of S is of type (i) above, any unknown maximal closed subgroup of S is primitive. Hence, much as in the ordinary O'Nan–Scott Theorem, we reduce to looking at primitive closed subgroups G of S, and it is not clear if the assumption 'G is maximal closed in S' gives anything extra. The difficulty is that the proof of the O'Nan–Scott Theorem crucially uses the existence of the socle, and hence of minimal normal subgroups, and our group G need not have a socle. However, one can get somewhere using the notions of *closed normal subgroup* and *minimal closed normal subgroup* (that is, minimal subject to being closed, normal and non-trivial). As a start, note the following.

Lemma 7.11 Let $\mathcal{M}$ be ω-categorical and $G := \mathrm{Aut}\ \mathcal{M}$. Then G has a minimal closed normal subgroup.

Proof. We shall use Zorn's Lemma, so it suffices to show that the intersection of any chain of non-trivial closed normal subgroups of G is non-trivial. Note that if N is a closed normal subgroup of G then for all $k < \omega$ the orbits of N on M^k are the classes of some O-definable equivalence relation on M^k. Let $(N_i : i \in I)$ be an infinite descending chain of non-trivial closed normal subgroups of G, and for each $i \in I$ let Υ_i be the set of O-definable equivalence relations on powers of M whose classes are fixed by N_i but not by G. A compactness argument shows that it is consistent that there is a non-trivial $g \in G$ which fixes setwise each class of each relation in each Υ_i, and by ω-categoricity such a g must lie in $\cap(N_i : i \in I)$. The result follows.

Now let $\mathcal{M}$ be an ω-categorical structure and suppose that $G = \mathrm{Aut}\ \mathcal{M}$ and is primitive. Since the intersection of two closed normal subgroups of G is closed and normal, any two minimal closed normal subgroups are equal or disjoint, and in the latter case they commute. Much as in the first step in the proof of the O'Nan–Scott Theorem we now have three cases.

(i) G has a unique minimal closed normal subgroup A, and A is abelian.

(ii) G has a unique minimal closed normal subgroup N, and N is non-abelian.

(iii) G has exactly two disjoint minimal closed normal subgroups N_1 and N_2, with $N_1 \neq N_2$, $C_G(N_1) = N_2$, $C_G(N_2) = N_1$.

On Peter Cameron's suggestion, it seems that a more natural case division is the following:

(a) any minimal closed normal subgroup of G is regular on M;

(b) any minimal closed normal subgroup of G acts oligomorphically on M;

(c) neither of the above.

Note that in case (i), A must act regularly on M. Similarly, in case (iii), each of N_1, N_2 acts regularly on M (as in the finite case). So both (i) and (iii) imply that (a) holds, and (a) can also hold in case (ii). Note too that (a), (b) and (c) are mutually exclusive.

In case (c) I have been unable to make any progress, even under the additional assumption that G is maximal closed in $\text{Sym}(M)$. The difficulty is that we know nothing of the structure of a minimal closed normal subgroup of G; in particular, this subgroup may itself not have a minimal closed normal subgroup (this happens, for example, in the group $X[\Lambda]$ discussed in §6 of [13], though admittedly $X[\Lambda]$ is not maximal closed in the symmetric group). If case (b) holds and N is a minimal closed normal subgroup of G, then by Lemma 7.11 N itself has a minimal closed normal subgroup K. It seems likely that quite a lot of the O'Nan–Scott Theorem should go through here. I have not checked this, and there may be difficulties involving the analogue of Schreier's Conjecture.

In case (a), let N be a regular minimal closed normal subgroup of G. Clearly N is a minimal normal subgroup of G, so is characteristically simple. We may identify M with N, and G is the semidirect product of N with G_1, the stabiliser of the identity. Now G_1 acts oligomorphically on N by conjugation. Hence N is an ω-categorical group. Since N is regular, it is infinite. By a theorem of John Wilson (see Apps [4]), any infinite ω-categorical, characteristically simple group is elementary abelian, or a Boolean power of a finite simple group, or a perfect p-group (and Wilson conjectures that the last case cannot occur). Thus, we have a good description of G in case (a). In particular, if N is abelian (i.e. if case (i) occurs), then $G \leq AGL(\aleph_0, p)$ (p a prime), with equality if and only if G is maximal closed in $\text{Sym}(M)$.

I remark that case (iii) can occur, and implies that (a) holds. We may identify M with N_1, with N_1 acting on M by its left regular representation, and N_2 on M by an action isomorphic to its right regular representation. This is an action analogous to the diagonal case in the O'Nan–Scott Theorem (see Theorem 4.1(d) with $k = 2$). An example of such an action is the following. Let N be a non-abelian infinite ω-categorical characteristically simple group and let

$$G := \{(a_1, a_2) : a_1, a_2 \in \text{Aut } N, a_1 \equiv a_2 \bmod \text{Inn } N\}.$$

Let G act on the right cosets of the subgroup

$$H := \{(a, a) : a \in \text{Aut } N\}.$$

Then in this permutation representation, G is primitive, closed and oligomorphic, of type (iii).

NOTE ADDED IN PROOF

Several topics touched on in this paper have been extended since the first draft was written. For completeness, I record some of these here.

First, J.L. Alperin pointed out that if $\kappa = \omega$ then the group S/B_ω admits an outer automorphism of infinite order (corresponding to "almost conjugation" by the function $n \mapsto n + 1$ on ω). A paper by Alperin, Covington and myself is in preparation, showing that when $\kappa = \omega$ the outer automorphism group of S/B_ω is infinite cyclic.

Second, substantial progress has been made on the small index property. In a paper by W.A. Hodges, I.M. Hodkinson, D. Lascar, S. Shelah ["The small index property for ω-stable ω-categorical structures and for the random graph", *J. London Math. Soc.*, to appear] it is shown that the automorphism group of the countable universal homogeneous graph has the small index property, as does the automorphism group of any ω-categorical, ω-stable structure. For a discussion of the small index property and reconstruction of a structure from its automorphism group, see Lascar's article in this book.

There has been further work on the question: for which ideals Π on $\mathcal{P}(\Omega)$ is $S_{\{\Pi\}}$ maximal in S? Various questions, mainly for uncountable Ω, are examined in two papers by Covington and Mekler ['Subgroups of infinite symmetric groups which are full for large sets', *J. Alg.*, to appear; 'Stabilisers of trivial ideals', *Bull. London Math. Soc.*, to appear]. In a paper by Covington, Macpherson and Mekler ['Stabilisers of ideals and infinite symmetric groups', preprint] the case when Ω is countable is investigated, and many combinatorially natural ideals are found whose stabilisers are maximal in S. In particular, Proposition 4.11 is extended.

Next, I should mention that some of the topics of this paper (maximal subgroups, supplements of normal subgroups, cofinality $c(G)$ of a group G) have recently been investigated for various other automorphism groups. In particular, Lascar and Gourion have proved $c(G) > \omega$ for various G, and Glass and McCleary have results on maximal subgroups and supplements for $\mathrm{Aut}(\mathbb{Q}, <)$.

Finally, the remarks on the O'Nan–Scott Theorem at the end of the paper have recently been extended by myself and Cheryl Praeger ["Infinitary versions of the O'Nan–Scott Theorem", preprint]. There are general versions of the O'Nan–Scott Theorem under assumptions of the existence of minimal closed normal subgroups which themselves have minimal closed normal subgroups, or of minimal normal subgroups with minimal normal subgroups. Since there is no infinitary analogue of the classification of the finite simple groups, we cannot expect nearly such powerful applications as in the finite case.

References

[1] S. Adeleke, P.M. Neumann, On infinite Jordan groups, handwritten manuscript.

[2] S. Adeleke, P.M. Neumann, Infinite bounded permutation groups, handwritten manuscript.

[3] G. Ahlbrandt, M. Ziegler, Quasi finitely axiomatizable totally categorical theories, *Ann. Pure Appl. Logic* **30** (1986), 63–82.

[4] A. Apps, On the structure of $\aleph_0$-categorical groups, *J. Alg.* **81** (1983), 320–339.

[5] M. Aschbacher, L. Scott, Maximal subgroups of finite groups, *J. Alg.* **92** (1985), 44–80.

[6] R. Baer, Die Kompositionsreihe der Gruppe aller eineindeutigen Abbildungen einer unendlichen Menge auf sich, *Studia Math.* **5** (1935), 15–17.

[7] J.T. Baldwin, *Fundamentals of stability theory*, Springer, Berlin, 1988.

[8] R.W. Ball, Maximal subgroups of symmetric groups, *Trans. Amer. Math. Soc.* **121** (1966), 393–407.

[9] J.E. Baumgartner, S. Shelah, S. Thomas, Maximal subgroups of infinite symmetric groups, preprint.

[10] M. Brazil, J.A. Covington, T. Penttila, C.E. Praeger, A. Woods, *Proc. London Math. Soc.*, to appear.

[11] P.J. Cameron, Transitivity of permutation groups on unordered sets, *Math. Z.* **148** (1976), 127–139.

[12] P.J. Cameron, Finite permutation groups and finite simple groups, *Bull. London Math. Soc.* **13** (1981), 1–22.

[13] P.J. Cameron, Some treelike objects, *Quart. J. Math. Oxford* (2) **38** (1987), 155–183.

[14] P.J. Cameron, Oligomorphic permutation groups, *London Math. Soc. Lecture Notes* **152**, CUP, Cambridge, 1990.

[15] P.J. Cameron, J.I. Hall, Some groups generated by transvection subgroups, *J. Alg.* **140** (1991), 184–209.

[16] C.C. Chang, H.J. Keisler, *Model Theory*, North-Holland, Amsterdam, 1973.

[17] G. Cherlin, L. Harrington, A.H. Lachlan, $\aleph_0$-categorical, $\aleph_0$-stable structures, *Ann. Pure Appl. Logic* **28** (1985), 103–135.

[18] J. Dieudonné, Les determinants sur un corps non commutatif, *Bull. Soc. Math. France* **71** (1943), 27–45.

[19] J.D. Dixon, P.M. Neumann, S. Thomas, Subgroups of small index in infinite symmetric groups, *Bull. London Math. Soc.* **18** (1986), 580–586.

[20] M. Droste, W.C. Holland, H.D. Macpherson, Automorphism groups of infinite semilinear orders (II), *Proc. London Math. Soc. (3)* **58** (1989), 479–494.

[21] E. Engeler, A characterisation of theories with isomorphic denumerable models, *Notices Amer. Math. Soc.* **6** (1959), 161.

[22] D.M. Evans, Subgroups of small index in infinite general linear groups, *Bull. London Math. Soc.* **18** (1986), 587–590.

[23] D.M. Evans, Infinite permutation groups and minimal sets, *Quart. J. Math. Oxford* (2) **38** (1987), 461–471.

[24] D.M. Evans, The small index property for infinite dimensional classical groups, *J. Alg.* **136** (1991), 248–264.

[25] D.M. Evans, P. Hewitt, Counterexamples to a conjecture on relative categoricity, *Ann. Pure Appl. Logic* **46** (1990), 201–209.

[26] D.M. Evans, W.A. Hodges, I.M. Hodkinson, Automorphisms of bounded abelian groups, *Forum Math.* **3** (1991), 523–541.

[27] A.M.W. Glass, Ordered permutation groups, *London Math. Soc. Lecture Notes* **5**, CUP, Cambridge, 1981.

[28] A. Hajnal, A remark on the homogeneity of infinite permutation groups, *Bull. London Math. Soc.* **22** (1990), 529–532.

[29] J.I. Hall, The number of trace-valued forms and extraspecial groups, *J. London Math. Soc. (2)* **37** (1988), 1–13.

[30] G. Higman, On infinite simple groups, *Publ. Math. Debrecen* **3** (1954), 221–226.

[31] W.A. Hodges, Categoricity and permutation groups, *Logic Colloquium '87* (ed. H.-D. Ebbinghaus *et al.*), Elsevier, Amsterdam, 1989, 53–72.

[32] W.A. Hodges, I.M. Hodkinson, H.D. Macpherson, Omega-categoricity, relative categoricity and coordinatisation, *Ann. Pure Appl. Logic* **46** (1990), 169–199.

[33] W.C. Holland, The lattice-ordered group of automorphisms of an ordered set, *Michigan Math. J.* **10** (1963), 399–408.

[34] T. Jech, *Set theory*, Academic Press, New York, 1978.

[35] W.M. Kantor, Homogeneous designs and geometric lattices, *J. Comb. Theory (A)* **38** (1985), 66–74.

[36] H.A. Kierstead, P. Nyikos, Hypergraphs with finitely many isomorphism subtypes, *Trans. Amer. Math. Soc.* **312** (1989), 699–718.

[37] D. Lascar, Autour de la propriété du petit indice, *Proc. London Math. Soc. (3)* **62** (1991), 25–53.

[38] D. Lascar, Les automorphismes d'un ensemble fortement minimal, *J. Symb. Logic* **57** (1992), 238–251.

[39] H. Läuchli, P.M. Neumann, On linearly ordered sets and permutation groups of countable degree, *Arch. Math. Logic* **27** (1988), 189–192.

[40] M.W. Liebeck, C.E. Praeger, J. Saxl, A classification of the maximal subgroups of the alternating and symmetric groups, *J. Alg.* **111** (1987), 365–383.

[41] M.W. Liebeck, C.E. Praeger, J. Saxl, On the O'Nan–Scott reduction theorem for finite primitive permutation groups, *J. Austral. Math. Soc.* **44** (1988), 389–396.

[42] A. Litman, S. Shelah, Models with few isomorphic expansions, *Israel J. Math* **28** (1977), 331–338.

[43] J.T. Lloyd, Lattice-ordered groups and o-permutation groups, Ph.D. thesis, Tulane University, New Orleans, Louisiana, 1964.

[44] H.D. Macpherson, Groups of automorphisms of $\aleph_0$-categorical structures, *Quart. J. Math. Oxford (2)* **37** (1986), 449–465.

[45] H.D. Macpherson, Orbits of infinite permutation groups, *Proc. London Math. Soc. (3)* **51** (1985), 246–284.

[46] H.D. Macpherson, Maximal subgroups of infinite dimensional general linear groups, *J. Austral. Math. Soc.*, to appear.

[47] H.D. Macpherson, P.M. Neumann, Subgroups of infinite symmetric groups, *J. London Math. Soc. (2)* **42** (1990), 64–84.

[48] H.D. Macpherson, C.E. Praeger, Maximal subgroups of infinite symmetric groups, *J. London Math. Soc. (2)* **42** (1990), 85–92.

[49] H.D. Macpherson, A.H. Mekler, S. Shelah, The number of infinite substructures, *Math. Proc. Cam. Phil. Soc.* **109** (1991), 193–209.

[50] R. Mansfield, G. Weitkamp, Recursive aspects of descriptive set theory, *Oxford Logic Guides* **11**, Oxford University Press, Oxford, 1985.

[51] J. van Mill, An introduction to $\beta\mathbb{N}$, *Handbook of Set-Theoretic Topology* (Eds. K. Kunen, J.E. Vaughan), North-Holland, Amsterdam, 1984, 503–567.

[52] R.G. Möller, The automorphism groups of regular trees, *J. London Math. Soc.*, to appear.

[53] G. Moran, Conjugacy classes whose square is an infinite symmetric group, *Trans. Amer. Math. Soc.* **316** (1989), 493–522.

[54] Y.N. Moschovakis, *Descriptive set theory*, North-Holland, Amsterdam, 1980.

[55] P.M. Neumann, Some primitive permutation groups, *Proc. London Math. Soc. (3)* **50** (1985), 265–281.

[56] P.M. Neumann, Homogeneity of infinite permutation groups, *Bull. London Math. Soc.* **20** (1988), 305–312.

[57] P. Nyikos, Transitivities of permutation groups, preprint.

[58] E.B. Rabinovich, A certain problem of Scott, *Dokl. Akad. Nauk. BSSR* **19** (1975), 583–584.

[59] D. Ramsay, On linearly ordered sets and permutation groups of uncountable degree, D.Phil. thesis, Oxford, 1990.

[60] F. Richman, Maximal subgroups of infinite symmetric groups, *Can. Math. Bull.* **10** (1967), 375–381.

[61] A. Rosenberg, The structure of the infinite general linear group, *Ann. of Math.* **68** (1958), 278–294.

[62] M. Rubin, On the reconstruction of $\aleph_0$-categorical structures from their automorphism groups, preprint.

[63] W. Rudin, Homogeneity properties in the theory of Cech compactifications, *Duke Math. J.* **23** (1956), 409–420.

[64] C. Ryll-Nardzewski, On categoricity in power $\leq \aleph_0$, *Bull. Acad. Pol. Sci. Ser. Math. Astr. Phys.* **7** (1959), 545–548.

[65] W.R. Scott, *Group theory*, Prentice-Hall, New Jersey, 1964.

[66] S.W. Semmes, Endomorphisms of infinite symmetric groups, *Abstracts Amer. Math. Soc.* **2** (1981), 426.

[67] S.W. Semmes, Infinite symmetric groups, maximal subgroups and filters, *Abstracts Amer. Math. Soc.* **3** (1982), 38.

[68] S. Shelah, *Classification theory and the number of non-isomorphic models*, North-Holland, Amsterdam, 1978.

[69] S. Shelah, S. Thomas, Subgroups of small index in infinite symmetric groups II, *J. Symb. Logic* **54** (1989), 95–99.

[70] S. Shelah, S. Thomas, Implausible subgroups of infinite symmetric groups, *Bull. London Math. Soc.* **20** (1988), 313–318.

[71] S. Shelah, S. Thomas, Homogeneity of infinite permutation groups, *Arch. Math. Logic* **28** (1989), 143–147.

[72] G. Stoller, Example of a subgroup of S_∞ which has a set-transitive property, *Bull. Amer. Math. Soc.* **69** (1963), 220–221.

[73] L. Svenonius, $\aleph_0$-categoricity in first-order predicate calculus, *Theoria* **25** (1959), 82–94.

[74] S. Thomas, Reducts of the random graph, *J. Symb. Logic* **56** (1991), 176–181.

[75] S. Thomas, Permutation groups which act transitively on moieties, preprint.

[76] J.K. Truss, Infinite permutation groups II. Subgroups of small index, *J. Alg.* **120** (1989), 494–515.

[77] H. Wielandt, Unendliche Permutationsgruppen, Universität Tubingen, 1959.

[78] E.L. Wimmers, The Shelah p-point independence theorem, *Isr. J. Math.* **43** (1982), 28–48.

[79] B.I. Zil'ber, Strongly minimal totally categorical theories, I, II, III, *Siberian J. Math.* **21** (1980), 219–230; **25** (1984), 396–412; **25** (1984), 559–572.

Homogeneous Partially Ordered Sets

ALAN H. MEKLER*

Department of Mathematics and Statistics
Simon Fraser University
Burnaby, B.C., V5A 1S6
Canada

In [9], Schmerl classified the countable partially ordered sets which are homogeneous. This classification was extended by Droste who studied partially ordered sets (henceforth to be known as posets) which are 2-homogeneous. In this paper we will survey their results and others on posets which are either homogeneous or partially homogeneous.

Recall that a relational structure is $(\leq)n$-*homogeneous* if every isomorphism between substructures of cardinality (at most) n extends to an automorphism. A structure is *homogeneous* if it is n-homogeneous for all finite n. The simplest example, other than a set with no relations at all, is the rational linear order, $\mathbb{Q}$. The proof that there is up to isomorphism only one countable dense linear order without endpoints shows that $\mathbb{Q}$ is homogeneous. In fact it is easy to see that $\mathbb{Q}$ is the only $\leq$ 2-homogeneous countable linear order other than the one element linear order.

The classification of posets is relative to that of linear orders, i.e., we attempt to describe the posets in terms of linear orders. Of course we have complete knowledge of the countable homogeneous linear order, but there are many complicated uncountable homogeneous linear orders. One fact that we will use later is that a $\leq$ 2-homogeneous linear order is homogeneous. The basic approach to the classification is to analyze the simple cases which occur when various small posets are forbidden to appear and then analyze the remaining case where all of these posets occur. Our analysis of these posets is close to Schmerl's original approach with some influence from Droste. Droste was also concerned with a weaker notion which might be called set-homogeneity, i.e., if two finite substructures are isomorphic then there is *some* automorphism carrying one to the other.

Before beginning the classification we recall some standard definitions. Suppose $\mathbb{P}$ is a poset. A *chain* is a linearly ordered subset and an *antichain* is a subset of pairwise incomparable elements. We use $a \parallel b$ to indicate that a and b are incomparable. If A is a subset of $\mathbb{P}$, we use $x < A$ to denote that for every $a \in A$, $x < a$, etc.

1 The Easy Cases

The first stage of the analysis is to consider posets which omit certain 2 or 3 element subposets (or types) (see figure 1). These types are (a) the 2 element chain, (b) a two element chain and a point incomparable to both, (c) two incomparable points with a common upper

*Research partially supported by NSERC grant A8948.

N.W. Sauer et al. (eds.), Finite and Infinite Combinatorics in Sets and Logic, 279–288.

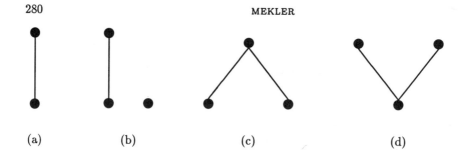

(a)   (b)   (c)   (d)

Figure 1: types which may be omitted

bound, and (d) two incomparable points with a common lower bound. (A poset *omits* a type if there is no subposet isomorphic to the type. The poset *realizes* a type if there is a subposet isomorphic to the type.)

The following theorem is meant to be self-evident.

Theorem 1 *Suppose that* $\mathbb{P}$ *is a* $\leq$ *2-homogeneous poset.*

1. *If* $\mathbb{P}$ *omits* (a) *then* $\mathbb{P}$ *is an antichain. Furthermore* $\mathbb{P}$ *is homogeneous.*

2. *Suppose* $\mathbb{P}$ *realizes* (a) *and omits* (b)*. Then there is a chain* C *and an antichain* A *so that* $\mathbb{P} = C \times A$, *where* $(c,a) > (d,b)$ *if and only if* $c > d$. *Furthermore* $\mathbb{P}$ *is homogeneous.*

3. *Suppose* $\mathbb{P}$ *realizes* (a) *and omits* (c) *and* (d)*. Then there is a chain* C *and an antichain* A *so that* $\mathbb{P} = C \times A$, *where* $(c,a) > (d,b)$ *if and only if* $c > d$ *and* $a = b$. *Furthermore* $\mathbb{P}$ *is homogeneous.*

4. *Suppose* $\mathbb{P}$ *realizes* (a)*,* (b) *and* (c) *but omits* (d)*. Then for any* $a \in \mathbb{P}$, $\{b : b > a\}$ *is a chain.*

5. *Suppose* $\mathbb{P}$ *realizes* (a)*,* (b) *and* (d) *but omits* (c)*. Then for any* $a \in \mathbb{P}$, $\{b : b < a\}$ *is a chain.*

In each of the last two cases the poset is called a *tree*. There is a very interesting and deep theory concerning the automorphism groups of trees, the most notable being Rubin's results on the reconstruction of trees from their automorphism groups ([7]). However they are a dead end in the search for homogeneous posets.

Lemma 2 *No tree is* $\leq$ *3-homogeneous.*

Proof. Suppose T is a counterexample. Consider the points a, b, c in Figure 2. Suppose φ is an automorphism so that $\varphi(a) = a$, $\varphi(b) = c$ and $\varphi(c) = b$. Consider $\varphi(d)$. Then $\varphi(d) < c, a$, since $d < b, a$. Similarly $\varphi(d) \parallel b$. But there is no such element. $\square$

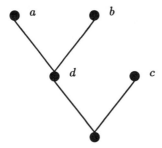

Figure 2: There is no $\leq$ 3-homogeneous tree.

2   Posets containing the pentagon

The study of $\leq$ 2-homogenous posets is reduced to studying those which realize types (b), (c), (d). This data can be summarized using the pentagon. We say that a poset realizes the pentagon if there are elements a, b, c, d, e so that $a < b < c < d$, $a < e < d$ and $e \parallel b, c$.

Lemma 3 *Suppose* $\mathbb{P}$ *is a* $\leq$ *2-homogeneous poset. Then* $\mathbb{P}$ *realizes* (b), (c), (d) *if and only if* $\mathbb{P}$ *realizes the pentagon.*

   Proof. Since the pentagon realizes (b), (c), (d), we only need to do one direction. The proof is most easily seen by a picture.

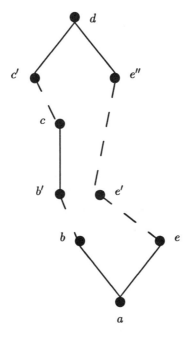

There is one other countable homogeneous poset which is known, the generic poset. The generic poset is the unique countable poset, $\mathbf{G}$, satisfying the axioms which say (in the presence of countability) that it is the universal homogeneous poset; namely if P is a finite subposet of $\mathbf{G}$ and $P \subseteq Q$ where Q is a finite poset then there is an embedding of Q into $\mathbf{G}$ which is the identity on P. One can restrict the axiom to the case where $|Q \backslash P| = 1$. Suppose $Q \backslash P = \{x\}$. Then x partitions P into three sets, $U = \{a \in P : x < a\}$, $D = \{a \in P : x > a\}$ and $A = \{a \in P : x \parallel a\}$. Let B be the antichain consisting of the maximal elements of D and C be the antichain consisting of the minimal elements of U. If $y \in \mathbf{G}$ is such that $B < y < C$ and $y \parallel A$, then $P \cup \{y\}$ is isomorphic to Q. We say that (A, B, C) is a (k, m, n)-configuration if $|A| = k$, $|B| = m$, $|C| = n$, and

1. B and C are antichains,

2. $B < C$,

3. for all $a \in A$ and $b \in B$, $a \not\leq b$, and for all $a \in A$ and $c \in C$, $c \not\leq a$.

Let $T_{k,m,n}$ be the first-order sentence which says that

> If (A, B, C) is a (k, m, n)-configuration then there is x so that $B < x < C$ and $x \parallel A$.

So we have the following theorem.

Theorem 4 *The theory of the generic poset is axiomatized by the poset axioms together with the axioms $T_{k,m,n}$, for $k, m, n < \omega$.*

The proof that these axioms are $\aleph_0$-categorical uses Fraïssé's characterization of universal homogeneous models in terms of amalgamation classes ([4]).

To complete the classification of the homogeneous posets, it suffices to show that a countable homogeneous poset containing the pentagon is the generic poset. We will take a slightly roundabout route to this result. First we prove some results about $\leq$ 2-homogeneous posets realizing the pentagon.

Lemma 5 *Suppose $\mathbb{P}$ is a $\leq$ 2-homogeneous poset realizing the pentagon. If X is a finite subset of $\mathbb{P}$ then there are points a, b, c so that $a < X < b$ and $c \parallel X$.*

Proof. Since $\mathbb{P}$ realizes type (c) and is 2-homogeneous every two points have an upper bound. By repeating this observation we can find a, b so that a is a lower bound for X and b is an upper bound for X. Since $\mathbb{P}$ realizes (b) there is c so that $c \parallel a, b$. $\square$

The following lemma is a variant of Schmerl's original argument that a homogeneous poset which contains the pentagon realizes all finite posets.

Lemma 6 *Suppose $\mathbb{P}$ is a $\leq$ 4-homogeneous poset which realizes the pentagon. Then $\mathbb{P}$ realizes every poset on at most five points.*

Proof. The proof is by induction on n, the size of the poset and then on k, the number of non-maximal elements. The case $n = 1$ is trivial. Also if there are no non-maximal elements, then the poset is an antichain and so we are done by Lemma 5. Suppose we know the induction hypothesis for $n-1$ and for n with $k-1$ non-maximal elements. Let Q be a poset of size n with k non-maximal elements. First suppose that Q has a unique minimal element, i.e., there is $a \in Q$ so that $a < Q \setminus \{a\}$. Then by the induction hypothesis we can identify $Q \setminus \{a\}$ with a subset of $\mathbb{P}$. Then by Lemma 5, this set has a lower bound.

Now let a, b be two minimal elements of Q. We can assume neither a nor b is maximal as otherwise we could finish by Lemma 5. There are two cases to consider. First assume $\{x \in Q: a < x\} = \{x \in Q: b < x\}$. Add a new point c to Q so that $a, b < c$ and for all $x \in Q \setminus \{a, b\}$, $c \parallel x$ if $a \parallel x$ and $c < x$ if $a < x$. Let Q' denote this new poset. Let U be the set of maximal elements of $\{x \in Q \setminus \{a, b\}: x \parallel a\}$. Let R be $Q' \setminus \{a, b\}$ and $S = \{a, b, c\} \cup U$. By the induction hypothesis there is an embedding $\varphi: R \to \mathbb{P}$. Also there is an embedding $\psi: S \to \mathbb{P}$, since S is isomorphic to the poset of type (c) together with an antichain. Let θ be an automorphism of $\mathbb{P}$ which extends $\varphi \circ \psi^{-1}|(\psi(S \setminus \{a, b\}))$. (Since the size of the domain of the partial function is at most 4, θ exists.) By replacing ψ by $\theta \circ \psi$, we can assume that φ and ψ agree on their common domain. One can then check that $\varphi(R) \cup \{\psi(a), \psi(b)\}$ is isomorphic to Q', which contains Q.

Now assume $\{x \in Q: a < x\} \neq \{x \in Q: b < x\}$. Without loss of generality we can assume there is d so that $a < d$ but $b \not\leq d$. Let Q' be $Q \cup \{c\}$ where $b < c$ and $c \parallel y$ if $y \neq b$. Let R be $Q' \setminus \{a\}$ and S be $Q' \setminus \{b\}$. Since R and S have one less non-maximal element than Q, there are embeddings φ and ψ of R and S respectively into $\mathbb{P}$. As above we can assume they agree on their common domain. We claim that $\varphi(R) \cup \{\psi(a)\}$ is isomorphic to Q. All that we need to check is that $\psi(a) \parallel \varphi(b)$. Suppose not. If $\psi(a) \leq \varphi(b)$, then $\psi(a) < \varphi(c) = \psi(c)$, which is a contradiction. Similarly if $\varphi(b) < \psi(a)$, then $\varphi(b) < \varphi(d)$, which again is a contradiction. $\square$

Since the theorem above is only about sets of size 5, it would be possible to give a brute force argument. However the same proof shows that if $\mathbb{P}$ is a homogeneous poset realizing the pentagon, then $\mathbb{P}$ realizes every finite poset. This fact together with Theorem 1 completes Schmerl's classification of the countable homogeneous posets.

Theorem 7 *Suppose $\mathbb{P}$ is a countable homogeneous poset. Then $\mathbb{P}$ is either isomorphic to the generic poset or $\mathbb{P}$ is one of the first three types of posets in Theorem 1, namely, an antichain, a chain of antichains or an antichain of chains.*

3 Finite axiomatization and partial homogeniety

In [1], Albert and Burris showed that the theory of the generic poset is finitely axiomatized. Recall the axioms $T_{k,m,n}$ from Theorem 4.

Theorem 8 [1] *The theory of the generic poset is axiomatized by the axioms for posets together with $T_{k,m,n}$ where $k \leq 1$ and $m, n \leq 2$ or $k = 2$ and $m, n \leq 1$.*

Proof. (Sketch) The proof heavily uses the transitivity of the partial order. As an example consider the case where (A, B, C) form a $(1, 3, 2)$-configuration. We want to find x so that $B < x < C$ and $x \parallel A$. Let $B = \{b_1, b_2, b_3\}$. By the axiom $T_{1,2,2}$ we can choose x_1 so that $b_1, b_2 < x_1 < C$ and $x_1 \parallel A$. If $x_1 > b_3$, we are done, otherwise we can apply the axiom again with $B' = \{x_1, b_3\}$. In general, one first proves by induction that $\mathbb{P}$ satisfies all the axioms $T_{1,m,n}$. Suppose now that (A, B, C) is a (k, m, n)-configuration. By repeatedly applying axioms of the form $T_{1,m,n}$, we can find y,z so that $y < A < z$ and $y \parallel B$, $z \parallel C$. Applying axioms of the form $T_{1,m,n}$ again, there is x_1, x_2 so that $B < x_1 < x_2 < C$, $y \parallel x_1$ and $z \parallel x_2$. Finally by $T_{2,1,1}$, there is x so that $x_1 < x < x_2$ and $y,z \parallel x$. $\square$

It is possible to eliminate several of the Albert–Burris axioms. In their proof they show the following lemma.

Lemma 9 *For all m,n the axiom $T_{0,m,n}$ follows from the axioms $T_{0,r,s}$ where $r,s \le 2$.*

We can now begin the elimination of some of the axioms.

Theorem 10 *The theory of the generic poset is axiomatized by the poset axioms together with $T_{0,m,n}$, $T_{k,m,0}$ and $T_{k,0,m}$ where $k,m,n \le 2$ and $k+m \le 3$.*

Let S denote this set of axioms.

Lemma 11 *The axioms $T_{1,m,n}$ follow from S where $m,n \le 2$ and $m+n \le 3$.*

Proof. We will only do two representative cases. We first verify that $T_{1,1,2}$ follows. (The axiom $T_{1,2,1}$ can be treated dually and the axiom $T_{1,1,1}$ is a simpler case.) Let a,b,c_0,c_1 be as in the hypothesis of the $T_{1,1,2}$. By applying $T_{1,0,2}$ find b' so that $b' < c_0, c_1$ and $b' \parallel a$. Next choose c' so that $c' > b, b'$ and $c' \parallel a$. Finally choose x so that $b,b' < x < c_0, c_1, c'$. Since $b' < x < c'$, $x \parallel a$.

Next we consider $T_{1,2,2}$. Suppose a, b_1,b_2 and c_1,c_2 are as in $T_{1,2,2}$. First choose d_1 so that $b_1 < d_1 < c_1,c_2$ and $d_1 \parallel a$. If $b_2 < d_1$, then we are in a simpler case, so assume that $d_1 \parallel b_2$. Next choose x' so that $d_1, b_2 < x' < c_1, c_2$. Note that $x' \not\le a$, as otherwise we would have $d_1 < a$. Finally choose x so that $d_1, b_2 < x < x'$ and $a \parallel x$. Suppose a, b_1,b_2 and c_1,c_2 are as in $T_{1,2,2}$. First choose d_1 so that $b_1 < d_1 < c_1,c_2$ and $d_1 \parallel a$. If $b_2 < d_1$, then we are in a simpler case, so assume that $d_1 \parallel b_2$. Next choose x' so that $d_1, b_2 < x' < c_1, c_2$. Note that $x' \not\le a$, as otherwise we would have $d_1 < a$. Finally choose x so that $d_1, b_2 < x < x'$ and $a \parallel x$. $\square$

Lemma 12 *The axiom $T_{2,1,1}$ follows from S.*

Proof. Let a_0,a_1,b,c be as in the hypothesis of $T_{2,1,1}$. Choose x_0 so that $a_0 \parallel x_0$ and $b < x_0 < c$ (such a choice can be made by the previous lemma). There are 3 cases. If $x_0 \parallel a_1$, we are done. Assume that $x_0 < a_1$ (the other case is dual). Now choose x_1 so that $x_0 < x_1$ and $x_1 \parallel a_0, a_1$. (This step is an application of $T_{2,1,0}$.) Now choose x so that $x_0 < x < x_1, c$ and $x \parallel a_1$. Since $b < x_0$ and $a \parallel x_0, x_1$, a is as desired. $\square$

This proposition finishes the proof of Theorem 10.

There are two important properties of this new set of axioms. The new axioms only require that we can find extensions for embeddings of posets of size at most 4, while the original axioms required that we could find an extension for a poset of size 5. Also $T_{0,2,2}$ is the only one of the reduced axioms which talks of a poset of size 4.

An immediate consequence of this refinement of the Albert and Burris theorem and Lemma 6 is the following theorem of Droste and Macpherson [3].

Theorem 13 *Any $\leq$ 4-homogeneous poset which realizes the pentagon satisfies the theory of the generic poset.*

Proof. Consider an instance of the axioms for some A, B, and C. By Lemma 6, we can find A', B', C' so that the poset $A \cup B \cup C$ and $A' \cup B' \cup C'$ are isomorphic via an isomorphism carrying A' to A, B' to B and C' to C and there is x' so that $B' < x' < C'$ and $x' \parallel A'$. We can let x be the image of x' under an automorphism which is guaranteed by $\leq$ 4-homogeneity. $\square$

As a consequence of this theorem they were able to deduce the following result.

Theorem 14 *If $\mathbb{P}$ is a countable poset which is $\leq$ 4-homogeneous, then $\mathbb{P}$ is homogeneous.*

Proof. If $\mathbb{P}$ omits the pentagon, then by our previous analysis it is homogeneous. (Note that this deduction did not require countability.) If it realizes the pentagon then it is the generic poset, which is homogeneous. $\square$

We also get information about posets which are $\leq$ 3-homogeneous.

Theorem 15 *If $\mathbb{P}$ is a $\leq$ 3-homogeneous poset which realizes all finite posets then $\mathbb{P}$ satisfies all the axioms of S except perhaps $T_{0,2,2}$.*

Proof. All the other axioms involve at most 3 points. $\square$

These theorems raise a number of questions: can the methods be applied to familiar structures such as graphs, can 4 be lowered to 3, and can the assumption that $\mathbb{P}$ is countable be omitted? The answers to all the questions are negative.

Macpherson [5] has shown that a finitely axiomatizable complete theory with trivial algebraic closure has the strict order property. In particular, the theory of none of the homogeneous graphs is finitely axiomatized. Droste and Macpherson [3] give examples to show that for all n there is a graph which is n-homogeneous but not $(n+1)$-homogeneous. Constructions of partially homogeneous posets will occupy us for the rest of this paper.

4 Partially homogeneous posets

In [3], Droste and Macpherson construct countable posets which are $\leq$ 3-homogeneous but not 4-homogeneous as well as countable posets which are $\leq$ 2-homogeneous but not 3-homogeneous. (The integers are 1-homogeneous but not 2-homogeneous. Also there are trees which are 2-homogeneous.) We will sketch the proof that there is a $\leq$ 3-homogeneous not 4-homogeneous poset which contains all finite posets. In constructing such a poset one of the axioms must be falsified. By Theorem 15, they had to falsify $T_{0,2,2}$. In order to do this they introduce a 4-ary predicate symbol, R, and add to the theory of posets axioms which say that if $R(a_1, a_2, b_1, b_2)$ holds then $\{a_1, a_2\}$ and $\{b_1, b_2\}$ witness the failure of the axiom $T_{0,2,2}$. In order to state the axioms precisely it is convenient to introduce some abbreviations. Let $\mathrm{Web}(x_1, x_2, y_1, y_2)$ abbreviate $x_1 \parallel x_2 \wedge y_1 \parallel y_2 \wedge x_1, x_2 < y_1, y_2$ (i.e., $(\emptyset, \{x_1, x_2\}, \{y_1, y_2\})$ is a $(0,2,2)$-configuration). Let $(u_1, u_2, v_1, v_2) \preceq (x_1, x_2, y_1, y_2)$ abbreviate $\mathrm{Web}(u_1, u_2, v_1, v_2) \wedge \mathrm{Web}(x_1, x_2, y_1, y_2) \wedge \bigwedge_j \bigvee_i x_i \leq u_j \wedge \bigwedge_j \bigvee_i v_j \leq y_i$. In a poset if $(c_1, c_2, d_1, d_2) \preceq (a_1, a_2, b_1, b_2)$ and $c_1, c_2 < z < d_1, d_2$ for some z, then $a_1, a_2 < z < b_1, b_2$.

Let T be the axioms of posets together with the following:

1. $R(x_1, x_2, y_1, y_2) \leftrightarrow R(x_2, x_1, y_1, y_2) \leftrightarrow R(x_1, x_2, y_2, y_1)$,

2. $R(x_1, x_2, y_1, y_2) \rightarrow \text{Web}(x_1, x_2, y_1, y_2)$,

3. $R(x_1, x_2, y_1, y_2) \rightarrow \neg \exists z (x_1, x_2 < z < y_1, y_2)$,

4. $(R(x_1, x_2, y_1, y_2) \wedge (u_1, u_2, v_1, v_2) \preceq (x_1, x_2, y_1, y_2)) \rightarrow R(u_1, u_2, v_1, v_2)$.

The first three axioms give the intended meaning of R. The last axiom is present so that the models of T form an amalgamation class and so there is a countable homogeneous model of T.

Theorem 16 *There is a poset which is ≤ 3-homogeneous but not 4-homogeneous. Further the poset contains all countable posets.*

Proof. Let $\mathbb{P}$ be the countable homogeneous model of T. Then it is not 4-homogeneous as a poset since there are two sorts of webs on 4 elements. But it is ≤ 3-homogeneous as a poset since the isomorphism type of any three elements of a model of T is determined by its isomorphism type as a poset. $\square$

Droste and Macpherson were able to construct countably many examples of ≤ 3-homogeneous posets which are not 4-homogeneous by varying the size of the webs which are not filled. Saracino and Wood [8] were able to construct $2^{\aleph_0}$ examples by adding predicates which pick out webs which are not filled and do not arise as in the fourth axiom. They also produce examples where the model is not $\aleph_0$-categorical. Examples are also given of posets which are 2-homogeneous, not 3-homogeneous but contain all finite posets. In some sense they are analogous and work by falsifying $T_{0,1,2}$. It doesn't seem to be easy to construct examples where we falsify a different axiom, say $T_{1,0,2}$.

One reason to be interested in the question of whether there exist $\leq n$-homogeneous but not $(n+1)$-homogeneous posets in uncountable cardinalities is that the existence of such posets show the somewhat roundabout proof of Theorem 14 is necessary. For example there is a silly proof that if a countable linear order is ≤ 2-homogeneous then it is homogeneous. We observe that any ≤ 2-homogeneous linear order is a dense linear order without endpoints. Then countability implies that it is isomorphic to $\mathbb{Q}$. This proof is silly since ≤ 2-homogeneity lets one piece together automorphisms and show that any ≤ 2-homogeneous linear order is homogeneous. The existence of say ≤ 5-homogeneous posets which are not homogeneous shows that we were not missing such a simple proof in the case of posets.

A detailed proof of the following theorem should appear in a paper by Droste and Mekler which is supposed to be in preparation.

Theorem 17 *For every $n < \omega$ there is a $\leq n$-homogeneous poset which is not $(n+1)$-homogeneous.*

Proof. (Sketch) The result is first established under certain set theoretic hypotheses and then turned into an absolute result by general methods. We can prove the existence of such a poset assuming CH, or $\diamondsuit(\aleph_1)$ or that $\aleph_1$ Cohen reals have been added to the universe.

It is probably easiest to describe the proof assuming $\Diamond(\aleph_1)$. Recall that $\Diamond(\aleph_1)$ says that there is a sequence of functions $\langle f_\alpha : \alpha \to \alpha : \alpha < \omega_1 \rangle$ so that for every function $f : \omega_1 \to \omega_1$ there is a (stationary set of) α such that $f|\alpha = f_\alpha$. Fix such a sequence. The poset we will construct will have universe ω_1. At each stage of the construction we will examine f_α and when appropriate take steps to stop it from extending to a homomorphism.

Form a new language L' by adding to the language of posets a new $(n+1)$-ary predicate R. Let T' be the theory of posets together with the axiom $R(x_0, \ldots, x_n) \to \bigwedge_{i \neq j} x_i \neq x_j$. The finite models of T' form an amalgamation class with the joint embedding property so there is a universal homogeneous countable model of T' (also called the generic model of T'). For L'-structures, we will use the word automorphism to denote automorphisms as posets and L'-automorphism to indicate automorphisms as L'-structures. The use of Aut and $\mathrm{Aut}_{L'}$ is analogous. Suppose that A is the generic model of T', and consider $\mathrm{Aut}_{L'}(A)$. This group is homogeneous on A viewed as an L'-structure. Since two subsets of size at most n are isomorphic as L'-structures if and only if they are isomorphic as L-structures, $\mathrm{Aut}_{L'}(A)$ is $\leq n$-homogeneous on A viewed as a poset. On the other hand since there are $A = \{a_0, \ldots, a_n\}$ and $B = \{b_0, \ldots, b_n\}$ which are isomorphic as posets and $A \models R(a_0, \ldots, a_n) \wedge \neg R(b_0, \ldots, b_n)$, $\mathrm{Aut}_{L'}(A)$ is not $(n+1)$-homogeneous. We are not done since A will be homogeneous as a poset.

The strategy is to define, for $\alpha < \omega_1$, an increasing continuous sequence (A_α, G_α) where A_α is a generic L'-structure and G_α is a countable $\leq n$-homogeneous subgroup of $\mathrm{Aut}_{L'}(A_\alpha)$. (The embedding of G_α into G_β for $\alpha < \beta$ is done so that every element of G_α extends to an element of $\mathrm{Aut}_{L'}(A_\beta)$.) The construction will be done so that if we set $A = \bigcup A_\alpha$ and $G = \bigcup G_\alpha$ then $\mathrm{Aut}(A) = G = \mathrm{Aut}_{L'}(A)$. So A will be the desired poset. To aid the bookkeeping we can assume the underlying set of A_α is $\omega \cdot \alpha$. To construct (A_0, G_0), let A_0 be a generic L'-structure and G_0 be any $\leq n$-homogeneous subset of $\mathrm{Aut}_{L'}(A_0)$. At limit ordinals the construction is determined.

At stage α, if f_α is an automorphism of A_α which is not an element of G_α then we will want to construct $A_{\alpha+1}$ so that f_α does not extend to an automorphism of $A_{\alpha+1}$. Of course we will have to do this in such a way that automorphisms we previously stopped from extending do not gain extensions. The idea is to choose an upward closed subset X of A_α and then adjoin a new point a so that $a < X$ and $a \parallel (A_\alpha \setminus X)$. Next we extend the action of G_α by adding new points a_g for $g \in G_\alpha$ where $a_g < g(X)$ and a_g is incomparable with everything else. We add no new R relations. This gives an L'-structure with an action of G_α on it. By alternately amalgamating one point extensions of finite subsets and ensuring that the action of G_α extends, we build $A_{\alpha+1}$ a generic L'-structure with G_α acting on $A_{\alpha+1}$. If X is chosen correctly there will be no element $b \in A_{\alpha+1}$ such that $b < f_\alpha(X)$ and $b \parallel (A_\alpha \setminus f_\alpha(X))$. Hence f_α does not extend to an automorphism of $A_{\alpha+1}$. (The choice of X is not too delicate as there is a comeagre set of choices which will work.)

We have completed the sketch of the proof if we assume $\Diamond$. For the general case, we will use a transfer theorem which Shelah assures me is true and follows along the lines of [6]. Assume that all models carry an underlying poset structure which satisfies the axioms for the generic poset. We enrich first order logic by a second order quantifier Q_{Po}, where the interpretation of $Q_{\mathrm{Po}} f \ldots$ is "for all automorphisms $f \ldots$". The logic $L(Q_{\mathrm{Po}})$ is fully compact but the property which interests us most is that its validities are absolute, e.g., if $\Diamond(\aleph_1)$ implies that a sentence has a model then the sentence has a model. Given this fact we are done since the sentence which expresses of a model that it is $\leq n$-homogeneous but not $(n+1)$-homogeneous is expressible in $L(Q_{\text{Po}})$. $\square$

There is a key property of posets which satisfy the generic poset axioms that is used both to choose the set X in the proof above and to prove the properties of $L(Q_{\text{Po}})$, namely that any automorphism is determined by its restriction to any infinite definable subset. This property allows us to deduce that two different automorphisms remain different even when restricted to an infinite definable subset and to extend local definability to global definability.

References

[1] Albert, M. and Burris, S., Finite axiomatizations for existentially closed posets and semilattices, *Order* 3 (1986) 169–178.

[2] Droste, M., Structure of partially ordered sets with transitive automorphism groups, *Memoirs Amer. Math. Soc.* 334 (1985).

[3] Droste, M. and Macpherson, H.D., On k-homogeneous posets and graphs, *J. Comb. Th. (Ser A)* 56 (1991) 1–15.

[4] Fraïssé, R., Sur l'extension aux relations de quelques propriétés des ordres, *Ann. Sci. École Norm. Sup.* 71 (1954) 361–388.

[5] Macpherson, H.D., Finite axiomatizability and theories with trivial algebraic closure, *Notre Dame J. Formal Logic* 32 (1991) 188–192.

[6] Mekler, A. and Shelah, S., Compactness results for isomorphism quantifiers, preprint.

[7] Rubin, M., The reconstruction of trees from their automorphism groups, preprint.

[8] Saracino, D. and Wood, C., Partially homogeneous partially ordered sets, *J. Comb. Th. (Ser. A)* (to appear).

[9] Schmerl, J., Countable homogeneous partially ordered sets, *Algebra Universalis* 9 (1979) 317–321.

Cardinal Representations

E. C. MILNER[*]

Department of Mathematics and Statistics,
University of Calgary,
2500 University Drive N.W.,
Calgary, Alberta, Canada
T2N 1N4

1   Introduction

In these lectures I shall describe some joint work with Fred Galvin and Maurice Pouzet which was done over several years but mostly when Fred and I were guests of Maurice at the Université Claude Bernard, Lyon in 1988. Fuller details will be found in [4].

A *preclosure* on a non-empty set E is a function $\varphi : \mathcal{P}(E) \to \mathcal{P}(E)$ such that (i) $X \subseteq \varphi(X) \subseteq \varphi(Y)$ whenever $X \subseteq Y \subseteq E$. Although not an essential part of the definition, it is convenient to assume that φ is normalized so that (ii) $\varphi(\emptyset) = \emptyset$. A subset $X \subseteq E$ is *closed* with respect to φ if $\varphi(X) = X$. The preclosure φ is a *closure* if it is idempotent, *i.e.* satisfies (iii) $\varphi(\varphi(X)) = \varphi(X)$ for every subset $X \subseteq E$. A *topological closure* is a closure φ which satisfies the remaining Kuratowski axiom *(iv)* $\varphi(X \cup Y) = \varphi(X) \cup \varphi(Y)$ for all $X, Y \subseteq E$. We also use the algebraic terminology and say that the preclosure φ has *character* n ($< \omega$) if, for every subset $X \subseteq E$, $\varphi(X) = \bigcup \{\varphi(Y) : Y \subseteq X, |Y| \leq n\}$; also, we say that φ has *finite character* or is *algebraic* if $\varphi(X) = \bigcup \{\varphi(Y) : Y \subseteq X, |Y| < \omega\}$.

Let φ be a preclosure on E. A subset $X \subseteq E$ is *independent* if $x \notin \varphi(X \setminus \{x\})$ for each element $x \in X$. A subset $X \subseteq E$ is *dependent* if it is not independent. In particular, the empty set is independent. The *independence number* of the preclosure φ is

$$\mathrm{ind}(\varphi) = \min\{\kappa : |X| < 1 + \kappa \text{ for every independent set } X \subseteq E\}.$$

In particular, for finite k, $\mathrm{ind}(\varphi) = k$ means that k is the maximum size of an independent set; and $\mathrm{ind}(\varphi) = \omega$ means that every independent set is finite, but there is no bound on the size. A subset $X \subseteq E$ is called a *generating set* for the preclosure φ if $\varphi(X) = E$, and we define the *dimension* of φ to be:

$$\dim(\varphi) = \min\{|X| : \varphi(X) = E\}.$$

A preclosure φ of character 1 on a set E is determined by its values on singletons and so corresponds to a (loop-free) directed graph $\mathcal{G}_\varphi$ on E in which there is a directed edge

---

[*]Research supported by NSERC Grant No. A5198.

N.W. Sauer et al. (eds.), Finite and Infinite Combinatorics in Sets and Logic, 289–298.

(x,y) from x to $y(\neq x)$ if and only if $y \in \varphi(\{x\})$. In this case, $\mathrm{ind}(\varphi)$ is the smallest cardinal κ such that every subset $X \subseteq E$ of cardinality $1 + \kappa$ contains an edge, and $\dim(\varphi)$ is the minimal size of a dominating set in $\mathcal{G}_\varphi$. In a similar way, a closure φ of character 1 corresponds to a quasi order (a reflexive, transitive binary relation) $\leq_\varphi$ on E, where $x \leq_\varphi y$ if and only if $x \in \varphi(\{y\})$. In this case, the independent subsets are the antichains, and the generating sets are the cofinal subsets, so that $\mathrm{ind}(\varphi)$ is the width of $\langle E, \leq_\varphi \rangle$ and $\dim(\varphi)$ is the cofinality $\mathrm{cf}(\langle E, \leq_\varphi \rangle)$. In fact, closures of character 1 or quasi-orders are just the topological closures which are algebraic.

We were interested in the structures of cofinal subsets of a partially ordered set, and this led us to consider the possibility of representing a family of generating sets of a closure or preclosure in the following uniform manner. This definition is suggested by the well-known elementary observation that any linearly ordered set contains a well-ordered cofinal subset which is order isomorphic to a regular cardinal.

Definition 1.1 *A family of sets $\mathcal{A} \subseteq \mathcal{P}(E)$ is called a* cardinal representation *of the pre-closure φ on E if, and only if, for any set $X \subseteq \bigcup \mathcal{A}$,*

$$\varphi(X) = E \Leftrightarrow \forall A \in \mathcal{A} \ (|X \cap A| = |A|).$$

Using this terminology, the above remark about linearly ordered sets is simply the observation that, if φ is the closure of character 1 corresponding to the linear order, then there is a cardinal representation for φ, $\mathcal{A} = \{A\}$ of size 1 (*i.e.* A is a cofinal subset and a set $B \subseteq A$ is cofinal if and only if $|B| = |A|$).

Of course, there are other uniform ways of generating a family of generating sets apart from the cardinal representations that we have chosen to consider. For example, call a family $\mathcal{A} \subseteq \mathcal{P}(E)$ a *transversal representation* for the preclosure φ on E if $\varphi(\bigcup \mathcal{A}) = E$ and, for every set $X \subseteq \bigcup \mathcal{A}$, we have

$$\varphi(X) = E \Leftrightarrow \forall A \in \mathcal{A} \ (X \cap A \neq \emptyset).$$

However, this is not so interesting since, for any preclosure φ, the family of sets

$$\mathcal{A} = \{A \subseteq E : A \cap S \neq \emptyset \text{ for every generating set } S \subseteq E\},$$

is easily seen to be a transversal representation of φ, and so the only real question here would be how small such a representation can be. For cardinal representations the situation is different. There are closures which do not have a cardinal representation, and so it is of interest to ask which ones do. Also, apart from giving rise to interesting questions which we have been unable to answer, the notion may actually be useful! We illustrate with two examples in §6.

2   Weak Cardinal Representations

It turned out to be convenient to consider the following one-sided version of Definition 1.1.

Definition 2.1 *If φ is a preclosure on E, then a family $\mathcal{A} \subseteq \mathcal{P}(E)$ is called a* weak cardinal representation *of φ if*

$$\forall X \subseteq \bigcup \mathcal{A} \; [\forall A \in \mathcal{A}(|X \cap A| = |A|) \; \Rightarrow \; \varphi(X) = E].$$

Of course, the set of one-element subsets of S is trivially a weak cardinal representation of φ if S is any generating set. Consequently, the only question of interest for weak cardinal representations is whether a closure or preclosure, φ, has such a representation of cardinality strictly smaller than the dimension $\dim(\varphi)$.

We begin with an outline proof of the following useful, but slightly surprising, result about preclosures which have a finite weak cardinal representation.

Theorem 2.2 *If the preclosure φ has a finite weak cardinal representation $\mathcal{A} = \{A_1, \ldots, A_k\}$ such that $\bigcup \mathcal{A}$ is countable, then it has a cardinal representation $\mathcal{A}' = \{A'_1, \ldots, A'_k\}$ such that $\bigcup \mathcal{A}' \subseteq \bigcup \mathcal{A}$ and $|A'_i| \leq |A_i|$ $(1 \leq i \leq k)$.*

Sketch Proof: Let $\mathcal{D}$ be the set of all sequences $(B_1, \ldots, B_k)$ such that $\{B_1, \ldots, B_k\}$ is a weak cardinal representation of φ, and define a quasi-order on $\mathcal{D}$ by the rule that $(B'_1, \ldots, B'_k) \leq (B_1, \ldots, B_k)$ if and only if $B'_1 \cup \ldots \cup B'_k \subseteq B_1 \cup \ldots \cup B_k$ and $|B'_i| \leq |B_i|$ for $1 \leq i \leq k$. For $(B_1, \ldots, B_k) \in \mathcal{D}$ define

$$\mathcal{D}(B_1, \ldots, B_k) = \{(B'_1, \ldots, B'_k) \in \mathcal{D} : (B'_1, \ldots, B'_k) \leq (B_1, \ldots, B_k)\}.$$

By hypothesis, $(A_1, \ldots, A_k) \in \mathcal{D}$ and $|A_1 \cup \ldots \cup A_k| \leq \omega$.

Call B_i an *inessential* term of $(B_1, \ldots, B_k)$ if there is a set $X \subseteq B_i$ such that $|X| < |B_i|$ and $\varphi(X \cup \bigcup\{B_j : j \neq i\}) = E$. Clearly, if there are no inessential terms of $(B_1, \ldots, B_k) \in \mathcal{D}$, then $\{B_1, \ldots, B_k\}$ is a cardinal representation of φ.

Without loss of generality we can assume that $(A_1, \ldots, A_k)$ is minimal in the sense that there is no $(A'_1, \ldots, A'_k) \in \mathcal{D}(A_1, \ldots, A_k)$ such that $|A'_i| < |A_i|$ for some i. We may also assume that $(A_1, \ldots, A_k)$ is minimal in the sense that there is no $(A'_1, \ldots, A'_k) \in \mathcal{D}(A_1, \ldots, A_k)$ having fewer inessential terms. The main part of the proof is to show that such a minimal sequence $(A_1, \ldots, A_k)$ in fact has no inessential terms, for the theorem follows from this. $\square$

Corollary 2.3 *A preclosure on ω has a finite cardinal representation if and only if it has a finite weak cardinal representation.*

The condition that the members of $\mathcal{A}$ be countable sets is essential for the theorem. For example, consider the (topological) closure φ defined on the real line $\mathbf{R}$ by setting $\varphi(X) = \mathbf{R}$ if X contains a subset order isomorphic to $\mathbf{Q}$, the set of rational numbers, and $\varphi(X) = X$ otherwise. Clearly, if A is any uncountable set of real numbers, then $\{A\}$ is a weak cardinal representation of φ of size 1. On the other hand, it is not difficult to show that every cardinal representation of φ has cardinality $2^{\aleph_0}$. (Note that φ does have cardinal representations, e.g. $\mathcal{A} = \{\mathbf{Q} \backslash X : \varphi(X) \neq \mathbf{R}\}$.)

The above example also shows that, in general, the members of a cardinal representation or weak cardinal representation of a preclosure may not be disjoint from one another. However, in the case when a preclosure has a finite cardinal or weak cardinal representation, then it can be shown that there is also such a representation in which the members are pairwise disjoint.

Theorem 2.4 *Suppose that $\{A_1, \ldots, A_k\}$ is a finite (weak) cardinal representation of a preclosure φ. Then there are pairwise disjoint sets $A'_i \subseteq A_i$ ($1 \leq i \leq k$) such that $\{A'_1, \ldots, A'_k\}$ is also a (weak) cardinal representation.*

3 Closures and preclosures of character 1

We can identify a directed graph or a quasi-order with the corresponding preclosure or closure of character 1, and so it makes sense to say that it has a cardinal representation. The two theorems of this section provide some motivation for our initial interest in the subject of cardinal representation.

First we will prove the following result about directed graphs. It says that a countable directed graph with no infinite independent set has a finite cardinal representation. Therefore, by Theorem 2.4 it follows that there is a finite number of pairwise disjoint sets of vertices $A_1, \ldots, A_k$ such that a set $B \subseteq \bigcup\{A_i : 1 \leq i \leq k\}$ is a dominating set of the graph if and only if $|A_i \cap B| = |A_i|$ for $1 \leq i \leq k$.

Theorem 3.1 *A countable directed graph with no infinite independent set has a finite cardinal representation.*

Let $\mathcal{G} = (E, \mathcal{E})$ be a directed graph with edge set $\mathcal{E} \subseteq E \times E$. For $X \subseteq E$, we define $\mathcal{G}(X) = X \cup \{y \in E : (y, x) \in \mathcal{E} \text{ for some } x \in X\}$. A set $X \subseteq E$ is *finitely compatible* if, for every finite set $F \subseteq X$, there is an element $x \in E$ such that $F \subseteq \mathcal{G}(\{x\})$; a set $X \subseteq E$ is *centered* if it satisfies the stronger condition that, for every finite set $F \subseteq X$, there is $x \in X$ such that $F \subseteq \mathcal{G}(\{x\})$. In order to prove Theorem 3.1 we use the following lemma which is a special case of a theorem of S. Todorčević [9].

Lemma 3.2 *If the directed graph $\mathcal{G} = (E, \mathcal{E})$ has no infinite independent set, then E is a finite union of centered sets.*

Proof. Suppose for a contradiction that $E = E_0$ is not a finite union of centered sets. Let $n < \omega$ and suppose that we have already chosen $x_i \in E(i < n)$ and $E_n \subseteq E$ so that E_n is not a finite union of centered sets. Then there is a finite set $X \subseteq E_n$ such that $X \nsubseteq \mathcal{G}(\{y\})$ for every $y \in E_n$. Hence E_n is the union of the sets $E_{n,x}$ ($x \in X$), where $E_{n,x} = \{y \in E_n : x \notin \mathcal{G}(\{y\})\}$. Therefore, for some $x = x_n \in X$, $E_{n+1} = E_{n,x} \backslash \mathcal{G}(\{x\})$ is not a finite union of centered sets. This defines $x_n (n < \omega)$. By the construction $\{x_n : n < \omega\}$ is an infinite independent set.   □

Proof of Theorem 3.1. Let $\mathcal{G}$ be a directed graph on ω. By the lemma we can assume that $\omega = \bigcup\{E_i : i < n\}$, where each $E_i(i < n)$ is finitely compatible. Then there are elements $x_{i,k} \in \omega$ ($i < n; k \in \omega$) so that $E_i \cap \{0, 1, \ldots, k\} \subseteq \mathcal{G}(\{x_{i,k}\})$. Put $A_i = \{x_{i,k} : k \in \omega\}$ ($i < n$). If A_i is finite, then $x_{i,k} = x_i$ for infinitely many k and so $E_i \subseteq \mathcal{G}(\{x_i\})$; in this case we may assume $A_i = \{x_i\}$. If A_i is infinite, then $E_i \subseteq \mathcal{G}(A_i')$ for any infinite set $A_i' \subseteq A_i$. It follows that $\mathcal{A} = \{A_i : i < n\}$ is a weak cardinal representation. By Theorem 2.2 there is a finite cardinal representation. $\square$

In view of this result it is natural to ask the following.

Problem 3.3 *Does every directed graph (preclosure of character 1) have a cardinal representation?*

The answer to this question is positive for closures (Theorem 3.4).

For a quasi-ordered set $\langle P, \leq \rangle$ which is either countable or contains no infinite antichain, there is an analogue of the Cantor theorem for linear orders; in this case there is a cofinal subset P' such that every cofinal subset of P contains a cofinal isomorphic copy of P' (see [6]). However, for general partial orders this is not true, there may be very many cofinal subsets none of which contains a cofinal subset which is isomorphic to a cofinal subset of another. While the next theorem is probably too general to be really useful, it does give some information about the cofinal subsets of a quasi-ordered set. For a quasi-ordered set $\langle P, \leq \rangle$ and $x \in P$, we define $P(\geq x) = \{y : y \geq x\}$. The *direct sum* of disjoint quasi-orders $\langle P_i, \leq_i \rangle$ $(i \in I)$ is the quasi-order $\langle P, \leq \rangle$, where $P = \bigcup\{P_i : i \in I\}$ and $\leq = \bigcup\{\leq_i : i \in I\}$.

Theorem 3.4 *Any quasi-ordered set $\langle P, \leq \rangle$ has a cardinal representation, $\mathcal{A}$, of cardinality $|\mathcal{A}| \leq \mathrm{cf}(\langle P, \leq \rangle)$.*

Proof: A quasi-order $\langle P, \leq \rangle$ is *cofinally regular* if $\mathrm{cf}(P(\geq x)) = \mathrm{cf}(P)$ $(\forall x \in P)$. Any quasi-order contains a final segment $P(\geq x)$ which is cofinally regular and hence there is a cofinal subset which is the direct sum of cofinally regular quasi-orders, say $P = \bigcup\{P_i : i \in I\}$, where each $\langle P_i, \leq \rangle$ is cofinally regular. Let $\mathrm{cf}(\langle P_i, \leq \rangle) = \kappa_i$ $(i \in I)$, so that $\mathrm{cf}(\langle P, \leq \rangle) = \sum_{i \in I} \kappa_i$. Let D_i be a cofinal subset of P_i of cardinality κ_i, and for each $x \in D_i$ let $D_{i,x} = D_i \cap P(\geq x)$. Then $|D_{i,x}| = \kappa_i$ $(i \in I, x \in D_i)$. We claim that $\mathcal{A} = \{D_{i,x} : i \in I, x \in D_i\}$ is a cardinal representation.

Let $B \subseteq \bigcup \mathcal{A}$. If $|B \cap D_{i,x}| = \kappa_i$ for all $i \in I$ and $x \in D_i$, then B is cofinal in P. Indeed, if $z \in P$, there are $i \in I$, $x \in D_i$, $b \in B \cap D_{i,x}$ such that $z \leq x \leq b$. Conversely, if B is cofinal, then $|B \cap D_{i,x}| = \kappa_i$ for all $i \in I, x \in D_i$; otherwise, there is $y \in D_{i,x}$ such that $B \cap P(\geq y) = \emptyset$. $\square$

In [4] we proved a little more.

Theorem 3.5 *A quasi-ordered set $\langle P, \leq \rangle$ has a finite cardinal representation if and only if there is a cofinal subset of P which is the direct sum of a finite number of chains.*

After proving Theorem 3.4 about quasi-orders we realized that, even more generally, every topology has a cardinal representation. By a cardinal representation of a topological space E we mean a cardinal representation for the corresponding closure φ, where $\varphi(X)$ is the topological closure of X in this space. A π-*base* is a set $\mathcal{B}$ of non-empty open sets such that every non-empty open set contains a member of $\mathcal{B}$, and $\pi(E) = \min\{|\mathcal{B}| : \mathcal{B} \text{ a } \pi\text{-base}\}$. The proof of the following theorem is essentially the same as the proof of Theorem 3.4.

Theorem 3.6 *A topological space E has a cardinal representation $\mathcal{A}$ of size $|\mathcal{A}| \leq \pi(E)$.*

4   A closure of character 3

As we already remarked, there are closures which have no cardinal representation at all. The following is a simple example from [4] due to Fred Galvin.

Theorem 4.1 *There is a closure of character 3 on ω which has no cardinal representation (and therefore has no finite weak cardinal representation).*

Proof. Let $P_n = \{2n, 2n+1\}\,(n < \omega)$. For $X \subseteq \omega$, define $\varphi(X) = X \cup \bigcup\{P_n : P_n \cap X \neq \emptyset$ and $P_m \subseteq X$ for some $m > n\}$. Then φ is a closure of character 3. Suppose for a contradiction that $\mathcal{A}$ is a cardinal representation for φ. Then $\bigcup \mathcal{A} \cap P_n \neq \emptyset$ for every $n < \omega$ and $P_n \subseteq \bigcup \mathcal{A}$ for infinitely many n. Choose $n < \omega$ so that $P_n \subseteq \bigcup \mathcal{A}$. Since $\varphi(\bigcup \mathcal{A} \backslash P_n) \neq \omega$, it follows that $|A \cap (\bigcup \mathcal{A} \backslash P_n)| < |A|$ for some $A \in \mathcal{A}$. Therefore, A is finite and there is $x \in P_n \cap A \neq \emptyset$. Now $\varphi(\bigcup \mathcal{A} \backslash \{x\}) = \omega$, but $|(\bigcup \mathcal{A} \backslash \{x\}) \cap A| < |A|$, and this contradicts the assumption that $\mathcal{A}$ is a cardinal representation. $\square$

By Theorems 3.1 and 3.4, any preclosure on ω of character one has a cardinal representation, and so does any closure of character one (on a set of arbitrary cardinality). It is natural to ask the following question.

Problem 4.2 *Is there a closure (or a preclosure on ω) of character two, which has no cardinal representation?*

5   Further results

One of our main results is the following theorem, and in the next section we shall give two applications of this.

Theorem 5.1 *Let φ be a closure with $ind(\varphi) = k < \omega$. Then φ has a cardinal representation, $\mathcal{A}$, such that $|\mathcal{A}| \leq k$ and for each $A \in \mathcal{A}$, $|A|$ is either 1 or an infinite regular cardinal.*

REMARK. As in Dilworth's theorem [2], the finiteness of $ind(\varphi)$ is an essential condition for Theorem 5.1. In fact, a slight variation of a well-known example used to show the failure of Dilworth's theorem in the infinite case, shows that this theorem also fails badly when $ind(\varphi) \geq \omega$. Consider the direct product $\kappa \otimes \kappa^+$, where κ is any regular infinite cardinal. In this case, there is no infinite independent set $(ind(\varphi) = \omega)$, but there is no weak cardinal representation of size less than κ. For suppose $\mathcal{A}$ is a family of subsets of cardinality $|\mathcal{A}| < \kappa$. For $A \in \mathcal{A}$, if $|A| < \kappa^+$ put $B(A) = A$, and if $|A| = \kappa^+$ let $B(A) = A \cap C$, where C is a chain of the form $\{(\xi, \beta) : \beta < \kappa^+\}$ such that $|A \cap C| = \kappa^+$. Thus $B = \bigcup\{B(A) : A \in \mathcal{A}\}$ meets each member of $\mathcal{A}$ in a subset of the full size, but B is not cofinal. Hence there is no weak cardinal representation of size less than κ. Note that by Theorem 3.4 there is a cardinal representation; in fact it is easy to see that $\{A_\alpha : \alpha < \kappa\}$ is a cardinal representation of size κ, where $A_\alpha = \{(\xi, \beta) : \alpha < \xi < \kappa, \beta < \kappa^+\}$.

This last example shows that there is no bound on the size of a cardinal representation of a closure φ if the independence number $ind(\varphi) = \omega$. When we wrote [4] we did not know if such a closure necessarily had a cardinal representation, although we could answer this question for closures on a countable set ([4], Theorem 2.5). Subsequently Pouzet (unpublished) has proved the more general result.

Theorem 5.2 *If φ is a closure such that $ind(\varphi) \leq \omega$ then φ has a cardinal representation.*

The condition that $ind(\varphi) \leq \omega$ (*i.e.* no infinite independent set) is needed in Theorem 5.2 in view of the example of Galvin in §3.

Nothing can be said about the size of the cardinal representation in Theorem 5.2 even for closures on a countable set. An example is given in [4] of a T_1 topological closure on ω with no infinite independent set and no countable cardinal representation, and it is consistent that there is no cardinal representation of size less that $2^{\aleph_0}$ (and $2^{\aleph_0}$ arbitrarily large). However, for algebraic closures there is a finite cardinal representation, but the proof is not easy.

Theorem 5.3 *If φ is an algebraic closure on ω and $ind(\varphi) \leq \omega$, then φ has a finite cardinal representation.*

The three theorems of this section have all been about closures, and it is natural to ask if there are analogous results for preclosures. Trying to generalize Theorem 5.1 to preclosures, we looked at the simplest case of a tournament (a complete graph in which each edge is given an orientation). Of course, a tournament corresponds to a preclosure of character 1 and independence number 1. In the tournaments we looked at, there was always a finite weak cardinal representation (even one of size 2!), and we thought this might always be the case. However, A. Hajnal [5], using the continuum hypothesis CH, gave an example of a tournament having no finite weak cardinal representation. Afterwards Z. Nagy and Z. Szentmiklóssy [8] showed, without any additional set-theoretic assumptions, that for any cardinal $\kappa \geq \omega$, there is a tournament on a set of size $(\kappa^{<\kappa})^+$ for which there is no weak cardinal representation, $\mathcal{A}$, of cardinality $|\mathcal{A}| < \kappa$. In other words, Theorem 5.1 fails badly for preclosures.

However, there is an analogue of Theorem 5.1 for preclosures on a countable set.

Theorem 5.4 *Let φ be a preclosure on ω such that $ind(\varphi) = k < \omega$. Then φ has a cardinal representation, $\mathcal{A}$, with $|\mathcal{A}| \leq k$.*

We do not know if there are analogues of Theorems 5.2 and 5.3 for preclosures.

Problem 5.5 *Is there a preclosure, φ, such that $ind(\varphi) = \omega$ and φ has no cardinal representation?*

Problem 5.6 *Is there an algebraic preclosure, φ, on ω such that $ind(\varphi) = \omega$ and φ has no (or no finite) cardinal representation?*

Of course, by Theorem 3.1, we do know the answer to this last question for preclosures of character 1.

6   Applications of Theorem 5.1

One easy application of Theorem 5.1 is an alternative proof of the following theorem of Milner and Pouzet [7].

Theorem 6.1 *If the dimension of the closure φ is a singular infinite cardinal, then* $\mathrm{ind}(\varphi) \geq \omega$.

Proof: Let $\dim(\varphi) = \kappa$, where κ is a singular cardinal, and let A be a generating set of cardinality $|A| = \kappa$. Consider the restriction, φ', of φ to A, i.e. $\varphi'(X) = \varphi(X) \cap A$ for $X \subseteq A$. If $\mathrm{ind}(\varphi) < \omega$, then $\mathrm{ind}(\varphi') < \omega$ also, say $\mathrm{ind}(\varphi') = k < \omega$. By Theorem 5.1 there is a cardinal representation $\mathcal{A} = \{A_1, \ldots, A_\ell\}$ for φ' where $\ell \leq k$ and each $|A_i|$ is either 1 or a regular infinite cardinal which is necessarily less than κ. Then $\dim(\varphi') \leq |\bigcup \mathcal{A}| < \kappa$, a contradiction.   □

The next application is about gaps in a partially ordered set of finite breadth. An immediate corollary of this is a theorem of D. Duffus and M. Pouzet [3] about gaps in a lattice of finite breadth.

Let $\mathcal{P} = \, <P, \leq>$ be a partially ordered set. For $X, Y \subseteq P$, we write $X \leq Y$ if and only if $x \leq y$ holds whenever $x \in X$ and $y \in Y$. Also, for $X \subseteq P$, let $X^+ = \{y \in P : X \leq \{y\}\}$ and $X^- = \{y \in P : \{y\} \leq X\}$. In particular, we have $\emptyset^+ = \emptyset^- = P$. A *gap* in $\mathcal{P}$ is a pair of subsets (A, B) of P such that $A \leq B$ and $A^+ \cap B^- = \emptyset$ (i.e. there is no element $x \in P$ such that $A \leq \{x\} \leq B$). A *subgap* of (A, B) is a gap (A', B') such that $A' \subseteq A$ and $B' \subseteq B$. The gap (A, B) is *irreducible* if $|A'| = |A|$ and $|B'| = |B|$ for every subgap (A', B'). A gap (A, B) is *regular* if $|A|$ and $|B|$ are infinite regular cardinals or zero.

According to [1] a lattice has finite *breadth* b if b is the least integer such that any join $x_1 \vee \ldots \vee x_n$ of $(n > b)$ elements is a join of some b of these elements (the same then holds for the meets). It is easy to see that this is equivalent to saying that b is the largest integer such that there is a set X of b elements so that $\bigvee Y < \bigvee X$ for every proper subset Y of X. This suggests that we define the *breadth* of a partially ordered set $\mathcal{P} = \,<P, \leq>$ to be $b(\mathcal{P}) = \sup\{|X| : X \subseteq P \text{ and } (\forall Y \subseteq X)(Y \neq X \Rightarrow Y^+ \neq X^+)\}$. Note that the one-sidedness of the definition is only apparent; we can replace $Y^+ \neq X^+$ by $Y^- \neq X^-$. For, if X is such that $\forall Y \subseteq X$ $(Y^+ \neq X^+)$, then $X_1 = \{f(x) : x \in X\}$ satisfies $|X_1| = |X|$ and $\forall Y \subseteq X_1$ $(Y^- \neq X_1^-)$, where $f(x) \in (X \backslash \{x\})^+ \backslash X^+$ $(x \in X)$. Therefore, the dual partial order $\mathcal{P}^d = \,<P, \geq>$ has the same breadth as $\mathcal{P}$.

The result of Duffus and Pouzet [3] referred to above is the following.

Theorem 6.2 *A gap in a lattice of finite breadth contains a regular, irreducible subgap.*

It is an immediate corollary of the following theorem.

Theorem 6.3 *Let (A, B) be a gap in the partially ordered set $\mathcal{P}$ of finite breadth k. Then there are nonnegative integers m, n and pairwise disjoint subsets $A_1, \ldots, A_m \subseteq A$ and $B_1, \ldots, B_n \subseteq B$ such that:*
(i) *each $|A_i|$, $|B_j|$ $(1 \leq i \leq m, 1 \leq j \leq n)$ is either an infinite regular cardinal or 1;*
(ii) *for $A' \subseteq A_1 \cup \ldots \cup A_m$ and $B' \subseteq B_1 \cup \ldots \cup B_n$, (A', B') is a gap if and only if, for all i, j, $|A' \cap A_i| = |A_i|$ and $|B' \cap B_j| = |B_j|$;*
(iii) *$m \leq k$ and $n \leq k$;*
(iv) *$m, n \leq 1$ if $\mathcal{P}$ is a lattice.*

Proof: Without loss of generality we can assume that the largest and least elements of P, if they exist, do not belong to $A \cup B$ (since $(A \backslash P^-, B \backslash P^+)$ is a subgap of (A, B)). Define functions ψ, φ on the subsets of $A \cup B$ by

$$\psi(X) = (X \cap A)^+ \cap (X \cap B)^-,$$

$$\varphi(X) = ((\psi(X))^- \cap A) \cup ((\psi(X))^+ \cap B).$$

It is easy to check that $X \subseteq \varphi(X) \subseteq \varphi(Y)$ and $\varphi(X) = \varphi(\varphi(X))$ for $X \subseteq Y \subseteq A \cup B$. Also, by our assumption, $\varphi(\emptyset) = \emptyset$, and so φ is a (normalized) closure on $A \cup B$.

For a subset $X \subseteq A \cup B$, $(X \cap A, X \cap B)$ is a gap if and only if $\psi(X) = \emptyset$. If $\psi(X) = \emptyset$, then $\varphi(X) = A \cup B$. On the other hand, if $x \in \psi(X) \neq \emptyset$, then either $A \not\leq \{x\}$ or $\{x\} \not\leq B$ and so $\varphi(X) \neq A \cup B$. Thus, for $X \subseteq A \cup B$, we have

$$(X \cap A, X \cap B) \text{ is a gap } \Leftrightarrow \varphi(X) = A \cup B.$$

Since the breadth of $\mathcal{P}$ is k, it follows that, whenever $Z \subseteq A$ and $|Z| > k$, there is some $z \in Z$ such that $(Z \backslash \{z\})^+ = Z^+$ and hence $z \in \varphi(Z \backslash \{z\})$. It follows that there is no φ-independent subset of A of size $k + 1$; similarly for subsets of B. Therefore $\operatorname{ind}(\varphi) \leq 2k$ and so, by Theorem 5.1, there are subsets $A_1, ..., A_m \subseteq A$, $B_1, ..., B_n \subseteq B$ such that $m + n \leq 2k$ and $\mathcal{A} = \{A_1, ..., A_m, B_1, ..., B_n\}$ is a cardinal representation for φ such that (i) holds and by Theorem 2.4 we may assume that the members of $\mathcal{A}$ are even pairwise disjoint. Also, (ii) is an immediate consequence of the above equivalence and the definition of a cardinal representation. By (ii), for each i ($1 \leq i \leq m$), $(\bigcup_{h \neq i} A_h, \bigcup_j B_j)$ is not a gap, and so there is $y_i \in P$ such that $\bigcup_{h \neq i} A_h \leq \{y_i\} \leq \bigcup_j B_j$. Therefore, $A_i \not\leq \{y_i\}$ and so there is $x_i \in A_i$ such that $x_i \not\leq y_i$. Let $X = \{x_i : 1 \leq i \leq m\}$; then $|X| = m$ and $(X')^+ \neq X^+$ for each proper subset X' of X, and therefore $m \leq k$. Similarly, $n \leq k$, and so (iii) holds.

Finally, to prove (iv) assume $\mathcal{P}$ is a lattice. Suppose for a contradiction that $m > 1$, and that A_1, A_2 are both non-empty. Since $(\bigcup_{h \neq 1} A_h, \bigcup_j B_j)$ is not a gap, it follows that there is $x_1 \in P$ such that $\bigcup_{h \neq 1} A_h \leq \{x_1\} \leq \bigcup_j B_j$. Similarly there is x_2 such that $\bigcup_{h \neq 2} A_h \leq \{x_2\} \leq \bigcup_j B_j$. Let x be the join of x_1 and x_2; then $\bigcup_h A_h \leq \{x\} \leq \bigcup_j B_j$ and this contradicts the fact that $(\bigcup_i A_i, \bigcup_j B_j)$ is a gap (by (ii)). Therefore $m \leq 1$, and similarly $n \leq 1$. $\square$

References

[1] G. Birkhoff, *Lattice Theory*, (3rd edition), Amer. Math. Soc., Providence, RI. 1967.

[2] R.P. Dilworth, A decomposition theorem for partially ordered sets, *Ann. Math.(2)* **51** (1950), 161–166.

[3] D. Duffus and M. Pouzet, Representing ordered sets by chains, *Annals of Discrete Mathematics* **23** (1984), 81–98.

[4] F. Galvin, E.C. Milner and M. Pouzet, Cardinal representations for closures and pre-closures, *Trans. Amer. Math. Soc.* **328** (1991), 667–693.

[5] A. Hajnal, Private communication.

[6] E.C. Milner and M. Pouzet, On the cofinality of partially ordered sets, *Ordered Sets* (Banff, 1981) (ed. by I. Rival), D. Reidel (1982), 279–298.

[7] E.C. Milner and M. Pouzet, On the independent subsets of a closure system with singular dimension, *Algebra Universalis* **21** (1985), 25–32.

[8] Z. Nagy & Z. Szentmiklóssy, On the representation of tournaments (preprint).

[9] S. Todorčević, Directed sets and cofinal types, *Trans. Amer. Math. Soc.* **290** (1985), 711–723.

A Vertex-to-Vertex Pursuit Game Played With Disjoint Sets of Edges

STEWART NEUFELD
RICHARD J. NOWAKOWSKI*
Dalhousie University

Abstract

The problem is to determine the number of 'cops' needed to capture a 'robber' where the game is played with perfect information, the different sides moving alternately. The 'cops' capture the 'robber' if one of them occupies the same vertex as the robber at any time in the game. Normally, both sides can move along any edge present in the graph. We investigate the game where the two sides move along disjoint sets of edges. Two natural situations occur. One, given a graph, the cops move along the edges of the graph and the robber moves along the complementary edges. Two, the adversaries can move along disjoint sets of edges present in a product of graphs.

1 Introduction and preliminaries

The game of 'cop and robber' was introduced by Quilliot [16] and, independently, by Nowakowski and Winkler [14]. The game rules were: given a graph G, the cop chooses a vertex of G, then the robber chooses a vertex. They then move alternately — each can move to an adjacent vertex or pass. The cop wins if he ever occupies the same vertex as the robber; the robber wins if this situation never occurs. In [16] and [14], the authors characterize those graphs in which the cop has a winning strategy. Aigner and Fromme [1] pose the question: given a graph G, determine the least number of cops required to capture a robber on G. This is called the *copnumber* of G and denoted $c(G)$. In [1], they show that if G is a finite, planar graph then $c(G) \leq 3$. Andreae [2, 3] pursued this further showing that excluded minors play a part in determining the copnumber. Since then several other papers have appeared with connections to Cayley graphs and girth [7, 8, 10] and bridged graphs [4]. Tošić and Šćekić [18] and Neufeld [12] investigated the game using variants of the passing option. (This idea first appeared in [1].) Maamoun and Meyniel [11] and Tošić [19] examined the copnumber of the Cartesian product. This was extended in [13] where the strong, Cartesian and categorical products are considered.

In this paper, we extend the investigation to the situation where the cops and robber do not have the same moves, specifically, when the robber and cops move along disjoint sets of edges. In Section 2, they use the same vertices but complementary sets of edges. In Section

*Partially supported by a grant from the NSERC.

N.W. Sauer et al. (eds.), Finite and Infinite Combinatorics in Sets and Logic, 299–312.
© 1993 *Kluwer Academic Publishers. Printed in the Netherlands.*

3, the underlying graph is a product graph and the adversaries are restricted to subsets of edges defined in terms of the Cartesian and categorical products.

Let $G_i = (V_i, E_i)$, $i = 1, 2, \ldots, n$ be irreflexive graphs. In all cases, the vertex set of the product graph is the Cartesian product of the vertex sets of the factors. To avoid confusion, we will denote the product of the vertex sets as $V_1 \cdot V_2$. In the product graph, let the vertices be given by $\underline{a} = (a_1, a_2, \ldots, a_n)$. In the *Cartesian* product, $\square_{i=1}^{n} G_i$, $\underline{a} = (a_1, a_2, \ldots, a_n)$ is adjacent to $\underline{b} = (b_1, b_2, \ldots, b_n)$ just if $\underline{a}$ and $\underline{b}$ differ in precisely one coordinate i and (a_i, b_i) is an edge in G_i. An example is the n-cube which can be considered the Cartesian product of n paths of length one. In the *categorical* product, $\times_{i=1}^{n} G_i$, $(\underline{a}, \underline{b})$ is an edge just if for all i, (a_i, b_i) is an edge in G_i. Note that if all the graphs are bipartite then the resulting product graph is not connected. In the *strong* product, $\boxtimes_{i=1}^{n} G_i$, $(\underline{a}, \underline{b})$ is an edge just if for each i, either (a_i, b_i) is an edge in G_i or $a_i = b_i$. (These product symbols are due to J. Nešetřil and represent the edges present when two edges are multiplied together.) Figure 1 gives examples of these products. Note that for two graphs G_1 and G_2, the edge set for $G_1 \boxtimes G_2$ is the disjoint union of the edge sets of $G_1 \square G_2$ and $G_1 \times G_2$ but this is no longer true if there are three or more graphs in the product.

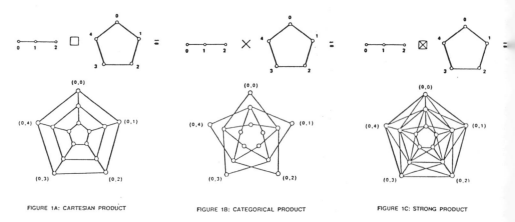

FIGURE 1A: CARTESIAN PRODUCT   FIGURE 1B: CATEGORICAL PRODUCT   FIGURE 1C: STRONG PRODUCT

We use the symbol $XY(\boxtimes_{i=1}^{n} G_i)$ to denote the least number of cops needed to capture a robber in $\boxtimes_{i=1}^{n} G_i$ where the cops are restricted to the set of edges X and the robber is restricted to the set of edges Y. Here, X and Y can be any of the following: S which denotes the set of edges in the Cartesian product (the Straight edges); $\overline{S}$ the set of edges in the strong product minus the set of edges in the Cartesian product; O the set of edges in the categorical product (the Oblique edges); and finally $\overline{O}$ the set of edges in the strong product minus the edges in the categorical product. Note that $S \subseteq \overline{O}$ and $O \subseteq \overline{S}$ so that

$$S\overline{S}(\boxtimes_{i=1}^{n} G) \geq SO(\boxtimes_{i=1}^{n} G) \geq \overline{O}O(\boxtimes_{i=1}^{n} G)$$

and

$$\overline{S}S(\boxtimes_{i=1}^{n} G) \leq OS(\boxtimes_{i=1}^{n} G) \leq O\overline{O}(\boxtimes_{i=1}^{n} G).$$

In all these products the size of the vertex set increases exponentially. In the categorical product, the number of edges grows exponentially, whereas the growth is only linear for the Cartesian product. For example, if G is regular with degree r, then the degree of a vertex in $\times_{i=1}^{n} G$ is r^n. In $\square_{i=1}^{n} G$, the degree is rn. In the strong product the degree would be $(r+1)^n - 1$, so the degree of a vertex in the complement of the categorical product is $(r+1)^n - r^n - 1$ and the degree in the complement of the Cartesian product is $(r+1)^n - rn - 1$.

In the proofs, we use a combination of strategies. Some are based on the strategies in the original graphs. These are introduced by considering the *shadow* or the projection of the robber onto one or more coordinates. However, the cops cannot always move in every coordinate on every move. In these cases, we use other strategies designed to catch the robber but which do not take into account the robber's position.

The main such strategy we call the *vertex separation strategy*. The *vertex separation number* of a graph G, denoted $vs(G)$, is defined in the following way. Arrange all the vertices of G in a horizontal row and draw all the edges as semi-circular arcs in the same plane. This is called a *linear layout*. Consider all vertical lines that intersect the layout. The separation number for a vertical line is the number of vertices that lie to the left of the line which are incident with an edge cut by the line. The vertex separation number for the layout is the maximum separation number taken over all the vertical lines and $vs(G)$ is the least vertex separation number taken over all linear layouts of G.

Take the linear layout $\mathcal{L} = \{a_1, a_2, \ldots, a_n\}$ that realizes the vertex separation number. The vertex separation strategy is this. We proceed inductively and we suppose that the $v + 1$ cops occupy a cutset between the vertices to the left and right of any vertical line passing between a_i and a_{i+1}. This is true initially if the cops occupy the first $vs(G) + 1$ vertices of the layout. By the definition of vertex separation, one of the $v + 1$ occupied vertices is not adjacent to any vertex to the right of this line. This cop is free to move and occupy a_{i+2}. These occupied vertices now form a cutset for the vertices to the right and left of a vertical line between a_{i+1} and a_{i+2}.

This strategy ignores the whereabouts of the robber but will eventually capture him. This is, in fact, the strategy used in a game considered in [15]. Here, the graph must be searched (both vertices and edges) to eliminate the possibility of an infinitely fast intruder being present. The searchers move from vertex to vertex along the edges. The problem is to determine the least number of searchers, $s(G)$, needed. Ellis *et al*, [6], showed $vs(G) \leq s(G) \leq vs(G) + 2$ and if the searchers only have to search the vertices then $vs(G) \leq s(G) \leq vs(G) + 1$.

In most cases, the lower bounds are derived from noticing how many cops are needed to dominate the neighbourhood of the robber given that they use different edge sets.

For any undefined terms, see [5].

2 Complementary sets of edges

In this section, we suppose that the robber moves along the edges of the graph G and the cops move along the edges of the complementary graph, $\overline{G}$. The least number of cops required to capture the robber will be denoted by $\overline{c}(G)$. In this case the game does not last very long and essentially $\gamma(G)$ — the domination number of G — many cops are needed.

Theorem 2.1 *If G is a finite, connected graph then*

$$\gamma(\overline{G}) - 1 \leq \overline{c}(G) \leq \gamma(\overline{G}).$$

Proof. If the cops occupy a dominating set in $\overline{G}$ then the robber cannot choose a safe vertex.

Place $\gamma(\overline{G}) - 2$ cops on $\overline{G}$. The robber chooses a vertex that is not adjacent, in $\overline{G}$, to any of the cops. Suppose that the cops move so that at least one of them is adjacent to the robber. There must be a vertex not adjacent, in $\overline{G}$, to any of the cops or to the vertex occupied by the robber, otherwise $\gamma(\overline{G}) - 1$ vertices would dominate $\overline{G}$. In G, this vertex is adjacent to the robber, consequently the robber will always have a safe move.   □

Both cases in Theorem 2.1 can occur. If G is a complete graph then it is easy to see that $\overline{c}(G) = \gamma(\overline{G})$. For the other equality consider a graph G where diam$(G) \geq 4$. Then there are two vertices, say a and b, such that $d(a, b) \geq 4$. Let the cop occupy vertex a. The robber must choose some vertex that is adjacent, in G, to a or else the cop will win on the next move. But now the cop will move to b and the robber cannot move to a vertex adjacent, in G, to b and so the cop wins on his second move.

3   Mixed products

First we need some results concerning *copwin* graphs, that is, graphs in which one cop suffices to catch the robber (under the original rules). If the robber tried to extend the game for as many moves as possible then on his last move he would find himself in a *corner*. That is, he would be adjacent to the cop and every vertex to which the robber could move would also be adjacent to the cop. The cop is on a *dominating* vertex for this corner. In this section, we extend this notion to say that the robber is *cornered* if he is in a corner with the cop on a dominating vertex for the corner, or if the robber occupies the same vertex as the cop.

Lemma 3.1 *Let G be a finite, copwin graph. If after some move the cop is adjacent to the robber then the cop can win by, on every turn, moving to some vertex adjacent (or equal) to that occupied by the robber.*

Proof. The result is certainly true in a singleton graph. In [14] it was shown that every copwin graph is dismantlable down to a singleton by removing corners. We now induct on the number of corners needed to reduce the graph. Note that since these graphs are copwin every winning strategy brings the cop to a vertex adjacent to that occupied by the robber. Suppose that G is the smallest copwin graph in which the robber has a means of avoiding capture if the cop follows a strategy of always staying adjacent (or equal) to the robber. Remove a corner a from G. In $G - \{a\}$ the cop has a winning strategy which allows him to stay adjacent to the robber, therefore the robber's strategy in G must use the vertex a. However, this strategy could work in $G - \{a\}$ since the robber can use a dominating vertex for a instead of a. This situation is impossible by the induction hypothesis, therefore to stop the robber moving to any dominating vertex for a the cop must be on a vertex adjacent to all the dominating vertices of a. But then when the robber moves to a the cop moves to a dominating vertex, remains adjacent to the robber and wins.   □

In $\boxtimes_{i=1}^{n} G_i$, for a given i, a *supercop* is a set of 2^{n-1} cops all of whom have the same ith coordinate but occupy all 2^{n-1} vertices of $\Pi_{j \neq i}\{a_j, b_j\}$, where a_j and b_j are adjacent vertices in G_j. This concept is useful when considering the product of more than two graphs.

3.1 Cartesian versus Categorical Edges

The robber, who moves on the categorical edges, has an increasing advantage as the number of terms in the product increases. In fact, as we see in Theorem 3.2, the copnumber increases exponentially. A cop can only move in one coordinate at a time and to obtain an upper bound for the number of cops we use the 'vertex separation' strategy.

For any graph G, let $\delta(G)$ be the minimum degree, $\beta(G)$ be the independence number and $\overline{\beta}(G) = |V(G)| - \beta(G)$. For the product of two graphs, we have the following results.

Theorem 3.1 *Let G and H be finite, connected graphs. Then*

$$SO(G \boxtimes H) \leq \min\{c(H) + vs(G), c(G) + vs(H), c(G) + \overline{\beta}(H) - 1, c(H) + \overline{\beta}(G) - 1\}$$

and

$$SO(G \boxtimes H) \geq \max\{c(G), c(H)\}.$$

Proof. The lower bound follows since the cops must win in the projections on both G and H.

Suppose that $c(H) + vs(G) \leq c(G) + vs(H)$. Take the linear layout of G that realizes $vs(G)$. Let the vertices of this layout be $\{a_1, a_2, \ldots, a_m\}$. The $c(H) + vs(G)$ cops are placed in the subgraph $\{a_1\} \cdot H$ and then they proceed to capture the robber's shadow. Since the cops are moving along the Cartesian edges they can capture the shadow and stay within $\{a_1\} \cdot H$. The cop on the shadow stays in $\{a_1\} \cdot H$ and the rest move onto $\{a_2\} \cdot H$. This procedure is repeated in $\{a_i\} \cdot H$, for $i = 2, 3, \ldots, vs(G) + 1$. The cops that are left behind in each $\{a_i\} \cdot H$, remain in that subgraph and follow the robber's shadow whenever it moves. Again, this is possible since they move along the Cartesian edges. The cops now move forward through the subgraphs $\{a_i\} \cdot H$, the next to be occupied coming from the vertex separation strategy. Each time, all the nonshadowing cops plus the cop from the subgraph $\{a_j\} \cdot H$ (where a_j is a vertex no longer needed in the cutset) move to the next subgraph and capture the shadow. At no time can the robber move past this block of $vs(G)$ subgraphs without moving into one of them (since the projection of the shadowing cops forms a cutset in H). Moreover, he must move to a vertex adjacent, by a Cartesian edge, to a shadowing cop who would then capture. Therefore, the robber cannot escape and is eventually caught in some $\{a_k\} \cdot H$.

Suppose that $c(G) + \overline{\beta}(H) - 1 \leq c(H) + \overline{\beta}(G) - 1$. Let $J = \{a_1, a_2, \ldots, a_j\}$ be the complement of a maximum sized independent set of H. First, place all the cops in $\{a_1\} \cdot H$ then capture the robber's shadow in this copy of H. Leave a capturing cop in this copy and move all the rest to $\{a_2\} \cdot H$ and repeat the procedure in this copy and, in turn, in $\{a_3\} \cdot H$ to $\{a_j\} \cdot H$. Whenever the robber moves, the shadowing cops follow the shadow in their copy of H. At this stage, either the robber has been captured or is in $\{x\} \cdot H$, $x \in H - J$. In the latter case the robber is under threat of capture from one of the shadowing cops. The robber can only move to vertices of the form $\{y\} \cdot H$, $y \in J$ where he will be adjacent, by a Cartesian edge, to a shadowing cop. $\square$

In most cases the upper bound of $c(G) + \beta(H) - 1$ is larger than $c(G) + vs(H)$. One exception is in the case of complete graphs. In this case, it is easy to show that $SO(K_n \boxtimes K_m) = \min\{n, m\} - 1$. The upper or lower bound of Theorem 3.1 can be improved upon by 1 if one of the graphs is a tree or a cycle.

Corollary 3.1 *Let C be a cycle and T a tree. Then*

(a) $SO(T \boxtimes H) \leq c(H) + 1$;

(b) *if $|C| \geq 5$ then $c(H) + 1 \leq SO(C \boxtimes H)$.*

Proof. (a) Let $x \in V(T)$. Place the cops on any vertex of $\{x\} \cdot H$ and capture the robber's shadow in this copy of H. The shadowing cop now follows the robber's shadow in $\{x\} \cdot H$. The robber cannot pass through $\{x\} \cdot H$ without being caught and so he is trapped in $T' \cdot H$ where T' is a proper subgraph of T. Move the other cops to $\{y\} \cdot H$ where $y \in V(T')$ and y is adjacent to x. Capture the shadow in $\{y\} \cdot H$. The robber is restricted to $T'' \cdot H$ where T'' is a proper subgraph of T'. Move the cop from $\{x\} \cdot H$ to $\{y\} \cdot H$. Repeat this procedure until the robber is trapped in $\{z\} \cdot H$ where z is a leaf of T. The robber cannot move and so only one cop need now go and capture the stationary robber.

(b) The robber can initially choose a copy of H at distance at least two from one of the cops. Also, he can maintain this distance. An additional $c(G)$ cops are then required to capture the robber. $\qquad \square$

If we take the product of more than two graphs then we have an exponential lower bound but only a generalization of the upper bounds given in Theorem 3.1. (For clarity, the following bounds are only given for the product of a graph with itself. These can be easily extended to products of different graphs.)

Theorem 3.2 *Let G be a finite, connected graph, $m = |V(G)|$ and suppose that $n > 2$. Then*

(a) $SO(\boxtimes_{i=1}^{n} G) \leq \min\{(c(G) + \overline{\beta}(G) - 1)m^{n-2}, (c(G) + vs(G))m^{n-2}\}$;

(b) $SO(\boxtimes_{i=1}^{n} G) \geq \delta(G)^{n-1}/n$.

Proof. (a) For the upper bound the cops are placed so as to play the strategy of the previous theorem in the first two coordinates but each cop is replicated m^{n-2} times so as to occupy all positions in the other $n - 2$ coordinates.

For the lower bound note the robber can move to at least $\delta(G)^n$ different vertices and a cop to $n\delta(G)$ vertices all of which could be in the robber's neighbourhood. Therefore at least $\delta(G)^{n-1}/n$ cops are needed to trap the robber. $\qquad \square$

3.2 Categorical versus Cartesian Edges

Since the number of categorical edges grows faster than the Cartesian edges, the cops should have the advantage as the number of factors in the product increases. The copnumber for this version should not grow as fast as in the previous game, but the upper bound in Theorem 3.4 is exponential and the best lower bound is only linear.

However, for two graphs we know the copnumber to within 1.

Theorem 3.3 *Let G and H be finite, connected graphs, both with at least two vertices.*

(a) *If $c(G) = c(H) = 1$, then $OS(G \boxtimes H) = 2$.*

(b) *If $c(G) \geq c(H) \geq 1$, then $c(G) \leq OS(G \boxtimes H) \leq c(G) + 1$.*

Proof. (a) Since one cop cannot dominate the neighbourhood of the robber then $2 \leq OS(G \boxtimes H)$.

Place two cops, S_1 and S_2, on adjacent vertices in a copy of G, say $G \cdot \{x\}$. Let them play in $G \cdot \{x, y\}$, y adjacent to x, until one of them captures the shadow of the robber, keeping both cops adjacent to each other in the projection on G. Now the cops play their winning strategy in H. Only one cop is needed so they can always have identical second coordinates. Also, they move so as to keep one of the cops on the shadow of the robber in G. This is accomplished by swopping first coordinates if the robber's move does not change his first coordinate, otherwise one of the cops follows the shadow while the other moves so as to change his first coordinate to that just vacated by the first cop. When the robber is cornered in the projection on H, he cannot move in G or he will be captured by the shadowing cop. As well, he cannot move in H or he will be captured by the second cop. Moreover, he must move since he is being attacked by this cop. Hence the robber is captured on the next move.

(b) Note that $c(G) \geq 2$. Since the cops must win in the projection on to G then $c(G) \leq OS(G \boxtimes H)$.

First, the cops capture the robber's shadow on a copy of G, say $G \cdot \{x\}$ with cop S_1. With the remaining $c(G)$ cops playing in $\{x, y\} \cdot G$, y adjacent to x, we capture a vertex in G adjacent to the shadow, say with cop S_2. This capture may occur in $G \cdot \{x\}$ or $G \cdot \{y\}$. In either case the robber cannot safely move into $G \cdot \{x\}$ or to $G \cdot \{z\}$ where z is adjacent to x. This follows since the robber is captured if S_2 is in $G \cdot \{x\}$. If S_2 is in $G \cdot \{y\}$, then he moves to the robber's shadow in $G \cdot \{x\}$ and S_1 moves adjacent to the robber in $G \cdot \{z\}$. The robber is now trapped and a third cop moves to capture him.

Now S_1 along with the remaining $c(G) - 1$ cops can successfully play in H. The cops S_1 and S_2 move together; they manage their first coordinates as the two cops in part (a). The second coordinate is more complicated. After the shadow and a vertex adjacent to the shadow have been captured, if S_1 and S_2 now have the same coordinate then they move so as to always have the same second coordinate. If the two are different (then necessarily adjacent) S_2 will always change its second coordinate to that just left by S_1.

Suppose that S_3 captures the shadow of the robber in H, say in $\{z\} \cdot H$. If S_3 is either of S_1 or S_2 then the robber is captured.

Suppose S_3 is not either of S_1 and S_2. If the robber moves to a different copy of G (i.e. moves in H) S_3 also moves to this copy and moves closer to the robber in G. Hence, the robber cannot indefinitely move in H. If the robber moves in G or passes, S_3 passes. At each turn, whatever the robber does, the unit of S_1 and S_2 moves toward the robber. If the robber moves in H, the distance between the robber and S_1 and S_2 does not increase. If the robber exercises his other two options this distance decreases. Eventually, we must have S_3 adjacent to the robber in G and either S_1 or S_2 adjacent in H. The robber now has no safe move and so is captured by one of these three cops.   □

For the product of more than two graphs we have

Theorem 3.4 Let G_i, $i = 1, 2, \ldots, n$ be finite graphs with $c(G_n) \geq c(G_{n-1}) \geq \ldots \geq c(G_1)$. Then

(a) $\max\{c(G_n), 2\} \leq OS(\boxtimes_{i=1}^{n} G_i) \leq c(G_n) 2^{n-1}$;

(b) if $\text{girth}(G_i) \geq 5$ for all $i = 1, 2, \ldots, n$, then $OS(\boxtimes_{i=1}^{n} G_i) \geq n + 1$.

Proof. (a) First suppose that $c(G_n) \geq 2$.

For each i, let x_i and y_i be adjacent vertices in G_i. Initially, place $\lceil c(G_n)/2 \rceil$ cops on one-half of the vertices of $\{x,y\}^n$ and $\lfloor c(G_n)/2 \rfloor$ on the others. In any one coordinate, depending on the initial placement, after a finite number of moves if the robber does not move in that coordinate on each turn then a single cop suffices to capture the robber — the strategy is for each cop to decrease the distance to the robber each time the robber passes. In the product, in a particular coordinate, call the cops starting on x_i as the leading cops and the others as the trailing cops for that coordinate. The trailing cops will always move to the last vertex (in that coordinate) occupied by the leading cops. The robber can only move in one coordinate at a time therefore the only way to avoid capture in every coordinate (and hence in the game) is to (eventually) keep moving in one coordinate only, say the ith coordinate. After the robber has been captured in the other $n-1$ coordinates, in each coordinate the leading and trailing cops keep switching vertices — denote these vertices by a_j, and b_j, $j \neq i$.

There are $c(G_n)$ ($\geq c(G_i)$) supercops in the ith coordinate, so by playing the winning strategy in G_i one supercop will capture the robber in that coordinate and one of the 2^{n-1} cops with the same ith coordinate will also capture in the product.

If $c(G_n) = 1$ then place the 2^{n-1} cops on $\Pi_{i=1}^{n-1}\{x_i,y_i\} \times \{x_n\}$. There is one supercop in G_n. The cops capture the robber's shadow in the first $n-1$ coordinates proceeding as in the previous paragraph. The supercop then captures the robber's shadow in G_n and the robber is captured in the product.

For the lower bounds, the cops must win in the projection of the game onto G_n. If $c(G_n) = 1$ then an extension of Theorem 3.3 gives that two cops are needed.

(b) If the girth of all the graphs is at least 5 then consider the situation immediately before the robber's last move. For each coordinate, there must be a cop equal to the robber in that coordinate and adjacent in all the rest. These will capture the robber if he moves. There must also be a cop adjacent to the robber in each coordinate in case the robber passes. □

The lower bound is tight: if $m > 2$ then $OS(\boxtimes_{i=1}^n K_m) = 2$. The strategy is to initially place the cops on two vertices that are different in every coordinate. The robber then chooses his vertex. Now, the cops can move so that the three have different values in all coordinates. Both cops threaten to capture and since the robber can only change the value of one coordinate he cannot escape from both cops.

3.3   Cartesian considerations

This section is concerned with the cases where the adversaries are on the Cartesian and complement (within the strong product) of the Cartesian edges. These are the most extreme cases of all the variants. All theorems concern the product of three or more graphs since in the case of two graphs the situation was covered in the previous sections.

In the case of the robber on the Cartesian edges and the cops on the rest, the copnumber is known in almost all cases.

Theorem 3.5 *Let G_i, $i = 1, 2, \ldots, n$ be finite graphs with $c(G_n) \geq c(G_{n-1}) \geq \ldots \geq c(G_1)$.*

(a) *If $c(G_n) = 1$, then $\overline{S}S(\boxtimes_{i=1}^n G_i) \leq 2$ for $n = 2, 3$ and $\overline{S}S(\boxtimes_{i=1}^n G_i) = 1$ for $n \geq 4$.*

(b) *If $c(G_n) > 1$, then $\overline{S}S(\boxtimes_{i=1}^n G_i) = c(G_n)$ for $n \geq 3$.*

Proof. Recall that the cops have to change at least two of their coordinates on each move while the robber changes only one.

We note first that if the robber confined himself to moving in the nth coordinate only, then $c(G_n)$ cops would be required. Hence, $c(G_n) \leq \overline{S}S(\boxtimes_{i=1}^n G_i)$.

(a) If $n = 2$, then the result follows from Theorem 3.3.

If $n = 3$, we place the two cops on the same vertex. Each cop plays the winning strategy in each of the coordinates. Eventually the cops will have two of their coordinates, say the first and second, equal to the robber's and in the third coordinate they will be adjacent to the robber. If the robber moves in either of the first two coordinates the cops, who have to move in at least two coordinates, will be adjacent to the robber in two, equal in the third and so will be able to capture. If the robber passes, the two cops move in the first two coordinates (to an adjacent vertex) but now the second cop also moves in the third coordinate to occupy the same vertex as the robber. Both cops threaten the robber. If he moves in the third coordinate the second cop will capture. If he moves in either of the other coordinates he cannot move to a vertex occupied by a cop, else again he would be captured by the first cop. Therefore he must run away in either the first or second coordinate. The cops follow the winning strategy in that coordinate and swop their third coordinates. Eventually, the robber will be cornered in both the first and second coordinates. Therefore he must pass or move in one coordinate to a vertex occupied by a cop. In either case one cop will be adjacent in at least two coordinates and equal on the rest and so can now capture.

If $n > 3$, we let one cop play his winning strategy in all coordinates simultaneously until his position is equal to the robber's in $n - 1$ coordinates and adjacent in the nth coordinate. It is clear that this can be readily accomplished. The robber now has three options: 1) he may pass; or he may move in the nth coordinate either 2) to a vertex adjacent to the cop's position (in this coordinate) or 3) to a vertex not adjacent (in this coordinate). He cannot move in any coordinate other than n or he will be captured.

1) If he passes, then the cop moves in two other coordinates, say $n - 2, n - 1$. The robber, now under attack, must move in one of these three coordinates. The cop now moves so that three of the four coordinates $n - 3$, $n - 2$, $n - 1$ and n are adjacent to the coordinates of the robber and is equal in the fourth. The robber cannot move in any of the other coordinates where he and the cop's coordinates coincide without being immediately captured. Neither can he move so as to occupy the same coordinate as the cop in the other three. Therefore, on each turn the robber must move to a vertex not adjacent to the cop in one of the three coordinates. The cop follows the robber using the winning strategy in that coordinate. By Lemma 3.1 the cop can always stay adjacent or equal to the robber. In the coordinate where there was equality he moves to the last previously occupied vertex. Since G is copwin, this can only happen finitely often in each coordinate. Therefore, the robber will be cornered in each coordinate, i.e. his only moves will leave him adjacent to the cop in at least two coordinates and equal in the rest. The cop will then win on the next move.

2) The second option for the robber is to move to a vertex adjacent to the cop in the nth coordinate. The cop then makes his nth coordinate identical to the robber's and three other coordinates, say $n-3, n-2$, and $n-1$, which is now the situation in 1) above.

3) If the robber moves in the nth coordinate to a vertex not adjacent to the cop, then the cop plays his winning strategy on the nth coordinate and makes the same moves as before in coordinates $n-2$ and $n-1$ which gives the same situation as 1) above.

(b) We begin by placing $c(G_n)$ cops on the same vertex. Since the cops can play on all of the coordinates simultaneously and the robber can only play on one at a time, with each move the cops move closer to the robber in $n-1$ coordinates and maintain their distance to the robber in the other coordinate. Hence, it is clear that eventually the cops' position will be the same as the robber's in $n-1$ coordinates and different in, say, the ith coordinate. On their next move, let one cop, S_1, move in his kth and ith coordinates, $k \neq i$. Let the remainder move in their jth and ith coordinates, $k \neq j \neq i$. Now let the cops play their winning strategy in the ith coordinate. Since the cops must change at least two coordinates at every turn, one of them changes his kth coordinate and the rest their jth coordinate in such a way that after their move their position in these coordinates is either equal or adjacent to the robber's.

Note that it does the robber no good to play on any coordinate other than i since the cops can maintain their positions relative (i.e. either adjacent or equal) to the robber on all these coordinates and at the same time move closer to the robber in the ith coordinate. Eventually, the robber will be captured in the ith coordinate (since $c(G_i) \leq c(G_n)$) by some cop, say S_2. Now either the robber is captured or he is adjacent to S_2 in either coordinate j or k. (If $S_1 = S_2$, then in the following argument, designate some other cop as S_1 and interchange j and k.) The robber can now only pass or move in coordinate j. If he passes, S_1 moves as before in the kth coordinate and toward the robber in the ith coordinate. The cop S_2 passes. If the robber moves in the jth coordinate, then S_2 moves in coordinates k and j so that he is adjacent to the robber in both of these coordinates. The cop, S_1, makes both of these coordinates equal to the robber and moves toward the robber in the ith coordinate. The robber may now move in either coordinate k or j but S_2 can remain adjacent to the robber in one or both of these coordinates and S_1 can maintain equality in coordinates k and j and continue to move towards the robber in the ith coordinate. Eventually, S_1 must be adjacent to the robber in coordinate i. The robber now has no safe move and so will be captured on the next turn.   □

As an example, we have $\overline{SS}(\boxtimes_{i=1}^{n} K_{m_i}) = 1$ for $n \geq 3$. The robber must choose an initial position with all coordinates equal to the cop except one or he will be caught immediately. The cop then moves to make all coordinates different from the robber. Since $n \geq 3$, after the robber moves he must have at least two of his coordinates different from the cop and so he will be caught on the next move.

If the edge sets are reversed then we have the following result. (For clarity we restrict our considerations to the product of a graph with itself, the appropriate generalizations are clear but messy.)

Theorem 3.6 *Let G be a finite graph, $\delta = \delta(G)$ and $m = |V(G)|$. Then*

(a) $S\overline{S}(\boxtimes_{i=1}^{n}G) \leq (c(G) + vs(G))m^{n-2}$;

(b) $S\overline{S}(\boxtimes_{i=1}^{n}G) \geq \dfrac{(\delta+1)^n - \delta n}{\delta n + 1}$;

(c) *if the girth of G is at least 5, then for $n > 2$,*

$$\frac{(\delta+1)^n - \delta n}{(n-1)\delta + 2} \leq S\overline{S}(\boxtimes_{i=1}^{n}G).$$

Proof. (a) The proof is identical to that given in Theorem 3.2.

The lower bound in (b) is based on the relative sizes of a cop's and robber's neighbourhood.

In (c), the robber has $(\delta + 1)^n - \delta n$ possible moves. Suppose that a cop is adjacent to the robber in k coordinates and equal in the rest. Then the intersection of the cop's neighbourhood with the robber's is $k + (n-k)\delta + 1 = f(k)$ where $1 \leq k \leq n$. Clearly, $f(k)$ is maximized when $k = 1$. Hence, each cop can be adjacent to at most $(n-1)\delta + 2$ vertices in the neighbourhood of the robber.   □

For products of some complete graphs, a better bound can be found. In particular, if $m_n > \Pi_{i=1}^{n-1}m_i$ then

$$\Pi_{i=1}^{n-1}m_i \geq S\overline{S}(\boxtimes_{i=1}^{n}K_{m_i}) \geq \Pi_{i=1}^{n-1}m_i - 1.$$

The upper bound comes from the theorem. If $m_n \geq \Pi_{i=1}^{n-1}m_i + 1$ and there are $\Pi_{i=1}^{n-1}m_i - 2$ cops then they cannot occupy the vertices of a dominating set. In particular there are at least two vertices of K_{m_n} and of $\boxtimes_{i=1}^{n-1}K_{m_i}$ that are not occupied by shadows of cops. Therefore, if the robber is threatened by capture there is at least one vertex in the product that differs from all the cops' positions in at least two coordinates. Consequently, $\Pi_{i=1}^{n-1}m_i - 2$ cops do not suffice.

3.4   Categorical Edges

In these cases, both edge sets grow exponentially, but the player on $\overline{O}$ has the advantage as the number of factors increases. The following results can be extended to a product of arbitrary graphs, but, for clarity we restrict our attention to the product of one graph.

In the first case, the copnumber is known almost exactly. Surprisingly, the copnumber increases linearly with n despite the fact that the cops seem to have a larger advantage as n increases.

Theorem 3.7 *Let G be a finite graph. Then for $n > 1$,*

(a) $\overline{O}\overline{O}(\boxtimes_{i=1}^{n}G) \leq (c(G) - 1)(n-1) + \min\{vs(G) + 1, \overline{\beta}(G)\}$;

(b) $\overline{O}\overline{O}(\boxtimes_{i=1}^{n}G) \geq (c(G) - 2)(n-1)$.

Proof. (a) With $c(G)$ cops we play in each of the first $n-1$ coordinates only. It is clear the robber will be caught in each coordinate and that at least one cop will have $\lceil \frac{(n-1)}{c(G)} \rceil$ coordinates equal to the robber. The addition of a new cop means that in each coordinate the robber can be caught again. By the pigeon-hole principle if there are $(n-1)(c(G)-1)$ cops, there is one cop who has caught the robber in $n-1$ coordinates.

Let $\mathcal{L} = \{a_1, a_2, \ldots, a_m\}$ be a linear layout of G which realizes the vertex separation number of G. Initially place $c(G) + vs(G) + 1$ cops so that their nth coordinate is a_1. Play the strategy given in the preceding paragraph so that one cop has captured the robber on all the first $n-1$ coordinates. That cop now follows the robber's shadow in the first $n-1$ coordinates and does not move in the last. The other cops change their last coordinate to a_2 and proceed to capture the robber (with one cop) in the first $n-1$ coordinates and, again, this cop keeps following the robber's shadow in the first $n-1$ coordinates and does not move in the last. This procedure is carried out for a_i, $i = 3, \ldots, vs(G) + 1$. The cops now use the vertex separation strategy — for the next vertex a_j, the released cop plus the non-shadowing cops move to a_j and capture the robber in the first $n-1$ coordinates. The robber is trapped in successively smaller subgraphs and so will be caught.

Note that if $c(G) = 1$ then only 1 cop is needed to capture the robber on the first $n-1$ coordinates.

Let $J = \{a_1, a_2, \ldots, a_j\}$ be the complement of a maximum independent set in G. Initially place $(c(G) - 1)(n - 1) + j$ cops so that their nth coordinate is a_1. Play the strategy given in the first paragraph so that one cop has captured the robber on all the first $n-1$ coordinates. Move the other cops so that their last coordinate is a_2, capture the robber's shadow, move the others so that their last coordinate is a_3, etc. For each i the capturing cop keeps his last coordinate as a_i but follows the robber in the other coordinates. After the robber's shadow has been caught by a cop with nth coordinate equal to a_j, either the robber has been caught or at least one of the j capturing cops is adjacent in the last coordinate. The robber is threatened with capture but his only moves are to vertices with last coordinate being a vertex of $J = \{a_1, a_2, \ldots, a_j\}$, but in each of these the capturing cop will capture on the next move.

(b) There are variants of cops and robber where the robber and/or cops are forced to move (i.e. cannot pass) on each move. The respective copnumbers are within one of $c(G)$ (see [12]). Therefore, in this game, where the robber must either pass or move in all coordinates simultaneously, the robber can avoid $c(G) - 2$ different cops in each of $n-1$ coordinates.□

With the edge sets reversed, the robber now has the advantage as n increases and this is reflected in the exponentially increasing lower bound.

Theorem 3.8 *Let G be a finite graph, $\delta = \delta(G) \geq 2$ and $c(G) \geq 2$. For $n > 2$,*
(a) $(1 + \frac{1}{\delta})^n - 1 \leq O\overline{O}(\boxtimes_{i=1}^n G) \leq 2^{n-1} c(G)^n$;
(b) *If* $\text{girth}(G) \geq 5$ *then* $O\overline{O}(\boxtimes_{i=1}^n G) \geq 2^n - 1$.

Proof. (a) In G, choose two adjacent vertices and designate one vertex as 'leading', the other as 'trailing'. Place $c(G)^n$ cops on each permutation of these two vertices in the first $n-1$ coordinates, $2^{n-1} c(G)^n$ cops in total. Now proceed to capture the robber in the first $n-1$ coordinates. Once the robber's shadow is captured in a coordinate the leading vertex remains on the robber's shadow at each turn. If the robber passes in this coordinate the roles of leading and trailing are interchanged. Eventually, this gives $c(G)$ cops on every permutation of the leading and trailing vertices where in each of the first $n-1$ coordinates one of these vertices is on the robber's shadow. There are $c(G)$ supercops in the nth coordinate. Now, the supercops proceed to capture the robber in the final coordinate which means he is now captured by some cop in all coordinates.

The lower bound arises from the consideration of the possible overlap of the cops' and robber's neighbourhoods.

(b) Consider the robber's last move. If he does not move then there must be a cop ready to capture him. If he moves in only one coordinate then he has $n\delta$ neighbours and any one cop can dominate at most δ of these. Similarly, if the robber moves in k coordinates then he has $\binom{n}{k}\delta^k$ neighbours and a cop can dominate at most δ^k of these. Hence, at least $\sum_{k=0}^{n-1}\binom{n}{k}$ cops are required. $\square$

Theorem 3.9 *If $m_i \geq n$ for each i, then $O\overline{O}(\boxtimes_{i=1}^{n} K_{m_i}) > n$.*

Proof. If $m_i > n$ then the strategy is to place the jth cop on the vertex $(j, j, \ldots, j)$, $j = 1, 2, \ldots, n$. To avoid capture, the robber must choose a vertex whose coordinates are a permutation of 1 through n. The cops then move so that for each i, all numbers from 1 to $n + 1$ occur in the ith coordinate except for the value that is in the robber's ith coordinate. Each cop then threatens the robber. To avoid capture the robber must change every coordinate which is not a legal move.

If there are only $n - 1$ cops then for each i, $1 \leq i \leq n - 1$, the robber can always move to have the same ith coordinate as the ith cop. The nth coordinate need never change. $\square$

4   Comments and open problems

1. The upper bound in Theorem 3.2 is obtained by using two coordinates only, not all n. In general, the upper and lower bounds are some distance apart. Is it possible that some more sophisticated strategy could use fewer cops?

One specific case where this is possible is: let $m_3 \geq m_2 \geq m_1$, then $SO(\boxtimes_{i=1}^{3} K_{m_i}) \leq m_1 m_2 + 1 - (m_1 + m_2)$. This can be accomplished by placing the cops on those vertices whose last coordinate is a 1 and that do not have precisely one 1 in the other coordinates. To avoid capture on the first move, the robber must occupy a vertex which has precisely one 1 in the first 2 coordinates and some value other than 1 in the 3rd coordinate. Say the first coordinate of the robber is not 1. Then, the cops that have the same first coordinate as the robber move so as to change that coordinate to a 1. The cop on $(1, 1, 1)$ moves so that his first coordinate is the same as the robber's. This cop threatens to capture the robber if he passes and the robber cannot move to a safe vertex and so he loses.

2. The bounds in Theorem 3.4 are very far apart. The upper bound is achieved in $OS(\boxtimes_{i=1}^{n} K_2)$. There, the cops are faced with a graph of 2^{n-1} disconnected edges. If we consider only graphs with chromatic number 3 or greater (so that the cops have a connected graph to play on), what are the correct bounds?

3. What is the value of $O\overline{O}(\boxtimes_{i=1}^{n} K_m)$ if $n > m$? Based on dominating the robber's neighbourhood, at least

$$\frac{m^n - (m-1)^n - 1}{(m-1)^n + 1}$$

cops are needed.

References

[1] M. Aigner, M. Fromme, A game of cops and robbers, *Discrete Appl. Math.* **8** (1984) 1–12.

[2] T. Andreae, Note on a pursuit game played on graphs, *Discrete Appl. Math.* **9** (1984), 111–115.

[3] T. Andreae, On a pursuit game played on graphs for which a minor is excluded, *J. Combin. Theory (Ser. B)* **41** (1986), 37–47.

[4] R. Anstee, M. Farber, On bridged graphs and cop-win graphs, *J. Combin. Theory (Ser. B)* **44** (1988), 22–28.

[5] G. Chartrand, L. Lesniak, *Graphs and Digraphs*, Wadsworth & Brooks/Cole, 1986.

[6] J.A. Ellis, I.H. Sudborough and J.S. Turner, *Graph separation and search number*, 21st Annual Allerton Conference on Communication Control and Computing (1983), 224–233.

[7] P. Frankl, Cops and robbers in graphs with large girth and Cayley graphs, *Discrete Appl. Math.* **17** (1987), 301–305.

[8] P. Frankl, On a pursuit game on Cayley graphs, *Combinatorica* **7** (1987), 67–70.

[9] G. Hahn, F. Laviolette, N. Sauer, R.E. Woodrow, *On cop-win graphs*, manuscript 1991.

[10] Y.O. Hamidoune, On a pursuit game of Cayley digraphs, *Europ. J. Combin.* **8** (1987), 289–295.

[11] M. Maamoun and H. Meyniel, On a game of policemen and robber, *Discrete Appl. Math.* **17** (1987), 307–309.

[12] S. Neufeld, M.Sc. Thesis, Dalhousie University, 1991.

[13] S. Neufeld, R. Nowakowski, Copnumbers of Cartesian, strong and categorical products of graphs, manuscript 1991.

[14] R. Nowakowski, P. Winkler, Vertex to vertex pursuit in a graph, *Discrete Mathematics* **43** (1983), 235–239

[15] T.D. Parsons, Pursuit-evasion in a graph, *Lecture Notes in Math.* **642** (Springer, Berlin 1978).

[16] A. Quilliot, *Thèse d'Etat*, Université de Paris VI, 1983.

[17] A. Quilliot, A short note about pursuit games played on a graph with a given genus, *J. Combin. Theory (Ser. B)* **38** (1985), 89–92.

[18] R. Tošić, S. Šćekić, An analysis of some partizan graph games, *Proc. 4th Yugoslav Seminar on Graph Theory*, Novi Sad (1983), pp. 311–319.

[19] R. Tošić, The search number of the Cartesian product of graphs, *Univ. u Novom Sabu Zb. Rad. Prirod.-Mat Fak. Ser. Mat* **17** (1987), 239–243.

Graphs and Posets with no Infinite Independent Set

MAURICE POUZET *

Institut de Mathématiques-Informatique
Université Claude Bernard
43 Bd du 11 Novembre 1918
69622 Villeurbanne Cédex France

Dedicated to Eric C. Milner at the occasion of this third coming of age.

Abstract

We consider the extension to graphs, directed or not, of the notion of better-quasi-ordering due to Nash–Williams and its specialization, the α-better-quasi-ordering. We find whether such graphs have, among their spanning subgraphs with the same property, some which are an ordering. As a typical result, we obtain that if a countable undirected graph has no infinite independent set, then it has a spanning subgraph which is the comparability graph of a poset with no infinite antichain, whereas this conclusion may fail if the graph is uncountable. We also solve a question of P. Ribenboim concerning compatible orderings on the rational numbers.

We discuss some specific instances of a general question: given a graph G which does not contain a certain pattern, does G contain, as a spanning subgraph, a comparability graph with the same property?

Forbidden patterns are first infinite independent sets, then the maps defined on some collections of finite sets of integers, leading to the definition of better-quasi-ordering introduced by Nash–Williams in 1965 [14]. Positive results are obtained by means of a very simple construction of a comparability graph from an arbitrary graph. Proofs follow from straightforward application of basic techniques of the theory of well-quasi-ordering and better-quasi-ordering. Negative results are illustrated by means of shift-graphs and their generalizations.

Results are presented in the next two sections. The first one is devoted to graphs with no infinite independent set. In the second, we present the extension to α-well-graphs and to better-graphs of results of section 1. Despite the fact that (unoriented) graphs with no infinite independent set are ω-well-graphs we prefer to do so in order to make the paper more accessible to the reader unfamiliar with better-quasi-ordering techniques. We conclude this section by discussing a question of P. Ribenboim [20] concerning compatible orderings

---

*Research supported by PRC Math–Info.
AMS subject classification (1991): 0.6 a10.
Key Words: poset, graph, well-quasi-ordering, better-quasi-ordering, independent, set, ordered groups.

N.W. Sauer et al. (eds.), Finite and Infinite Combinatorics in Sets and Logic, 313–335.
© 1993 *Kluwer Academic Publishers. Printed in the Netherlands.*

on the rational numbers, which motivated this research. We solve it by means of an old result of Erdős and Tarski [5].

We present our results in an informal way, leaving details to sections 3 and 4. We use the language of graph theory, and most of our results can be understood in terms of the commonly accepted terminology about graphs. However, the graphs we have in mind are rather binary relations, sometimes reflexive. More specifically, in the sequel a *graph* is a pair $G = (V, \mathcal{E})$ where $\mathcal{E}$ is a subset of $V \times V$, we write $x \sim y \pmod{G}$ instead of $(x, y) \in \mathcal{E}$. If $\mathcal{E}$ contains $\Delta_V := \{(x, x) : x \in V\}$, the graph is *reflexive*. The *symmetric hull* of $G = (V, \mathcal{E})$ is the graph $(V, \mathcal{E} \cup \mathcal{E}^{-1})$; we say that such a graph is *undirected*. A *comparability graph* is the symmetric hull of a poset; hence, in principle, such a graph is reflexive. The *complement* of G is the graph $G^c = (V, V \times V \backslash \mathcal{E})$. Hence, the complement of a graph, in the usual sense, is reflexive. A graph $G' = (V', \mathcal{E}')$ is a *spanning subgraph* of $G = (V, \mathcal{E})$ if $V' = V$ and $\mathcal{E}' \subseteq \mathcal{E}$. Other unexplained terminology is borrowed from set theory. Ordinal numbers are denoted by greek letters, ω denotes the first infinite ordinal; the cardinality of a set A is denoted $|A|$, the cofinality of a cardinal x is denoted $\mathrm{cf}(x)$.

Our construction consists of enumerating the vertices of a graph, ordering them from the left to the right and keeping as many vertices as possible. More precisely let $G = (V, \mathcal{E})$ be a graph (either oriented or unoriented), let $\mathcal{L}$ be a well ordering of V and let $x_0, x_1, \ldots, x_\alpha, \ldots$ be an enumeration of $V(G)$ according to this well ordering. We define the ordering, that we denote $\leq$, by a double induction. Suppose that the set $\{x_\beta : \beta < \alpha\}$ has already been ordered. We compare its elements to x_α as follows: suppose the comparisons between elements of $\{x_\gamma : \gamma < \beta\}$ and x_α already defined, then we set $x_\beta \leq x_\alpha$ if the following conditions hold:

$$x_\beta \sim x_\alpha (\mathrm{mod}\ G) \tag{1}$$

$$x_\gamma \leq x_\beta \quad \text{implies} \quad x_\gamma \leq x_\alpha \quad \text{for all } \gamma, \gamma < \beta. \tag{2}$$

This clearly defines an ordering with no infinite descending chain. By construction, each pair of distinct and comparable elements form an edge of G. We denote $F(G, \mathcal{L})$ the resulting poset.

Let $(L, \leq)$ be a chain; the *L-shift-graph* is the undirected graph S_L whose vertices are all pairs $u = (i, j) \in L^2$ such that $i < j$, and whose edges are pairs $\{u, v\}$ such that $u = (i, j), v = (j, k)$ for some $i < j < k$. Such a graph contains no triangle, thus its complement contains no independent set on three elements.

Part of this work was done at the University of Calgary in May and August, 1991. We thank members of the Department of Mathematics and Statistics for their hospitality and the NSERC of Canada for support. We thank A. Marcone for helpful discussions and E.C. Milner for his help in writing this paper.

1 Graphs containing no infinite independent set

Theorem 1 *Let G be an undirected graph on a countable set of vertices V and $\mathcal{L}$ be a well ordering of V with type ω. If G contains no infinite independent set then $F(G, \mathcal{L})$ does not either.*

Corollary *Let G be an undirected reflexive graph with countably many vertices. If no infinite subset of the vertex set $V(G)$ is formed of pairwise independent vertices, then $V(G)$ can be ordered in such a way that each comparable pair is an edge of G, and there is neither an infinite antichain, nor an infinite descending chain in this ordering.*

Our second result shows that this conclusion fails badly if the vertex set is uncountable:

Theorem 2 *If L is an uncountable chain, then every spanning subgraph of the complement of S_L which is a comparability graph contains an infinite independent set.*

Theorem 2 is a direct consequence of the following result about power sets of comparability graphs.

Given a graph $G := (V, \mathcal{E})$ let $\mathcal{P}(G)$ be the graph whose vertex set is made of subsets of V, two vertices X_1, X_2 being joined by an edge if for every $i \neq j \in \{1, 2\}$, and every $x \in X_i$, there is some $y \in X_j$ such that $(x, y) \in \mathcal{E}$. If G is an undirected reflexive graph, then $\mathcal{P}(G)$ is the discrete analog of the power set of a metric space endowed with the Hausdorff metric.

Theorem 3 *Let G be a graph having a spanning subgraph which is a comparability graph with no infinite independent set. Then every collection $\mathcal{C}$ of vertices of $\mathcal{P}(G)$ such that $\mathrm{cf}(|\mathcal{C}|) > \omega$ contains a complete subgraph of size $|\mathcal{C}|$.*

If G is the complement of the shift graph S_L of a chain L, then the collection of vertices of $\mathcal{P}(G)$ made of sets of the form $X_i = \{(i', j') \in L \times L : i = i' < j'\}$ is an independent set of $\mathcal{P}(G)$. Indeed for $i < j$, there is no vertex in X_j joined to the vertex (i, j) of X_i. Consequently, if L is uncountable then $\mathcal{P}(G)$ contains an independent set of size $\aleph_1$. From theorem 3 it follows that G has no spanning subgraph which is a comparability graph with no infinite independent set.

We deduce theorem 3 from the following result about posets, obtained jointly with Eric Milner ([12], theorem 6).

Theorem 4 *Let P be an ordered set with no infinite antichain, then every set J of ideals of P such that $\mathrm{cf}(|J|) > \omega$ contains a chain (ordered by inclusion) of size $|J|$.*

The ordering in theorem 1 is obtained as a sub-ordering of the linear ordering induced on the vertex set V by an arbitrary ordering of type ω. It makes sense to ask for which well-orderings $\mathcal{L}$ on V, the poset $F(G, \mathcal{L})$ has no infinite antichain, and also for which countable linear orderings $\mathcal{L}$ on V one can find a sub-ordering with no infinite antichain whose comparabilities consist only of edges of G.

The answers are different.

Theorem 5 *Let G be the complement of the shift-graph $S_{\omega+1}$ associated with the chain $\omega + 1$, and $\mathcal{L}$ be the lexicographic ordering defined on $V := [\omega + 1]^2$ according to the second difference (i.e. $(i, j) \leq (i', j')$ if either $j < j'$ or $j = j'$ and $i \leq i'$). Then $\mathcal{L}$ has order type $\omega.2$, however $F(G, \mathcal{L})$ contains an infinite antichain (namely, the set of (i, ω) such that $i < \omega$).*

On the other hand, with the help of theorems 1 and 2 we get:

Theorem 6 *Let $\mathcal{L}$ be a linear ordering on a set V. The following conditions are equivalent:*

(i) *For every reflexive undirected graph $G = (V, \mathcal{E})$ with no infinite independent set, there is an ordering $\mathcal{P}$ contained in $\mathcal{L} \cap \mathcal{E}$ which has no infinite independent set;*

(ii) *V is countable and the chains $\omega.\omega^*$ and $\omega^*.\omega$ do not embed in $(V, \mathcal{L})$.*

2 Better-graphs and better-quasi-ordering

Let us recall the basic ingredients of the notion of better-quasi- ordering, as introduced by C.St.J.A. Nash–Williams in 1965 [14], cf. also [15].

We start with his notion of *barrier*. First, we identify finite subsets of ω and strictly increasing finite sequences of non negative integers. We denote such an object by the list of its elements, e.g., $\{3, 4, 8\}$. Let $[\omega]^{<\omega}$ be the collection of finite subsets of ω.

Given $s \in [\omega]^{<\omega}$, let s^* be $s \backslash \text{Min } s$; given s and t, we denote $s \ll t$ the fact that s is an initial segment of t, and $s \lhd t$ the fact that $s^* \ll t$. This latter relation will be called the $\lhd$-relation. For example, $\{1, 2, 7\} \ll \{1, 2, 7, 8, 10\}$, and $\{1, 2, 7\} < \{2, 7, 8, 10\}$.

A *barrier* is any subset B of $[\omega]^{<\omega}$ such that:

B1 *Members of B are pairwise incomparable for inclusion;*

B2 *The support of B, defined by $\bar{B} = \cup\{s : s \in B\}$, is infinite;*

B3 *Every infinite subset X of $\bar{B}$ has an initial segment s (with respect to the ordering induced by ω on X) which belongs to B.*

As observed in [16], each barrier, once lexicographically ordered, is well-ordered. Order-types of barriers are of the form $\alpha.n$, where α is a countably infinite indecomposable ordinal, n a positive integer, subject to the restriction that $n = 1$ if $\alpha < \omega^\omega$ [1]. For example, $[\omega]^n$ is a barrier for each positive integer, its order type is the ordinal ω^n; for $n = 2$, the $\lhd$-relation on $[\omega]^2$ defines the oriented shift-graph on ω. The subset $B = \{s \in B : l(s) = \text{Min } s + 1\}$ is a barrier of order type ω^ω.

Given a countable (infinite) ordinal number α, a (directed) graph $G = (V, \mathcal{E})$ is α-*well* if every map $f : (B, \lhd) \rightarrow G$, where B is a barrier of type at most α, is *good* (see 3.2) — that is there are s and t in B such that $s \lhd t$ and $f(s) \sim f(t)$ mod G. One can observe that it is enough to consider barriers B of type $\text{ind}(\alpha)$, the largest indecomposable ordinal below α. The graph G is *better* if it is α-well for every countably infinite ordinal number α. In the case $\mathcal{E}$ is a quasi-ordering (that is $\mathcal{E}$ is reflexive and transitive) this latter definition of this is *better- quasi-ordering* (b.q.o.) due to C.St.J.A. Nash–Williams. For $\alpha = \omega$, the ω-well notion amounts to the usual notion of *well-quasi-ordering* (w.q.o.): every infinite sequence $x_0, \ldots, x_n, \ldots$ contains some increasing pair, i.e. $x_n \leq x_m$ for some $n < m$. Notice that if a graph is α-well for some α then G is necessarily reflexive.

Theorem 7 *Let G be a countable directed graph, $\mathcal{L}$ be an enumeration of its vertices, with order type ω and α be a countably infinite ordinal. If G is α-well, then the ordered set $F(G, \mathcal{L})$ is α-well.*

If α is the order-type of a barrier B, then the graph G, formed with B as a set of vertices and the complement of the $\lhd$-relation as edge set is not α-well.

Theorem 8 (A. Marcone, 1992) [10]. *For each denumberable indecomposable ordinal α, there is a barrier B with order type α such that the graph G is β-well for all $\beta < \alpha$.*

From these two results it follows that uncountably many conditions are needed in order to define better-quasi-ordering; precisely:

Corollary *For each denumberable indecomposable ordinal α there is a denumerable ordered set which is β-well-ordered for all $\beta < \alpha$ and not α-well-ordered.*

We have obtained this result jointly with A. Marcone (see also [10]). It was previously announced in [16] but our hint given there was incorrect.

On the negative side we have

Theorem 9 *Let α be a denumberable indecomposable ordinal; if $\alpha < \omega^\omega$, then there exists an α-well unoriented graph such that each spanning subgraph which is a comparability graph has an infinite independent set.*

Problem 1 Does this hold without the condition $\alpha < \omega^\omega$?

In surprising contrast with the negative theorems 2, 6 and 9, we have the following result.

Theorem 10 *Let $G := (V, \mathcal{E})$ be a graph, $\mathcal{L}$ be a linear well-ordering of its vertices. If G is a better-graph, then $F(G, \mathcal{L})$ is b.q.o. in the Nash–Williams sense.*

Corollary 1 *Every better-graph, possibly uncountable, contains, as a spanning subgraph, an ordering which is a b.q.o. In particular every better-unoriented-graph contains, as a spanning subgraph, the comparability graph of a b.q.o.*

Corollary 2 *Let $(E_\alpha)_{\alpha < \kappa}$ be an increasing sequence of sets, and on each E_α let $\mathcal{P}_\alpha$ be an ordering. If $\mathrm{cf}(\kappa) \neq \omega$ and each $\mathcal{P}_\alpha$ is b.q.o., then on $E = \bigcup_{\alpha < \kappa} E_\alpha$ there is a b.q.o. $\mathcal{P}'$ which is contained in $\mathcal{P} := \bigcup_{\alpha < \kappa} \mathcal{P}_\alpha$.*

On the other hand, from theorems 2 and 6 we have

Proposition. *On each denumberable ordinal α there is a w.q.o. $\mathcal{P}_\alpha$ such that there is no w.q.o. $\mathcal{C}'$ on ω_1 whose union is contained in $\mathcal{P} := \bigcup_{\alpha < \omega_1} \mathcal{P}_\alpha$.*

Indeed, let $S_{\omega_1}^c$ be the complement of the ω_1-shift graph. Enumerate the vertex set, namely $[\omega_1]^2$, into an ω_1-sequence. According to theorem 6, for each countable α, the graph induced on the first α terms contains, as a spanning subgraph a w.q.o. $\mathcal{P}_\alpha$. From theorem 2, every ordering contained in $\mathcal{P} := \bigcup_{\alpha < \omega_1} \mathcal{P}_\alpha$ has an infinite antichain (while the graph $\alpha < \omega_1$ having $\mathcal{P}$ as set of edges has no infinite independent set).

A very special instance of Corollary 2 whose formulation does not involve the b.q.o. notion is this:

Proposition. *For each countable ordinal α, let f_α be an injective map from α into ω and let $\mathcal{P}_\alpha$ be the ordering defined on α by $\alpha' \leq \alpha'' \bmod \mathcal{P}_\alpha$ if $\alpha' \leq \alpha''$ and $f_\alpha(\alpha') \leq f_\alpha(\alpha'')$. Then there is a w.q.o. $\mathcal{P}'$ on ω_1 which is contained in $\mathcal{P} := \bigcup_{\alpha < \omega_1} \mathcal{P}_\alpha$.*

To derive this from Theorem 10, it suffices to know that ordinals are b.q.o., finite intersections of b.q.o. are bqo, b.q.o. are w.q.o. Indeed, from these first two facts each $\mathcal{P}_\alpha$ is b.q.o.; theorem 10 then applies, the $\mathcal{P}'$ it gives is w.q.o. In fact, a much simpler approach gives a stronger conclusion.

Lemma. *Let $(E_\alpha)_{\alpha<\kappa}$ be an increasing sequence of sets, with $cf(\kappa)$:*

> *If $\mathcal{P}_\alpha$ is a well-ordering on E_α for $\alpha < \kappa$ then on $E = \cup_{\alpha<\kappa}E_\alpha$ there is a well-ordering $\mathcal{P}'$ contained in $\mathcal{P} := \cup_{\alpha<\kappa}\mathcal{P}_\alpha$.*

Proof. Instead of building the ordering by comparing pairs we enumerate its elements in order. Suppose $x_0,\ldots,x_\beta,\ldots$ $(\beta < \alpha)$ defined in such a way that, $(x_\beta,y) \in \mathcal{P}$ for all $\beta < \alpha$, $y \in F := E\backslash\{x_\beta : \beta < \alpha\}$. If $F \neq \emptyset$ then select x_α in F such that $(x_\alpha,y) \in \mathcal{P}$, for all $y \in F$. If no such an element exists, then for each $x \in F$, there is some $y_x \in F$ such that $(y_x,x) \in \mathcal{P}$, meaning $y_x < x$ (mod $\mathcal{P}_\gamma$) whenever $\{x,y_x\} \subseteq F_\gamma := E_\gamma \cap F$. Then define a sequence $x_0,x_1,\ldots,x_n,\ldots$ $(n < \omega)$ starting with an arbitrary $x_0 \in F$, and $x_{n+1} = y_{x_n}$. Since $cf(\kappa) \neq \omega$, then $\{x_n : n < \omega\}$ is contained in some E_γ. But then the sequence is strictly decreasing with respect to $\mathcal{P}_\gamma$ which is impossible.

Corollary. *If each $\mathcal{P}_\alpha$ is the intersection of n well-orderings $n < \omega$ then $\mathcal{P} := \cup\mathcal{P}_\alpha$ contains the intersection $\mathcal{P}'$ of n well-orderings, each defined on $E := \cup_{\alpha<\kappa}E_\alpha$, hence $\mathcal{P}'$ has no infinite antichain.*

This study was motivated by the following question asked by P. Ribenboim [20]: Is there some subset P of $\mathbb{Q}^+$, the set of non negative rational numbers, such that

(1) $0 \in P$, $P + P \subseteq P$,

(2) $1 \notin P$,

(3) the ordering defined by $x \leq_P y$ if $y - x \in P$ has no infinite antichain.

Clearly, if a subset $P \subseteq \mathbb{Q}^+$ satisfies (1) and (2) then the comparability graph of P is contained in the set $\mathcal{P}$ of pairs (x,y) such that $x - y \neq 1/n$ for every integer n. It is easy to see that the graph $G = (\mathbb{Q},\mathcal{P})$ has no infinite independent set. (In fact, for every pair $\{x,y\} \in [\mathbb{Q}]^2$, the subset of z such that neither x nor y is joined to z if finite.) Hence, prior to the Ribenboim question was to prove that G has a spanning ordering with no infinite antichains. Theorem 1 asserts that this holds; but theorem 6 shows that the fact that G has no infinite independent set is not enough to insure that there is such an ordering in the natural ordering on $\mathbb{Q}$. This then suggests to look for some conditions on a graph $G = (V,\mathcal{E})$, stronger than forbidding infinite antichains, insuring that for every linear ordering $\mathcal{L}$ on the set of vertices, the graph $\mathcal{E} \cap \mathcal{L}$ contains an ordering with no infinite antichain. Theorem 5 shows that they have to be quite strong (e.g. conditions of b.q.o. type will not suffice). While this approach to the Ribenboim problem was unsuccessful, it turned out, during a visit of A. Marcone in the spring of 1992, that it could be applied to build obstructions to better-quasi-ordering. We discovered recently that known properties of posets with no infinite antichain lead to a negative solution of Ribenboim's problem.

Theorem 11 *The natural ordering on the rational numbers, and the reverse ordering, are the only orders which are compatible with the addition and have no infinite antichains.*

3   Proofs of the results of Section 1

3.1   Proof of Theorem 1

Our proof relies on basic properties of posets which contain no infinite antichain and no infinite descending sequence. Posets satisfying such properties have been intensively studied — see e.g. [8] for a survey – and are known under the name of *well-quasi-ordered sets* (w.q.o.) (this term covering quasi-ordered sets and ordered sets as well). Supposing that there is some infinite antichain, the "minimal bad sequence" technique due to C.St.J.A. Nash–Williams [13] provides some infinite antichain Y_0 such that the set $\check{Y}_0 := \{z : z < y \text{ for some } y \in Y_0\}$ is w.q.o. Then, we derive a contradiction from a theorem, due to G. Higman [7], asserting that if P is w.q.o. then the collection of its finite subsets is w.q.o. once ordered by "majorization" (A is majorized by A' if every member of A is majorized by some member of A'). Here are the details.

We apply the following observation.

Lemma *Let P be a poset; if P is not w.q.o., then there is some final segment F of P such that*

1) *F is not finitely generated, that is there is no finite subset X of F such that $F = \{y : y \geq x \text{ for some } x \in X\}$.*

2) *$P \backslash F$ is w.q.o.*

Hence if P is well-founded then $L = \operatorname{Min} F$, the set of minimal elements of F is an infinite antichain, and the set $\check{F} := \{z : z < y \text{ for some } y \in L\}$ is w.q.o.

Proof. If P is not w.q.o., then some final segment i of P is not finitely generated. Zorn's lemma provides a maximal one. It has the required property.   □

The minimal bad sequence technique of Nash–Williams is an effective translation of this fact, avoiding Zorn's lemma.

Suppose that $F(G, \omega)$ is not w.q.o. According to the lemma above, it contains some infinite antichain, say Y_0 such that the set $Y_0 := \{z : z < y \text{ for some } y \in Y_0\}$ (where $<$ is the ordering defined by (1) and (2)) contains no infinite antichain. Write $Y_0 = \{y_0, y_1, \ldots, y_n, \ldots\}$ in an increasing order with respect to the enumeration of $V(G)$.

For each n, the set $Z_n = \{y : y < y_n\}$ is a finite initial segment of $\check{Y}_0$. Since this latter set is w.q.o., then, according to the Higman theorem mentioned above, the collection of its finitely generated initial segments, once ordered by inclusion, is also w.q.o. Since, every infinite sequence of elements of a w.q.o. set contains an infinite increasing subsequence, from the sequence of Z_n we get an infinite increasing subsequence, say $Z_{n_0} \subseteq Z_{n_1} \subseteq \ldots \subseteq Z_{n_j} \ldots$. Since G has no infinite independent set, the set $\{y_{n_0}, y_{n_1}, \ldots, y_{n_j}, \ldots\}$ contains two distinct vertices y_{n_i}, y_{n_j}, such that $n_i < n_j$ and $y_{n_i} \sim y_{n_j}$. These two vertices satisfy condition (1). Since $Y_{n_i} \subseteq Y_{n_j}$, they satisfy (2). Hence $y_{n_i} \leq y_{n_j}$ contradicting the fact that these vertices are incomparable.

Remark Let E be a set. Let us consider the collection of subsets X of $[E]^2$ such that $X \cap [Y]^2 \neq \emptyset$ for every infinite $Y \subseteq E$. From Ramsey's Theorem it follows that this collection forms a filter $\mathcal{F}$. It is not difficult to see that if E is countable then $\mathcal{F}$ is the intersection of the collection of $U_0 U$ where U is a non-principal ultrafilter on E, and

$$U_0 U := \{x \subseteq [E]^2 : \{x \in E : \{y \in E : \{x, y\} \in X\} \in U\} \in U\}.$$

Our corollary of Theorem 1 says that if E is countable, the sub-collection of all irreflexive comparability graphs is coinitial in $\mathcal{F}$.

Even on $E := \omega$, we do not know the coinitial structure of this filter.

3.2   Proof of Theorem 3

Theorems 3 and 4 are about a collection $\mathcal{C}$ of vertices. We prefer to consider a family of vertices, (in fact indexed by a chain), thus allowing possible repetitions of vertices. The following lemma proves that this modification does not change the validity of the results. In the sequel Theorems 3 and 4 will refer to this modification.

Lemma 3.2.1 *Lef f be a map from a set A onto a set B. Then there is a subset $B' \subseteq B$ such that $\mathrm{cf}(|B'|) \in \{1, \mathrm{cf}(|A|)\}$ and $|f^{-1}(B'')| = |A|$ for all $B'' \subseteq B'$ such that $|B''| = |B'|$.*

Proof. We prove that B contains a subset B' such that $|f^{-1}(B')| = |A|$ and either $|B'| \in \{1, |A|\}$ or $|B'|$ is regular, $|A|$ is singular, and the cardinals $|f^{-1}(x)|$ (for $x \in B'$) are all distinct and distinct from $|A|$. Such a B' satisfies the required property. To show that there is such a subset, consider all subsets C of B such that $|f^{-1}(C)| = |A|$ and select one, say C_1, with minimum cardinality. If $|C_1| \in \{1, |A|\}$ set $B' = C_1$. If not, let $\kappa = \mathrm{cf}(|C_1|)$. From the minimality of $|C_1|$, we have $|f^{-1}(C)| < |A|$ for all $C \subseteq C_1$ such that $|C| < |C_1|$. Since $1 < |C_1| < |A|$, it follows that A is singular and $\mathrm{cf}(|A|) = \kappa$. Now for each $\lambda < |A|$ there is some $x \in C_1$ such that $|f^{-1}(x)| \geq \lambda$, otherwise $|f^{-1}(C_1)| = \cup\{f^{-1}(X) : x \in C_1\} \leq \lambda$. $|C_1| < |A|$, thus we may inductively define a sequence $(x_\alpha)_{\alpha < \kappa}$ in C_1 such that the sequence $(\lambda_\alpha)_{\alpha \in \kappa}$ where $\lambda_\alpha := |f^{-1}(x_\alpha)|$ is strictly increasing and $\sum_{\alpha < \kappa} \lambda_\alpha = |A|$. We set $B' = \{x_\alpha : \alpha < \kappa\}$. $\square$

For the rest of the discussion we make the following conventions.

Let $G := (V, \mathcal{E})$, $G' := (V', \mathcal{E}')$ be two graphs; a *map* f *from* G *to* G' is any map $f : V \to V'$. Such a map f is *good* if there are vertices $u, v \in V$ such that:

$$(u, v) \in \mathcal{E} \quad \text{and} \quad (f(u), f(v)) \in \mathcal{E}' \tag{3}$$

otherwise f is *bad*. The map f is *perfect* if

$$(u, v) \in \mathcal{E} \Rightarrow (f(u), f(v)) \in \mathcal{E}' \tag{4}$$

Perfectness is another name for *graph-homomorphisms* from G to G'. Clearly, f is bad, as a map from G to G', if and only if f is perfect from G into G^c.

Given a graph $G := (V, \mathcal{E})$, let $\mathcal{P}(G)$ be the graph $(\mathcal{P}(V), \underline{\mathcal{E}})$ where $(X, Y) \in \underline{\mathcal{E}}$ if for every $x \in X$ there is some $y \in Y$ such that $(x, y) \in \mathcal{E}$. Clearly the graph $P(G)$ defined in section 1 verifies $\mathcal{P}(G) = (\mathcal{P}(V), \underline{\mathcal{E}} \cap \underline{\mathcal{E}}^{-1})$.

In order to prove theorem 3, it suffices to show:

Lemma 3.2.2 *Let G be a graph having a spanning subgraph which is a comparability graph with no infinite antichain. Let L be a chain. If $\mathrm{cf}(|L|) > \omega$ then every $f : (L,<) \to \mathcal{P}(G)$ is perfect on a subchain L' of size $|L'| = |L|$.*

Indeed, applying this lemma to $f|_{L'} : (L',>) \to \mathcal{P}(G)$ gives a subchain L'' for which $f|_{L''} : (L'',<) \to P(G)$ is perfect. This is theorem 3.

Proof. Let $G := (V, \mathcal{E})$ be a graph containing a comparability graph $G' = (V, \mathcal{E}')$ as a spanning subgraph; let $(L, <)$ be a chain and $f : L \to V$.

Claim 1. We may suppose that G is a comparability graph.

Indeed, since $\mathcal{P}(G')$ is a subgraph of $\mathcal{P}(G)$, the fact that a restriction of f is perfect with respect to $\mathcal{P}(G')$ ensures that it is perfect with respect to $\mathcal{P}(G)$.

Then let $\leq$ be an ordering on V such that $\mathcal{E}$ is the set of (x,y) such that $x \leq y$ or $y \leq x$, and let $P := (V, \leq)$. A subset X of V is *convex* if:

$$x \leq z \leq y \quad \text{and} \quad x,y \in X \quad \text{implies} \quad z \in X. \tag{5}$$

The *convex hull* of a subset X is the least convex set $\hat{X}$ containing X; in fact:

$$\hat{X} = \{z \in V : x \leq z \leq y \text{ for some } x,y \in X\}. \tag{6}$$

Lemma 3.2.3 *Let X, Y be subsets of V;*

$$X \sim \hat{Y} \bmod \mathcal{P}(G) \Rightarrow X \sim Y \bmod \mathcal{P}(G). \tag{7}$$

Proof. Suppose that $X \sim Y$ does not hold. Let $x \in X$ be a witness of this fact, that is x is incomparable to y, (which we denote $x \| y$), for all $y \in Y$. Let $z \in \hat{Y}$. Then from (5), there are $u, v \in Y$ such that $u \leq z \leq v$.

Since $u \in Y$ we have $u \not\leq x$ and, since $u \leq z$, it follows $z \not\leq x$; similarly, we obtain $z \not\leq x$. This shows that $x \| z$. Since this holds for all $z \in \hat{Y}$, the element x witnesses the fact that $X \sim \hat{Y}$ does not hold. $\qquad\square$

Claim 2. We may suppose that each $f(i)$ is a convex subset of V.

Indeed, let $g : L \to \mathcal{P}(V)$ be defined by setting $g(i) = \hat{f(i)}$.
If g is perfect on a subset L' of L, then from the previous lemma it follows that f is perfect on L'. $\qquad\square$

From now on, the ordering has no infinite antichain and the cofinality of L is uncountable.

Claim 3. We may suppose that each $f(i)$ is of the form $f(i) := I_i \cap F_i$, where $I_i := I_{i,1} \cup \ldots \cup I_{i,r}$, $F_i := F_{i,1} \cup \ldots \cup F_{i,r}$ and all $I_{i,l}$, (resp. $F_{i,k}$), are ideals, (resp. filters).

Indeed, each $f(i) = I_i \cap F_i$ where $I_i = \downarrow f(i)$, $F_i = \uparrow f(i)$.
Now, since $(P, \leq)$ has no infinite antichain, it follows from a result of Erdős and Tarski [5] that *each initial segment is a finite union of ideals* (cf. [6] 5.3 (2), p. 112 for details), hence $I_i := I_{i,1} \cup \ldots \cup I_{i,r_i^1}$; similarly F_i is a finite union of filters $F_i := F_{i,1} \cup \ldots \cup F_{i,r_i^2}$.

We may suppose that r_i^1, r_i^2 has a common value r_i. Now, since $\mathrm{cf}(\kappa) \neq \omega$, it follows that r_i is constant, $r_i = r$, on a subchain L' of size κ. It suffices to replace L by L'.   □

Let $\mathrm{comp}\,(\mathcal{J}(P))$, (resp. $\mathrm{comp}\,(\mathcal{F}(P))$), be the comparability graph of the collection $\mathcal{J}(P)$ of ideals, (resp. $\mathcal{F}(P)$ of filters), of P, ordered by inclusion.

Claim 4. We may suppose that for each l, $1 \leq l \leq r$ (resp. each k, $1 \leq k \leq r$) the map $i \rightarrow I_{i,l}$ from $(L,<)$ into $\mathrm{comp}\,(\mathcal{J}(P))$, (resp. $i \rightarrow F_{i,k}$ from $(L,<)$ into $\mathrm{comp}\,(\mathcal{F}(P))$ is perfect.

Indeed it suffices to apply theorem 4 successively $2r$ times.

Claim 5. There is no infinite subset L' of L such that $i < j \Rightarrow X_i \not\prec X_j$.

Suppose the contrary. Let L' be such a set. For each pair $(i,j) \in [L']^2$, let $\theta(i,j) = \{(k,l) : x \| X_j \text{ for some } x \in I_{i,l} \cap F_{i,k}\}$. Since $\theta(i,j) \subseteq R = [\{1,\ldots,r\}]^2$, there is some subset θ of R, and some infinite subset $L'' \subseteq L'$ such that $\theta(i,j) = \theta$ for all $(i,j) \in [L'']^2$. Let $(k,l) \in \theta$, and let $L_1'' := L'' \backslash \mathrm{Max}\, L''$, (O if L'' has a largest element).

Sub-Claim 1.

$$I_{i,l} \cap F_{i,k} \not\subseteq I_{j,l} \cup F_{j,k} \tag{8}$$

and

$$I_{i,l} \supset I_{j,l} \quad \text{and} \quad F_{i,k} \supseteq F_{j,k} \tag{9}$$

for all $(i < j) \in [L_1'']^2$.

Indeed, let $(i,j) \in [L_1'']^2$. Since $\theta(i,j) = \theta$ and $(k,l) \in \theta$, there is some $x \in I_{i,l} \cap F_{i,k}$ such that x is incomparable to each element of X_j. We have $x \notin I_{j,l} \cup F_{j,k}$. Indeed, suppose for an example that $x \in I_{j,l}$. Let $y \in I_{j,l} \cap F_{j,k}$ (there is such an element since j is not the largest element of L''). Since $I_{j,l}$ is an ideal, there is some $z \in I_{j,l}$ such that $x, y \leq z$. Since $F_{j,k}$ is a final segment $z \in F_{j,k}$ hence $z \in I_{j,l} \cap F_{j,k}$. But $I_{j,l} \cap F_{j,k} \subseteq \hat{X}$ thus x is comparable to some element of $\hat{X}$, which is impossible. (9) follows from (8).   □

From the Ramsey Theorem [18], L'' contains either an ω chain, or an ω^* chain, hence an infinite subset D of L_1'' such that each $i \in D \backslash \mathrm{Max}\, D$ has a successor i^+ in D.

$$\text{For each } i \in D_1 \text{ select } x_i \in I_{i,l} \cap F_{j,k} \backslash (I_{i^+,l} \cup F_{i^+,k}) \tag{10}$$

Sub-Claim 2. The set of all x_i forms an infinite antichain.

Indeed, let $i < j$. If $x_i \leq x_j$, then since $x_j \in I_{j,l}$ we get $x_i \in I_{j,l}$. Since $I_{i,l} \subseteq I_{i^+,l}$ we get $x_i \in I_{i^+,l}$ contradicting (9). If $x_j \leq x_i$, then, similarly, since $x_j \in F_{j,k}$, we get $x_i \in F_{j,k}$, and since $F_{j,k} \subseteq F_{l^+,k}$ we get $x_i \in F_{l^+,k}$ contradicting (10).

This sub-claim contradicts the fact that the ordering has no infinite antichain, proving our claim.   □

Let $A := \{\{i,j\} : i < j \text{ and } X_i \sim X_j\}$, let $B := \{\{i,j\} : i < j \text{ and } X_i \not\sim X_j\}$. From our claim, there is no infinite subset L' of L such that $[L']^2 \subseteq B$, hence, from the Erdős–Dushnik–Miller Theorem [2] (cf. also [4]), there is a subset L' of L with $|L'| = |L|$ such that $[L']^2 \subseteq A$. This is exactly the conclusion of Lemma 3.2.2. $\qquad\square$

We have seen that theorem 2 follows from theorem 3. How do their strengths compare? As we have shown, theorem 3 follows from lemma 3.2.2, which in turn is a special case. We now show that theorem 2 can be translated into the following lemma, which is very close to lemma 3.2.2.

Lemma 3.2.4 *Let G be a graph having a spanning subgraph which is a comparability graph with no infinite independent set.*

If L is an uncountable chain, then every $f : (L, <) \to \mathcal{P}(G)$ is good.

Problem 2 Let G be a graph with no infinite independent set. Suppose that the conclusion of lemma 3.2.4 holds; does the conclusion of lemma 3.2.2 hold too?

Our translation is based upon the next lemma:

Let $G = (V, \mathcal{E})$ be a graph and let $S(G)$ be the graph whose vertex set is $\mathcal{E}$ and edge set is the set of pairs (u, v) such that $u := (i, j) \in \mathcal{E}$, $v := (j, k) \in \mathcal{E}$ for some $i, j, k \in V$. This is the *directed-shift-graph*. The ordinary shift-graph is obtained by taking its symmetric hull.

Lemma 3.2.5 *Let $G := (V, \mathcal{E})$, $G' = (V', \mathcal{E}')$ be two graphs. The following properties are equivalent:*

(i) every map from $S(G')$ into G is good;

(ii) every map from G' into $\mathcal{P}(G)$ is good.

Proof. (i) $\Rightarrow$ (ii). Let $g : G' \to \mathcal{P}(G)$. Suppose g is bad. Let $u = (i, j) \in \mathcal{E}'$. Since g is bad, $(g(i), g(j)) \notin \mathcal{E}$. From the definition of $\mathcal{E}$, this means that there is some $x \in g(i)$ such that $(x, y) \notin \mathcal{E}$ for all $y \in g(i)$. Let $f(i, j)$ be such an element x, this defines a map f from $S(G')$ into G. This map f is bad. Indeed, let $u, v \in V(S(G'))$ such that $(u, v) \in E(S(G'))$. From the definition of $S(G')$, this means that $u = (i, j) \in \mathcal{E}'$, $v = (j, k) \in \mathcal{E}'$ for some i, j, k. According to our definition of f, we have $f(u) \in g(i)$ and $f(v) \in g(j)$. Since $(f(u), y) \notin \mathcal{E}$ for all $y \in g(j)$ we get $(f(u), f(v)) \notin \mathcal{E}$ proving that f is bad.

(ii) $\Rightarrow$ (i). Let $f : S(G') \to G$. Define $g : G' \to \mathcal{P}(G)$, setting $g(i) := \{f(i, j) : (i, j) \in \mathcal{E}'\}$. Suppose (ii) holds, then g is good, that is $(g(i), g(j)) \notin \mathcal{E}$ for some $(i, j) \in \mathcal{E}'$. Since $f(i, j) \in g(i)$ there is some $y \in g(j)$ such that $(f(i, j), y) \in \mathcal{E}$. Since $y := f(i, k)$ for some k such that $(i, k) \in \mathcal{E}'$, and $((i, j), (j, k)) \in E(S(G'))$, it follows that f is good, thus (i) holds. $\qquad\square$

N.B. The substance of this result is basic, both in the theory of ordered sets and the theory of graphs. For $G' := (\omega, <)$ and G a poset, this result, due to R. Rado [17], was the starting point of the notion of better-quasi- ordering. For $G := (V, =)$ and $G' := (\kappa, <)$, it was employed in order to show that the shift graph of κ has chromatic number $\lceil \log_2 \kappa \rceil$ ($=$ the least λ such that $\kappa \leq 2^\lambda$) (Erdős–Hajnal [3]). We will see other uses e.g. remarks 3.2.7, 4.2.3 and proof of theorem 10.

If the graph G in lemma 3.2.5 is symmetric, we can perfectly replace $S(G')$ by its symmetric hull, hence the conclusion of lemma 3.2.4 amounts to this one:

If L is an uncountable chain, then every map from S_L into G is good.   (11)

If this property holds we get theorem 2 as follows.

Let L be an uncountable chain and S_L be the L-shift-graph. Let G be a graph on the same set of vertices. If G is a comparability graph with no infinite antichain then from the property above applied to the identity map it follows that S_L and G have a common edge, hence G is not a subgraph of the complement of S_L. To derive (11) from theorem 2, it suffices to observe that if f is a map from a set V' to the vertex set V of a graph $G = (V, \mathcal{E})$ then the graph $G' = (V', \mathcal{E}')$ where $\mathcal{E}' = \{\{i, j\} : i \neq j$ and $f(i) \simeq f(j) \bmod G\}$ is a comparability graph (and has no infinite independent set), provided that G is a comparability graph (and has no infinite independent set).   □

From this discussion we have.

Theorem 3.2.6 *Let $G := (V, \mathcal{E})$ be a (reflexive, undirected) graph with no infinite independent set. The properties listed below are related by the following implications:*

$$(iv) \to (iii) \to (ii) \to (i)$$

(i) *Every spanning subgraph of G which is a comparability graph has an infinite independent set;*

(ii) *There is a chain $(L, <)$, with $\mathrm{cf}(|L|) > \omega$, and a map $f : (L, <) \to \mathcal{P}(G)$ for which no restriction to a subchain of the same size is perfect;*

(iii) *There is an uncountable chain $(L, <)$ and a bad map $f : (L, <) \to \mathcal{P}(G)$;*

(iv) *G has an induced subgraph isomorphic to a spanning subgraph of the complement of the shift graph S_L over some uncountable chain.*

Note that (ii) $\to$ (i) is theorem 3, whereas (iv) $\to$ (i) is theorem 2.

Remark 3.2.7 From theorem 9 it follows that (i) does not imply (ii), hence (i) does not imply (iv). Indeed, let G be a ω^2-well unoriented graph satisfying the conclusion of theorem 9. Let f be a map from an infinite chain L into $\mathcal{P}(G)$; if f is bad on some infinite sub-chain L', then, according to lemma 3.2.5, there is a bad map from $[L']^2$ into G, hence a bad map from $[\omega]^2$ into G, which is impossible. Hence, from the Erdős– Dushnik–Miller theorem, f is perfect on a subchain of size $|L|$.

Another example of a graph satisfying the conclusion of theorem 2 can be obtained from Milner–Pouzet [12]. Indeed, that paper shows an example of a topological graph $G = (V, \mathcal{E})$ containing an independent set of size $\aleph_1$ and a dense set A with no infinite independent subset ([12] theorem 3, p. 251). It is also proved there that no such graph exists if we impose the additional requirement that the graph $G|_A$ induced on A is a comparability graph ([12] theorem 4, p. 291). From this fact, $G|_A$ cannot have a spanning subgraph which is a comparability graph with no infinite independent set. Indeed, let G'_A be a subgraph of A

which is a comparability graph on A. Since we may suppose that the topology induced on A is the discrete topology (if not, add all subsets of A to the open sets, A remains dense and the graph remains a topological graph), the graph G'_A coincides with the graph $G'|_A$ induced on A by its closure G'. By definition G' is a topological graph. Since this is a spanning subgraph of G it also contains independent sets of size $\aleph_1$. From the result we mention G'_A contains an infinite independent set. $\square$

3.3 Proof of Theorem 5

According to the lexicographical ordering $\mathcal{L}$, elements of $V := [\omega + 1]^2$ form the following $\omega 2$-sequence:

$$(0,1),\ (0,2),\ (1,2),\ (0,3),\ (1,3),\ (2,3),\ \ldots,\ (0,j+1),\ (1,j+1),\ \ldots,\ (j,j+1),$$
$$\ldots,\ (0,\omega),\ (1,\omega),\ \ldots,\ (k,\omega).$$

With that in mind, it is not hard to check that if $\mathcal{E}$ is the complement of the shift-graph, then the ordering on $F(G,\mathcal{L})$ consists of pairs $((i,j),(i',j'))$ such that

$$\begin{aligned}
&\text{either} && i' = 0 \text{ and } j < j'; \\
&\text{or} && i = i' \text{ and } j < j'; \\
&\text{or} && j < i'.
\end{aligned}$$

Hence, $\{(i,\omega) : i < \omega\}$ forms an infinite antichain. $\square$

Notice that the restriction of this ordering to the set $[\omega_*]^2 := \{(i,j) : 0 < i < j < \omega\}$ is the prototypal example of w.q.o. for which the collection of initial segments is not w.q.o. Defined by Rado, this poset R, with lemma 3.2.5, leads to the discovery of b.q.o. Laver proved that every w.q.o. for which the collection of initial segments is not w.q.o. contains an isomorphic copy of R.

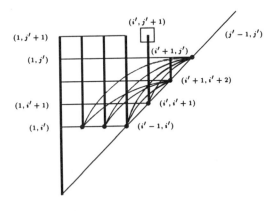

3.4   Proof of Theorem 6

(ii) → (i): Observe first that (ii) minus the countability condition amounts to the fact that $(V, \mathcal{L})$ decomposes into a finite ordinal sum of chains $(V_i, \mathcal{L}_i)$, each being well ordered or dually well ordered. Next, if two linear orderings $\mathcal{L}_1$, $\mathcal{L}_2$ (on two disjoint sets V_1, V_2 respectively) satisfy condition (i) then the ordinal sum $(V_1, \mathcal{L}_1) + (V_2, \mathcal{L}_2)$ and the dual $(V_1, \mathcal{L}_1)^*$ satisfy (i) too. Consequently, we may suppose that $(V, \mathcal{L})$ is well ordered. Let α be the order type of $(V, \mathcal{L})$. If α is finite, (ii) holds with any ordering $\mathcal{P}$ contained in $\mathcal{L} \cap \mathcal{E}$. If $\alpha = \omega$, it holds by theorem 1. If $\alpha > \omega$, select a linear ordering $\mathcal{L}'$ on V with type ω. From theorem 1, obtain $\mathcal{P}' \subseteq \mathcal{L}' \cap \mathcal{E}$ such that $\mathcal{P}'$ has no infinite antichain. Let $\mathcal{P} = \mathcal{P}' \cap \mathcal{L}$. Since $\mathcal{P}'$ and $\mathcal{L}$ are w.q.o. the intersection is w.q.o.   □

(i) → (ii): The fact that V must be countable follows from theorem 2. Observe that for every $V' \subseteq V$, the ordering $\mathcal{L}'$ induced by $\mathcal{L}$ on V' satisfies (i) too. Indeed, given a graph $G' = (V', \mathcal{E}')$ with no infinite independent set, we may extend it (as an induced subgraph) to a graph $G = (V, \mathcal{E})$ having no infinite independent subset, (e.g., set $\mathcal{E} = \mathcal{E}' \cup \{(x, y) : x \notin V', y \in V\}$). From (i) applied to G we get some $\mathcal{P} \subseteq \mathcal{E}$ with no infinite antichain. The ordering $\mathcal{P}' := \mathcal{P} \cap V' \times V'$ has the same properties. Because of that it suffices to show that $\mathcal{L}$ cannot have order-type $\omega.\omega^*$ or $\omega^*.\omega$. Clearly the first case is enough. We deduce it from the following lemma:

Lemma. *Let G be the comparability graph of $[\omega]^2$, the set of pairs (i, j) such that $i < j < \omega$, componentwise ordered (i.e., $(i, j) \leq (i', j')$ if $i \leq i'$ and $j \leq j'$). This graph has no infinite independent set. The linear ordering $\mathcal{L}$ on $[\omega]^2$ defined by $(i, j) \leq (i', j')$ if $i > i'$ or $i = i'$ and $j \leq j'$ has order-type $\omega.\omega^*$, however, every ordering contained in $\mathcal{E} \cap \mathcal{L}$ has an infinite antichain.*

Proof. G is the comparability graph of a w.q.o., hence it has no infinite independent set; trivially $\mathcal{L}$ has type $\omega.\omega^*$. Suppose that some ordering $\mathcal{P} \subseteq \mathcal{L} \cap \mathcal{E}$ contains no infinite antichain. In this case, each set of the form $X_i = \{(i, j) : i < j < \omega\}$ contains some ω-chain C_i. For each i, pick $x_i \in C_i$. Again, since $\mathcal{P}$ has no infinite antichain then there are at least two integers i, j such that $i < j$ and $x_i > x_j \pmod{\mathcal{P}}$. But since infinitely many elements of X_i dominate x_i, these elements dominate x_j. Since in the graph G no element belonging to X_j is joined to infinitely many elements of X_i, this gives a contradiction.   □

B.q.o. theory has three basic ingredients, namely the extension of Ramsey theorem to barriers, the extension of the "minimal bad sequence technique", the "forerunning technique" all due to Nash-Williams. The first one occurs in the proofs of theorems 7 and 10, under the form indicated below. Given a graph $G = (V, \mathcal{E})$, a *G-array* is a map f whose domain dom f is a barrier B and the range is a subset of V. Looking at f as a map from the graph $(B, \lhd)$ into G, the extension of the Ramsey theorem to barriers gives the following result of Nash-Williams.

A G-array has a sub-G-array which is either bad or perfect.

4   Proofs of remaining results

4.1   Proof of Theorem 7

The "minimal bad sequence technique", on which relies the proof of theorem 1, was extended by Nash–Williams to maps defined on barriers as follows:

Let P be a poset. Bad P-arrays (if there are any) are compared with respect to an auxiliary ordering $\preceq$ on P which is a subordering of the original one (that is $x \preceq y$ implies $x \leq y \bmod P$). If f and g are two bad P-arrays, we write $f \preceq g$ if:

$$\text{dom } f \text{ is a sub-barrier of dom } g \tag{12}$$

$$f(x) \preceq g(x) \text{ for all } x \in \text{dom } f \tag{13}$$

A bad map g is *irreducibly bad* if every f smaller than g in this ordering is the restriction of g to its domain.

Lemma 4.1.1 [14] *If the auxiliary ordering on P is well-founded, then every bad map dominates some irreducibly bad map.*

Let P be the collection of finite sequences of elements of P equipped with the Higman ordering:

$$\text{if } s = (s_{(0)}, \ldots, s_{(n)}), \ t = (t_{(0)}, \ldots, t_{(n)}) \text{ are two members of } P^{<\omega}$$

then $s \leq t$ if there is some subsequence t' of t such that s is smaller than t' componentwise, that is there is some order-preserving map $h : \text{dom } s \to \text{dom } t$ such that $s(i) \leq t(h(i))$ for all $i \in \text{dom } s$. For example, the empty sequence is the least element of P.

Identifying each element a of P to the one element sequence taking value a, then P identifies to a subposet of $P^{<\omega}$. Defining the auxiliary ordering $\preceq$ on $P^{<\omega}$ by $s \preceq t$ if either $s = t$ or $s \leq t$ and $|\text{dom } s| < |\text{dom } t|$, one gets a well-founded ordering. One obtains the following lemma of Nash–Williams proving that if P is α-well then $P^{<\omega}$, too.

Lemma 4.1.2 *Let P be a poset. For every bad $P^{<\omega}$-array g, there is some bad P-array g' such that $g' \leq g$.*

Let $I_{\text{fin}}(P)$ be the collection consisting of finite initial segments of P, ordered by inclusion. From lemma 4.1.2 above follows directly this:

Lemma 4.1.3 *For every bad map $g : C \to I_{\text{fin}}(P)$ there is some sub-barrier $C' \subseteq C$ and a bad map $g' : C' \to P$ such that $g'(s) \in g(s)$ for all $s \in C'$.*

We deduce theorem 7 from lemma 4.1.2 and 4.1.3. Let G be an α-well graph on ω and let $P := F(G, \omega)$. Suppose that P is not α-well. Let $f : B \to P$ be a bad map, where B is a barrier of type α. Since the ordering $\leq$ defined on P is well-founded, hence, choosing $\prec$ equals to $\leq$, lemma 4.1.2 provides a minimal bad map $f' : B' \to P$ defined on a sub-barrier B' of B such that $f'(s) \leq f(s)$ for all $s \in B'$.

<u>Claim 1.</u> There is a sub-barrier C of B' such that if $s \lhd t$ then:

$$f'(s) < f'(t) \quad (\text{mod } \omega) \tag{14}$$

$$f'(s) \sim f'(t) \quad (\text{mod } G). \tag{15}$$

Indeed, since ω is a b.q.o., f' considered as a map from B' into ω is perfect on a sub-barrier B''. Since f', as a map into P is bad, $s \lhd t$ implies $f(s) \neq f(t)$, hence (14) holds for all s, t in B''. Now, since G is α-good, and B'' has type at most α, the map $f' : B'' \to G$ is perfect on a sub-barrier C, that is (15) holds. $\square$

For each $s \in C$, let $g(s) := \{y : y < f'(s) \ (\text{mod } P)\}$. This defines a map $g : C \to I_{\text{fin}}(P)$.

<u>Claim 2.</u> The map g is bad. Otherwise, there are s, t in C such that $s \lhd t$ and

$$g(s) \subseteq g(t). \tag{16}$$

For such a pair s, t, conditions (14) and (15) hold. According to the definition of the ordering on P, conditions (14), (15) and (16) amount to $f'(s) \leq f'(t) \ (\text{mod } P)$. This contradicts the assumption that f' is bad. $\square$

From lemma 4.1.3 above there is some bad map $g' : C' \to P$ such that $g'(s) \in g(s)$ for all $s \in C'$. Such a map satisfies $g' \leq f'$. Since f' is minimal bad, g' must be the restriction of f' to C'. But this is impossible, since from the definition of $g(s)$ we have $g(s) < f'(s)$ for all $s \in C'$. This contradiction proves theorem 7.

4.2 Proof of Theorem 9

Let L be a chain, and $[L]^{<\omega}$ be the collection of finite subsets of L. We define the $\lhd$ relation on $[L]^{<\omega}$ as we did for $[\omega]^{<\omega}$: if $s, t \in [L]^{<\omega}$ we set $s \lhd t$ whenever $s_* \ll t$, meaning that $s_* := s \backslash \text{Min} \, s$ is an initial segment of t. Given an integer m, $m \geq 2$, the *unoriented L-shift graph of order* m, that we denote $S_m(L)$, is the symmetric hull of $([L]^{<\omega}, \lhd)$. Let G be its complement.

Lemma 4.2.1 *If L is infinite then G is ω^{m-1}-well and not ω^m-well.*

Proof. (a) G is not ω^m-well. Since L is infinite, then, either ω or ω^* embeds into L. Suppose that ω embeds into L; let φ be such an embedding, and $\bar{\varphi}$ be its natural extension from $[\omega]^m$ into $[L]^m$. Clearly we have

$$s \lhd t \leftrightarrow \bar{\varphi}(s) \lhd \bar{\varphi}(t)$$

hence $\bar{\varphi} : ([\omega]^m, \lhd) \to G$ is bad.

If ω^* embeds into L, then ω embeds into L^*. Due to the fact that we consider subsets of L of the same size, the canonical map from $[L]^{<\omega}$ onto $[L^*]^{<\omega}$ (which associates to every sequence $(s_0, s_1, \ldots s_{m-1})$ the sequence $(s_{m-1}, \ldots, s_0)$) preserves the $\lhd$ relation, hence G is the symmetric hull of $S_m(L^*)$ and, by the above result, $\bar{\varphi}$ is bad.

(b) G is ω^{m-1}-well.

Suppose that some $f : ([\omega]^{m-1}, \lhd) \to G$ is bad that is

$$\text{either } s \lhd t \to f(s) \lhd f(t) \tag{17}$$

$$\text{or } f(t) \lhd f(s). \tag{18}$$

for all $s, t \in [\omega]^{m-1}$ such that $s \lhd t$.

This amounts to saying that f is a graph-homomorphism from $S_{m-1}(\omega)$ into $S_m(L)$. Since the girthof $S_{m-1}(\omega)$ is $2m-1$ and the girth of $S_m(L)$ is $2m+1$, such an f cannot exist. Here is a direct proof:

Proof. Divide $[\omega]^m$ into two parts, A_1, A_2, an element u of $[\omega]^m$ belonging to A_1 if $u = s \cup t$, with $s \lhd t$ and $f(s) \lhd f(t)$, and to A_2 otherwise. By the Ramsey theorem [18], there is some infinite subset I of ω such that either $[I]^m \subseteq A_1$ that is (17) holds for all $s, t \in [I]^{m-1}$ or $[I]^M \subseteq A_2$, that is (18) holds for all $s, t \in [I]^{m-1}$. Each case is impossible. Indeed, for this first one, observe that for each $k \geq 0$, the graph $([I]^{m-1}, \lhd)$ contains two paths of length $m-1$ and $m+k$ having the same extremities:

$$s = s_0 \lhd s_1 \lhd s_2 \ldots \lhd s_{m-2} \lhd s_{m-1} = t \atop t = t_0 \lhd t_1 \lhd t_2 \ldots \lhd t_{m-1} \lhd t_m \lhd \ldots t_{m+k} = t \quad \text{and}$$

Trivially this does not hold in L^m, even for $k = 0$. The second case is similar. $\square$

Lemma 4.2.2 *If L is large enough then every ordering on $[L]^m$ such that $x \lhd y$ or $y \lhd x$ implies $x \not\leq y$ contains some infinite antichain.*

Let $\leq$ be a (partial) ordering on $[L]^m$. Divide $[L]^{2m}$ into finitely many classes as follows:

F' and F'' belong to the same class if the map $\bar{\varphi}$ from $[F']^m$ onto $[F'']^m$ which extends the canonical isomorphism of the chain F' onto the chain F'' is an order-isomorphism with respect to the ordering $\leq$.

The number of classes is bounded by the number of orders on $\binom{2m}{m}$ elements. From the Erdős–Rado Theorem, see [14], it follows that if $|L|$ is large enough then there is a subset L' of L whose order type is either $\omega.m$ or ω^*m and such that all the $2m$-element subsets belong to the same class. Suppose this holds.

Let U be the set of pairs (x, y) such that:

$$x := (x_0, \ldots x_{m-1}) \in [L']^m, \quad y := (y_0, \ldots y_{m-1}) \in [L']^m$$

and

$$x_0 < y_0 < x_1 < y_1 < \ldots < x_{m-1} < y_{m-1}.$$

Now, suppose that the ordering on $[L]^m$ has no comparable pair (x, y) satisfying $x \lhd y$.

Claim 1. U consists of incomparable pairs.

Suppose on the contrary that some pair $(x, y) \in U$ satisfies $x \leq y$ or $y \leq x$, and for an example $x \leq y$. In this case, every other pair $(x', y') \in U$ also satisfies $x' \leq y'$. Indeed, the $2m$ element subsets $\{x_0, y_0, \ldots x_{m-1}, y_{m-1}\}$ and $\{x'_0, y'_0, \ldots x'_{m-1}, y'_{m-1}\}$ belong to the same class. Because of that, and the fact that L' is infinite, we may suppose that x and y are such that y_{m-1} is not the largest element of L'. Choose $t \in L'$ such that $t > y_{m-1}$. Let $z = (x_1, x_2, \ldots x_{m-1}, t)$. We have $(y, z) \in U$, hence $y \leq z$. From $x \leq y$ we deduce $x \leq z$. But, from the definition of z, we have $x \lhd z$, contradicting our hypothesis. The case $y \leq x$ is similar. $\Box$

With no loss of generality, we may suppose that L' has type $\omega.m$. Let $a_0, a_1, \ldots a_\alpha, \ldots$ $(\alpha < \omega.m)$ be the list of elements of L' according to this ordering. Let $t_0, t_1, \ldots, t_n, \ldots$ $(n < \omega)$ be the sequence of members of $[L']^m$ defined by:

$$t_n = (a_n, a_{\omega+n}, \ldots, a_{\omega k+n}, \ldots, a_{\omega(m-1)+n})$$

Claim 2. The t_n's form an infinite antichain.

Indeed, from the construction of these elements we have $(t_n, t_m) \in U$ for all $n < m < \omega$. $\Box$

Problem 3 Which is the least cardinal $\kappa(m)$ (depending on m) such that the conclusion of Lemma 4.2.2 holds for every chain L satisfying $|L| > \kappa(m)$?

Remark 4.2.3 A similar proof as above shows that $\kappa(m) \leq \beth_{2m-1}(\omega)$. (We recall that $\beth_0(\lambda) = \lambda$, $\beth_{m+1}(\lambda) = 2^{\beth_m(\lambda)}$.) Indeed, add a well ordering on L and classify the $2m$-element subsets of L in such a way that the original ordering and this well ordering are preserved. The following instance of the Erdős–Rado Theorem:

$$\beth_{2m-1}(\omega)^+ \rightarrow (\omega_1)^{2m}_\omega$$

provides a monochromatic L' of type ω_1 or ω_1^*. With much effort, we proved that $\kappa(m) \leq \beth_m(\omega)$. On the other hand, it is very easy to see that $\kappa(m) \geq \beth_{m-2}(\omega)$. For that, let R be the "Rado ordering" defined on $[\omega]^2$ (see the proof of Theorem 5) and let $\kappa = \beth_{m-2}(\omega)$. Observe that there is a bad f from $[\kappa]^m$ into R. Indeed, define $\beth_k(R)$ by setting $\beth_0(R) = R$, $\beth_{k+1} = \mathcal{P}(\beth_k(R))$. From the fact that $\beth_1(R)$ contains an independent set of size $\aleph_0$ it follows that $\beth_k(R)$ contains an independent set of size $\beth_{k-1}(\omega)$, hence $\beth_{m-1}(R)$ contains an independent set of size κ. This provides a bad g from κ into $\beth_{m-1}(R)$. Lemma 3.2.5 applied m times gives a bad map f from $[\kappa]^m$ into R. Let $\mathcal{L}$ be an order on $[\kappa]^m$ which is a w.q.o. and contains the $\lhd$-relation (e.g. the product ordering, the lexicographical ordering). For $s, t \in [\kappa]^m$ set:

$$s \leq t \text{ if } f(s) \leq f(t) \pmod{R} \text{ and } s \leq t \pmod{\mathcal{L}}.$$

This defines an ordering. Since R and $\mathcal{L}$ are w.q.o., this ordering is a w.q.o. Since f, as a map from $[\kappa]^m$ into R, is bad and since $\mathcal{L}$ contains $\lhd$, the comparability graph of this ordering is included in the complement of the symmetric hull of the $\lhd$-relation. That is the conclusion of Lemma 4.2.2 fails.

We conjecture that $\kappa(m) = \beth_{m-2}(\omega)$. For $m = 2$ this is theorem 2. We do not know whether $\kappa(3) = \beth(\omega)$, $\beth_2(\omega)$ or $\beth_3(\omega)$.

4.3 Proof of Theorem 10

The proof is based on the "forerunning technique" of C.St.J.A. Nash–Williams [14]. Our presentation follows Milner's paper [11].

Let B, B' be two barriers. We say that B *foreruns* B', and denote $B' \sqsubseteq B$, if $\bar{B}' \subseteq \bar{B}$ and for every $b' \in B'$ there is some $b \in B$ such that $b \ll b'$. We say that B *strictly foreruns* B' and denote $B' \sqsubset B$ if in addition $B' \subseteq B$. For an example we have $[\omega]^{n+1} \sqsubset [\omega]^n$. Let P be a poset and $\prec$ be an ordering on P such that $x \prec y$ implies $x \leq y$ (mod P). If $f : B \to P$ and $f' : B' \to P$ are two P-arrays, we say that f *foreruns* f' (resp. f *strictly foreruns* f') if B foreruns B' (resp. B *strictly foreruns* B') and

$$f'(b) = f(b) \quad \text{if} \quad b \in B' \cap B \tag{19}$$

$$f'(b') \prec f(b) \quad \text{if} \quad b \in B,\ b' \in B' \backslash B \text{ and } b \ll b'. \tag{20}$$

A bad P-array f if *strictly minimal bad* if there is no bad P-array f' such that f strictly foreruns f. We recall the following fact (cf. [11], Theorem 2.19).

Lemma 4.3.1 *If $\prec$ is a well-founded ordering on P, then for every bad P-array f there is a strictly minimal bad P-array f' such that f foreruns f'.*

We prove theorem 10 as follows. Let $G := (V, \mathcal{E})$ be a better-graph, $\mathcal{L}$ be a well-ordering on V and $P := F(G, \mathcal{L})$. Suppose that P is not a better-quasi-ordering, that is there is some bad P-array. Let $f : B \to P$ be such a map. Since $\mathcal{L}$ is a strengthening of the order $\leq$ defined on P, this order is well-founded. Hence, choosing $\preceq$ equal to $\leq$, we get from lemma 4.3.1 a strictly minimal bad $f' : B' \to P$ such that f foreruns f'. As in the proof of theorem 7 there is a sub-barrier C of B' such that (14) and (15) of claim 1 holds for all $s,\ t$ in C such that $s \triangleleft t$. For each $s \in C$, let $g(s) := \{y : y < f'(s) \pmod{P}\}$. This defines a map g from C into $I(P)$ the set of initial segments of P, ordered by inclusion. As in claim 2 of the proof of Theorem 7, this map is bad. Let $B'' = \{s \cup t : s \triangleleft t \text{ and } s, t \in C\}$. This is a barrier and in fact $(B'', \triangleleft)$ is the shift-graph associated to $(C, \triangleleft)$. Hence, as in the proof of lemma 3.2.5, we may define a bad map $f'' : B'' \to P$. Indeed, let $u \in B''$. There is a unique pair (s, t) of elements of C such that $u = s \cup t$ and $s \triangleleft t$. Since g is bad, $g(s) \not\subseteq g(t)$, we select $f''(u)$ arbitrary in $g(s) \backslash g(t)$. If u, u' are two elements of B'' such that $u \triangleleft u'$, then the pairs (s, t) and (s', t') as above respectively associated to u and u' satisfy $t = s'$. Since $f''(u) \not\subseteq g(t)$ and $f''(u') \in g(t)$ it follows that $f''(u) \not\leq f''(u')$ amounting to the fact that f'' is bad. Clearly B' strictly foreruns B''. Since $f''(s \cup t) < f'(s)'$, the map f' strictly foreruns f''. Since f foreruns f', it follows that f strictly foreruns f'', contradicting the fact that f is strictly minimal bad. □

4.4 Proof of Theorem 11

Let $\preceq$ be a partial order on $\mathbb{Q}$ which is compatible with the addition. We start with an observation we owe to M. Giraudet.

Claim 1 The natural ordering on $\mathbb{Q}$ contains either $\preceq$ or its reverse as a subordering.

Proof. Let P be the positive cone associated with our given ordering: $P := \{x \in \mathbb{Q} : 0 \preceq x\}$. Clearly the binary relation defined by $x \leq_P y$ if $y - x \in P$ coincides with $\preceq$. Hence our claim reduces to prove that either $P \subseteq \mathbb{Q}^+$ or $P \subseteq \mathbb{Q}^-$. Set $\mathbb{Q}_*^+ = \mathbb{Q}^+ \backslash \{0\}$.

CASE 1. $\mathbf{Q}_*^+ \cap P \neq \emptyset$. In this case $P \subseteq \mathbf{Q}^+$. Indeed, supposing the contrary there are $y \in \mathbf{Q}_*^+ \cap P$ and $y' \in P \backslash \mathbf{Q}^+$. Let p, q, p', q' be integers such that $y = \frac{p}{q}$, $y' = -\frac{p'}{q'}$. Since $y \in P$ we have $p'qy \in P$ hence $p'p \in P$. Since $y' \in P$ we have $pq'y' \in P$ hence $-pp' \in P$. Since $\leq_P$ is an ordering, this imposes $pp' = 0$, that is $y = 0$ or $y' = 0$ contradicting our choice of such elements.

CASE 2. $\mathbf{Q}_*^+ \cap P = \emptyset$. Set $P' = -P$. If $\mathbf{Q}_*^+ \cap P' = \emptyset$ then $P = \{0\}$, in which case our ordering is the equality relation, and our claim trivially holds. If not, then from Case 1, we have $P' \subseteq \mathbf{Q}^+$ giving $P \subseteq \mathbf{Q}^-$.

Suppose that $\preceq$ is not a linear ordering. We prove that there are infinite antichains. According to Claim 1, we may suppose that the natural ordering on $\mathbf{Q}$ contains $\preceq$, that is $\mathbf{Q}^+$ contains P.

CASE 1. There are some $r \in \mathbf{Q}$ and some non-empty interval $I =]u, v[$ of $\mathbf{Q}$ such that r is incomparable (mod P) to every $y \in I$.

For such an r and I, choose an integer a, large enough, and $b \in \mathbf{Q}$ so that the affine transformation

$$\varphi(x) = ax + b$$

sends r and I into I. Define inductively an infinite sequence, setting $x_0 = r$, $x_{n+1} = \varphi(x_n)$. Since a is a positive integer, the map φ preserves the incomparability relation, hence the x_n's form an infinite antichain.

CASE 2. Case 1 does not hold. Observe that if $d \in \mathbf{Q}_*^+ \backslash P$ then for every $r \in \mathbf{Q}$, and every non-empty interval $I =]u, v[$ included in $[r, r + d]$ there is some $y \in I$ such that r and y are incomparable (mod P) (otherwise, $r + d$ is incomparable to every element of I contradicting the fact that Case 1 does not hold).

SUB-CASE 1. $\mathbf{Q}_*^+$ is a filter (mod P). Since $\mathbf{Q}_*^+$ is a final segment (mod P), this amounts to say that $\mathbf{Q}_*^+$ is down-directed meaning that every pair of members of $\mathbf{Q}_*^+$ has a lower bound (mod P) in $\mathbf{Q}_*^+$. Let $d \in Q_*^+ \backslash P$.

<u>Claim 2.</u> For every $\alpha \in [0, d[$, every finite and non-empty subset x of $[0, \alpha]$, the interval $]\alpha, d[$ is not included in $\downarrow X := \{y : y \leq x \pmod{P} \text{ for some } x \in X\}$. Indeed, suppose the contrary, then $]\alpha, +\infty[\subseteq \cup \mathcal{F}$ where $\mathcal{F} = \{\downarrow x : x \in X\} \cup \{[d, +\infty[\}$. Since $\mathbf{Q}_*^+$ is a filter, $]\alpha, +\infty[$ is a filter too; since $\mathcal{F}$ is a finite family of initial segments (mod P) it follows that $]\alpha, +\infty[$ is included into some member of $\mathcal{F}$. Since $]\alpha, +\infty[\subseteq [d, +\infty[$ is impossible, we have $]\alpha, +\infty[\subseteq \downarrow x$ for some $x \in X$. It follows that x is comparable to every element of $]\alpha, x + d[$ contradicting the observation above. This proves our claim.

From this, define inductively an infinite sequence $x_0, \ldots x_n, \ldots$ such that

$$]x_n, d[\not\subseteq \downarrow \{x_i : i \leq n\}$$

(set $x_0 = 0$ and $x_0, \ldots, x_n$ being defined, choose for x_{n+1} any element of $]x_n, d[\backslash \downarrow \{x_i : i \leq n\}$). The x_n's form an infinite antichain.

SUB-CASE 2. $\mathbf{Q}_*^+$ is not a filter (mod P).

<u>Claim 3.</u> $\mathbf{Q}_*^+$ contains incompatible subsets of arbitrarily large (finite) size. (A subset $A \subseteq \mathbf{Q}_*^+$ is *incompatible* if no pair of distinct elements of A has a common lower bound in $\mathbf{Q}_*^+$).

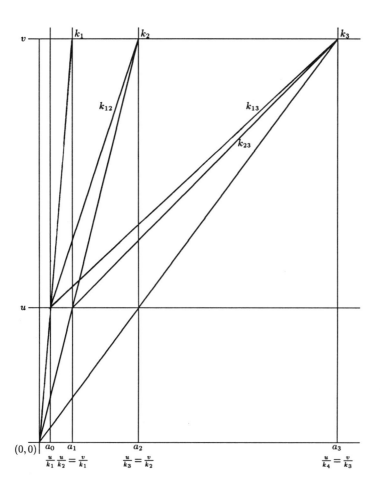

Proof. Since $\mathbb{Q}_*^+$ is not a filter, it contains two elements with no common lower bound in $\mathbb{Q}_*^+$. Given two such elements, say u and v, with $u < v \pmod{\mathbb{Q}}$, an $n+1$-element subset A of $\mathbb{Q}_*^+$ is incompatible if for every pair u', v' in A such that $u' < v' \pmod{\mathbb{Q}}$, the affine map φ such that $\varphi(u') < u$ and $\varphi(v') = v$ writes:

$$\varphi(x) = kx + \alpha$$

with k a positive integer and $\alpha \in \mathbb{Q}^+$.

Indeed, if u' and v' have a common lower bound w in $\mathbb{Q}_*^+$, then, as a straightforward computation shows, $\varphi(w)$ is a lower bound of u and v which belongs to $\mathbb{Q}_*^+$, and this is impossible.

Fix a positive integer n (i.e. $n \geq 1$). Define A as follows (see picture). Its elements, enumerated in an increasing order, are $a_0, \ldots, a_n$. For each integer i, $i < n$, the affine map

φ which transforms a_i into u and a_{i+1} into v writes $\varphi(x) = k_{i+1}x$, with $k_{i+1} \in \mathbb{Q}_*^+$, that is:

$$a_0 = \frac{u}{k_1}, \quad a_1 = \frac{v}{k_1} = \frac{u}{k_2}, \ldots, \quad a_i = \frac{v}{k_i} = \frac{u}{k_{i+1}}, \ldots \quad a_n = \frac{v}{k_n}$$

giving $k_{i+1} = k_1 \cdot 1/(\frac{v}{u})^i$ and $a_{i+1} = \frac{1}{k_1} \times \frac{v^{i+1}}{u^i}$. Given i, j with $i + 1 < j \leq w$, let $\theta(x) = k_{i+1,j}x + \alpha_{i+1,j}$ be the affine map transforming a_i into u, a_j into v.
Clearly $\alpha_{i+1,j} \in \mathbb{Q}_*^+$. An elementary computation gives:

$$k_{i+1,j} = k_1 \cdot 1/\left(\frac{v}{u}\right)^i \left(1 + \frac{v}{u} + \cdots \left(\frac{v}{u}\right)^{j-i-1}\right).$$

Hence, all coefficients k_{i+1}, $k_{i+1,j}$ write k_1 times a rational expression of u and v. Choosing for k_1 a large enough integer (e.g. for $n = 3$, $\frac{v}{u} = 3$, $k_1 = 368$) we obtain that all these coefficients are integers. By the above observation A is incompatible. According to a result of Erdős–Tarski [5], if a poset contains incompatible subsets of arbitrarily large finite size, then it contains an infinite incompatible subset, hence an infinite antichain. Consequently, in our case, $\mathbb{Q}_*^+$ contains an infinite antichain. This proves theorem 11.

Problem Is subcase 1 possible?

References

[1] R. Assous, Caractérisation du type d'ordre des barrières de Nash–Williams, *Publ. Dep. Math. Univ. Claude-Bernard (Lyon 1)* **11** (1974), 89–106.

[2] B. Dushnik– E.W. Miller, Partially ordered sets, *Amer. J. of Math.* **63** (1941), 600–610.

[3] P. Erdős and A. Hajnal, On chromatic number of infinite graphs.

[4] P. Erdős, A. Hajnal, A. Mate and R. Rado, Combinatorial set theory: partition relations for cardinals, *Akadémiai Kiado*, Budapest (1984), 347pp.

[5] P. Erdős and A. Tarski, On families of mutually exclusive sets, *Annals of Math.*, vol. 44 (1943), 315–329.

[6] R. Fraisse, Theory of relations, *Studies in Logic*, vol. 118, North–Holland (1986), 397pp.

[7] G. Higman, Ordering by divisibility in abstract algebras, *London Math. Soc.* **(3)2** (1952), 326–336.

[8] J.B. Kruskal, The theory of well-quasi-ordering: a frequently discovered concept, *J. Comb. Th. A* **13** (1972), 297–305.

[9] R. Laver, Better-quasi-orderings and a class of trees, in *Studies in foundations and combinatorics: Advances in Mathematics*, Supplementary Series 1, New–York, Academic Press, (1978), 31–48.

[10] A. Marcone, Foundations of bqo theory, preprint, Oct. 1992, 17pp.

[11] E.C. Milner, Basic wqo- and bqo- theory, in *Graphs and Order* (I. Rival, ed.), NATO ASI Series C vol. 147 (1985), 487–502, Reidel Pub., Dordrecht.

[12] E.C. Milner and M. Pouzet, The Erdős–Dushnik–Miller Theorem for topological graphs and orders, *ORDER* 1 (1985), 249–257.

[13] C.St.J.A. Nash–Williams, On well-quasi-ordering finite trees, *Proc. Cambridge Phil. Soc.* 59 (1963), 833–835.

[14] C.St.J.A. Nash–Williams, On well-quasi-ordering infinite trees, *Proc. Cambridge Phil Soc.* 61 (1965), 697–720.

[15] C.St.J.A. Nash–Williams, On better-quasi-ordering transfinite sequences, *Proc. Cambridge Phil. Soc.* 64 (1968), 273–290.

[16] M. Pouzet, Sur les prémeilleurordres, *Annales Inst. Fourier* 22 (1972), 1–20.

[17] R. Rado, Partial ordering of sets of vectors, *Mathematika* 1 (1954), 89–95.

[18] F.P. Ramsey, On a problem in formal logic, *Proc. London Math. Soc.* 30 (1930), 264–286.

[19] J.G. Rosenstein, *Linear orderings*, Academic Press (1982), New York.

[20] P. Ribenboim, *Problem 5 in Problem sessions and Abstracts of the First Meeting on Ordered Groups and Infinite Permutation Groups*, (CIRM Luming, July 2–6, 1990), edited by M. Giraudet, Univ. Paris 7, July 1992.

Problems About Planar Orders

IVAN RIVAL

Department of Computer Science
The University of Ottawa
Ottawa K1N 6N5 Canada

Abstract

Quite apart from their relevance, in modern theoretical computer science, to efficient graphical data structures devised especially for decision-making problems in which choices must be made from among ranked alternatives, *upward drawings* of ordered sets are interesting mathematical objects. The theory of *planar* upward drawings is situated somewhere on the common ground of combinatorial optimization, topological graph theory, and combinatorial geometry. This is a survey of current research directions, highlighting the principal unsolved problems.

Chief among graphical data structures for ordered sets is the *upward drawing*, which we occasionally shorten to *drawing*—alias, *diagram, Hasse diagram*, or *directed (acyclic) covering graph*— according to which the elements of an ordered set P are drawn on a surface (usually the plane) as disjoint small circles, arranged in such a way that, for $a, b \in P$, the circle corresponding to a is higher than the circle corresponding to b whenever $a > b$ and a monotonic arc is drawn to join them just if *a covers b* (that is, for each $x \in P, a > x \geq b$ implies $x = b$). These arcs are drawn to avoid incidence with unwanted circles. In symbols, we write $a \succ b$ and say that a is an *upper cover* of b or b is a *lower cover* of a. A monotonic arc joining a to b may be drawn as a monotonic polygonal path and, wherever convenient, we use a straight segment for the monotonic arc itself. As is the custom we identify an ordered set with (all or) any of its drawings. The *covering graph* stands for the undirected companion of the upward drawing (see Figures 1,2).

Loosely speaking, graphical data structures must be *drawn* in order that they may easily be *read*. This makes them useful as data structures to code and store ordered sets. For references to these and other applied aspects of upward drawings, see [25], [27], [29], [28].

Among the criteria for a "good" drawing, perhaps the most obvious is "planarity". An ordered set is *planar* if it has an upward drawing in which no arcs cross. Thus, planarity is a property of the order and a planar ordered set may have upward drawings which are not planar, and in this respect, it is an apparent abuse of language to identify a planar ordered set with one of its nonplanar upward drawings. (See Figures 3,4.)

Planarity is by now a classical theme in combinatorics and graph algorithm research. For (undirected) graphs there are well-known and elegant combinatorial characterizations of the graphs with a planar representation and efficient algorithms for testing whether a graph has such a one at all [9], [16]. For ordered sets the situation seems much more complicated.

Problem 1 (Planarity Testing) *Is there an effective procedure to test whether an ordered set has a planar upward drawing?*

N.W. Sauer et al. (eds.), Finite and Infinite Combinatorics in Sets and Logic, 337–347.
© 1993 *Kluwer Academic Publishers. Printed in the Netherlands.*

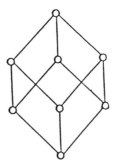

Figure 1: An upward drawing of Q_3

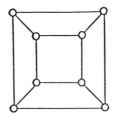

Figure 2: The covering graph of Q_3

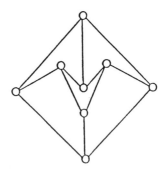

Figure 3: Planar upward drawing of an ordered set with the same covering graph as Q_3

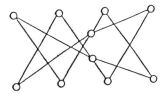

Figure 4: A nonplanar upward drawing of a planar ordered set

As an indication of the difficulty, consider this—almost provocatively—weaker version of it: *is there any procedure at all—that is a finite one—to test whether an n-element ordered set is planar?* Indeed, although there are finite procedures, none seems to be at all obvious (cf. [12], [29]). For lattices, at least, planarity is well understood [14]. The reason, perhaps, stems from this striking connection between graphs and lattice drawings.

Theorem 1 ([22]) *A lattice is planar if and only if the graph obtained from its drawing by adjoining a single edge, from the top to the bottom, is itself planar—as an undirected graph.*

Not only does this lead to an effective planarity testing procedure for lattices, based on graphs, it also readily produces an effective procedure for testing whether an n-element ordered set has (order) dimension at most two, that is, whether it is order embeddable in the direct product of two chains, each with at most n elements. For, if an ordered set P has dimension at most two, then its completion, which is an order sublattice of the direct product of two chains, each with at most n elements, must be a planar lattice with at most n^2 elements.

It is also easy to decide whether an ordered set has a planar drawing in which all maximal elements and all minimal elements appear on the exterior face of its drawing, for, adjoining a top and bottom, in this case would necessarily produce a planar lattice ([14]).

These are two important recent results.

Theorem 2 ([4]) *For any ordered set of width two with n vertices there is a decision procedure to test in time $O(n^{\frac{3}{2}})$ whether it is planar.*

The construction approach provides a minimal list of width two ordered sets (containing, for instance, the order illustrated in Figure 5) such that an arbitrary ordered set P of width two is planar if and only if no member of this list is a *homeomorph* of a *subdrawing* of P. Generalizing this to width three is fraught with new complications (cf. Figure 6). Indeed, it is not even known whether there is a finite minimal list of nonplanar ordered sets of width three with this property. Moreover, in the width three case there may even be a decomposition of a planar ordered set into three chains which, if placed along vertical lines, cannot produce a planar representation (cf. Figure 7).

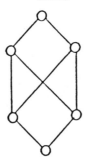

Figure 5: A nonplanar ordered set of width two

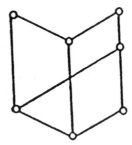

Figure 6: A nonplanar ordered set of width three which contains no subdrawing homeomorphic to any width two nonplanar ordered set

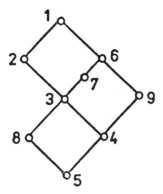

Figure 7: The three chains $A = \{1 > 2 > 3 > 4 > 5\}, B = \{6 > 7 > 8\}, C = \{9\}$ cannot be
placed along three verticals to produce a planar representation

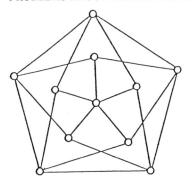

Figure 8: The smallest, triangle-free, noncovering graph [18] , [7]

Theorem 3 ([10], [32]) *There is an efficient planarity testing procedure for an ordered set, provided that it has a bottom.*

What is at stake for a better understanding of planarity?

Problem 2 (Characterization) *Which graphs are covering graphs?*

There is little hope for a computationally satisfying characterization.

Theorem 4 ([19]) *The decision problem, whether an undirected graph is a covering graph, is NP-complete.*

The problem is, nevertheless, of considerable interest [21], [26].

An *orientation* of a graph is an order (or upward drawing) whose covering graph is this same graph. Thus, a graph is a covering graph just if it has an orientation. The triangle is the smallest "noncovering" graph, for any attempt to "orient" it yields a "nonessential" edge. The smallest triangle-free noncovering graph has eleven vertices (see Figure 8). The scarcity of "small" examples of noncovering graphs suggests this more astute approach.

Problem 3 (Noncovering Graphs) *Which triangle-free graphs are not covering graphs?*

A triangle-free graph which is planar has a vertex three-colouring [7]. Thus, if its vertices are coloured with positive integers $1, 2, 3$, the relation $a < b$, just if there is a path from a to b in the graph, of strictly increasing colours , is an orientation of the graph.

Theorem 5 (Folklore) *Any triangle-free planar graph is a covering graph.*

A *bipartite* orientation of a graph is an orientation of it in which every element is either minimal or maximal.

Theorem 6 ([2]) *A bipartite graph has a bipartite planar orientation if and only if the undirected graph itself is planar.*

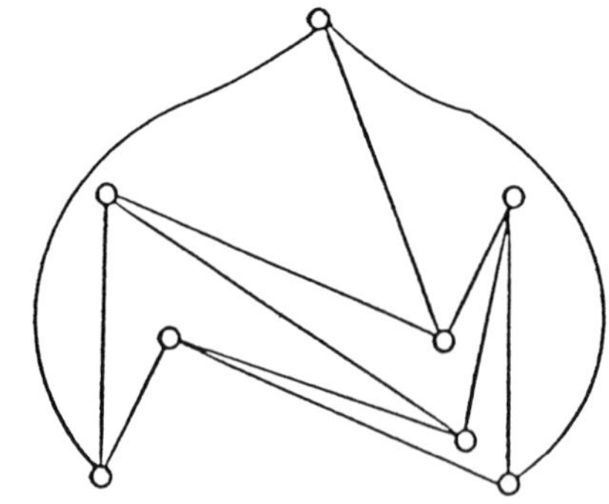

Figure 9: Planar bipartite orientation of the planar bipartite covering graph of Q_3

Every planar n-element covering graph has at least $2^{\frac{n}{3}}$ orientations [17]. Is one of them planar? It is an innocent question of surprising difficulty.

Problem 4 (Planar Orientation) *Does every triangle-free planar graph have a planar orientation?*[1]

Planarity, for lattices, serves even as an algorithm for "lattice testing" (cf. [14]).

Theorem 7 ([3]) *A planar ordered set with top and bottom is a lattice.*

The keypoint is that a *four-cycle* $\{a_i < c_i | i = 1, 2\}$ with no element between the a_i's and the b_i's, together with top and bottom, has no planar upward drawing.

A *doubly irreducible* element is an element with at most one upper cover and at most one lower cover. The starting point for the theory of planar lattices is this simple and far-reaching observation.

Theorem 8 ([1]) *Every planar lattice (with at least three elements) has a doubly irreducible element on the left boundary of any planar upward drawing.*

An important generalization is a *dismantlable ordered set*, that is, an ordered set P whose elements can be labelled $a_1, a_2, \ldots, a_n$ such that each $a_i, i = 1, 2, \ldots, n-2$ is doubly irreducible in $P \setminus \{a_1, a_2, \ldots, a_{i-1}\}$. Every planar lattice is dismantlable. More general, though, than planar lattices, they are, not surprisingly, simpler to characterize. A subset $\{a_1, a_2, \ldots, a_n, c_1, c_2, \ldots, c_n\}, n \geq 3$, is a *cycle* if, $c_i > a_i, c_i > a_{i+1}, i = 1, 2, \ldots, n-1$, and $c_n > a_n, c_n > a_1$ are the only comparabilities.

Theorem 9 ([13]) *A lattice is dismantlable if and only if it contains no cycle.*

Dismantlable lattices have numerous combinatorial connections (cf. [29]). Several problems have resisted solution during the last two decades [24] [14].

Problem 5 (Irreducible Width) *In any dismantlable lattice, is width $\geq$ dimension?*

---

[1]A. Kisielewicz and I. Rival have just announced a positive solution.

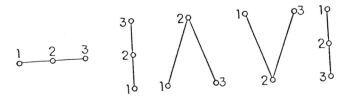

Figure 10: All (four) orientations of the three-element path

Not apparently of direct combinatorial significance, this next problem, if positively settled, would imply a positive solution to the last one. Its two parts are essentially identical.

Problem 6 (Dismantlable Embedding)   (i) *Is every dismantlable lattice order embeddable in a modular dismantlable lattice?*

(ii) *Is every dismantlable lattice order embeddable in the subgroup lattice of the direct product of two cyclic groups, each of prime power order?*

We must search far before we come upon a nontrivial property—different, for example, from the number of elements, edges, etc.—invariant with respect to all orientations of a covering graph. None of the familiar combinatorial parameters *height*, *width* or *dimension*, is an example (see Figure 10).

Problem 7 (Orientation Invariants) *What are nontrivial order theoretical properties common to all orientations of the covering graph?*

Planarity is not such an invariant either ...

Theorem 10 ([11]) *In every lattice orientation of the covering graph of a planar lattice, there is always a doubly irreducible element.*

Recent progress in computational geometry has led to the first genuine instance of a nontrivial orientation invariant. It originates in a computational metaphor for iceflow analysis based on order (cf. [31], [6], [30]). Given a collection of disjoint convex figures, each assigned a fixed direction of motion, say that a figure *A* *blocks* a figure *B* if there is a line joining a point of *B* to a point of *A* along the direction assigned to *B*. The transitive closure of this blocking relation is an order—provided this relation contains no directed cycle. In this case we call it an *m-directional blocking order*, where *m* is the number of different directions used among the convex figures (one for each figure) (see Figure 11).

Theorem 11 ([31]) *There is a one-to-one correspondence between one-directional blocking orders and planar lattices.*

Blocking relations are constructed in terms of convex figures. One-directional blocking relations (that is, planar lattices) can always be represented using only line segments, in fact using just horizontal line segments with the upward perpendicular direction of motion for each figure [31], cf. [20].

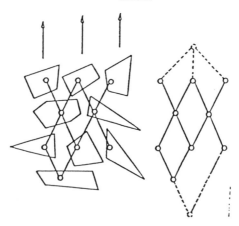

Figure 11: A one-directional blocking relation

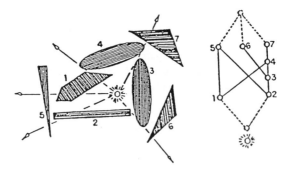

Figure 12: A one-light source order

Problem 8 (Line Segment Blocking) *Every m-directional blocking relation can be represented using only line segments.*

A companion metaphor to one-directional blocking order is "one-light source order". Consider a light source located on a plane, disjoint from a collection of disjoint connected figures. A figure *A obstructs* a figure *B* if a ray from the light source passes through *B* before it passes through *A*. The transitive closure is a *one-light source order*—provided there are no directed cycles (see Figure 12). If the light source is far away from the collection of disjoint convex figures, then the light rays are, in effect, parallel and the one-light source order becomes a one-directional blocking order [6].

A *spherical ordered set* is an ordered set with bottom and top elements having an upward drawing on the surface of a sphere such that the bottom is located at the South Pole, the top at the North Pole, all arcs are strictly increasing northward on the sphere, and no pair of arcs cross. For instance, $\mathbf{Q}_3$ is spherical (see Figure 12) while the ordered set illustrated in Figure 14 is not.

Here is the basic result about spherical ordered sets.

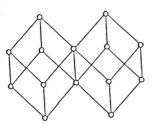 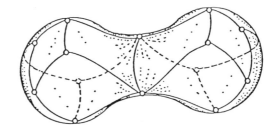

Figure 13: An ordered set with an upward drawing on a "peanut surface" (topologically equivalent to a sphere), and yet with no upward drawing on a sphere

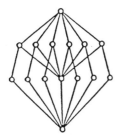 

Figure 14: A lattice with *order genus one*

Theorem 12 ([6])   (i) *An ordered set is spherical if and only if it has a bottom, a top, and its covering graph is planar.*

(ii) *An ordered set is a one-light source order if and only if it can be obtained from a spherical order by removing its bottom and its top.*

There are ordered sets with upward drawings on topological spheres and yet none on spheres [5], [23] (see Figure 13).

These results and examples set the stage for the first nontrivial example of a long sought after orientation invariant. This example comes from topological graph theory. Every surface is topologically equivalent to a sphere with handles; its *genus* is the number of handles that must be added to obtain its homeomorphism type. The *graph genus* of an (undirected) graph is the smallest number g such that the graph can be drawn, without edge crossings, on a surface with genus g, that is, on a sphere with g handles [8]. The *order genus* of an ordered set P is the smallest integer g such that it can be drawn, without edge crossings, on a surface with genus g, in such a way that, whenever $a < b$ in P, the z-coordinate of a is smaller than the z-coordinate of b, and all edges of P are monotonic with respect to the z-coordinate (see Figure 14).

Such drawings of ordered sets on surfaces, without edge crossings, obviously constitute a higher-dimensional analogue of "planarity". This extension of upward drawings to the realm of topological graph theory has substance, too.

Theorem 13 ([5]) *The order genus of an upward drawing equals the graph genus of its covering graph.*

As the order genus depends only on the covering graph, every orientation of a fixed (covering) graph has the same order genus.

Theorem 14 ([5]) *Order genus is an orientation invariant.*

References

[1] K. A. Baker. P. Fishburn and F. S. Roberts (1971) Partial orders of dimension 2, *Networks* **2**, 11–28.

[2] G. di Battista, W.-P. Liu and I. Rival (1990) Bipartite graphs, upward drawings, and planarity, *Inform. Proc. Letters* **36**, 317–322.

[3] G. Birkhoff (1967) *Lattice Theory*, 3rd edition, American Mathematical Society, page 32, ex. 7(a).

[4] J. Czyzowicz. A. Pelc and I. Rival (1990) Planar ordered sets of width two, *Math. Slovaca* **40** (4), 375–388.

[5] K. Ewacha, W. Li, and I. Rival (1991) Order, genus and diagram invariance, *Order* **8**.

[6] S. Foldes, I. Rival and J. Urrutia, Light sources, obstructions, and spherical orders, *Discrete Math.*, to appear.

[7] H. Grötsch (1958) Ein Dreifarbensatz für dreikreisfreie Netze auf der Kugel, *Wiss. Z. Martin-Luther-Univ. Halle (Math. Natur. Reihe)* **8**, 109–119.

[8] M. Henle (1979) *A Combinatorial Introduction to Topology*, Freeman.

[9] J. Hopcroft and R. E. Tarjan (1974) Efficient planarity testing, *J. Assoc. Comp. Mach.* **21** (4) 549–568.

[10] M. Hutton and A. Lubiw (1990) *Symp. Discrete Appl. Math.*

[11] R. Jégou, R. Nowakowski, and I. Rival (1987) The diagram invariant problem for planar lattices, *Acta. Sci. Math. (Szeged)* **51**, 103–121.

[12] D. Kelly (1987) Fundamentals of planar ordered sets, *Discrete Math.* **63** 197–216.

[13] D. Kelly and I. Rival (1974) Crowns, fences and dismantlable lattices, *Canad. J. Math.* **26**, 1257–1271.

[14] D. Kelly and I. Rival (1975) Planar lattices, *Canad. J. Math.* **27** 636–665.

[15] J. G. Lee, W.-P. Liu, R. Nowakowski, and I. Rival (1988) Dimension invariance of subdivision, *Technical Report, University of Ottawa* **TR-88-30**, pp. 1–14.

[16] A. Lempel, S. Even and I. Cederbaum (1967) An algorithm for planarity testing of graphs, in *Theory of Graphs, International Symposium*, Rome (1966) (P. Rosenstiehl ed.), Gordon and Breach, pp. 215–232.

[17] W.-P. Liu and I. Rival (1991) Enumerating orientations of ordered sets, *Discrete Math.*.

[18] J. Mycielski (1955) Sur le colorage des graphes, *Colloq. Math.* **3**, 161–162.

[19] J. Nešetřil and V. Rödl (1987) The complexity of diagrams, *Order* **3**, 321–330.

[20] R. J. Nowakowski, I. Rival and J. Urrutia (1990) Representing orders on the plane by translating points and lines, *Discrete Appl. Math.* **27**, 147–156.

[21] O. Ore (1962) *Theory of Graphs*, Amer. Math. Soc., Providence.

[22] C. R. Platt (1976) Planar lattices and planar graphs, *J. Comb. Th. Ser. B* **21** 30–39.

[23] K. Reuter and I. Rival (1991) Genus of orders and lattices, in *Graph-Theoretic Concepts in Computer Science* (R. Möhring ed.), Lect. Notes Comp. Sci. **484**, pp. 260–275.

[24] I. Rival (1976) Combinatorial inequalities for semimodular lattices of breadth two, *Algebra Univ.* **6**, 303–311.

[25] I. Rival (1985) The diagram, in *Graphs and Order* (I. Rival ed.), Reidel, pp. 103–133.

[26] I. Rival (1985) The diagram, *Order* **2**, 101–104.

[27] I. Rival (1989) Graphical data structures for ordered sets, in *Algorithms and Order* (I. Rival ed.), Kluwer, pp. 3–31.

[28] I. Rival (1993) *Introducing Ordered Sets*, to appear.

[29] I. Rival (1993) Reading, drawing, and order, in *Algebras and Orders* (I. G. Rosenberg ed.), Kluwer.

[30] I. Rival (1993) Order, iceflow and surfaces, in *The Birkhoff Symposium* (R. Wille ed.)

[31] I. Rival and J. Urrutia (1988) Representing orders by translating convex figures in the plane, *Order* **4**, 319–339.

[32] C. Thomassen (1989) Planar acyclic oriented graphs, *Order* **5**, 349–361.

Superstable and Unstable Theories of Order

TODD J. SCHNEIDER

Department of Mathematics
Providence College
Providence, RI 02919

Abstract

The main theorem in section one of this note is a sufficient condition on a theory of an order of finite height that ensures superstability. A characterization of the types over a model of such a theory is also given. In the second section examples of finite height orders of dimension two with unstable theories are given. As a corollary, we define a set of finite orders which provide a sufficient condition for the instability of an order of finite height.

Most of the terminology is standard; all models are infinite with finite relational similarity type and only theories having such models are considered. $S_1(T)$ is the set of 1-types of the theory T and $S_1(A)$ is the set of 1-types over the model A. If φ is a formula of the language, then $qr(\varphi)$ denotes its quantifier rank. Bold face $\mathbf{a}, \mathbf{b}$'s denote finite sequences.

An *order* is a structure $A = (A, <)$ with a binary antisymmetric and transitive relation, $a|b$ is an abbreviation for $\neg(a < b \lor b < a \lor a = b)$, a (strict) *path* is a finite sequence $(a_i : i < n)$ so that

$$\bigwedge_{i<n} (a_i < a_{i+1} \lor a_i > a_{i+1}) \land \bigwedge_{i<n-2} a_i | a_{i+2},$$

and the *length* of this path between a_0, a_{n-1} is $n-2$. For $k \in \omega$, let $\lambda(x, y : k)$ be the formula that asserts the existence of a path between x and y having length k. If $\mathbf{a}, \mathbf{b} \in A$, we will write $\lambda(\mathbf{a}, \mathbf{b} : k)$ for $\bigvee_{i \in |\mathbf{a}|, j \in |\mathbf{b}|} \lambda(a_i, b_j : k)$.

An order is *connected* if there is a path between any two points and a *component* is a maximal connected set. The *height* of an order, $ht(A)$, is the supremum of lengths of chains in the order, and the *height of a point a*, $ht(a)$, is the supremum of lengths of chains from a minimal element to a. From : *x is a dead-end off y* iff x is comparable only to y and to no other points of the order. For $a \in A$ define subsets $S(a)$ and $P(a)$ as follows:

$$S(a) = \{b \in A : \exists x[(x \le b < a \lor a < b \le x) \land \exists y(y|a \land y \nmid x)]\},$$

$$P(a) = \{c \in A : c|a \land \exists b \in S(a)(c \nmid b)\}.$$

1 Superstability: A Sufficient Condition

An order, or theory thereof, which has arbitrarily long chains is, by compactness, unstable. So for the remainder we assume that $T = Th(A)$ is a theory of an order of finite height.

N.W. Sauer et al. (eds.), Finite and Infinite Combinatorics in Sets and Logic, 349–353.
© *1993 Kluwer Academic Publishers. Printed in the Netherlands.*

An order $\mathcal{A}$ or its theory T is *uniformly locally finite* if there is $m \in \omega$ such that

$$T \models \forall x[|S(x)| < m \vee |P(x)| < m].$$

A 1-type $p(v) \in S_1(A)$ *is in* $\mathcal{A}$, if $p(v) \vdash v = a$, for some $a \in A$. If a 1-type $p(v)$ is in the model, then it is in some component.

Lemma 0 *Let $\mathcal{A}$ be uniformly locally finite. If $p(v) \in S_1(A)$ is not in $\mathcal{A}$ and for all $a \in A$, $p(v) \vdash v|a$, then $p(v) \vdash \neg \lambda(v, a : k)$, for all $a \in A$ and all $k \in \omega$.*

Proof. Let $\mathcal{B} \succ \mathcal{A}$ realizing p by $b \in B - A$. Suppose b is connected to some point of A. Then there are $a \in A$, $n \in \omega$, and $c_1, \ldots, c_n \in B$ forming a path from b to a of minimal length with $a \not| c_n$ and $b \not| c_1$. By hypothesis, the length $n > 0$.

If the path length n equals 1, then $b \in P(a)$. If $P(a)$ is finite, then every point of it is an element of A, so that $b \in A$. Therefore, we assume $P(a)$ is infinite. Thus $S(a)$ is finite and there is some $a' \in A$ which witnesses $b \in P(a)$, $b|a$ and $b \not| a'$. By hypothesis b is incomparable with every point of A. In either case there is a contradiction.

Suppose the length n is greater than 1, then c_{n-1} is in $P(a)$. If $P(a)$ is finite, then $c_{n-1} \in A$ and we have a path of length $n - 1$ from b to a point of A. Thus, we may assume that $P(a)$ is infinite and $S(a)$ is finite. It follows that $c_n \in S(a)$ and hence is an element of A. In either case we may find a path of length $n - 1$ from b to a point of A. This contradicts the minimality of the original path and point a.

Lemma 1 *Let $p(v) \in S_1(A)$ be not in $\mathcal{A}$ an order of finite height. If $\mathcal{A}$ is uniformly locally finite, then there are at most finitely many $a_i \in A$ such that $p(v) \vdash v \not| a_i$.*

Proof. Supposing otherwise, let $\{a_i : i \in \omega\} \subset A$ be such that $p(v) \vdash v \not| a_i$. We may assume for all $i, j \in \omega$, $ht(a_i) = ht(a_j) \neq ht(v)$. It follows that $p(v) \vdash a_i \in S(v)$ for all $i \in \omega$, and that, since $p(v)$ is not in $\mathcal{A}$, $S(a_i)$ is infinite for each $i \in \omega$. From the uniform local finiteness, $P(a_i)$ must be uniformly finite for all $i \in \omega$. But for each $i \neq j$, $a_i \in P(a_j)$, contradicting its finiteness.

Lemma 2 *Let T be a theory of an order of finite height that is uniformly locally finite. Let $\mathcal{B} \succ \mathcal{A} \models T$ with $b \in B - A$ and let $\Gamma(v)$ be the union of the sets*

(i) $Th(A, <, a)_{a \in A}$,

(ii) $\{\varphi(v) : \mathcal{B} \models \varphi(b)\}$,

(iii) $\{v < a : a \in A \wedge \mathcal{B} \models b < a\} \cup \{v > a : a \in A \wedge \mathcal{B} \models b > a\}$,

(iv) $\{v|a : a \in A \wedge \mathcal{B} \models b|a\}$,

(v) $\{\exists \mathbf{y} \psi(v, a, \mathbf{y}) : qr(\psi) = 0 \wedge a \in A \wedge \exists \mathbf{c} \in B \wedge \mathcal{B} \models \psi(b, a, \mathbf{c})\}$,

(vi) $\{v \neq a : a \in A\}$.

Then $\Gamma(v)$ has a unique extension to a complete 1-type over A.

Proof. Suppose there are $p(v), q(v) \in S_1(A)$ with $\Gamma(v) \subset p(v), q(v)$ and $p(v) \neq q(v)$.

We may assume that for some $a \in A$, $\Gamma \vdash v \nmid a$, for otherwise $\Gamma(v) \cup \{v | a : a \in A\}$ has an extension to a 1-type over A, say $p(v)$, and by Lemma 0 $p(v)$ proves that v is not connected to any point of A. It follows that $p(v)$ will be unique.

If $\Gamma \vdash v \nmid a$, for $a \in A$, then $S(a)$ is infinite in $\mathcal{A}$, since $\Gamma(v)$ is not in A, and $P(a)$ is finite, containing only points of A.

Let $(A, <, a)_{a \in A} \prec (C, <, a)_{a \in A} \models p(b), q(c)$ for $b, c \in C$. Note that $b | c$, for $\Gamma(v) \vdash ht(v) = k$, for some k less than or equal to the height of the order. We wish to show that b and c are ω-equivalent over A, hence $p(v) = q(v)$. To do this we show the existence of a winning strategy for player II in the Ehrenfeucht game of length n.

First some simplifications. It follows from parts (iii), (iv) and (ii) of the definition of Γ that b, c are 0-equivalent over A and both b, c have the same number and distribution of dead-ends. We may also assume that b, c are not dead-ends off a, and hence $b, c \in S(a)$. Finally observe that, if in the course of play of a game a path from either b or c is chosen, say $c_0, \ldots, c_k$, then either $c_0 \nmid a$ or $c_0 \in P(a)$. In either case, $b < a < c_0 \leftrightarrow c < a < c_0$ or $b < c_0 \leftrightarrow c < c_0$, and $c_0, \ldots, c_k$ is a path for both b and c to c_k.

Therefore, we may assume that in any play of a game no points of A comparable to b, c or dead-ends off either b, c are played, since II has a winning response.

The game begins and player I picks a point, d_0. II's response depends on where d_0 lies in relation to any points of A comparable to b and hence c. We suppose that $a \in A$ is an arbitrary point comparable to both b and c.

If $d_0 > a$, then II responds with the same point and calls it e_0. If either $b < d_0 < a$ or $c < d_0 < a$, then from part (v) of the definition of $\Gamma(v)$ there is an e_0 0-equivalent to d_0 over b, a.

If $d_0 | a$ and either $d_0 \nmid c$ or $d_0 \nmid b$, then $d_0 \in P(a)$ and is an element of A. Hence, II may respond by also choosing $e_0 = d_0$.

If d_0 is incomparable to each of a, b, c, then player II may choose this same element $e_0 = d_0$.

Inductively, we assume that player II has a winning strategy for any Ehrenfeucht game of length $< n$.

Suppose sequences $\mathbf{d} = (c, d_0, \ldots, d_{k-1})$ and $\mathbf{e} = (b, e_0, \ldots, e_{k-1})$, $k < n$, have been chosen which are 0-equivalent over A. From the simplifications above, we may assume that these sequences do not form paths from b, c to e_{k-1}, d_{k-1}, respectively.

Player I's next move is d_k. If $\neg \lambda(d_k, \mathbf{d} : j)$, for each $j \leq n - k$, then d_k is sufficiently far away so as to pose no potential threat and player II may choose this same element, $e_k = d_k$.

If $\lambda(d_k, \mathbf{d} : j)$ for some $j \leq n - k$, then let j be the least such and let $\mathbf{d}' \subset \mathbf{d}$ be that subsequence such that $\lambda(d_k, \mathbf{d}'(i) : j)$, for each $i < |\mathbf{d}'|$. If $\mathbf{d}'$ contains points of A let these be $a_0, \ldots, a_l$. If $\mathbf{d}' \cap A$ is empty, then $a_0 \in A$ can be an arbitrary point sufficiently far away from d_k, and the sequences $\mathbf{d}, \mathbf{e}$, and $l = 0$. Consider the quantifier free formulas $\psi(v, a_i, \mathbf{u}, x)$, for $i \leq l$, formed by the atomic diagram of the sequence $(c, a_i, \mathbf{d}', d_k)$. The formulas $\exists \mathbf{u} \exists x \psi(v, a_i, \mathbf{u}, x)$ are in $\Gamma \subset p(v)$. It follows that player II can pick e_k such that $(b, \mathbf{e}, e_k)$ is 0-equivalent to $(c, \mathbf{d}, d_k)$.

Therefore, player II has a winning strategy in the n-game and it follows that $p(v) = q(v)$.

Lemma 3 Let $p(v) \in S_1(A)$ and not in A, an order of finite height which is uniformly locally finite. Then there are at most $|A| \cdot |S(T)|$ many such 1-types.

Proof. Given such a 1-type $p(v)$, either v is incomparable with every point of A, in which case $p(v)$ generates a new component, of which there are at most $|S_1(T)|$-many; or, by Lemma 1, v is comparable to finitely many points of A.

There are at most $|S_2(T)|$ many sets of Σ-formulas with two free variables consistent with T. By Lemma 2, there are at most $|A^{<\omega}| \cdot |S_2(T)|$ many 1-types $p(v)$ not in A comparable with finitely many points of A.

In either case the number of 1-types over A is at most

$$|S_1(T)| + |A^{<\omega}| \cdot |S_2(T)| \leq |A| \cdot |S(T)|.$$

Theorem 4 *Let T be a theory of an order of finite height which is uniformly locally finite. Then T is superstable. Moreover, if $|S(T)| \leq \aleph_0$, then T is ω-stable.*

Proof. For $p \in S_1(T)$, either p is in A, or it isn't. There are at most $|A|$-many 1-types in A, while the latter possibilities are enumerated in Lemma 3. In all cases

$$|S_1(A)| \leq |A| + |A| \cdot |S(T)| \leq |A| + 2^\omega.$$

If $|S(T)| \leq \aleph_0$, then $|S_1(A)| \leq |A|$.

2   Unstable Orders of Dimension 2

It had been known [1] that the least dimension of an unstable order of finite height is at least 3. We present an example whose dimension is 2, contradicting the notion that the class of dimension 2 finite height orders might have only stable theories.

Let $A = \{x_i | i \in \omega\} \cup \{y_j | j \in \omega\}$. Order A as

$$x_i < y_j \text{ iff } i \leq j, \ x_i | x_j \text{ and } y_i | y_j, \text{ for all } i, j \in \omega.$$

Then $(A, <)$ has height 2: the x_i's are the minimal elements and the y_j's are the maximal elements. In particular, x_0 is below all y_j's and y_j has exactly $j + 1$ predecessors.

DIMENSION 2

The dimension of $(A, <)$ is 2. We exhibit two linear extensions on A which realize the order.

$$L_0 : \ x_0 \ L_0 \ y_0 \ L_0 \ x_1 \ L_0 \ y_1 \ldots x_n \ L_0 \ y_n \ L_0 \ x_{n+1} \ L_0 \ y_{n+1} \ldots .$$

$$L_1 : \ \ldots x_{n+1} \ L_1 \ x_n \ L_1 \ldots x_1 \ L_1 \ x_0 \ldots y_{m+1} \ L_1 \ y_m \ L_1 \ldots y_1 \ L_1 \ y_0.$$

Clearly the intersection of these linear orders forces the x_i's and the y_j's into pairwise incomparable sets, respectively. And $x_i | y_j$ iff $j < i$.

INSTABILITY

Let $\Psi(u, v)$ be the formula $\exists z \exists w [z < u, v \wedge z | w \wedge w < v \wedge w | u]$.

Observe that when $i < j$, then for all x, $x < y_i \rightarrow x < y_j$, so that on the y_j's Ψ is antisymmetric. Since $x_j < y_j \wedge x_j | y_i$ for all $i < j$, Ψ is transitive on the y_j's.

Thus $\Psi(y_i, y_j)$ iff $i < j$. Hence $Th(A, <)$ is unstable.

Theorem 5 *There is an order of height 2 and dimension 2 whose theory is unstable.*

Corollary 6 *For each $n \in \omega$, $n > 2$, there exists an order of height n and dimension 2 whose theory is unstable.*

Proof. Extend the height 2 example above by adding a chain (or chains) of length $n-2$ below some minimal element(s): $x_i < z_{i,0} < z_{i,1} < \ldots < z_{i,n-3} < x_i$. To see that these orders have dimension 2, the linear extensions shown above may be modified as follows.

$$L_0 : \ z_{0,0} \ L_0 \ldots z_{0,n-2} \ L_0 \ x_0 \ L_0 \ y_0 \ L_0 \ z_{1,0} \ L_0 \ldots z_{1,n-2} \ L_0$$
$$x_1 \ L_0 \ y_1 \ldots z_{m,0} \ L_0 \ldots z_{m,n-2} \ L_0 \ x_m \ L_0 \ y_m \ldots$$

$$L_1 : \ \ldots z_{k+1,0} \ L_1 \ldots z_{k+1,n-2} \ L_1 \ x_{k+1} \ L_1 \ z_{k,0} \ L_1 \ldots z_{k,n-2} \ L_1$$
$$x_k \ L_1 \ldots z_{0,0} \ L_1 \ldots z_{0,n-2} \ L_1 \ x_0 \ldots y_{m+1} \ L_1 \ y_m \ L_1 \ldots$$

The y_i's remain orderable by the formula Φ above.

Define finite orders $P_n = \{x_i : i \in n\} \cup \{y_j : j \in n\}$ as $x_i|x_j$, $y_i|y_j$, for $i < j < n$ and $x_i < y_j$ iff $i < j < n$. Then from compactness we have the following theorem.

Proposition 7 *If T is an order of finite height and some model embeds P_n (as suborders) for arbitrarily large $n \in \omega$, then T is unstable.*

Reference

[1] Kenneth Smith, Stability and categoricity of lattices, *Can. J. Math.* **33** (1981), No. 6, pp. 1380–1419.

Advances in Cardinal Arithmetic

SAHARON SHELAH[*][†]

The Hebrew University
Jerusalem, Israel

Department of Mathematics
Rutgers University
New Brunswick, NJ USA

Abstract

If $\mathrm{cf}\kappa = \kappa$, $\kappa^+ < \mathrm{cf}\lambda = \lambda$ then there is a stationary subset S of $\{\delta < \lambda : \mathrm{cf}(\delta) = \kappa\}$ in $I[\lambda]$. Moreover, we can find $\overline{C} = \langle C_\delta : \delta \in S \rangle$, C_δ a club of λ, $\mathrm{otp}(C_\delta) = \kappa$, guessing clubs and for each $\alpha < \lambda$ we have: $\{C_\delta \cap \alpha : \alpha \in \mathrm{nacc}C_\delta\}$ has cardinality $< \lambda$.

We prove that, for example, there is a stationary subset of $S_{<\aleph_1}(\lambda)$ of cardinality $\mathrm{cf}(S_{<\aleph_1}(\lambda), \subseteq)$.

We prove the existence of nice filters where instead of being normal filters on ω_1 they are normal filters with larger domains, which can increase during a play. They can help us transfer the situation on $\aleph_1$-complete filters to normal ones.

We consider ranks and niceness of normal filters, such that we can pass, say, from $\mathrm{pp}_{\Gamma(\aleph_1)}(\mu)$ (where $\mathrm{cf}\mu = \aleph_1$) to $\mathrm{pp}_{\mathrm{normal}}(\mu)$.

We consider some weakenings of G.C.H. and their consequences. Most have not been proved independent of ZFC.

1 $I[\lambda]$ is Quite Large and Guessing Clubs

On $I[\lambda]$ see [6], [5], [7, §4] (but this section is self-contained; see Definition 1.1 and Claim 1.2 below). We shall prove that for regular κ, λ, such that $\kappa^+ < \lambda$, there is a stationary $S \subseteq \{\delta < \lambda : \mathrm{cf}\delta = \kappa\}$ in $I[\lambda]$. We then investigate "guessing clubs" in (ZFC).

Definition 1.1 For a regular uncountable cardinal λ, $I[\lambda]$ is the family of $A \subseteq \lambda$ such that $\{\delta \in A : \delta = \mathrm{cf}\delta\}$ is not stationary and for some $\langle \mathcal{P}_\alpha : \alpha < \lambda \rangle$ we have:

(a) $\mathcal{P}_\alpha$ is a family of $< \lambda$ subsets of α;

(b) for every limit $\alpha \in A$ such that $\mathrm{cf}(\alpha) < \alpha$ there is $x \subseteq \alpha$, $\mathrm{otp}(x) < \alpha = \sup x$ such that
$$\bigwedge_{\beta < \alpha} x \cap \beta \in \bigcup_{\gamma < \alpha} \mathcal{P}_\gamma.$$

[*]Partially supported by the BSF, Publ. 420.
[†]I thank Alice Leonhardt for typing (and retyping) the manuscript so nicely and accurately.

355

N.W. Sauer et al. (eds.), *Finite and Infinite Combinatorics in Sets and Logic*, 355–383.
© 1993 *Kluwer Academic Publishers. Printed in the Netherlands.*

We know (see [6], [5] or below)

Claim 1.2 *Let $\lambda > \aleph_0$ be regular.*

(1) $A \in I[\lambda]$ *iff (note: by (c) below the set of inaccessibles in A is not stationary and) there is $\langle C_\beta : \beta < \lambda \rangle$ such that:*

    (a) C_β *is a closed subset of β;*

    (b) *if $\alpha \in \mathrm{nacc} C_\beta$ then $C_\alpha = C_\beta \cap \beta$ (nacc stands for "non-accumulation");*

    (c) *for some club E of λ, for every $\delta \in A \cap E$: $\mathrm{cf}\delta < \delta$ and $\delta = \sup C_\delta$, and $\mathrm{cf}(\delta) = \mathrm{otp}(C_\delta)$;*

    (d) $\mathrm{nacc}(C_\beta)$ *is a set of successor ordinals.*

(2) $I[\lambda]$ *is a normal ideal.*

Proof.

1) THE "IF" PART:

Assume $\langle C_\beta : \beta < \lambda \rangle$ satisfy (a), (b), (c) with a club E for (c). For each limit $\alpha < \lambda$ choose a club e_α of order type $\mathrm{cf}(\alpha)$. We define, for $\alpha < \lambda$:

$$\mathcal{P}_\alpha =: \{C_\beta : \beta \leq \alpha\} \cup \{e_\beta : \beta \leq \alpha\} \cup \{e_\gamma \cap \alpha : \gamma \leq \min(E \setminus (\alpha + 1))\}.$$

It is easy to check that $\langle \mathcal{P}_\alpha : \alpha < \lambda \rangle$ exemplify "$A \in I[\lambda]$".

THE "ONLY IF" PART:

Let $\langle \mathcal{P}_\alpha : \alpha < \lambda \rangle$ exemplify "$A \in I[\lambda]$" (by Definition 1.1). Without loss of generality if $C \in \mathcal{P}_\alpha$, and $\zeta \in C$ then $C \backslash \zeta \in \mathcal{P}_\alpha$ and $C \cap \zeta \in \mathcal{P}_\alpha$.

For each limit $\beta < \lambda$ let e_β be a club of β, $\mathrm{otp}(e_\beta) = \mathrm{cf}(\beta)$ and $\mathrm{cf}\beta < \beta \Rightarrow \mathrm{cf}\beta < \min(e_\beta)$. Let $\langle \gamma_i : i < \lambda \rangle$ be strictly increasing continuous, each γ_i a non-successor ordinal $< \lambda$, $\gamma_0 = 0$, and $\gamma_{i+1} - \gamma_i \geq \aleph_0 + |\bigcup_{\alpha \leq \gamma_i} \mathcal{P}_\alpha| + |\gamma_i|$ and $\gamma_i \in A \Rightarrow \mathrm{cf}(\gamma_i) < \gamma_i$.

Let F_i be a one to one function from $\left(\bigcup_{\alpha \leq \gamma_i} \mathcal{P}_\alpha\right) \times \gamma_i$ into $\{\zeta + 1 : \gamma_i < \zeta + 1 < \gamma_{i+1}\}$. Now we define $C_\alpha \subseteq \alpha$ as follows.

Assume α is a successor ordinal, and let $i(\alpha)$ be such that $\gamma_{i(\alpha)} < \alpha < \gamma_{i(\alpha)+1}$. If $\alpha \notin \mathrm{Rang} F_{i(\alpha)}$, let $C_\alpha = \emptyset$. If $\alpha = F_{i(\alpha)}(x, \beta)$ (so $x \in \bigcup_{\varepsilon \leq \gamma_{i(\alpha)}} \mathcal{P}_\varepsilon$, $\beta < \gamma_{i(\alpha)}$), let C_α be the closure (in the order topology on α) of:

$$F_j(x \cap \zeta, \beta) : \begin{cases} \text{(i)} & \zeta \in x, \\ \text{(ii)} & \mathrm{otp}(x \cap \zeta) \in e_\beta, \\ \text{(iii)} & j < i(\alpha) \text{ is minimal such that } x \cap \zeta \in \bigcup_{\varepsilon \leq \gamma_j} \mathcal{P}_\varepsilon, \\ \text{(iv)} & \text{if } \xi \in x \cap \zeta, \mathrm{otp}(x \cap \xi) \in e_\beta \text{ then} \\ & \qquad (\exists j(1) < j)[x \cap \xi \in \bigcup_{\varepsilon \leq \gamma_{j(1)}} \mathcal{P}_\varepsilon], \\ \text{(v)} & \beta < \min x. \end{cases}$$

Now for $\alpha < \lambda$ limit, choose C_α: if possible, $\mathrm{nacc} C_\alpha$ is a set of successor ordinals, C_α is a club of α, $[\beta \in \mathrm{nacc} C_\alpha \Rightarrow C_\beta = \beta \cap C_\alpha]$; if this is impossible, let $C_\delta = \emptyset$. Let $E =: \{\gamma_i : i \text{ limit} < \lambda\}$. Now we can check the condition in 1.2(1).

2) By Definition 1.1 $I[\lambda]$ is an ideal; by 1.2(1) $I[\lambda]$ includes the ideal of non-stationary subsets of λ. By the last phrase and Definition 1.1, clearly $I[\lambda]$ is normal. $\square_{1.2}$

Claim 1.3 *If κ, λ are regular, $S \subseteq \{\delta < \lambda : \mathrm{cf}\,\delta = \kappa\}$, $S \in I[\lambda]$, S stationary, $\kappa^+ < \lambda$ then we can find $\overline{\mathcal{P}} = \langle \mathcal{P}_\alpha : \alpha < \lambda \rangle$ such that for $\delta(*) =: \kappa$ we have:*

$$\bigoplus_{\overline{\mathcal{P}},S}^{\lambda,\delta(*)} : \begin{cases} \text{(i)} & \mathcal{P}_\alpha \text{ is a family of closed subsets of } \alpha, |\mathcal{P}_\alpha| < \lambda; \\ \text{(ii)} & \mathrm{otp}\,C \leq \delta(*) \text{ for } C \in \cup_\alpha \mathcal{P}_\alpha; \\ \text{(iii)} & \text{for some club } E \text{ of } \lambda, \text{ we have:} \\ & \quad [\alpha \notin E \Rightarrow \mathcal{P}_\alpha = \emptyset] \text{ and} \\ & \quad [\alpha \in E \Rightarrow (\forall C \in \mathcal{P}_\alpha)(\mathrm{otp}\,C \leq \delta(*))]; \\ & \quad [\alpha \in E \backslash (S \cap \mathrm{acc}E) \Rightarrow (\forall C \in \mathcal{P}_\alpha)(\mathrm{otp}\,C < \delta(*))]; \\ & \quad [\alpha \in S \cap \mathrm{acc}E \Rightarrow (\exists! C \in \mathcal{P}_\alpha)(\mathrm{otp}\,C = \delta(*))]; \\ & \quad [\alpha \in S \cap \mathrm{acc}E \ \& \ C \in \mathcal{P}_\alpha \ \& \ \mathrm{otp}\,C = \delta(*) \Rightarrow \alpha = \sup C)]; \\ \text{(iv)} & C \in \mathcal{P}_\alpha \ \& \ \beta \in \mathrm{nacc}C \Rightarrow \beta \cap C \in \mathcal{P}_\beta; \\ \text{(v)} & \text{for any club } E \text{ of } \lambda, \text{ for some } \delta \in S \cap E \text{ and } C \in \mathcal{P}_\delta \text{ we have} \\ & \quad C \subseteq E \ \& \ \mathrm{otp}\,C = \delta(*). \end{cases}$$

Proof. Let $\langle C_\alpha : \alpha < \lambda \rangle$ witness "$S \in I[\lambda]$" as in 1.2(1); without loss of generality $\mathrm{otp}\,C_\alpha \leq \delta(*)$. For any club E let us define $\mathcal{P}_E^\alpha$ by induction on $\alpha < \lambda$:

$$\mathcal{P}_E^\alpha =: \{\alpha \cap g\ell(C_\beta, E) : \alpha \in E \text{ and } \alpha \leq \beta < \min[E \backslash (\alpha + 1)]\}$$
$$\cup \{C \cup \{\beta\} : \text{ for some } \beta \in E, \beta < \alpha, C \in \mathcal{P}_E^\beta \text{ and } \mathrm{otp}(C) < \delta(*)\}$$

where

$$g\ell(C_\beta, E) =: \{\sup(E \cap (\gamma + 1)) : \gamma \in C_\beta \text{ and } \gamma > \min E\}.$$

Note that $|\mathcal{P}_E^\alpha| \leq |\min(E \backslash (\alpha + 1))| < \lambda$. We can prove that for some club E of λ $\langle \mathcal{P}_E^\alpha : \alpha < \lambda \rangle$ is as required except (v) which can be corrected (just by trying successively κ^+ clubs $E_\zeta(\zeta < \kappa^+)$ decreasing with ζ, see [13]) and (iv) which is guaranteed by demanding E to consist of limit ordinals only and the second set in the union defining $\mathcal{P}_E^\alpha$. $\square_{1.3}$

The following lemma gives a sufficient condition for the existence of "quite large" stationary sets in $I[\lambda]$ of almost any fixed cofinality.

Lemma 1.4 *Suppose*

(i) $\lambda > \kappa > \aleph_0$, λ and κ are regular,

(ii) $\overline{\mathcal{P}} = \langle \mathcal{P}_\alpha : \alpha < \kappa \rangle$, $\mathcal{P}_\alpha$ a family of $< \lambda$ closed subsets of α,

(iii) $I_{\overline{\mathcal{P}}} =: \{S \subseteq \kappa : \text{for some club } E \text{ of } \kappa, \text{ for no } \delta \in S \cap E \text{ is there a club } C \text{ of } \delta, \text{ such that } C \subseteq E \text{ and } [\alpha \in \mathrm{nacc}C \Rightarrow C \cap \alpha \in \cup_{\beta < \alpha} \mathcal{P}_\beta]\}$ *is a proper ideal on κ.*

Then there is $S^* \in I[\lambda]$ such that for stationarily many $\delta < \lambda$ of cofinality κ, $S^* \cap \delta$ is stationary in δ; moreover for some club E of δ of order type κ,

$$\{\mathrm{otp}(\alpha \cap E) : \alpha \in E \backslash S^*\} \in I_{\overline{\mathcal{P}}}.$$

Remark 1.4A: The "for stationarily many" in the conclusion can be strengthened to: a set whose complement is in the ideal defined in [13], §2.

Proof. Let χ be regular large enough, N^* be an elementary submodel of $(H(\chi), \in, <^*_\chi)$ of cardinality λ such that $(\lambda+1) \subseteq N^*, \overline{P} \in N$. Let $\overline{C} = \langle C_i : i < \lambda \rangle$ list $N^* \cap \{A \subseteq \lambda : |A| < \kappa\}$ and let

$$S^* = \{\delta < \lambda : \mathrm{cf}(\delta) < \kappa \text{ and for some } A \subseteq \delta,\ \delta = \sup A,$$
$$\mathrm{otp} A < \kappa \text{ and } (\forall \alpha < \delta)[A \cap \alpha \in \{C_i : i < \delta\}]\}.$$

Clearly $S^* \in I[\lambda]$; so we should only find enough $\delta < \lambda$ of cofinality κ as required. So let E_0^a be a club of λ. We can choose inductively $M_\zeta (\zeta \leq \kappa)$ such that:

(a) $M_\zeta \prec (H(\chi), \in, <^*_\chi)$,

(b) $\|M_\zeta\| < \lambda$, $M_\zeta \cap \lambda$ an ordinal,

(c) M_ζ is increasing continuous,

(d) $N, \kappa, \overline{P}, \overline{C}, E_0^a$ belongs to M_0,

(e) $\langle M_\varepsilon : \varepsilon \leq \zeta \rangle \in M_{\zeta+1}$.

Let $\delta_\zeta = \sup(M_\zeta \cap \lambda)$, so $\langle \delta_\zeta : \zeta \leq \kappa \rangle$ is strictly increasing continuous, so $\delta =: \delta_\kappa$ has cofinality κ. Hence there is a strictly increasing continuous sequence $\langle \alpha_\zeta : \zeta < \kappa \rangle \in N^*$ with limit δ, and clearly $E = \{\zeta < \kappa : \alpha_\zeta = \delta_\zeta\}$ is a club of κ. We know that

$$T =: \{\zeta < \kappa : \zeta \text{ limit and for some club } C \text{ of } \zeta,\ C \subseteq E \text{ and } \bigwedge_{\varepsilon < \zeta} [C \cap \varepsilon \in \bigcup_{\xi < \zeta} P_\xi]\}$$

is stationary; moreover, $\kappa \backslash T \in I_{\overline{P}}$ (see assumption (iii)) and clearly $T \subseteq E$. Clearly it suffices to show

(∗) $\zeta \in T \Rightarrow \delta_\zeta \in S^*$.

Suppose $\zeta \in T$, so there is C, a club of ζ such that $C \subseteq E$ and $\bigwedge_{\varepsilon < \zeta} [C \cap \varepsilon \in \bigcup_{\xi < \zeta} P_\xi]$. Let $C^* = \{\delta_\varepsilon : \varepsilon \in C\}$, so C^* is a club of δ_ζ of order type $\leq \zeta < \kappa$ (which is $< \delta_0 \leq \delta_\zeta$). It suffices to show for $\xi \in C$ that $\{\delta_\varepsilon : \varepsilon \in \xi \cap C\} \in \{C_i : i < \delta_\zeta\}$. For this end we shall show

(α) $\{\delta_\varepsilon : \varepsilon \in C \cap \xi\} \in \{C_i : i < \lambda\}$,

(β) $\{\delta_\varepsilon : \varepsilon \in C \cap \xi\} \in M_{\xi+1}$.

This suffices as $\langle C_i : i < \lambda \rangle \in M_0 \prec M_{\xi+1}$ and $M_{\xi+1} \cap \{C_i : i < \lambda\} = \{C_i : i \in \lambda \cap M_{\xi+1}\} = \{C_i : i < \delta_{\xi+1}\}$.

PROOF OF (α): Remember $\langle \alpha_\varepsilon : \varepsilon < \kappa \rangle \in N^*$. Also $\langle P_\varepsilon : \varepsilon < \kappa \rangle \in N^*$ hence $\bigcup_{\varepsilon < \kappa} P_\varepsilon \subseteq N^*$ (as $\kappa < \lambda$, $|P_\varepsilon| < \lambda, \mathrm{cf}\lambda = \lambda$) and $C \cap \xi \in \bigcup_{\varepsilon < \kappa} P_\varepsilon$; hence $C \cap \xi \in N^*$. Together $\{\alpha_\varepsilon : \varepsilon \in \xi \cap C\} \in N^*$; as $\varepsilon \in C \Rightarrow \varepsilon \in E \Rightarrow \alpha_\varepsilon = \delta_\varepsilon$ (from $C \subseteq E$ and the definition of E), and from the definition of $\langle C_i : i < \lambda \rangle$, we finish.

PROOF OF (β): We know $\overline{P} \in M_0$; as $|P_\varepsilon| < \lambda$, $\kappa < \lambda$ and $M_\varepsilon \cap \lambda$ is an ordinal, clearly $\bigcup_{\varepsilon < \kappa} P_\varepsilon \subseteq M_0$ (remember $|P_\varepsilon| < \lambda$, $\kappa < \lambda$). So for $\varepsilon < \zeta$, $C \cap \varepsilon \in \bigcup_{\gamma < \zeta} P_\gamma \subseteq M_0 \subseteq M_{\xi+1}$.

As $\langle M_i : i \leq \xi \rangle \in M_{\xi+1}$ clearly $\langle \delta_i : i \leq \xi \rangle \in M_{\xi+1}$ hence by the previous sentence $\langle \delta_i : i \in C \cap \xi \rangle \in M_{\xi+1}$, as required. $\qquad\qquad\qquad\qquad\qquad\qquad\qquad\qquad\square_{1.4}$

Conclusion 1.5 *If κ, λ are regular, $\kappa^+ < \lambda$ then there is a stationary $S \subseteq \{\delta < \lambda : \mathrm{cf}\delta = \kappa\}$ in $I[\lambda]$.*

Proof. If $\lambda = \kappa^{++}$ — use [9], 4.1. So assume $\lambda > \kappa^{++}$. By [9], 4.1 the pair (κ, κ^{++}) satisfies the assumption of 1.3 for $S = \{\delta < \kappa^{++} : \mathrm{cf}\delta = \kappa\}$; (i.e. κ, λ there stands for κ, κ^{++} here). Hence the conclusion of 1.3 holds for some $\overline{\mathcal{P}} = \langle \mathcal{P}_\alpha : \alpha < \kappa^{++} \rangle$, $|\mathcal{P}_\alpha| < |\kappa^{++}|$. Now apply 1.4 with (κ^{++}, λ) here standing for (κ, λ) there (we have just proved $I_{\overline{\mathcal{P}}}$ is a proper ideal, so assumption (ii) holds). Note:

$$(*) \quad \{\delta < \kappa^{++} : \mathrm{cf}\delta = \kappa\} \notin I_{\overline{\mathcal{P}}}.$$

Now the conclusion of 1.4 (see the "moreover" and choice of $\overline{\mathcal{P}}$, i.e.$(*)$) gives the desired conclusion. $\qquad\qquad\qquad\qquad\qquad\qquad\qquad\qquad\qquad\qquad\qquad\qquad\quad\square_{1.5}$

Conclusion 1.6 *If $\lambda > \kappa$ are uncountable regular, $\kappa^+ < \lambda$, then for some stationary $S \subseteq \{\delta < \lambda : \mathrm{cf}\delta = \kappa\}$ and some $\overline{\mathcal{P}} = \langle \mathcal{P}_\alpha : \alpha < \lambda \rangle$ we have: $\oplus_{\overline{\mathcal{P}},S}^{\lambda,\kappa}$ from the conclusion of 1.3 holds.*

Proof. As κ is regular apply 1.5 and then 1.3. $\qquad\qquad\qquad\qquad\qquad\square_{1.6}$

Now 1.6 was a statement I have long wanted to know, still sometimes we want to have "$C_\delta \subseteq E, \mathrm{otp}C = \delta(*)$", $\delta(*)$ not a regular cardinal. We shall deal with such problems.

Claim 1.7 *Suppose*

(i) $\lambda > \kappa > \aleph_0$, λ and κ are regular cardinals,

(ii) $\overline{\mathcal{P}}_\ell = \langle \mathcal{P}_{\ell,\alpha} : \alpha < \kappa \rangle$ for $\ell = 1, 2$, where $\mathcal{P}_{1,\alpha}$ is a family of $< \lambda$ closed subsets of α, $\mathcal{P}_{2,\alpha}$ is a family of $\leq \lambda$ clubs of α and $[C \in \mathcal{P}_{2,\alpha}$ & $\beta \in C \Rightarrow C \cap \beta \in \bigcup_{\gamma < \alpha} \mathcal{P}_{1,\gamma}]$,

(iii) $I_{\overline{\mathcal{P}}_1, \overline{\mathcal{P}}_2} =: \{S \subseteq \kappa :$ for some club E of κ, for no $\delta \in S \cap E$ is there $C \in \mathcal{P}_{2,\alpha}$, $C \subseteq E\}$ is a proper ideal on κ.

Then we can find $\overline{\mathcal{P}}_\ell^* = \langle \mathcal{P}_{\ell,\alpha}^* : \alpha < \lambda \rangle$ for $\ell = 1, 2$ such that:

(A) $\mathcal{P}_{1,\alpha}^*$ is a family of $< \lambda$ closed subsets of α;

(B) $\beta \in \mathrm{nacc}C$ & $C \in \mathcal{P}_{1,\alpha}^* \Rightarrow C \cap \beta \in \mathcal{P}_{1,\beta}^*$;

(C) $\mathcal{P}_{2,\delta}^*$ is a family of $\leq \lambda$ clubs of δ (for δ limit $< \lambda$) $[\beta \in \mathrm{nacc}C$ & $C \in \mathcal{P}_{2,\delta}^* \Rightarrow C \cap \beta \in \mathcal{P}_{1,\beta}^*]$;

(D) for every club E of λ, for some strictly increasing continuous sequence $\langle \delta_\zeta : \zeta \leq \kappa \rangle$ of ordinals $< \lambda$ we have

$$\{\zeta < \kappa : \zeta \text{ limit, and for some } C \in \mathcal{P}_{2,\zeta} \text{ we have:}$$

$$\{\delta_\varepsilon : \varepsilon \in C\} \in \mathcal{P}_{2,\delta_\zeta}^* \text{ (hence } [\xi \in \mathrm{nacc}C \Rightarrow \{\delta_\varepsilon : \varepsilon \in C \cap \xi\} \in \mathcal{P}_{1,\delta_\xi}^*])\}$$

$$\equiv \kappa \bmod I_{\overline{\mathcal{P}}_1, \overline{\mathcal{P}}_2};$$

(E) *we have e_δ a club of δ of order type $\mathrm{cf}(\delta)$ for any limit $\delta < \lambda$; such that for any $C \in \bigcup_{\alpha<\lambda} P^*_{2,\alpha}$ for some $\delta < \lambda$, $\mathrm{cf}\delta = \kappa$ and $C' \in \bigcup_{\beta<\kappa} P_{2,\beta}$ we have $C = \{\gamma \in e_\delta : \mathrm{otp}(e_\delta \cap \gamma) \in C'\}$.*

Proof. Same proof as 1.4. (Note that without loss of generality $[C \in P_{1,\alpha}$ & $\beta < \alpha < \kappa \Rightarrow C \cap \beta \in P_{1,\beta}]$). $\square_{1.7}$

Conclusion 1.8: *If $\delta(*)$ is a limit ordinal and $\lambda = \mathrm{cf}\lambda > |\delta(*)|^+$ then we can find $\overline{P}^*_\ell = \langle P^*_{\ell,\alpha} : \alpha < \lambda \rangle$ for $\ell = 1,2$ and stationary $S \subseteq \{\delta < \lambda : \mathrm{cf}\delta = \mathrm{cf}\delta(*)\}$ such that:*

$$\bigoplus^{\lambda,\delta(*)}_{\overline{P}^*_1,\overline{P}^*_2,S} : \begin{cases}
\text{(A)} & P^*_{1,\alpha} \text{ is a family of } < \lambda \text{ closed subsets of } \alpha \text{ each of order type } < \delta(*); \\
\text{(B)} & \beta \in \mathrm{nacc}C \ \& \ C \in P^*_{1,\alpha} \Rightarrow C \cap \beta \in P^*_{1,\beta}; \\
\text{(C)} & P^*_{2,\delta} \text{ is a family of } \leq \lambda \text{ clubs of } \delta \text{ (yes, maybe } = \lambda) \text{ of order type } \delta(*), \\
& \text{and } [\beta \in \mathrm{nacc}C \ \& \ C \in P^*_{2,\delta} \Rightarrow C \cap \beta \in P^*_{1,\beta}]; \\
\text{(D)} & \text{for every club } E \text{ of } \lambda, \text{ for some } \delta \in E \cap S, \ \mathrm{cf}\delta = \mathrm{cf}(\delta(*)) \text{ and there is} \\
& C \in P^*_{2,\beta} \text{ such that } C \subseteq E.
\end{cases}$$

Proof. If $\lambda = |\delta(*)|^{++}$ (or any successor of regulars) use [3], III, 6.4(2) or [13], 2.14(2) (c)&(d)). If $\lambda > |\delta(*)|^{++}$ let $\kappa = |\delta(*)|^{++}$ and let $S_1 = \{\delta < \kappa^{++} : \mathrm{cf}\delta = \mathrm{cf}\delta(*)\}$; applying the previous sentence we get $\overline{P}^*_1, \overline{P}^*_2$ satisfying $\bigoplus^{\kappa^{++},\delta(*)}_{\overline{P}^*_1,\overline{P}^*_2,S_1}$, hence satisfying the assumption of 1.7 so we can apply 1.7. $\square_{1.8}$

Definition 1.9 $^+\bigoplus^{\lambda,\delta(*)}_{\overline{P}_1,\overline{P}_2,S}$ is defined as in 1.8 except that we replace (C) by:

(C)$^+$ $P^*_{2,\delta}$ is a family of $< \lambda$ clubs of δ of order type $\delta(*)$.

Remark 1.9A Note that if $P_\alpha = P_{1,\alpha} \cup P_{2,\alpha}$, $|P_{2,\alpha}| \leq 1$, $P_{1,\alpha} = \{C \in P_\alpha : \mathrm{otp}C < \delta(*)\}$, $P_{2,\alpha} = \{C \in P_\alpha : \mathrm{otp}C = \delta(*)\}$ then $^+\bigoplus^{\lambda,\delta(*)}_{\overline{P}_1,\overline{P}_2,S} \Leftrightarrow \bigoplus^{\lambda,\delta(*)}_{\overline{P},S}$.

Claim 1.10 *Suppose $\lambda = \mathrm{cf}\lambda > |\delta(*)|^+$, $\delta(*)$ a limit ordinal, additively indecomposable (i.e. $\alpha < \delta(*) \Rightarrow \alpha + \alpha < \delta(*)$), $\bigoplus^{\lambda,\delta(*)}_{\overline{P}_1,\overline{P}_2,S}$ from 1.8 and*

$(*)$ $\alpha \in S \Rightarrow |P_{2,\alpha}| \leq |\alpha|$.

(Note: a non-stationary subset of S does not count; e.g. for λ successor cardinal the α with $|\alpha|^+ < \lambda$. Note: $^+\bigoplus^{\lambda,\delta(*)}_{\overline{P}_1,\overline{P}_2,S}$ holds by $(*)$ and if λ is successor then $^+\bigoplus^{\lambda,\delta(*)}_{\overline{P}_1,\overline{P}_2,S}$ suffices).

Then for some stationary $S_1 \subseteq S$ and $\overline{P} = \langle P_\alpha : \alpha < \lambda \rangle$ we have: $P_\alpha \subseteq P_{1,\alpha} \cup P_{2,\alpha}$ and:

$$^*\bigotimes^{\lambda,\delta(*)}_{\overline{P},S_1} : \begin{cases}
\text{(i)} & P_\alpha \text{ is a family of closed subsets of } \alpha, \ |P_\alpha| < \lambda; \\
\text{(ii)} & \mathrm{otp}C < \delta(*) \text{ if } C \in P_\alpha, \ \alpha \notin S_1; \\
\text{(iii)} & \text{if } \alpha \in S_1 \text{ then: } P_\alpha = \{C_\alpha\}, \ \mathrm{otp}C_\alpha = \delta(*), \ C_\alpha \text{ a club of } \alpha \text{ disjoint to } S_1; \\
\text{(iv)} & C \in P_\alpha \ \& \ \beta \in \mathrm{nacc}C \Rightarrow \beta \cap C \in P_\beta; \\
\text{(v)} & \text{for any club } E \text{ of } \lambda \text{ for some } \delta \in S_1 \text{ we have } C_\delta \subseteq E.
\end{cases}$$

Remark: Note there are two points we gain: for $\alpha \in S_1$, P_α is a singleton (as in 1.3), and an ordinal α cannot have a double role — C_α a guess (i.e. $\alpha \in S_1$) and C_α is a proper initial segment of such C_δ. When $\delta(*)$ is a regular cardinal this is easier.

Proof. Let $\mathcal{P}_{2,\alpha} = \{C_{\alpha,i} : i < \alpha\}$ (such a list exists as we have assumed $|\mathcal{P}_{2,\alpha}| \leq |\alpha|$, we ignore the case $\overline{\mathcal{P}}_{2,\alpha} = \emptyset$). Now

$(*)_0$ for some $i < \lambda$ for every club E of λ for some $\delta \in S \cap E$ we have $C_{\delta,i} \setminus E$ is bounded in α. [Why? If not, for every $i < \lambda$ there is a club E_i of λ such that for no $\delta \in S \cap E$ is $C_{\delta,i} \setminus E$ bounded in α . Let $E^* = \{j < \lambda : j$ a limit ordinal, $j \in \bigcap_{i<j} E_i\}$, it is a club of λ, hence for some $\delta \in S \cap E^*$ and $C \in \mathcal{P}_{2,\delta}$ we have $C \subseteq E^*$. So for some $i < \alpha$, $C = C_{\delta,i}$, so $C \subseteq E^* \subseteq E_i \cup i$ hence $C_{\delta,i} \setminus i \subseteq E_i$, contradicting the choice of E_i.]

$(*)_1$ for some $i < \lambda$ and $\gamma < \delta(*)$, letting $C_\delta =: C_{\delta,i} \setminus \{\zeta \in C_{\delta,i} : \text{otp}(\zeta \cap C_{\delta,i}) < \gamma\}$ we have: for every club E of λ, for some $\delta \in S \cap E$ we have: $C_\delta \subseteq E$. [Why? Let $i(*)$ be as in $(*)_0$, and for each $\gamma < \delta(*)$ suppose E_γ exemplify the failure of $(*)_1$ for $i(*)$ and γ, now $\bigcap_{\gamma<\delta(*)} E_\gamma$ is a club of λ exemplifying the failure of $(*)_0$ for $i(*)$, contradiction. So for some $\gamma < \delta(*)$ we succeed.]

$(*)_2$ Without loss of generality $|\mathcal{P}_{2,\alpha}| \leq 1$, so let $\mathcal{P}_{2,\alpha} = \{C_\alpha\}$. [Why? Let i, γ and C_δ (for $\delta \in S$) be as in $(*)_1$ and use $\mathcal{P}'_{1,\alpha} = \{C \setminus \{\zeta \in C : \text{otp}(\zeta \cap C) < \gamma\} : C \in \mathcal{P}_{1,\alpha}\}$, $\mathcal{P}'_{2,i} = \{C_\delta\}$.]

$(*)_3$ for some $h : \lambda \to |\delta(*)|^+$, for every $\alpha \in S$ we have $h(\alpha) \notin \{h(\beta) : \beta \in C_\alpha\}$. [Why? Choose $h(\alpha)$ by induction on α.]

$(*)_4$ for some $\beta < |\delta(*)|^+$, for every club E of λ, for some $\delta \in S \cap h^{-1}(\{\beta\})$, $C_\delta \subseteq E$. [Why? If for each β there is a counterexample E_β then $\bigcap\{E_\beta : \beta < |\delta(*)|^+\}$ is a counterexample for $(*)_2$.]

Now we have gotten the desired conclusion. $\square_{1.10}$

Claim 1.11 *If* $S \subseteq \{\delta < \lambda : \text{cf}\delta = \kappa\}$, $S \in I[\lambda]$, $\kappa^+ < \lambda = \text{cf}\lambda$, *then for some stationary* $S_1 \subseteq S$ *and* $\overline{\mathcal{P}}_1$ *we have* $^*\oplus_{\overline{\mathcal{P}}_1,S_1}^{\lambda,\delta(*)}$.

Proof. Same proof as 1.3 (plus $(*)_3$, $(*)_4$ in the proof of 1.8). $\square_{1.11}$

Claim 1.12 *Assume* $\lambda = \mu^+$, $|\delta(*)| < \mu$, $\text{cf}(\delta(*)) \neq \text{cf}\mu$. *Then we can find stationary* $S \subseteq \{\delta < \lambda : \text{cf}\delta = \text{cf}\delta(*)\}$ *and* $\overline{\mathcal{P}}$ *such that* $^*\otimes_{\overline{\mathcal{P}},S}^{\lambda,\delta(*)}$.

Remark: This strengthens 1.8.

Proof.
CASE (α): μ REGULAR.
By [3], III, 6.4(2)] or [13], 2.14(2) ((c)&(d)).

CASE β: μ SINGULAR.
Let $\theta =: \text{cf}\mu$, $\sigma =: |\delta(*)|^+ + \theta^+$ and $\mu = \sum_{\zeta<\theta} \mu_\zeta$, $\langle \mu_\zeta : \zeta < \theta \rangle$ strictly increasing, $\mu_0 > \sigma$ and for each $\alpha < \lambda$ let $\alpha = \bigcup_{\zeta<\theta} A_{\alpha,\zeta}$, $\langle A_{\alpha,\zeta} : \zeta < \theta \rangle$ increasing, $|A_{\alpha,\zeta}| \leq \mu_\zeta$.
By 1.6 there is a sequence $\overline{\mathcal{P}} = \langle \mathcal{P}_\alpha : \alpha < \lambda \rangle$ and stationary $S_1 \subseteq \{\delta < \lambda : \text{cf}(\delta) = \sigma\}$ such that $\oplus_{\overline{\mathcal{P}},S_1}^{\lambda,\sigma}$ of 1.3 holds. Let $\bigcup\{\mathcal{P}_\alpha : \alpha < \lambda\} \cup \{\emptyset\}$ be $\{C_\alpha : \alpha < \lambda\}$ such that $C_\alpha \subseteq \alpha$,

$[\alpha \in S_1 \Rightarrow \alpha = \sup C_\alpha \And C_\alpha \in \mathcal{P}_\alpha \And \text{otp} C_\alpha = \sigma]$ and $[\alpha \notin S_1 \Rightarrow \text{otp} C_\alpha < \sigma]$. For some club E_1^* of λ, $[\alpha \in E_1^* \Rightarrow \bigcup_{\beta<\alpha} \mathcal{P}_\beta = \{C_\beta : \beta < \alpha\}]$.

Looking again at $\bigoplus_{\bar{\mathcal{P}},S_1}^{\lambda,\sigma}$, we can assume $S_1 \subseteq E_1^*$ & $(\forall\delta)[\delta \in S_1 \Rightarrow C_\delta \subseteq E_1^*]\}$, hence because we can replace every C_α by $\{\beta \in C_\alpha : \text{otp}(\beta \cap C_\alpha)$ is even $\}$, without loss of generality

$$(*)\ [\delta \in S_1\ \And\ \alpha \in \text{nacc} C_\delta \Rightarrow \alpha \cap C_\delta \in \{C_\beta : \beta < \text{Min}(C_\delta \cap \alpha)\}].$$

Without loss of generality $[\beta \in A_{\alpha,\zeta} \Rightarrow C_\beta \subseteq A_{\alpha,\zeta}]$ (just note $|C_\beta| \le \sigma < \mu_\zeta$) and $\alpha \in A_{\beta,\zeta} \Rightarrow A_{\alpha,\zeta} \subseteq A_{\beta,\zeta}$. For $\alpha \in S_1$ let $C_\alpha = \{\beta_{\alpha,\varepsilon} : \varepsilon < \sigma\}$ ($\beta_{\alpha,\varepsilon}$ increasing in ε) and let $\beta_{\alpha,\varepsilon}^* \in [\beta_{\alpha,\varepsilon},\beta_{\alpha,\varepsilon+1})$ be minimal such that $C_\alpha \cap \beta_{\alpha,\varepsilon+1} = C_{\beta_{\alpha,\varepsilon}^*}$ (exists by $(*)$ above). Without loss of generality every C_α is an initial segment of some C_β, $\beta \in S_1$ (if not, we redefine it as $\emptyset$).

$(*)_1$ there are $\gamma = \gamma(*) < \theta$ and stationary $S_2 \subseteq S_1$ such that for every club E of λ, for some $\delta \in S_2$ we have: $C_\delta \subseteq E$, and for arbitrarily large $\varepsilon < \sigma$, $\beta_{\delta,\varepsilon}^* \in A_{\beta_{\delta,\varepsilon+1},\gamma}$. [Why? If not, for every $\gamma < \theta$ (by trying $\gamma(*) = \gamma$) there is a club E_γ of λ exemplifying the failure of $(*)_1$ for γ. Let $E = \bigcap_{\gamma<\theta} E_\gamma \cap E_1^*$, so E is a club of λ, hence

$$S' =: \{\delta : \delta < \lambda, \delta \in S_1 (\text{so cf}\delta = \sigma) \text{ and } C_\delta \subseteq E\}$$

is a stationary subset of λ. For each $\delta \in S'$ and $\varepsilon < \sigma$, for some $\gamma = \gamma(\delta,\varepsilon) < \theta$ we have $\beta_{\delta,\varepsilon}^* \in A_{\beta_{\delta,\varepsilon+1},\gamma}$, but as $\sigma = \text{cf}\sigma \ne \text{cf}\theta = \theta$ for some $\gamma(\delta)$, $\{\varepsilon < \sigma : \varepsilon\gamma(\delta,\varepsilon) = \gamma(\delta)\}$ is unbounded in σ. But $\delta \in E_{\gamma(\delta)}$, contradiction].

$(*)_2$ Without loss of generality: if $\beta \in \text{nacc} C_\alpha$, $\alpha < \lambda$ then $(\exists\xi \in A_{\beta,\gamma(*)})[\beta > \xi > \sup(\beta \cap C_\alpha)\ \And\ \beta \cap C_\alpha = C_\xi]$. [Why? Define C_α' for $\alpha < \lambda$:

$$C_\alpha^0 = \{\beta : \beta \in \text{nacc} C_\alpha \text{ and } (\exists\xi \in A_{\beta,\gamma(*)})[\beta > \xi \ge \sup(\beta \cap C_\alpha)\ \And\ \beta \cap C_\alpha = C_\xi]\}.$$

$$C_\alpha' \text{ is: } \begin{cases} \emptyset & \text{if } \alpha \in S_2, \alpha > \sup C_\alpha^0, \\ \alpha \cap \text{closure of } C_\alpha^0 & \text{otherwise.} \end{cases}$$

Now $\langle C_\alpha : \alpha < \lambda \rangle$ can be replaced by $\langle C_\alpha' : \alpha < \lambda \rangle$].

$(*)_3$ For some $\gamma_1 = \gamma_1(*) < \theta$, for every club E of λ, for some $\delta \in E$: $\text{cf}(\delta) = \text{cf}(\delta(*))$, and there is a club e of δ satisfying: $e \subseteq E$, $\text{otp}(e)$ is $\delta(*)$, and for arbitrarily large $\beta \in \text{nacc}(e)$ we have $e \cap \beta \in \{C_\zeta : \zeta \in A_{\delta,\gamma_1}\}$. [Why? If not, for each $\gamma_1 < \theta$ there is a club E_{γ_1} of λ for which there is no δ as required. Let $E =: \bigcap_{\gamma_1<\theta} E_{\gamma_1}$, so E is a club of λ, hence for some $\alpha \in \text{acc}(E) \cap S_2$, $C_\alpha \subseteq E$. Letting again $C_\alpha = \{\beta_{\alpha,\varepsilon} : \varepsilon < \sigma\}$ (increasing), $C_\alpha \cap \beta_{\alpha,\varepsilon} = C_{\delta,\beta_{\delta,\varepsilon}^*}$ where $\beta_{\delta,\varepsilon}^* \in A_{\beta_{\delta,\varepsilon+1},\gamma(*)}$ clearly $\delta =: \beta_{\alpha,\delta(*)}$, $e = \{\beta_{\delta,\varepsilon} : \varepsilon < \delta(*)\}$ satisfies the requirements except the last. As $\text{cf}(\delta(*)) \ne \text{cf}(\mu)$, for some $\gamma_1(*) < \theta, \gamma_1(*) \ge \gamma(*)$ and $\{\varepsilon < \delta(*) : \beta_{\delta,\varepsilon}^* \in A_{\beta_{\delta,\delta(*)},\gamma_1(*)}\}$ is unbounded in $\delta(*)$. Clearly $\delta =: \beta_{\alpha,\delta(*)}$, $e =: C_\alpha \cap \delta$ satisfies the requirement. Now this contradicts the choice of $E_{\gamma_1(*)}$].

$(*)_4$ For some club E^a of λ, for every club $E^b \subseteq E^a$ of λ, for some $\delta \in E^b$ we have:

(a) $\text{cf}(\delta) = \text{cf}(\delta(*))$;

(b) for some club e of δ : $e \subseteq E^b$, $\mathrm{otp}(e) = \delta(*)$, and for arbitrarily large $\beta \in \mathrm{nacc}(e)$ we have $e \cap \beta \in \{C_\xi : \varepsilon \in A_{\delta,\gamma_1(*)}\}$;

(c) for every $\beta \in A_{\delta,\gamma_1(*)}$ we have: $C_\beta \subseteq E^a \Rightarrow C_\beta \subseteq E^b$ (we could have demanded $C_\beta \cap E^a = C_\beta \cap E^b$). [Why? If not we choose E_i for $i < \mu^+_{\gamma_1(*)}$ by induction on i, $[j < i \Rightarrow E_i \subseteq E_j]$, E_i a club of λ, and E_{i+1} exemplify the failure of E_i as a candidate for E^a. So $\bigcap_i E_i$ is a club of λ hence by $(*)_3$ there are δ and e as there. Now $\langle \{\beta \in A_{\delta,\gamma_1(*)} : C_\beta \subseteq E_i\} : i < \mu^+_{\gamma_1(*)}\rangle$ is a decreasing sequence of subsets of $A_{\delta,\gamma_1(*)}$ of length $\mu^+_{\gamma_1(*)}$, and $|A_{\delta,\gamma_1(*)}| \le \mu_{\gamma_1(*)}$, hence it is eventually constant. So for every i large enough, δ contradicts the choice of E_{i+1}].

Let $S = \{\delta < \lambda : \mathrm{cf}(\delta) = \mathrm{cf}(\delta(*))$, and there is a club $e = e_\delta$ of δ satisfying: $e \subseteq E^a$, $\mathrm{otp}(e) = \delta(*)$, $\alpha \in \mathrm{nacce} \Rightarrow e \cap \alpha \in A_{\alpha,\gamma(*)}$ and for arbitrarily large $\beta \in \mathrm{nacc}(e)$ we have $e \cap \beta \in \{C_\xi : \xi \in A_{\delta,\gamma(*)}\}\}$. So S is stationary, let for $\delta \in S$, C^*_δ be an e as above. For $\alpha < \lambda$ let $\mathcal{P}_{1,\alpha} = \{C_\beta : \beta \le \alpha, \beta \in A_{\alpha,\gamma_2(*)}\}$.

$(*)_5$

(a) for every club E of λ, for some $\delta \in S$, $C^*_\delta \subseteq E$;

(b) C^*_δ is a club of δ, $\mathrm{otp}(C^*_\delta) = \delta(*)$;

(c) if $\beta \in \mathrm{nacc}C^*_\delta (\delta \in S)$ then $C^*_\delta \cap \beta \in \mathcal{P}_{1,\beta}$;

(d) $|\mathcal{P}_{1,\beta}| \le \mu_{\gamma(*)}$, $\mathcal{P}_{1,\beta}$ is a family of closed subsets of β of order type $< \delta(*)$.

[Why? This is what we have proved.]

Now repeating $(*)_3$, $(*)_4$ of the proof of 1.10, and we finish. $\qquad\square_{1.12}$

Claim 1.13

(1) *Assume* $\lambda = \mu^+$, $|\delta(*)| < \mu$, $\aleph_0 < \mathrm{cf}(\delta(*)) = \mathrm{cf}(\mu)(< \mu)$; *then we can find stationary* $S \subseteq \{\delta < \lambda : \mathrm{cf}\delta = \mathrm{cf}(\delta(*))\}$ *and* $\overline{P}$ *such that* $^*\otimes^{\lambda,\delta(*)}_{\overline{P},S}$, *except when:*

$\oplus$ *for every regular* $\sigma < \mu$, *we can find* $h : \sigma \to \mathrm{cf}(\mu)$ *such that for no* δ, ϵ *do we have: if* $\delta < \sigma$, $\mathrm{cf}(\delta) = \mathrm{cf}(\mu)$, $\epsilon < \mathrm{cf}\mu$ *then* $\{\alpha < \delta : h(\alpha) < \epsilon\}$ *is not a stationary subset of* δ.

(2) *In 1.12 and 1.13(1) we can have* $\mu > \sup_{\alpha<\lambda} |\mathcal{P}_\alpha|$.

(3) *If 1.13(2), if* μ *is strong limit we can have* $|\mathcal{P}_\alpha| \le 1$ *for each* α.

Remark Compare with [7], §3.

Proof. Left to the reader (reread the proof of 1.12 and [7], §3).

Claim 1.14 *Let* κ *be regular uncountable. We can choose for each regular* λ, $\overline{\mathcal{P}}^\lambda = \langle \mathcal{P}^\lambda_\alpha : \alpha < \lambda\rangle$ *(assuming global choice) such that:*

(a) *for each* λ, $\mathcal{P}^\lambda_\alpha$ *is a family of* $\le \lambda$ *of closed subsets of* α *of order type* $< \kappa$.

(b) *if χ is regular, F is the function $\lambda \mapsto \overline{\mathcal{P}}^\lambda$ (for λ regular $< \chi$), $\aleph_0 < \kappa = \mathrm{cf}\kappa$, $\kappa^{++} < \chi$, $x \in H(\chi)$ then we can find $\overline{N} = \langle N_i : i \leq \kappa \rangle$, an increasing continuous chain of elementary submodels of $(H(\chi), \in, <_\chi^*, F)$, $\langle N_j : j \leq i \rangle \in N_{i+1}$, $\|N_i\| = \aleph_0 + |i|$, $x \in N_0$ such that:*

    (*) *if $\kappa^+ < \theta = \mathrm{cf}\theta \in N_i$, then for some club C of $\sup(N_\kappa \cap \theta)$ of order type κ, for any $j_1^i < j < \kappa$ we have:*

$$C \cap \sup(N_j \cap \theta) \in N_{j+1} \text{ and } \mathrm{otp}(C \cap \sup(N_j \cap \theta)) = j.$$

Proof. Let $\langle C_\alpha : \alpha \in S \rangle$ be such that $S \subseteq \{\alpha \leq \kappa^{++} : \mathrm{cf}\alpha \leq \kappa\}$ is stationary, $\mathrm{otp}C_\alpha \leq \kappa$, $[\beta \in C_\alpha \Rightarrow C_\beta = \beta \cap C_\alpha]$, C_α a closed subset of α, $[\alpha \text{ limit} \Rightarrow \alpha = \sup C_\alpha]$, $\{\alpha \in S : \mathrm{cf}\alpha = \kappa\}$ stationary, and for every club E of κ^{++} there is $\delta \in S$, $\mathrm{cf}(\delta) = \kappa$, $C_\delta \subseteq E$. For $i \in \kappa^{++} \backslash S$ let $C_i = \emptyset$. Now for every regular $\lambda > \kappa^+$ and $\alpha \leq \lambda$, let $e_\alpha^\lambda \subseteq \alpha$ be a club of α for $\alpha \leq \lambda$ limit and let

$$\overline{\mathcal{P}}_\alpha^\lambda = \{\{i \in e_\delta : i < \alpha, \mathrm{otp}(e_\delta \cap i) \in C_\beta\} : \delta < \lambda \text{ has cofinality } \kappa^{++}, \text{ and } \beta \in S\}.$$

Given $x \in H(\chi)$, we choose by induction on $i < \kappa^{++}$, M_i, N_i such that:

$N_i \prec M_i \prec (H(\chi), \in, <_\chi^*, F)$,

$\|M_i\| = |i| + \aleph_0$,

$\|N_i\| = |C_i| + \aleph_0$,

$M_i(i < \kappa^{++})$ is increasing continuous,

$x \in M_0$,

$\langle M_j : j \leq i \rangle \in M_{i+1}$,

N_i is the Skolem Hull of $\{\langle N_j : j \in C_\zeta \rangle : \zeta \in C_i\}$.

We leave the checking to the reader. $\square_{1.14}$

2 Measuring $S_{<\kappa}(\lambda)$

We prove that two natural ways to measure $S_{<\kappa}(\lambda)$ (κ regular uncountable) give the same cardinal: the minimal cardinality of a cofinal subset; i.e. its cofinality (i.e. $\mathrm{cov}(\lambda, \kappa, \kappa, 2)$) and the minimal cardinality of a stationary subset. The theorem is really somewhat stronger: for appropriate normal ideal on $S_{<\kappa}(\lambda)$, some member of the dual filter has the right cardinality.

The problem is natural and I did not trace its origin, but until recent years it seems (at least to me) it surely is independent, and I find it gratifying we get a clean answer. I thank P. Matet and M. Gitik for reminding me of the problem.

We then find applications to Δ-systems and largeness of $I[\lambda]$.

Definition 2.1

(1) $(\overline{C}, \overline{\mathcal{P}}) \in \mathcal{T}^*[\theta, \kappa]$ if

    (i) $\aleph_0 < \kappa = \mathrm{cf}\kappa < \theta = \mathrm{cf}\theta$,

(ii) $\overline{C} = \langle C_\delta : \delta \in S \rangle$, $\overline{\mathcal{P}} = \langle \mathcal{P}_\delta : \delta \in S \rangle$,

(iii) $S \subseteq \theta$, S is stationary (we shall write $S = S(\overline{C})$),

(iv) C_δ is an unbounded subset of δ (not necessarily closed),

(v) $id^a(\overline{C})$ is a proper ideal (i.e. for every club E of θ for some $\delta \in S$, $C_\delta \subseteq E$),

(vi) $\bigwedge_{\delta \in S} \text{otp} C_\delta < \kappa$ (hence $[\delta \in S \Rightarrow \text{cf}(\delta) < \kappa]$),

(vii) $\mathcal{P}_\delta$ is a directed family of bounded subsets of C_δ, $\bigcup_{x \in \mathcal{P}_\delta} x = C_\delta$, and $|\mathcal{P}_\delta| < \kappa$,

(viii) for every $\alpha < \theta$ the set

$$\mathcal{P}_\alpha^* =: \{a \cap \alpha : \text{for some } \delta \in S \text{ we have } \alpha < \delta \in S, a \in \mathcal{P}_\delta \text{ and } \alpha \in C_\delta\}$$

has cardinality $< \theta$ or at least

(viii)$^-$ for some list $\langle a_i : i < \theta \rangle$ of $\bigcup_\alpha \mathcal{P}_\alpha$ we have: $\mathcal{P}_\alpha \subseteq \{a_j : j < \alpha\}$,

(ix) for $x \in \bigcup_{\delta \in S} \mathcal{P}_\delta$, $|\{y \in \bigcup_{\delta \in S} \mathcal{P}_\delta : y \subseteq x\}| < \kappa$.

(2) $\overline{C} \in \mathcal{T}^0[\theta, \kappa]$ if $(\overline{C}, \overline{\mathcal{P}}) \in \mathcal{T}^*[\theta, \kappa]$ with $\mathcal{P}_\delta = \{C_\delta \cap \alpha : \alpha \in C_\delta\}$ or at least $\mathcal{P}_\delta = \{C_\delta \cap \alpha : C_\delta \cap \alpha \text{ has a least element}\}$.

(3) $\overline{C} \in \mathcal{T}^1[\theta, \kappa]$ if $(\overline{C}, \overline{\mathcal{P}}) \in \mathcal{T}^*[\theta, \kappa]$ with $\mathcal{P}_\delta = S_{<\aleph_0}(C_\delta)$.

Note that:

Claim 2.2

(1) If $\theta = \text{cf}\theta > \kappa = \text{cf}\kappa > \sigma = \text{cf}\sigma$, then there is $\overline{C} \in \mathcal{T}^1[\theta, \kappa]$ such that:

$$\{\delta \in S(\overline{C}) : \text{cf}\delta = \sigma\} \neq \emptyset \mod id^a(\overline{C}).$$

(2) If $S \subseteq \{\delta < \theta : \text{cf}\delta < \kappa\}$ is stationary, $\overline{C}$ an S-club system, $|C_\delta| < \kappa$, and $id^a(\overline{C})$ a proper ideal, then $\overline{C} \in \mathcal{T}^1[\theta, \kappa]$.

(3) In (2) if in addition $|\{C_\delta \cap \alpha : \alpha \in C_\delta, \delta \in S\}| < \theta$ then $\overline{C} \in \mathcal{T}^0[\theta, \kappa]$.

(4) In part (1) if θ is a successor of regular then we can demand $\overline{C} \in \mathcal{T}^0[\theta, \kappa]$ each C_δ closed.

(5) In part (1) if $\theta = \text{cf}\theta > \kappa = \text{cf}\kappa > \sigma = \text{cf}\sigma$ then there is $\overline{C} \in \mathcal{T}^0[\theta, \kappa]$ such that: $\{\delta \in S(\overline{C}) : \text{cf}\delta = \sigma\} \neq \emptyset \mod id^a(\overline{C})$.

Proof.

(1) By [13], §2 and then part (2).

(2) Check.

(3) Check.

(4) By [3], III, 6.4(2) (or [13], 2.14(2) ((c)&(d))).

(5) By 1.5 and 1.11 (so we use the non-accumulation points).

Remember (see [Sh52], §3)

Definition 2.3 (1) $\mathcal{D}^{\kappa}_{<\kappa}(\lambda)$ is the filter on $\mathcal{S}_{<\kappa}(\lambda)$ defined by:
for $X \subseteq \mathcal{S}_{<\kappa}(\lambda)$:
$X \in \mathcal{D}^{\kappa}_{<\kappa}(\lambda)$ iff there is a function F from $\bigcup_{\zeta<\kappa} {}^{\zeta}[\mathcal{S}_{<\kappa}(\lambda)]$ to $\mathcal{S}_{<\kappa}(\lambda)$ such that: if $a_{\zeta} \in \mathcal{S}_{<\kappa}(\lambda)$ for $\zeta < \kappa$, is increasing continuous and for each $\zeta < \kappa$ we have $F(\langle \ldots, a_{\xi}, \ldots \rangle)_{\xi \leq \zeta} \subseteq a_{\zeta+1}$ then $\{\zeta < \kappa : a_{\zeta} \in X\} \in \mathcal{D}_{\kappa}$ ($\mathcal{D}_{\kappa}$ the filter generated by the family of clubs of κ).

Similarly

Definition 2.4 For $\lambda \geq \theta = \mathrm{cf}\theta > \kappa = \mathrm{cf}\kappa > \aleph_0$, $(\overline{C}, \overline{\mathcal{P}}) \in \mathcal{T}^*[\theta, \kappa]$ we define a filter $\mathcal{D}_{(\overline{C},\overline{\mathcal{P}})}(\lambda)$ on $\mathcal{S}_{<\kappa}(\lambda)$; (let $\chi = \beth_{\omega+1}(\lambda)$):
$Y \in \mathcal{D}_{(\overline{C},\overline{\mathcal{P}})}(\lambda)$ iff $Y \subseteq \mathcal{S}_{<\kappa}(\lambda)$ and for some $x \in H(\chi)$, for every $\langle N_{\alpha}, N^*_a : \alpha < \theta, a \in \bigcup_{\delta \in S} \mathcal{P}_{\delta} \rangle$ satisfying $\otimes$ below, and also $[a \in \mathcal{P}_{\delta} \,\&\, \delta \in S \,\&\, \alpha < \kappa \Rightarrow x \in N^*_a \,\&\, x \in N_{\alpha}]$, there is $A \in \mathrm{id}^a(\overline{C})$ such that: $\delta \in S(\overline{C}) \backslash A \Rightarrow \bigcup_{a \in \mathcal{P}_{\delta}} N^*_a \cap \lambda \in Y$, where

$$\otimes : \begin{cases}
\text{(i)} & N_{\alpha} \prec (H(\chi), \in, <^*_{\chi}); \\
\text{(ii)} & \|N_{\alpha}\| < \theta, \; N_{\alpha} \cap \theta \text{ an initial segment}; \\
\text{(iii)} & \langle N_{\beta} : \beta \leq \alpha \rangle \in N_{\alpha+1}; \\
\text{(iv)} & N_{\alpha} \text{ increasing continuous}; \\
\text{(v)} & N^*_a \prec (H(\chi), \in, <^*_{\chi}) \text{ for } a \in \bigcup_{\delta \in S} \mathcal{P}_{\delta}; \\
\text{(vi)} & \|N^*_a\| < \kappa, \; N^*_a \cap \kappa \text{ an initial segment}; \\
\text{(vii)} & b \subseteq a \text{ (both in } \bigcup_{\delta \in S} \mathcal{P}_{\delta}) \text{ implies } N^*_b \prec N^*_a; \\
\text{(viii)} & \text{if } \alpha \in a \in \bigcup_{\delta \in S} \mathcal{P}_{\delta} \text{ then } \langle N_{\beta}, N^*_b : \beta \leq \alpha, b \subseteq \alpha, b \in \bigcup_{\delta \in S} \mathcal{P}_{\delta} \rangle \\
& \text{belongs to } N^*_a; \\
\text{(ix)} & \langle N_{\beta}, N^*_b : \beta \leq \alpha, b \subseteq \alpha + 1, b \in \bigcup_{\delta \in S} \mathcal{P}_{\delta} \rangle \text{ belongs to } N_{\alpha+1}; \\
\text{(x)} & a \subseteq N^*_a \text{ and } \alpha \in a \Rightarrow \alpha \cap a \in N^*_a; \\
\text{(xi)} & a \subseteq \alpha, a \in \bigcup_{\delta \in S} \mathcal{P}_{\delta} \text{ implies } N^*_a \in N_{\alpha+1} \text{ (remember (viii) of 2.1)}.
\end{cases}$$

Clearly

Claim 2.5

(1) Any $\chi \geq 2^{\lambda}$ can serve, and $x = (Y, \lambda, \overline{C}, \overline{\mathcal{P}})$ is enough.

(2) $\mathcal{D}_{(\overline{C},\overline{\mathcal{P}})}(\lambda)$ is a fine normal filter on $\mathcal{S}_{<\kappa}(\lambda)$ when $(\overline{C}, \overline{\mathcal{P}}) \in \mathcal{T}^*[\theta, \kappa]$, $\lambda \geq \theta$, hence it extends $\mathcal{D}_{<\kappa}(\lambda)$. (Remember $\mathrm{id}^a(\overline{C})$ is a proper ideal.)

Theorem 2.6 Suppose $\lambda > \kappa = \mathrm{cf}\kappa > \aleph_0$. Then the following three cardinals are equal for $(\overline{C}, \overline{\mathcal{P}}) \in \mathcal{T}^*[\kappa^+, \kappa]$:

$\mu(0) = \mathrm{cf}(\mathcal{S}_{<\kappa}(\lambda), \subseteq)$,

$\mu(1) = \mathrm{cov}(\lambda, \kappa, \kappa, 2) = \min\{|\mathcal{P}| : \mathcal{P} \subseteq \mathcal{S}_{<\kappa}(\lambda), \text{ and for every } a \subseteq \lambda, |a| < \kappa, \text{ there is } b \in \mathcal{P}, a \subseteq b\}$,

$$\mu(2) = \min\{|S| : S \subseteq \mathcal{S}_{<\kappa}(\lambda) \ \text{is stationary}\},$$

$$\mu(3) = \mu_{(\overline{C},\overline{\mathcal{P}})} = \min\{|Y| : Y \in \mathcal{D}_{(\overline{C},\overline{\mathcal{P}})}(\lambda)\}.$$

Remark 2.6A

(1) It is well known that if $\lambda > 2^{<\kappa}$ then the equality holds.

(2) This is close to "strong covering".

(3) In the proof we may replace "$\theta = \kappa^+$" by "$\lambda > \theta = \mathrm{cf}\theta > \kappa$" if $\alpha < \theta \Rightarrow \mathrm{cov}(\alpha, \kappa, \kappa, 2) < \theta$.

(4) Note if $\lambda = \kappa$, then $\mu(1) = \mu(2)$ trivially.

(5) Note that only $\mu(3)$ has $(\overline{C}, \overline{\mathcal{P}})$ in its definition, so actually $\mu(3)$ does not depend on $(\overline{C}, \overline{\mathcal{P}})$.

Remark 2.6B

(1) We can weaken in Definition 2.1(1) demand (ix) as follows:

(ix) there is a sequence $\langle a_i, \mathcal{P}_i^* : i < \lambda \rangle$ such that

   (a) $|a_i| < \kappa$, $\mathcal{P}_i^*$ is a family of $< \kappa$ subsets of a_i;
   (b) for every δ and $x \in \mathcal{P}_\delta$, for some $i < \delta$, $a_i = x$ and

$$(\forall b)[b \in \mathcal{P}_\delta \ \& \ b \subseteq a \Rightarrow b \in \mathcal{P}_i^*].$$

In this case 2.6, 2.6A(3) (and 2.5) remain true and we can strengthen 2.2.

(2) We can even use another order on $\mathcal{P}_\delta$ (not $\subseteq$).

Proof. Clearly $\lambda \leq \mu(0) = \mu(1) \leq \mu(2) \leq \mu(3)$ (the last by 2.5(2)). So we shall prove $\mu(3) \leq \mu(1)$, (suffices by 2.2(1)) and let $\mathcal{P}$ exemplify $\mu(1) = \mathrm{cov}(\lambda, \kappa, \kappa, 2)$.

Let χ be e.g. $\beth_3(\lambda)^+$, M_λ^* be the model with universe $\lambda + 1$ and all functions definable in $(H(\chi), \in, <_\chi^*, \lambda, \kappa, \mu(1))$. Let M^* be an elementary submodel of $(H(\chi), \in, <_\chi^*)$ of cardinality $\mu(1)$, $\mathcal{P} \in M^*$, $M_\lambda^* \in M^*$, $(\overline{C}, \overline{\mathcal{P}}) \in M^*$ and $\mu(1) + 1 \subseteq M^*$ hence $\mathcal{P} \subseteq M^*$. It is enough to prove that $M^* \cap \mathcal{S}_{<\kappa}(\lambda)$ belongs to $\mathcal{D}_{(\overline{C},\overline{\mathcal{P}})}(\lambda)$.

So let N_i (for $i < \kappa^+$), N_a^* (for $a \in \bigcup_{\delta \in S} \mathcal{P}_\delta$) be such that: they satisfy $\otimes$ of 2.4 and M_λ^*, M^*, $\mathcal{P}$, λ, κ, $\overline{C}$, $\overline{\mathcal{P}}$ belong to every N_α, N_a^*. It is enough to prove that $\{\delta < \kappa^+ : \lambda \cap \bigcup_{a \in \mathcal{P}_\delta} N_x^* \in M^*\} = \kappa^+ \bmod \mathrm{id}^a(\overline{C})$.

For each $i \in S$ there is a set a_i such that $(\bigcup_{y \in \mathcal{P}_i} N_y^*) \cap \lambda \subseteq a_i \in \mathcal{P}$; so without loss of generality $a_i \in N_{i+1}$. Let $\mathfrak{a}_i =: \mathrm{Reg} \cap a_i \cap \lambda^+ \backslash \kappa^{++}$, so $\mathfrak{a}_i$ is a set of $< \kappa$ regular cardinals $> \kappa^+$ and $\mathfrak{a}_i \in N_{i+1}$ too, so there is $\langle \mathfrak{b}_\lambda[\mathfrak{a}_i] : \lambda \in \mathrm{pcf}\mathfrak{a}_i \rangle$ as in [13], 2.6, without loss of generality it is definable from $\mathfrak{a}_i$ (in $(H(\chi), \in, <_\chi^*)$). Also $a \in \mathcal{P} \subseteq M^*$ so $a \in M^*$, so $\mathfrak{a} \in M^*$. Hence $\langle \mathfrak{b}_\lambda[\mathfrak{a}_i] : \lambda \in \mathrm{pcf}\mathfrak{a}_i \rangle \in N_{i+1} \cap M^*$, and also there is $\langle f_{\theta,\alpha}^{\mathfrak{a}_i} : \alpha < \theta, \theta \in \mathrm{pcf}\mathfrak{a}_i \rangle$ as in [13], 1.2, and again without loss of generality it belongs to $N_{i+1} \cap M^*$. As $\max \mathrm{pcf}\mathfrak{a}_i \leq \mathrm{cov}(\lambda, \kappa, \kappa, 2) \leq \mu(1)$ (first inequality by [10], 5.4), clearly each $f_{\theta,\alpha}^{\mathfrak{a}_i} \in M^*$. Let h be the function with domain $\bigcup_{i \in S} \mathfrak{a}_i$, $h(\theta) = \sup(\theta \cap \bigcup_{i < \kappa^+} N_i)$. So by [13], 2.3(1) each $h\lceil \mathfrak{a}_i$ has the form $\max\{f_{\theta_\ell,\alpha_\ell}^{\mathfrak{a}_i} : \ell < n\}$ hence belongs to M^*. Let e be a definable function in $(H(\chi), \in, <_\chi^*, \lambda, \kappa)$, Dom $e = \lambda + 1$, e_α is a club of α of order type cfα, enumerated as $\langle e_\alpha(\zeta) : \zeta < \text{cf}\alpha \rangle$. Now for each $\theta \in \bigcup_{i<\kappa^+} \mathfrak{a}_i$,

$$E_\theta =: \{ i < \kappa^+ : (\forall \zeta < \kappa^+)[[e_{h(\theta)}(\zeta) \in N_i \Leftrightarrow \zeta < i], \ i \text{ is limit,}$$

$$\theta \in \bigcup_{j<i} \mathfrak{a}_j \text{ and } \sup(N_i \cap \theta) = \sup\{e_{h(\theta)}(\zeta) : \zeta < i\}]\}$$

is a club of κ^+, hence

$$E = \{\delta < \kappa^+ : \delta \text{ limit and } [\theta \in (\bigcup\{N_y : \text{ for some } \alpha \in S \text{ and } \zeta \in C_\alpha \cap \delta$$

we have $\sup y < \zeta, \ y \in \mathcal{P}_\alpha\}) \ \& \ \theta \in \text{Reg} \cap \lambda^+\backslash\kappa^{++} \Rightarrow \delta \in E_\theta]$ and $N_\delta \cap \kappa^+ = \delta\}$

is a club of κ^+ (note: we use (viii) of Definition 2.1(1)). For each $\delta \in E \cap S$ such that $C_\delta \subseteq E$, let $\delta^* = \sup(\kappa \cap \bigcup_{y \in \mathcal{P}_\delta} N_y^*)$ so $\delta^* < \kappa$, and we define by induction on n models $M_{y,\delta,n}$ for every $y \in \mathcal{P}_\delta$ (really, they do not depend on δ). Now $M_{y,\delta,0}$ is the Skolem Hull in M_λ^* of $\{i : i \in y\} \cup \{j : j < \delta^*\}$. $M_{y,\delta,n+1}$ is the Skolem Hull in M_λ^* of

$$M_{y,\delta,n} \cup \{e_\theta(\zeta) : \theta \in (\text{Reg} \cap \lambda^+\backslash\kappa^{++}) \cap M_{y,\delta,n} \text{ and } \zeta \in y\}.$$

Now (A), (B), (C), (D), (E) below suffice to finish.

(A) We can easily prove by induction on n that:

(a) for $y \in \mathcal{P}_\delta$ we have $M_{y,\delta,n} \subseteq \bigcup_{y \in \mathcal{P}_\delta} N_y^*$;

(b) for $z \subseteq y$ in $\mathcal{P}_\delta$ we have $M_{z,\delta,n} \subseteq M_{y,\delta,n}$;

(c) for $y \in \mathcal{P}_\delta$ and $m < n$ we have $M_{y,\delta,m} \subseteq M_{y,\delta,n}$;

(d) assume $i \in y$ (hence $i \in E$), $\{y, z\} \subseteq \mathcal{P}_\delta$, $\sup z < i$, $z \subseteq y$ and $\theta \in \bigcup_{i<\kappa^+} N_i \cap \text{Reg} \cap \lambda^+\backslash\kappa^{++}$; then:
$\theta \in N_z^* \cap N_i \Rightarrow e_{h(\theta)}(i) = \sup(\theta \cap N_i) \in N_y^*$ and
$\theta \in M_{z,\delta,n} \cap N_i \Rightarrow \sup(M_{z,\delta,n} \cap \theta) \leq e_{h(\theta)}(i) \leq \sup(M_{y,\delta,n} \cap \theta)$.

(B) We can also prove that $\langle M_{y,\delta,n} : n < \omega, y \in \mathcal{P}_\delta \rangle$ is definable in $(H(\chi), \in, <_\chi^*)$ from the parameters δ, M_λ^*, $(\overline{C}, \overline{\mathcal{P}})$ and $h\restriction\mathfrak{a}_i$, all of them belong to M^*, hence the sequence, and $\bigcup_{n<\omega, y \in \mathcal{P}_\delta} M_{y,\delta,n}$ belongs to M^*.

(C) $(\bigcup_{n<\omega, y \in \mathcal{P}_\delta} M_{y,\delta,n}) \cap \text{Reg} \cap (\kappa^+, \lambda^+)$ is a subset of $\mathfrak{a}_i$ (use (A)(a) and definition of $a_i, \mathfrak{a}_i$).

(D) if $\sigma \in \bigcup_{n<\omega, y \in \mathcal{P}_\delta} M_{y,\delta,n}$, $\sigma \in \text{Reg} \cap \lambda^+\backslash\kappa$ then $\sigma \cap \bigcup_{n<\omega} M_{y,\delta,n}$ is unbounded in $\sigma \cap \bigcup_{y \in \mathcal{P}_\delta} N_\delta^*$ [when $\sigma > \kappa^+$ use $(*)$, for $\sigma = \kappa^+$ as C_δ is equal to $\bigcup_{y \in \mathcal{P}_\delta} y$ and $\delta = \sup C_\delta$, for $\sigma = \kappa$ see (d), choice of $M_{y,\delta,0}$].

(E) $\bigcup_{n<\omega, y \in \mathcal{P}_\delta} M_{y,\delta,n} \cap \lambda = \bigcup_{y \in \mathcal{P}_\delta} N_y^* \cap \lambda$. (See [14], 3.3A, 5.1A). $\square_{2.6}$

Conclusion 2.7 Suppose $\lambda > \kappa > \aleph_0$ are regular cardinals and $(\forall \mu < \lambda)[\text{cov}(\mu, \kappa, \kappa, 2) < \lambda]$. If for $\alpha < \lambda$, a_α is a subset of λ of cardinality $< \kappa$ and $S \in \mathcal{D}_{<\kappa}(\lambda)$ (or just $S \neq \emptyset$

mod $\mathcal{D}^\kappa_{\leq\kappa}(\lambda))$ *then we can find a stationary* $T \subseteq \{\delta < \lambda : \mathrm{cf}\delta = \kappa\}$, $c \subseteq \lambda$ *and* $\langle b_\delta : \delta \in T \rangle$ *such that:*

$$a_\delta \subseteq b_\delta \in S \ for \ \delta \in T$$

and

$$b_\delta \cap \delta = c \ for \ \delta \in T.$$

Remark: See on this and on 2.9 Rubin Shelah [2] and [12], §6.

Conclusion 2.8 *If* $\lambda > \kappa > \aleph_0$, λ *and* κ *are regular cardinals and* $[\kappa < \mu < \lambda \Rightarrow \mathrm{cov}(\mu,\kappa,\kappa,2) < \lambda]$ *then* $\{\delta < \lambda : \mathrm{cf}(\delta) < \kappa\} \in I[\lambda]$.

Proof. Use $\mu(3)$ of 2.6.

Claim 2.9 *Let* $(*)_{\mu,\lambda,\kappa}$ *mean: if* $a_i \in S_{<\kappa}(\lambda)$ *for* $i \in S$, $S \subseteq \{\delta < \mu : \mathrm{cf}\delta = \kappa\}$ *is stationary, then for some* $b \in S_{<\kappa}(\lambda)$, $\{i \in S : a_i \cap i \subseteq b\}$ *is stationary. Let* $(*)^-_{\mu,\lambda,\kappa}$ *be defined similarly but* $\{i \in S : a_i \subseteq b\}$ *only unbounded. Then for* $\aleph_0 < \kappa < \lambda < \mu$ *regular we have:*

$$\mathrm{cov}(\lambda,\kappa,\kappa,2) < \mu \Rightarrow (*)_{\mu,\lambda,\kappa} \Rightarrow (*)^-_{\mu,\lambda,\kappa}$$
$$\Rightarrow (\forall \lambda' \leq \lambda)[\kappa < \lambda' \leq \lambda \ \& \ \mathrm{cf}\lambda' < \kappa \Rightarrow \mathrm{pp}_{<\kappa}\lambda' < \mu].$$

Remark So it is conceivable that the $\Rightarrow$ are $\Leftrightarrow$. See [12], §3.

Proof. Straightforward.

3 Nice Filters Revisited

This generalizes [11] (and see there).
  See [15], §5 on this generalization of normal filters.

Conventions 3.1

  (1) We use $\aleph_1$ rather than an uncountable regular κ for simplicity.

  (2) Let μ^* be $> \aleph_1$ and $\mathcal{Y}_i = \{i\} \times (\bigcup_{\mu<\mu^*} \mu)$, $\mathcal{Y} = \bigcup_{i<\omega_1} \mathcal{Y}_i$, $\iota(y) = i$ when $y \in \mathcal{Y}_i$.

  (3) Let Eq denote a set of equivalence relations e on $\mathcal{Y}$ refining $\bigcup_{i<\omega_1} \mathcal{Y}_i \times \mathcal{Y}_i$ with $< \mu^*$ equivalence classes, each class of cardinality $|\mathcal{Y}|$. We say $e_1 \leq e_2$ if e_2 refines e_1. If not said otherwise, every e is in Eq. Let Eq_μ be the set of all such equivalence relations with $< \mu$ equivalence classes. Let $\iota(x/e) = \iota(x)$.

Definition 3.2

  (1) Let $\mathrm{FIL}(e) = \mathrm{FIL}(e,\mathcal{Y})$ denote the set of D such that:

  (a) D is a filter on $\mathcal{Y}/e$,
  (b) for any club C of ω_1, $\bigcup_{i\in C} \mathcal{Y}_i/e \in D$,

(c) (*normality*) if $X_i \in D$ for $i < \omega_1$ then $\{(\delta, j)/e : (\delta, j) \in \mathcal{Y}, \delta \text{ limit and}$
$i < \delta \Rightarrow (\delta, j) \in X_i\}$ belongs to D.

(2) $\mathrm{FIL}(\mathcal{Y}) = \mathrm{FIL}(Eq, \mathcal{Y})$ is $\bigcup_{e \in Eq} \mathrm{FIL}(e, \mathcal{Y})$. For $D \in \mathrm{FIL}(\mathcal{Y})$, let $e = e[D]$ be such that $D \in \mathrm{FIL}(e, \mathcal{Y})$.

(3) For $D \in \mathrm{FIL}(e)$ let $D^{[*]} = \{X \subseteq \mathcal{Y} : \{y/e : y/e \subseteq X\} \in D\}$.

(4) For $D \in \mathrm{FIL}(\mathcal{Y})$ and $e(1) \geq e(D)$, let $D^{[e(1)]} = \{X \subseteq \mathcal{Y}/e(1) : X^{[*]} \in D^{[*]}\}$.

(5) For $A \subseteq \mathcal{Y}/e$, $A^{[*]} = \bigcup_{(x/e) \in A} x/e$, and for $e(1) \geq e$ let

$$A^{[e(1)]} = \{y/e(1) : y/e \in A\}.$$

Definition 3.2A For $D \in \mathrm{FIL}(e, \mathcal{Y})$, let D^+ be $\{Y \subseteq \mathcal{Y}/e : Y \neq \emptyset \bmod D\}$.

Definition 3.3

(0) For $f : \mathcal{Y}/e \to X$ let $f^{[*]} : \mathcal{Y} \to X$ be $f^{[*]}(x) = f(x/e)$. We say $f : \mathcal{Y} \to X$ is supported by e if it has the form $g^{[*]}$ for some $g : \mathcal{Y}/e \to X$. Let $e_1, e_2 \in Eq$, $f_\ell : \mathcal{Y}/e_\ell \to X$; we say $f_1 = f_2^{[e_1]}$ if $f_1^{[*]} = f_2^{[*]}$.

(1) Let $F_c({}^\omega\omega, e) = F_c({}^\omega\omega, e, \mathcal{Y})$ be the family of $\bar{g}$, a sequence of the form $\langle g_\eta : \eta \in u \rangle$, $u \in f_c({}^\omega\omega) = $ the family of non-empty finite subsets of ${}^{\omega>}\omega$ closed under initial segment, and for each $\eta \in u$ we have $g_\eta \in {}^\mathcal{Y}\mathrm{Ord}$ is supported by e. Let $\mathrm{Dom}\ \bar{g} = u$, $\mathrm{Range}\ \bar{g} = \{g_\eta : \eta \in u\}$. We let $e = e(\bar{g})$, an abuse of notation.

(2) We say $\bar{g}$ is decreasing for D or D-decreasing (for $D \in \mathrm{FIL}(e, I)$) if $\eta \triangleleft \nu \Rightarrow g_\nu <_D g_\eta$.

(3) If $u = \{<>\}$, $g = g_{<>}$ we write g instead of $\langle g_\eta : \eta \in u \rangle$.

Definition 3.4

(1) For $e \in Eq$, $D \in \mathrm{FIL}(e, \mathcal{Y})$ and D-decreasing $\bar{g} \in F_c({}^\omega\omega, e)$ we define a game $G^*(D, \bar{g}, e, \mathcal{Y})$ (we may omit $\mathcal{Y}$). In the nth move (stipulating $e_{-1} = e$, $D_{-1} = D$, $\bar{g}_{-1} = \bar{g}$):

> player I chooses $e_n \geq e_{n-1}$ and $A_n \subseteq \mathcal{Y}/e_n$, $A_n \neq \emptyset \bmod D_{n-1}^{[e_n]}$ and he chooses $\bar{g}^n \in F_c({}^\omega\omega, e_n)$ extending $\bar{g}_{n-1}$ (i.e. $\bar{g}^{n-1} = \bar{g}^n \lceil \mathrm{Dom}\ \bar{g}_{n-1})$, $\bar{g}^n$ supported by e_n and $\bar{g}^n$ is $(D_n^{[e_n]} + A_n)$-decreasing, player II chooses $D_n \in \mathrm{FIL}(e_n, \mathcal{Y})$ extending $D_{n-1}^{[e_n]} + A_n$.

In the end, the second player wins if $\bigcup_{n<\omega} \mathrm{Dom}\ \bar{g}^n$ has no infinite branch.

(2) $G^{\bar{\gamma}}(D, \bar{g}, e, \mathcal{Y})$ is defined similarly to $G^*(D, \bar{g}, e, \mathcal{Y})$ ($\mathrm{Dom}\ \bar{\gamma} = \mathrm{Dom}\ \bar{g}$) but the second player has, in addition, to choose an ordinal α_η for $\eta \in \mathrm{Dom}\ \bar{g}^n \setminus \bigcup_{\ell<n} \mathrm{Dom}\ \bar{g}^\ell$ such that $[\eta \triangleleft \nu\ \&\ \nu \in \mathrm{Dom}\ \bar{g}^{n-1} \Rightarrow \alpha_\nu < \alpha_\eta]$ and $\alpha_\eta = \gamma_\eta$ for $\eta \in \mathrm{Dom}\ \bar{g}$.

(3) $wG^*(D, \bar{g}, e, \mathcal{Y})$ and $wG^{\bar{\gamma}}(D, \bar{g}, e, \mathcal{Y})$ are defined similarly but e is not changed during a play.

(4) If $\bar{\gamma} = \langle \gamma_{<>} \rangle$, $\bar{g} = \langle g_{<>} \rangle$ we write $\gamma_{<>}$ instead of $\bar{\gamma}$, $g_{<>}$ instead of $\bar{g}$.

(5) If $E \subseteq \text{FIL}(\mathcal{Y})$ the games G_E^*, $G_E^{\bar{\gamma}}$ are defined similarly, but player II can choose filters only from E (so we like to have $A \in D^+$, $D \in E \Rightarrow D + A \in E$).

Remark 3.4A Denote the above games $G_0^*, G_0^{\bar{\gamma}}$. Another variant is

(3) For $e \in Eq$, $D \in \text{FIL}(e, \mathcal{Y})$ and D-decreasing $\bar{g} \in F_c(^\omega\omega)$ we define a game $G_1^*(D, \bar{g}, e, \mathcal{Y})$. We stipulate $e_{-1} = e$, $D_{-1} = D$.

In the nth move first player chooses $e_n, e_{n-1} \leq e_n \in Eq$ and $D'_n \in \text{FIL}(e_n, \mathcal{Y})$ such that:

> (*) for some $A_n \subseteq \mathcal{Y}/e_{n-1}$, $A_n \neq \emptyset \bmod D_{n-1}$ we have:
>> (i) $(D_{n-1} + A_n)^{[e_n]} \subseteq D_n$;
>> (ii) D'_n is the normal filter on $\mathcal{Y}/e_n$ generated by $(D_{n-1} + A_n)^{[e_n]} \cup \{A_\zeta^n : \zeta < \zeta_n^*\}$ where for some $\langle C_\zeta : \zeta < \zeta_n \rangle$ we have:
>>> (a) each C_ζ is a club of ω_1,
>>> (b) if $\zeta_\ell < \zeta_n^*$ for $\ell < \omega$, $i \in \bigcap_{\ell < \omega} C_{\zeta_\ell}$, $x \in \mathcal{Y}/e_{n-1}$, and $\iota(x) = i$, then for some $x' \in \mathcal{Y}/e_n$, we have $x' \subseteq x$, $x' \in \bigcap_{\ell < \omega} A_{\zeta_\ell}^n$.
>
> First player also chooses $\bar{g}^n$ extending $\bar{g}^{n-1}$ D'_n-decreasing and the second player chooses D_n, $D'_n \subseteq D_n \in \text{FIL}(e_n, \mathcal{Y}_n)$.

(4) We define $G_1^{\bar{\gamma}}(D, \bar{g}, e, \mathcal{Y})$ as in (2) using G_1^* instead of G_0^*.

(5) If player II wins, e.g. $G_E^{\bar{\gamma}}(D, \bar{f}, e, \mathcal{Y})$ this is true for $E' =: \{D' \in G : \text{player II wins } G_{E^*}^{\bar{\gamma}}(D', \bar{f}, e, Y)\}$.

Definition 3.5

(1) We say $D \in \text{FIL}(\mathcal{Y})$ is nice to $\bar{g} \in F_c(^\omega\omega, e, \mathcal{Y})$, $e = e(D)$, if player II wins the game $G^*(D, \bar{g}, e)$ (so in particular $\bar{g}$ is D-decreasing, $\bar{g}$ supported by e).

(2) We say $D \in \text{FIL}(\mathcal{Y})$ is nice if it is nice to $\bar{g}$ for every $\bar{g} \in F_c(^\omega\omega, e, \mathcal{Y})$.

(3) We say D is nice to α if it is nice to the constant function α. We say D is nice to $g \in {}^{\aleph_1}\text{Ord}$ if it is nice to $g^{[e(D)]}$.

(4) "Weakly nice" is defined similarly but e is not changed.

Remark "Nice" in [11] is the weakly nice here, but formally they act on different objects; but if $x \mathrel{e} y \Leftrightarrow \iota(x) = \iota(y)$ we get a situation isomorphic to the old one.

Claim 3.6 *Let $D \in \text{FIL}(\mathcal{Y})$ and $e = e(D)$.*

(1) If D is nice to f, $f \in F_c(^\omega\omega, e, \mathcal{Y})$, $g \in F_c(^\omega\omega, e, \mathcal{Y})$ and $g \leq f$ then D is nice to f.

(2) *If D is nice to f, $e = e(D) \leq e(1) \in Eq$ then $D^{[e(1)]}$ is nice to $f^{[e(1)]}$.*

(3) *The games from 3.4(2) are determined and winning strategies do not need memory.*

(4) *D is nice to $\bar{g}$ iff D is nice to $g_{<>}$ (when $\bar{g} \in F_c({}^{\omega}w, e, \mathcal{Y})$ is D-decreasing).*

(5) *If $Eq' \subseteq Eq$ and for simplicity $\bigcup_{i<\omega_1}\{i\} \times \mathcal{Y}_i \in Eq'$ and for every $e \in Eq'$, $e \leq e(1) \in Eq$ for some permutation π of $\mathcal{Y}$, $\pi(e) = e$, $\pi(e(1)) \leq e(2) \in Eq'$ then we can replace Eq by Eq'.*

(6) *For $Eq = Eq_\mu$ (where $\mu \leq \mu^*$) there is Eq' as above with: $|Eq'|$ countable if μ is a successor cardinal $(> \aleph_1)$, $|Eq'| = cf\mu$ if μ is a limit cardinal.*

Proof. Left to the reader. (For part (4) use 3.7(2) below.) $\square_{3.6}$

Claim 3.7

(1) *Second player wins $G^*(D,\bar{g},e)$ iff for some $\bar{\gamma}$ second player wins $G^{\bar{\gamma}}(D,\bar{g},e)$.*

(2) *If second player wins $G^\gamma(D,f,e)$ then for any D-decreasing $\bar{g} \in F_c({}^\omega w, e, \mathcal{Y})$, $\bar{g}$ supported by e and $\bigwedge_{\eta,x} g_\eta(x) \leq f(x)$, the second player wins in $G^{\bar{\gamma}}(D,\bar{g},e)$, when we let*

$$\gamma_\eta = \gamma \times [\max_{\eta \trianglelefteq \nu \in \mathrm{Dom}\,\bar{g}} (\ell g(\nu) - \ell g(\nu) + 1)].$$

(3) *If $u_1, u_2 \in f_c({}^{\omega>}\omega)$, $h : u_1 \to u_2$ satisfies $[\eta \triangleleft \nu \Leftrightarrow h(\eta) \triangleleft h(\nu)]$ and for $\ell = 1, 2$ we have $\bar{g}^\ell \in F_2({}^{\omega>}\omega, e_2, \mathcal{Y})$, $g_\eta^1 = g_{h(\eta)}^2$ (for $\eta \in u_1$), $\bar{\gamma}^\ell = \langle \gamma_\eta^\ell : \eta \in u_2 \rangle$ is $\triangleleft$-decreasing sequence of ordinals, $\gamma_\eta^1 \geq \gamma_{h(\eta)}^2$ and the second player wins in $G^{\bar{\gamma}^2}(D,\bar{g}^2,e,\mathcal{Y})$ then the second player wins in $G^{\bar{\gamma}^1}(D,\bar{g}^1,e,\mathcal{Y})$.*

Proof.

(1) The "if" part is trivial, the "only if" as in [11].

The following is a consequence of a theorem of Dodd and Jensen [DoJ]:

Theorem 3.8 *If λ is a cardinal, $S \subseteq \lambda$ then:*

(1) *$K[S]$, the core model, is a model of ZFC $+ (\forall \mu \geq \lambda)2^\mu = \mu^+$.*

(2) *If in $K[S]$ there is no Ramsey cardinal $\mu > \lambda$ (or much weaker condition holds) then $(K[S], V)$ satisfies the μ-covering lemma for $\mu \geq \lambda + \aleph_1$; i.e. if $B \in V$ is a set of ordinals of power $\leq \mu$ then there is $B' \in K[S]$, $B \subseteq B'$, $V \models |B'| \leq \mu$.*

(3) *If $V \models (\exists\mu \geq \lambda)(\exists\kappa)[\mu^\kappa > \mu^+ > 2^\kappa]$ then in $K[S]$ there is a Ramsey cardinal $\mu > \lambda$.*

Lemma 3.9 *Suppose $f \in {}^{\aleph_1}\mathrm{Ord}$, $\lambda > \lambda_0 =: \sum_{\alpha<\mu^*} 2^{|\alpha|^{\aleph_0}} + \prod_{i<\omega_1}|f(i) + 1| + |Eq|$, and for every $A \subseteq \lambda_0$, in $K[A]$ there is a Ramsey cardinal $> \lambda_0$, then for every normal filter $D \in \mathrm{FIL}(e, \mathcal{Y})$, D is nice to f.*

Remark: The point in the proof is that via forcing we translate the filters from $\text{FIL}(e, \mathcal{Y})$ to normal filters on ω_1 [for higher κ's cardinal restrictions are better].

Proof. Without loss of generality $(\forall i) f(i) \geq 2$.

Let $S \subseteq \lambda_0$ be such that $[\alpha < \mu^* \; \& \; A \subseteq 2^{|\alpha|^{\aleph_0}} \Rightarrow A \in L[S]]$, $Eq \in L[S]$ and: if $g \in {}^{\aleph_1}\text{Ord}$, $(\forall i < \omega_1) g(i) \leq f(i)$ then $g \in L[S]$ (possible as $\prod_{i < \omega_1} |f(i) + 1| \leq \lambda_0$). We work for awhile in $K[S]$. In $K[S]$ there is a Ramsey cardinal $\mu > \lambda_0$ (see 3.8(3)). Let, in $K[S]$,

$$I = \{X : X \subseteq \mu, X \cap \omega_1 \text{ a countable ordinal } > 0, \{\omega_1, \mu\} \subseteq X,$$
$$\text{moreover } X \cap \lambda_0 \text{ is countable}\}.$$

Let

$$J = \{X \in I : X \text{ has order type } \geq f(X \cap \omega_1)\}.$$

Now for $g \in {}^{\aleph_1}\text{Ord}$ such that $\bigwedge_{i < \omega_1} g(i) < f(i)$ let $\hat{g}$ be the function with domain J, $\hat{g}(X) = $ the $g(X \cap \omega_1)$-th member of X.

Let $D = \{A_i : \omega_1 \leq i \leq 2^{|\mathcal{Y}/e|}\}$ and we arrange $\langle A_i : \omega_1 \leq i < 2^{|\mathcal{Y}/e|}\rangle \in L[S]$, (as $\mathcal{Y}/e$ has cardinality $< \mu^*$, so $2^{|\mathcal{Y}/e|} \leq \lambda_0$).

Let F be the minimal fine normal filter on I (in $K[S]$) to which J_D belongs where

$$J_D = \{X : X \in J \text{ and } i \in (\omega_1, 2^{|\mathcal{Y}/e|}) \cap X \Rightarrow X \cap \omega_1 \in A_i\}.$$

Clearly it is a proper filter as $K[S] \models$ "μ is a Ramsey cardinal".

Observation 3.9A [in $K[S]$]. Assume P is a proper forcing notion of cardinality $\leq |\alpha|^{\aleph_0}$ for some $\alpha < \mu^*$ (or just P, $\text{MAC}(P) \in K[S]$ and $\{X \in I : X \cap |\text{MAC}(P)| \text{ is countable}\} \in F$ where $\text{MAC}(P)$ is the set of maximal antichains of P) and let F^P be the normal fine filter which F generates in V^P. Then

(1) F-positiveness is preserved; i.e. if $X \in V$, $X \subseteq I$, $F \in \text{FIL}(\mathcal{Y})$ and $V \models$ "$X \neq \emptyset \bmod F$" then $\Vdash_P$ "$X \neq \emptyset \bmod F^P$".

(2) Moreover, if $Q \lessdot P$, (Q proper and) P/Q is proper then forcing with P/Q preserve F^Q-positiveness.

Let $\mathcal{P}(\mathcal{Y}/e) = \{A_\zeta^e : \zeta < 2^{|\mathcal{Y}/e|}\}$.

Now we describe a winning strategy for the second player. In the side we choose also (p_n, Γ_n, f_n), $\bar{\gamma}^n$, W_n such that[1] (where e_n, A_n are chosen by the second player):

(A) (i) $P_n = \prod_{\ell \leq n} Q_\ell$, Q_ℓ is $\text{Levy}(\aleph_1, \mathcal{Y}/e_n)$ (we could use iterations, too, here it does not matter);

(ii) $p_n \in P_n$;

(iii) p_n increasing in n;

(iv) f_n is a P_n-name of a function from ω_1 to $\mathcal{Y}/e_n$;

(v) $p_n \Vdash_{P_n}$ "$f_n(i) \in \mathcal{Y}_i/e_n$";

---

[1] By the homogeneity of the forcing notion the value of p_n is immaterial.

(vi) $p_{n+1} \Vdash$ "$f_{n+1}(i) \subseteq f_n(i)$ for every $i < \omega_1$";

(vii) f_n is given naturally — it can be interpreted as the generic object of Q_n except trivialities.

(B) (i) $\bar{\gamma}^n, \bar{g}^n$ has the same domain, $\gamma^n_\eta < \mu$;

(ii) $p_n \Vdash_{P_n}$ "$W_n \subseteq J_D$, $W_{n+1} \subseteq W_n$";

(iii) $\bar{\gamma}^n = \gamma^{n+1} \restriction \operatorname{Dom} \bar{\gamma}^n$, $\operatorname{Dom} \bar{\gamma}^n = \operatorname{Dom} \bar{g}^n$;

(iv) $p_n \Vdash_{P_n}$ "$\{X \in J_D : \text{for } \ell \in \{0,...,n\}, f_\ell(X \cap \omega_1) \in A_\ell \text{ and } \bigwedge_{\eta \in \operatorname{Dom} \bar{g}^n} \hat{g}_\eta(X) = \gamma_\eta$ and for $\ell \in \{-1,0,...,n-1\}, \zeta \in X \cap 2^{|\mathcal{Y}/e_\ell|}$ we have: $A^{e_\ell}_\zeta \in D_\ell \Rightarrow f_\ell(X \cap \omega_1) \in A^{e_\ell}_\zeta\} \supseteq W_n \neq \emptyset \bmod F^{P_n}$"

(C) $D_n = \{Z \subseteq \mathcal{Y}/e_n : p_n \Vdash_{P_n}$ "$\{X \in J_D : f_n(X \cap \omega_1) \notin Z\} = \emptyset \bmod D^{P_n}_n + W_n$"$\}$.

Note that $D_n \in K[S]$, so every initial segment of the play (in which the second player uses this strategy) belongs to $K[S]$. $\square_{3.9}$

Remark 3.9B

(1) From the proof, instead $K[S] \models$ "λ is Ramsey", $K[S] \models$ "$\mu \to (\alpha)^{<\omega}_2$ for $\alpha < \lambda_0$" is enough for showing 3.9.

(2) Also if $\prod(|f(i)| + 1) < \mu_0$, $[\alpha < \mu_0 \Rightarrow |\alpha|^{\aleph_0} < \mu_0]$, it is enough: $S \subseteq \alpha < \mu_0 \Rightarrow$ in $K[S]$ there is $\mu \to (\alpha)^{<\omega}_2$.

Theorem 3.10 *Let* $D^* \in \operatorname{FIL}(e, \mathcal{Y})$ *be a normal ideal on* $\aleph_1$. *If for every* $f : \aleph_1 \to (\sum_{\chi < \mu^\bullet} \chi^{\aleph_1})^+$, D^* *is nice to* f, *then for every* $f \in {}^{\aleph_1}\operatorname{Ord}$, D *is nice to* f.

Proof. As in [11], 1.7.

Remark 3.10A So, the existence of μ, $\mu \to (\alpha)^{<\omega}_{\aleph_0}$ for every $\alpha < (\sum_{\chi < \mu^\bullet} \chi^{\aleph_1})^+$, is enough for "$D^*$ is nice".

Conclusion 3.11 *Let* $\lambda_0 = \sum_{\chi < \mu^\bullet} 2^{\chi^{\aleph_0}} + |Eq|$, $\mu^* \geq \aleph_2$; *if for every* $S \subseteq \lambda_0$ *there is a Ramsey cardinal in* $K[S]$ *above* λ_0 *then every* $D \in \operatorname{FIL}(\mathcal{Y})$ *is nice.*

Proof. By 3.9, 3.10.

Concluding Remark 3.12

(1) We could have used other forcing notions, not $\operatorname{Levy}(\aleph_1, \mathcal{Y}/e_n)$. E.g. if $\mu = \aleph_2$ we could use finite iterations of the forcing of Baumgartner to add a club of ω_1, by finite conditions. (So this forcing notion has cardinality $\aleph_1$.) Then in 3.9 we can weaken the demands on $\lambda_0 : \lambda_0 = \sum_{\chi < \mu_0} 2^\chi + \prod_{i < \omega_1} |1 + f(i)| + |Eq|$, hence also in 3.11, $\lambda_0 = \sum_{\chi < \mu^\bullet} 2^\chi$ is O.K.

(2) Concerning $|Eq|$ remember 3.6(5), (6).

(3) Similarly to (1). If $\bigwedge_{\theta < \mu} \operatorname{cov}(\theta, \aleph_1, \aleph_1, 2) < \mu$ then by 2.6 we can use forcing notions of Todorčevič for collapsing $\theta < \mu$ which has cardinality $< \mu$.

(4) If we want to have $\lambda_0 =: \prod_{i < \omega_1} |f(i) + 2|$ (or even $T_D(f+2)$), we can get this by weakening further the first player letting him choose only A_n which are easily definable from the $\bar{g}^{n-1}$, we shall return to it in a subsequent paper.

4 Ranks

Convention 4.1 Like 3.1 and: $\bar{g} \in F_c({}^\omega w, e^*, \mathcal{Y})$, $\eta^* \in \operatorname{Dom} \bar{g}^*$, ν^* an immediate successor of η^* not in Dom g^*, $D^* \in \operatorname{FIL}(e^*, \mathcal{Y})$ is such that in $G^{\bar{\gamma}^*}(D^*, \bar{g}^*, e^*)$ second player wins (all constant). $\operatorname{FIL}^*(e, \mathcal{Y})$ will be the set of $D \in \operatorname{FIL}(e, \mathcal{Y})$ such that $e \geq e^*$, $(D^*)^{[e]} \subseteq D$ and in $G^{\bar{\gamma}^*}(D^*, \bar{g}^*, e^*)$ second player wins. (So actually $\operatorname{FIL}(e^*, \mathcal{Y})$ depends on $D^*, \bar{g}^*, e^*$, too.)

Definition 4.2

(1) $rk_D^5(f)$ for $D \in \operatorname{FIL}^*(e, \mathcal{Y})$, $f \in {}^{\mathcal{Y}/\ell}\operatorname{Ord}$, $f <_D \bar{g}_{\eta^*}^*$ will be: the minimal ordinal α such that for some $D_1, e_1, \bar{\gamma}^1$ we have $D^{[e_1]} \subseteq D_1 \in \operatorname{FIL}(e_1, \mathcal{Y})$, $\bar{\gamma}^1 = \bar{\gamma}^* {}^\wedge \langle \nu^*, \alpha \rangle$ (i.e. Dom $\bar{\gamma}^1 = (\operatorname{Dom} \bar{\gamma}^*) \cup \{\nu^*\}$, $\bar{\gamma}^1 \lceil \operatorname{Dom} \bar{\gamma}^* = \bar{\gamma}^*$, $\gamma_{\nu^*}^1 = \alpha$) and in $G^{\bar{\gamma}^1}(D, \bar{g}^* {}^\wedge \langle \nu^*, f \rangle)$ second player wins and ∞ if there is no such α .

(2) $rk_D^4(f)$ is $\sup\{rk_{D+A}^5(f) : A \in D^+\}$.

Claim 4.3

(1) $rk_D^5(f)$ is (under the circumstances of 4.1, 4.2) an ordinal $< \gamma_{\eta^*}^*$.

(2) $rk_D^4(f)$ is an ordinal $\leq \gamma_{\eta^*}^*$.

Claim 4.4 If $D \in \operatorname{FIL}^*(e, \mathcal{Y})$, $h <_D f <_D g_{\eta^*}^*$ then $rk_D^5(h) < rk_D^5(f)$.

Proof. Let e_1, D_1 witness $rk_D^5(f) = \alpha$ so $e(D) \leq e_1$, $D \subseteq D_1 \in \operatorname{FIL}^*(e_1, \mathcal{Y})$ and in $G^{\bar{\gamma}^* {}^\wedge \langle \nu^*, \alpha \rangle}(D_1, \bar{g}^* {}^\wedge \langle \nu^*, f \rangle, e)$ second player wins. We play for the first player: $e = e_1$, $A_0 = \mathcal{Y}/e_1$, $\bar{g}^0 = \bar{g}^* {}^\wedge \langle \nu^*, f \rangle {}^\wedge \langle \nu^* {}^\wedge \langle 0 \rangle, g \rangle$, now the first player should be able to answer say e_2, D_2, $\bar{\gamma}^2$. So $\gamma_{\nu^* {}^\wedge \langle 0 \rangle}^2 < \gamma_{\nu^*}^2 = \alpha$, and by 3.7(3), we know that in $G^{\bar{\gamma}'}(D_2, \bar{g}^* {}^\wedge \langle \nu^*, g \rangle, e_2)$ where $\bar{\gamma}' = \bar{\gamma}^{\wedge} \langle \nu^*, \gamma_{\nu^* {}^\wedge \langle 0 \rangle}^2 \rangle$, second player wins. $\square_{4.4}$

Claim 4.5 Let $e \geq e^*$, $D \in \operatorname{FIL}^*(e, \mathcal{Y})$.

(1) For $e \geq e(D)$, $A \in (D^{[e]})^+$, $f \in {}^{\mathcal{Y}/\ell}\operatorname{Ord}$, $f <_D g_{\eta^*}^*$ we have:

$$rk_D^5(f) \leq rk_{D^{[e]}+A}^5(f) \leq rk_{D^{[e]}+A}^4(f) \leq rk_D^4(f).$$

(2) If $e_2 \geq e_1 \geq e(D)$, $f_\ell \in {}^{\mathcal{Y}}\operatorname{Ord}$ is supported by e_ℓ, $f_1 \leq_D f_2 <_D g_{\eta^*}^*$ then $rk_D^\ell(f_1) \leq rk_D^\ell(f_2)$ for $\ell = 4, 5$.

5   More on Ranks and Higher Objects

Convention 5.1

(a) μ^* is a cardinal $> \aleph_1$ (using $\aleph_1$ rather than an uncountable regular κ is to save parameters).

(b) $\mathcal{Y}$ is a set of cardinality $\sum_{\kappa < \mu^*} \kappa$.

(c) ι is a function from $\mathcal{Y}$ onto ω_1, $|\iota^{-1}(\{\alpha\})| = |\mathcal{Y}|$ for $\alpha < \omega$.

(d) Eq is the set of equivalence relations e on $\mathcal{Y}$ such that:

   (α) $y \, e \, z \Rightarrow \iota(y) = \iota(z)$,
   (β) each equivalence class has cardinality $|\mathcal{Y}|$,
   (γ) e has $< \mu^*$ equivalence classes.

(e) D denotes a normal filter on some $\mathcal{Y}/e$ ($e \in Eq$), we write $e = e(D)$. The set of such D's is $\mathrm{FIL}(\mathcal{Y})$.

(f) E denotes a set of D's as above, such that:

   (α) for some $D = \min E \in E$,

$$(\forall D')[D' \in E \Rightarrow (e, D) \le (e(D'), D')],$$

   (β) if $D \in E$, $A \subseteq \mathcal{Y}/e_1$, $e_1 \ge e(D)$, $A \ne \emptyset \bmod D$ then $D^{[e_1]} + A \in E$.

(g) $E^{[e]} =: \{D \in E : e(D) = e\}$.

(h) $\mathcal{E}$ denotes a set of E's as above, such that:

   (α) there is $E = \mathrm{Min}\, \mathcal{E} \in \mathcal{E}$ satisfying

$$(\forall E')(E' \in E \Rightarrow E' \subseteq E),$$

   (β) if $D \in E \in \mathcal{E}$ then

$$E_{[D]} = \{D' : D' \in E \text{ and } (e(D), D) \le (e(D'), D')\} \in \mathcal{E}.$$

Definition 5.2

(1) We say E is λ-divisible when: for every $D \in E$, and Z a set of cardinality $< \lambda$, there are D', j such that:

   (α) $D' \in E$;
   (β) $(e(D), D) \le (e(D'), D')$;
   (γ) $j : \mathcal{Y}/e(D') \to Z$;
   (δ) for every function $h : \mathcal{Y}/e(D) \to Z$,

$$\{y/e(D') : h(y/e(D)) = j\,(y/e(D'))\} \ne \emptyset \bmod D'.$$

(2) We say E has λ-sums when: for every $D \in E \in \mathcal{E}$ and sequence $\langle Z_\zeta : \zeta < \zeta^* < \lambda \rangle$ of subsets of $\mathcal{Y}/e(D)$ there is $Z^* \subseteq \mathcal{Y}/e(D)$ such that: $Z^* \cap Z_\zeta = \emptyset \mod D$ and: [if $(e(D), D) \leq (e', D')$, $e' = e(D')$, $D' \in E_{[D]}$ and $\bigwedge_\zeta Z_\zeta^{[e']} = \emptyset \mod D'$, then $Z^* \in D'$].

(3) We say E has weak λ-sums if for every $D \in E \in \mathcal{E}$ and sequence $\langle Z_\zeta : \zeta < \zeta^* < \lambda \rangle$ of subsets of $\mathcal{Y}/e(D)$ there is D^*, $D^* \in E_{[D]}$ such that:

   (α) if $(e(D), D) \leq (e', D')$, $D' \in E_{[D]}$ and $Z_\zeta = \emptyset \mod D'$ for $\zeta < \zeta^*$, $e(D^*) \leq e(D')$, then $D^* \subseteq D'$, and

   (β) $Z_\zeta = \emptyset \mod D^*$ for $\zeta < \zeta^*$.

(4) If $\lambda = \mu^*$ we omit it. We say $\mathcal{E}$ is λ-divisible if every $E \in \mathcal{E}$ is. Similarly we define "$\mathcal{E}$ has [weak] λ-sums" by modifying clause [(3)] (2), replacing E by $\mathcal{E}$ and D by E.

We now define variants of the games from §3.

Definition 5.3 For a given $\mathcal{E}$, for every $E \in \mathcal{E}$:

(1) We define a game $G_2^*(E, \bar{g})$.

   In the n-th move first player chooses $D_n \in E_{n-1}$ (stipulating $E_{-1} = E$) and choose $\bar{g}_n \in F_c({}^\omega \omega, e(D_n), \mathcal{Y})$ extending $\bar{g}_{n-1}$ (stipulating $\bar{g}_{-1} = \bar{g}$) such that $\bar{g}_n$ is D_n-decreasing. Then the second player chooses E_n, $(E_{n-1})_{[D_n]} \subseteq E_n \in \mathcal{E}$.

   In the end the second player wins if $\bigcup_{n<\omega} \mathrm{Dom}\, \bar{g}_n$ has no infinite branch.

(2) We define a game $G_2^{\bar{\gamma}}(E, \bar{g})$ where $\mathrm{Dom}\, \bar{\gamma} = \mathrm{Dom}\, \bar{g}$, each γ_n an ordinal, $[\eta \vartriangleleft \nu \Rightarrow \gamma_\eta > \gamma_\nu]$ similarly to $G_2^*(D, \bar{g})$ but the second player in addition chooses an indexed set $\bar{\gamma}_n$ of ordinals, $\mathrm{Dom}\, \bar{\gamma}_n = \mathrm{Dom}\, \bar{g}_n$, $\bar{\gamma}_n | \mathrm{Dom}\, \bar{\gamma}_{n-1} = \bar{\gamma}_{n-1}$ and $[\eta \vartriangleleft \nu \Rightarrow \gamma_{n,\eta} > \gamma_{n,\nu}]$.

Definition 5.4

(1) We say $\mathcal{E}$ is nice to $\bar{g} \in F_c({}^\omega \omega, e, \mathcal{Y})$ if for every $E \in \mathcal{E}$ with $e \leq e(E)$ the second player wins the game $G_2^*(E, \bar{g})$.

(2) We say $\mathcal{E}$ is nice if it is nice to $\bar{g}$ whenever $E \in \mathcal{E}$, $e \leq e(E)$, $\bar{g} \in F_c({}^\omega \omega, e, \mathcal{Y})$, $\bar{g}$ is $(\min E)$-decreasing, we have: $\mathcal{E}_{[E]}$ is nice to $\bar{g}$.

(3) If $\mathrm{Dom}\, \bar{g} = \{<>\}$ we write $g_{<>}$ instead of $\bar{g}$.

(4) We say $\mathcal{E}$ is nice to α if it is nice to the constant function α.

Claim 5.5

(1) *If $\mathcal{E}$ is nice to f, $f \in F_c({}^\omega \omega, e, \mathcal{Y})$, $g \in F_c({}^\omega \omega, e, \mathcal{Y})$, $g \leq f$ then $\mathcal{E}$ is nice to f.*

(2) *The games from 5.4 are determined, and the winning side has winning strategy which does not need memory.*

(3) *The second player wins $G_2^*(E, \bar{g})$ iff for some $\bar{\gamma}$ second player wins $G_2^{\bar{\gamma}}(E, g)$.*

(4) *If the second player wins $G_2^{\bar{\gamma}}(E, f)$, $\bar{g} \in F_e({}^\omega \omega, e(E))$, $g_\eta \leq f$ for $\eta \in \text{Dom}(\bar{g})$ then the second player wins in $G_2^{\bar{\gamma}}(E, \bar{g})$ when we let*

$$\gamma_\eta = \gamma + [\max_{\eta \trianglelefteq \nu \in \text{Dom}\, \bar{g}} (\ell g \nu - \ell g \eta + 1)].$$

Lemma 5.6 *Suppose $f_0 \in {}^{(\mathcal{Y}/e)}\text{Ord}$, $e \in Eq$, $\lambda_0 =: \sup_{e_0 \leq e \in Eq} \prod_{x \in \mathcal{Y}/e} (f_0^{[e]}(x) + 1)$.*

(1) *If there is a Ramsey cardinal $\geq \bigcup \{f(x) + 1 : x \in \text{Dom}\, f_0\}$ then there is a μ^*-divisible $\mathcal{E}$ nice to f_0 having weak μ^*-sums.*

(2) *If for every $A \subseteq \lambda_0$ there is in $K[A_0]$ a Ramsey cardinal $> \lambda_0$, then there is a μ^*-divisible $\mathcal{E}$ which has weak μ^*-sums and is nice to f.*

(3) *In part 2 if $\lambda_0 = 2^{<\mu_0}$ then there is a μ^*-divisible nice $\mathcal{E}$ which has weak μ^*-sums.*

Remark: This enables us to pass from "$pp_{\Gamma(\theta, \aleph_1)}$ large" to "pp_{normal} is large".

Proof. (1) Define $f_1 \in {}^{(\aleph_1)}\text{Ord}$, $f_1(i) = \sup\{f_0(y/e) : \iota(y) = i\}$, let λ be such that: $\lambda \to (\sup_{i < \aleph_1} f_1(i))_2^{\leq \omega}$ (or just $\emptyset \notin D_n^*$ — see below), let $\lambda_n = (\lambda^{\mu^*})^{+n}$,

$I_n = \{s : s \subseteq \lambda_n, s \cap \omega_1$ a countable ordinal$\}$,

$J_n = \{s \in I_n : s \cap \lambda$ has order type $\geq f_0(s \cap \omega_1)\}$.

Let D_n^* be the minimal fine normal filter on J_n.

Let for $n < \omega$ and $e \in Eq$, $H_{n,e} = \{h : h$ a function from J_n into $\mathcal{Y}/e$ such that $\iota(h(s)) = s \cap \omega_1\}$.

Let $P_n = \{p : p \subseteq J_n, p \neq \emptyset \bmod D_n^*\}$, $P = \bigcup_{n < \omega} P_n$ and for $p \in P$ let $n(p)$ be the unique n such that $p \in P_n$.

Let $p \leq q$ (in P) if $n(p) \leq n(q)$ and $\{s \cap \lambda_{n(p)} : s \in q\} \subseteq p$. Now for every $e \in Eq$, $n < \omega$, $p \in P_n$, $h \in H_{n,e}$ we let:

$$D_p^{n,e,h} = \{A \subseteq \mathcal{Y}/e : h^{-1}(A) \supseteq p \bmod D_{n(p)}^*\},$$

$$E_p^{n,e,h} = \{D_q^{n^1,e^1,h^1} : p \leq q \in P, n^1 = n(q), \text{ and } (n^1, e^1, h^1) \geq (n, e, h)\},$$

where $(n^1, e^1, h^1) \geq (n, e, h)$ means: $n \leq n^1 < \omega$, $e \leq e^1 \in Eq$, $h^1 \in H_{n^1, e^1}$ and for $s \in J_{(n^1)}$, $h^1(s)^{[e]} = h(s \cap \lambda_n)$. We define $(p^1, n^1, e^1, h^1) \geq (p, n, e, h)$ similarly and let

$$\mathcal{E}_p^{n,e,h} = \{E_q^{n^1,e^1,h^1} : p \leq q \in P, n^1 = n(q), (n^1, e^1, h^1) \geq (n, e, h)\}.$$

[Note: $(p^1, n^1, e^1, h^1) \geq (p, n, e, h)$ implies $D_{p^1}^{n^1,e^1,h^1} \supseteq D_p^{n,e,h}$, $E_{p^1}^{n^1,e^1,h^1} \subseteq E_p^{n,e,h}$ and $\mathcal{E}_{p^1}^{n^1,e^1,h^1} \subseteq \mathcal{E}_p^{n,e,h}$.] Now any $\mathcal{E} = \mathcal{E}_p^{n,e,h}$ ($p \in P$) is as required.

A new point is "$\mathcal{E}$ is μ-divisible". So suppose $E \in \mathcal{E} = \mathcal{E}_p^{n,e,h}$ so $E = E_q^{n^1,e^1,h^1}$ for some $(q, n^1, e^1, h^1) \geq (p, n, e, h)$. Let Z be a set of cardinality $< \mu^*$, so $(\lambda_{n^1})^{|Z|} = \lambda_{n_1}$; let $\{h_\zeta : \zeta < \zeta^* = |\mathcal{Y}/e_1|^{|Z|} \leq 2^\mu \leq \lambda_{n^1}\}$ list all functions h from $\mathcal{Y}/e_1$ to Z. Let $\langle S_\zeta : \zeta <$

$|\mathcal{Y}/e_1|^{|Z|}\rangle$ list a sequence of pairwise disjoint stationary subsets of $\{\delta < \lambda_{n^1+1} : \mathrm{cf}\delta = \aleph_0\}$. Let $e_2 \in Eq$ be such that $e_1 \leq e_2$ and for every $y \in \mathcal{Y}$, $\{z/e_2 : z\, e_1\, y\} = \{x(y/e,t) : t \in Z\}$; we let q_2, $q \leq q_2 \in P$ be: $q_2 = \{s \in J_{n^1+1} : s \cap \lambda_{n^1} \in q$ and $\sup s \in \bigcup_\zeta S_\zeta\}$; lastly we define $h^2 : J_{n^1+1} \to \mathcal{Y}/e_1$ by: $h^2(s) = x(h^1(s \cap \lambda_{n^1}), h_\zeta(s \cap \lambda_{n^1}))$ if $s \in q_2$, $\sup s \in S_\zeta$ (for $s \in J_{n^1+1}\backslash q_2$ it does not matter).

The proof that q_2, e_2, h^2 are as required is as in [2] and more specifically [8].

As for proving "$\mathcal{E}_p^{n,e,h}$ has weak μ^*-sums" the point is that the family of fine normal filters on J_n has μ^*-sum.

(2) Similar to 3.9 (and 3.6(5),(6)).

(3) Similar to [11], 1.7. $\qquad\qquad\qquad\qquad\qquad\qquad\qquad\qquad\qquad\qquad\square_{5.6}$

6 Hypotheses: Weakening of GCH

We define some hypotheses; except for the first we do not know now whether their negations are consistent with ZFC.

Hypothesis 6.1

(A) $pp(\lambda) = \lambda^+$ for every singular λ .

(B) If $\mathfrak{a}$ is a set of regular cardinals, $|\mathfrak{a}| < \mathrm{Min}\,\mathfrak{a}$ then $|\mathrm{pcf}\mathfrak{a}| \leq |\mathfrak{a}|$.

(C) If $\mathfrak{a}$ is a set of regular cardinals, $|\mathfrak{a}| < \mathrm{Min}\,\mathfrak{a}$ then $\mathrm{pcf}\mathfrak{a}$ has no accumulation point which is inaccessible (i.e.: λ inaccessible $\Rightarrow \sup(\lambda \cap \mathrm{pcf}\mathfrak{a}) < \lambda$).

(D) For every λ, $\{\mu < \lambda : \mu$ singular and $pp\mu \geq \lambda\}$ is countable.

(E) For every λ, $\{\mu < \lambda : \mu$ singular and $\mathrm{cf}\mu = \aleph_0$ and $pp\mu \geq \lambda\}$ is countable.

(F) For every λ, $\{\mu < \lambda : \mu$ singular of uncountable cofinality, $pp_{\Gamma(\mathrm{cf}\mu)}(\mu) \geq \lambda\}$ is finite.

(D)$_{\theta,\sigma,\kappa}$ For every λ, $\{\mu < \lambda : \mu > \mathrm{cf}\mu \in [\sigma,\theta)$ and $pp_{\Gamma(\theta,\sigma)}(\mu) \geq \lambda\}$ has cardinality $< \kappa$.

(A)$_\Gamma$ If $\mu > \mathrm{cf}\mu$ then $pp_\Gamma(\mu) = \mu^+$ (or in the definition of $pp_\Gamma(\mu)$ the supremum is on the empty set).

(B)$_\Gamma$, (C)$_\Gamma$ Similar versions (i.e. use pcf_Γ).

We concentrate on the parameter free case.

Claim 6.2 *In 6.1, we have:*

(1) (A) $\Rightarrow$ (B) $\Rightarrow$ (C);

(2) (A) $\Rightarrow$ (D) $\Rightarrow$ (E), (A) $\Rightarrow$ (F);

(3) (E) + (F) $\Rightarrow$ (D) $\Rightarrow$ (B). *[Last implication — by the localization theorem [13], §2.]*

Theorem 6.3 *Assume Hypothesis 6.1A.*

(1) *For every $\lambda > \kappa$, $cov(\lambda, \kappa^+, \kappa^+, 2) = \begin{cases} \lambda^+ & \text{if } cf(\lambda) \leq \kappa, \\ \lambda & \text{if } cf(\lambda) > \kappa. \end{cases}$*

(2) *For every $\lambda > \kappa = cf\kappa > \aleph_0$, there is a stationary $S \subseteq S_{\leq\kappa}(\lambda)$, $|S| = \lambda^+$ if $cf(\lambda) \leq \kappa$ and $|S| = \lambda$ if $cf(\lambda) > \kappa$.*

(3) *For μ singular, there is a tree with $cf\mu$ levels, each level of cardinality $< \mu$, and with $\geq \mu^+$ $(cf(\mu))$-branches.*

(4) *If $\kappa \leq cf\mu < \mu \leq 2^\kappa$ then there is an entangled linear order $\mathcal{T}$ of cardinality μ^+.*

Proof.

(1) By [14], §1.

(2) By part (1) and 2.6.

(3), (4) By [10], §4.

Theorem 6.4 [Hypothesis 6.1(D)]. *If $\lambda > 2^{\aleph_0}$, and $\lambda > \theta \geq cf\lambda + 2^{\aleph_0}$ then $cov(\lambda, \lambda, \theta^+, 2) =^+ pp_\theta(\lambda)$.*

Remark See [14], §3, §5 on earlier results; [16] for later results.

Proof. We prove by induction on $pp_\theta(\lambda)$ (not on λ!) for fixed θ. For a given λ, let

$$\Theta_1 =: \{\mu : \lambda \leq \mu < pp_\theta^+(\lambda), cf\mu \leq \theta, pp_\theta^+(\mu) = pp^+(\lambda)\},$$

$$\Theta_2 =: \{\mu : \lambda \leq \mu < pp_\theta^+(\lambda), cf\mu \leq \theta \text{ and } cov(\mu, \mu, \theta^+, 2) \geq pp_\theta^+(\mu)\}.$$

As we know that $[\lambda \leq \mu < pp_\theta^+(\lambda) \ \& \ cf\mu \leq \theta \Rightarrow pp_\theta^+(\mu) \leq pp_\theta^+(\lambda)]$ (by [10], 2.3) and by the induction hypothesis clearly $\Theta_2 \subseteq \Theta_1$. But by Hypothesis 6.1(D) we have Θ_1 countable hence Θ_2 is countable (really $|\Theta_1| \leq \theta$ suffices). By [10], 5.3(10) Θ_2 is closed hence it has a last element σ. By [10], proof of 5.4(1)—first part $cov(\alpha, \sigma^+, \theta^+, 2) < pp^+(\lambda)$ for $\alpha < pp^+(\lambda)$ (and as said above $\sigma \in \Theta_1$). Now apply 6.5 below (we have Hypothesis 6.1(C) by 6.2(3) + 6.2(1) with $\lambda, \chi, \theta, \kappa$ there standing for $\sigma, pp^+(\lambda), \theta, cf\lambda$ here). $\square_{6.4}$

Claim 6.5 *Suppose*

(a) $\lambda > cf\lambda = \kappa$, $\lambda > \theta \geq \kappa$,

(b) $\chi = cf\chi > \lambda$ *and* $cov(\alpha, \lambda^+, \theta^+, 2) < \chi$ *for* $\alpha < \chi$,

(c) $pp_\theta^+(\lambda) \leq \chi$,

(d) $\lambda > 2^{\aleph_0}$ *if* $\kappa = \aleph_0$,

(e) *if χ is inaccessible then Hypothesis 6.1(C).*

Then

(α) $\text{cov}(\lambda, \lambda, \theta^+, 2) < \chi$;

(β) *moreover, for some* $\lambda_0 < \lambda$, $\text{cov}(\chi, \lambda_0^+, \theta^+, 2) = \chi$.

Proof. We concentrate on the case $\text{cf}\lambda = \aleph_0$, which is harder [if $\text{cf}\lambda > \aleph_0$ it suffices to choose f_ξ for $\xi < \omega$]. Note that in the conclusion, (β) follows from (α) (by [10], 5.3 (10). Let $\chi^* = \beth_3(\lambda)^+$, and choose by induction on $\zeta \le (2^{\aleph_0})^+$ a model $M_\zeta^* \prec (H(\chi)^*, \in, <_{\chi^*}^*)$, $\|M_\zeta^*\| < \chi$, $M_\zeta^* \cap \chi$ an ordinal, M_ζ^* increasing continuous in ζ, $\{\kappa, \chi, \lambda, \theta\} \in M_0^*$ and $\langle M_\xi^* : \xi \le \zeta \rangle \in M_{\zeta+1}^*$. Let $M^* = M_{(2^{\aleph_0})^+}^*$. Let $\mathcal{P}_\zeta =: S_{<\lambda}(\lambda) \cap M_\zeta^*$, and $\mathcal{P} = \mathcal{P}_{(2^{\aleph_0})^+}$. Clearly $\mathcal{P}$ is a family of $< \chi$ subsets of λ each of cardinality $< \lambda$, so it suffices to prove:

(∗) if $a \subseteq \lambda$, $|a| \le \theta$ then for some $A \in \mathcal{P}$, $a \subseteq A$.

Given $a \subseteq \lambda$, $|a| \le \theta$, we define by induction on $\zeta < \omega_1$, f_ζ such that:

(a) $f_\zeta \in M^*$, f_ζ belongs to $\prod(\lambda \cap \text{Reg})$.

(b) For $w \subseteq \zeta$ satisfying $(\exists A \in M^*)[\{f_\xi : \xi \in w\} \subseteq A \,\&\, |A| < \lambda]$, let A_w be the $<_{\chi^*}^*$-first such A of minimal cardinality and we let N_w^a be the Skolem Hull of $\{f_\xi : \xi \in w\}$ in $(H(\chi^*), \in, <_{\chi^*}^*)$ and N_w^b be the Skolem Hull of $A_w = A_w \cup \{f_\xi : \xi \in w\}$ in $(H(\chi^*), \in, <_{\chi^*}^*)$. We demand for every such w that: for every large enough $\sigma \in \lambda \cap \text{Reg} \cap N_w^a$ we have $\sup(\sigma \cap N_w^b) < f_\zeta(\sigma)$.

For defining f_ζ, let $W_\zeta = \{w \subseteq \zeta : A_w \text{ well defined}\}$ so $W_\zeta \subseteq M^*$, $|W_\zeta| \le 2^{\aleph_0}$ hence for some $\xi(\zeta) < (2^{\aleph_0})^+$, $W_\zeta \subseteq M_{\xi(\zeta)}^*$. For $w \in W_\zeta$, let N_w^+ be the Skolem Hull of A_w in $(H(\chi^*), \in, <_{\chi^*}^*)$, so $N_w^+ \in M_{\xi(\zeta)+1}^*$ (see its definition) and $\|N_w^+\| = |A_w|$ hence

$$\mathfrak{a}_w = \{\sigma : \sigma \in N_w^+ \cap \lambda \cap \text{Reg} \cap N_w^+ \backslash |A_w|^+\}$$

belongs to $M_{\xi(\zeta)+1}^*$, and it includes an end segment of $\lambda \cap \text{Reg} \cap N_w^+$. Now by [14], 3.2, $\text{cf}_{\le\theta}(\prod \mathfrak{a}_w / J_\lambda^{bd}) < \chi$ (we use Hypothesis 6.1(C) if χ is inaccessible). As $\mathfrak{a}_w \in M_{\xi(\zeta)+1}$ there is $f_w^\zeta \in (\prod \mathfrak{a}_w) \cap M_{\xi(\zeta)+1}^*$ such that:

(∗) for every large enough $\sigma \in N_w^+ \cap \text{Reg} \cap \lambda$ we have $\sup(\sigma \cap N_w^+) < f_w^\zeta(\sigma)$,

but $N_w^a \subseteq N_w^+$ hence

(∗)′ for every large enough $\sigma \in N_w^a \cap \text{Reg} \cap \lambda$ we have $\sup(\sigma \cap N_w^+) < f_w^\zeta(\sigma)$.

Now $M_{\xi(\zeta)+1} \in M_{\xi(\zeta)+2}$, $\|M_{\xi(\zeta)+1}\| < \chi$ hence there is a cofinal $\mathcal{P}' \subseteq S_{\le\lambda}(|M_{\xi(\zeta)+1}|)$ of cardinality $< \chi$ in $M_{\xi(\zeta)+2}$; as $M_{\xi(\zeta)+2} \cap \chi$ is an ordinal, necessarily $\mathcal{P}' \subseteq M_{\xi(\zeta)+2}$ hence there is $A^\zeta \in M_{\xi(\zeta)+2}$ such that $\bigwedge_{w \in W_\zeta} f_w^\zeta \in A^\zeta$ and $|A^\zeta| \le \lambda$. So there is $f_\zeta \in \prod(\text{Reg} \cap \lambda)$ in $M_{\xi(\zeta)+2}$ satisfying $(\forall f)[f \in A^\zeta \,\&\, (\exists\theta)[\theta < \lambda \,\&\, f\restriction (\text{Reg} \cap \lambda\backslash\theta) < f_\zeta]]$.

Now there is $A \in M^*$, $|A| \le \lambda$, $\{f_\xi : \xi < \omega_1\} \subseteq A$ (by assumption (b) of the claim), hence for some $A \in M^*$, $|A| < \lambda$ and $w^* = \{\xi < \omega_1 : f_\xi \in A\}$ is uncountable. For each $\xi \in w$, for some $\lambda_\xi < \lambda$,

$$\lambda_\xi < \sigma \in \lambda \cap \text{Reg} \cap N_{w^* \cap \xi}^a \Rightarrow \sup(N_{w^* \cap \xi}^b \cap \sigma) < f_\xi(\sigma).$$

As we assume $\text{cf}\lambda = \aleph_0$, for some $\lambda(*) < \lambda$, there are $\xi_0 < \xi_1 < \ldots < \xi_n < \ldots$ in w^* such that $\lambda_{\xi_n} \le \lambda(*)$.

Let $N^* =$ Skolem Hull of $A \cup (\lambda(*) + 1)$ in $(H(\chi^*), \in, <^*_{\chi^*})$; it belongs to M^*, hence $N^* \cap \lambda \in \mathcal{P}$. So it suffices to show that $N^b_{\{\xi_n : n < \omega\}}$ is a subset of N^*, which is done as in [14], 3.3A, 5.1A.

$\square_{6.5}$

Remark 6.5A

(1) We may want to omit the "$\lambda > 2^{\aleph_0}$ and $\theta \geq \mathrm{cf}\lambda + 2^{\aleph_0}$" in 6.4, 6.5. Of course, this is used only in 6.5, and we may replace it by: for some $\lambda_0 < \lambda$

$(*)_{\lambda_0}$ if c is a two place function from λ_0 to κ such that $[\alpha < \beta < \gamma \Rightarrow c(\alpha, \gamma) \leq \max\{c(\alpha, \beta), c(\beta, \gamma)\}$, then for some $n_0 < \omega$ and infinite $w \subseteq \lambda_0$ we have $\alpha \in w \ \& \ \beta \in w \ \& \ \alpha < \beta \Rightarrow c(\alpha, \beta) \leq n_0$.

Unfortunately, this is equivalent to

$(*)'_{\lambda_0}$ there are functions $f_\alpha \in {}^\omega \mathrm{Ord}$ for $\alpha < \lambda_0$ such that: $\alpha < \beta \Rightarrow f_\beta <_{J^{bd}} f_\alpha$

[why? $(*)'_{\lambda_0} \Rightarrow (*)_{\lambda_0}$ using $c(\alpha, \beta) = \min\{n : (\forall m)[m \geq n \Rightarrow f_\alpha(m) > f_\beta(m)]\}$.
$(*)_{\lambda_0} \Rightarrow (*)'_{\lambda_0}$ as for each $\alpha < \lambda_0$ and n we define when $f_\alpha(n) \geq \zeta$:

$$f_\alpha(n) \geq \zeta \Leftrightarrow \bigwedge_{\xi < \zeta} (\exists \beta)[\alpha < \beta \ \& \ c(\alpha, \beta) \leq n \ \& \ f_\beta(n) \geq \xi].$$

Now $f_\alpha(n)$ is the minimal value; if it is ∞ we get contradiction to the choice of c, and $[\alpha < \beta \ \& \ c(\alpha, \beta) = n \leq m \Rightarrow f_\alpha(m) > f_\beta(m)]$ is as required.

Claim 6.6 *Assume* (E) *(or just* $(D)_{\theta, \aleph_0, \theta}$*).*
If $\kappa \leq \theta = \mathrm{cf}\mu < \mu < 2^\mu$ *then there is an entangled linear order of cardinality* μ^+.

Proof. By [10], 2.1 for some strictly increasing continuous $\langle \mu_i : i < \theta \rangle$, $\mu = \bigcup_{i < \kappa} \mu_i$ and $\mu^+ = \mathrm{tcf} \prod \mu_i^+ / J^{bd}_\theta$. Now note

$(*)$ for some $i < \kappa$, for every $j \in (i, \kappa)$, $\mu^+ \notin \mathrm{pcf}\{\mu_\alpha^+ : i < \alpha < j\}$.

Now we can choose by induction on $\zeta < \theta$, $i(\zeta) < \theta$ such that $i(\zeta)$ strictly increasing and $\mu_{i(\zeta)} > \max \mathrm{pcf}\{\lambda_j : i < j < \bigcup_{\xi < \zeta} i(\xi)\}$. Now to $\langle \mu_{i(\zeta)}^+ : \zeta < \theta \rangle$ apply [10], 4.12.

References

[1] T. Dodd and R.B. Jensen, *The covering lemma for K*, Ann. of Math Logic **22** (1982), 1–30. [DoJo]

[2] M. Rubin and S. Shelah, *Combinatorial problems on trees: partitions, Δ-systems and large free subsets*, Annals of Pure and Applied Logic **33** (1987), 43–82. [RuSh117]

[3] S. Shelah, *Universal Classes*, (with III, IV, revised from 300, V, VI, VII completed, and 322), O.U.P. (submitted). [Sh-e]

[4] S. Shelah, *Cardinal Arithmetic*, Oxford University Press (most Chapters listed individually), accepted. [Sh-g]

[5] S. Shelah, Appendix: Stationary Sets, Abstract elementary classes, *Proc. of the USA–Israel Conference on Classification Theory*, Chicago 12/85; ed. J. Baldwin, Springer Lecture Notes 1292 (1987) 483–497. [Sh88a]

[6] S. Shelah, On successor of singular cardinals, *Proc. of the ASL Meeting*, Mons, Aug. 1978, *Logic Colloquium 78*, ed. M. Boffa, D. van Dalen and K. McAloon; *Studies in Logic and the Foundation of Math.*, Vol. 97, North Holland Publ. Co., Amsterdam (1979), 357–380. [Sh108]

[7] S. Shelah, Diamonds and Uniformization, *J. of Symb. Logic* [sb 4/82, Ne, acc 10/82, pr 1/84, NSF+BSF] **49** (1984), 1022–1033. [Sh186]

[8] S. Shelah, *The Existence of Coding Sets*, Springer Verlag Lecture Notes Volume 1182 (1986), 188–202. [Sh212]

[9] S. Shelah, Reflection of stationary sets and successor of singulars, *Archiv für Math. Logic* **31** (1991) 25–34. [Sh351]

[10] S. Shelah, $\aleph_{\omega+1}$ has a Jonsson Algebra, [dn 5-7/88, dn Nov 18, 1988]. [Sh355]

[11] S. Shelah, Bounding $pp(\mu)$ when $\mu > cf(\mu) > \aleph_0$, *Cardinal Arithmetic Ch. V*, Oxford Univ. Press, in press. [Sh386]

[12] S. Shelah, Advanced: Cofinalities of Small Reduced Products, *Cardinal Arithmetic*, Oxford Univ. Press, in press. [Sh371]

[13] S. Shelah, There are Jonsson algebras in many inaccessible cardinals, *Cardinal Arithmetic Ch. III*, Oxford Univ. Press, in press. [Sh365]

[14] S. Shelah, Cardinal arithmetic, *Cardinal Arithmetic Ch IX*, Oxford Univ. Press, in press. [Sh400]

[15] S. Shelah, More on cardinal arithmetic, *Archiv für Math. Logic*. [Sh410]

[16] S. Shelah, Further cardinal arithmetic, *Israel J. Math.*, accepted. [Sh430]

Conjectures of Rado and Chang and Cardinal Arithmetic

STEVO TODORČEVIĆ

Matematicki Institut

Kneza Mihaila 35

11001 Beograd, p.p. 367

Jugoslavija

Abstract

We study the following conjecture of Richard Rado: a family of intervals of a linearly ordered set is the union of countably many disjoint subfamilies iff every subfamily of size $\aleph_1$ has this property. We connect it with a well-known two-cardinal transfer principle of model theory known as Chang's Conjecture and show that it solves the Singular Cardinals Problem.

The countable case of a general conjecture proposed by Richard Rado in [11] is the following statement (RC): *a family of intervals (or rather convex sets) of a linearly ordered set is σ-disjoint (i.e., the union of countably many disjoint subfamilies) iff each of its subfamilies of size $\aleph_1$ is σ-disjoint.* The conjecture was based on his result from [10] which says that for an integer k a family of intervals is k-disjoint iff every subfamily of size $k + 1$ has this property. In [17] we proved that RC is equivalent to the following familiar statement about trees: *a tree T is special (i.e., the union of countably many antichains) iff every subtree of T of size $\aleph_1$ is special.* Since every tree T has a conjugate poset T^c (on the same underlying set and with the property that antichains of T are chains of T^c), we see that Rado's Conjecture is a special case of the following statement considered by Fred Galvin (unpublished) as a generalization of Dilworth's chain decomposition theorem ([3]): a partially ordered set is the union of countably many chains iff each of its subposets of size $\aleph_1$ has this property. In [17] we proved that RC is a consistent statement, while the corresponding result for the statement of Galvin is still open. Searching further through the literature for more statements similar to RC one gets the impression that RC might be the weakest statement of its sort ever considered. This explains another side of our interest in RC as possibly the weakest (and therefore more likely well-chosen) compactness principle at the level where almost any other kind of compactness fails badly. In this note we shall prove that RC has many interesting consequences. They will be obtained as applications of our combinatorics of nonspecial trees developed in [16] and [18] which generalizes the usual combinatorics of the uncountable. For example we shall frequently use the following result ([16], [18]) which corresponds to (and strengthens) the classical Pressing Down Lemma of ω_1.

Pressing Down Lemma for Trees. Every regressive mapping defined on a nonspecial tree must be constant on a nonspecial subtree.

N.W. Sauer et al. (eds.), Finite and Infinite Combinatorics in Sets and Logic, 385–398.

We shall also consider the following well-known conjecture from model theory due to C.C. Chang: every model of the form $\langle A, U, ... \rangle$, where A has size $\aleph_2$ and U is a distinguished unary relation of size $\aleph_1$ has an elementary submodel $\langle B, B \cap U, ... \rangle$ such that B is uncountable but $B \cap U$ is countable. This is a well studied statement whose set-theoretic analysis reveals a large cardinal strength that has been exactly determined (see e.g. [6]). In this note we shall show that RC implies a strengthening of Chang's Conjecture strong enough to give a restriction on the size of the continuum.

The results of this note can also be interpreted as results about a large cardinal axiom that seems much weaker than strong compactness (see [6]). To illustrate this, let r_0 be the minimal cardinal θ with the property that every non-σ-disjoint family of intervals of a linearly ordered set contains a subfamily of size $< \theta$ with the same property. Clearly, every strongly compact cardinal is bigger than or equal to r_0. Then, for example, the result of §3 says that the Singular Cardinals Hypothesis holds above r_0, which is a considerable improvement over a result of Solovay [15] which says that SCH holds above a strongly compact cardinal. To get the impression of this improvement consider, for example, the minimal cardinal r_1 which has the property that every graph of uncountable chromatic number has an uncountably chromatic subgraph of size $< r_1$. Or, the minimal cardinal r_2 with the property that every open cover of a Tychonoff cube of the form $\mathbf{N}^A$ has a subcover of size $< r_2$. Or, the minimal cardinal r_3 that gives the compactness to the logic $L_{\omega_1\omega}$, the smallest and the most important of the infinitary extensions of the usual logic $L_{\omega\omega}$ that allows countable disjunctions and conjunctions but still only finite strings of quantifiers. Clearly $r_0 \leq r_1 \leq r_2 \leq r_3 \leq$ the first strongly compact cardinal, simply because an uncountable cardinal κ is by definition strongly compact iff the logic $L_{\kappa\kappa}$ is κ-compact, i.e., a set S of sentences of $L_{\kappa\kappa}$ has a model iff every subset of S of size $< \kappa$ has a model. On the other hand $r_0 = \aleph_2$ is a consistent statement being equivalent to Rado's Conjecture while it is well-known (and easily proved) that r_1, r_2 and r_3 are much bigger than $\aleph_2$.

This paper was written during the academic year 1990–91 when we were receiving financial support from NSERC grants of A. Dow, C. Laflamme, J. Steprans and W. Weiss. We would like to thank these mathematicians as well as F. Tall, S. Watson and other members of the Toronto Set Theory Seminar for their support.

1   Rado's Conjecture and the Continuum

In this section we shall see that RC has a strong influence on the set of reals. In particular, we shall prove the following.

Theorem 1. *Rado's Conjecture implies that the continuum has size at most $\aleph_2$.*

This will be done by analyzing a single mapping

$$e : [\omega_2]^2 \to \omega_1$$

such that $e(\alpha, \gamma) \neq e(\beta, \gamma)$ for $\alpha < \beta < \gamma < \omega_2$. We shall work in the structure $[\omega_2]^\omega$ of all countable infinite subsets of ω_2 and use the corresponding notion of a closed unbounded set and the notion of a stationary set. We shall say that $A \subseteq \omega_2$ is *closed under e* if the following conditions are satisfied:

(a) A is closed under taking predecessors and successors and $A \cap \omega_1$ is an ordinal,

(b) if $\alpha < \beta$ are in A, then $e(\alpha, \beta)$ is also in A,

(c) if $\nu = e(\alpha, \beta)$ and if ν and β are in A, then α is also in A.

For $\nu < \omega_1$ and $\alpha < \omega_2$ let

$$F_\nu(\alpha) = \{\xi < \alpha : e(\xi, \alpha) < \nu\}.$$

Let C_e be the set of all A in $[\omega_2]^\omega$ that are closed under e. Then C_e is a closed and unbounded subset of $[\omega_2]^\omega$. Let

$$S_e = \{A \in C_e : A \neq F_\nu(\alpha) \text{ for all } \nu < \omega_1 \text{ and } \alpha < \omega_2\}.$$

This is the first object that we associate to e. We first mention a simple but useful property of S_e:

Lemma 1. *If $A \subset B$ are two distinct elements of C_e such that $A \cap \omega_1 = B \cap \omega_1$, then A is not an element of S_e.*

Lemma 2. *$S_e \cap [Y]^\omega$ is nonstationary in $[Y]^\omega$ for every uncountable proper subset Y of ω_2.*

Proof. Suppose $S_e \cap [Y]^\omega$ is stationary and work for a contradiction. Note that in this case Y is also closed under e.
CASE 1. $Y \cap \omega_1 \in \omega_1$. Then we can find $A \subset B$ in $S_e \cap [Y]^\omega$ such that $Y \cap \omega_1 \subseteq A$. Then $A \cap \omega_1 = B \cap \omega_1$ contradicting Lemma 1.
CASE 2. $\omega_1 \subseteq Y$. Then $Y = \alpha$ for some α and S_e is disjoint from the set $\{F_\nu(\alpha) : \nu < \omega_1\}$ which is closed and unbounded in $[Y]^\omega$, a contradiction.

The second object that we associate to e is the tree T_e of all countable continuous chains of elements of S_e whose unions are also elements of S_e. Note that if $A \subset B$ are elements of some t from T_e, then by Lemma 1 we must have $A \cap \omega_1 < B \cap \omega_1$. Hence every t in T_e is a well-ordered chain of elements of S_e. The ordering of T_e is, of course, end-extension.

Lemma 3. *$T_e^\delta = \{t \in T_e : \bigcup t \subseteq \delta\}$ is a special subtree of T_e for all $\delta < \omega_2$.*

Proof. The proof is by induction on δ. Consider a t in T_e^δ of a limit length. Let $\nu = (\bigcup t) \cap \omega_1$. Since $\bigcup t \neq F_\nu(\delta)$ there are two cases to consider. If there is $\alpha_t < \delta$ such that $e(\alpha_t, \delta) = \nu_t < \nu$ but α_t is not in $\bigcup t$, let $H_0(t)$ be any proper initial segment s of t such that $\nu_t < (\bigcup s) \cap \omega_1$. In the other case there exist ξ_t in $\bigcup t$ such that $e(\xi_t, \delta) \geq \nu$. Let $H_1(t)$ be any proper initial part s of t such that $\xi_t \in \bigcup s$. This defines a regressive mapping on T_e^δ, so by the PDL for trees it suffices to show that for every s both $H_0^{-1}(s)$ and $H_1^{-1}(s)$ are special trees.

$H_0^{-1}(s)$ *is special.* It suffices to show that for a given $\mu < (\bigcup s) \cap \omega_1$ the set W_μ of all t in $H_0^{-1}(s)$ with property $\nu_t = \mu$ is special. So consider one such μ and let $\alpha < \delta$ be unique such that $e(\alpha, \delta) = \mu$. Then $\alpha \notin \bigcup t$ for all t in W_μ. By the induction hypothesis T_e^α is special so removing some elements from W_μ we may assume that $\bigcup t$ has some elements above α for all t in W_μ. If W_μ is nonspecial then there must be w in W_μ such that the set X_w of all t in W_t extending w is also nonspecial. But consider a β in $\bigcup w$ above α. Since $\bigcup t$ for t in W_w are closed under e we must have that $(\bigcup t) \cap \omega_1 \le e(\alpha, \beta)$ for any such t. This means that X_w does not have elements of length $> e(\alpha, \beta)$ contradicting the fact that W_w is not special.

$H_1^{-1}(s)$ *is special.* It suffices to show that for a given $\xi \in \bigcup s$ the set W_ξ of all t in $H_1^{-1}(s)$ such that $\xi_t = \xi$ is special. Note that for every t in W_ξ, $(\bigcup t) \cap \omega_1 \le e(\xi, \delta)$, so W_ξ does not have elements of height $> e(\xi, \delta)$, and therefore, must be a special subtree. This finishes the proof of Lemma 3.

Lemma 4. T_e *is special iff* S_e *is nonstationary.*

Proof. If S_e is stationary, then T_e is not only nonspecial, it is, in fact, a Baire tree, i.e., the intersection of every sequence $\{D_n : n < \omega\}$ of dense open subsets of T_e is dense. To see this choose a countable elementary submodel M of some large enough H_θ such that $M \cap \omega_2$ is in S_e and M contains all the relevant objects. Now construct an increasing sequence $\{t_n\}$ of elements of $T_e \cap M$ such that $t_n \in D_n \cap M$ and such that if $t_\omega = \bigcup_n t_n$ then $\bigcup t_\omega = M \cap \omega_2$. Then t_ω is an element of $\bigcap_n D_n$.

Suppose now S_e is not stationary and choose an $f : [\omega_2]^{<\omega} \to \omega_2$ such that the club C_f of all elements of $[\omega_2]^\omega$ that are closed under f is a subset of $C_e \setminus S_e$. Consider a t in T_e of limit length. Since $\bigcup t$ is not closed under f there exists a finite subset a_t of $\bigcup t$ such that $f(a_t)$ is not in $\bigcup t$. Let $H(t)$ be any proper initial part s of t such that $a_t \subseteq \bigcup s$. This defines a regressive mapping H on T_e. By the PDL for trees it suffices to show that for every s, $H^{-1}(s)$ is special. For this it suffices to show that for every finite subset a of $\bigcup s$ the set W_a of all t in $H^{-1}(s)$ such that $a_t = a$ is special. Let $\alpha = f(a)$. By Lemma 3 we may remove the elements of T_e^α from W_a, i.e., we may assume that for every t in W_a, the set $\bigcup t$ has ordinals above α. If W_a is not special we can find a w in W_a so that the set X_w of all t from W_a extending w is also nonspecial. Choose β in $\bigcup w$ above α. Then for every t in X_w, since α is not in $\bigcup t$ and since $\bigcup t$ is closed under e, we must have that $(\bigcup t) \cap \omega_1 \le e(\alpha, \beta)$. Hence X_w does not have elements of length $> e(\alpha, \beta)$ and, therefore, must be special, a contradiction. This finishes the proof of Lemma 4.

We have already pointed out the result of [17] which says that RC is equivalent to the statement that a nonspecial tree must contain a nonspecial subtree of size $\aleph_1$. So by Lemmas 3 and 4, RC implies that T_e is a special tree and that S_e is a nonstationary set. To finish the proof of Theorem 1 all we need now is a result of Baumgartner and Taylor [1] that every closed and unbounded subset of $[\omega_2]^\omega$ has size at least $\mathfrak{c}$ together with the observation that the complement of S_e in C_e has size $\aleph_2$.

We have kept our promise by giving a proof of Theorem 1 based on an analysis of a single mapping $e : [\omega_2]^2 \to \omega_1$. Of course, RC has similar influence on S_f for any other such mapping f. It turns out that the statement that S_f is nonstationary for every f gives a severe restriction on the set of reals. Probably the simplest and at the same time the most informative result in this direction is the following strengthening of Theorem 1.

Theorem 2. *If RC holds then every transitive model M satisfying a sufficiently large fragment of ZFC and having the property $\omega_2^M = \omega_2$ must contain all reals.*

Proof. Suppose $\mathbf{R} \not\subseteq M$. Then by a result of Gitik [5], $S = [\omega_2]^\omega \setminus M$ is stationary in $[\omega_2]^\omega$. Choose an $f : [\omega_2]^2 \to \omega_1$ such that $f \in M$ and $f(\alpha, \gamma) \neq f(\beta, \gamma)$ for all $\alpha < \beta < \gamma < \omega_2$. Note that $S \cap C_f$ is a subset of S_f. Hence S_f is also stationary. But this contradicts RC.

Thus S_e is stationary (and, therefore, RC is false) in any forcing extension that adds a new real and preserves ω_1 and ω_2. Rado's Conjecture is not strong enough to decide which of the two remaining values is equal to the continuum. In [17] we have proved the consistency of RC and CH while the consistency of RC and $\underline{c} = \aleph_2$ is obtained by applying the lemmas of [17] to the model of Mitchell [7].

2   Chang's Conjecture

Recall that Chang's Conjecture is the statement that every structure (with countable similarity type) of the form $\langle A, U, ... \rangle$, where A has size $\aleph_2$ and U is a distinguished unary relation of size $\aleph_1$, has an elementary substructure of the form $\langle B, B \cap U, ... \rangle$ such that B is uncountable but $B \cap U$ is countable. Statements of this form are well studied both in model theory and set theory and are closely related to some large cardinal axioms (see [6]). In this section we shall present the following connection between RC and CC.

Theorem 3. *Rado's Conjecture implies Chang's Conjecture.*

The proof will actually give a stronger result that will also include Theorem 1 of §1. It involves the following version of the cut and choose game of Mycielski and Ulam ([8], [21]) due to Galvin [4]. The game $G_\omega(\omega_2, \omega_1)$ has two players I and II. Player I starts by splitting ω_2 into ω_1 pieces, i.e., choosing an $f_0 : \omega_2 \to \omega_1$. II responds by choosing countably many of the pieces, i.e., by choosing a countable ordinal δ_0. Then player I chooses another $f_1 : \omega_2 \to \omega_1$ and II responds by choosing $\delta_1 < \omega_1,...$ and so on for ω steps. Player II wins a play $f_0, \delta_0, f_1, \delta_1, ...$ iff the set

$$\{\alpha < \omega_2 : \ f_n(\alpha) < \sup_i \delta_i \text{ for all } n < \omega\}$$

is unbounded in ω_2; otherwise I wins. The following result is strong enough to give Theorem 3 as well as the results of §1.

Theorem 4. *RC implies that II has a winning strategy in the game $G_\omega(\omega_2, \omega_1)$.*

To deduce Theorem 3 from Theorem 4 we consider the following strengthening of Chang's Conjecture which we denote by CC*: for every large enough regular cardinal θ and a countable elementary submodel M of $\langle H_\theta, \in, < \rangle$, where $<$ is a well-ordering of H_θ, there exists a countable elementary submodel M^* of $\langle H_\theta, \in, < \rangle$ such that $M \subseteq M^*$, $M^* \cap \omega_1 =$

$M \cap \omega_1$ but $M^* \cap \omega_2 \neq M \cap \omega_2$. Now, Theorem 3 follows from Theorem 4 and the following result of Shelah [12; p. 398].

Lemma 5. *If* II *has a winning strategy in the game* $G_\omega(\omega_2, \omega_1)$, *then* CC* *holds.*

Proof. Take a countable elementary submodel M of $\langle H_\theta, \in, < \rangle$. Then M contains a winning strategy σ for II in $G_\omega(\omega_2, \omega_1)$. Let $\{f_n : n < \omega\}$ enumerate all functions from ω_2 into ω_1 that are elements of M. Playing $f_0, f_1, \dots$ against σ we conclude that there is an ordinal $\alpha < \omega_2$ above $M \cap \omega_2$ such that $f_n(\alpha) \in M \cap \omega_1$ for all $n < \omega$. Let M^* be the Skolem closure of $M \cup \{\alpha\}$ in $\langle H_\theta, \in, < \rangle$. Then M^* is as required.

To give another proof of Theorem 1 all we need is the following fact.

Lemma 6. CC* *implies that for every stationary* $S \subseteq [\omega_2]^\omega$ *there exists* $\alpha < \omega_2$ *such that* $S \cap [\alpha]^\omega$ *is stationary.*

Proof. Suppose $S \cap [\alpha]^\omega$ is not stationary for all $\alpha < \omega_2$. Then for each $\alpha < \omega_2$ we can choose one-to-one $e_\alpha : \alpha \to \omega_1$ such that for every limit $\nu < \omega_1$ the set $A_{\nu\alpha}$ of all $\xi < \alpha$ such that $e_\alpha(\xi) < \nu$ is not in S. If S is stationary then there is a countable elementary submodel M of some large enough $\langle H_\theta, \in, < \rangle$ such that S and $\{e_\alpha : \alpha < \omega_2\}$ are elements of M and $M \cap \omega_2$ is in S. Let M^* be the submodel guaranteed by CC*. Then every ordinal of $M^* \cap \omega_2$ that is not in $M \cap \omega_2$ is above every ordinal of $M^* \cap \omega_2$. So let α be the minimal ordinal from the difference of these two sets. Then e_α is an element of M^* and by the elementarity of M^*, $M \cap \omega_2$ is equal to $A_{\nu\alpha}$, where $\nu = M^* \cap \omega_1 = M \cap \omega_1$. But this contradicts the fact that $M \cap \omega_2$ is an element of S. This finishes the proof.

Hence CC* implies that the sets considered in §1 are all nonstationary. It follows that CC* implies the conclusions of Theorems 1 and 2 of §1. Note that Chang's Conjecture itself does not have anything to do with the size of the continuum. This follows from the well-known (and easily checked) fact that CC is preserved by *ccc* forcings. Thus, in particular, CC does not imply CC* nor RC.

Now we concentrate on the proof of Theorem 4 that involves an analysis of the game $G = G_\omega(\omega_2, \omega_1)$. Let $F(\omega_2, \omega_1)$ denote the union of ω_1 and the set of all mappings from ω_2 into ω_1. For A in $[F(\omega_2, \omega_1)]^\omega$, let

$$D_A = \{\alpha < \omega_2 : f(\alpha) \in A \text{ for all } f \text{ in } A\}.$$

Finally, set

$$S_G = \{A \in [F(\omega_2, \omega_1)]^\omega : A \cap \omega_1 \in \omega_1 \text{ and } D_A \text{ is bounded in } \omega_2\}.$$

Our interest in S_G is based on the following simple fact.

Lemma 7. II *has a winning strategy in* G *iff* S_G *is nonstationary.*

For A and B in S_G we shall say that B *strongly includes* A if $A \subset B$ and $A \cap \omega_1 < B \cap \omega_1$. Let T_G be the set of all countable continuous strong-inclusion-chains t of elements of S_G

such that $\bigcup t$ is also an element of S_G. The ordering of T_G is end-extension. Then T_G is a tree that is very closely related to the game G as the next two lemmas show.

Lemma 8. *Every subtree of T_G of size $\aleph_1$ is special.*

Proof. Let U be a subtree of T_G of size $\aleph_1$ that is closed under taking initial parts. Let $\delta < \omega_2$ be a bound to all D_A for A appearing in an element of U. Consider a t in U that has a limit length. Since δ is a bound of D_A for $A = \bigcup t$ there exists f_t in $\bigcup t$ such that $f_t(\delta) \geq (\bigcup t) \cap \omega_1$. Let $H(t)$ be any proper initial part s of t such that f_t is an element of $\bigcup s$. This defines a regressive mapping H on U. By the PDL for trees it suffices to show that $H^{-1}(s)$ is special for every s in U. For this it suffices to show that for every f in $\bigcup s$ the set W_f of all t in $H^{-1}(s)$ such that $f_t = f$ is a special tree. To see this note that for every t in W_f, $f(\delta) \geq (\bigcup t) \cap \omega_1$ so t cannot have length $> f(\delta)$ (since its elements are increasing under the strong inclusion). Hence W_f must be special. This finishes the proof.

By Lemma 8, RC implies that the whole tree T_G must be special so the proof of Theorem 3 is finished once we show the following.

Lemma 9. *If T_G is special, II has a winning strategy in the game G.*

Proof. By Lemma 7 it suffices to show that if S_G is stationary, then T_G is not special. In fact, in this case T_G is a Baire tree. This is proved by an argument from the proof of Lemma 4.

3   Cardinal Arithmetic

In this section we show that RC solves the Singular Cardinals Problem, i.e., that 2^θ for singular cardinals θ has the least possible value depending on the values for regular cardinals less than θ. This will be deduced from the following generalization of Theorem 1 of §1.

Theorem 5. RC *implies $\theta^{\aleph_0} = \theta$ for regular $\theta \geq \aleph_2$.*

The proof is by induction on θ. The result of Theorem 1 is of course the case $\theta = \aleph_2$. Note that the first difficulty occurs for $\theta = \kappa^+$ when κ is a singular cardinal of cofinality ω. The general structure of the proof is very similar to that of §1 but the real difficulty lies in finding the analogues of set S_e and the corresponding T_e.

We start by choosing an

$$e : [\kappa^+]^2 \to \kappa$$

such that $e(\alpha, \gamma) \neq e(\beta, \gamma)$ for $\alpha < \beta < \gamma$. We also fix an increasing sequence $\{\kappa_i\}$ of uncountable regular cardinals that converges to κ and consider the following two versions of the ordering of eventual dominance in the product $\prod_i \kappa_i$: $a <^* b$ (resp. $a \leq^* b$) iff $a(i) < b(i)$ (resp. $a(i) \leq b(i)$) for all but finitely many integers i. The second object which we choose is a sequence

$$A = \{a_\alpha : \alpha < \kappa^+\} \subseteq \prod_i \kappa_i,$$

which is closed under finite changes of its elements, which is $\leq^*$-increasing, i.e., $a_\alpha \leq^* a_\beta$ for $\alpha < \beta$, and such that for every α there is β such that $a_\alpha <^* a_\beta$. We shall also consider the ordering $<$ of everywhere dominance between the elements of the product $\prod_i \kappa_i$ defined by: $a < b$ iff $a(i) < b(i)$ for all i. For a countable $<$-increasing sequence $t = \{t_\xi : \xi < \lambda\}$ of elements of A we let $\sup t$ be the element b of $\prod_i \kappa_i$ defined by

$$b(i) = \sup\{t_\xi(i) : \xi < \lambda\}.$$

Note that $\sup t' = \sup t$ for every subsequence t' of t which is cofinal in t. For $\delta < \kappa^+$ and $i < \omega$, set

$$A_i(\delta) = \{a_\alpha : \alpha < \delta \text{ and } e(\alpha, \delta) < \kappa_i\}.$$

Then $A_i(\delta)$ has size at most κ_i, so the set $A_i^*(\delta)$ of all $\sup t$ for t a countable $<$-increasing sequence of elements of $A_i(\delta)$ also has size $\leq \kappa_i$. This follows from our inductive hypothesis $\kappa_i^{\aleph_0} = \kappa_i$. Let

$$A^* = \bigcup_{\delta < \kappa^+} \bigcup_{i < \omega} A_i^*(\delta).$$

Then A^* has size κ^+ and it is the set that is going to play the role of complement of the set of S_e of §1. The associated tree T_A we choose to be the set of all $<$-increasing countable sequences t of elements of A such that t has a successor length (i.e., $t = \{t_\xi : \xi \leq \mu\}$ for some μ) and satisfies the following restriction:

(∗) $\sup\{t_\xi : \xi < \lambda\} \notin A^*$ for every limit $\lambda \leq \mu$.

The ordering of T_A is of course the end-extension. The following lemma contains the crucial point of the proof of Theorem 5.

Lemma 10. *Every subtree of T_A of size $\leq \kappa$ is special.*

Proof. Let U be a subtree of T_A of size κ and let $\delta < \kappa^+$ be such that if an a_α occurs in a sequence of U, then $\alpha < \delta$. For $t \in T_A$ let a_t denote the last element of t. Then for every i we can define U_i to be the set of all t in U such that a_t is an element of $A_i(\delta)$. It suffices to show that U_i is special for every integer i. In fact we shall show that no element of U_i has infinitely many predecessors in U_i. For suppose there is u in U_i with infinitely many predecessors $t_0 \subset t_1 \subset \dots$ in U_i. For $n < \omega$, let $\lambda_n + 1$ be the length of t_n, and let $\lambda = \sup_n \lambda_n$. Then λ is a limit ordinal less than or equal to the length of u. Since u is a $<$-increasing sequence it follows that

$$\sup\{u_\xi : \xi < \lambda\} = \sup\{u_{\lambda_n} : n < \omega\}.$$

Since $u_{\lambda_n} = a_{t_n}$ are all elements of $A_i(\delta)$ it follows that

$$\sup\{u_\xi : \xi < \lambda\} \in A_i^*(\delta),$$

contradicting (∗) for u. This finishes the proof.

Now, RC implies that T_A must be a special tree (see [17]), so the following lemma completes the inductive step $\theta = \kappa^+$ in the proof of Theorem 5.

Lemma 11. *If T_A is special, then $\kappa^{\aleph_0} = \kappa^+$.*

Proof. Choose a $\leq^*$-decreasing sequence $\{b_\xi : \xi < \lambda\}$ of eventually different elements of $\prod_i(\kappa_i + 1)$ which are $<^*$-bounds of A and such that $\{b_\xi : \xi < \lambda\}$ cannot be properly end-extended to a sequence satisfying these conditions. An argument of Shelah [13; p. 62] (see also [20; pp. 359–360]) shows that λ must be a successor ordinal. [First observe that $\lambda < \underline{c}^+$ follows from $\underline{c}^+ \to (\omega)^2_\omega$. Let $B_i = \{b_\xi(i) : \xi < \lambda\}$ for $i < \omega$. If λ is a limit ordinal, the product $\prod_i B_i$ would contain κ^+ different elements contradicting the fact that κ is bigger than $\underline{c}$.] So, let $b = b_{\lambda-1}$, and

$$A_b = \{a \in A : a < b\}.$$

Claim 1. Every element of $\prod_i b(i)$ is $<$-dominated by an element of A_b.

Proof. Since A_b is also closed under finite changes, it suffices to show that A_b is $<^*$-cofinal in $\prod_i b(i)$. Suppose there is an x in the product that is not dominated by any element of A_b. Then since κ^+ is bigger than the continuum there exists $<^*$-cofinal $B \subseteq A_b$ and infinite $I \subseteq \omega$ such that $a(i) < x(i)$ for all a in B and i in I. Let b_λ be equal to x on I and to b $(= b_{\lambda-1})$ on the complement of I. Then b_λ can be added on the top of the sequence $\{b_\xi : \xi < \lambda\}$ contradicting its maximality.

Let λ_i be the cofinality of $b(i)$. Then for the same cardinality reasons the sequence $\{\lambda_i\}$ must converge to κ. Working below with a subsequence of $\{\lambda_i\}$ we may assume that the λ_i's are increasing and uncountable. Let T_{A_b} be the subtree of T_A consisting only of sequences of elements of A_b. The conclusion of the lemma will be deduced assuming only that T_{A_b} is special and, in fact, that it is not Baire. By working above a fixed element of the tree we may assume that there is a sequence $\{D_i\}$ of upwards closed cofinal subsets of T_{A_b} such that $\bigcap_i D_i = \emptyset$.

Choose a sequence $\{U_i\}$ of antichains of T_{A_b} such that:

(a) $U_0 = \{\emptyset\}$, $U_{i+1} \subseteq D_i$ and $|U_{i+1}| = \lambda_i$,

(b) every element of U_i has λ_i successors in U_{i+1} and every element of U_{i+1} has a predecessor in U_i,

(c) for every t in U_i the set of all ordinals of the form $a_u(i)$ for u in U_{i+1} extending t is cofinal in $b(i)$ and has order type λ_i.

Then $U = \bigcup_i U_i$ is a subtree of T_{A_b} of height ω in which every branch (=maximal chain) has type ω. Let $[U]$ denote the set of all branches of U. Then every $\{t_i\}$ of $[U]$ will be identified with the sequence $t = \bigcup_i t_i$ of elements of A_b. A subtree V of U will be called *properly splitting* if for every i and t in $V \cap U_i$, the set of all u in $V \cap U_{i+1}$ extending t has size λ_i. Fixing such a V and a t in $V \cap U_i$, we see that the set of all branches of V containing t is covered by λ_i sets of the form

$$C_\alpha = \{v \in [V] : [\sup v](i) \leq \alpha\},$$

where $\alpha < b(i)$. Since C_α's are closed in $[V]$, we can apply a lemma of Namba and Prikry [9] and get a properly splitting subtree $W_t \subseteq V$ containing t (as a minimal element) and an

$\alpha < b(i)$ such that $[W_t] \subseteq C_\alpha.$[1] Starting with U and its minimal element $\emptyset$ and successively applying this procedure, we can find a properly splitting subtree $W \subseteq U$ containing $\emptyset$ such that for all i,

(d) if s and t are two distinct elements of $W \cap U_i$ and if u and v are elements of $[W]$ containing s and t, respectively, then $[\sup u](i) \neq [\sup v](i)$.

It follows that if $\{u_i\}$ and $\{v_i\}$ are two distinct ω-branches of W and if $u = \bigcup_i u_i$ and $v = \bigcup_i v_i$, then

$$[\sup u](i) \neq [\sup v](i)$$

for almost all i. Note that by Claim 1 for every ω-branch $\{t_i\}$ of W the sequence $t = \bigcup_i t_i$ can be extended to a sequence t' by adding to it a single element of A_b that $<$-dominates everything in t. Since t' cannot be an element of T_{A_b} it means that it does not satisfy the requirement (∗). There can be only one reason for this and this is that the function $\sup t$ is an element of A^*. Since A^* has size κ^+ and since W has $|\prod_i \lambda_i| = \kappa^{\aleph_0}$ many ω-branches, the proof of Lemma 1 is finished.

Theorem 6. RC *implies* $\theta^\lambda = \theta$ *for regular* θ *and* λ *such that* $\theta \geq 2^\lambda$.

Proof. The proof is by induction on θ and λ with Theorem 5 as the case $\lambda = \aleph_0$. The only difficulty occurs when $\theta = \kappa^+$ for some singular κ. In this case $\theta^\lambda = \theta$ follows either from the induction hypothesis on λ (if $\mathrm{cf}\kappa < \lambda$) or from the induction hypothesis on θ and a theorem of Silver [14] (if $\mathrm{cf}\kappa \geq \lambda$).

Corollary 7. RC *implies* $\theta^{\mathrm{cf}\theta} = \theta^+ \cdot 2^{\mathrm{cf}\theta}$ *for every infinite cardinal* θ.

The conclusion of Corollary 7 is usually called the Singular Cardinals Hypothesis, SCH. Note that it implies $2^\theta = \theta^+$ for every strong limit singular cardinal θ. This paper can be considered not only as a study of a certain hypothesis RC but also as a study of a certain large cardinal axiom. To see this, let $\mathbf{r}_0$ be the minimal cardinal θ (if there is one) such that every non-σ-disjoint family of intervals of a linearly ordered set must contain a subfamily of size less than θ with the same property. Clearly $\mathbf{r}_0$ is not bigger than the first strongly compact cardinal if such a cardinal exists (see [6]). Note also that RC is equivalent to the statement that $\mathbf{r}_0 = \aleph_2$. It should be clear that the proof of Theorem 5 also gives the following fact about $\mathbf{r}_0$.

Theorem 8. $\theta^\lambda = \theta$ *for regular* θ *and* λ *such that* $\theta \geq \mathbf{r}_0^\lambda$.

Corollary 9. $\theta^{\mathrm{cf}\theta} = \theta^+ \cdot 2^{\mathrm{cf}\theta}$ *for every* $\theta \geq \mathbf{r}_0$.

---

[1]At the suggestion of the referee we give an explanation of this point. Let $\mathcal{P}$ be the set of all properly splitting subtrees of the tree of all sequences appearing in one of the finite subproducts $\prod_{i<n} \lambda_i$ ($n < \omega$). We let $\mathcal{P}$ be ordered by inclusion. This is Namba's version of a poset of K. Prikry ("On models using perfect sets", a preprint originally planned to appear in the Proceedings of the 1967 U.C.L.A. Summer Institute of Set Theory). What Namba and Prikry prove ([9]) is that the corresponding Boolean algebra is (ω, λ_i)-distributive for all $i < \omega$ which immediately gives the statement about closed sets C_α that we need. These types of arguments are by now quite frequent in the literature and we refer the reader to [2] for a detailed historical discussion.

Hence, the Singular Cardinal Hypothesis holds above the cardinal r_0. Note that this is a considerable improvement over a result of Solovay [15] which says that SCH holds above the first strongly compact cardinal.

The construction of §1 suggests another route to the SCH. To see this, let RP denote the statement that for every uncountable θ and a stationary $S \subseteq [\theta]^\omega$ there exists $X \subseteq \theta$ of size $\aleph_1$ such that $S \cap [X]^\omega$ is stationary in $[X]^\omega$. Note that §1 in particular shows that RP for $\theta = \aleph_2$ implies $\aleph_2^{\aleph_0} = \aleph_2$. This led us to conjecture (several years ago) that RP implies $\theta^{\aleph_0} = \theta$ for every regular cardinal $\theta \geq \omega_2$. This has been essentially proved by Velickovic [22] using a strengthening of RP, call it RP*, which requires the set X to be closed under a given $f : S \to \theta$ at least when restricted to a stationary subset of $S \cap [X]^\omega$. We don't know whether there is any connection between RC and RP (or RP*) but we do know that the following weakening of both RC and RP* (and possibly of RP?), denoted by RC^l, is still sufficient to give us the SCH: *if a linearly ordered set is not the union of countably many well-ordered subsets then it must contain a subset of size $\aleph_1$ with the same property.* To see the connection, the reader is reminded of our result from [17] which says that RC is equivalent to the statement that if a partially ordered set is not the union of countably many well-founded subsets then it must contain a subset of size $\aleph_1$ with the same property. We finish this section with a remark which might be useful in applications of these ideas. Note that all consequences of RC given in this paper are in fact deduced from the following weakening of RC (which can naturally be denoted by RC^b): *every nontrivial Baire tree contains a nonspecial subtree of size $\aleph_1$.* Thus, it is expected that a Baire tree T, all of whose small subtrees are special, might be a more useful object in applications than such a tree T which is merely nonspecial.

4 Square Sequences

In this section we study the effect of RC on square sequences (see [6] for definitions). It turns out that it is very similar to the effect (discovered in [19]) of the Proper Forcing Axiom on such sequences.

Theorem 10. *Assume RC. Let $\theta > \omega_1$ be a regular cardinal and let $\Gamma \subseteq \theta$ be a set of limit ordinals that contains all ordinals less than θ of cofinality ω_1. Then for every sequence $\langle C_\alpha : \alpha \in \Gamma \rangle$ with properties*

(i) *C_α is a closed and unbounded subset of α,*

(ii) *if α is a limit point of C_β, then $\alpha \in \Gamma$ and $C_\alpha = C_\beta \cap \alpha$,*

there is a closed and unbounded $C \subseteq \theta$ such that $\alpha \in \Gamma$ and $C_\alpha = C \cap \alpha$ for every limit point α of C.

Proof. The key idea of the proof is the same as that of [19] though we shall use its variation given by Velickovic [23]. More precisely, the tree that we associate to $\langle C_\alpha : \alpha \in \Gamma \rangle$ is essentially the variation of our original poset given in [23], while the fact that its speciality implies the desired conclusion is an immediate consequence of the main Claim of [19]. For the convenience of the reader we shall reproduce here the argument of that Claim. Let T

be the set of all closed countable subsets of θ such that if α is a limit point of t and if α is in Γ, then $C_\alpha \cap t$ must be bounded in α. The ordering of T is end-extension.

Claim 1. *Every subtree U of T of size $\aleph_1$ is special.*

Proof. Let δ be the minimal ordinal such that $t \subseteq \delta$ for all t in U. The proof is by induction on δ. We may assume that $\mathrm{cf}\delta = \omega_1$ and so δ is a member of Γ. Let t be an element of U of limit height. If $\alpha = \max(t)$ is not a limit point of C_δ we let $H_0(t)$ be a proper initial part s of t such that the interval $(\max(s), \max(t))$ has no points of C_δ. If α is a limit point of C_δ, then by (ii) α is an element of Γ and $C_\alpha = C_\delta \cap \alpha$. By the definition of T, $t \cap C_\alpha$ is bounded in α, so let $H_1(t)$ be any proper initial part s of t such that $t \setminus s$ has no points in C_δ except $\alpha = \max(t)$. To show that U is special by the Pressing Down Lemma for trees, it suffices to show that $H_0^{-1}(s)$ and $H_1^{-1}(s)$ are special for all s in U. (We are assuming here that U is closed under taking initial parts of its elements.) Since clearly $H_1^{-1}(s)$ is an antichain for all s, we may concentrate only on $H_0^{-1}(s)$. Let γ be the minimal point of C_δ above $\max(s)$. Then $\max(t) \leq \gamma$ for all t in $H_0^{-1}(s)$. So $H_0^{-1}(s)$ is special by the inductive hypothesis on δ.

By RC, the whole tree must be special. It turns out that the speciality of T is equivalent to the conclusion of Theorem 10, though we shall prove only that if the conclusion fails then the tree T is Baire. So let $\{D_i\}$ be a given sequence of cofinal subsets of T closed upwards, and let t_0 be a given element of T. Choose a continuous $\in$-chain $\{N_\xi : \xi < \theta\}$ of elementary submodels of $H_{(2^\theta)^+}$ of size $< \theta$ such that N_0 contains all relevant objects and such that $\alpha_\xi = N_\xi \cap \theta$ is an ordinal in θ for all ξ. Then $C = \{\alpha_\xi : \xi < \theta\}$ is closed and unbounded in θ. Choose an elementary submodel M of $H_{(2^\theta)^+}$ containing all these objects such that $\delta = M \cap \theta$ is an ordinal of cofinality ω_1. Then $\delta \in C \cap \Gamma$, and so C_δ is defined. By the elementarity of M and the assumption that the conclusion of Theorem 10 fails, for every $\gamma < \delta$ there must be $\xi < \delta$ such that $\alpha_\xi > \gamma$ and $\alpha_\xi \notin C_\delta$. So we can choose an increasing sequence $\{\xi_i\}$ of ordinals $< \delta$ such that $\alpha_{\xi_i} \notin C_\delta$ and $(\alpha_{\xi_i}, \alpha_{\xi_{i+1}}) \cap C_\delta \neq \emptyset$ for all i. Let $\alpha = \sup_i \alpha_{\xi_i}$. Then α is a limit point of C_δ, so $\alpha \in \Gamma$ and $C_\alpha = C_\delta \cap \alpha$. Inductively on i, starting from t_0, we construct a strictly increasing sequence $\{t_i\}$ of elements of T such that for all i, t_{i+1} is an element of $D_i \cap N_{\xi_i}$ and $t_{i+1} \setminus t_i$ is a subset of the interval

$$(\max(C_\alpha \cap \alpha_{\xi_i}), \alpha_{\xi_i}).$$

Then the $\max(t_i)$'s converge to α, so if we let t be the union of the t_i's together with the point α, we get an element of T. Since t extends t_0 and since it is an element of $\bigcap_i D_i$, we are done. This finishes the proof.

Corollary 11. *If RC holds, then $\square_\kappa$ fails for every uncountable cardinal κ.*

Note that if the class Γ of Theorem 10 contains every ordinal $(< \theta)$ of uncountable cofinality less than or equal to some fixed cardinal κ, the proof shows that the associated tree T has all subtrees of size $\leq \kappa$ special. So we have the following result about the cardinal r_0 defined at the end of §3.

Theorem 12. *$\square_\kappa$ fails for all $\kappa \geq \mathrm{r}_0$.*

This should be compared with a result of Solovay [15] which says that $\square_\kappa$ fails above a strongly compact cardinal.

References

[1] J. Baumgartner and A. Taylor, Saturation properties of ideals in generic extensions, II., *Trans. Amer. Math. Soc.* **271** (1982), 587–609.

[2] L. Bukovsky and E. Coplakova, Minimal collapsing extension of models of ZFC, *Ann. Pure Appl. Logic* **46** (1990), 265–298.

[3] R.P. Dilworth, A decomposition theorem for partially ordered sets, *Ann. of Math* **51** (1950), 161–166.

[4] F. Galvin, Unpublished Notes on Games. $\approx$ 1980.

[5] M. Gitik, Nonsplitting stationary subsets of $\wp_\kappa(\kappa^+)$, *J. Symbolic Logic* **50** (1985), 881–894.

[6] A. Kanamori and M. Magidor, *The evolution of large cardinal axioms in set theory*, Lecture Notes in Math. **669** (1978), Springer–Verlag, Berlin, Heidelberg, New York.

[7] W.J. Mitchell, Aronszajn trees and the independence of the transfer property, *Ann. Math. Logic* **5** (1972), 21–46.

[8] J. Mycielski, On the axiom of determinateness (II), *Fund. Math.* **59** (1966), 203–212.

[9] K. Namba, Independence proof of (ω, ω_α)–distributive law in complete Boolean algebras, *Comment. Math. Univ. St. Pauli.* **XIX** (1970), 1–12.

[10] R. Rado, Covering theorems for ordered sets, *Proc. London Math. Soc. (2)* **50** (1946), 509–535.

[11] R. Rado, Theorems on intervals of ordered sets, *Discrete Math.* **35** (1981), 199–201.

[12] S. Shelah, *Proper Forcing*, Lecture Notes in Math. **940** (1982), Springer–Verlag, Berlin, Heidelberg, New York.

[13] S. Shelah, Jonsson algebras in successor cardinals, *Israel J. Math.* **30** (1978), 57–64.

[14] J.H. Silver, On the singular cardinal problem, *Proc. Int. Congr. Math.*, Vancouver 1974.

[15] R.M. Solovay, Strongly compact cardinals and the GCH, in: *Proc. Symp. Pure Math. Vol 25.*, American Math. Soc., Providence, 1974.

[16] S. Todorčević, Stationary sets, trees and continua, *Publ. Inst. Math. Beograd* **43** (1981), 249–262.

[17] S. Todorčević, On a conjecture of R. Rado, *J. London Math. Soc. (2)* **27** (1983), 1–8.

[18] S. Todorčević, Partition relations for partially ordered sets, *Acta Mathematica* **155** (1985), 1–25.

[19] S. Todorčević, A note on the proper forcing axiom, *Contemporary Math.* **31** (1984), 209.

[20] S. Todorčević, Remarks on cellularity in products, *Compositio Math.* **57** (1986), 357–372.

[21] S. Ulam, Combinatorial analysis in infinite sets and some physical theories, *SIAM Rev.* **6** (1964), 343–355.

[22] B. Veličković, Forcing axioms and stationary sets, preprint, 1986.

[23] B. Veličković, Jensen's □ principle and the Novak number of partially ordered sets, *J. Symbolic Logic* **51** (1986), 47–58.

Random Structures and Zero–One Laws

PETER WINKLER

Bell Communications Research
445 South St.
Morristown, NJ 07962-1910 USA

Abstract

Random structures (such as the classical random graphs of Erdős and Rényi) are playing an increasingly large role in the theory of computing, as well as in discrete mathematics. The surprising and useful fact that random structures can have properties not found in known constructions—indeed, that they can "have properties" at all—rests on the phenomenon of 0-1 laws; that is, on the fact that for many properties P the probability that a random structure satisfies P is guaranteed to approach either 0 or 1.

We will make a tour of various spaces of random structures, noting when 0-1 laws hold and when they do not; proofs or sketches of proofs are given for the major results. We will also discuss progress in two directions, characterizing structures where the 0-1 law holds for first-order logic, and extending the 0-1 law as far as possible to more expressive logics.

Ultimately, we wish to improve both our own and our readers' intuition about how random structures behave.

1 Introduction

One hundred gentlemen check their hats at a fancy-dress ball, but the hat-check person loses all the slips and hands the hats back at random after the event. What is the probability that at least one man gets his own hat back?

This classic question, no doubt familiar to most readers, has been known as the "problème des rencontres" in France, the "derangement problem" in England and "the hat-check girl" (now "person") in America. An easy inclusion-exclusion argument shows that for n men the probability is precisely

$$1 - \frac{1}{2!} + \frac{1}{3!} - \frac{1}{4!} + \cdots - \frac{(-1)^n}{n!}$$

which rapidly approaches $1 - 1/e \approx 63\%$, so rapidly in fact that the answer is pretty much the same for 7 men as for a billion.

If the problem is posed to a class of bright, eager undergraduates (or even the more usual kind) all or nearly all will guess either that the probability approaches zero or that it approaches one, as the number of men grows. After they see the answer they will concede that their intuition was lousy, but in fact it can be argued that their intuition is excellent; they were faced with an atypical example. (Later we shall see an example involving random intervals which is in one respect even more bizarre.)

399

N.W. Sauer et al. (eds.), Finite and Infinite Combinatorics in Sets and Logic, 399–420.

In mathematics we are accustomed to dealing with properties that are quite special and cannot be expected to hold in a "random" case; or, less commonly, properties that nearly always hold. (A useful clue is that if a property is called "regular" or "normal" it is rare indeed!) Frequently we imagine a "generic case" in which nothing special happens; thus a generic finite set of points in the plane contains no three that are collinear; a generic topological space won't be Hausdorff; a generic Hausdorff space won't be locally compact.

Since *constructed* objects tend to have unusual properties, it may happen that generic objects, despite their numerical dominance, are hard to come by. Thus, we may wish to find generic objects by random selection; indeed, there are now myriad examples, in discrete mathematics and theoretical computer science especially, of useful properties that are obtainable from random structures but not via known deterministic constructions.

But can randomly chosen structures be said to "have" properties? Our intuition, especially since the inception of the study of random graphs by Erdős and Rényi, has been that if we find a "space" of random structures and look at larger and larger structures from that space, they will begin to look alike—that properties in which we are interested will either hold in nearly all cases or fail in nearly all cases.

Beginning with the work of Glebskii et al. [11] and Fagin [10], we now have theorems to back our intuition. The theorems are called "zero-one laws" and assert that, for a given space of random structures and a given class of properties, the limit of the probability that any property in the class holds in increasingly large structures from the space is either 0 or 1. Basically, this work is a survey devoted to the question: when do 0-1 laws hold, and when do they not?

In §2, we take a tour through various spaces of random structures including models of random graphs and random partially ordered sets, giving a few examples of applications and techniques.

§3 presents the classical 0-1 law for random graphs, with (our version of) Fagin's proof, plus commentary. A brief description of first-order logic and its main properties is included.

In §4, we examine the 0-1 law situation for random (partial) orders and see that the choice of model is critical. Some counterexamples to 0-1 laws are presented, with explanations of the contexts in which they arose and the techniques for dealing with them.

§5 returns to random graphs, but to the much more challenging case where the edge probability is a function of the number of vertices. The results of Shelah and Spencer are described, together with an attempt to de-mystify their proofs.

In a brief §6, we describe the curious case of random intervals.

In §7, we give a brief description of Compton's results, which classify spaces of structures according to growth rate with the idea of predicting whether 0-1 laws will hold, or even whether first-order sentences will have limiting (i.e., converging) probabilities.

Finally, in §8, we outline the (mostly very recent) progress in extending 0-1 laws to the most expressive logics possible.

2   Some Examples of Spaces of Random Structures

By a "structure" we will always mean a set with various functions and relations defined upon it. The "size" or "cardinality" of a structure is the number of elements in its underlying set.

A *space of random structures* is actually a sequence of probability spaces. A class **C** of finite structures is specified, and for each positive integer n a probability distribution is given on the class $\mathbf{C}_n$ of structures from **C** of size n. Typically, the structures in $\mathbf{C}_n$ are all considered to have a fixed, labeled underlying set (e.g. the numbers from 1 to n); but we may also consider $\mathbf{C}_n$ to be a set of isomorphism classes of structures, in which case we get a space of "unlabeled" random structures.

Often the probability distribution on $\mathbf{C}_n$ is taken to be uniform, but by no means always. It is important to have a natural way to build (perhaps "select" is a better word) random structures of size n, preferably one with a substantial degree of independence so that calculations can easily be made. Sometimes ease of construction is compatible with uniformity; at other times, compromises must be made.

Following are some spaces of random structures. The list is by no means exhaustive but it attempts, within the confines of the author's personal preferences, to be representative.

2.1 PERMUTATIONS—uniform

We begin with the space of random permutations Π_n of a fixed n-element set, each permutation taken with the same probability $1/n!$. It is on this space that the derangement problem of the introduction takes place. We may also consider permutations of an unlabeled set, which amounts to considering each possible "cycle structure" with equal probability; see the comment at the end of §7.

2.2 RANDOM GRAPHS—uniform

Study of the now-classical random graph construction, initiated primarily by Erdős and Rényi [9], became the paradigm for analysis of random structures of all kinds.

Fix a set V of n vertices, where we imagine n to be a large number. For convenience we take V to be $\{1, 2, \ldots, n\}$. For each pair $\{i, j\}$ of vertices, let $\mathbf{e}_{i,j}$ be a Bernoulli random variable which takes the value 1 with probability p and zero otherwise; and we let the $\mathbf{e}_{i,j}$'s be independent for different pairs $\{i, j\}$.

Now a random graph $\mathbf{G}_{n,p}$ is defined on V by $E(G) = \{\{i, j\} : \mathbf{e}_{i,j} = 1\}$.

In the case of $p = 1/2$, it is apparent that each possible graph G on V will occur with probability precisely

$$\Pr(G) = 2^{-\binom{n}{2}}$$

and therefore $\mathbf{G}_{n,1/2}$ duplicates the uniform distribution among graphs on a labeled set of n vertices.

Let us look at an application of this construction, which illustrates also a standard technique for dealing with this and other random structures. The *Ramsey number* $r(k)$ is defined to be the least integer n so that *any* graph on n vertices contains either a clique (that is, a copy of the complete graph) on k vertices, or an independent set of size k. An inductive argument shows that $r(k) < 2^{2k}$, but other than for small values of k, we do not know $r(k)$ exactly.

To bound $r(k)$ from below, we need large graphs with no large cliques or independent sets; amazingly, random graphs drawn from the uniform distribution are better for this purpose than any construction so far discovered. The following argument is classical Erdős:

Fix n and let $\mathbf{G} = \mathbf{G}_{n,1/2}$. Let X be a particular set of vertices of size k, and denote by $\langle X \rangle$ the subgraph generated by X. The probability that $\langle X \rangle$ is a clique is of course

$$2^{-\binom{k}{2}}$$

and therefore the expected number of cliques of size k in $\mathbf{G}$ is

$$\binom{n}{k} \cdot 2^{-\binom{k}{2}}$$

which tends to zero for $k = c \log n$, when $c < 2$. Hence, as $n \to \infty$, there will almost surely be no cliques (or, similarly, independent sets) of size $c \log n$ in G, for $c > 2$; we have bounded $r(k)$ from below asymptotically by $2^{k/2}$.

There have been slight improvements in both upper and lower bounds for $r(k)$ but if indeed it is the case that $r(k)^{1/k}$ approaches a constant, that constant could still be anywhere in the interval $[\sqrt{2}, 4]$.

A variation on the construction of $\mathbf{G}_{n,p}$ is obtained by fixing the number of edges instead of the edge-probability; specifically, we let $\mathbf{G}_{n,m}$ be given by taking $E(\mathbf{G}_{n,m})$ to be a random subset of $\binom{V}{2}$ of size m. Since the number of edges in the previous construction will almost always be near $m = p\binom{n}{2}$, $\mathbf{G}_{n,m}$ will be very closely related to $\mathbf{G}_{n,p}$ in this case. In practice, $\mathbf{G}_{n,p}$ has proved to be easier to work with.

Another variation on the uniform space $\mathbf{G}_{n,1/2}$ is obtained by de-labeling the vertex set, that is, by taking with uniform probability any isomorphism class of graphs on n vertices. Since for n large most instances of $\mathbf{G}_{n,1/2}$ will be rigid, i.e. will have no non-trivial automorphism, this space also corresponds closely to $\mathbf{G}_{n,1/2}$ on an $n!$-to-1 basis.

2.3 RANDOM GRAPHS—varying edge probability

As we shall see, setting p in $\mathbf{G}_{n,p}$ to a constant value (strictly between 0 and 1) independent of n produces a rather limited kind of random graph, and denies us the opportunity to witness threshold phenomena. Of greater interest, therefore, is letting $p = p(n)$ vary with n, typically by being a smooth decreasing function such as $n^{-\alpha}$ with $0 < \alpha < 2$.

These more general $\mathbf{G}_{n,p(n)}$ help to model "sparse" graphs, which arise frequently in applications, as opposed to "dense" graphs whose numbers of edges are proportional to $\binom{n}{2}$. Further, by slowly changing $p(n)$ from n^{-2} to $1/2$, we may witness the "evolution" of a random graph—a mathematician's version of childbirth and development.

2.4 RANDOM ORDERS—uniform

If one tries to mimic the construction of $\mathbf{G}_{n,1/2}$ for partially ordered sets (which we shall simply call "ordered sets" or even "orders"), one runs afoul either of independence or transitivity. If a random variable is used to determine, for a pair $\{i, j\}$, whether $i < j$ or $i > j$ or i is incomparable to j, then of course the decisions to put $i < j$ and $j < k$ force $i < k$ destroying independence. We give three models for random orders which nonetheless manage to incorporate some forms of independence.

Naturally we can obtain a uniform space by direct definition, as in Kleitman and Rothschild [14]. Accordingly, let us fix a ground set $V = \{1, 2, \ldots, n\}$, and let $\mathbf{P}_n$ be a partial ordering of V chosen uniformly from all possibilities. Interestingly, $\mathbf{P}_n$ can after all be described—asymptotically, at least—by an independent process. Roughly speaking, almost all orders drawn from the uniform space take the following form: of the n elements of $\mathbf{P}_n$, $n/4$ are minimal and the rest form an antichain of "middle" elements. For each maximal element i and each middle element j, we put i over j independently with probability $1/2$; similarly for middle elements j and minimal elements k. Automatically, the maximals will lie above the minimals by transitivity.

On account of this characterization, the uniform space fulfilled its original purpose of helping to obtain an asymptotic count of partial orderings of a labeled set. From the point of view of ordered-set theorists looking for interesting models, however, it was naturally a bit of a disappointment.

2.5 RANDOM ORDERS—fixed linear extension

Another model for random orders, studied by Albert Frieze [1], avoids the transitivity problem by orienting a random graph acyclicly and only later taking the transitive closure. We may begin with a fixed ground set $\{1, 2, \ldots, n\}$, where now the numbering is important, and as in the Erdős–Rényi construction put an edge between each i and j independently with probability $1/2$. We now put i less than j in the "random graph order" $\mathbf{O}_n$ if there are vertices $h_1, \ldots, h_k$, $k \geq 2$, such that

$$i = h_1 < h_2 < \ldots < h_k = j$$

as numbers, and $\{h_m, h_{m+1}\}$ is an edge for every m, $1 \leq m \leq k - 1$. In this model the natural ordering $1, 2, \ldots, n$ of the ground set is automatically a linear extension; that is, $\mathbf{O}_n$ is forced to be consistent with the numerical ordering.

This construction suffers somewhat from lack of symmetry, in that the ground vertices do not all play the same role; but it does turn out to be quite interesting to try to determine what effect closure under transitivity has after random edges have been selected.

2.6 RANDOM ORDERS—fixed dimension

The *dimension* of an ordered set P, as defined by Dushnik and Miller [8], is the least k such that the binary order relation of P can be described as the intersection of k linear orders $L_1, \ldots, L_k$. Each L_i is then perforce a linear extension of P. Since the intersection of *all* the linear extensions of an ordered set P is necessarily P itself, every finite ordered set has finite dimension. Equivalently, if we (partially) order the Euclidean space $\mathbf{R}^d$ coordinate-wise—that is, by $(x_1, \ldots, x_d) \leq (y_1, \ldots, y_d)$ iff $x_i \leq y_i$ for each i—then the dimension of a finite ordered set is the least k for which it can be embedded in $\mathbf{R}^k$.

The dimension parameter lends itself rather nicely to a construction for random orders which exhibits several forms of independence. Let us fix a (usually small) number k and a (usually large) number n, and construct a random order $\mathbf{P}_k(n)$ as follows: choose independently k random linear orderings $\mathbf{L}_1, \ldots, \mathbf{L}_k$ of the ground set $\{1, 2, \ldots, n\}$, and put $i < j$ in $\mathbf{P}_k(n)$ iff $i < j$ in every $\mathbf{L}_h$, $1 \leq h \leq k$. Of course, this might result in an ordered set of dimension strictly less than k, but it is easy to see that this rapidly becomes unlikely for large n.

The independence of the k choices of linear orders can be replaced by independence of n choices of points, or even nk choices of real numbers, in the following equivalent (up to measure 0) construction: let n points be chosen independently from the Lebesgue (uniform) distribution on the unit cube $[0,1]^k$ in Euclidean k-space, and order them coordinate-wise (i.e. take the partial ordering induced by the coordinate order described above).

A third construction is almost equivalent to the other two, in fact it bears the same sort of relation to them that $\mathbf{G}_{n,p}$ does to $\mathbf{G}_{n,m}$. Again we choose points from the Lebesgue distribution on $[0,1]^k$, but this time the points are given by a Poisson process of density n so that the number of points in the random order may not be n exactly. The gain is that now the event of having a point appear in some subset of the cube is independent of any similar event involving a disjoint subset.

Let us again look at an application, just to get an idea of typical technique. A (finite) ordered set P is said to be *ranked* if there is a rank function $r : P \to \mathbb{N}$ such that the minimal elements have rank 0, the maximal rank h for some fixed h, and all covering pairs (x,y) in P satisfy $r(y) = r(x) + 1$. (We say that y "covers" x, or that (x,y) is a covering pair, if $x < y$ but there is no z such that $x < z < y$.) A simpler (but less intuitive) characterization of ranked ordered sets is given by noting that all maximal chains are of the same length.

Ranked ordered sets occur very frequently in mathematics (e.g. the Boolean algebra of subsets of a set, the partitions of a number, etc.) Given a non-ranked ordered set, what is the largest ranked *suborder* that one may be assured of finding inside? Well, trivially every order on $1 + ab$ elements contains either a chain of length $a + 1$ or an antichain of size $b + 1$, and chains and antichains are ranked, so we are assured of finding in any order on n elements a ranked suborder of size $\geq \sqrt{n}$. Random orders of dimension 2 can be used to show that up to a constant, we can do no better in general:

Theorem 2.1 (Linial, Saks and Shor [17]). *There exist ordered sets on n elements which contain no ranked suborder of size greater than $4e\sqrt{n}$.*

Proof. The first step is to show that the probability that $\mathbf{P}_2(m)$ is *itself* ranked is no more than $16^m/m!$; we omit this (admittedly critical) part since it is not germane to our discussion. Then, however, the expected number of ranked suborders of cardinality $4e\sqrt{n}$ in $\mathbf{P}_2(n)$ is at most

$$\binom{n}{4e\sqrt{n}} 16^{4e\sqrt{n}} / (4e\sqrt{n})!$$

which goes to zero as $n \to \infty$. It follows that almost every $\mathbf{P}_2(n)$ has no ranked suborder of cardinality greater than $4e\sqrt{n}$.

It would be interesting to know whether there is a *construction* with even smaller maximum ranked suborders than $\mathbf{P}_2(n)$, but at the moment—as is often the case—the best result is obtained via random structures.

Given two (closed) intervals I and J on the real line, we say that I is *adjacent* to J if they intersect, and that I *precedes* J if I lies entirely to the left of J. Thereby any set of n intervals determines an n-vertex "interval graph" and an n-point "interval order".

To obtain a random interval graph or random interval order we obviously need a way to construct random intervals, and the simplest and most natural was suggested by Scheinerman [21]. Simply select $2n$ points $x_1, y_1, x_2, y_2, \ldots, x_n, y_n$ independently from *any* continuous distribution on the real line, and let the ith interval I_i be $[x_i, y_i]$ or $[y_i, x_i]$ according to which of x_i, y_i is the smaller.

For the purpose of generating an interval graph or interval order all continuous distributions are equivalent, since all sets of $2n$ distinct points on the line are "isomorphic"; only the order in which they are chosen matters. One might equally well simply pair up the integers from 1 to $2n$ at random to obtain the intervals. Thus, the construction has a pleasing robustness; whether this robustness leads to good behavior will be seen later.

2.7 RANDOM STRUCTURES for a LANGUAGE

Suppose a list $\bar{a} = a_1, a_2, \ldots, a_r$ of positive integers and a list $\bar{b} = b_1, b_2, \ldots, b_s$ of non-negative integers are given. Suppose further that a set S is endowed with r relations R_i and s functions F_j such that R_i has "arity" a_i (i.e. is an a_i-ary relation) and F_j has arity b_j. Then S is said to be a "structure of similarity type $(\bar{a}, \bar{b})$". (Nullary functions correspond to distinguished elements.)

Structures of cardinality n of a fixed similarity type comprise, under the uniform distribution, a class of random structures of special interest in model theory. Here is one example: take $\bar{a} = (2), \bar{b} = ()$ so that we are talking about sets with one binary relation. Such sets might be thought of as directed graphs, with loops allowed.

One variation of this class, of considerable interest (although we will not dwell on it in this work), is obtained by adding the random relations and functions not to a set but to a set with some "built-in" relation or function already on it. The built-in relation is most likely to be a linear ordering, which is accessed in the language either by an order relation or by a successor function.

3 The Basic Zero-One Law for Graphs

The 0-1 law for the first-order theory of random graphs (example 2.2: uniform) was proved by Glebskii et al. [11] in 1969, but only became known in the West from Fagin's quite different proof [10] in 1976. These two proofs became paradigms for subsequent proofs of zero-one laws; we will present (our version of) the more elegant Fagin proof below.

Let us first review the (few) needed facts about first-order logic, with special attention to graphs. (For greater detail, the reader is referred to Schoenfield [23] or Chang and Keisler [4].) The first-order language of graphs, which we will denote $L_1(G)$, consists of *formulas*, which are finite strings of symbols built inductively from the following elements:

variable symbols x, y, z, x_1 etc;

equality symbol $=$;

binary relation symbol $\sim$, to denote adjacency;

left and right parentheses (,);

logical connectives $\wedge$, $\vee$ and $\neg$;

quantifiers $\forall$ (for all) and $\exists$ (there exists).

Only the $\sim$ symbol is special to graphs; generally, there will be some finite or countably infinite list of relation and function symbols appropriate to a given type of structure. Any set equipped with relation and functions matching a language L will be called an "L-structure."

We now proceed to build the class of formulas by induction; for convenience, we consider only the language of graphs. We begin by placing "$x = y$" and "$x \sim y$" in the class of formulas, for all variables x and y; in these x and y are said to be "free" variables. They take on a truth value in a particular $L_1(G)$-structure (e.g. a graph) provided elements of that structure have been assigned to the free variables, and of course that value will be "TRUE" just if the elements are equal or, respectively, satisfy the binary relation on the structure.

If α and β are formulas then $(\alpha) \wedge (\beta)$, $(\alpha) \vee (\beta)$, and $\neg(\alpha)$ are also formulas (with the obvious truth values in a given structure). Further, $\exists x(\alpha)$ and $\forall x(\alpha)$ are formulas provided x is a free variable of α (but x becomes a "bound" variable in the new formula). To improve the readability, we omit parentheses when no confusion will result.

For a given structure and a given assignment A of the *rest* of the free variables of α, $\exists x(\alpha)$ will be TRUE if α is true for some assignment to x which extends A. The truth value of $\forall x(\alpha)$ may be defined by noting its equivalence to $\neg \exists x(\neg \alpha)$.

A *sentence* ϕ is a formula with no free variables; then no assignment of variables is needed to obtain a truth value for ϕ relative to a given structure. If ϕ is true for S we write $S \models \phi$ and say that "S satisfies ϕ". A collection T of sentences is called a "theory"; if S satisfies every sentence in T then we write $S \models T$ and say that "S satisfies T" or that "S is a model of T".

The following two sentences constitute the "theory of graphs" T_G, in that a structure with binary relation satisfies the sentences iff it is a (simple, loopless) graph:

"$\forall x \forall y((x \sim y \wedge y \sim x) \vee (\neg x \sim y \wedge \neg y \sim x))$" ($\sim$ is symmetric).

"$\forall x(\neg x \sim x)$" ($\sim$ is irreflexive).

Here are some other examples of sentences in $L_1(G)$, with their meanings in terms of graphs:

"$\exists x \forall y(\neg x \sim y)$" ($G$ contains an isolated vertex).

"$\exists x \exists y \exists z \exists w(x \sim y \wedge y \sim z \wedge z \sim w \wedge w \sim x \wedge \neg x \sim z \wedge \neg y \sim w)$" ($G$ contains an induced 4-cycle).

"$(\exists x \exists y(\neg x \sim y)) \wedge (\forall x \forall y(x \sim y \vee (\exists z(x \sim z \wedge z \sim y))))$" ($G$ has diameter 2).

On the other hand, the property "G is connected" seems not to be expressible in first-order logic (and indeed it is not), since we cannot employ an infinite disjunction of sentences saying "G has diameter k". The property "G is three-colorable" is also inexpressible in first-order logic; here the problem is that we cannot quantify over "second-order" objects such as coloring functions.

If T is a theory and ϕ a sentence, we write $T \vdash \phi$ and say that "ϕ is a consequence of T" if there is a finite list of sentences, ending with ϕ, such that each member of the list is deducible from previous members and sentences in T by means of the usual rules of logical deduction. It is immediate from these rules that if $S \models T$ and $T \vdash \phi$ then $S \models \phi$.

T is said to be *inconsistent* if $T \vdash$ "$\exists x (\neg x = x)$" and *consistent* otherwise; thus if T has a model, it must be consistent. The converse, known often as "Gödel's Completeness Theorem", is also true and is the crucial theorem asserting adequacy of the standard laws of deduction for first-order logic.

In fact, even more is true:

Theorem 3.1. (Löwenheim–Skolem—see e.g. [23]): *If T is consistent then it has a countable (that is, finite or countably infinite) model.*

Let us make two easy deductions from these results. The first is "compactness":

Theorem 3.2. *If every finite subset of T has a model then so does T.*

Proof. Otherwise T is inconsistent; but the proof of "$\exists x (\neg x = x)$" from T can use only finitely many sentences of T, therefore some finite part of T is already inconsistent and has no model.

A theory T is said to be *complete* (for a language L) if for any sentence ϕ of L, either $T \vdash \phi$ or $T \vdash \neg \phi$. Evidently if S is any structure for the language then the set of all sentences satisfied by S (known as the "theory of S" or $\mathrm{Th}(S)$) is a complete theory.

Theorem 3.3. *Suppose a consistent theory T has only one countable model, that is, all countable models of T are isomorphic. Then T is complete.*

Proof. Otherwise there is a sentence ϕ such that neither ϕ nor $\neg \phi$ is a consequence of T, from which we may deduce that both $T \cup \{\phi\}$ and $T \cup \{\neg \phi\}$ are consistent. These both have countable models, on account of the Löwenheim–Skolem Theorem, which cannot be isomorphic, a contradiction.

Now let us suppose that a class of random structures has been fixed, together with an appropriate first-order language L. Let $\mathbf{S}_n$ be a random structure of size n in our class; then we may define, for any sentence ϕ of L,

$$\mathrm{Pr}_n(\phi) = \mathrm{Pr}(\mathbf{S}_n \models \phi)$$

and

$$\mathrm{Pr}(\phi) = \lim_{n \to \infty} \mathrm{Pr}_n(\phi)$$

when the limit exists.

The result of Glebskii et al. and Fagin may then be stated as follows:

Theorem 3.4. *Let ϕ be any first-order sentence in the language of graphs, and let random graphs be chosen uniformly (e.g. $\mathbf{G} = \mathbf{G}_{n,1/2}$). Then $\mathrm{Pr}(\phi)$ exists and is equal either to 0 or 1.*

Proof (Fagin [10]). We begin by defining a sequence of sentences which we call "Alice's Restaurant axioms" because they assert that you can get anything you want. The kth

Alice's Restaurant axiom Ψ_k is given by:

$$\forall x_1 \forall y_1 \forall x_2 \forall y_2 \cdots \forall x_k \forall y_k \exists z((z \sim x_1 \ \wedge \ z \sim x_2 \ \wedge \cdots \wedge \cdots \wedge \ z \sim x_k \ \wedge$$

$$\neg z \sim y_1 \ \wedge \ \neg z = y_1 \ \wedge \ \neg z \sim y_2 \ \wedge \ \neg z = y_2 \ \wedge \cdots \wedge \ \neg z \sim y_k \ \wedge \ \neg z = y_k) \ \vee$$

$$(x_1 = y_1 \ \vee \ x_1 = y_2 \ \vee \cdots \vee \ x_k = y_1 \ \vee \cdots \vee \ x_k = y_k)).$$

We may translate as follows: if X and Y are two disjoint sets of vertices each of cardinality at most k, then there is a vertex z outside $X \cup Y$ which is adjacent to every vertex in X but not to any vertex in Y.

To show that $\Pr(\Psi_k) = 1$ for each k we employ a typical expectation argument. Fix two disjoint vertex sets X and Y in $\{1, 2, \ldots, n\}$ with $|X|, |Y| \le k$. For a vertex z not in $X \cup Y$, let W_z be the event that z "works", i.e. is adjacent to the members of X and not to the members of Y. Then the events W_z for various z are independent, since different edge-variables $e_{i,j}$ are involved. Hence

$$\Pr(X \text{ and } Y \text{ witness the failure of } \Psi_k)$$

$$= \Pr(W_z \text{ fails for every } z)$$

$$= \prod_z \Pr(W_z \text{ fails})$$

$$\le (1 - 2^{-2k})^{n-2k}.$$

It follows that the expected number of bad pairs X, Y is at most

$$\binom{n}{2k}\binom{n-k}{k}(1 - 2^{-2k})^{n-2k}$$

which approaches zero as $n \to \infty$. Therefore $\Pr(\text{a bad pair } X, Y \text{ exists in } \mathbf{G}_{n,1/2})$ goes to 0, meaning that $\Pr(\Psi_k) = 1$ as required.

We now need to show that the set $\Psi = \{\Psi_k : k = 1, 2, \ldots\}$ is a complete theory, using Theorem 3.3. To this end, suppose that G and H are two countable (necessarily infinite) graphs which satisfy Ψ; we show that G and H are isomorphic.

The technique is a back-and-forth argument, familiar to those readers who recall the textbook proof that any two countable dense linear orderings without endpoints are isomorphic. We define a sequence of nested 1-to-1 partial mappings $f_1, f_2, \ldots$ from $V(G)$ to $V(H)$, each an isomorphism between subgraphs of order i.

Let $V(G) = \{g_1, g_2, \ldots\}$ and $V(H) = \{h_1, h_2, \ldots\}$. For i odd, let i' be the least index such that $g_{i'}$ is not in the domain of F_{i-1}; let $X = \{f_{i-1}(g_j) : g_j \sim g_{i'}\}$ and $Y = \{f_{i-1}(g_j) : \neg(g_j \sim g_{i'})\}$. Then axiom Ψ_{i-1}, applied to H, is enough to ensure that there is a vertex $h_{i''}$ such that putting $f_i(g_{i'}) = h_{i''}$ extends the isomorphism.

For i even we instead find the least index i' such that $h_{i'}$ is not in the range of f_{i-1}, and use Ψ_{i-1} on G to find a suitable $G_{i''}$, setting $f_i(g_{i''}) = h_{i'}$.

This construction ensures that for all j both g_j and h_j eventually are covered by some f_i, and therefore that the union of the f_i's is indeed an isomorphism between G and H. We have shown that Ψ has only one countable model and is therefore complete.

Before proceeding, we make some observations. First, notice that if a random graph $\mathbf{G}_{\omega,1/2}$ is constructed as usual but on a countably infinite set of vertices, then it will with probability 1 satisfy all the Ψ_k's; hence, any two such graphs will (with probability 1) be isomorphic. This even holds if one of them is a $\mathbf{G}_{\omega,p}$ and the other a $\mathbf{G}_{\omega,q}$ for $p \neq q$. The unique result is sometimes called the "Rado graph" (see [20]), and can be constructed deterministically without much difficulty by just inductively ensuring that the Ψ_k's are satisfied.

Second, notice that in order merely to embed G isomorphically *into* H, we need only that H satisfy the theory Ψ. Thus the Rado graph is "universal" for all countable (including finite) graphs; that is, any finite or countable graph is an induced subgraph of the Rado graph.

Now let us conclude Fagin's proof of the 0-1 law. Let ϕ be any sentence in the first-order language of graphs; then since Ψ is complete, either ϕ or its negation is a consequence of Ψ. If the former, then by finiteness of proofs (compactness) ϕ is already a consequence of $\Psi_1, \ldots, \Psi_k$ for some k, but the conjunction of these (equivalent, actually, to just Ψ_k) has limiting probability 1 among random graphs. Hence $\Pr(\phi) = 1$ and in the opposite case, where $\neg(\phi)$ is a consequence of Ψ, we get $\Pr(\phi) = 0$.

Let us make a few remarks about Fagin's proof. First, it does more than necessary; as we shall see, when we move to other structures the "unique countable random structure" of Fagin's proof will not always exist. However, we will be able to get around that shortcoming in some cases.

Second, the proof applies to $\mathbf{G}_{n,p}$ for any constant p, and even to functions $p(n)$ suitably near constant; it suffices, in fact, for $p(n)$ to satisfy $n^{-\varepsilon} < p(n) < n^\varepsilon$ asymptotically, for all $\varepsilon > 0$.

Thirdly, and importantly, the proof applies (and was applied by Fagin) to the uniform space of structures for any language with only relational symbols, that is, with no function symbols. The Alice's Restaurant axioms must be amended to ensure the existence of an element bearing any possible combination of relations to a given finite set of elements. For example, uniform random digraphs (with loops) are equivalent to unrestricted structures for a language with one binary relation (say, R); in that case the kth Alice's Restaurant axiom would say that for any three disjoint sets U, V, X, and Y each of size at most k there is a new element w such that wRu and uRw for each $u \in U$, wRv but $\neg vRw$ for each $v \in V$, etc., *and wRw* (and of course there is another, similar w' with $\neg w'Rw'$). Note that here even Ψ_0 has content.

This version is therefore called the "0-1 law for first-order logic with relation symbols".

It is easy to see why Fagin's proof fails if there is a function symbol in the language, even a nullary function (that is, a constant symbol). In fact, the 0-1 law flops miserably in a language with just one constant symbol (say, c) and a unary relation U: the sentence $c \in U$ already has probability $1/2$. Fagin himself noticed that in the language with just one unary function symbol F, the sentence "$\exists x (F(x) = x)$" has limiting probability $1/e$ (as in the case mentioned in the introduction, where F is constrained to be a permutation).

There is, however, a reasonable recourse when the 0-1 law fails; we might at least hope that for each first-order sentence ϕ, $\Pr(\phi)$ *exists*. This is indeed the case if L contains just a unary function, or even many unary functions; here Lynch [18] showed that in fact every sentence then has a limiting probability which can be expressed by integer constants with arithmetic operators, including e^x.

4   The Zero-One Law for Ordered Sets

The 0-1 law for the random orders of Kleitman and Rothschild (uniform) was proved by Compton [6], using Fagin's approach. The language of partial orders has one binary relation symbol "<" (we could as well use "≤") which is constrained to be transitive, antisymmetric and irreflexive. Compton defines the following theory, which is seen via a typical back-and-forth argument to be complete by virtue of having only one countable model.

First, we constrain the order to have height at most 3; formally the sentence can be written

$$\neg \exists w \exists x \exists y \exists z (w < x \;\wedge\; x < y \;\wedge\; y < z).$$

Now the minimal, middle and maximal elements have easy first-order descriptions, by means of which we can give two types of Alice's Restaurant axioms.

The sentences α_k will say that for any two disjoint sets X and Y of middle elements, each of size at most k, there is a minimal element which is below the elements in X but not the ones in Y, and a maximal element which is above the elements in X but not those in Y.

The sentences β_k will say that for any two disjoint sets X' and Y' of maximal elements and any two disjoint sets X'' and Y'' of minimal elements, all of size at most k, there is a middle element which lies between the elements of X' and the elements of X'' but is incomparable to any element of $Y' \cup Y''$. (Note that β_1 already insures that every maximal element lies above every minimal.)

Now, the reader will perhaps already have noted that Compton's theory is not universal for ordered sets, nor even for ordered sets of height three—it does, finally, embed every countable *ranked* order of height three. There is a countable, universal ordered set, however, and its theory *does* have just one countable model. The Alice's Restaurant axioms for the theory of this ordered set say that for any three disjoint sets X, Y and Z, each of size at most k, such that no element of Y lies above any element of X or below any element of Z and no element of X lies below an element of Z, there is an element w which lies below every element of X and above every element of Z while being incomparable to everything in Y.

This theory looks like a much better candidate to help prove the 0-1 law for ordered sets than does Compton's theory, but the ugly duckling is the winner here; only Compton's axioms are satisfied by (sufficiently large versions of) the ordered sets described by Kleitman and Rothschild, and these are the ones which almost all orders grow up to look like.

(NOTE: Compton [6] actually proved much more; the 0-1 law for uniform random orders hold even in least-fixed-point logic, and for unlabeled orders. He was also able to establish the computational complexity of deciding whether $\Pr(\phi) = 1$ in each case.)

What about the random orders of fixed dimension, i.e. $\mathbf{P}_k(n)$? The random-points-in-an-n-cube construction gives a natural interpretation to $\mathbf{P}_k(\omega)$, since a countably infinite set of points can be chosen. It does turn out that for fixed k, the theory of $\mathbf{P}_k(\omega)$ indeed has just one countable model—a nice generalization of the set of rational numbers, to which the linearly ordered set $\mathbf{P}_1(\omega)$ is isomorphic. Setting up the Alice's Restaurant axioms for these theories is a bit trickier than in the cases we have thus far seen, because in some sense the natural language for dealing with random orders of dimension k contains k linear order symbols $<_1, \ldots, <_k$ rather than the partial order $<$ obtained as their intersection.

The Alice's Restaurant axioms need to be translated to the weaker language with just the partial order relation; this can be done ([25]).

So, do we then get the 0-1 law for random orders of fixed dimension? No, the models described above—while they are universal for orders of dimension k—are bad news for Fagin's proof. For example, their axioms immediately imply that there is no minimal element, yet *every* finite ordered set has a minimal element.

Well, in dimension 1 (the case of linear orders) we were looking at the wrong model. The theory of the linear order $\omega + \omega^*$, that is, a copy of the non-negative integers followed by an inverted copy, is a complete theory (though it has more than one countable model—e.g. $\omega + \omega^* + \omega + \omega^*$); each sentence of this theory is indeed true for sufficiently large finite linear orders. Thus, we do get a 0-1 law for linear orders, but this is hardly a great surprise since there is only one linear order of cardinality n, for each n.

This author was at one time quite excited about the prospect of finding the right analog to $\omega + \omega^*$ in dimensions 2, 3, etc. but soon discovered that no such analog exists: the 0-1 law fails in dimensions 2 and higher!

Probably the simplest example with limiting probability other than 1 or 0 for $\mathbf{P}_2(n)$ is the following:

$$\exists x \exists y \forall z (\neg z > x \wedge \neg z < y \wedge \neg y < x).$$

This sentence (let us call it χ) asserts the existence of an incomparable pair x, y such that x is maximal and y is minimal. The context in which χ arose is quite interesting. At first it was not clear how to prove that the random orders of fixed dimension are (almost always) connected, but later [26] an application of a correlation lemma and some ordinary calculus established that the comparability graph of $P_k(n)$ has diameter at most 3. (The *comparability graph* $G(P)$ of an ordered set P has as its vertices the elements of P, with $x \sim y$ in G just when $x < y$ or $y < x$ in P.)

The easiest way to show that the diameter of $G(\mathbf{P}_k(n))$ is precisely 3 would have been to find an incomparable max-min pair, since such a pair cannot be connected by a path of length less than 3. Such pairs exist in abundance when k is 3 or more, but it turns out that in $\mathbf{P}_2(n)$ they exist only 3/4 of the time! (Eventually, a much more difficult argument was found to show that even in the remaining 25% of the cases, $G(\mathbf{P}_2(n))$ still has diameter 3.)

To see that $\Pr(\chi) = 3/4$ in dimension 2, let $\mathbf{L}_1$ and $\mathbf{L}_2$ be the generating random linear orderings, and let t_i and b_i be the top and bottom elements of $\mathbf{L}_i$, $i = 1, 2$. Perforce t_1 and t_2 are maximal, and b_1 and b_2 are minimal; if $t_1 < b_1$ in $\mathbf{L}_2$, or $t_2 < b_2$ in $\mathbf{L}_1$, we have an incomparable max-min pair. Each of these events has probability 1/2 and in the limit (when four elements are almost surely distinct) they are independent; hence

$$\Pr((t_1, b_1) \text{ or } (t_2, b_2)) \text{ is an incomparable max-min pair}) = 3/4.$$

If neither event occurs, then we claim that no incomparable max-min pair can exist. For, every maximal element must lie above t_1 in $\mathbf{L}_2$ (else t_1 would be larger) and above t_2 in $\mathbf{L}_1$; every minimal element must lie below b_1 in $\mathbf{L}_2$ and below b_2 in $\mathbf{L}_1$, hence below every maximal element.

Another 0-1 law counterexample arises in the problem of counting the 2-dimensional orderings of a (labeled) n-element set [28]. Let Ψ be the sentence

$$\exists x \exists y \forall z (\neg x = y \wedge \neg x < y \wedge \neg y < x \wedge (z < x \leftrightarrow z < y) \wedge (x < z \leftrightarrow y < z)),$$

which says that there exists an incomparable pair (x, y) such that the remaining elements of the ordered set are either above both x and y, below both, or incomparable to both. It is not difficult to see that Ψ can hold in $\mathbf{P}_2(n)$ only if x and y are consecutive in both $\mathbf{L}_1$ and $\mathbf{L}_2$, but with $x < y$ in one and $y < x$ in the other. If this does happen, let $\mathbf{L}_i'$ be the result of switching x and y in $\mathbf{L}_i$; then $(\mathbf{L}_1, \mathbf{L}_2)$, $(\mathbf{L}_1', \mathbf{L}_2')$, $(\mathbf{L}_2, \mathbf{L}_1)$ and $(\mathbf{L}_2', \mathbf{L}_1')$ all yield the same ordered set. It turns out that this is the only *likely* way to get more than 2 representations of the same 2-dimensional ordered set; in general, if there are s pairs $(x_1, y_1), \ldots, (x_s, y_s)$ in a given $\mathbf{P}_2(n)$ witnessing Ψ, then almost always the number of representations of $\mathbf{P}_2(n)$ as the intersection of two linear orderings is precisely 2^{s+1}.

Now the number of such pairs in $\mathbf{P}_2(n)$ takes on a Poisson distribution in the limit, much like the number of fixed points in a random permutation; in particular $\Pr(\Psi) = 1 - 1/e$. It follows that, asymptotically, the number of 2-dimensional orderings of an n-element set is

$$\sum_{s=0}^{\infty} \frac{n!^2}{es!} \cdot \frac{1}{2^{s+1}} = \frac{n!^2}{2\sqrt{e}}.$$

It proved oddly difficult to find counterexamples to the 0-1 law for $\mathbf{P}_k(n)$ with $k > 2$, so much so that this author was moved (foolishly) to conjecture that the 0-1 law might hold again in dimensions 3 and higher. However, he was finally able [27] to construct counterexamples in higher dimensions.

The last event in this chain of discovery concerned the question of whether each first-order sentence in $\mathbf{P}_k(n)$ does at least have a limiting probability. This did not seem like too much to ask (at least to this author) but Spencer [24] was able, by interpreting parts of arithmetic (a standard but by no means easy technique), to obtain a sentence with no limiting probability for two-dimensional random orders.

5 Thresholds and 0-1 Laws for Sparse Random Graphs

In Erdős and Rényi's seminal paper "On the evolution of random graphs" [9] the approach was to begin with a large set of vertices, with edge-probability zero, and then gradually increase the edge-probability $p = p(n)$, watching as various graph properties appear and disappear. The striking observation, made many times, was that properties seem to appear and disappear quite suddenly, at certain "thresholds". (In what follows, p will always be a function of n.)

Suppose, for instance, that P is a monotone-upward property; that is, the result of adding an edge to a graph with property P is another graph with property P. A function f is said to be a *threshold* for property P if:

(1) $\Pr(G_{n,p} \text{ has } P) \to 1$ when $p/f \to \infty$ and

(2) $\Pr(G_{n,p} \text{ has } P) \to 0$ when $p/f \to 0$.

As an example, we consider for any fixed graph H the property P_H defined as follows: G has P_H if G contains a copy of H as a (not necessarily induced) subgraph. First, we need a definition:

The *maximum density* $\delta(H)$ of a graph H is the maximum over all non-empty subgraphs J of H of $|E(J)|/|V(J)|$.

Now if a graph J has e edges and v vertices, then the expected number of occurrences of J in $\mathbf{G}_{n,p}$ is

$$\binom{n}{v}\frac{v!}{|\text{Aut}(J)|}p^e$$

which approaches zero when $pn^{v/e} \to 0$ but grows without bound when $pn^{v/e} \to \infty$. (Here $\text{Aut}(J)$ is the automorphism group of J, so that $v!/|\text{Aut}(J)|$ is the number of different graphs on a labeled set of k vertices which are isomorphic to J.) With a little help from the second moment method (see e.g. [3]) and a bit of extra work when H is not itself its densest subgraph, we deduce that $n^{-1/\delta(H)}$ is the threshold for property P_H.

Now P_H is a first-order property of graphs, so we see that we cannot expect a 0-1 law to hold through the range $p \sim n^{-1/\delta(H)}$. By the same token, we cannot expect any first-order property to have a constant threshold function, because we know that for any first-order sentence ϕ, $\Pr(\mathbf{G}_{n,c} \models \phi)$ is stuck at 0 or at 1 for all constants $c \in (0,1)$. Thus we see that the threshold function p of n should either be a threshold function for some first-order property, or should be the edge-probability for some 0-1 law.

Before proceeding, though, let us note that many interesting graph properties (such as connectivity and 3-colorability, mentioned earlier) are not expressible in first-order logic. Nonetheless, they too have sharp thresholds; and they too have limiting probability 1 or 0 when edge-probability is constant. Thus, we may expect that 0-1 laws exist for larger classes of properties than the first-order properties. We will take up this matter again in §6.

In [22], Shelah and Spencer go a long way toward characterizing the functions p for which a first-order 0-1 law applies, in fact all the way as far as p of the form n^{-a} are concerned. Let us consider first the range $p \leq n^{-1}$, and for any function $f(n)$ let us say that $p \ll f(n)$ (resp. $p \gg f(n)$) if $\lim_{n\to\infty} p(n)/f(n) = 0$ (resp ∞).

Theorem 5.1. *If $p \ll n^{-2}$, or if $n^{-1-1/k} \ll p \ll n^{-1-1/(k+1)}$ for some positive integer k, or if $n^{-1-\varepsilon} \ll p \ll n^{-1}$ for all $\varepsilon > 0$, or if $n^{-1} \ll p \ll n^{-1}\log n$, or if $n^{-1}\log n \ll p \ll n^{-1+\varepsilon}$ for all $\varepsilon > 0$, then the 0-1 law holds for $\mathbf{G}_{n,p}$.*

Proof. We are not foolish enough to try to prove this in detail here, but in fact it is not hard to say enough to put the reader on track. In each case an application of Fagin's technique, or a slight variation, does the trick.

Notice that $p = n^{-1-1/k}$ is the threshold for containment of any fixed tree T_{k+1} on $k+1$ vertices, thus with k edges. Hence, for p between $n^{-1-1/k}$ and $n^{-1-1/(k+1)}$, we would expect $\mathbf{G}_{n,p}$ to contain arbitrarily many copies of all trees on $k+1$ vertices but no larger trees, nor any graphs of density ≥ 1. An infinite graph with these properties satisfies a first-order theory which says that (1) for each t, there are t copies of every tree of order $k+1$ or less; (2) there are no trees of size $k+2$; and (3) there are no cycles. This theory is easily seen to have just one countable model, and Fagin's proof goes through without a hitch.

The case where $p \ll n^{-2}$ is trivial since then there are almost always no edges. When p is less than n^{-1} but more than $n^{-1-\varepsilon}$ for any ε, then the axioms will ensure infinitely many trees of all sizes and isomorphism types, but still no cycles; this theory now has more than one countable model, because there may be one or many *infinite* tree-components; but it can still be shown to be a complete theory.

Between n^{-1} and $n^{-1}\log n$, we need all trees and all cycles as well, but cycles with a vertex of finite degree r (for any r) and all graphs of density > 1 are forbidden. Here a model will have infinitely many tree components as before, plus infinitely many components containing a unique cycle all of whose vertices have infinite degree, plus maybe some infinite tree components. Again, a complete theory.

The final case—between $n^{-1}\log n$ and n^{-1}-to-n^0 range is neater to state but much tougher to prove.

Theorem 5.2. *If* $p = n^{-a}$, *for* $a \in (0,1)$ *and irrational, then* $\mathbf{G}_{n,p}$ *satisfies the 0-1 law.*

Proof. Here the method of "elimination of quantifiers" (used by Glebskii et al. [11] in their original proof of the basic 0-1 law for constant p) is employed. We give only a *very* rough sketch.

The basic device is a *pair* of graphs $H \subset H'$; such a pair is said to be of type (v, e) if there are v vertices in $V(H') - V(H)$ and e edges in $E(H') - E(H)$. (Edges within H are, in fact, totally ignored.) We are interested in whether a particular list $x_1, x_2, \ldots, x_{V(H)}$ of vertices of $\mathbf{G}_{n,p}$ can be extended so as to play the role of H in H'; this will be possible in infinitely many ways when $v/e < 1/a$ (provided no graphs between H and H' are too dense) but in only boundedly many ways otherwise. The "closure" of the x_i's (relative to formulas of a particular length) consists of those points plus the finitely many other vertices which appear in the dense extensions.

Now, one shows by induction on the length of the formula ϕ (the interesting case being when a quantifier is added) that whether a set of x_i's satisfies $\phi(\bar{x})$ is determined by the closure, the behavior of the rest of the graph being predictable. But it turns out that the closure of the empty set is always empty, therefore the truth value of a *sentence* doesn't depend on anything at all—it is (almost) always true or false.

Spencer and Shelah also succeed in constructing (using the word loosely) a formula which, for a particular smooth function p, has the rather frightening property that its probability in $\mathbf{G}_{n,p}$ goes to 1 if $n \to \infty$ in such a way that $\log^* n \equiv 25 \bmod 100$ but to 0 if the "25" is replaced by a "75". (Here $\log^* n$ is the number of applications of the log function needed to reduce n to 1.) Thus, we cannot hope to show that limits exist in all the cases where the 0-1 law fails.

6   Random Intervals and a Curious Counterexample

The constructions in §2 of random interval graphs and random interval orders behave, in many respects, as spaces of random structures ought to. For example, the maximum number of disjoint intervals (corresponding to the independence number of the interval graph, and the height of the interval order) is a well-behaved random variable which is concentrated about $2\sqrt{n}/\sqrt{\pi}$ (as is proved in [13] using subadditivity and an integral equation). When the authors of that paper asked themselves about the *diameter* of a random interval graph, however, the initial result was only that the diameter was almost always 2 or 3.

Let us note that for an interval graph to have diameter 2 it is necessary (and of course sufficient) to have a vertex which is adjacent (thus incomparable, in the interval order) to all other vertices—that is, an interval which meets all the others. What is the probability that, in a random collection of intervals, there is one which meets all the others?

The initial analysis assumed uniform distribution of endpoints in the unit interval (recall that all continuous distributions are equivalent for this purpose), which meant that the intervals could be represented as random points from the uniform distribution on $[0,1]^2$. Approximating this distribution by a Poisson distribution revealed that the probability in question approaches $2/3$.

Startling as it was to learn that a third of the random interval graphs would have different diameter from their sisters, the authors were even more stunned to learn from a more careful analysis that the probability does not merely *approach* $2/3$—it actually *is* $2/3$!

Here is a sketch of a combinatorial proof. Let us use the discrete model, where integers from 1 to $2n$ inclusive are paired at random, but let us number the intervals in the following peculiar way.

Let I_1 be the interval one of whose endpoints is $n+1$; let us say that I_1 "goes left" if its other endpoint (which we call the "outside" endpoint) is less than $n+1$, otherwise it "goes right". Once $I_1, \ldots, I_k$ are labeled, I_{k+1} is labeled as follows: if I_k went left, that is if its outside endpoint is less than its inside (initial) endpoint, then the inside endpoint of I_{k+1} will be the least number above $n+1$ which is not already an endpoint of a labeled interval. Otherwise we go to the *largest* number *below* $n+1$ which is not already an endpoint of a labeled interval.

In this way the intervals are labeled from inside out, as it were, and the left-right rule ensures that the points on either side of $n+1$ are used about equally. We continue the process until there are only four remaining unused points, say $a < b < c < d$; it is easy to show that a is below $n+1$ and c and d are above, but b may be either above or below $n+1$ depending on whether the last interval went left or right.

Suppose first that b is left of $n+1$. There are three ways to hook up these last four points, and we claim that two of these ways produce an interval (actually two intervals, in this case) which intersects all others; but the third way (a to b and c to d) ensures that no such interval exists. To see this, just note that every previous interval had an inside endpoint between b and c, else b or c would instead have been chosen.

The case where only a lies to the left of $n+1$ is a tad trickier, and we leave it to the reader, but again 2 of the 3 remaining ways to hook up a, b, c and d lead to the "big" interval and the third prevents one from existing. So, regardless of the respective probabilities of the two cases, the probability that there will be an interval which intersects all others will be exactly $2/3$.

7 Growth Rate and General 0-1 Laws

It would be wonderful if someone came along to *characterize* the spaces of random structures for which 0-1 laws hold. Obviously, if this had occurred we would not have wasted the reader's time by considering the many cases above separately.

However, some serious progress has been made in determining which *uniform* spaces of random structures have 0-1 laws, when the numbers of such structures do not grow too rapidly. We give a brief sketch of this work of Kevin Compton [5].

The spaces **C** considered by Compton are uniform, considered either as labeled or unlabeled structures, and closed under disjoint unions and components. (Connectivity is definable for sets with relations and functions in a natural way, which extends the notion for graphs.) Thus we require that if two structures in **C** are merely put side-by-side the result is again a structure of **C**; and that a connected component of any structure in **C** is again in **C**. These requirements are very reasonable and indeed apply in every uniform space we have considered.

Let us fix such a space **C** and for each n, let a_n be the number of **C**-structures on the set $\{1, 2, \ldots, n\}$ and let b_n the number of isomorphism classes of **C**-structures of size n. It is often the case that most structures of **C** are *rigid*, that is, have no non-trivial automorphisms, in which case a_n will tend to be larger than b_n by a factor of $n!$. This helps explain why we use exponential generating functions for the a_n's and ordinary generating functions for the b_n's, as follows:

$$a(n) = \sum_{n=0}^{\infty} \frac{a_n}{n!} x^n$$

and

$$b(n) = \sum_{n=0}^{\infty} b_n x^n.$$

Let us now use **L** to stand for the uniform space of labeled **C**-structures, **U** for unlabeled. We will say that **L** is *slow-growing* if $a(x)$ has non-zero radius of convergence, and similarly for **U** and $b(x)$. Compton's results apply to the slow-growing cases, of which there are a number of good examples; one is the space of random permutations, mentioned above in §2.1. However, the uniform graphs and ordered sets that we have looked at above are all fast-growing.

Among the slow-growing structures, it turns out that the behavior of 0-1 laws depends on the limit of the ratios of successive coefficients for the two generating functions. Specifically:

Theorem 7.1 [5]. *Let* **L** *be a slow-growing uniform labeled space of random structures, closed under disjoint union and components. Then* **L** *has a 0-1 law for first-order logic if and only if, for any* m,

$$\lim_{n \to \infty} \frac{a_{n-m}/(n-m)!}{a_n/n!} = \infty.$$

Theorem 7.2 [5]. *Let* **U** *be a slow-growing uniform unlabeled space of structures, closed under disjoint union and components. Then* **U** *has a 0-1 law for the first-order logic if and only if, for any* m,

$$\lim_{n \to \infty} \frac{b_{n-m}}{b_n} = 1.$$

One interesting consequence of the second theorem is that the *unlabeled* space of random permutations actually does satisfy a 0-1 law. Of course, the sentence

$$\exists x (f(x) = x),$$

which expresses the property that a fixed point exists, has non-trivial limiting probability among labeled random permutations; but in the non-labeled world, the students' intuition must be correct—or half of the students' intuitions, anyway. We leave the question to the reader: if the hat-check person first selects an *unlabeled* permutation from the uniform distribution, then returns the hats in accordance with a random labeling of that permutation, is it likely that someone will get his own hat back, or not?

This question brings up a point which we have not hitherto addressed. When a 0-1 law does hold, how hard is it to figure out for a given sentence ϕ whether $\Pr(\phi) = 0$ or $\Pr(\phi) = 1$? In the case of random graphs, the theory Ψ used in Fagin's proof is necessarily recursively axiomatizable and complete, hence decidable. This means that there is an algorithmic procedure for deciding whether a particular sentence ϕ is a consequence of Ψ or not; such a procedure exists, since we can simultaneously search for a proof of ϕ and a proof of $\neg\phi$ from the axioms of Ψ. One or the other search must eventually terminate, and the result will tell us whether $\Pr(\phi)$ is 0 or 1.

Computer theorists will want to know more, though, and the answer was supplied by Grandjean [12]: the set of sentences with limiting probability 1 for uniform random graphs is PSPACE-complete. This implies that unless there is unexpected collapse of the complexity hierarchy, there is no *efficient* algorithm to solve this problem. [Note: in really bizarre cases where the 0-1 law holds, but not with a Fagin-type proof, the decision problem may even be recursively unsolvable.]

Before moving on, we wish to remark that Compton's results go far beyond those stated above. He has extended Theorems 7.1 and 7.2, for example, to monadic second-order logic—but higher logics are the subject of §8.

8 Stronger Logics

We saw earlier that, despite the great interest in establishing 0-1 laws for first-order logic, properties of interest to mathematicians (e.g. in graph theory) are often not expressible by first-order sentences. There are basically two shortcomings: (1) sentences of first-order logic are required to be finite in length; and (2) no quantification is permitted over sets or relations. It is the first shortcoming which prevents us from expressing connectedness (although connectedness can also be formulated as a finite statement with a set quantifier), and the second which prevents us from expressing 3-colorability.

The bad news is that the obvious strengthenings of first-order logic to cover these shortcomings are too strong: they flunk the 0-1 test, even for uniform structures over the language (example 2.8). For example, each of the following two sentences holds only for structures with an even number of elements, and therefore has no limiting probability:

$$\bigvee_{n=1}^{\infty}\exists x_1\exists y_1\cdots\exists x_n\exists y_n(\neg x_1 = y_1 \wedge \neg x_1 = y_2 \wedge \cdots \wedge \neg x_n = y_n \wedge$$

$$\forall z(z = x_1 \vee z = y_1 \vee \cdots \vee z = x_n \vee z = y_n))$$

and

$$\exists R\forall x\exists y\forall z(xRy \wedge yRx \wedge (z = y \vee (\neg zRx \wedge \neg xRz))).$$

The good news is that there are logics less strong than infinite first-order logic and existential second-order logic but still strong enough to cover many interesting properties, for which the 0-1 law does hold (for uniform structures over a language). In fact, we are fairly close to characterizing, in some senses, the reasonably natural languages for which the 0-1 law holds.

In one attempt to introduce some "infinite" power into the language, Blass, Gurevich and Kozen [2] extended the 0-1 law to finite "positive" sentences with a fixed-point operator. Kolaitis and Vardi were able to widen that class to cover "partial" fixed-point operators on general sentences, and now ([16]) have hugely improved these results by showing that the 0-1 law holds for infinite sentences *as long as they have only finitely many variables*. This delightful result more than subsumes all previous results in this direction.

Since, as we have seen, even one existential second-order quantifier is enough to scrap the 0-1 law, logicians were led to study the idea of allowing such a quantifier, but restricting the form of the remaining first-order part of the sentence. The most natural way to impose such restrictions is to limit the form of the initial first-order quantifiers, e.g. forbidding more than one universal quantifier (yielding what is known as the "Ackermann class" of sentences for the first-order part) or requiring all existential quantifiers to be in front (this forms the "Bernays–Schönfinkel class").

Now, it turns out that back in the first-order world, the Ackermann class and the Bernays–Schönfinkel class are precisely those "prefix classes" for which the finite satisfiability problem is solvable. The "finitely satisfiability problem" for a class of sentences is the problem of deciding for a given sentence from that class whether there exists some finite structure which satisfies the sentence.

With the help of the negative results of Pacholski and Szwast [19], Kolaitis and Vardi [15] showed that the Ackermann and Bernays–Schönfinkel classes are also precisely the prefix classes which, when preceded by a second-order existential quantifier, yield a 0-1 law! Interestingly, the "fragments" of second-order logic created in this way cover both connectivity and k-colorability for graphs.

In their proofs Kolaitis and Vardi use the "elimination of quantifiers" method of Glebskii et al. for their infinitary logic results, but the Fagin technique presented here for their second-order results. Kolaitis and Vardi have also made an intriguing observation which could nail down the 0-1 law even further. When considering more general restrictions on the first-order part of an NP sentence, they noticed that there is a tendency for the 0-1 law to hold when the restricted first-order sentences are "finitely controllable", that is, have the property that satisfiability implies satisfiability by some finite model. If this correspondence holds then one might expect some deep underlying relationship between finite controllability and the 0-1 law.

References

[1]   M.H. Albert & A.M. Frieze, Random graph orders, *Order* **6** (1989), 19–30.

[2]   A. Blass, Y.Gurevich & D. Kozen, A zero-one law for logic with a fixed point operator, *Information and Control* **67** (1985), 70–90.

[3]   B. Bollobás, *Random Graphs*, Academic Press, New York, 1985.

[4]   C.C. Chang & H.J. Keisler, *Model Theory*, North-Holland, 1990.

[5]   K.J. Compton, A logical approach to asymptotic combinatorics I: First-order properties, *Advances in Mathematics* **78** (1987), 65–96.

[6] K.J. Compton, The computational complexity of asymptotic problems I: partial orders, *Inform. and Comput.* **78** (1988), 108–123.

[7] K.J. Compton, 0-1 laws in logic and combinatorics, in *Algorithms and Order*, I. Rival, ed., NATO ASI Series, Kluwer Academic Publishers, Dordrecht, (1989) 353–383.

[8] B. Dushnik & E.W. Miller, Partially ordered sets, *Amer. J. Math.* **63** (1941), 600–610.

[9] P. Erdős & A. Rényi, On the evolution of random graphs, *Publ. Math. Inst. Hungar. Acad. Sci.* **5** (1960), 17–61.

[10] R. Fagin, Probabilities on finite models, *J. Symbolic Logic* **41** (1) (1976), 50–58.

[11] Y.V. Glebskii, D.I. Kogan, M.I. Liogon'kii & V.A. Talanov, Range and degree of realizability of formulas in the restricted predicate calculus, *Kibernetika* **2** (1969), 17–28.

[12] E. Grandjean, Complexity of the first-order theory of almost all structures, *Information and Control* **52** (1983), 180–204.

[13] J. Justicz, E. Scheinerman, & P. Winkler, Random intervals, *Amer. Math. Monthly*, **97** (10) (1989), 881–889.

[14] D.J. Kleitman & B.L. Rothschild, Asymptotic enumeration of partial orders on a finite set, *Trans. Amer. Math. Soc.* **205** (1975), 205–210.

[15] Ph.G. Kolaitis & M.Y. Vardi, Zero-one laws and decision problems for fragments of second-order logic, *Information and Computation* **87** (1990), 302–338.

[16] Ph.G. Kolaitis & M.Y. Vardi, Zero-one laws for infinitary logics, *Proc. 5th IEEE Symp. on Logic in Computer Science* (1990), 156–167.

[17] N. Linial, M. Saks & P. Shor, Largest induced suborders satisfying the chain condition, *Order* **2** (1985), 265–268.

[18] J. Lynch, Probabilities of first-order sentences about unary functions, *Trans. Amer. Math. Soc.* **287** (1985), 543–568.

[19] L. Pacholski & W. Szwast, The 0-1 law fails for the class of existential second-order Gödel sentences with equality, *Proc. 30th IEEE Symp. on Foundations of Computer Science* (1989), 160–163.

[20] R. Rado, Universal graphs and universal functions, *Acta Arith.* **9** (1964), 331–340.

[21] E.R. Scheinerman, Random interval graphs, *Combinatorica* **8** (1988), 357–371.

[22] S. Shelah and J. Spencer, Zero-one laws for sparse random graphs, *J. Amer. Math. Soc.* **1** (1988), 97–115.

[23] J.R. Shoenfield, *Mathematical Logic*, Addison-Wesley, 1967.

[24] J. Spencer, Nonconvergence in the theory of random orders, *Order* **7** (1991), 341–348.

[25] P. Winkler, Random orders, *Order* **1** (1985), 317–331.

[26] P. Winkler, Connectedness and diameter for random orders of fixed dimension, *Order* **2** (1985), 165–171.

[27] P. Winkler, A counterexample in the theory of random orders, *Order* **5** (1989), 363–368.

[28] P. Winkler, Random orders of dimension 2, *Order* **7** (1991), 329–339.

Isotone Maps: Enumeration and Structure

NEJIB ZAGUIA

Computer Science Department
University of Ottawa, 34 George Glinski
Ottawa, Ontario, Canada. K1N 6N5

Abstract

In the last decades "isotone" maps have attracted attention. Let P and Q be two ordered sets. A map $f : P \to Q$ is *isotone* (or *order-preserving, monotone*) if $f(x) \leq f(y)$ in Q whenever $x \leq y$ in P. Despite the simple idea of "order-preserving" not much progress has been made in most of the questions. The objective here is to impart some of the flavor of recent investigations, with no claim of completeness. We highlight a few ideas and results concerning two different problems on isotone maps. First we expose new approaches and questions related to the enumeration problem of isotone maps. Second, we investigate the new concept of a pair of perpendicular orders, that is, a pair of orders with only the trivial isotone self-maps in common. Although pairs of perpendicular orders are relevant to the study of the structure of the lattice of clones on a set, this concept is interesting in its own right, since its investigation has led to attractive questions in the theory of ordered sets.

The best known studies on isotone maps are mainly concerned with the fixed point property. An ordered set P is said to satisfy the *fixed point property*, f.p.p., if each isotone map f of P to itself has a fixed point, that is $f(x) = x$ for some x; otherwise P is *fixed point free*. The first and best known example of orders with the f.p.p. is due to Tarski [30]:

Every complete lattice has the f.p.p.

Two important questions about the f.p.p. in the theory of ordered sets are still unsettled:

What are the ordered sets with the f.p.p.?

if P and Q have the f.p.p. then does the direct product? (See I. Rival [22],[23].)

A lot of work has been done on these two questions, which created a fruitful research area. Other examples of finite ordered sets with the f.p.p. which have been useful are the dismantlable orders. A finite ordered set P is *dismantlable* if the elements can be labelled $a_1, \ldots, a_n$ such that a_i is *irreducible* [that is, a_i has either precisely one upper cover or precisely one lower cover] in $P - \{a_1, \ldots, a_{i-1}\}$, $i = 1, \ldots, n - 1$. It is interesting to notice that dismantlability of a finite order P is related to the structure of P^P. We denote by P^Q the set of all isotone maps of Q to P, ordered by $f \leq g$ if $f(x) \leq g(x)$ for each x in Q. Indeed Duffus and Rival [9] showed that *a finite ordered set P is dismantlable if and only if P^P is connected.*

Problems in isotone maps are difficult and the enumeration problem is no exception. In fact, the arithmetic of ordered structures tells us very little about the cardinalities of the

N.W. Sauer et al. (eds.), Finite and Infinite Combinatorics in Sets and Logic, 421–430.
© 1993 *Kluwer Academic Publishers. Printed in the Netherlands.*

structures. In most of the cases a compact formula that gives the exact answer is beyond hope. Usually, we only seek an approximate answer or an efficient way to compute the answer. We review recent results on enumeration of isotone maps. Also the problems of enumerating special classes of isotone maps as linear extensions and automorphisms are discussed.

1   The number of isotone maps

A very special instance of the enumeration problem of isotone maps is the famous problem of Dedekind:

> What is the number $D(n)$ of elements of $FBD(n)$, the n-generated free distributive lattice?

It is the same as the number of isotone maps of the Boolean lattice 2^n [the set of all subsets of $\{1, \ldots, n\}$, ordered by inclusion] to a 2-element chain. The exact values of $D(n)$ are known for $n \leq 8$. Notice that $D(7)$ has been known since 1965 (Church [6]), and only recently Wiedemann [31] computed $D(8)$. This computation used about 200 hours of time on one Cray-2 processor.

The most important result, considered by many as a solution to Dedekind's problem, is the asymptotic estimate by Korshunov [14]. Using the idea that almost all antichains in 2^n are contained in the union of the middle 3 levels, he has proved the following.

For n even,

$$D(n) \sim 2^{E(n)} \exp\{C(n, n/2 - 1)(2^{-n/2} + n^2 2^{-n-5} - n2^{-n-4})\};$$

For n odd,

$$D(n) \sim 2^{E(n)+1} \exp\{C(n, (n-3)/2)(2^{-(n+3)/2} + n^2 2^{-n-6} - n2^{-n-3})$$

$$+ (C(n, (n-1)/2)(2^{-(n+1)/2} + n^2 2^{-n-4})\}.$$

Here $E(n) = C(n, [n/2])$ is the size of the middle level of 2^n. [$C(n, k)$ stands for the number of subsets of size k taken from a set of size n.]

The exact value of $D(8)$ confirms that, in this case, the estimate of Korshunov is very accurate. Notice that Kisieliwicz [13] gave a precise yet computationally useless formula for $D(n)$. A fuller discussion on Dedekind's problem and general spectral problems can be found in the survey paper of Quackenbush [20]. (See also [21]).

There are few classes of orders for which we have a compact formula or an efficient way to compute the number of isotone maps. Clearly an antichain of size n has n^n self-maps and all of them are isotone. A chain of size n has $C(2n - 1, n)$ isotone self-maps. Although it is trivial that an ordered set P of size n has at most n^n isotone self-maps, it is not in general true that $|P^P| \geq C(2n - 1, n)$. In fact the number of isotone self-maps is not an increasing function with respect to order extensions. [See the examples in Figure 1.]

31   36   275   234

Examples of ordered sets and the number of their isotone self-maps.

Figure 1.

The enumeration problem is hard even for orders with simple structures such as zigzags or cycles. [See Figure 2.] Currie and Visentin [7] and Duffus, Rödl, Sands and Woodrow [10] gave independently a very accurate estimate for the number of isotone self-maps f_n and c_n of zigzags and cycles respectively, with n elements. They proved that

$$f_n \sim (n/2\sqrt{2})(1+\sqrt{2})^n \text{ and } c_n \sim [\sqrt{n}/(2^{1/4}\sqrt{\pi})](1+\sqrt{2})^n.$$

In their proof, Currie and Visentin showed that there is a one-to-one correspondence between the isotone self-maps of the zigzag and the lattice paths on the plane [with unit steps of the form $(i,j) \to (i+1,j+1)$, $(i,j) \to (i+1,j)$ or $(i,j) \to (i+1,j-1)$] on $n-1$ steps. Actually the lattice path corresponding to an isotone self-map f is $(1,f(1))$, $(2,f(2))$, ..., $(n,f(n))$. Using the result for zigzags with an even number of vertices and generating functions, they obtain the estimate for cycles.

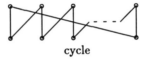

zigzag   cycle

Figure 2.

By only counting isotone self-maps whose images have a simple structure, such as a chain, it is possible to show that there are exponentially many isotone self-maps for any ordered set. For instance an easy argument shows that $|P^P| \geq 2^{n(k-1)/k}$, where $|P| = n$ and k is the size of a longest chain in P (Duffus, Rödl, Sands, and Woodrow [10]). To see this, suppose that $C = \{c_1, \ldots, c_k\}$ is a maximum-sized chain and let $S_i = \{x \in P : \text{height}(x) = i\}$, where height$(x)$ stands for the size of the largest chain with top $= x$. Let $m_i = |S_i|$ and let $m_j = \min\{m_i : i = 1, \ldots, k\}$. Now for each partition of the levels S_i, $i \neq j$, into two levels $A_i \cup B_i$, we associate an isotone self-map f of P defined by:

$$f(x) = \begin{cases} c_i & \text{if } x \in A_i; \\ c_{i+1} & \text{if } x \in B_i \text{ and } i < j; \\ c_{i-1} & \text{if } x \in B_i \text{ and } i \geq j; \\ c_j & \text{if } x \in S_j. \end{cases}$$

There are $\prod_{i \neq j} 2^{m_i} = 2^{\sum_{i \neq j} m_i} = 2^{n-m_j}$ such partitions. Since $m_j \leq n/k$, then we have at least $2^{n(k-1)/k}$ isotone self-maps of P whose images are the chain C. Using as images other structures than chains, namely subsets of the form $1 \oplus 3$ [linear sum of a singleton and a 3-element antichain] when they exist, Duffus, Rödl, Sands and Woodrow [10] showed that the number of isotone self-maps for a bipartite ordered set is at least 2^n. For instance let $P = L_1 \cup L_2$ be a bipartite ordered set and suppose that $|L_1| \geq n/2$. Clearly if L_2 has an element of degree at least three, say $x \in L_2$ and $a, b, c \in L_1$ cover x, then any map that sends L_2 to x and maps L_1 into $\{x, a, b, c\}$ without restrictions, is isotone, and there are $4^{n/2} = 2^n$ such maps. For the other cases they use the enumeration results for cycles and zigzags to complete the proof. It should be possible to improve this bound to 2^n for all ordered sets.

There are special classes of isotone maps for which the enumeration problem has been studied. The best known example is the set of linear extensions. Let P be an ordered set of size n. A *linear extension* of P is an isotone map of P onto an n-element chain. The number of linear extensions $e(P)$ of an ordered set P is an important parameter to measure the complexity of P. Moreover, $e(P)$ appears in the "information-theoretic" lower bound for the sorting problem of the ordered set P, and also in the context of searching the order P for some element. (See Atkinson [1].) The recurrence formula $e(P) = \sum_{x \in \min(P)} e(P - \{x\})$ is easy to see. A non-obvious generalization has been proven by Sidorenko [28]. He showed that $e(P) = \sum_{x \in A} e(P - \{x\})$, where A is an antichain which intersects every maximal chain in P. However, these formulas do not lead to an efficient way to calculate $e(P)$. For many classes of orders, there are efficient methods to compute $e(P)$. For instance Atkinson and Chang [2] gave an algorithm with time complexity $O(mn)$ to compute $e(P)$ for ordered sets of width two, where m and n are the respective sizes of two chains whose union is P. More generally, and for ordered sets of width at most k, Steiner [29] gave an algorithm whose running time is $O(n^{k+1})$. It uses dynamic programming on the set of order ideals of P. For ordered sets whose diagrams are trees, there is an algorithm that runs in $O(n^5)$, (Atkinson [3]). But, in general, the problem is hard. A very interesting result has been obtained by Brightwell and Winkler [5] who showed that counting linear extensions for arbitrary ordered sets is #P-complete. (That is, roughly, as difficult as counting the number of satisfying assignments of a Boolean formula.)

Automorphisms constitute another interesting class of isotone self-maps. An *automorphism* of an ordered set P is a one-to-one and onto map $\alpha : P \to P$ such that both α and α^{-1} are isotone self-maps of P. Let $\mathrm{Aut} P$ stand for all automorphisms of P. There are no substantial classes of orders for which we know how to compute $|\mathrm{Aut} P|$. For instance, every group is isomorphic to the group $\mathrm{Aut} P$ for some ordered set P (Birkhoff [4]). But this result does not tell much about the enumeration of $\mathrm{Aut} P$, and the orders used have a very special structure.

Recently Rival and Rutkowski [24] proposed the following conjecture:

$$\lim_{|P| \to \infty} |\mathrm{Aut} P| / |P^P| = 0.$$

There are known facts that give the impression that proving this conjecture should be easy. For instance, almost every order is *rigid*, that is, has only one automorphism, the identity (Prömel [19]), and as we saw there are exponentially many isotone self-maps for every ordered set. However, the only isotone self-maps which we know best are the ones whose images are chains (or at most complete lattices). But these isotone self-maps are not enough to settle the conjecture. For an antichain of size n, there are $n!$ automorphisms. But there are only n isotone self-maps whose images are chains. And, even if we restrict ourselves to isotone self-maps whose images are bounded by some fixed value k, we have $\sum_{1 \le i \le k} C(n,i)i^n$ such maps, which is less than $n!$ for large enough n. Another difficulty in settling the conjecture is that $|\mathrm{Aut}P|$ is not easy to compute.

Cycle-free orders is a special class for which the conjecture is true (Liu, Rival and Zaguia [16]). A subset $C_{2n} = \{x_1, y_1, \ldots, x_n, y_n\}$, $n \ge 2$, of an ordered set P is a *cycle* if $x_i \le y_i$, $x_{i+1} \le y_i$ (modulo n), are the only comparabilities among these elements and, for the case $n = 2$, there is no element z in P such that $z > x_1, x_2$ and $z < y_1, y_2$. An ordered set P is *cycle-free* if it contains no cycles. Note that every ordered set whose undirected covering graph contains no (graph) cycle is cycle-free. However the undirected covering graph of a cycle-free order may contain (graph) cycles. [See Figure 3.] Cycle-free orders with a top and a bottom are exactly the dismantlable lattices. Let $I(P)$ be the set of irreducible elements in P with a top and a bottom. The subdiagram of the diagram of a cycle-free ordered set P consisting of $I(P)$, will be an ordered subset P consisting of disjoint chains. Let $c(P)$ be the number of these chains. A first step in the proof is to show that every automorphism of a cycle-free ordered set P is determined by its action on the irreducible elements of P, and since every automorphism induces a permutation of the chain of $I(P)$, it follows that $|\mathrm{Aut}P| \le c(P)!$. A second step is to show that $|P^P| > 2^{c(P)} \cdot |\mathrm{Aut}P|$, whenever P is a cycle-free order of size at least 6. Now if $c(P) \ge \log n$ then $|P^P| > n \cdot |\mathrm{Aut}P|$, otherwise $|\mathrm{Aut}P| \le (\log n)!$ and since $|P^P| \ge 2^{cn}$ for some $c > 0$, the ratio $|\mathrm{Aut}P|/|P^P|$ converges to zero, as $|P|$ grows large. Notice that from the proof there follows an efficient procedure with time complexity $O(n^3)$ to compute the number of automorphisms of cycle-free orders, [16].

Counting linear extensions is #P-complete. However little is known about the complexity status of the counting problems of isotone maps and automorphisms.

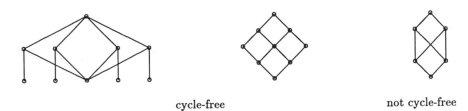

cycle-free   not cycle-free

Figure 3.

One of the difficulties in enumerating isotone maps is that there is no well-defined procedure to generate all of them. In the case of lattices, the (algebraic) polynomials built from the two binary lattice operations and involving constants of the lattice are mechanical devices to produce isotone self-maps. In order to give an estimate of the number of isotone self-maps in case of lattices, it may be a useful approach to consider the enumeration of isotone self-maps which are polynomials. A related question is to find the right concept that generalizes to ordered sets. (See [26].)

It is still an open question to find the largest constant c such that every ordered set P of size n has at least c^n isotone self-maps. Notice that c must be at least $2^{2/3}$ and is at most $1 + \sqrt{2}$. (See [7], [10])

2 Perpendicular orders

Recently Demetrovics, Miyakawa, Rosenberg, Simovici and Stojmenovic [8] initiated the study of pairs of "perpendicular orders". Let P and Q be orders on the same underlying set. Obviously P^P and Q^Q have common elements, namely, the identity map ($f(x) = x$ for each x) and the constant maps which, for any element a in the underlying set, send every x to a. We say that P and Q are *perpendicular* and write $P \perp Q$ if P^P and Q^Q have only these trivial elements in common. In introducing this concept, Demetrovics et al. [8] were interested in the problem to describe the structure of the lattice of "clones" on a set. A *clone* on a set A is a set of n-ary operations, or functions on A (i.e. maps $f : A^n \to A$), which contains all of the projection maps e_n^i where $e_n^i(a_1, \ldots, a_n) = a_i$ for all $a_1, \ldots, a_n \in A$. The set of all clones on a set A, when ordered by inclusion, forms a lattice L_A. Notice that the meet of two clones in L_A is the set theoretic intersection. The structure of the lattice L_A has been well-studied during the last decades. (See [27].) For instance, the *maximal clones* (i.e. the coatoms or the elements covered by the clone O_A of all operations on A) are explicitly known. It is also known, that the intersection of all maximal clones is the bottom element Q_A of L_A, where Q_A is the set of all projections. A special class of clones, called *isotone*, is the class of all clones which preserve a given ordering relation on the set A. For instance, *any perpendicular pair of orders yields a pair of isotone clones whose intersection (that is their meet in L_A) is precisely the clone consisting of all projections and all constant maps* (Länger and Pöschel [15], Pálfy [18]).

The first examples of pairs of perpendicular orders, for each size $n \geq 4$, has been given by Demetrovics et al. [8]. These examples are bipartite and have a very simple structure. A natural question is,

> given an ordered set P, how to construct an ordered set Q which is perpendicular to P?

A subset S of P with at least two elements is called *autonomous* if for every x, y in S and $z \in P \setminus S$, $x \leq z$, $x \geq z$ if and only if $y \leq z$, $y \geq z$, respectively. An autonomous set is an important concept in the structure theory of comparability graphs. (See Kelly [12].)

A necessary condition for a pair of orders on the same underlying set to be perpendicular is that they contain no common autonomous subset. Let P and Q be two orders on the same set with a common autonomous subset S, and let $a \in S$. Then the self-map f defined by $f(x) = a$ if $x \in S$ and $f(x) = x$ otherwise, is a common isotone self-map for P and Q which is neither the identity nor a constant map, that is P and Q are not perpendicular. It is, in general, not a sufficient condition.

"Complementary orders" are natural candidates to consider, in order to construct pairs of perpendicular orders. A *complement* of an ordered set P, denoted comp(P), is an order on the same underlying set as P, such that x is comparable to y in comp(P) if and only if x is noncomparable to y in P. Notice that an order may have more than one complement, or none at all. Actually, *an order has a complement if and only if it has (order) dimension at most two.* (Dushnik and Miller [11].) Rival and Zaguia [25] proved the following:

> An ordered set P is perpendicular to comp(P) for any complement of P if and only if P has no autonomous subsets.

In their proof they showed that if P has an autonomous subset then P and comp(P) have a common autonomous subset for any complement of P.

As an obvious consequence, every ordered set of dimension at most 2 and with no autonomous subset has a perpendicular. It is an open question to show *whether or not every order with no autonomous subset has a perpendicular.* [See Figure 4 for orders with no perpendiculars.] In case of linear orders, autonomous subsets correspond to intervals. A characterization of pairs of perpendicular linear orders has been given by Demetrovics et al. [8].

Two linear orders on the same set are perpendicular if and only if they contain no common interval.

To see this consider two linear orders P and Q on the same underlying set which are not perpendicular, and let $f \in P^P \cap Q^Q$ which is neither the identity map nor a constant map. Clearly f cannot be an automorphism for otherwise $P = Q$. Let $a \neq b$ such that $f(a) = f(b) = u$. The set $I = \{x \in P : f(x) = u\}$ is a common interval for P and Q.

Orders with no perpendiculars.

Figure 4.

Other constructions of pairs of bipartite perpendicular orders are deduced easily from the following, although technical, result (Rival and Zaguia [25]).

Let P and Q be bipartite orders on the same underlying set with $L_1 = \min Q$ and $L_2 = \max Q$ such that

(i) *Q has no autonomous subsets,*

(ii) *degree$(x) < |L_i|$ for each $x \in Q \setminus L_i$, $i = 1, 2$,*

(iii) *for every $x \in L_1$, $y \in L_2$, $x < y$ in Q if and only if x is noncomparable to y in P.*

Then every common isotone self-map for P and Q is either a constant or an automorphism.

For instance this result may be used to construct bipartite perpendicular orders for cycles with at least 6 elements. To see this let $\{x_1 < y_1 > x_2 < y_2 > \ldots < y_{n-1} > x_n < y_n > x_1\}$, $n \geq 3$, be a cycle. If n is even, let $L_1 = \{x_i, y_i : i \text{ odd}\}$ and $L_2 = \{x_i, y_i : i \text{ even}\}$. If n is odd, let $L_1 = \{x_i, y_i : i \text{ odd and } i < n\} \cup \{x_n\}$ and let $L_2 = \{x_i, y_i : i \text{ is even and } i < n\} \cup \{y_n\}$. Now the cycle and $Q = L_1 \cup L_2$ satisfy the condition of the theorem above and therefore the only nontrivial common isotone self-maps are automorphisms. [See Figure 5.] But an automorphism cannot interchange elements from different levels and thus they are perpendicular. A similar construction can be done for even zigzags.

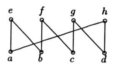 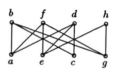

A cycle and its perpendicular.

Figure 5.

The concept of perpendicularity may be naturally extended to graphs, where isotone self-maps are replace by adjacency-preserving self-maps. (That is x adjacent to y implies $f(x)$ adjacent to $f(y)$ or $f(x) = f(y)$.) For instance if f is a non-constant adjacency-preserving self-map of a graph G and for its complement, then f is an automorphism of G if and only if G contains no autonomous subset. Notice too that if two graphs on the same set of vertices have a common edge then they cannot be perpendicular. Thus the only possible perpendiculars to a graph G must be subgraphs of the complement of G. A natural question is this: *What are the graphs that have a perpendicular?*

Given an ordered set P, let $q(P)$ be the number of orderings on the same set as P and which are perpendicular to P. It is interesting to study the function q. The only known result gives a recurrence formula to compute $q(\mathbf{k})$ when $\mathbf{k}$ is a linear ordering of size k. Moreover the ratio $2q(\mathbf{k})/k!$ tends to e^{-2} as k goes to ∞. [Nozaki, Miyakawa and Rosenberg [17]).

It is interesting to study the concept of perpendicularity for infinite ordered sets.

References

[1] M.D. Atkinson, The complexity of orders, in *Algorithms and Orders* (I. Rival ed.), NATO ASI. **255**, Kluwer Acad. Publishers (1989), Dordrecht, 195–230.

[2] M.D. Atkinson and H.W. Chang, Computing the number of mergings with constraints, *Information Processing Letters* **24** (1987), 289–292.

[3] M.D. Atkinson, On computing the number of linear extensions of a tree, *Order* **7** (1990), 23–25.

[4] G. Birkhoff, Sobre los grupos de automorfismos, *Revista Unión Mat. Argentina* **11** (1946), 155–157.

[5] G. Brightwell and P. Winkler, Counting linear extensions is #P-complete, *DIMACS technical report* **90-49** (1990).

[6] R. Church, Enumeration by rank of the free distributive lattice with 7 generators, *Notices Amer. Math. Soc.* **11** (1965), 724.

[7] J.D. Currie and T.I. Visentin, The number of order-preserving maps of fences and crowns, *Order* **8** (1991), 133–142.

[8]   J. Demetrovics, M. Miyakawa, I.G. Rosenberg, D.A. Simovici and I. Stojmenovic, In-
      tersections of isotone clones on a finite set, *Proc. 20th Internat. Symp. Multiple-Valued
      Logic*, Charlotte, 248–253.

[9]   D. Duffus and I. Rival, A structure theory for ordered sets, *Discrete Math.* **35** (1981),
      53–118.

[10] D. Duffus, V. Rödl, B. Sands and R.E. Woodrow, Enumeration of order preserving
      maps, to appear in *Order* **9** (1992).

[11] B. Dushnik and E.W. Miller, Partially ordered sets, *Amer. J. Math* **63** (1941), 600–610.

[12] D. Kelly, Comparability graphs, in *Graphs and Orders* (I. Rival ed.) NATO ASI C
      147, D. Reidel Publ. Co. (1985), 3–40.

[13] A. Kisielewicz, A solution of Dedekind's problem on the number of isotone Boolean
      functions, preprint.

[14] A.D. Korshunov, On the number of monotone Boolean functions, *Probl. Kibern.* **38**
      (1981), 5–108, (in Russian).

[15] F. Länger and R. Pöschel, Relational systems with trivial endomorphisms and poly-
      morphisms, *J. Pure Appl. Algebra* **2** (1984), 129–142.

[16] W.P. Liu, I. Rival and N. Zaguia, Automorphisms, isotone self-maps and cycle-free
      orders, *Technical report* **-91-07** (1991), Computer Science Department, University of
      Ottawa.

[17] A. Nozaki, M. Miyakawa, I.G. Rosenberg and G. Pogosyan, The number of orthogonal
      permutations, *E.T.L. technical report* **91-5** (1991).

[18] P.P. Pálfy, Unary polynomial in algebra I, *Algebra Universalis* **18** (1984), 162–273.

[19] H.J. Prömel, Counting unlabeled structures, *J. Comb. Theor.* **A (44)** (1987), 83–93.

[20] R. Quackenbush, Enumeration in classes of ordered structures, in *Ordered Sets* (I. Rival
      ed.), D. Reidel, Dordrecht (1982), 523–554.

[21] R. Quackenbush, Dedekind's problem, *Order* **2** (1986), 415–417.

[22] I. Rival, Fixed points of direct products, *Order* **1** (1984), 103–105.

[23] I. Rival, The fixed point property, *Order* **2** (1985), 219–221.

[24] I. Rival and A. Rutkowski, Does almost every isotone self-map have a fixed point?
      Technical report **90-21** (1990), Computer Science Department, University of Ottawa.

[25] I. Rival and N. Zaguia, Perpendicular orders, *Technical report* **91-10** (1991), Computer
      Science Department, University of Ottawa.

[26] I. Rival and N. Zaguia, Images of simple lattice polynomials, *Technical report* **91-24**
      (1991), Computer Science Department, University of Ottawa.

[27] I.G. Rosenberg, Completeness properties of multiple-valued logic algebra, in *Computer Science and Multiple-valued Logic, Theory and Applications* (D.C. Rine ed.), North-Holland, 2nd revised ed. (1984), 144–186.

[28] A.F. Sidorenko, The number of linear extensions of a partially ordered set as a function of its incomparability graph, *Math Notes* **29** (1981), no. 1–2, 40–44.

[29] G. Steiner, On computing the information theoretic bound for sorting: counting the linear extensions of a poset, preprint (1990).

[30] A. Tarski, A lattice-theoretical fixpoint theorem and its applications, *Pacific J. Math.* **5** (1955), 285–309.

[31] D. Wiedemann, A computation of the eighth Dedekind number, *Order* **8** (1991), 5–6.

PROBLEM SESSIONS

Introduction

A Banff conference would not be the same without problem sessions. This time there were four, held (as in the previous two Banff meetings) in the evenings in the Solarium. In our introduction to the sessions of the 1984 meeting (*Graphs and Order*), we had gloomily reported that the Solarium may be eliminated in the process of renovations to the Banff Centre; happily, that did not occur, and the old familiar surroundings — small tables, comfy chairs, shuffleboard, well-stocked bar — were all there. A couple of portable blackboards, and a tape recorder, and we were all set. What follows are the proceedings of these sessions, gleaned from the recordings, notes, and written and oral versions of the problems given to us by the presenters. As before, we have tried to reproduce, within reason, the informal conversational spirit of the sessions as they actually happened, complete with some audience remarks. Sometimes information about a problem was obtained later in the conference or after it ended, and these have been added in as parenthetical notes. Our thanks to all participants for their cooperation and help in reconstructing these proceedings. We would also like to thank David Gunderson for taking charge of the tape recorder.

Mathematics is always the main, but never the sole, attraction during a Banff problem session. The bar was again a popular feature, with Springbok a best-seller among beers (and no doubt a revelation to most non-Albertans). Musically we were entertained on more than one occasion by the pianistic talents and vast memory of Adrian Mathias, who displayed a repertoire ranging from Beethoven to Lehrer (Tom, that is).

Time passes and customs change, and what was once accepted without thought becomes a problem. So it was that the question of smoking in the Solarium came up, with participants lined up on both sides. After the first two problem sessions, to which several people could not attend in comfort because of the presence of smoke, we made the next two sessions non-smoking ones, resulting in the other faction not attending. And unfortunately, the unscientific conclusion one might derive from the whole affair is that smokers propose more problems than non-smokers! This conundrum was one reason the number of problem sessions was lower than at other Banff meetings. Perhaps by the next meeting (whenever that may be) a solution to this problem will have turned up. Meanwhile, if anyone out there can tell us how to satisfy everybody, we'd like to hear from you!

We trust that the following record is correct, and that it brings back fond memories to those who were present, arouses feelings of envy in those who weren't (soon replaced by a resolve not to miss the next Banff meeting!), and inspires new ideas and solutions in everyone. Enjoy!

Bill Sands
Robert E. Woodrow

N.W. Sauer et al. (eds.), Finite and Infinite Combinatorics in Sets and Logic, 431–453.
© 1993 *Kluwer Academic Publishers. Printed in the Netherlands.*

Problem Session 1: Infinite Graphs

Chairman: M. POUZET

P. ERDŐS: This first problem is really very old.

The notation $\alpha \to (\alpha, 3)^2$ means that if α is an order type, and G is a graph on a set of order type α, without triangle, then there is an independent set in G of order type α. For example, Specker showed $\omega^2 \to (\omega^2, 3)^2$.

1.1 *The conjecture would be that then*

$$\alpha \to (\alpha, n)^2,$$

i.e. if there is no complete subgraph of size n, there must be an independent set of type α.

Nothing much is known, but no counter-examples have been found. Another problem of Hajnal and myself is the following:

1.2 *If G has sufficiently large chromatic number show it contains a triangle-free subgraph of large chromatic number.*

It is not clear that the chromatic number would equal that of the graph. In fact a stronger conjecture would be that any graph of sufficiently large chromatic number contains a subgraph of large chromatic number with no small odd cycle. You cannot extend this to even cycles because we have a theorem that states that any graph of chromatic number $\aleph_1$ must contain all finite bipartite graphs, indeed it must contain $K_{n,\aleph_0}$ for each $n < \omega$.

There is another and more elementary problem of Hajnal and myself.

1.3 *Let G be a graph of infinite chromatic number. Let $2l_i + 1$, $l_1 < l_2 < \ldots$ be the lengths of the cycles of odd order occurring in G. Is it true that*

$$\sum_i \frac{1}{l_i} = \infty?$$

In fact, perhaps the upper density of the l_i is positive, in which case one can give upper and lower bounds for this density. Actually I should offer \$100 for this problem, and also for the problem on subgraphs without small odd cycles.

A. HAJNAL: For that one you should have chromatic number bigger than $\aleph_0$ because Rödl did it for triangles.

ERDŐS: Yes. I meant it should be infinite.

HAJNAL: I was just trying to save you some money.

ERDŐS: Once I stated the problem about the sum of reciprocals a bit incorrectly and it was solved by Gyárfás, Komjáth and Szemerédi: if a graph has n vertices and kn edges

433

then the sum of the reciprocals of the distinct cycles is at least $c \log k$. This is I feel much easier than the chromatic number and odd cycles problem.

*   *   *   *   *

ROBERT BONNET: I will give two problems in topology. Let $\mathcal{X}$ be an infinite compact Hausdorff space. Say that $\mathcal{X}$ is c.o. if for each closed subspace F there is a clopen subspace U which is homeomorphic to F.

Let us remark that: in $\mathcal{X}$ there are infinitely many isolated points, and if the set of isolated points is dense then $\mathcal{X}$ is scattered.

1.4 *Is it true that if $\mathcal{X}$ is a c.o. space then $\mathcal{X}$ must be scattered?*

The answer is yes if the topology is determined by a linear order (Bekkali, Rubin, and myself).

S. Shelah has shown a consistency result for a related problem.

1.5 *Is it consistent that for compact $\mathcal{X}$ such that each closed $F \subset \mathcal{X}$ is a c.o. space, we have each F homeomorphic to $\alpha + 1$ for some ordinal α?*

(This would imply that $\mathcal{X}$ is hereditarily c.o.)

We know that the answer is yes if the Cantor–Bendixson rank of $\mathcal{X}$ is less than ω_1. For rank ω_1, assuming $\Diamond$, the answer is no.

S. SHELAH: Can you translate this in terms of Boolean algebras?

BONNET: B is a Boolean algebra (corresponding to the compact set). For every ideal I of B the Boolean algebra $A = B/I$ is such that for each J, an ideal of A, $A \cong (A/J) \times C$ for some C.

*   *   *   *   *

REINHARD DIESTEL: Let G be a graph. A ray is a one way infinite path of G. Two rays are equivalent if there are infinitely many paths joining them. The equivalence classes of rays are the ends. A simple example is that of a tree: if you fix a root the ends are in 1–1 correspondence with the rays beginning at the root. A forest is end faithful if it contains (for a given root) one ray from each end.

1.6 *Does every graph contain an end faithful forest?*

In 1964, Halin originally asked whether every connected infinite graph contains an end-faithful spanning tree. That was recently disproved independently by Paul Seymour and Robin Thomas, and by Carsten Thomassen. However, the examples have only one end.

*   *   *   *   *

GENA HAHN: This is a short problem from some work with C. La Viollette, Norbert Sauer, and Robert Woodrow. Say a graph G is bridged if each cycle of length at least 4 has a bridge, i.e. two vertices whose distance in G is less than the distance in the cycle.

1.7 *Suppose G is an infinite bridged graph. Is it true that any finite set of vertices lies in a finite, induced, and bridged subgraph?*

It is true if the diameter is two.

*   *   *   *   *

A. HAJNAL: This is a problem of myself and Szentmiklossy.

1.8 *Is it consistent that there exists a sequence F_α, $\alpha < \omega_2$, a subset of $^{\omega_1}\omega_1$ such that for $\alpha < \beta < \omega_2$, $\{\eta : F_\beta(\eta) \leq F_\alpha(\eta)\}$ is finite?*

Of course this would imply $2^{\aleph_0} \geq \aleph_2$. As far as I know, no one knows whether this is consistent. The problem has a twin.

1.9 *Is it consistent that there exists a sequence $A_\alpha : \alpha < \omega_2$, $A_\alpha \subset \omega_1$, such that for $\alpha < \beta < \omega_2$, $A_\alpha \backslash A_\beta$ is finite and $A_{\alpha+1} \backslash A_\alpha$ is of cardinality $\aleph_1$?*

This implies there are $\aleph_2$ subsets of ω_1 of size ω_1, which pairwise have finite intersection. This was proved consistent by J. Baumgartner but the forcing he used does not give the sequence.

*   *   *   *   *

CLAUDE LAFLAMME: This problem is concerned with the difficulty of finding a homogeneous set for the infinite Ramsey theorem. One way to do this is to consider the minimum size of a family which contains a homogeneous set for each partition. Specifically, consider partitions of the k-element subsets of ω into two classes. Say $\mathcal{X}$ is good if it contains a homogeneous set for each partition, and set h_k equal to the minimum cardinality of a good family $\mathcal{X}$. We have $h_1 \leq h_2 \leq \dots$.

I know essentially nothing about these except it is consistent that $h_1 < h_2$, and also it is consistent that they are all equal and less than the continuum.

1.10 *Can $h_2 < h_3$?*

One further comment: changing the number of colours doesn't affect the cardinals.

*   *   *   *   *

ERIC MILNER: I want to mention a problem that arose in connection with discussions with a Chinese colleague, Wang Shang Shi, who is visiting at Calgary.

If $(P, \leq)$ is a partially ordered set, $\mathrm{cf}(P) = \min\{|Q| : Q \leq P, Q$ cofinal in $P\}$.

1.11 *Does every partially ordered set have a cofinal subset in which every maximal chain* *has regular cardinality?*

I think the answer must be no.

[*Editor's note*: At the end of the session Hajnal presented a partial solution, found during the session. We summarize this at the end of this section.]

*   *   *   *   *

MAURICE POUZET: An âge $\mathcal{A}$ is a collection of finite relational structures of the same similarity type, such that

(i) $G \in \mathcal{A}$ & $H \leq G \Rightarrow H \in \mathcal{A}$

(ii) $\forall G, H \in \mathcal{A}$ $\exists K \in \mathcal{A}$ $G, H \leq K$,

where $R \leq R'$ denotes that R is embeddable in R'.

1.12 *Are the following equivalent?*

(i) $\mathcal{A}$ *contains no infinite antichain for* $\leq$.

(ii) *The collection* $\mathcal{D}(\mathcal{A})$ *of âges contained in* $\mathcal{A}$ *is well-founded when ordered by inclusion.*

(iii) $\mathcal{D}(\mathcal{A})$ *is finite or countable.*

Some of the implications ((i) $\Rightarrow$ (iii) $\Rightarrow$ (ii))are standard from well-quasi-ordering theory. If $\mathcal{D}(\mathcal{A})$ is well-founded one can define a height function by

$$h(\mathcal{A}') = \sup\{h(\mathcal{A}'') + 1 : \mathcal{A}'' \subset \mathcal{A}'\} \text{ for } \mathcal{A}' \subset \mathcal{A}.$$

1.13 *What values for* $h(\mathcal{A})$ *can be attained?*

*   *   *   *   *

NORBERT SAUER: Let me first tell you a problem from Universal Algebra that originated with Kaiser. Given a lattice L, any polynomial (in $x, \vee, \wedge$) with constants from L, is order preserving.

1.14 *Are there lattices for which the polynomials are the only order preserving mappings?*

It is trivial to observe that L can have no non-trivial congruence relation. R. Wille has classified the finite possibilities. From elementary counting L cannot be countable. Both Maurice Pouzet and I have shown that L must have both a 0 and 1.

Generalizing Kaiser's problem to ordered sets one obtains the following.

1.15 *Are there infinite partial orders for which all order preserving maps are definable (with parameters)?*

PARTIAL SOLUTION:

* * * * *

A. HAJNAL: We have two partial answers to Milner's problem [1.11].

First, Stevo Todorčević points out that an $\aleph_{\omega+1}$ Suslin Tree is a counter-example.

The second uses GCH. Take an ordered set $(P, \leq)$ which contains neither an $\aleph_{\omega+1}$ increasing or decreasing sequence and consider its "Sierpinskization", $(Q, \leq)$. (That is, take a well-ordering $\leq_1$ of P, as a cardinal, and let $(Q, \leq)$ be the two dimensional order which is the intersection of $\leq_1$ and $\leq$.) Now it is obvious that the cofinality is at least $\aleph_{\omega+1}$. But any set of size $\aleph_{\omega+1}$ contains an $\aleph_\omega$-chain, which is a contradiction.

Problem Session 2: Finite Combinatorics

Chairman: H. LEFMANN

J. SCHÖNHEIM: I have some very finite problems. I will start with one which is the most finite. It is a problem about a set S having exactly 10 elements. Then a system T of 45 triples of elements of S can have the property that every quadruple contains at least one of the triples. And 45, by Turán, is the minimum number with which you can do that. This is known. Now the number of pairs of elements of S is also 45. So my problem is:

2.1 *Can a set of 45 triples of elements of S have simultaneously the two properties: to cover all the quadruples, and to have distinct representatives which are all the pairs of S?*

I don't know if this is possible or not.

The second problem is a little more algebraic. Consider an abelian group G having $k^2 + k + 1$ elements, and consider a complete difference set D of G which contains 0. Define $D^* = D \setminus \{0\}$.

2.2 *Is $4D^* = G$?*

Why 4? Because for 3 it's not true, for 5 it's true, but with the method to prove 5 we cannot prove 4.

Let S be a hereditary system. Then Chvátal conjectures that among the maximal intersecting subsystems of S, not hereditary of course, you have a *star*: this means sets each containing the same element. My question is:

2.3 *Is this true in the particular case that S is not only hereditary but also a matroid?*

For my fourth question, let $1 < m_1 < m_2 < \ldots < m_s$ be odd integers, and consider the group

$$G = Z_{m_1} \times Z_{m_2} \times \cdots \times Z_{m_s}.$$

Consider also the subgroup $H = \langle (1, 1, \ldots, 1) \rangle$.

2.4 *Prove that in every coset of H there is a vector having all its components different from 0.*

I believe that if you solve this you can win a prize from Professor Erdős, because this is one of his questions in another form.

* * * * *

I. RIVAL: My question is about orientations.

2.5 *Conjecture: Every planar triangle-free graph has a planar orientation.*

We know that if the graph is triangle-free and planar then the chromatic number is at most 3, so there is always *some* orientation (i.e. a diagram of an ordered set), in fact in which there are three levels corresponding to the three colour classes. The trouble is that this orientation won't necessarily be planar. In fact, in the bipartite case, the answer is always yes, but it's not at all obvious: for instance, for the orientation which works is

[di Battista, Liu and Rival (1990), *Inform. Proc. Letters*].

*   *   *   *   *

M. POUZET: I ask two questions, the first with I. Rival:

2.6 *Look at the number of labelled posets with n vertices and k comparabilities, and move k from 0 to $\binom{n}{2}$; is this sequence unimodal (i.e., does it grow and then decrease)?*

The same question for the number of unlabelled posets (i.e., up to isomorphism) is right now considered by R. Fraïssé.

The second question is

2.7 *Given a collection $\mathcal{E}$ of equivalence relations on a finite set E, how can we decide whether the first and second projections are the only maps $f : E \times E \longrightarrow E$ such that $f(x,x) = x$ and f preserves each equivalence relation in $\mathcal{E}$ [i.e. for each $\theta \in \mathcal{E}$, if $x \overset{\theta}{\sim} x'$ and $y \overset{\theta}{\sim} y'$, then $f(x,y) \overset{\theta}{\sim} f(x',y')$]? Classify these $\mathcal{E}$'s.*

R. BONNET: For the first question, this is true for $n = 1, 2, \ldots, 7$. [A calculation of N. Lygeros.]

*   *   *   *   *

D. GUNDERSON: I have two problems. The first one may be trivial or already known.

2.8 *For which finite projective planes is the dual (interchanging points and lines) isomorphic to the original?*

If the plane is Desarguesian apparently this is known in some cases.

The second problem is a little closer to home. Suppose that $F \to (G)_2^1$ (this is the reduced Ramsey arrow), and F is minimal with respect to number of vertices first and then number of edges.

2.9 *Conjecture: If G^* is a graph with $V(G^*) = V(G)$ and $E(G^*) \supset E(G)$, then $G^* \not\subseteq F$.*

Perhaps the problem will be easier in the ordered case, that is, we fix a linear ordering of the vertices of G. For example, if $G = \, \vert$, the only candidate for G^* is the triangle; what I say is that in the minimal graph F which will give me a monochromatic $\vert$, there is no triangle. The minimal graph for $\vert$ in the ordered case is

,

which doesn't have a triangle.

*   *   *   *   *

N. SAUER: We looked with Mike Stone at identities of the full semigroup on a finite set, and the following quite nice thing happened. Let S be a finite set. If you have a self-map $f : S \longrightarrow S$, call the *deficiency* of f the number

$$\operatorname{def} f = |S| - |f(S)|.$$

For example, a permutation has deficiency 0. Now let S be a collection of functions from S to S; what is the smallest deficiency of any element in the semigroup generated by S, and is there a way of calculating it? For example, if $S = \{A, B, C\}$ and you wanted to test for deficiency 0, you multiply the three functions together; if the product ABC has deficiency 0 then certainly no element in the semigroup generated by S has deficiency larger than 0. If you look at two generators ($S = \{A, B\}$), then the word ABA^2B^2AB has the following property: if the deficiency of this word is not larger than 1 then no word in the semigroup generated by S has deficiency larger than 1. Both of these examples have the property that they are independent of the size of S. We can prove ["Composing functions to reduce image size", *Ars Combinatoria*, to appear] that for any size $n = |S|$ of alphabet S and any deficiency k, there exists a word W_k^n with exactly this property: if you want to test whether any word in the semigroup generated by S has deficiency at most k, you test W_k^n. This looks very effective, except that we can't get a good bound on the size of W_k^n — we get n^{4^k}. The question is obvious:

2.10 *Can we say something reasonable about the smallest size of such words W_k^n ?*

*   *   *   *   *

P. WINKLER: The subject is partitions of the edges of a graph into cliques. We define a *greedy* partition as one in which we just remove *maximal* cliques: given a graph, pick any edge, pick a few more edges until you have a maximal clique, remove it, repeat the operation; eventually you have a clique partition of the graph. Now a classical theorem of Erdős, Goodman, and Posá states that *any graph with n vertices has a partition into at most $n^2/4$ cliques.* The unique extremal case is a complete balanced bipartite graph.

2.11 *Conjecture: For any graph G, any greedy partition of the edges of G into cliques has at most $n^2/4$ cliques.*

Now the Erdős–Goodman–Posá theorem was improved independently by Chung and by Györi and Kostochka in the following way: *for any graph G there exists a partition such that*

$$\sum (no. \ of \ vertices \ in \ the \ cliques) \leq \frac{n^2}{2}.$$

Both results are proved by induction. Of course we make the corresponding

2.12 *Conjecture: For any graph G, and any greedy partition of the edges of G,*

$$\sum (no. \ of \ vertices \ in \ the \ cliques) \leq \frac{n^2}{2}.$$

The Chung and Györi–Kostochka theorem, like Erdős–Goodman–Posá, has the property that the only extremal case is the complete balanced bipartite graph. However, this last conjecture, if true, has many extremal examples. For example if you take a complete balanced tripartite graph, and you decompose it greedily in the worst possible manner, then the above sum is exactly $n^2/2$.

[It is reported that Sean McGuinness has recently proved the first conjecture.]

$$* \qquad * \qquad * \qquad * \qquad *$$

J. ŠIRÁŇ: I have two very finite problems; let's hope they are not infinitely difficult! The first is with Peter Horák. A finite graph G is called *maximally non-hamiltonian* (MNH) if it is non-hamiltonian but if you add any new edge it becomes hamiltonian. So it means that every pair of nonadjacent vertices in G is connected by a hamiltonian path. At first glance you might say that such graphs must contain many edges; this is not true because there are infinitely many examples of cubic MNH graphs, one of them being the Petersen graph. Another way to look at the "density" of MNH graphs is to consider *girth*. There are infinitely many MNH graphs that are triangle-free, in fact infinitely many of girth 4, infinitely many of girth 5, 6, and 7, but no MNH graph of girth higher than 7 is known. So my question:

2.13 *Are there MNH graphs of arbitrarily large girth?*

The second problem (a very simple question) I have asked with Herbert Fleischner.

2.14 *Does there exist a simple 4-regular graph which has precisely one hamiltonian cycle?*

There are examples of 4-regular multigraphs which have this property. If you consider 3-regular graphs, it is known that if such a graph is hamiltonian it contains at least 3 hamiltonian cycles.

P. ERDŐS: Can you show a MNH graph of girth 7?

ŠIRÁŇ: It's the Coxeter graph on 28 vertices. And starting from this graph we are able to build infinitely many of them.

ERDŐS: Can you show a MNH graph which has the smallest number of edges? Do you know how many edges it would have?

ŠIRÁŇ: Bondy has proved that if you have a MNH graph on n vertices, then the lower bound for the number of edges is $3n/2$, and there are infinitely many such examples for cubic graphs.

$$\ast \qquad \ast \qquad \ast \qquad \ast \qquad \ast$$

N. ZAGUIA: The first problem is proposed jointly with Ivan Rival. Two orders P and Q on the same set are called *perpendicular* if the only common order-preserving self-maps for P and Q (i.e. the elements of $P^P \cap Q^Q$) are the identity and the constant maps.

2.15 *Is it true that every order which contains no nontrivial autonomous set has a perpendicular?*

[*Editor's note.* A (nontrivial) *autonomous set* is a proper subset A of an order P, A with at least two elements, such that for all $x \in P \setminus A$, if $x < a$ for some $a \in A$ then $x < a$ for all $a \in A$, and the same for $>$. This definition was not given during the problem session. The above problem has been altered slightly since that session.]

The second problem concerns the complexity status of two problems about ordered sets.

2.16 *What is the complexity status of*
(a) *Counting the number of order-preserving self-maps?*
(b) *Counting the number of automorphisms?*

The third problem is related to a kind of one-machine scheduling problem.

2.17 *Given an ordered set, is there an effective procedure for finding a suborder of maximum size having width k?*

If $k = 1$, this is just finding a chain of maximum size, which is obviously very simple to find. For $k = 2$ the problem becomes difficult.

$$\ast \qquad \ast \qquad \ast \qquad \ast \qquad \ast$$

G. HAHN: I have two related problems. For one next to nothing is known, the other may be known.

2.18 *Pick a group G and a set S of generators for it, not containing the identity, and construct the Cayley graph of G with symbol S; can anything "intelligent" be said about the stability number (the size of the largest independent set) of that graph?*

2.19 *Is it NP-complete to find the stability number of the Cayley graph?*

*   *   *   *   *

J.-M. BROCHET: There are two maps $f, g : \omega \longrightarrow \omega$ such that if $n < \omega$ and G is an undirected graph with no independent set of size $n + 1$, then

(i) if each subset of $V(G)$ of size at most $f(n)$ is contained in a cycle, then G has a hamiltonian cycle;

(ii) there are at most $g(n)$ subsets of $V(G)$ which are maximal with respect to being covered by a cycle.

The question is:

2.20 *Find good estimations of the minimal functions f and g.*

In the proof that f and g exist I got a very bad upper bound.

A remark concerning (i). You can show that for such a graph G, if a subset W of $V(G)$ is contained in a cycle, then it is contained in a cycle of length at most $2m$, where $m = |W|$, so this length may be bounded by a function of f. This provides a polynomial algorithm to decide whether G has a hamiltonian cycle (if you bound the independence number of G). The degree of the polynomial is related to f.

*   *   *   *   *

H. LEFMANN: The problem is the following:

2.21 *Is it true that for every positive integer c and for every coloring*

$$\Delta : \{1, 2, 3, \ldots\} \longrightarrow \{1, 2, \ldots, c\}$$

there exist positive integers x, y, z such that $\Delta(x) = \Delta(y) = \Delta(z)$ and $x^2 + y^2 = z^2$?

If we consider the equation $x + y = z$ instead, the result is true, as was shown by Schur. Surprisingly, the equations

$$\frac{1}{x} + \frac{1}{y} = \frac{1}{z} \quad \text{and} \quad \frac{1}{\sqrt{x}} + \frac{1}{\sqrt{y}} = \frac{1}{\sqrt{z}}$$

also have this property. If we consider $x^3 + y^3 = z^3$, we know the result cannot be true because this equation has no solution in positive integers. This is the only argument I know for this case!

P. ERDŐS: Did you try it with two colours?

LEFMANN: Of course, and in the case $c = 2$ the situation is completely unclear. On the other hand for example the equation $x - y = z^2$ has this property; this was proved by Bergelson and Fürstenberg. Another equation, $x + y = z^2$, doesn't have this property. The linear case is solved, essentially, and Walter Deuber will speak about this I think.

Problem Session 3: Homogeneous Structures and Permutation Groups

Chairman: D. EVANS

M. POUZET: Posets and graphs can be viewed as a kind of metric space. In this setting the set of values is a residuated ordered monoid H equipped with an involution. It is known that the category of finite metric spaces over H has the amalgamation property with respect to isometric embeddings.

3.1 *Assuming H countable, describe in terms of the monoid structure or the metric structure of H, the homogeneous countable metric space associated to H.*

Such a description would include some well known examples of homogeneous structures: the Jónsson construction of the countable homogeneous poset, the Rado graph, the countable homogeneous metric space; and some lesser known examples: a graph which isometrically embeds all finite graphs and is homogeneous with respect to isometries, and a transitive system where distances are rational languages.

*   *   *   *   *

ALAN MEKLER: I am giving this problem in proxy for Greg Cherlin. In model theoretic terms:

3.2 *Does the generic (i.e. homogeneous universal) triangle-free graph have the finite model property? Is any sentence true in this countable graph true in some finite graph?*

For triangle-free graphs the question reduces to the following. Let Γ be a triangle-free graph. Γ has the *n-extension property* if, for every independent set X of size at most m and every Y of size k such that $m + k \leq n$ and $X \cap Y = \emptyset$, there is a vertex z such that z is adjacent to every vertex in X and no vertex in Y. The problem is equivalent to whether for all n there is a finite triangle-free graph with the n-extension property.

Michael Albert has produced an infinite family of triangle-free graphs with the 3-extension property. (The graphs are obtained by stretching a graph with 16 vertices.)

It is still open whether there is a triangle-free graph with the 4-extension property. Preliminary estimates by Cherlin indicate the minimal valence would be at least 64.

*   *   *   *   *

DUGALD MACPHERSON: A countable structure M is set-homogeneous if, whenever U and V are isomorphic finite substructures of M, there is $g \in \operatorname{Aut} M$ with $Ug = V$. Cherlin's myopic local order $S(3)$, with the orientation on arcs forgotten, is a set-homogeneous graph which is not homogeneous. Allowing twisted amalgamation, one can develop a construction following Fraïssé producing interesting examples of relational structures. It is fun to play with this in a simple language.

3.3 *Classify the countable set homogeneous graphs.*

M. POUZET: Beside twisted amalgamation what do you know about such structures?

MACPHERSON: Not much. They are model complete, for example.

POUZET: What about adding extra predicates?

MACPHERSON: One could hope they become homogeneous on adding finitely many relations.

S. SHELAH: You want an explicit list?

MACPHERSON: Some examples would help.

$$* \qquad * \qquad * \qquad * \qquad *$$

JACINTA COVINGTON: My question is very short.

3.4 *For which filters F on Ω is the stabilizer $S_{\{F\}}$ maximal in the full symmetric group,* $\operatorname{Sym}(\Omega)$?

We know two necessary conditions (which aren't sufficient):

(1) F is *closed*, i.e. $\Sigma \in F$, $\Sigma \overset{\lambda}{=} \Gamma \Rightarrow \Gamma \in F$ where $\lambda = \min\{|\Delta| : \Delta \in F\}$ is the *depth* of F. ($\Sigma \overset{\lambda}{=} \Gamma$ iff the symmetric difference $\Gamma \triangle \Sigma$ has cardinality less than λ.)

(2) F is *superclosed* (or almost principal), i.e. F is closed and has no almost intersection of cardinality λ. (Φ is an almost intersection if $\Phi \overset{\lambda}{\subseteq} \Delta$ for all $\Delta \in F$.)

The stabilizer of a filter sends almost intersections to almost intersections.

J. LARSON: What can λ be? Finite?

COVINGTON: It is only interesting when λ is infinite.

J. STEPRANS: Are there any examples of these filters which can be generated by less than $\mathfrak{c}$ sets?

COVINGTON: Al Woods has done a study of the case where there is a chain as a filter base, where there are only two orbits, and his work may answer your question.

*   *   *   *   *

DAVID EVANS: These problems come from joint work with Alan Camina.

Suppose $|\Omega| = \aleph_0$ and $G \leq \mathrm{Sym}\,(\Omega)$ is oligomorphic (i.e. G has finitely many orbits on k-sets, $\forall k \in \mathbb{N}$). (In other words we have an $\aleph_0$-categorical structure M with $G = \mathrm{Aut}\,M$.)

Given an enumeration $(\Omega, <)$ of Ω of type ω, define a partial order $\ll$ on Ω by: $a \ll b$ if $a < b$ and there exists $g \in G$ such that $ag = b$ and if $c < a$ then $cg < b$. Call $(\Omega, <)$ a *nice enumeration* if $(\Omega, \ll)$ has no infinite set of pairwise incomparable elements.

3.5 *For every pair* (G, Ω), *is there a nice enumeration of* Ω?

The answer is yes for $(\mathrm{Sym}\,(X), [X]^k)$, $(GL(\aleph_0, q), [V(\aleph_0, q)]^k)$ (Albrecht–Ziegler) and any enumeration of the rationals [Evans].

This problem is due to Dugald Macpherson:

3.6 *When is every enumeration nice?*

For another problem again let $|\Omega| = \aleph_0$ and $G \subseteq \mathrm{Sym}\,(\Omega)$. Pick a field K and consider the vector space $K\Omega$ with basis Ω as a KG-module. There are many things one could ask (but only a few things one can say!) about these permutation modules. Here's one:

3.7 *If* (G, Ω) *is oligomorphic, can* $K\Omega$ *have an infinite ascending chain of submodules?*

If the answer to my first problem is "yes", then the answer here is "no".

*   *   *   *   *

JOHN TRUSS: A *permutation group* is a pair (G, Ω) where G is a subgroup of $\mathrm{Sym}\,\Omega$. We shall assume $|\Omega| = \aleph_0$. If (G_1, Ω_1), (G_2, Ω_2) are permutation groups, an *embedding* of (G_1, Ω_1) into (G_2, Ω_2) is a bijection $\alpha : \Omega_1 \to \Omega_2$ such that $\alpha^{-1} G_1 \alpha \leq G_2$.

$H(\mathbb{Q})$ is the group of all auto-homeomorphisms of the rationals $\mathbb{Q}$.

Let Γ be the random graph. A permutation g of Γ is *almost an automorphism* if the set of edges or non-edges destroyed by g is finite. Aaut (Γ), the group of almost automorphisms of Γ, is a highly transitive extension of $\mathrm{Aut}\,\Gamma$, the automorphism group of Γ. ($\mathrm{Aut}\,\Gamma$ is not even 2-transitive!)

Consider the following conditions on (G, Ω):

(*) if $g_1, g_2, \ldots, g_n \in G$, $\displaystyle\bigcap_{i=1}^{n} \mathrm{supp}\, g_i$ is empty or infinite;

(**) if $g_1, g_2, \ldots, g_n \in G \backslash \{1\}$, $\displaystyle\bigcap_{i=1}^{n} \mathrm{supp}\, (g_i)$ is infinite.

Resolving a problem of Peter Neumann, A. Mekler showed

(1) if $|G| \leq \aleph_0$ then (G, Ω) can be embedded in $(H(\mathbf{Q}), \mathbf{Q})$ if and only if it fulfills $(*)$.

(2) there is (G, Ω) (with $|G| > \aleph_0$) fulfilling $(*)$ which can not be embedded in $(H(\mathbf{Q}), \mathbf{Q})$.

3.8 *For which closed (or more generally Borel) permutation groups (G, Ω) does $(*)$ suffice for embeddability into $(H(\mathbf{Q}), \mathbf{Q})$? In particular, can* Aaut (Γ) *be so embedded?*

Aaut (Γ) is $F\sigma$ and it fulfills $(**)$. Note too that Aut (Γ) satisfies $(**)$ and it can be embedded.

[*Editor's note.* During the conference S. Shelah showed that there is no embedding of Aaut (Γ) into $H(\mathbf{Q})$. Subsequently this problem was the focus of an article by A. Mekler, R. Schipperus, S. Shelah, and J.K. Truss, "The random graph and automorphisms of the rational world," to appear in the *Bulletin of the London Mathematical Society.*]

Problem Session 4: Miscellaneous

Chairman: H. HARBORTH

W. DEUBER: I am speaking about matrices with entries in **Z** and with n columns. For such matrices I talk about the statement

> $\neg\phi$: there exists a least integer $r(A)$ and a colouring $\Delta : \mathbb{N} \to \{0, \ldots, r-1\}$ such that no solution $x_1, \ldots, x_n$ of

$$A \begin{pmatrix} x_1 \\ \vdots \\ x_n \end{pmatrix} = 0$$

is contained in one class.

(For its negation, ϕ, you can go to my talk tomorrow!) I call $r(A)$ the *partition rank* of A. Now the problem.

4.1 *Does* $\max r(A)$, *over all A satisfying* $\neg\phi$ *and having n columns, exist?*

It is known that for $n = 2$, $r(A) \leq 2$. The problem goes back to Richard Rado in 1933.

$$* \qquad * \qquad * \qquad * \qquad *$$

R. DIESTEL: There's a problem that arose from the "bounded graph" concept. Call a graph G *dominating* if there exists a labelling $f : V(G) \to \mathbb{N}$ of its vertices so that every sequence of natural numbers is dominated by some ray eventually; i.e., for every $\sigma : \mathbb{N} \to \mathbb{N}$ there exists a ray $x_0 x_1 \ldots$ in G and $n \in \mathbb{N}$ such that for all $m \geq n$, $f(x_m) \geq \sigma(m)$. Examples: I_{2^ω}, i.e. just 2^ω disjoint rays, is dominating (we just use every possible sequence to label the rays); the complete graph on countably many vertices is dominating; T_ω, the ω-regular tree (which is contained in K_ω), is dominating. Subdivisions of T_ω, such that every edge is subdivided only a bounded number of times (the bound may depend on the vertices where the edge is incident), are also bounded. We say that a subdivision T of T_ω is *locally bounded* if for every vertex v of T_ω there exists $k \in \mathbb{N}$ such that every edge of T_ω incident with v is (in T) subdivided at most k times. It's not difficult to prove that these trees are also dominating.

Two more examples of non-locally bounded subdivisions of T_ω. Enumerate the entire edge set of T_ω, and subdivide the kth edge k times, for each k; then one can show that this graph is *not* dominating. The example that seems to cause problems at the moment is:

4.2 *For each vertex, enumerate the edges above it, and subdivide the first of these edges once, the second twice, the third three times, and so on, so that for any k, every vertex has*

451

exactly one upper edge which is subdivided k times; is this graph dominating?

This should *not* be dominating.

A. MEKLER: I guess, more generally, you could give them the "dominating graph conjecture".

DIESTEL: Yeah, okay, so let's be ridiculously bold:

4.3 *Dominating graph conjecture: any dominating graph contains either uncountably many disjoint rays or contains a locally bounded subdivision of T_ω.*

MEKLER: For people who are less bold, you could try a compromise, and define a *dominating tree* to be one that happened to have a labelling that works (leaving to somebody else the problem of characterizing those), and then try to prove that a graph is dominating if it contains either many independent rays or a dominating tree.

DIESTEL: — whatever these may be. That's nice.

*   *   *   *   *

M. POUZET: The problem is this:

4.4 *What can be said about the structure of posets with singular cofinality; in particular, which posets are necessarily embeddable into every poset with singular cofinality?*

For a partially ordered set P you look for the least κ such that there is a cofinal subset of size κ; κ is the cofinality $\mathrm{cf}(P)$ of P. We observed with Eric Milner that if κ is singular and $\kappa^{<\mathrm{cf}(\kappa)} = \kappa$ then P contains a direct sum $\sqcup_{i<\mathrm{cf}(\kappa)} P_i$ where $\vee_{i<\mathrm{cf}(\kappa)} \mathrm{cf}(P_i) = \mathrm{cf}(P) = \kappa$. In particular P contains an infinite antichain. One other observation: if P is up-directed and $\mathrm{cf}(P)$ is singular then P embeds the binary tree. The next step could be: describe the structure of cofinal subsets for posets not containing a direct sum of finite chains of unbounded size.

*   *   *   *   *

P. ERDŐS: This is a very old problem of Faber, Lovász and myself.

4.5 *Suppose you have n complete graphs of size n, and suppose they are edge disjoint. Is it then true that their union has chromatic number n?*

The best result in this direction is due to Jeff Kahn. He proved that the union has chromatic number $(1 + o(1))n$. This problem is nearly 20 years old. It is known for $n \leq 10$.

*   *   *   *   *

H. HARBORTH: Consider each n-digit binary number and the triangle of all its binary *derivatives*. For example for the 4-digit number 0100 we take additions mod 2 of consecutive digits and we have

$$0\ 1\ 0\ 0$$
$$1\ 1\ 0$$
$$0\ 1$$
$$1$$

Then count the number of 1's in each such triangle. You get 2^n numbers.

4.6 *Which number of 1's occurs most often, and how often?*

Many years ago I could prove that it always is possible to find an n-digit binary number so that half of the digits in the triangle are 1's, but I don't know if this is the number of 1's which occurs most often. (The maximum possible number of 1's is nearly 2/3 of the entries in the triangle.)